Power Sys

Second Edition

Manuel

BAD ~~Manuel~~
Cut
1-21-26

ABOUT THE AUTHORS

Badri Ram, formerly Professor and Head of P.G. Electrical Engineering Department and Dean of Faculty of Engineering, Patna University, Patna, holds a B.E. in Electrical Engineering with distinction from Ranchi University, M.E. from Indian Institute of Science, Bangalore; and Ph.D. from Patna University. He has about forty years of teaching and research experience. An author of four books, he has a number of papers in various national and international journals to his credit. He received the K.F. Antia Award, 1969–70, for the best paper from the Institution of Engineers (India). He was awarded honorary fellowship by the National Association of Computer Educators and Trainers in the year 1996.

D N Vishwakarma is currently Professor of Electrical Engineering at the Institute of Technology, Banaras Hindu University, Varanasi. He obtained his B.Sc.(Engineering), M.Sc.(Engineering) and Ph.D. from Patna University, Patna. He had earlier served as Associate Professor of Electrical Engineering at the Bihar College of Engineering, Patna (presently, National Institute of Technology Patna). He has a teaching and research experience of over 35 years and has contributed about 60 papers to various national and international journals and conferences. He is a senior member of IEEE (USA), a fellow of the Institution of Engineers (India), and a fellow of the Institution of Electronics and Telecommunication Engineers.

Power System Protection and Switchgear
Second Edition

Badri Ram
Former Professor and Head
PG Department of Electrical Engineering
Bihar College of Engineering, Patna
and
Former Dean, Faculty of Engineering
Patna University

D N Vishwakarma
Professor
Department of Electrical Engineering
Institute of Technology—Banaras Hindu University
Varanasi

McGraw Hill Education (India) Private Limited
NEW DELHI

McGraw Hill Education Offices
New Delhi New York St Louis San Francisco Auckland Bogotá Caracas
Kuala Lumpur Lisbon London Madrid Mexico City Milan Montreal
San Juan Santiago Singapore Sydney Tokyo Toronto

McGraw Hill Education (India) Private Limited

Published by McGraw Hill Education (India) Private Limited
P-24, Green Park Extension, New Delhi 110 016

Power System Protection and Switchgear, 2/e

Copyright © 2011, by McGraw Hill Education (India) Private Limited.

Eighth reprint 2013
RQZLCRBORAACY

No part of this publication may be reproduced or distributed in any form or by any means, electronic, mechanical, photocopying, recording, or otherwise or stored in a database or retrieval system without the prior written permission of the publishers. The program listings (if any) may be entered, stored and executed in a computer system, but they may not be reproduced for publication.

This edition can be exported from India only by the publishers,
McGraw Hill Education (India) Private Limited.

ISBN (13 digit): 978-0-07-107774-3
ISBN (10 digit): 0-07-107774-X

Vice President and Managing Director: *Ajay Shukla*
Head—Higher Education (Publishing and Marketing): *Vibha Mahajan*
Publishing Manager—SEM & Tech. Ed.: *Shalini Jha*
Senior Editorial Researcher: *Koyel Ghosh*
Executive—Editorial Services: *Sohini Mukherjee*
Sr Production Manager: *Satinder S Baveja*
Production Executive: *Anuj Kumar Shriwastava*
Marketing Manager—Higher Education: *Vijay S Jaganathan*

General Manager—Production: *Rajender P Ghansela*

Information contained in this work has been obtained by McGraw Hill Education (India), from sources believed to be reliable. However, neither McGraw Hill Education (India) nor its authors guarantee the accuracy or completeness of any information published herein, and neither McGraw Hill Education (India) nor its authors shall be responsible for any errors, omissions, or damages arising out of use of this information. This work is published with the understanding that McGraw Hill Education (India) and its authors are supplying information but are not attempting to render engineering or other professional services. If such services are required, the assistance of an appropriate professional should be sought.

Typeset at Bharati Composers, D-6/159, Sector-VI, Rohini, Delhi 110 085, and printed at A. P Offset, M-135, Panchsheel Garden Naveen Shahdara, Delhi 110 032

Contents

Preface *xiii*

1. Introduction 1
    1.1 Need for Protective Systems *1*
    1.2 Nature and Causes of Faults *2*
    1.3 Types of Faults *3*
    1.4 Effects of Faults *4*
    1.5 Fault Statistics *5*
    1.6 Evolution of Protective Relays *6*
    1.7 Zones of Protection *9*
    1.8 Primary and Back-up Protection *10*
    1.9 Essential Qualities of Protection *11*
    1.10 Performance of Protective Relays *12*
    1.11 Classification of Protective Relays *14*
    1.12 Components of a Protection System *18*
    1.13 Classification of Protective Schemes *19*
    1.14 Automatic Reclosing *20*
    1.15 Current Transformers (CTs) for Protection *20*
    1.16 Voltage Transformers (VTs) *22*
    1.17 Basic Relay Terminology *23*
    Exercises *31*

2. Relay Construction and Operating Principles 32
    2.1 Introduction *32*
    2.2 Electromechanical Relays *33*
    2.3 Static Relays *50*
    2.4 Numerical Relays *61*
    2.5 Comparison between Electromechanical Relays and Numerical Relays *63*
    Exercises *64*

3. Current and Voltage Transformers   65
3.1 Introduction   65
3.2 Current transformers (CTs)   65
3.3 Voltage Transformers (VTs)   89
3.4 Summation transformer   93
3.5 Phase-sequence Current-segregating Network   94
   Exercises   95

4. Fault Analysis   96
4.1 Introduction   96
4.2 Per-Unit System   98
4.3 One-Line Diagram   102
4.4 Impedance and Reactance Diagrams   103
4.5 Symmetrical Fault Analysis   107
4.6 Symmetrical Components   146
4.7 Unsymmetrical Fault Analysis   168
4.8 Grounding (Earthing)   198
   Exercises   215

5. Overcurrent Protection   221
5.1 Introduction   221
5.2 Time-current Characteristics   221
5.3 Current Setting   225
5.4 Time Setting   226
5.5 Overcurrent Protective Schemes   228
5.6 Reverse Power or Directional Relay   234
5.7 Protection of Parallel Feeders   237
5.8 Protection of Ring Mains   237
5.9 Earth Fault and Phase Fault Protection   238
5.10 Combined Earth Fault and Phase Fault Protective Scheme   240
5.11 Phase Fault Protective Scheme   240
5.12 Directional Earth Fault Relay   240
5.13 Static Overcurrent Relays   241
5.14 Numerical Overcurrent Relays   245
   Exercises   246

6. Distance Protection   249
6.1 Introduction   249
6.2 Impedance Relay   250
6.3 Reactance Relay   258
6.4 MHO (Admittance or Angle Admittance) Relay   261

6.5 Angle Impedance (OHM) Relay *267*
6.6 Input Quantities for Various Types of Distance Relays *269*
6.7 Sampling Comparator *270*
6.8 Effect of arc Resistance on the Performance of Distance Relays *271*
6.9 Reach of Distance Relays *274*
6.10 Effect of Power Surges (Power Swings) on the Performance of Distance Relays *277*
6.11 Effect of Line Length and Source Impedance on Distance Relays *281*
6.12 Selection of Distance Relays *283*
6.13 MHO Relay with Blinders *283*
6.14 Quadrilateral Relay *284*
6.15 Elliptical Relay *286*
6.16 Restricted MHO Relay *287*
6.17 Restricted Impedance Relay *289*
6.18 Restricted Directional Relay *289*
6.19 Restricted Reactance Relay *290*
6.20 Some Other Distance Relay Characteristics *290*
6.21 Swivelling Characteristics *293*
6.22 Choice of Characteristics for Different Zones of Protection *294*
6.23 Compensation for Correct Distance Measurement *295*
6.24 Reduction of Measuring Units *298*
6.25 Switched Schemes *299*
6.26 Auto-reclosing *301*
 Appendix *304*
 Exercises *305*

7. Pilot Relaying Schemes   310

7.1 Introduction *310*
7.2 Wire Pilot Protection *310*
7.3 Carrier Current Protection *315*
 Exercises *326*

8. Differential Protection   327

8.1 Introduction *327*
8.2 Differential Relays *328*
8.3 Simple (Basic) Differential Protection *329*
8.4 Percentage or Biased Differential Relay *339*

8.5 Differential Protection of 3-Phase Circuits *346*
8.6 Balanced (Opposed) Voltage Differential Protection *346*
   Exercises *347*

9. Rotating Machines Protection   349
9.1 Introduction *349*
9.2 Protection of Generators *349*
   Exercises *361*

10. Transformer and Buszone Protection   364
10.1 Introduction *364*
10.2 Transformer Protection *364*
10.3 Buszone Protection *375*
10.4 Frame Leakage Protection *376*
   Exercises *377*

11. Numerical Protection   379
11.1 Introduction *379*
11.2 Numerical Relay *381*
11.3 Data Acquisition System (DAS) *386*
11.4 Numerical Relaying Algorithms *390*
11.5 Mann–Morrison Technique *390*
11.6 Differential Equation Technique *391*
11.7 Discrete Fourier Transform Technique *393*
11.8 Walsh–Hadamard Transform Technique *402*
11.9 Rationalised Haar Transform Technique *416*
11.10 Block Pulse Functions Technique *427*
11.11 Wavelet Transform Technique *433*
11.12 Removal of the dc Offset *441*
11.13 Numerical Overcurrent Protection *442*
11.14 Numerical Distance Protection *444*
11.15 Numerical Differential Protection *445*
   Exercises *450*

12. Microprocessor-Based Numerical Protective Relays   452
12.1 Introduction *452*
12.2 IC Elements and Circuits for Interfaces *453*
12.3 A/D Converter, Analog Multiplexer, S/H Circuit *457*
12.4 Overcurrent Relays *471*
12.5 Impedance Relay *475*
12.6 Directional Relay *481*

12.7 Reactance Relay *484*
12.8 Generalised Mathematical Expression for Distance Relays *487*
12.9 Measurement of R and X *489*
12.10 Mho and Offset Mho Relays *492*
12.11 Quadrilateral Relay *498*
12.12 Generalised Interface for Distance Relays *501*
12.13 Microprocessor Implementation of Digital Distance Relaying Algorithms *502*
  Exercises 504

13. Artificial Intelligence Based Numerical Protection 506

13.1 Introduction *506*
13.2 Artificial Neural Network (ANN) *507*
13.3 Fuzzy Logic *518*
13.4 Application of Artificial Intelligence to Power System Protection *520*
13.5 Application of ANN to Overcurrent Protection *522*
13.6 Application of ANN to Transmission Line Protection *522*
13.7 Neural Network Based Directional Relay *523*
13.8 ANN Modular Approach for Fault Detection, Classification and Location *523*
13.9 Wavelet Fuzzy Combined Approach for Fault Classification *525*
13.10 Application of ANN to Power Transformer Protection *526*
13.11 Power Transformer Protection Based on Neural Network and Fuzzy Logic *528*
13.12 Power Transformer Protection Based Upon Combined Wavelet Transform and Neural Network *529*
13.13 Application of ANN to Generator Protection *530*
  Exercises 531

14. Circuit Breakers 533

14.1 Introduction *533*
14.2 Fault Clearing Time of a Circuit Breaker *534*
14.3 Arc Voltage *535*
14.4 Arc Interruption *535*
14.5 Restriking Voltage and Recovery Voltage *537*
14.6 Resistance Switching *541*
14.7 Current Chopping *544*

14.8 Interruption of Capacitive Current 545
14.9 Classification of Circuit Breakers 546
14.10 Air-Break Circuit Breakers 547
14.11 Oil Circuit Breakers 548
14.12 Air Blast Circuit Breakers 553
14.13 SF_6 Circuit Breakers 555
14.14 Vacuum Circuit Breakers 562
14.15 Operating Mechanism 564
14.16 Selection of Circuit Breakers 565
14.17 High Voltage DC (HVDC) Circuit Breakers 566
14.18 Rating of Circuit Breakers 567
14.19 Testing of Circuit Breakers 570
Exercises 575

15. Fuses 578

15.1 Introduction 578
15.2 Definitions 578
15.3 Fuse Characteristics 582
15.4 Types of Fuses 582
15.5 Applications of HRC Fuses 588
15.6 Selection of Fuses 588
15.7 Discrimination 589
Exercises 590

16. Protection Against Overvoltages 592

16.1 Causes of Overvoltages 592
16.2 Lightning Phenomena 593
16.3 Wave Shape of Voltage Due to Lightning 595
16.4 Overvoltages Due to Lightning 596
16.5 Klydonograph and Magnetic Link 598
16.6 Protection of Transmission Lines against Direct Lightning Strokes 599
16.7 Protection of Stations and Sub-stations from Direct Strokes 603
16.8 Protection Against Travelling Waves 605
16.9 Peterson Coil 615
16.10 Insulation Coordination 617
16.11 Basic Impulse Insulation Level (BIIL) 618
Exercises 619

17. Modern Trends in Power System Protection   621

17.1 Introduction   *621*
17.2 Gas Insulated Substation/Switchgear (GIS)   *623*
17.3 Frequency Relays and Load-shedding   *628*
17.4 Field Programmable Gate Arrays (FPGA) Based Relays   *631*
17.5 Adaptive Protection   *637*
17.6 Integrated Protection and Control   *639*
17.7 Relay Reliability   *640*
17.8 Advantages of Fast Fault Clearing   *641*
Exercises   641

Appendix A: *8086 Assembly Language Programming   643*
Appendix B: *Orthogonal and Orthonormal Functions   650*
Appendix C: *Gray-code to Binary Conversion   651*
Appendix D: *Kronecker (or Direct) Product of Matrices   652*

Index   653

Preface to the Second Edition

The authors are thankful to the readers including teachers, students and engineers who have given overwhelming response to the first edition of the book. The response to the first edition of the book has been so encouraging that there have been about 40 reprints of this edition.

The proper operation of a power system requires an efficient, reliable and fast-acting protection scheme, which consists of protective relays and switching devices. There has been continuous improvement in the design of relaying schemes. In order to provide a foolproof protection system to the power system, innovative technology is complementing the conventional protection system. With revolutionary changes in the power system and tremendous developments in computer hardware technology, numerical relays based on microprocessors or microcontrollers are fast replacing the conventional protective relays.

Revision Rationale

Keeping in view the rapid technological advancement in power system protection, we have revised the present second edition of the book thoroughly and have added many new chapters. This edition of the book will be very useful to undergraduate and postgraduate students of electrical engineering, research scholars, students preparing for professional examinations and practicing engineers engaged in system-protection work. In revising this edition, we gratefully acknowledge the constructive suggestions, critical comments and inspiration received from readers and well-wishers for the improvement of the book.

New Additions

The present second edition of the book contains seven additional chapters, viz., *Chapter 3* on Current and Voltage Transformers, *Chapter 4* on Fault Analysis, *Chapter 8* on Differential Protection, *Chapter 9* on Rotating Machines Protection, *Chapter 11* on Numerical Protection, *Chapter 13* on Artificial Intelligence Based Numerical Protection, and *Chapter 17* on Modern Trends in Power System Protection. With the addition of these new chapters, the present edition of the book has become considerably up-to-date.

Chapter Organisation

Chapter 1 on Introduction now stands completely rewritten with addition of new sections on Performance of Protective Relays and Components of a Protection System and many new relay terminologies.

Chapter 2 on Relay Construction and Operating Principles has also undergone a thorough revision.

Chapter 3, a new chapter on Current and Voltage Transformers, describes the construction and operating principles of various types of current and voltage transformers (transducers) required in protection systems.

Chapter 4, a new chapter on Fault Analysis, discusses the per unit system, symmetrical fault analysis, symmetrical components, unsymmetrical fault analysis and grounding. In order to ensure adequate protection, the student must clearly understand the conditions existing on a power system during faults. These abnormal conditions provide the discriminating means for relay operation. Hence, fault analysis is very essential for the design of a suitable protection system.

Chapter 5 on Overcurrent Protection is the revised and renumbered Chapter 3 of the first edition. This chapter deals with overcurrent protection detailing electromagnetic and static overcurrent relays, directional relays, and overcurrent protective schemes.

Similarly, **Chapter 6** on Distance Protection is the revised and renumbered Chapter 4 of the first edition. It discusses different types of electromagnetic and static distance relays, arc resistance and power swings on performance of distance relays, realization of different types of operating characteristics of distance relays and autoreclosing.

Chapter 7 on Pilot Relaying Schemes is the same as Chapter 5 of first edition. It deals with pilot protection, wire pilot relaying and carrier pilot relaying.

Chapter 8, a new chapter on Differential Protection, discusses construction, operating principles and performance of various differential relays for differential protection.

Chapter 9, a new chapter on Rotating Machines Protection, deals with protection of generators and motors. We developed it by splitting Chapter 6 of the first edition and adding motor protection. We then used the remaining parts of Chapter 6 to develop **Chapter 10**, a new chapter on Transformer and Bus Zone Protection.

Chapter 11, a new chapter on Numerical Protection, discusses the numerical relay, data acquisition system, numerical relaying algorithms, numerical overcurrent protection, numerical distance protection and numerical differential protection of generator and power transformer. Numerical relay, which is the main component of the numerical protection scheme, is the latest development in the area of protection. The basis of this topic is numerical (digital) devices, e.g., microprocessors, microcontrollers, Digital Signal Processors (DSPs), etc., developed because of tremendous advancement in VLSI and computer hardware technology.

Chapter 12, a new chapter on Microprocessor Based Numerical Protective Relays, comprises some portions of Chapters 7 and 8 of the first edition and some new additions. It describes microprocessor-based overcurrent and distance relays and digital filtering algorithms for removal of dc offset from relaying signals and computation of R and X of a transmission line.

Chapter 13, a new chapter on Artificial Intelligence Based Numerical Protection, deals with application of Artificial Neural Networks (ANNs) and Fuzzy Logic to numerical protection of power system.

Chapter 14 on Circuit Breakers is the thoroughly revised and renumbered Chapter 9 of the first edition. It covers circuit breakers explaining the principle of circuit interruption and different types of circuit breakers.

Chapter 15 on Fuses is the same as Chapter 10 of the first edition. It describes the construction and operating principle of different types of fuses and gives the definitions of different terminologies related to a fuse. Again, **Chapter 16** is the same as Chapter 11 of the first edition, and is on Protection Against Overvoltages. **Chapter 17**, a new chapter on Modern Trends in Power System Protection, discusses the modern topics of power system protection such as Gas Insulated Substation (GIS), frequency relays and loadshedding, Field Programmable Gate Arrays (FPGA) based relays, adaptive protection, integrated protection and control, Supervisory Control And Data Acquisition (SCADA), travelling wave relay, etc.

Acknowledgements

We owe a special note of thanks to Mr. Harshit Nath who helped us a lot in the preparation of the manuscript of this edition. We also express our gratitude to Ms. Amrita Sinha and Ms. Kiran Srivastava for their constructive suggestions and cooperation during the manuscript-preparation stage of this edition. At this stage, we would also like to thank the many reviewers who took out time from their busy schedules and sent us their suggestions, all of which went a long way in helping us revise the book. We mention their names below.

Zakir Hussain	National Institute of Technology Hamirpur, Himachal Pradesh
Rakesh Kumar	Global Institute of Technology, Greater Noida, Uttar Pradesh
Sardindu Ghosh	National Institute of Technology Rourkela, Orissa
R C Jha	Birla Institute of Technology Mesra, Ranchi, Jharkhand
V B Babaria	Lalbhai Dalpatbhai College of Engineering, Ahmadabad, Gujarat
R Senthil Kumar	Bannari Amman Institute of Technology, Sathyamangalam, Erode, Tamil Nadu
K Lakshmi	K S Rangasamy College of Engineering, Tiruchengode, Tamil Nadu
S P Rajkumar	Sri Ramakrishna Engineering College, Coimbatore, Tamil Nadu
C V K Bhanu	Gayatri Vidya Parishad College of Engineering, Vishakhapatnam, Andhra Pradesh
M Narendra Kumar	Guru Nanak Engineering College, Hyderabad, Andhra Pradesh
S Ram Mohan Rao	College of Engineering, Andhra University, Vishakhapatnam, Andhra Pradesh

We wish to convey our feelings of indebtedness to our colleagues and students for the encouragement and useful suggestions they gave us while we were revising the book. Finally, we state our heartfelt gratitude to our respective families for the love, patience and inspiration, which they are never short of showering on us.

We welcome constructive criticism and suggestions from all concerned for further improvement of the future editions of the book. Please feel free to email your views at the publisher's website mentioned below.

Badri Ram
D N Vishwakarma

Feedback

We invite comments and suggestions from all the readers. You may e-mail your feedback to tmh.elefeedback@gmail.com *(kindly mention the title and author name in the subject line)*. Also, please report any piracy of the book spotted by you.

1 Introduction

1.1 NEED FOR PROTECTIVE SYSTEMS

An electrical power system consists of generators, transformers, transmission and distribution lines, etc. Short circuits and other abnormal conditions often occur on a power system. The heavy current associated with short circuits is likely to cause damage to equipment if suitable protective relays and circuit breakers are not provided for the protection of each section of the power system. Short circuits are usually called faults by power engineers. Strictly speaking, the term 'fault' simply means a 'defect'. Some defects, other than short circuits, are also termed as faults. For example, the failure of conducting path due to a break in a conductor is a type of fault.

If a fault occurs in an element of a power system, an automatic protective device is needed to isolate the faulty element as quickly as possible to keep the healthy section of the system in normal operation. The fault must be cleared within a fraction of a second. If a short circuit persists on a system for a longer, it may cause damage to some important sections of the system. A heavy short circuit current may cause a fire. It may spread in the system and damage a part of it. The system voltage may reduce to a low level and individual generators in a power station or groups of generators in different power stations may lose synchronism. Thus, an uncleared heavy short circuit may cause the total failure of the system.

A protective system includes circuit breakers, transducers (CTs and VTs), and protective relays to isolate the faulty section of the power system from the healthy sections. A circuit breaker can disconnect the faulty element of the system when it is called upon to do so by the protective relay. Transducers (CTs and VTs) are used to reduce currents and voltages to lower values and to isolate protective relays from the high voltages of the power system. The function of a protective relay is to detect and locate a fault and issue a command to the circuit breaker to disconnect the faulty element. It is a device which senses abnormal conditions on a power system by constantly monitoring electrical quantities of the systems, which differ under normal and abnormal conditions. The basic electrical quantities which are likely to change during abnormal conditions are current, voltage, phase-angle (direction) and frequency. Protective relays utilise one or more of these quantities to detect abnormal conditions on a power system.

Protection is needed not only against short circuits but also against any other abnormal conditions which may arise on a power system. A few examples of other abnormal conditions are overspeed of generators and motors, overvoltage, underfrequency, loss of excitation, overheating of stator and rotor of an alternator etc. Protective relays are also provided to detect such abnormal conditions and issue alarm signals to alert operators or trip circuit breaker.

A protective relay does not anticipate or prevent the occurrence of a fault, rather it takes action only after a fault has occurred. However, one exception to this is the Buchholz relay, a gas actuated relay, which is used for the protection of power transformers. Sometimes, a slow breakdown of insulation due to a minor arc may take place in a transformer, resulting in the generation of heat and decomposition of the transformer's oil and solid insulation. Such a condition produces a gas which is collected in a gas chamber of the Buchholz relay. When a specified amount of gas is accumulated, the Buchholz relay operates an alarm. This gives an early warning of incipient faults. The transformer is taken out of service for repair before the incipient fault grows into a serious one. Thus, the occurrence of a major fault is prevented. If the gas evolves rapidly, the Buchholz relay trips the circuit breaker instantly.

The cost of the protective equipment generally works out to be about 5% of the total cost of the system.

1.2 NATURE AND CAUSES OF FAULTS

Faults are caused either by insulation failures or by conducting path failures. The failure of insulation results in short circuits which are very harmful as they may damage some equipment of the power system. Most of the faults on transmission and distribution lines are caused by overvoltages due to lightning or switching surges, or by external conducting objects falling on overhead lines. Overvoltages due to lighting or switching surges cause flashover on the surface of insulators resulting in short circuits. Sometimes, insulators get punctured or break. Sometimes, certain foreign particles, such as fine cement dust or soot in industrial areas or salt in coastal areas or any dirt in general accumulates on the surface of string and pin insulators. This reduces their insulation strength and causes flashovers. Short circuits are also caused by tree branches or other conducting objects falling on the overhead lines.

Birds also may cause faults on overhead lines if their bodies touch one of the phases and the earth wire (or the metallic supporting structure which is at earth potential). If the conductors are broken, there is a failure of the conducting path and the conductor becomes open-circuited. If the broken conductor falls to the ground, it results in a short circuit. Joint failures on cables or overhead lines are also a cause of failure of the conducting path. The opening of one or two of the three phases makes the system unbalanced. Unbalanced currents flowing in rotating machines set up harmonics, thereby heating the machines in short periods of time. Therefore, unbalancing of the lines is not allowed in the normal operation of a power system. Other causes of faults on overhead lines are: direct lightning strokes, aircraft, snakes, ice and snow loading, abnormal loading, storms, earthquakes, creepers, etc. In the case of cables, transformers, generators and other equipment, the causes of faults are: failure of the solid insulation due to aging, heat, moisture or overvoltage, mechanical damage, accidental contact with earth or earthed screens, flashover due to overvoltages, etc.

Sometimes, circuit breakers may trip due to errors in the switching operation, testing or maintenance work, wrong connections, defects in protective devices, etc.

Certain faults occur due to the poor quality of system components or because of a faulty system design. Hence, the occurrence of such faults can be reduced by improving the system design, by using components and materials of good quality and by better operation and maintenance.

1.3 TYPES OF FAULTS

Two broad classifications of faults are
  (i) Symmetrical faults
  (ii) Unsymmetrical faults

1.3.1 Symmetrical Faults

A three-phase (3-ϕ) fault is called a symmetrical type of fault. In a 3-ϕ fault, all the three phases are short circuited. There may be two situations—all the three phases may be short circuited to the ground or they may be short-circuited without involving the ground. A 3-ϕ short circuit is generally treated as a standard fault to determine the system fault level.

1.3.2 Unsymmetrical Faults

Single-phase to ground, two-phase to ground, phase-to-phase short circuits; single-phase open circuit and two-phase open circuit are unsymmetrical types of faults.

Single-phase to Ground (L-G) Fault

A short circuit between any one of the phase conductors and earth is called a single phase to ground fault. It may be due to the failure of the insulation between a phase conductor and the earth, or due to phase conductor breaking and falling to the ground.

Two-phase to Ground (2L-G) Fault

A short circuit between any two phases and the earth is called a double line to ground or a two-phase to ground fault.

Phase-to-Phase (L-L) Fault

A short circuit between any two phases is called a line to line or phase-to-phase fault.

Open-circuited Phases

This type of fault is caused by a break in the conducting path. Such faults occur when one or more phase conductors break or a cable joint or a joint on the overhead lines fails. Such situations may also arise when circuit breakers or isolators open but fail to close one or more phases. Due to the opening of one or two phases, unbalanced currents flow in the system, thereby heating rotating machines. Protective schemes must be provided to deal with such abnormal situations.

Winding Faults

All types of faults discussed above also occur on the alternator, motor and transformer windings. In addition to these types of faults, there is one more type of fault, namely the short circuiting of turns which occurs on machine windings.

1.3.3 Simultaneous Faults

Two or more faults occurring simultaneously on a system are known as multiple or simultaneous faults. In simultaneous faults, the same or different types of faults may occur at the same or different points of the system. An example of two different types of faults occurring at the same point is a single line to ground fault on one phase and breaking of the conductor of another phase, both simultaneously present at the same point. The simultaneous presence of an L-G fault at one point and a second L-G fault on another phase at some other point is an example of two faults of the same type at two different points. If these two L-G faults are on the same section of the line, they are treated as a double line to ground fault. If they occur in different line sections, it is known as a *cross-country earth fault*. Cross-country faults are common on systems grounded through high impedance or Peterson coil but they are rare on solidly grounded systems.

1.4 EFFECTS OF FAULTS

The most dangerous type of fault is a short circuit as it may have the following effects on a power system, if it remains uncleared.

(i) Heavy short circuit current may cause damage to equipment or any other element of the system due to overheating and high mechanical forces set up due to heavy current.

(ii) Arcs associated with short circuits may cause fire hazards. Such fires, resulting from arcing, may destroy the faulty element of the system. There is also a possibility of the fire spreading to other parts of the system if the fault is not isolated quickly.

(iii) There may be reduction in the supply voltage of the healthy feeders, resulting in the loss of industrial loads.

(iv) Short circuits may cause the unbalancing of supply voltages and currents, thereby heating rotating machines.

(v) There may be a loss of system stability. Individual generators in a power station may lose synchronism, resulting in a complete shutdown of the system. Loss of stability of interconnected systems may also result. Subsystems may maintain supply for their individual zones but load shedding would have to be resorted in the sub-system which was receiving power from the other subsystem before the occurrence of the fault.

(vi) The above faults may cause an interruption of supply to consumers, thereby causing a loss of revenue.

High grade, high speed, reliable protective devices are the essential requirements of a power system to minimise the effects of faults and other abnormalities.

1.5 FAULT STATISTICS

For the design and application of protective scheme, it is very useful to have an idea of the frequency of occurrence of faults on various elements of power system. Usually the power stations are situated far away from the load centres, resulting in hundreds of kilometers' length of overhead lines being exposed to atmospheric conditions. The chances of faults occurring due to storms, falling of external objects on the lines, flashovers resulting from dirt deposits on insulators, etc., are greater for overhead lines than for other parts of the power system. Table 1.1 gives an approximate idea of the fault statistics.

Table 1.1 Percentage Distribution of Faults in Various Elements of a Power System

Element	% of Total Faults
Overhead Lines	50
Underground Cables	9
Transformers	10
Generators	7
Switchgears	12
CTs, VTs, Relays Control Equipment, etc.	12

From Table 1.1, it is evident that 50% of the total faults occur on overhead lines. Hence it is overhead lines that require more attention while planning and designing protective schemes for a power system.

Table 1.2 shows the frequency of occurrence of different types of faults (mainly the different types of short circuits) on overhead lines. From the table it is evident that the frequency of line to ground faults is more than any other type of fault, and hence the protection against L-G fault requires greater attention in planning and design of protective schemes for overhead lines.

Table 1.2 Frequency of Occurrence of Different Types of Faults on Overhead Lines

Types of Faults	Fault Symbol	% of Total Faults
Line to Ground	L-G	85
Line to Line	L-L	8
Double Line to Ground	2L-G	5
Three Phase	$3\text{-}\phi$	2

In the case of cables, 50% of the faults occur in cables and 50% at end junctions. Cable faults are usually of a permanent nature and hence, automatic reclosures are not recommended for cables.

1.6 EVOLUTION OF PROTECTIVE RELAYS

In the very early days of the power industry, small generators were used to supply local loads and fuses were the automatic devices to isolate the faulty equipment. They were effective and their performance was quite satisfactory for small systems.

However, they suffered from the disadvantage of requiring replacement before the supply could be restored. For important lines, frequent interruption in power supply is undesirable. This inconvenience was overcome with the introduction of circuit breakers and protective relays. Attracted armatured-type electromagnetic relays were first introduced. They were fast, simple and economical. As auxiliary relays their use will continue even in future, due to their simplicity and low cost. This type of relays operate through an armature which is attracted to an electromagnet or thought a plunger drawn into a solenoid. Plunger-type electromagnetic relays formed instantaneous units for detecting overcurrent or over-voltage conditions. Attracted armature-type electromagnetic relays work on both ac and dc. Later on, induction-type electromagnetic relays were developed. These relays use the electromagnetic induction principle for their operation and hence work on ac only. Since both attracted armature and induction-type electromagnetic relays operate by mechanical forces generated on moving parts due to electromagnetic forces created by the input quantities, these relays were called electromechanical relays. Induction disc-type inverse time-current relays were developed in the early 1920s to meet the selectivity requirement. They were used for overcurrent protection. For directional and distance relays, induction-cup-type units were widely used throughout the world. An induction-cup-type unit was fast and accurate due to its higher torque/inertia ratio. For greater sensitivity and accuracy, polarised dc relays are being used since 1939.

Attracted armature-type balanced-beam relays provided differential protection, distance protection as well as low burden overcurrent units. These relays operated when the magnitude of an operating signal was larger than the magnitude of the restraining signal. These relays were classified as *amplitude comparators*.

Single-input induction-type relays provided operations with time delays. Two-input induction type relays provided directional protection. Two- and three-input induction-type relays also provided distance protection. The operation of these relays was dependant on the phase displacement between the applied electrical inputs. These relays were classified as *phase comparators*.

In 1947, rectifier bridge-type comparators were developed in Norway and Germany. Polarised dc relays, energised from rectifier bridge comparators, challenged the position of induction-cup-type relays. They are widely used for the realisation of distance relay characteristics.

Electronic relays using vacuum tubes first appeared in the literature in 1928 and continued up to 1956. They were not accepted because of their complexity, short life of vacuum tubes and incorrect operation under transient conditions. But electronic valves were used in carrier equipment. There was automatic checking of the carrier channel. An alarm was sounded if any tube became defective, and it was replaced immediately.

Magnetic amplifiers were also used in protective relays in the past. A magnetic amplifier consists of a transformer and a separate dc winding. As the transformer action is controlled by the dc winding, the device is also known as *transductor*. This type of relay is rugged but slow in action. At present, such relays are not used.

Hall crystals were also used to construct phase comparators. Because of their low output and high-temperature errors, such relays have not been widely adopted except in Russian countries.

The first transistorised relay was developed in 1949, soon after the innovation of the transistor. Various kinds of static relays using solid-state devices were developed in the fifties. Multi-input comparators giving quadrilateral characteristics were developed in the sixties. Static relays possess the advantages of low burden on the CT and VT, fast operation, absence of mechanical inertia and contact troubles, long life and less maintenance. As static relays proved to be superior to electromechanical relays, they were used for the protection of important lines, power stations and sub-stations. But they did not replace electromechanical relays. Static relays were treated as an addition to the family of relays. In most static relays, the output or slave relay is a polarised dc relay which is an electromechanical relay. This can be replaced by a thyristor circuit, but it is used because of its low cost. Electromechanical relays have continued to be used because of their simplicity and low cost. Their maintenance can be done by less qualified personnel, whereas the maintenance and repair of static relays requires personnel trained in solid-state devices. Static relays using digital techniques have also been developed.

Static relays appeared to be the technology poised to replace the electromechanical counterparts in the late sixties when researchers ventured into the use of computers for power system protection. Their attempts and the advances in the Very Large Scale Integrated (VLSI) technology and software techniques in the seventies led to the development of microprocessor-based relays that were first offered as commercial devices in 1979. Early designs of these relays used the fundamental approaches that were previously used in the electromechanical and static relays.

In spite of the developments of complex algorithms for implementing protection functions, the microprocessor-based relays marketed in the eighties did not incorporate them. These relays performed basic functions, took advantage of the hybrid analog and digital techniques, and offered a good economical solution. At present, in microprocessor-based relays, different relaying algorithms are used to process the acquired information. Microprocessor/Microntroller-based relays are called numerical relays specifically if they calculate the algorithm numerically.

The modern power networks which have grown both in size and complexity require fast accurate and reliable protective schemes to protect major equipment and to maintain system stability. Increasing interest is being shown in the use of on-line digital computers for protection. The concept of numerical protection employing computers which shows much promise in providing improved performance has evolved during the past three decades. In the beginning, the numerical protection (also known as digital protection) philosophy was to use a large computer system for the total protection of the power system. This protection system proved to be very costly and required large space. If a computer is required to perform other control functions in addition to protection, it can prove to be economical. With the tremendous developments in VLSI and computer hardware technology, microprocessors that appeared in the seventies have evolved and made remarkable progress in recent years. The latest fascinating innovation in the field of computer technology is the development of microprocessors, microcontrollers, Digital Signal Processors (DSPs) and Field Programmable Gate Arrays (FPGAs) which are making in-roads in every activity of mankind. With the development of fast, economical, powerful and sophisticated microprocessors, microcontrollers, DSPs and FPGAs, there is a growing trend to develop numerical relays based on these devices.

The conventional relays of electromechanical and static types had no significant drawbacks in their protection functions, but the additional features offered by microprocessor technologies encouraged the evolution of relays that introduced many changes to the power industry. Economics and additionally, functionality were probably the main factors that forced the power industry to accept and cope with the changes brought by microprocessor/microcontroller-based numerical relays.

Multifunction numerical relays were introduced in the market in the late eighties. These devices reduced the product and installation costs drastically. This trend has continued until now and has converted microprocessor/microcontroller based numerical relays to powerful tools in the modern substations.

The inherent advantage of microprocessor/microcontroller-based protective schemes, over the existing static relays with one or very limited range of applications, is their flexibility. The application of microprocessors and microcontrollers to protective relays also result in the availability of faster, more accurate and reliable relaying units. A microprocessor or a microcontroller increases the flexibility of a relay due to its programmable approach. It provides protection at low cost and competes with conventional relays. A number of relaying characteristics can be realised using the same interface. Using a multiplexer, the microprocessor/microcontroller can obtain the required input signals for the realisation of a particular relaying characteristic. Different programs can be used for different characteristics. Individual types and number of relaying units are reduced to a great extent, resulting in a very compact protective scheme. Field tests have demonstrated their feasibility and some schemes are under investigation. A number of schemes have been put into service and their performances have been found to be satisfactory. Microprocessor/microcontroller-based numerical protective schemes are being widely used in the field.

At present, many trends are emerging. These include common hardware platforms, configuring the software to perform different functions, integrating protection with substation control and substituting cables carrying voltages and currents with optical fibre cables carrying signals in the form of polarized light.

On the software side, Artificial Intelligence (AI) techniques, such as Artificial Neural Networks (ANNs) and Fuzzy Logic Systems have attracted the attention of researchers and protection engineers and they are being applied to power system protection. ANN and Fuzzy Logic based intelligent numerical relays for overcurrent protection, distance protection of transmission lines and differential protection of transformers and generators are presently under active research and development stage. Adaptive protection is also being applied to protection practices. Recent work in this area includes feedback systems in which relays continuously monitor the operating state of the power system and automatically reconfigure themselves for providing optimal protection.

1.7 ZONES OF PROTECTION

A power system contains generators, transformers, bus bars, transmission and distribution lines, etc. There is a separate protective scheme for each piece of equipment or element of the power system, such as generator protection, transformer protection, transmission line protection, bus bar protection, etc. Thus, a power system is divided into a number of zones for protection. A protective zone covers one or at the most two elements of a power system. The protective zones are planned in such a way that the entire power system is collectively covered by them, and thus, no part of the system is left unprotected. The various protective zones of a typical power system are shown in Fig. 1.1. Adjacent protective zones must overlap each other, failing which a fault on the boundary of the zones may not lie in any of the zones (this may be due to errors in the measurement of actuating quantities, etc.), and hence no circuit breaker would trip. Thus, the overlapping between the adjacent zones is unavoidable. If a fault occurs in the overlapping zone in a properly protected scheme, more circuit breakers than the minimum necessary to isolate the faulty element of the system would trip. A relatively low extent of overlap reduces the probability of faults in this region and consequently, tripping of too many breakers does not occur frequently.

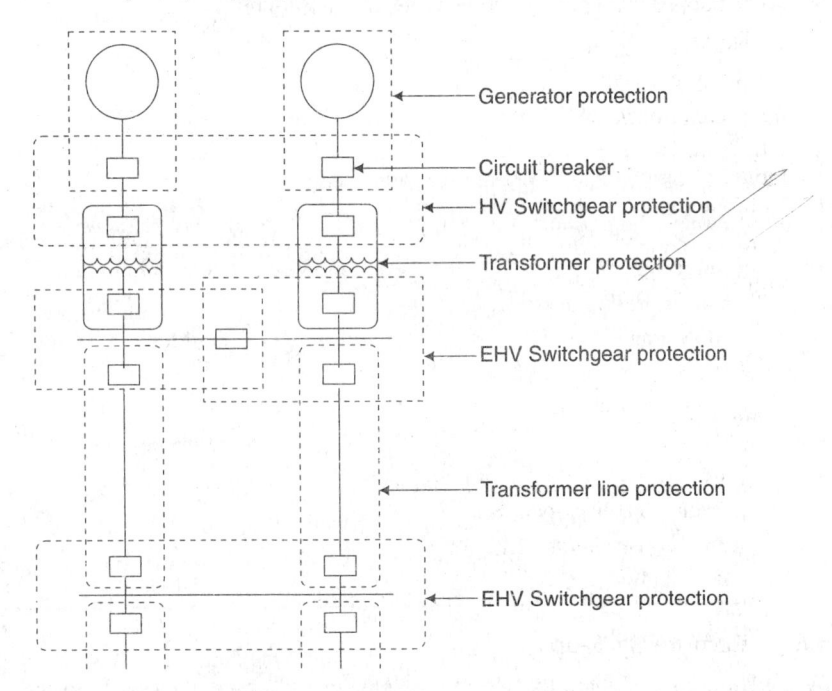

Fig. 1.1 *Zones of protection*

1.8 PRIMARY AND BACK-UP PROTECTION

It has already been explained that a power system is divided into various zones for its protection. There is a suitable protective scheme for each zone. If a fault occurs in a particular zone, it is the duty of the primary relays of that zone to isolate the faulty element. The primary relay is the first line of defense. If due to any reason, the primary relay fails to operate, there is a back-up protective scheme to clear the fault as a second line of defence.

The causes of failures of protective scheme may be due to the failure of various elements, as mentioned in Table 1.3. The probability of failures is shown against each item.

The reliability of protective scheme should at least be 95%. With proper design, installation and maintenance of the relays, circuit breakers, trip mechanisms, ac and dc wiring, etc. a very high degree of reliability can be achieved.

The back-up relays are made independent of those factors which might cause primary relays to fail. A back-up relay operates after a time delay to give the primary relay sufficient time to operate. When a back-up relay operates, a larger part of the power system is disconnected from the power source, but this is unavoidable. As far as possible, a back-up relay should be placed at a different station. Sometimes, a local back-up is also used. It should be located in such a way that it does not employ components (VT, CT, measuring unit, etc.) common with the primary relays which are to be backed up. There are three types of back-up relays:

(i) Remote back-up
(ii) Relay back-up
(iii) Breaker back-up

Table 1.3 Percentage failure rate of various equipment

Name of Equipment	% of Total Failures
Relays	44
Circuit breaker interrupters	14
AC wiring	12
Breaker trip mechanisms	8
Current transformers	7
DC wiring	5
VT	3
Breaker auxiliary switches	3
Breaker tripcoils	3
DC supply	1

1.8.1 Remote Back-up

When back-up relays are located at a neighbouring station, they back-up the entire primary protective scheme which includes the relay, circuit breaker, VT, CT and other elements, in case of a failure of the primary protective scheme. It is the cheapest and he simplest form of back-up protection and is a widely used back-up protection for transmission lines. It is most desirable because of the fact that it will not fail due to the factors causing the failure of the primary protection.

1.8.2 Relay Back-up

This is a kind of a local back-up in which an additional relay is provided for back-up protection. It trips the same circuit breaker if the primary relay fails and this operation takes place without delay. Though such a back-up is costly, it can be recommended where a remote back-up is not possible. For back-up relays, principles of operation that are different from those of the primary protection as desirable. They should be supplied from separate current and potential transformers.

1.8.3 Breaker Back-up

This is also a kind of a local back-up. This type of a back-up is necessary for a bus bar system where a number of circuit breakers are connected to it. When a protective relay operates in response to a fault but the circuit breaker fails to trip, the fault is treated as a bus bar fault. In such a situation, it becomes necessary that all other circuit breakers on that bus bar should trip. After a time-delay, the main relay closes the contact of a back-up relay which trips all other circuit breakers on the bus if the proper breaker does not trip within a specified time after its trip coil is energised.

1.9 ESSENTIAL QUALITIES OF PROTECTION

The basic requirements of a protective system are as follows:
  (i) Selectivity or discrimination
 (ii) Reliability
(iii) Sensitivity
(iv) Stability
 (v) Fast operation

1.9.1 Selectivity or Discrimination

Selectivity, is the quality of protective relay by which it is able to discriminate between a fault in the protected section and the normal condition. Also, it should be able to distinguish whether a fault lies within its zone of protection or outside the zone. Sometimes, this quality of the relay is also called discrimination. When a fault occurs on a power system, only the faulty part of the system should be isolated. No healthy part of the system should be deprived of electric supply and hence should be left intact. The relay should also be able to discriminate between a fault and transient conditions like power surges or inrush of a transformer's magnetising current. The magnetising current of a large transformer is comparable to a fault current, which may be 5 to 7 times the full load current. When generators of two interconnected power plants lose synchronism because of disturbances, heavy currents flow through the equipment and lines. This condition is like a short circuit. The flow of heavy currents is known as a power surge. The protective relay should be able to distinguish between a fault or power surge either by its inherent characteristic or with the help of an auxiliary relay. Thus, we see that a protective relay must be able to discriminate between those conditions for which instantaneous tripping is required and those for which no operation or a time-delay operation is required.

1.9.2 Reliability

A protective system must operate reliably when a fault occurs in its zone of protection. The failure of a protective system may be due to the failure of any one or more elements of the protective system. Its important elements are the protective relay, circuit breaker, VT, CT, wiring, battery, etc. To achieve a high degree of reliability, greater attention should be given to the design, installation, maintenance and testing of the various elements of the protective system. Robustness and simplicity of the relaying equipment also contribute to reliability. The contact pressure, the contact material of the relay, and the prevention of contact contamination are also very important from the reliability point of view. A typical value of reliability of a protective scheme is 95%.

1.9.3 Sensitivity

A protective relay should operate when the magnitude of the current exceeds the preset value. This value is called the pick-up current. The relay should not operate when the current is below its pick-up value. A relay should be sufficiently sensitive to operate when the operating current just exceeds its pick-up value.

1.9.4 Stability

A protective system should remain stable even when a large current is flowing through its protective zone due to an external fault, which does not lie in its zone. The concerned circuit breaker is supposed to clear the fault. But the protective system will not wait indefinitely if the protective scheme of the zone in which fault has occurred fails to operate. After a preset delay the relay will operate to trip the circuit breaker.

1.9.5 Fast Operation

A protective system should be fast enough to isolate the faulty element of the system as quickly as possible to minimise damage to the equipment and to maintain the system stability. For a modern power system, the stability criterion is very important and hence, the operating time of the protective system should not exceed the critical clearing time to avoid the loss of synchronism. Other points under consideration for quick operation are protection of the equipment from burning due to heavy fault currents, interruption of supply to consumers and the fall in system voltage which may result in the loss of industrial loads. The operating time of a protective relay is usually one cycle. Half-cycle relays are also available. For distribution systems the operating time may be more than one cycle.

1.10 PERFORMANCE OF PROTECTIVE RELAYS

When a fault occurs in a particular zone of the power system, the primary relays of that zone are expected to operate and initiate isolation of the faulty element. However, back-up relays surrounding that area are also alerted by the fault and begin to operate. They do not initiate tripping if the primary relays operate correctly. Information regarding operation of these back-up relays is not available when they do not trip. The back-up relay operates, if due to any reason the primary relay fails to operate. Though several primary relays are employed in many protection systems, but frequently only one of the relays actually initiates tripping of the circuit breaker. There may be no direct evidence regarding the other relays being in a correct operating mode.

The performance of the protective relay is documented by those relays that provide direct or specific evidence of operation. Relay performance is generally classified as

(i) Correct operation

(ii) Incorrect operation

(iv) No conclusion

1.10.1 Correct Operation

Correct operation of the relay can be either wanted or unwanted. The correct operation gives indication about (i) correct operation of atleast one of the primary relays, (ii) operation of none of the back-up relays to trip for the fault, and (iii) proper isolation of the trouble area in the expected time.

Almost all relay operations are corrected and wanted, i.e., the operation is as per plan and programme. There are the few cases of the correct but unwanted operation. If all relays and associated equipment perform correctly when their operation is not desired or anticipated, it is called "correct but unwanted operation."

1.10.2 Incorrect Operation

A failure, a malfunction, or an unplanned or unanticipated operation of the protective system results in incorrect operation of the relay. The incorrect operation of the relay can cause either incorrect isolation of an unfaulted area, or a failure to isolate a faulted area. The reasons for incorrect operation can be any one or a combination of (i) misapplication of relays, (ii) incorrect settings, (iii) personnel errors, and (iv) equipment malfunctions. Equipment that can cause an incorrect operation includes CTs, VTs, relays, breakers, cable and wiring, pilot channels, station batteries, etc.

1.10.3 No Conclusion

When one or more relays have or appear to have operated, such as the tripping of the circuit breaker, but no cause of operation can be found, it is the case of 'no conclusion'. Neither any evidence of a power system fault or trouble, nor apparent failure of the equipment, causes and extremely frustrating situation. Thus the cases of no conclusion involves considerable concern and thorough investigation. It is suspected that many of the cases of 'no conclusion' may be the result of personnel involvement which is not reported. Modern oscillographs and data-recording equipment which are being used nowaday in many power systems often provide direct evidence or clues regarding the problem, as well as indicating possibilities that could not have occurred.

1.11 CLASSIFICATION OF PROTECTIVE RELAYS

Protective relays can be classified in various ways depending on the technology used for their construction, their speed of operation, their generation of development, function, etc., and will be discussed in more details in the following chapters.

1.11.1 Classification of Protective Relays Based on Technology

Protective relays can be broadly classified into the following three categories, depending on the technology they use for their construction and operation.
  (i) Electromechanical relays
  (ii) Static relays
  (iii) Numerical relays

Electromechanical Relays

Electromechanical relays are further classified into two categories, i.e., (i) electromagnetic relays, and (ii) thermal relays. *Electromagnetic relays* work on the principle of either electromagnetic attraction or electromagnetic induction. *Thermal relays* utilise the electrothermal effect of the actuating current for their operation.

First of all, electromagnetic relays working on the principle of electromagnetic attraction were developed. These relays were called attracted armature-type electromagnetic relays. This type of relay operates through an armature which is attracted to an electromagnet or through a plunger drawn into a solenoid. Plunger type electromagnetic relays are used for instantaneous units for detecting over current or overvoltage conditions.

Attracted armature-type electromagnetic relays are the simplest type which respond to ac as well as dc. Initially attracted armature-type relays were called electromagnetic relays. Later on, induction type electromagnetic relays were developed. These relays use electromagnetic induction principle for their operation, and hence work with ac quantities only. Electromagnetic relays contain an electromagnet (or a permanent magnet) and a moving part. When the actuating quantity exceeds a certain predetermined value, an operating torque is developed which is applied on the moving part. This causes the moving part to travel and to finally close a contact to energise the trip coil of the circuit breaker.

Since both attracted armature and induction type electromagnetic relays operate by mechanical forces generated on moving parts due to electromagnetic forces created by the input quantities, these relays were called electromechanical relays. The term 'electromechanical relays' has been used to designate all the electromagnetic relays which use either electromagnetic attraction or electromagnetic induction principle for their operation and thermal relays which operate as a result of electrothermic forces created by the input quantities. Sometimes both the terms, i.e., electromagnetic relays and electromechanical relays are used in parallel.

Static Relays

Static relays contain electronic circuitry which may include transistors, ICs, diodes and other electronic components. There is a comparator circuit in the relay, which compares two or more currents or voltages and gives an output which is applied to either a slave relay or a thyristor circuit. The slave relay is an electromagnetic relay which finally closes the contact. A static relay containing a slave relay is a semi-static relay. A relay using a thyristor circuit is a wholly static relay. Static relays possess the advantages of having low burden on the CT and VT, fast operation, absence of mechanical inertia and contact trouble, long life and less maintenance. Static relays have proved to be superior to electromechanical relays and they are being used for the protection of important lines, power stations and sub-stations. Yet they have not completely replaced electromechanical relays. Static relays are treated as an addition to the family of relays. Electromechanical relays continue to be in use because of their simplicity and low cost. Their maintenance can be done by less qualified personnel, whereas the maintenance and repair of static relays requires personnel trained in solid state devices.

Numerical Relays

Numerical relays are the latest development in this area. These relays acquire the sequential samples of the ac quantities in numeric (digital) data form through the data acquisition system, and process the data numerically using an algorithm to calculate the fault discriminants and make trip decisions. Numerical relays have been developed because of tremendous advancement in VLSI and computer hardware technology. They are based on numerical (digital) devices, e.g., microprocessors, microcontrollers, Digital Signal Processors (DSPs), etc. At present microprocessor/microcontroller-based numerical relays are widely used. These relays use different relaying algorithms to process the acquired information. Microprocessor/microcontroller-based relays are called numerical relays specifically if they calculate the algorithm numerically. The term 'digital relay' was originally used to designate a previous-generation relay with analog measurement circuits and digital coincidence time measurement (angle measurement) using microprocessors. Now a days the term 'numerical relay' is widely used in place of 'digital relay'. Sometimes, both terms are used in parallel. Similarly, the term 'numerical protection' is widely used in place of 'digital protection'. Sometimes both these terms are also used in parallel.

The present downward trend in the cost of Very Large Scale Integrated (VLSI) circuits has encouraged wide application of numerical relays for the protection of modern complex power networks. Economical, powerful and sophisticated numerical devices (e.g., microprocessors, microcontrollers, DSPs, etc) are available today because of tremendous advancement in computer hardware technology. Various efficient and fast relaying algorithms which form a part of the software and are used to process the acquired information are also available today. Hence, there is a growing trend to develop and use numerical relays for the protection of various components of the modern complex power system. Numerical relaying has become a viable alternative to the traditional relaying systems employing electromechanical and static relays. Intelligent numerical relays using artificial Intelligence techniques such as Artificial Neural Networks (ANNs) and Fuzzy Logic Systems are presently under active research and development stage.

The main features of numerical relays are their economy, compactness, flexibility reliability, self-monitoring and self-checking capability, multiple functions, low burden on instruments transformers and improved performance over conventional relays of electromechanical and static types.

1.11.2 Classification of Protective Relays Based on Speed of Operation

Protective relays can be generally classified by their speed of operation as follows:
  (i) Instantaneous relays
  (ii) Time-delay relays
  (iii) High-speed relays
  (iv) Ultra high-speed relays

Instantaneous Relays

In these relays, no intentional time delay is introduced to slow down their response. These relays operate as soon as a secure decision is made.

Time-delay Relays

In these relays, an intentional time delay is introduced between the relay decision time and the initiation of the trip action.

High-speed Relays

These relays operate in less than a specified time. The specified time in present practice is 60 milliseconds (3 cycles on a 50 Hz system).

Ultra High-speed Relays

Though this term is not included in the relay standard but these relays are commonly considered to operate within 5 milliseconds.

1.11.3 Classification of Protective Relays Based on their Generation of Development

Relays can be classified into the following categories, depending on generation of their development.
  (i) *First-generation relays:* Electromechanical relays
  (ii) *Second-generation relays:* Static relays
  (iii) *Third-generation relays:* Numerical relays.

1.11.4 Classification of Protective Relays Based on their Function

Protective relays can be classified into the following categories, depending on the duty they are required to perform:
  (i) Overcurrent relays
  (ii) Undervoltage relays
  (iii) Impedance relays
  (iv) Underfrequency relays
  (v) Directional relays

These are some important relays. Many other relays specifying their duty they perform can be put under this type of classification. The duty which a relay performs is evident from its name. For example, an overcurrent relay operates when the current exceeds a certain limit, an impedance relay measures the line impedance between the relay location and the point of fault and operates if the point of fault lies within the protected section. Directional relays check whether the point of fault lies in the forward or reverse direction.

The above relays may be electromechanical, static or numerical.

1.11.5 Classification of Protective Relays as Comparators

Protective relays are basically comparators which must be able to carry out addition, subtraction, multiplication or division of some scalar or some phasor quantities and make comparisons of the input quantities as desired. Based upon this principle, the protective relays can be classified as comparators into the following categories.

(i) Single-input comparator
(ii) Dual-input comparator
(iii) Multi-input comparator

Single-input Comparator

These relays have only one input signal and are also known as *level detectors*. Such relays continuously monitor one electrical quantity and compare it with certain constant quantity, i.e., a reference or a base quantity which may be the pull of a spring or gravitational force. An example of this type of relay is an over current relay which measures the current of a circuit and compares it with a certain preset value of the reference or base quantities. Though these relays are simple in construction and operation, they have several drawbacks such as (i) they are non-directional, (ii) they are not reliable because their action depends upon a single quantity, and (iii) they fail to attain the desired reliability.

Dual-input Comparator

These relays have two input signals. Such relays measure one quantity and compare it with another quantity. The typical examples of such type of relays are distance relays and differential relays. The distance relay measures the current entering the circuit and compares it in magnitude or in phase angle with the local bus voltage. The differential relay measures the current entering the circuit and compares it with the current leaving the circuit at the other end. Dual-input comparators are of two types, i.e., amplitude comparator and phase comparator. The amplitude comparator compares only the amplitude of the two input signals irrespective of phase angle between them, whereas the phase comparator compares only the phase angle between the two input signals irrespective of their magnitudes. There is duality between amplitude and phase comparators, i.e., an amplitude comparator can be converted to a phase comparator and vice-versa if the input quantities to the comparator are modified. The modified input quantities are the sum and difference of the original two input quantities. These relays have several advantages and wide applications.

Multi-input Comparator

Multi-input comparators have more than two input signals and are used for the relaisation of special characteristics other than straight lines or circle. These comparators are also of two types, i.e., (i) multi-input phase comparator, and (ii) multi-input amplitude comparator. Multi-input phase comparator is used for realisation of quadrilateral characteristic whereas multi-input amplitude comparator is used for realisation of conic characteristics such as elliptical or hyperbolic characteristics.

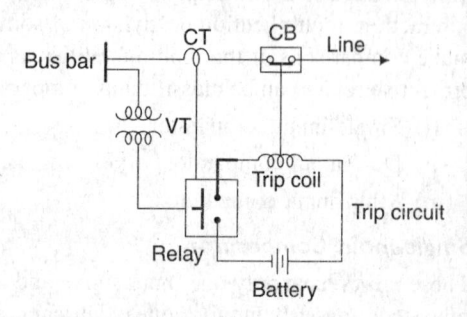

1.12 COMPONENTS OF A PROTECTION SYSTEM

A protection system consists of many other subsystems which contribute to the detection and removal of faults. As shown in Fig. 1.2, the main subsystems of the protection system are the transducers, relays, circuit breakers and trip circuit containing trip coil and battery.

The transducers, i.e., the current and voltage transformers (CTs and VTs) are used to reduce currents and voltages to standard lower values and to isolate protective relays from the high voltages of the power system. They constitute a major component of the protection system, and are discussed in detail in Chapter 3. Protective relays detect and locate the fault and issue a command to the circuit breaker (CB) to disconnect the faulty element. When a fault occurs in the protected circuit (i.e., the line in this case), the relay connected to the CT and VT actuates and closes its contacts to complete the trip circuit. Current flows from the battery in the trip circuit. As the trip coil of the circuit breaker is energized, the circuit breaker operating mechanism is actuated and it operates for the opening operation to disconnect the faulty element.

Fig. 1.2 Components of a protection system

A circuit breaker is a mechanical switching device capable of making, carrying and breaking currents under normal circuit conditions and also making, carrying for a specified time, and automatically breaking currents under specified abnormal circuit condition such as those of short circuits, i.e., faults. It is used to isolate the faulty part of the power system under abnormal conditions. A protective relay detects abnormal conditions and sends a tripping signal to the circuit breaker. A circuit breaker has two contacts—a fixed contact and a moving contact. Under normal conditions, these two contacts remain in closed position. When the circuit breaker is required to isolate the faulty part, the moving contact moves to interrupt the circuit. On the separation of the contacts, the flow of current is interrupted, resulting in the formation of an arc between the contacts. The medium in which circuit interruption is performed is designated by a suitable prefix, such as oil circuit breaker, air-break circuit breaker, air blast circuit breaker, sulphure hexafluoride circuit breaker, vacuum circuit breaker.

Since the primary function of a protection system is to remove a fault, the ability to trip a circuit breaker through a relay must not be compromised during a fault, when the ac voltage available at the substation may not be of sufficient magnitude. In case of a close-in three-phase fault, the ac voltage at the substation can be zero. Therefore the tripping power, as well as the power required by the relays cannot be obtained from the ac system, and is usually provided by the station battery. The battery which is permanently connected through a charger to the station ac service floats on the charger during normal steady-state conditions. The charger is of a sufficient VA capacity to provide all steady-state loads powered by the battery. The battery should also be rated to maintain adequate dc power for 8-12 hours following a station blackout. For better reliability EHV stations have duplicate batteries, each feeding from its charger, and connected to its own complement of relays. Since the severe transients produced by the electromechanical relays on the battery loads during their operation may cause maloperation of other sensitive relays in the substation, or may even damage them, it is common practice, as far as practicable, to separate electromechanical and static equipment by connecting them to different batteries.

1.13 CLASSIFICATION OF PROTECTIVE SCHEMES

A protective scheme is used to protect an equipment or a section of the line. It includes one or more relays of the same or different types. The following are the most common protective schemes which are usually used for the protection of a modern power system.

(i) Overcurrent protection

(ii) Distance protection

(iii) Carrier-current protection

(iv) Differential protection

1.13.1 Overcurrent Protection

This scheme of protection is used for the protection of distribution lines, large motors, equipment, etc. It includes one or more overcurrent relays. An overcurrent relay operates when the current exceeds its pick-up value.

1.13.2 Distance Protection

Distance protection is used for the protection of transmission or sub-transmission lines; usually 33 kV, 66 kV and 132 kV lines. It includes a number of distance relays of the same or different types. A distance relay measures the distance between the relay location and the point of fault in terms of impedance, reactance, etc. The relay operates if the point of fault lies within the protected section of the line. There are various kinds of distance relays. The important types are impedance, reactance and mho type. An impedance relay measure the line impedance between the fault point and relay location; a reactance relay measures reactance, and a mho relay measures a component of admittance.

1.13.3 Carrier-Current Protection

This scheme of protection is used for the protection of EHV and UHV lines, generally 132 kV and above. A carrier signal in the range of 50-500 kc/s is generated for the purpose. A transmitter and receiver are installed at each end of a transmission line to be protected. Information regarding the direction of the fault current is transmitted from one end of the line section to the other. Depending on the information, relays placed at each end trip if the fault lies within their protected section. Relays do not trip in case of external faults. The relays are of distance type and their tripping operation is controlled by the carrier signal.

1.13.4 Differential Protection

This scheme of protection is used for the protection of generators, transformers, motors of very large size, bus zones, etc. CTs are placed on both sides of each winding of a machine. The outputs of their secondaries are applied to the relay coils. The relay compares the current entering a machine winding and leaving the same. Under normal conditions or during any external fault, the current entering the winding is equal to the current leaving the winding. But in the case of an internal fault on the winding, these are not equal. This difference in the current actuates the relay. Thus, the relay operates for internal faults and remains inoperative under normal conditions or during external faults. In case of bus zone protection, CTs are placed on the both sides of the bus bar.

1.14 AUTOMATIC RECLOSING

About 90% of faults on overhead lines are of transient nature. Transient faults are caused by lightning or external bodies falling on the lines. Such faults are always associated with arcs. If the line is disconnected from the system for a short time, the arc is extinguished and the fault disappears. Immediately after this, the circuit breaker can be reclosed automatically to restore the supply.

Most faults on EHV lines are caused by lightning. Flashover across insulators takes place due to overvoltages caused by lightning and exists for a short time. Hence only on instantaneous reclosure is used in the case of EHV lines. There is no need for more than one reclosure for such a situation. For EHV lines, one reclosure in 12 cycles is recommended. A fast reclosure is desired from the stability point of view. More details have been given in Ch. 6.

On lines up 33 kV, most faults are caused by external objects such as tree branches etc. falling on the overhead lines. This is due to the fact that the support height is less than that of the trees. The external objects may not be burnt clear at the first reclosure and may require additional reclosures. Usually three reclosures at 15-120 seconds intervals are made to clear the fault. Statistical reports show that over 80% faults are cleared after the first reclosure, 10% require the second reclosure and 2% need the third reclosure, while the remaining 8% are permanent faults. If the fault is not cleared after 3 reclosures, it indicates that the fault is of permanent nature. Automatic reclosure are not used on cables as the breakdown of insulation in cables causes a permanent fault.

1.15 CURRENT TRANSFORMERS (CTs) FOR PROTECTION

Current transformers (CTs) are used to obtain reduced current signals from the power systems for the purpose of measurement, control and protection. They reduce the heavy currents of the power system to lower values that are suitable for the operation of relays and other instruments connected to their secondary windings. Besides reducing the current levels, CTs also isolate the relays and instruments circuit from the primary circuit which is a high voltage power circuit and allow the use of standardized current ratings for relays and meters. The current ratings of the secondary windings of the CTs have been standardized so that a degree of interchangeability among relays and meters of different manufacturers can be achieved. Since the standard current ratings of the secondary windings of the CTs are 5 or 1 ampere the protective relays also have the same current rating. The current transformers are designed to withstand fault currents (which may be as high as 50 times the full load current) for a few seconds. Protective relays require reasonably accurate reproduction of the normal and abnormal conditions in the power system for correct sensing and operation. Hence the current transformers should be able to provide current signals to the relays which are faithful reproductions of the primary currents. The measure of a current transformer performance is its ability to accurately reproduce the primary current in secondary.

The requirements of CTs used for relaying are quite different from those of metering (CTs for measuring instruments). CTs used for instrumentation are required to be accurate over the normal working range of currents, whereas CTs used for relaying are required to give a correct ratio up to several times the rated primary current. The CTs used for metering may have very significant errors during fault conditions, when the currents may be several times their normal value for a very short time. Since metering functions are not required during faults, this is not significant. CTs used for relaying are designed to have small errors during fault conditions, whereas their performance during normal steady state condition, when the relay is not required to operate, may not be as accurate.

The accuracy of a current transformer is expressed in terms of the departure of its ratio form its true ratio. This is called the ratio error, and is expressed as:

$$\text{Percent error} = \left[\frac{NI_s - I_p}{I_p}\right] \times 100$$

where, N = Nominal CT ratio

$$= \frac{\text{Rated primary current}}{\text{Rated secondary current}}$$

$$= \frac{\text{Number of secondary turns}}{\text{Number of primary turns}}$$

I_s = Secondary current, and

I_p = Primary current

The ratio error of a CT depends on its exciting current.

Current transformers are of electromagnetic, opto-electronic and Rogowski coil types. The electromagnetic type CTs which are magnetically coupled, multi-winding transformers can be classified into two categories: toroidal or bar primary CTs and wound primary CTs. Toroidal or bar-primary-type CTs, donot contain a primary winding and instead a straight conductor, (wire) which is a part of the power system and carries the current, acts as the primary. The conductor (wire) that carries the currents is encircled by a ring-type iron core on which the secondary winding is wound over the entire periphery. Wound-type CTs consist of a primary winding of fewer turns wound on the iron core and inserted in series with the conductor that carries the measured-current.

An opto-electronic or optical CT uses two light beams travelling through an optical fiber to measure the magnetic field around a current-carrying conductor, which gives a measure of the current flowing in the conductor. The phase displacement between the two beams is proportional to the level of current in the conductor.

The Rogowski Coil (RC) is a helical coil of wire with the lead from one end returning through the centre of the coil to the other end, so that both are at the same end of the coil. The whole assembly is then wrapped around the straight conductor whose current is to be measured. Since the voltage that is induced in the coil is proportional to the rate of change (derivative) of current in the conductor, the output of the Rogowski coil is usually connected to an electrical (or electronic) integrator circuit to provide an output signal that is proportional to the current.

Current transformers are described in detail in Chapter 3.

1.16 VOLTAGE TRANSFORMERS (VTs)

Voltage transformers (VTs) were previously known as potential transformers (PTs). They are used to reduce the power system voltages to lower values and to provide isolation between the high-voltage power network and the relays and other instruments connected to their secondaries. The voltage ratings of the secondary windings of the VTs have been standardized, so that a degree of interchangeability among relays and meters of different manufacturers can be achieved. The secondary windings of the voltage transformers are rated at 110 V line to line. Therefore, the voltage ratings of the voltage (pressure) coils of protective relays and measuring instruments (meters) are also 110 V line to line. The voltage transformers should be able to provide voltage signals to the relays (and meters) which are faithful reproductions of the primary voltages.

The accuracy of voltage transformers is expressed in terms of the departure of its ratio from its true ratio.

The percentage ratio error is given by

$$\text{Percent ratio error} = \left(\frac{KV_s - V_p}{V_p}\right) \times 100$$

where, K = Nominal voltage ratio

$$= \frac{\text{Rated primary voltage}}{\text{Rated secondary voltage}}$$

$$= \frac{\text{Number of primary turns}}{\text{Number of secondary turns}}$$

V_s = Secondary voltage, and

V_p = Primary voltage

Ideally a VT should produce a secondary voltage which is exactly proportional and in phase opposition to the primary voltage. But, in practice, this cannot be achieved owing to the voltage drops in the primary and secondary windings due to magnitude and power factor of the secondary burden. Thus, ratio and phase angle errors are introduced.

There are three types of voltage transformers:

(i) electromagnetic type, (ii) capacitor type, and (iii) opto-electronic type. The electromagnetic type of VT is similar to a conventional wound type transformer with additional feature to minimise errors. This type of VT is conveniently used for voltages up to 132 KV. Capacitor-type voltage transformer has a capacitance voltage divider and is used at higher system voltages, i.e., 132 KV and above. This type of voltage transformer is also known as Coupling Capacitor Voltage Transformer (CCVT). In an opto-electronic VT, a circular polarized light beam traveling through a fibre optic up the column is used to determine the voltage difference between the conductor and the ground. This type of voltage transformer is also known as electronic voltage transformer.

The voltage transformers (VTs) are described in detail in Chapter 3.

1.17 BASIC RELAY TERMINOLOGY

Relay: A relay is an automatic device by means of which an electrical circuit is indirectly controlled (opened or closed) and is governed by a change in the same or another electrical circuit.

Protective relay: A protective relay is an automatic device which detects an abnormal condition in an electrical circuit and causes a circuit breaker to isolate the faulty element of the system. In some cases it may give an alarm or visible indication to alert operator.

Operating force or torque: A force or torque which tends to close the contacts of the relay.

Restraining force or torque: A force or torque which opposes the operating force/torque.

Actuating quantity: An electrical quantity (current, voltage, etc) to which relay responds.

Pick-up (level): The threshold value of the actuating quantity (current, voltage, etc.) above which the relay operates.

Reset on drop-out (level): The threshold value of the actuating quantity (current, voltage, etc.) below which the relay is de-energised and returns to its normal position or state. Consider a situation where a relay has closed its contacts and the actuating current is still flowing. Now, due to some reason, the abnormal condition is over and the current starts decreasing. At some maximum value of the current the contacts will start opening. This condition is called reset or drop-out. The maximum value of the actuating quantity below which contacts are opened is called the reset or drop-out value.

Operating time: It is the time which elapses from the instant at which the actuating quantity exceeds the relays pick-up value to the instant at which the relay closes its contacts.

Reset time: It is the time which elapses from the moment the actuating quantity falls below its reset value to the instant when the relay comes back to its normal (initial) position.

Setting: The value of the actuating quantity at which the relay is set to operate.

Seal-in relay: This is a kind of an auxiliary relay. It is energised by the contacts of the main relay. Its contacts are placed in parallel with those of the main relay and is designed to relieve the contacts of the main relay from their current carrying duty. It remains in the circuit until the circuit breaker trips. The seal-in contacts are usually heavier than those of the main relay.

Reinforcing relay: This is a kind of an auxiliary relay. It is energised from the contacts of the main relay. Its contacts are placed in parallel with those of the main relay and it is also designed to relieve the main relay contacts from their current carrying duty. The difference between a reinforcing relay and a seal-in relay is that the latter is designed to remain in the circuit till the circuit breaker operates. But this is not so with the reinforcing relay. The reinforcing relay is used to hold a signal from the initiating relay (main relay) for a longer period. As the contacts of the main relay are not robust, they are closed for a short time.

Back-up relay: A back-up relay operates after a slight delay, if the main relay fails to operate.

Back-up protection: The back-up protection is designed to clear the fault if the primary protection fails. It acts as a second line of defence.

Primary protection: If a fault occurs, it is the duty of the primary protective scheme to clear the fault. It acts as a first line of defence. If it fails, the back-up protection clears the fault.

Measuring relay: It is the main protective relay of the protective scheme, to which energising quantities are applied. It performs measurements to detect abnormal conditions in the system to be protected.

Auxiliary relays: Auxiliary relays assist protective relays. They repeat the operations of protective relays, control switches, etc. They relieve the protective relays of duties like tripping, time lag, sounding and alarm, etc. They may be instantaneous or may have a time delay.

Electromagnetic relay: A relay which operates on the electromagnetic principle, i.e., an electromagnet attracts magnetic moving parts (e.g.,) plunger type moving iron type, attracted armature type). Such a relay operates principally by action of an electromagnetic element which is energized by the input quantity.

Electromachanical relay: An electrical relay in which the designed response is developed by the relative movement of mechanical elements under the action of a current in the input circuit. Such relay operates by physical movement of mechanical parts resulting from electromagnetic or electrothermic forces created by the input quantities.

Electrodynamic relay: A relay which has two or more coils and- operates due to interaction of fluxes produced by the individual coils

Ferrodynamic relay: A relay in which the electrodynamic action is reinforced by pieces of ferromagnetic material placed in the path of magnetic lines of force.

Static relays: These are solid state relays and employ semiconductor diodes, transistors, thyristors, logic gates, ICs, etc. The measuring circuit is a static circuit and there are no moving parts. In some static relays, a slave relay which is a dc potarised relay is used as the tripping device.

Analog relay: An analog relay is that in which the measured quantities are converted into lower voltage but similar signals, which are then combined or compared directly to reference values in level detectors to produce the desired output.

Digital relay: A digital relay is that in which the measured ac quantities are manipulated in analog form and subsequently converted into either square-wave voltages or digital form. Logic circuits or microprocessors compare either the phase relationships of the square waves or the magnitudes of the quantities in digital form to make a trip decision.

Numerical relay: A numerical relay is that in which the measured ac quantities are sequentially sampled and converted into numerical (digital) data form. A microprocessor or a microcontroller processes the data numerically (i.e., performs mathematical and/or logical operations on the data) using an algorithm to calculate the fault discriminants and make trip decisions.

Microprocessor-based relay: A microprocessor is used to perform all functions of a relay. It measures electrical quantities, makes comparisons, performs computations, and sends tripping signals. It can realise all sorts of relaying characteristics, even irregular curves which cannot be reslised by electromechanical or static relays easily.

Microcontroller-based relay: A microcontroller is used for performing all the function of the relay. It measures the electrical quantities by acquiring them in digital form through a data acquisition system, makes comparisons, processes the digital data to calculate the fault discriminants and make trip decisions. It can realise all sorts of relaying characteristics.

DSP-based relay: A Digital Signal Processor (DSP) is used to perform all the functions of a relay.

FPGA-based relay: A Field Programmable Gate Array (FPGA) is used to perform all the functions of a relay. It acquires the signals, processes them to calculate the fault discriminants and make trip decisions.

ANN-based relay: An Artificial Neural Network (ANN) is used for processing the relaying signals (current and voltage signals) and making trip decisions.

Overcurrent relay: A relay which operates when the actuating current exceeds a certain preset value (its pick-up value).

Undervoltage relay: A relay which operates when the system voltage falls below a certain preset value.

Directional or reverse power relay: A directional relay is able to detect whether the point of fault lies in the forward or reverse direction with respect to the relay location. It is able to sense the direction of power flow, i.e. whether the power is flowing in the normal direction or the reverse direction.

Polarised relay: A relay whose operation depends on the direction of current or voltage.

Flag or target: Flag is a device which gives visual indication whether a relay has operated or not.

Time-lag relay: A time-lag relay operates after a certain preset time lag. The time lag may be due to its inherent design feature or may be due to the presence of a time-delay component. Such relays are used in protection schemes as a means of time discrimination. They are frequently used in control and alarm schemes.

Instantaneous relay: An instantaneous relay has no intentional time delay in its operation. It operates in 0.1 second. Sometimes the terms high set or high speed relays are also used for the relays which have operating times less than 0.1 second.

Inverse time relay: A relay in which the operating time is inversely proportional to the magnitude of the operating current.

Definite time relay: A relay in which the operating time is independent of the magnitude of the actuating current.

Inverse Definite Minimum Time (IDMT) Relay: A relay which gives an inverse time characteristic at lower values of the operating current and definite time characteristic at higher values of the operating current.

Induction relay: A relay which operates on the principle of induction. Examples are induction disc relays, induction cup relays etc.

Moving coil relay: This type of a relay has a permanent magnet and a moving coil. It is also called a permanent magnet d.c. moving coil relay. The actuating current flows in the moving coil.

Moving iron relay: This is a dc polarised, moving iron type relay. There is an electromagnet, permanent magnet and a moving armature in its construction.

Printed disc relay: This relay operates on the principle of a dynamometer. There is a permanent magnet or an electromagnet and a printed disc. Direct current is fed to the printed circuit of the disc.

Thermal relay: This relay utilises the electrothermal effect of the actuating current for its operation.

Distance relay: A relay which measures impedance or a component of the impedance at the relay location is known as a distance relay. It is used for the protection of a transmission line. As the impedance of a line is proportional to the length of the line, a relay which measures impedance or its component is called a distance relay.

Impedance relay: A relay which measures impedance at the relay location is called an impedance relay. It is a kind of a distance relay.

Modified impedance relay: It is an impedance relay having shifted characteristics. The voltage coil includes some current biasing.

Reactance relay: A relay which measures reactance at the relay location is called a reactance relay. It is a kind of a distance relay.

MHO relay (admittance or angle admittance): This is a kind of a distance relay. It measures a particular component of the impedance, i.e. $\dfrac{Z}{\cos(\phi-\theta)}$, where ϕ is the power factor angle and θ is the design angle to shift MHO characteristic on the R-X diagram. Its characteristic on the R-X diagram is a circle passing through the origin. It is a directional relay. It is also known as an admittance or angle admittance relay.

Conduction relay: This is a MHO relay whose diameter (passing through the origin) lies on the R-axis.

Offset MHO characteristic: In an offset MHO relay, the MHO characteristic is shifted on the R-X diagram to include the origin.

Angle impedance relay (ohm relay): The characteristic of this relay on the R-X diagram is a straight line passing at an angle and cutting both the axes. It is a kind of a distance relay and is also called an Ohm relay.

Elliptical relay: The characteristic of an elliptical relay on the R-X diagram is an ellipse. This is also a kind of a distance relay.

Quadrilateral relay: The characteristic of a quadrilateral relay on the R-X diagram is a quadrilateral. This too is a kind of a distance relay.

Frequency sensitive relay: This is a relay which operates at a predetermined value of the system frequency. It may be an under-frequency relay or an over-frequency relay. An under-frequency relay operates when the system frequency falls below a certain value. An over-frequency relay will operate when the system frequency exceeds a certain preset value of the frequency.

Differential relay: A relay which operate in response to the difference of two actuating quantities.

Earth fault relay: A relay used for the protection of an element of a power system against earth faults is known as an earth fault relay.

Phase fault relay: A relay used for the protection of an element of a power system against phase faults is called a phase fault relay.

Negative sequence relay: A relay for which the actuating quantity is the negative sequence current. When the negative sequence current exceeds a certain value, the relay operates. This type of a relay is used to protect electrical machines against overheating due to unbalanced currents.

Zero sequence relay: A relay for which the actuating quantity is the zero sequence current. This type of a relay is used for earth fault protection.

Starting relay or fault detector: This is a relay which detects abnormal conditions and initiates the operation of other elements of the protective scheme.

Notching relay: A relay which switches in response to a specific number of applied impulses is called a notching relay.

Regulating relay: A regulating relay is activated when an operating parameter deviates from predetermined limits. This relay functions through supplementary equipment to restore the quantity to the prescribed limits.

Monitoring relay: A monitoring relay verifies conditions on the power system or in the protection system. This relay includes fault detectors, alarm units, channel-monitoring relays, synchronism verification, and network phasing. Power system conditions that do not involve opening circuit breakers during faults can be monitored by monitoring (verification) relays.

Synchronizing (or synchronism check) relay: This relay assures that proper conditions exist for interconnecting two sections of power system.

Biased relay: A relay in which the characteristics are modified by the introduction of some quantity other than the actuating quantity, and which is usually in opposition to the actuating quantity. The setting of this relay is modified by additional windings the amount of bias being dependent upon conditions in the protected circuit.

Primary relay: A relay which is directly connected to the protected circuit without interposing instrument transformers or shunt.

Secondary relay: A relay connected to the protected circuit through current and voltage transformers.

Sequential relay: A relay which instantaneously transfers its contact position from a particular combination of 'off' and 'on' position to another combination, every time it picks up or drops off, according to a predetermined programme which may or may not be adjustable and repetitive

Indicating relay: A measuring or auxiliary relay which displays a signal on energisation

Reclosing relay: This relay establishes a closing sequence for a circuit breaker following tripping by protective relays.

Supervisory relay: A measuring relay or a combination of measuring and auxiliary relays in a unit with a definite purpose of supervision.

Change over relay: An auxiliary relay with two positions either of which cannot be designated as 'on' and 'off' but which transfers the contact circuits from one connection to the other.

Two-step relay: A relay with two sets of contacts, one of them operates at a certain value of the characteristic quantity and the other after a further change of the quantity.

Current unbalance relay: This relay operates when the currents in a polyphase system are unbalanced by a predetermined amount.

Voltage unbalanced relay: This relay operates when the voltages in a poly phase system are unbalanced by a predetermined amount.

Ferraris relay: This relay moves by the interaction of the magnetic field of a coil and the currents induced in a metal body (disc or cup)

Protective zone: A power system is divided into a number of zones from the protection point of view. Each element of the power system has a separate protective scheme for its protection. The elements which come under a protective scheme are said to be in the zone of protection of that particular scheme. Similarly, a protective relay has its own zone of protection.

Reach: This term is mostly used in connection with distance relays. A distance relay operates when the impedance (or a component of the impedance) as seen by the relay is less than a preset value. This preset impedance (or a component of impedance) or corresponding distance is called the reach of the relay. In other words, it is the maximum length of the line up to which the relay can protect.

Overreach: Sometimes a relay may operate even when a fault point is beyond its present reach (i.e. its protected length).

Underreach: Sometimes a relay may fail to operate even when the fault point is within its reach, but it is at the far end of the protected line. This phenomenon is called underreach.

Selectivity or discrimination: It is the ability of a relay to discriminate between faulty conditions and normal conditions (or between a fault within the protected section and outside the protected section). In other words, it is the quality of the protective system by which it distinguishes between those conditions for which it should operate and those for which it should not.

Reliability: A protective relay must operate reliably when a fault occurs. The reliability of a protective relay should be very high, a typical value being 95%.

Sensitivity: A protective relay should be sensitive enough to operate when the magnitude of the actuating quantity exceeds its pick-up value.

Stability: This is the ability of the protective system to remain inoperative under all load conditions, and also in case of external faults. The relay should remain stable when a heavy current due to an external fault is flowing through it.

Fast operation: A protective relay should be fast enough to cause the isolation of the faulty section as quickly as possible to minimise the damage and to maintain the stability.

Burden: The power consumed by the relay circuitry at the rated current is known as its burden.

Blocking: The prevention of tripping of the relay is called blocking. It may be due to the operation of an additional relay or due to its own characteristic.

Energizing quantity: The electrical quantity, i.e., either current or voltage, which alone or in combination with other electrical quantities, must be applied to the relay for its functioning.

Characteristic quantity: A quantity, the value of which characterizes the operation of the relay for example, current for an overcurrent relay, voltage for a voltage relay, phase angle for a directional relay, time for an independent time delay relay, impedance for an impedance relay, frequency for a frequency relay.

Characteristic angle: The angle between the phasors representing two of the energizing quantities applied to a relay and used for the declaration of the performance of the relay.

System Impedance Ratio (SIR): The ratio of the power system source impedance to the impedance of the protected zone.

Characteristic Impedance Ratio (CIR): The maximum value of the system impedance ratio up to which the relay performance remains within the prescribed limits of accuracy.

Power swing: Oscillation between groups of synchronous machines caused by an abrupt change in load conditions.

Through fault current: The current flowing through a protected zone to a fault beyond that zone.

Unit system of protection: A unit system of protection is one which is able to detect and respond to faults occurring only within its own zone of protection. It is said to have absolute discrimination. Its zone of protection is well defined. It does not respond to the faults occurring beyond its own zone of protection. Examples of unit protection are differential protection of alternators, transformers or bus bars, frame leakage protection, pilot wire and carrier current protection.

Non-unit system of protection: A non-unit system of protection does not have absolute discrimination (selectivity). It has dependent or relative discrimination. The discrimination is obtained by time grading, current grading or a combination of current and time grading of the relays of several zones. In this situation, all relays may respond to a given fault. Examples of non-unit system of protection are distance protection and time graded, current graded or both time and current graded protection.

Restricted earth fault protection: This is an English term which may be misunderstood in other countries. It is used in the context of transformer or alternator. It refers to the differential protection of transformers or alternators against ground faults. It is called restricted because its zone of protection is restricted only to the winding of the alternator or transformer. The scheme responds to the faults occurring within its zone of protection. It does not respond to faults beyond its zone of protection.

Unrestricted protection: A protection system which has no clearly defined zone of operation and which achieves selective operation only by time grading.

Protective gear or equipment: It includes transducers (CTs and VTs), protective relays, circuit breakers and ancillary equipment to be used in a protective system.

Protective system: It is a combination of Protective gear equipment to secure isolation of the faulty element under predetermined conditions, usually abnormal or to give an alarm signal or both.

Protective scheme: A protective scheme may consist of several protective systems. It is designed to protect one or more elements of a power system.

Residual current: It is the algebraic sum of all currents in a multiphase system. It is denoted by I_{res}. In a 3-phase system $I_{res} = I_A + I_B + I_C$.

Transducers or instrument transformers: Current and voltage transformers (CTs and VTs) are collectively known as transducers or instrument transformers. They are used to reduce currents and voltages to standard lower values and to isolate protective relays and measuring instruments from the high voltages of the power system.

Switchgear: It is a general term covering switching and interrupting devices and their combination with associated control, metering, protective and regulating devices, also assemblies of these devices with associated inter-connections, accessories, enclosures and supporting structures used primarily in connection with generation, transmission, distribution, and conversion of electric power.

Circuit breaker: It is a mechanical switching device capable of making, carrying and breaking currents under normal circuit conditions and also making, carrying for a specified time, and automatically breaking currents under specified abnormal circuit conditions such as those of short circuit. The medium in which circuit interruption is performed may be designated by suitable prefix, such as, oil-circuit breaker, air-blast circuit breaker, air-break circuit breaker, sulphur hexafluoride circuit breaker, vacuum circuit breaker, etc.

Adaptive relaying: Adaptive relaying is defined as the protection system whose settings can be changed automatically so that it is attuned to the prevailing power system conditions.

Pilot wire: An auxiliary conductor used in connection with remote measuring devices or for operating apparatus at a distant point.

Pilot protection: A form of line protection that uses a communication channel as a means to compare electrical conditions at the terminals of a line.

Pilot wire protection: Pilot protection in which a metallic circuit is used for the communicating means between relays at the circuit terminals.

EXERCISES

1. Explain the nature and causes of faults. Discuss the consequences of faults on a power system.
2. What are the different types of faults? Which type of fault is most dangerous?
3. Discuss briefly the role of protective relays in a modern power system.
4. What do you understand by a zone of protection? Discuss various zones of protection for a modern power system.
5. Explain what you understand by primary and back-up protection. What is the role of back-up protection? What are the various methods of providing back-up protection?
6. Explain what you understand by pick-up and reset value of the actuating quantity.
7. Discuss what you understand by selectivity and stability of protective relay.
8. Discuss the essential qualities of a protective relay.
9. How is the relay performance classified? What indication the correct operation of the relay gives? What do you mean by "correct but unwanted operation"?
10. What do you understand by incorrect operation of the protective relay? What are the reasons of incorrect operation?
11. Discuss the classification of protective relays based on their speed of operation.
12. Differentiate between a digital relay and a numerical relay.
13. What is a numerical relay? Discuss its advantages over conventional relays of electromechanical and static types. How can an intelligent numerical relay be developed.
14. What are the various components of a protection system? Briefly describe their functions with the help of an schematic diagram.
15. Differentiate between a protective system and a protective scheme.
16. What do you understand by adaptive relaying? How can a relay be made adaptive?
17. Briefly describe the following types of relay.
    (i) Monitoring relay (ii) Regulating relay (iii) Auxiliary relay (iv) Synchronizing relay, and (v) Biased relay.

2 Relay Construction and Operating Principles

2.1 INTRODUCTION

The proper operation of the power system requires an efficient, reliable and fast-acting protection scheme, which basically consists of protective relays and switching devices. A protective relay, acting as a brain behind the whole system, senses the fault, locates it, and sends a command to appropriate circuit breaker to isolate only the faulty section, thus keeping the rest of the healthy system functional. It detects abnormal conditions on a power system by constantly monitoring the electrical quantities of the system, which are different under normal and abnormal (fault) conditions. The basic electrical quantities which are likely to change during abnormal conditions are current, voltage, phase angle (direction) and frequency. Protective relays utilize one or more of these quantities to detect abnormal conditions on a power system.

The basic relay circuit is illustrated in Fig. 2.1. There are two ways in which the circuit breaker trip coil is energized. In one method, the station battery is used to supply the current in the trip coil after the relay contacts are closed by the operation of the relay. In the second method, as soon as the relay operates, the CT secondary current flows through the trip coil and energizes it. This method does not require a station battery and it is used for the protection of feeders.

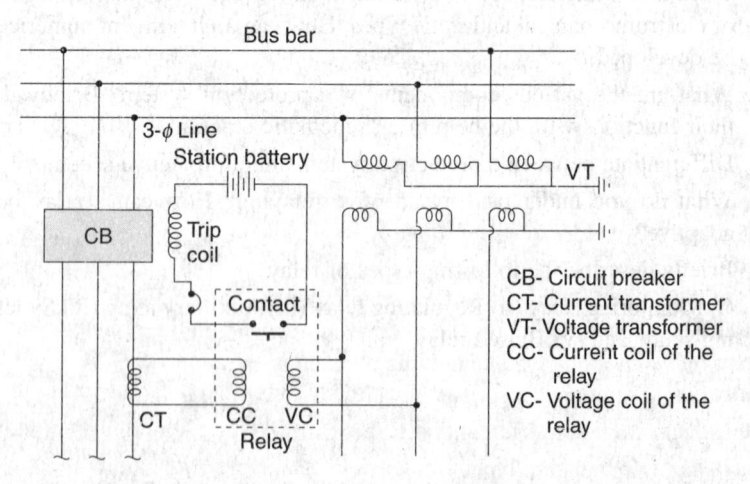

Fig. 2.1 *Relay connection*

CB- Circuit breaker
CT- Current transformer
VT- Voltage transformer
CC- Current coil of the relay
VC- Voltage coil of the relay

Protective relays are broadly classified into the following three categories depending on the technologies they use for their construction and operation.

(i) Electromechanical relays
(ii) Static relays
(iii) Numerical relays

There are various types of protective relays in each category, depending on the operating principle and application.

2.2 ELECTROMECHANICAL RELAYS

Electromechanical relays operate by mechanical forces generated on moving parts due to electromagnetic or electrothermic forces created by the input quantities. The mechanical force results in physical movement of the moving part which closes the contacts of the relay for its operation. The operation of the contact arrangement is used for relaying the operated condition to the desired circuit in order to achieve the required function. Since the mechanical force is generated due to the flow of an electric current, the term 'electromechanical relay' is used.

Most electromechanical relays use either electromagnetic attraction or electromagnetic induction principle for their operation. Such relays are called *electromagnetic relays*. Depending on the principle of operation, the electromagnetic relays are of two types, i.e., (i) attracted armature relays, and (ii) induction relays. Some electromechanical relays also use electrothermic principle for their operation and are based upon the forces created by expansion of metals caused by temperature rise due to flow of current. Such relays are called *thermal relays*. Most of the present day electromechanical relays are of either induction disc type or induction cup type.

The following are the principal types of electromechanical relays:

1. Electromagnetic relays
   (i) Attracted armature relays, and
   (ii) Induction relays
2. Thermal relays

2.2.1 Attracted Armature Relays

Attracted armature relays are the simplest type which respond to ac as well as dc. These relays operate through an armature which is attracted to an electromagnet or through a plunger which is drawn into a solenoid. All these relays use the same electromagnetic attraction principle for their operation. The electromagnetic force exerted on the moving element, i.e., the armature or plunger, is proportional to the square of the flux in the air gap or the square of the current. In dc relays this force is constant. In case of ac relays, the total electromagnetic force pulsates at double the frequency. The motion of the moving element is controlled by an opposing force generally due to gravity or a spring.

The following are the different types of construction of attracted armature relays.

(i) Hinged armature type
(ii) Plunger type

(iii) Balanced beam type
(iv) Moving-coil type
(v) Polarised moving-iron type
(vi) Reed type

Hinged Armature-Type Relays

Figure 2.2(a) shows a hinged armature-type construction. The coil is energised by an operating quantity proportional to the system current or voltage. The operating quantity produces a magnetic flux which in turn produces an electromagnetic force. The electromagnetic force is proportional to the square of the flux in the air gap or the square of the current. The attractive force increases as the armature approaches the pole of the electromagnet. This type of a relay is used for the protection of small machines, equipment, etc. It is also used for auxiliary relays, such as indicating flags, slave relays, alarm relays, annunciators, semaphores, etc.

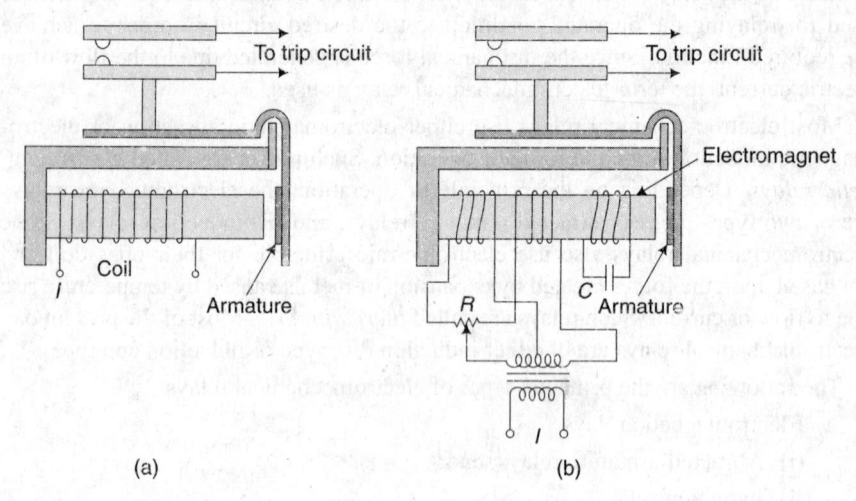

Fig. 2.2 *(a) Hinged armature-type relay (b) Modified hinged armature-type relay*

The actuating quantity of the relay may be either ac or dc. In dc relay, the electromagnetic force of attraction is constant. In the case of ac relays, sinusoidal current flows through the coil and hence the force of attraction is given by

$$F = K I^2 = K (I_{max} \sin \omega t)^2 = \frac{1}{2} K (I_{max}^2 - I_{max}^2 \cos 2\omega t)$$

From the above expression, it is evident that the electromagnetic force consists of two components. One component is constant and is equal to ½ $K I_{max}^2$. The other component is time dependent and pulsates at double the frequency of the applied ac quantity. Its magnitude is ½ $K I_{max}^2 \cos 2\omega t$. The total force is a double frequency pulsating force. This may cause the armature to vibrate at double the frequency. Consequently, the relay produces a humming sound and becomes noisy. This difficulty can be overcome by making the pole of the electromagnet of shaded construction. Alternatively, the electromagnet may be provided with two coils. One coil is energised with the actuating quantity. The other coil gets its supply through a phase shifting circuit.

The restraining force is provided by a spring. The reset to pick-up ratio for attracted armature type relays is 0.5 to 0.9. For this type of a relay, the ratio for ac relays is higher as compared to dc relays. The VA burden is low, which is 0.08 W at pick-up for the relay with one contact, 0.2 W for the relay with four contacts. The relay is an instantaneous relay. The operating speed is very high. For a modern relay, the operation time is about 5 ms. It is faster than the induction disc and cup type relays. Attracted armature relays are compact, robust and reliable. They are affected by transients as they are fast and operate on both dc and ac. The fault current contains a dc component in the beginning for a few cycles. Due to the presence of dc transient, the relay may operate though the steady state value of the fault current may be less than its pick-up. A modified construction as shown in Fig. 2.2(b) reduces the effect of dc transients.

Plunger-Type Relays

Figure 2.3 shows a plunger-type relay. In this type of a relay, there is a solenoid and an iron plunger which moves in and out of the solenoid to make and break the contact. The movement of the plunger is controlled by a spring. This type of construction has however become obsolete as it draws more current.

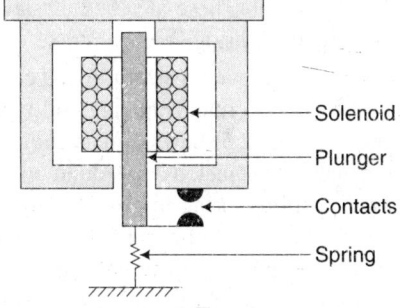

Fig. 2.3 Plunger-type relay

Balanced Beam Relays

Figure 2.4 shows a balanced beam relay which is also a kind of attracted armature type relay. As its name indicates, it consists of a beam carrying two electromagnets at its ends. One gives operating torque while the other retraining torque. The beam is supported at the middle and it remains horizontal under normal conditions. When the operating torque exceeds the restraining torque, an armature fitted at one end of the beam is pulled and its contacts are closed. Though now obsolete, this type of a relay was popular in the past for constructing impedance and differential relays. It has been superseded by rectifier bridge comparators and permanent magnet moving coil relays. The beam type relay is robust and fast in operation, usually requiring only 1 cycle, but is not accurate as it is affected by dc transients.

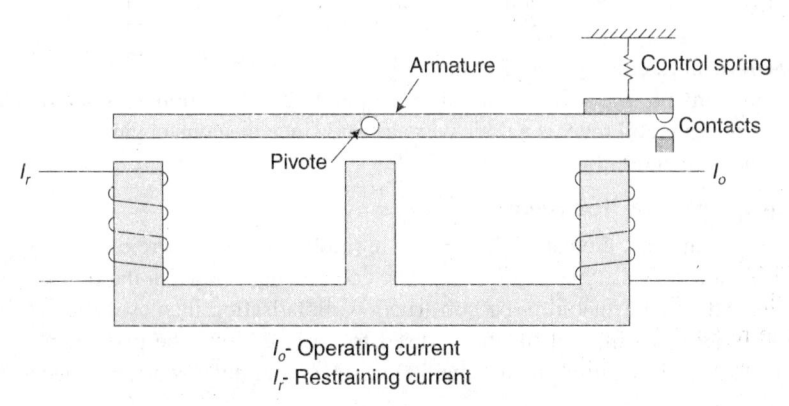

I_o- Operating current
I_r- Restraining current

Fig. 2.4 Balanced beam relay

Moving Coil Relays

Figure 2.5 shows a permanent magnet moving coil relay. It is also called a polarised dc moving coil relay. It responds to only dc actuating quantities. It can be used with ac actuating quantities in conjunction with rectifiers. Moving coil relays are most sensitivity type electromagnetic relays. Modern relays have a sensitivity of 0.1 mW. These relays are costlier than induction cup or moving iron type relays. The VA burden of moving coil relays is very small. These are used as slave relays with rectifier bridge comparators.

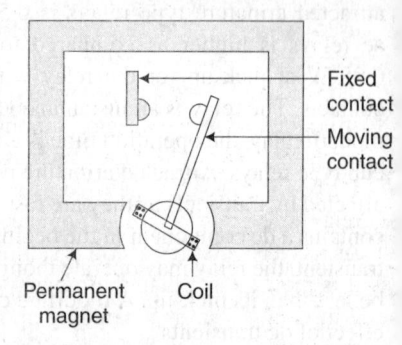

Fig. 2.5 *Rotating moving coil relays*

There are two types of moving coil relays: rotary moving coil and axially moving coil type. The rotary moving coil type is similar to a moving coil indicating instrument. Figure 2.5 shows a rotary moving coil type construction. The components are: a permanent magnet, a coil wound on a non-magnetic former, an iron core, a phosphor bronze spiral spring to provide resetting torque, jewelled bearing, spindle, etc. The moving coil assembly carries an arm which closes the contact. Damping is provided by an aluminum former. The operating time is about 2 cycles. A copper former can be used for heavier damping and slower operation. The operating torque is produced owing to the interaction between the field of the permanent magnet and that of the coil. The operating torque is proprtional to the current carried by the coil. The torque exerted by the spring is proportional to deflection. The relay has an inverse operating time/current characteristic.

Figure 2.6 shows an axially moving coil type construction. As this type has only one air gap, it is more sensitive than the rotary moving coil relay. It is faster than the rotary moving coil relay because of light parts. An operating time of the order of 30 msec can be obtained. Sensitivities as low as 0.1 mW can be obtained. Its coils are wound on a cylindrical former which is suspended horizontally. The coil has only axial movement. The relay has an inverse operating time/current characteristic. The axially moving coil relay is a delicate relay and since the contact gap is small, it has to be handled carefully.

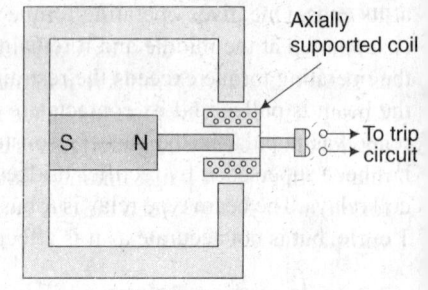

Fig. 2.6 *Axial moving coil relay*

Polarised Moving Iron Relays

Figure 2.7 shows a typical polarised moving iron relay. There are different types of constructions of this type (see Ref. 1). The construction shown in the figure is a flux shifting attracted armature type construction. Polarisation increases the sensitivity of the relay. A permanent magnet is used for polarisation. The permanent magnet produces flux in addition to the main flux. It is a dc polarised relay, meant to be used with dc only. However, it can be used with ac with rectifiers. Modern relays have sensitivities in the range of 0.03 to 1mW, depending on their construction. Using transistor amplifiers, a relay's sensitivity can be increased to 1 μW for pick-up. It is used as a slave relay with rectifier bridge comparators. As its current carrying coil is stationary, it is more robust than the moving coil type dc polarised relay. Its operating time is 2 msec to 15 msec depending upon the type of construction. An ordinary attracted armature type relay is not sensitive to the polarity of the actuating quantity whereas a dc polarised relay will only operate when the input is of the correct polarity.

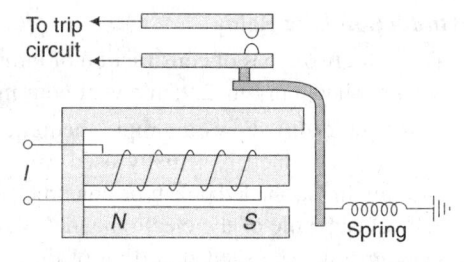

Fig. 2.7 Polarised moving iron relay

Reed Relays

A reed relay consists of a coil and nickel-iron strips (reeds) sealed in a closed glass capsule, as shown in Fig. 2.8. The coil surrounds the reed contact. When the coil is energised, a magnetic field is produced which causes the reeds to come together and close the contact. Reed relays are very reliable and are maintenance free. As far as their construction is concerned, they are electromagnetic relays. But from the service point of view, they serve as static relays. They are used for control and other purposes.

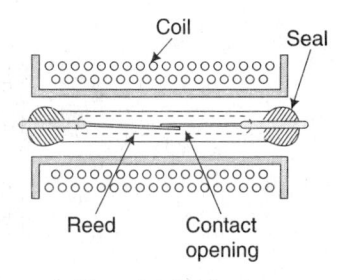

Fig. 2.8 Reed relay

They can also be used as a protective relays. They are quite suitable to be used as slave relays. Their input requirement is 1 W to 3 W and they have speed of 1 or 2 msec. They are completely bounce free and are more suitable for normally-closed applications. Heavy duty reed relays can close contacts carrying 2 kW at 30 A maximum current or at a maximum of 300 V dc supply. The voltage withstand capacity for the insulation between the coil and contacts is about 2 kV. The open contacts can withstand 500 V to 1 kV.

2.2.2 Induction Relays

Induction relays use electromagnetic induction principle for their operation. Their principle of operation is same as that of a single-phase induction motor. Hence they can be used for ac currents only. Two types of construction of these Relays are fairly standard: one with an induction disc and the other with an induction cup. In both types of relays, the moving element (disc or cup) is equivalent to the rotor of the induction motor. There is one contrast from the induction motor, i.e., the iron associated with the rotor in the relay is stationary. The moving element acts as a carrier of rotor currents, whereas the magnetic circuit is completed through stationary magnetic elements. Two sources of alternating magnetic flux in which the moving element may turn are required for the operation of induction-type relays. In order to produce an operating torque, the two fluxes must have a phase difference between them.

Induction Disc Relay

There are two types of construction of induction disc relays, namely the shaded pole type, as shown in Fig. 2.9; and watt hour meter type, as shown in Fig. 2.10.

Figure 2.9(a) shows a simple theoretical figure, whereas Fig. 2.9(b) shows the construction which is actually used in practice. The rotating disc is made of aluminium. In the shaded pole type construction, a C-shaped electromagnet is used. One half of each pole of the electromagnet is surrounded by a copper band know as the shading ring. The shaded portion of the pole produces a flux which is displaced in space and time with respect to the flux produced by the unshaded portion of the pole. Thus two alternating fluxes displaced in space and time cut the disc and produce eddy currents in it. Torques are produced by the interaction of each flux with the eddy current produced by the other flux. The resultant torque causes the disc to rotate.

(a) Simple construction

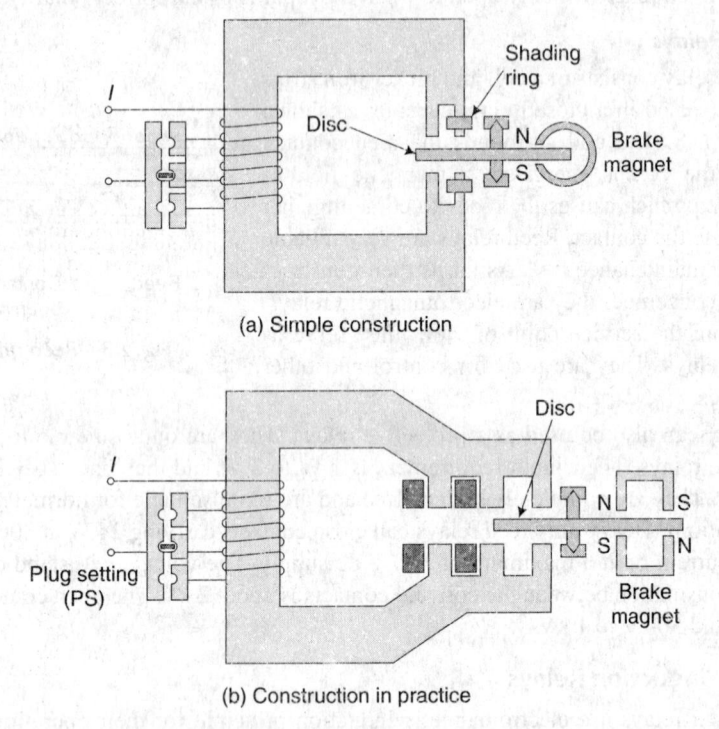

(b) Construction in practice

Fig. 2.9 *Shaded pole type induction disc relay*

In wattmetric type of construction, two electromagnets are used: upper and lower one. Each magnet produces an alternating flux which cuts the disc. To obtain a phase displacement between two fluxes produced by upper and lower electromagnets, their coils may be energised by two different sources. If they are energised by the same source, the resistances and reactances of the two circuits are made different so that there will be sufficient phase difference between the two fluxes.

Induction disc type construction is robust and reliable. It is used for overcurrent protection. Disc type units gives an inverse time current characteristic and are slow compared to the induction cup and attracted armature type relays. The induction disc type is used for slow-speed relays. Its operating time is adjustable and is employed where a time-delay is required. Its reset/pick-up ratio is high, above 95% because its operation does not involve any change in the air gap. The VA burden depends on its application, and is generally of the order of 2.5 VA. The torque is proportional to the square of the actuating current if single actuating quantity is used.

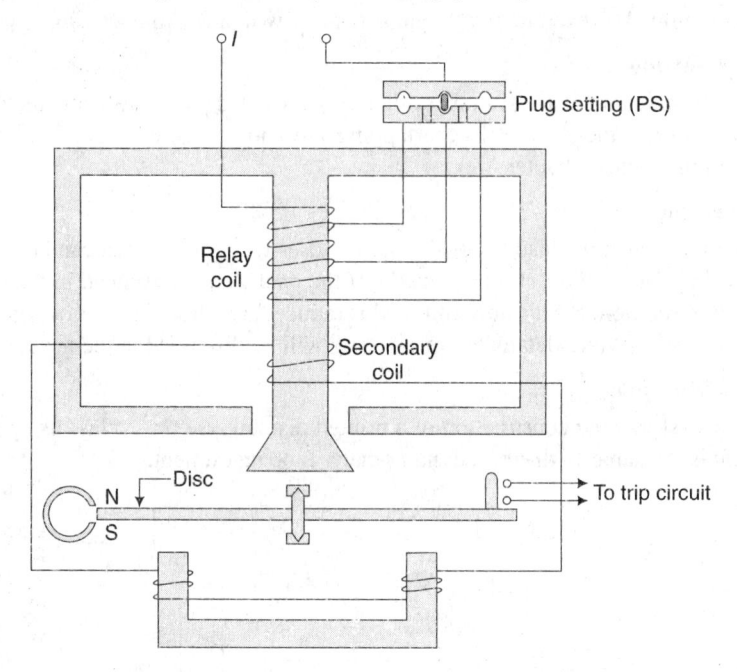

Fig. 2.10 *Wattmetric type induction-disc relay*

A spring is used to supply the resetting torque. A permanent magnet is employed to produce eddy current braking to the disc. The magnets should remain stable with age so that its accuracy will not be affected. Magnets of high coercive force are used for the purpose. The braking torque is proportional to the speed of the disc. When the operating current exceeds pick-up value, driving torque is produced and the disc accelerates to a speed where the braking torque balances the driving torque. The disc rotates at a speed proportional to the driving torque.

It rotates at a constant speed for a given current. The disc inertia should be as small as possible, so that it should stop rotating as soon as the fault current disappears when circuit breaker operates at any other location or fault current is for a short moment (i.e. transient in nature). After the cessation of the fault current, the disc will travel to some distance due to inertia. This distance should be minimum. It is called the over-run of the disc. A brake magnet is used to minimise over-run. The over-run is usually not more than 2 cycles on the interruption of a current which is 20 times the current setting.

At a current below pick-up value, the disc remains stationary by the tension of the control spring acting against the normal direction of disc rotation. The disc rests against a backstop. The position of the backstop is adjustable and therefore, the distance by which the moving contact of the relay travels before it closes contacts, can be varied. The distance of travel is adjusted for the time setting of the relay.

The rotor (disc) carries an arm which is attached to its spindle. The spindle is supported by jewelled bearings. The arm bridges the relay contacts. In earlier constructions, there were two contacts which were bridged when the relay operated. In modern units however, there is a single contact with a flexible lead-in.

Current Setting

In disc type units, there are a number of tapping provided on coil to select the desired pick-up value of the current. These tapping are shown in Fig. 2.10. This will be discussed in the next chapter.

Time Setting

The distance which the disc travels before it closes the relay contact can be adjusted by adjusting the position of the backstop. If the backstop is advanced in the normal direction of rotation, the distance of travel is reduced, resulting in a shorter operating time of the relay. More details on time-setting will be discussed in the next chapter.

Printed Disc Relay

Figure 2.11 shows the construction of a printed disc inverse time relay. Its operating principle is the same as that of a dynamometer type instrument.

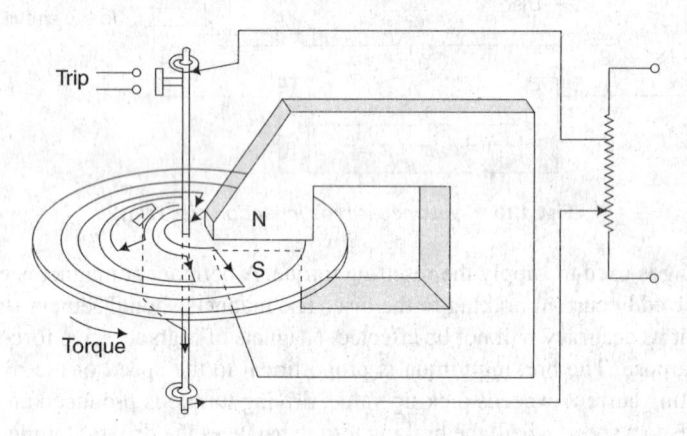

Fig. 2.11 *Printed disc inverse time relay*

There is a permanent magnet to produce a magnetic field. The current from the CT is fed to the printed disc through a rectifier. When a current carrying conductor is placed in a magnetic field, a force is developed, thereby a torque is exerted on it. On this very principle, torque is produced in a printed disc relay.

Figure 2.12 shows the construction of a printed disc extremely inverse time relay ($I^2t = K$ relay). To obtain $I^2t = K$ characteristic, an electromagnet and a printed disc are used. The electromagnet is energised from the CT through a rectifier.

Printed disc relays give a much more accurate time characteristic. They are also very efficient. A printed disc relay is 50 to 100 times more efficient than the induction disc type. The maximum efficiency that an induction disc relay can have is only about 0.05%, which is extremely poor. Characteristics other than inverse time-current characteristic can be obtained by including a non-linear network in between the printed circuit of the disc and the rectified current input.

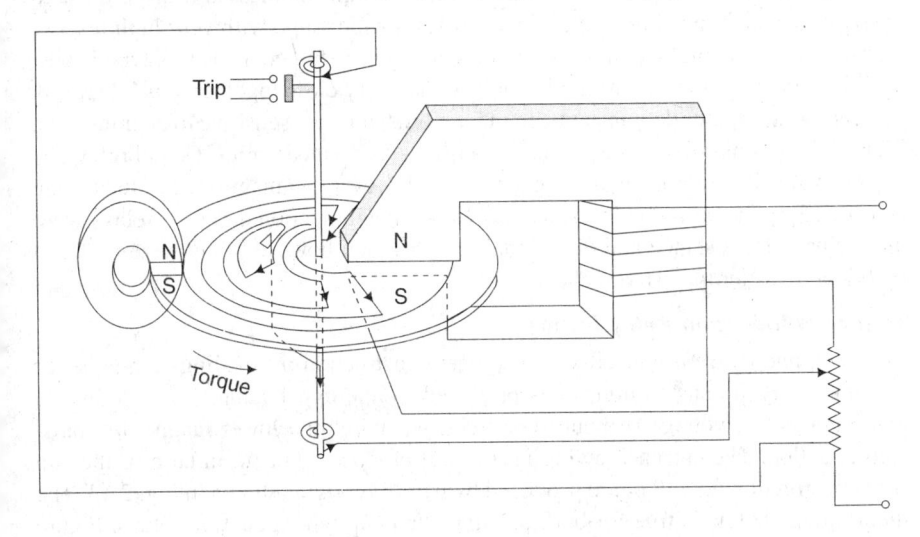

Fig. 2.12 *Printed disc extremely inverse time relay*

Induction Cup Relay

Figure 2.13 shows an induction cup relay. A stationary iron core is placed inside the rotating cup to decrease the air gap without increasing inertia. The spindle of the cup carries an arm which closes contacts. A spring is employed to provide a resetting torque. When two actuating quantities are applied, one may produce an operating torque while the other may produce restraining torque. Brake magnets are not used with induction cup type relays. It operates on the same principle as that of an induction motor. It employs a 4 or 8-pole structure.

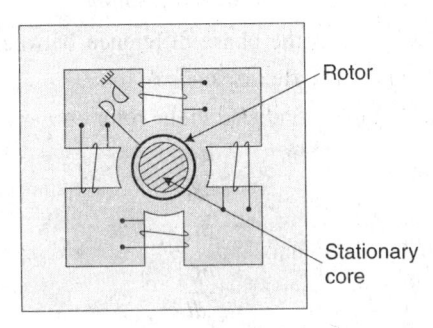

Fig. 2.13 *Induction cup relay*

The rotor is a hollow cylinder (inverted cup). Two pairs of coils, as shown in the figure, produce a rotating field which induces current in the rotor. A torque is produced due to the interaction between the rotating flux and the induced current, which causes rotation. The inertia of the cup is much less than that of a disc. The magnetic system is more efficient and hence the magnetic leakage in the magnetic circuit is minimum. This type of a magnetic system also reduces the resistance of the induced current path in the rotor. Due to the low weight of the rotor and efficient magnetic system its torque per VA is about three times that of an induction disc type construction. Thus, its VA burden is greatly reduced. It possesses high sensitivity, high speed and produces a steady non-vibrating torque. Its parasitic torques due to current or voltage alone are small. Its operating time is to the order of 0.01 second. Thus with its high torque/inertia ratio, it is quite suitable for higher speeds of operation.

Magnetic saturation can be avoided by proper design and the relay can be made to have its characteristics linear and accurate over a wide range with very high reset to pick-up ratio. The pick-up and reset values are close together. Thus this type is best suited where normal and abnormal conditions are very close together. It is inherently self compensating for dc transients. In other words, it is less sensitive to dc transients. The other system transients as well as transients associated with CTs and relay circuits can also be minimised by proper design. However, the magnitude of the torque is affected by the variation in the system frequency. Induction cup type relays were widely used for distance and directional relays. Later, however, they were replaced by bridge rectifier type static relays.

Theory of Induction Relay Torque

Fluxes ϕ_1 and ϕ_2 are produced in a disc type construction by shading technique. In watt-metric type construction, ϕ_1 is produced by the upper magnet and ϕ_2 by the lower magnet. A voltage is induced in a coil wound on the lower magnet by transformer action. The current flowing in this coil produces flux ϕ_2. In case of the cup type construction, ϕ_1 and ϕ_2 are produced by pairs of coils, as shown in Fig. 2.13. The theory given below is true for both disc type and cup type induction relays. Figure 2.14 shows how force is produced in a rotor which is cut by ϕ_1 and ϕ_2. These fluxes are alternating quantities and can be expressed as follows.

$$\phi_1 = \phi_{1m} \sin \omega t \qquad \phi_2 = \phi_{2m} \sin (\omega t + \theta)$$

where θ is the phase difference between ϕ_1 and ϕ_2. The flux ϕ_2 leads ϕ_1 by θ.

Voltages induced in the rotor are:

$$e_1 \propto \frac{d\phi_1}{dt}$$
$$\propto \phi_{1m} \cos \omega t$$
$$e_2 \propto \frac{d\phi_2}{dt}$$
$$\propto \phi_{2m} \cos (\omega t + \theta)$$

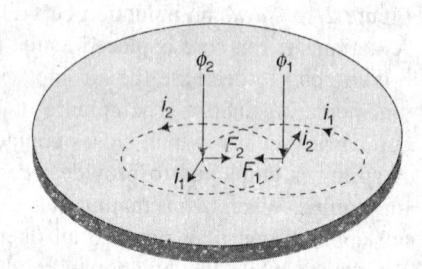

Fig. 2.14 Torque produced in an induction relay

As the path of eddy currents in the rotor has negligible self-inductance, with negligible error it may be assumed that the induced eddy currents in the rotor are in phase with their voltages.

$$i_1 \propto \phi_{1m} \cos \omega t$$
$$i_2 \propto \phi_{2m} \cos (\omega t + \theta)$$

The current produced by the flux interacts with other flux and vice versa. The forces produced are:

$$F_1 \propto \phi_1 \, i_2$$
$$\propto \phi_{1m} \sin \omega t \cdot \phi_{2m} \cos (\omega t + \theta)$$
$$\propto \phi_{1m} \, \phi_{2m} \cos (\omega t + \theta) \cdot \sin \omega t$$

$$F_2 \propto \phi_2 i_1$$
$$\propto \phi_{2m} \sin(\omega t + \theta) \cdot \phi_{1m} \cos \omega t$$
$$\propto \phi_{1m} \phi_{2m} \sin(\omega t + \theta) \cdot \cos \omega t$$

As these forces are in opposition, the resultant force is

$$F = (F_2 - F_1)$$
$$\propto \phi_{1m} \phi_{2m} [\sin(\omega t + \theta) \cos \omega t - \cos(\omega t + \theta) \cdot \sin \omega t]$$
$$\propto \phi_{1m} \phi_{2m} \sin \theta$$

The suffix m is usually dropped and the expression is written in the form of $F = K\phi_1 \phi_2 \sin \theta$. In this expression, ϕ_1 and ϕ_2 are rms values.

If the same current produces ϕ_1 and ϕ_2 the force produced is given by

$$F = K I^2 \sin \theta$$

where θ is the angle between ϕ_1 and ϕ_2. If two actuating currants M and N produce ϕ_1 and ϕ_2, the force produced is

$$F = KMN \sin \theta$$

2.2.3 Thermal Relays

These relays utilise the electro-thermal effect of the actuating current for their operation. They are widely used for the protection of small motors against overloading and unbalanced currents. The thermal element is a bimetallic strip, usually wound into a spiral to obtain a greater length, resulting in a greater sensitivity. A bimetallic element consists of two metal strips of different coefficients of thermal expansion, joined together. When it heats up one strip expands more than the other. This results in the bending of the bimetallic strip. The thermal element can be heated directly by passing the actuating current through the strip, but usually a heater coil is employed. When the bimetallic element heats up, it bends and deflects, thereby closing the relay contacts. For the ambient temperature compensation, a dummy bimetallic element *shielded* from the heater coil and designed to oppose the bending of the main bimetallic strip is employed. When the strip is in a spiral form, the unequal expansions of the two metals causes the unwinding of the spiral, which results in the closure of the contacts. Fig. 2.15(a) shows a simple arrangement to indicate the operating principle. Figure 2.15(b) shows a spiral form. Unimetallic strips are also used as thermal elements in a hair-pin like shape, as shown in Fig. 2.15(c). When the strip gets heated it expands and closes the contacts.

For the protection of 3-phase motors, three bimetallic strips are used. They are energised by currents from the three phases. Their contacts are arranged in such a way that if any one of the spirals moves differently from the other, due to an unbalance exceeding 12%, their contacts meet and cause the circuit breakers to trip. These spirals also protect the motor against overloading.

Thermocouples and resistance temperature detectors are also used in protection. In the protection of a large generator, such elements are placed in the stator slots. The element forms an arm of a balancing bridge. In normal condition, the bridge is balanced. When the temperature exceeds a certain limit, the bridge becomes unbalanced. The out-of-balance current energises a relay which trips a circuit breaker. This will be discussed in detail in the chapter dealing with protection of machines.

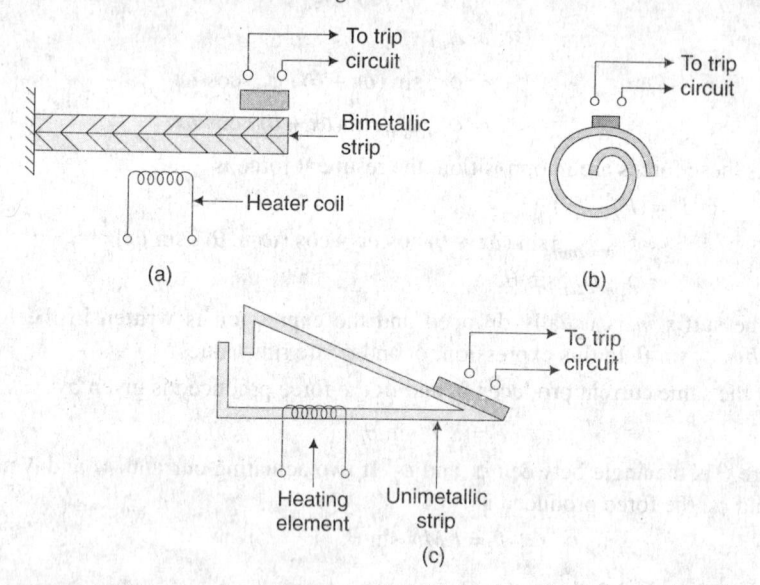

Fig. 2.15 (a) Bimetallic thermal relay (b) Bimetallic spiral type thermal relay (c) Unimetallic thermal relay

2.2.4 Auxiliary Relay, Auxiliary Switch and Flag

A protective relay is assisted by auxiliary relays for a number of important operations. A protective relay performs the task of measurement and under the required condition, it closes its contacts. It is relieved of other duties such as tripping, time lag, breaking of trip circuit current, giving alarm, showing flags etc. These duties are performed by auxiliary relays. Auxiliary relays repeat the operations of protective relays, control switches, etc. Repeat contact and auxiliary switches are also used to assist protective relays. The reasons for employing auxiliary relays, repeat contactors and auxiliary switches are:

(i) Protective relay contacts are delicate and light in weight. They are not capable of carrying large amount of current for a long period.

(ii) The protective relays do not have enough contacts to perform all duties required in a protective scheme.

The commonly used auxiliary relays have been described below.

Seal-in Relay

A seal-in relay is an auxiliary relay which is employed to protect the contacts of a protective relay. Once the protective relay closes its contacts, the seal-in relay is energized. Its contacts bypass the contacts of the protective relay, close and seal the circuit while the tripping current flows. It may also give an indication by showing a flag (target). It is an instantaneous relay, operates on attracted armature principle.

Time-lag Relays

Time-lag relays operate after a preset time-lag. They are used in protection schemes as a means of time discrimination, for example, time graded schemes which will be discussed in Chapter 5. They are also used in control and alarm circuits to allow time for the required sequence of operations to take place. The principle of producing time delays will be discussed later (see sec. 2.2.6).

Alarm Relays

An alarm relay gives both an audible and a visual indication. At a substation, it is sufficient to provide a trip alarm and one non-trip alarm, which is common to the whole substation. In the control room of a generating station, the trip alarm and non-trip alarm should be separate for each primary circuit. There is an arrangement for alarm cancellation by pressing a button. The alarm circuit is interrupted on pushing this button. When the relay is de-energised, the initiating contact of the cancellation mechanism is reset so that it can receive another alarm.

Repeat Contactors

A repeat contactor repeats the operation of a protective relay. It is sometimes needed because a protective relay may not have a sufficient number of contacts. It may also be required to take over the operation from the initiating relay if the contacts of the latter are not designed for carrying current for long periods. Its most important requirements are that it should be fast and absolutely reliable. It should also be robust and compact. It is usually mounted in the same case as the relay for which it is required to repeat the operation.

Repeat contactors operate on the attracted armature principle. It may be connected either in series or in parallel with the relay. It contains a number of contacts which are placed in parallel. However, having more than three contacts in parallel is usually not pratical.

Flag or Target

When a relay operates, a flag is indicated to show its operation. When on a relay panel there are several relays, it is the flag that indicates, the relay that has operated. This helps the operator to know the cause of the tripping of the circuit breaker. It is also called the target or indicator. Its coil is connected in series with the trip coil of the circuit breaker, as shown in Fig. 2.16. The resetting of a flag indicator is usually manual. There is a button or knob outside the relay case to reset the flag indicator. A flag indicator may either be electrical or mechanical. In a mechanical flag indicator, the movement of the armature of the relay pushes a small shutter to expose the flag. In an electrically operated flag indicator there is a solenoid which is energised when relay contacts are closed. Electrical flags being more reliable are preferred.

Auxiliary Switch

An auxiliary switch is connected in series with the trip-coil circuit, as shown in Fig. 2.16. It is mechanically interlocked with the operating mechanism of the circuit breaker so that the auxiliary switch opens when the circuit breaker opens. The opening of the auxiliary switch prevents unnecessary drainage of the battery.

When the trip-coil of the circuit breaker is energised, it actuates a mechanism of the circuit breaker, which causes the operating force to come into action to open the circuit breaker.

2.2.5 Connections for Seal-in Relay, Auxiliary Switch and Circuit Breaker Trip-Coil

Figure 2.16(a) shows the connection for a seal-in relay, circuit breaker trip-coil and auxiliary switch. In order to protect the contacts of the protective relay, a seal-in relay is employed. Its contacts bypass the contacts of the protective relay and seal the circuit closed, while the tripping current flows. Some relays employ a simple holding coil in series with the relay contacts, as shown in Fig. 2.16(b). The holding coil is wound on a small soft iron core which acts on a small armature on the moving-contacts assembly to hold the contacts tightly closed, once they have established the flow of current through the trip coil. The holding coils are used to protect the relay contacts against damage which may be caused due to the make and break action of the contacts.

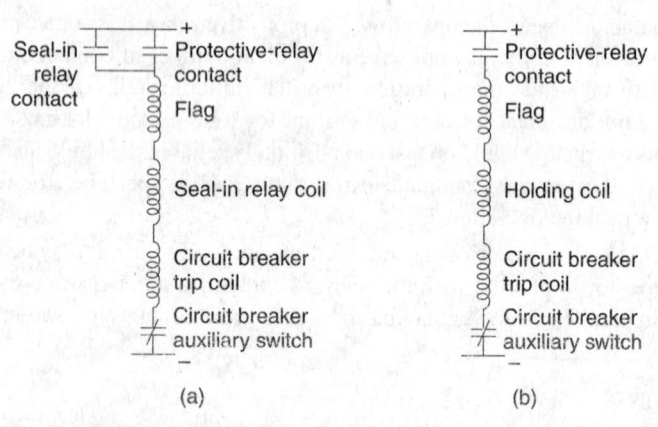

Fig. 2.16 *Circuit breaker trip coil circuit (a) With seal-in relay (b) With holding coil*

2.2.6 Techniques to Produce Time-delays

Sometimes, a protective relay is required to operate after a preset time delay. Intentional time delays are necessary for such relays. The intentional time delay may be caused by inherent design features of the relay or by a delay producing component of the relay. Sometimes, a starting relay or instantaneous relay is used in conjunction with a timing relay to perform certain operations after a preset time. A time-lag relay (timing relay) is an auxiliary relay designed to operate after a preset time-delay.

Mechanical Time-delay

The time delay may be produced by mechanical, electrical or electronic components. Oil dashpots, pneumatic damping, toothed gears, cams, mercury switches, etc. are some examples of mechanical devices which are used to produce a time delay. In an oil dashpot, there is a mechanically operated plunger. Oil flowing through an orifice in the cylinder retards the relay movement. The pneumatic timer contains a metal chamber, a diaphragm and solenoid. These mechanical devices are crude devices and do not produce accurate delays. The mercury switch, however, gives an accurate delay. The mercury tube has two sections. One section contains mercury and the other section contains contacts. The tube is tilted so that mercury flows from one section to the other and bridges the contacts. The flow of mercury is impeded by a construction between the two sections of the tube. The time setting is fixed by the design of the tube. It is not possible to have a range of time settings on a particular mercury switch. Toothed gears or cams are also used to produce time delay.

Thermal Time-delay
Thermal devices employing expansion of bimetal strip or spiral, unimetal (brass) strips, etc. are also used to produce time-delay.

Electrical Time-delay
Time delay can be produced by employing a short-circuiting ring around a solenoid pole; a circuit containing reactance, capacitance or non-linear resistance; a resonant circuit, etc.

Short-circuiting Ring
A short-circuiting copper band is fitted around the pole piece of an attracted armature hinged-type relay. This arrangement provides eddy current path for damping. To obtain a time-delay on pick-up, the band is placed at the armature end of the core. By this technique, a delay of about 0.1 s can be produced in pick-up with a large armature gap, and a stiff restraining spring. To obtain a delayed reset, the band is placed at the frame end of the solenoid. A delay up to 0.5 s in drop out can be obtained with a short lever arm and a light spring load. Time-delay can also be produced by employing a copper tube inside the coil.

Capacitance
A capacitor which is connected in parallel with the relay coil is changed through a resistor, as shown in Fig. 2.17(a). A longer time-delay is obtained by this technique. A delay of up to 0.5 s can be obtained on pick-up with a capacitor of reasonable size. For ac applications, a rectifier should be included.

Resonant Circuit
A resonant circuit, as shown in Fig. 2.17(b) can be employed to produce a delay of up to 3 cycles.

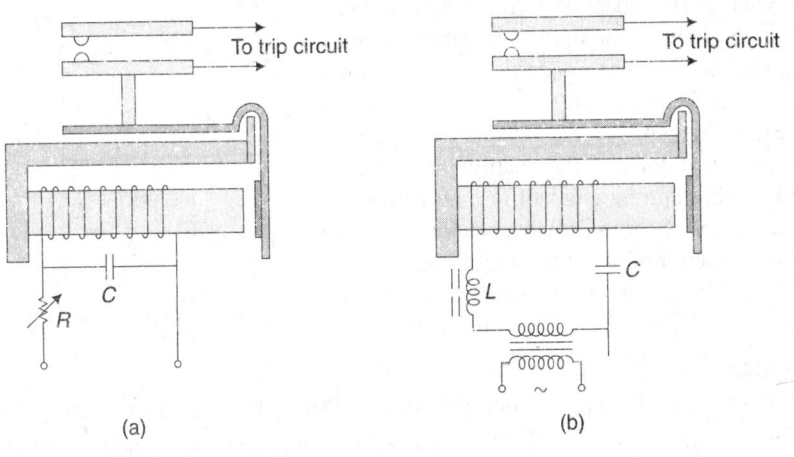

Fig. 2.17 *(a) Capacitor charge delay (b) Resonance build-up delay*

Ballistic Resistance

This technique is based on the principle of delaying the build-up of operating voltage. This includes thermistors or filament lamps. Figure 2.17(c) shows a metal filament lamp connected across the relay coil. A resistance is also placed in the circuit as shown in the figure. The hot resistance of the lamp filament is 10 times its cold resistance. The relay coil is short-circuited by the lamp, thereby keeping the magnetic flux to zero for a short time until the filament becomes incandescent.

Alternatively, a thermistor or a carbon filament can be placed in series with the relay coil. The thermistor resistance being high at room temperature limits the coil current. As the current drawn by the relay coil heats the thermistor, its resistance decreases until the relay current becomes sufficient for pick-up.

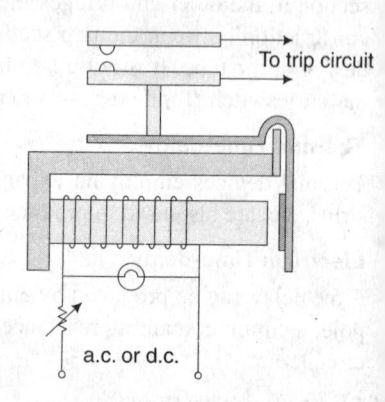

Fig. 2.17 (c) Lamp filament heating delay

Synchronous Motor

A synchronous motor, geared to a reduced speed can be also used on ac to produce more precise, long time delays.

Electronic Time Delay

Longer time delays are obtained with R-C circuits. R-C circuits are also used with electromagnetic relays, as shown in Fig. 2.17(a). Longer time delays can be obtained with R-C circuits when used with electronic relays rather than with electromagnetic relays. This is due to the face that a smaller current is needed with electronic relays, which in turn takes longer time to charge the capacitor.

Figure 2.18 shows a time-delay circuit employing a transistor. A constant dc voltage is applied to an R-C network to charge the capacitor C through resistor R. When the voltage of capacitor C reaches a suitable value, the transistor starts conducting. A realy is placed in the collector circuit. This relay operates when the transistor starts conducting. The time delay depends on the value of the capacitor and the magnitude of the charging current. As the charging current is small in a transistor circuit, a delay of several minutes can be obtained with a capacitor of only few microfarads. Delays of several hours can be obtained with tantalum capacitors of a few hundred microfarads.

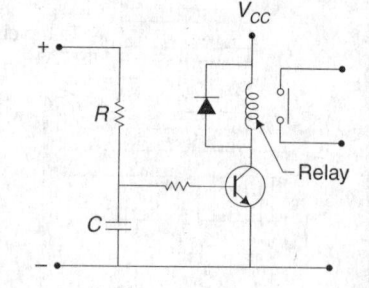

Fig. 2.18 Static time delay circuit

Counter

For obtaining even more accurate time-delays, electronic counters are used. A crystal oscillator or some suitable electronic circuitry is employed to generate a train of high frequency pulses. Counters are used as frequency dividers. A number of counters may be connected in cascade and different time-lags may be obtained from different stages of the cascaded counters.

2.2.7 Bearings

The pivot and jewel bearing is commonly used for precision relays. Spring-mounted jewels are used in modern relays. The design is such that shocks are taken on a shoulder and not on a jewel.

For high sensitivity and low friction, a single ball bearing between two cup-shaped sapphire jewels is used.

Multi-ball bearings provide friction as low as the jewel bearings and have greater resistance to shock. They are also capable of combining side-thrust and end-thrust in a single bearing. Miniature bearings less than 1.6 mm in diameter are now available.

Knife-edge bearings, pin bearings or resilient strips are used in hinged armature relays.

2.2.8 Backstops

When the moving part of the relay is stationary, it rests against a backstop. The material of a backstop should be chosen carefully so that it should not be sticky. To avoid magnetic adherence, the material should be non-metallic. The molecular adherence can be overcome using a hard surface rounded to a large radius. Smooth backstops made of agate or nylon are used.

2.2.9 Contacts

The reliability of protective relays depends on their contact performance. The following are the requirement of good contacts.

(i) Low contact resistance
(ii) High contact pressure
(iii) Freedom from corrosion
(iv) Bounce free
(v) Self-clearing action
(vi) Freedom from sparking
(vii) Dust proof

Silver is the most commonly used material for relay contacts. It has the lowest resistance. Copper is not used because of its higher resistance. Silver cadmium oxide is used where high currents are to be handled. It has low resistance like silver but does not weld or become sticky. An alloy consisting of 67% gold, 26% silver and 7% platinum is used for small currents and very light contact pressure. Non-corroding materials like gold, palladium or rhodium can be used for sensitive relays where the contact pressure needs to be very low. These materials are not recommended for protective relays where high contact pressure is required.

The most reliable relay contact are cylindrical contacts at right angles, as they give the optimum high pressure, bounce-proof contacts, hard smooth contact surfaces and dust proof relay cases minimise the maintenance of protective relays.

On silver contacts silver oxide does not form readily. Even when formed, its thickness does not exceed 10 Å and hence can easily be moved aside by high pressure or wiping action. Humid sulphurous and high temperature atmosphere causes corrosion. In polluted atmospheres where coal fires are used, silver sulphide is formed readily, especially in the presence of heat and humidity. It is not breakable like oxides but it is soft and thus can be squeezed aside by high pressure. A thin petrolatum coating can reduce corrosion of contacts without increasing their resistance. It is helpful in polluted atmospheres.

A dust-proof casing is usually used for modern relays. A filter is provided at the back to trap any dust and to allow the relay to breathe. A relay with such a casing is quite suitable for a dusty or otherwise dirty atmosphere. In a relay with poor ventilation, particularly in a sealed relay, high resistance polymers may appear on contacts. This is due to organic emanations from the coil insulation material. All insulating materials, except teflon give off organic vapour to a certain degree. Phenolic resin gives off organic vapours more than others. Polyester and epoxide varnishes now available have good performance and are quite satisfactory for coil insulation. Relay casings with good ventilation and having dust filters minimise the collection of high resistance polymers on the contacts. Encapsulated contacts as in the case of reed relays provide the best solution to the problem. Alternatively, the relay coil can also be encapsulated.

An electromechanical relay used with comparators is usually of a small rating. When such relays control auxiliary relays and timing units, they are to be protected with spark quenching circuits. A series resistor and capacitor connected across the contacts is a simple spark quenching circuit.

2.3 STATIC RELAYS

In a static relay, the comparison or measurement of electrical quantities is performed by a static circuit which gives an output signal for the tripping of a circuit breaker. Most of the present day static relays include a dc polarised relay as a slave relay. The slave relay is an output device and does not perform the function of comparison or measurement. It simply closes contacts. It is used because of its low cost. In a fully static relay, a thyristor is used in place of the electromagnetic slave relay. The electromechanical relay used as a slave relay provides a number of output contacts at low cost. Electromagnetic multicontact tripping arrangements are much simpler than an equivalent group of thyristor circuits.

A static relay (or solid state relay) employs semiconductor diodes, transistors, zener diodes, thyristors, logic gates, etc. as its components. Now-a-days, integrated circuits are being used in place of transistors. They are more reliable and compact.

Earlier, induction cup units were widely used for distance and directional relays. Later these were replaced by rectifier bridge type static relays which employed dc polarised relays as slave relays. Where overcurrent relays are needed, induction disc relays are in universal use throughout the world. But ultimately static relays will supersede all electromagnetic relays, except the attracted armature relays and dc polarised relays as these relays can control many circuit at low costs.

2.3.1 Merits and Demerits of Static Relays

The advantages of static relays over electromechanical relays are as follows.

(i) Low burden on CTs and VTs. The static relays consume less power and in most of the cases they draw power from the auxiliary dc supply
(ii) Fast response
(iii) Long life
(iv) High resistance to shock and vibration
(v) Less maintenance due to the absence of moving parts and bearings
(vi) Frequent operations cause no deterioration
(vii) Quick resetting and absence of overshoot
(viii) Compact size
(ix) Greater sensitivity as amplification can be provided easily
(x) Complex relaying characteristics can easily be obtained
(xi) Logic circuits can be used for complex protective schemes

The logic circuit may take decisions to operate under certain conditions and not to operate under other conditions.

The demerits of static relays are as follows:

(i) Static relays are temperature sensitive. Their characteristics may vary with the variation of temperature. Temperature compensation can be made by using thermistors and by using digital techniques for measurements, etc.
(ii) Static relays are sensitive to voltage transients. The semiconductor components may get damaged due to voltage spikes. Filters and shielding can be used for their protection against voltage spikes.
(iii) Static relays need an auxiliary power supply. This can however be easily supplied by a battery or a stabilized power supply.

2.3.2 Comparators

When faults occur on a system, the magnitude of voltage and current and phase angle between voltage and current may change. These quantities during faulty conditions are different from those under healthy conditions. The static relay circuitry is designed to recognise the changes and to distinguish between healthy and faulty conditions. Either magnitudes of voltage/current (or corresponding derived quantities) are compared or phase angle between voltage and current (or corresponding derived quantities) are measured by the static relay circuitry and a trip signal is sent to the circuit breaker when a fault occurs. The part of the circuitry which compares the two actuating quantities either in amplitude or phase is known as the *comparator*. There are two types of comparators—amplitude comparator and phase comparator.

Amplitude Comparator

An amplitude comparator compares the magnitudes of two input quantities, irrespective of the angle between them. One of the input quantities is an operating quantity and the other a restraining quantity. When the amplitude of the operating quantity exceeds the amplitude of the restraining quantity, the relay sends a tripping signal. The actual circuits for comparators will be discussed in subsequent chapters.

Phase Comparator

A phase comparator compares two input quantities in phase angle, irrespective of their magnitudes and operates if the phase angle between them is ≤ 90°.

2.3.3 Duality between Amplitude and Phase Comparators

An amplitude comparator can be converted to a phase comparator and vice versa if the input quantities to the comparator are modified. The modified input quantities are the sum and difference of the original two input quantities. To understand this fact, consider the operation of an amplitude comparator which has two input signals M and N as shown in Fig. 2.19(a). It operates when $|M| > |N|$. Now change the input quantities to $(M + N)$ and $(M - N)$ as shown in Fig. 2.19(b). As its circuit is designed for amplitude comparison, now with the changed input, it will operate when $|M + N| > |M - N|$. This condition will be satisfied only when the phase angle between M and N is less than 90°. This has been illustrated with the phasor diagram shown in Fig. 2.20. It means that the comparator with the modified inputs has now become a phase comparator for the original input signals M and N.

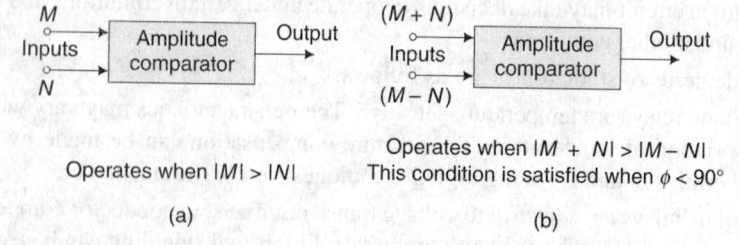

Fig. 2.19 *(a) Amplitude comparator (b) Amplitude comparator used for phase comparison*

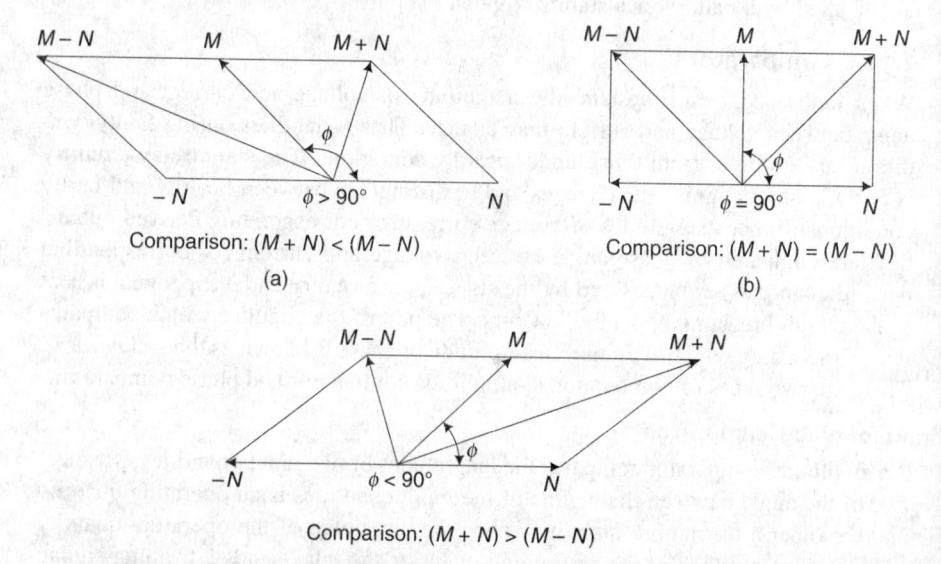

Fig. 2.20 *Phasor diagram for amplitude comparator used for phase comparison*

Relay Construction and Operating Principles 53

Similarly, consider a phase comparator shown in Fig. 2.21(a). It compares the phases of input signals M and N. If the phase angle between M and N, i.e. angle ϕ is less than 90°, the comparator operates. Now change the input, signals to $(M + N)$ and $(M - N)$, as in Fig. 2.21(b). With these changed inputs the comparator will operate when phase angle between $(M + N)$ and $(M - N)$, i.e. angle λ is less than 90°. This condition will be satisfied only when $|M| > |N|$. In other words, the phase comparator with changed inputs has now become an amplitude comparator for the original input signals M and N. This has been illustrated with phasor diagrams as shown in Fig. 2.22.

Operates when $\phi < 90°$
(a)

Operates when $\lambda < 90°$
This condition is satisfied when $|M| > |N|$
(b)

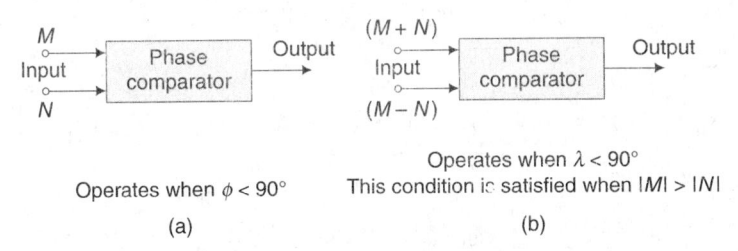

Fig. 2.21 (a) Phase comparator (b) Phase comparator used for amplitude comparison

Figure 2.20 shows three phasor diagrams for an amplitude comparator. The phase angle between the original inputs M and N is ϕ. Now the inputs to the amplitude comparator are changed to $(M + N)$ and $(M - N)$ and its behaviour is examined with the help of three phasor diagrams. The three phasor diagrams are with phase angle ϕ (i) greater than 90°, (ii) equal to 90° and (iii) less than 90°, respectively. When ϕ is less than 90°, $|M + N|$ become greater than $|M - N|$ and the relay operates with the modified inputs. When ϕ is equal to 90° or greater than 90°, the relay does not operate.

The phasor diagrams show that $|M + N|$ becomes greater than $|M - N|$ only when ϕ is less than 90°. This will be true irrespective of the magnitude of M and N. In other words, this will be true whether $|M| = |N|$ or $|M| > |N|$ or $|M| < |N|$. The figures have been drawn with $|M| = |N|$. The reader can draw phasor diagrams with $|M| < |N|$ or $|M| > |N|$. The results will remain the same. This shows that with changed inputs, the amplitude comparator is converted to a phase comparator for the original inputs.

Figure 2.22 shows three phasor diagrams for a phase comparator. The original inputs are M and N. Now the inputs of the phase comparators are changed to $(M + N)$ and $(M - N)$, and its behavior is examined with the help of three phasor diagrams drawn for (i) $|M| < |N|$, (ii) $|M| = |N|$ and (iii) $|M| > |N|$. The angle between $(M + N)$ and $(M - N)$ is λ. The angle λ becomes less than 90° only when $|M| > |N|$. As the comparator under consideration is a phase comparator, the relay will trip. But for the original inputs M and N, the comparator behaves as an amplitude comparator. This will be true irrespective of the phase angle ϕ between M and N. The figure has been drawn with ϕ less than 90°. The reader can check it by drawing phasors with $\phi = 90°$ or $\phi > 90°$. The result will remain the same.

2.3.4 Types of Amplitude Comparators

As the ratio of the instantaneous values of sinusoidal inputs varies during the cycle, instantaneous comparison of two inputs is not possible unless at least one of the

Comparision: $\lambda > 90°$
(a)

Comparision: $\lambda > 90°$
(b)

Comparison: $\lambda < 90°$
(c)

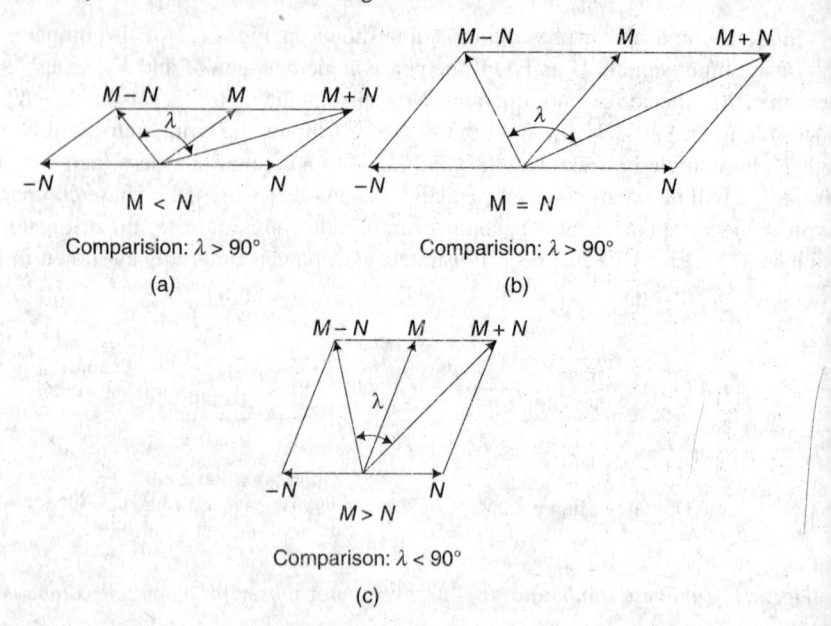

Fig. 2.22 *Phase comparator used for amplitude comparison*

signals is rectified. There are various techniques to achieve instantaneous comparison. In some techniques both inputs are rectified, while in some methods, only one of the inputs is rectified. When only one input signal is rectified, the rectified quantity is compared with the value of the other input at a particular moment of the cycle. Besides instantaneous (or direct) comparison, the integrating technique is also used.

The amplitude comparison can be done in a number of different ways. The following are some important methods which will be described to illustrate the principle.

  (i) Circulating current type rectifier bridge comparators
 (ii) Phase splitting type comparators
(iii) Sampling comparators.

Rectifier Bridge Type Amplitude Comparator

The rectifier bridge type comparators are widely used for the realisation of overcurrent and distance relay characteristics. The operating and restraining quantities are rectified and then applied to a slave relay or thyristor circuit. Figure 2.23(a) shown a rectifier bridge type amplitude comparator. There are two full wave rectifiers, one for the operating quantity and the other for the restraining quantity. The outputs of these bridges are applied to a dc polarised relay. When the operating quantity exceeds the restraining quantity, the relay operates. Figure 2.23(b) shows a rectifier bridge type amplitude comparator with the thyristor circuit as an output device.

To get more accurate results the bridge rectifier can be replaced by a precision rectifier employing an operational amplifier. The circuit for the precision rectifier has been discussed while describing microprocessor based relays.

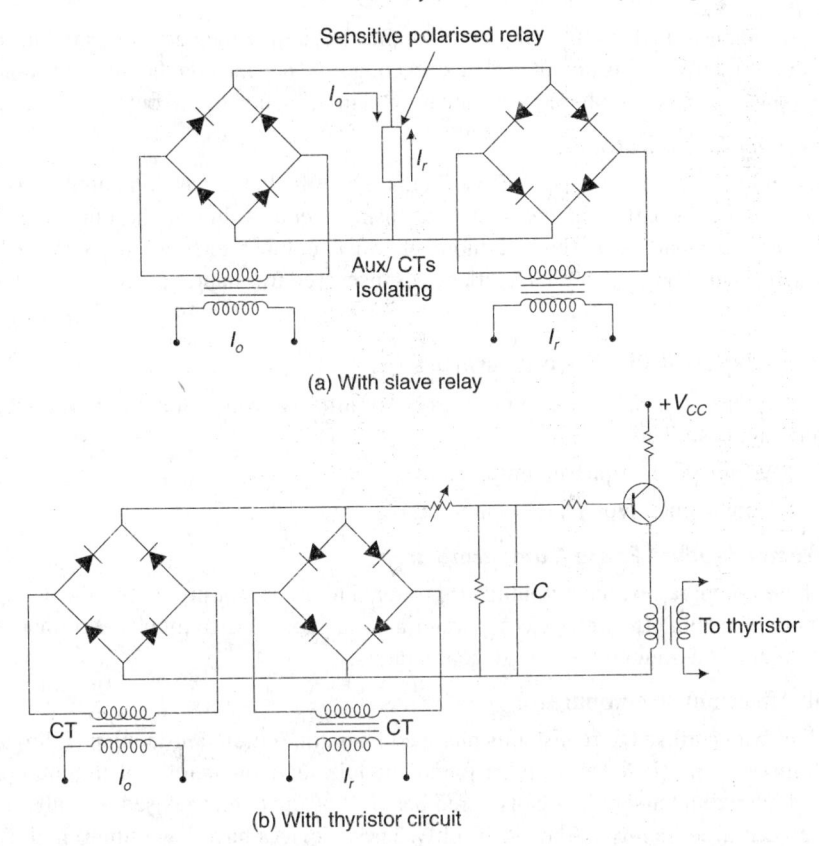

Fig. 2.23 Rectifier bridge type amplitude comparator

(a) With slave relay

(b) With thyristor circuit

Phase Splitting Type Amplitude Comparators

Figure 2.24 shows a phase splitting of inputs before rectification. The input is split into six components 60° apart, so that output after rectification is smoothed within 5%.

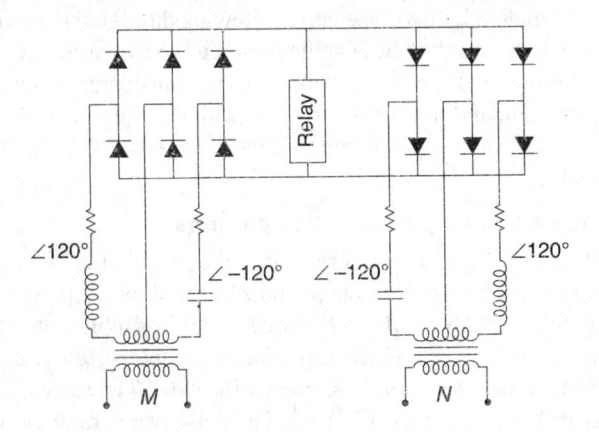

Fig. 2.24 Phase splitting type amplitude comparator

As both input signals to the relay are smoothed out before they are compared, a continuous output signal is obtained. The operating time depends on the time constant of the slowest arm of the phase-splitting circuit and the speed of the output device.

Sampling Comparators

In sampling comparators, one of the inputs is rectified and it is compared with the other input at a particular moment. The instantaneous value of the other input is sampled at a particular desired moment. Such comparators are used to realise reactance and MHO relay characteristics as discussed in Chapter 6 Section 6.7.1 and 6.7.2.

2.3.5 Types of Phase Comparators

Phase comparison can be made in a number of different ways. Some important techniques are described below.

(i) Vector product phase comparators

(ii) Coincidence type phase comparators.

(i) Vector Product Phase Comparators

In these comparators, the output is proportional to the vector product of the ac input signals. The Hall effect phase comparator and magneto-resistivity phase comparator come under this category of phase comparators.

Hall effect phase comparator

Hall effect is utilised to realise this phase comparator. Indium antimonide (InSb) and indium arsenide (InAs) have been found suitable semiconductors for this purpose. Of which indium arsenide is considered better. Protective relays based on Hall effect have been used mainly in the USSR only. These devices have low output, high cost and they can cause errors due to rising temperatures.

Magneto-resistivity

Some semiconductors exhibit a resistance variation property when subjected to a magnetic field. Suppose two input signals are V_1 and V_2. V_1 is applied to produce a magnetic field through a semiconductor disc. V_2 sends a current through the disc at a right angle to the magnetic filed. The current flowing through the disc is proportional to $V_1 V_2 \cos \phi$, where ϕ is the phase angle between the two voltages. Therefore, this can be used as a phase comparator. This device is considered to be better than the Hall effect type comparator because it gives a higher output, its construction and circuitry are simpler and no polarising current is required. This device is also used only in the USSR.

(ii) Coincidence Circuit Type Phase Comparators

In a coincidence circuit type phase comparator, the period of coincidence of positive polarity of two input signals is measured and compared with a predetermined angle, usually 90°. Figure 2.25 shows the period of coincidence represented by an angle ψ. If the two input signals have a phase difference of ϕ, the period of coincidence $\psi = 180 - \phi$. If ϕ is less than 90°, ψ will be greater than 90°. The relay is required to trip when ϕ is less than 90°, i.e. $\psi > 90°$. Thus, the phase comparator circuit is designed to trip signal when ψ exceeds 90°.

Various techniques have been developed to measure the period of coincidence. The following are some important ones which will be described to illustrate the principle.
(a) Phase-splitting type phase comparator
(b) Integrating type phase comparator
(c) Rectifier bridge type phase comparator
(d) Time-bias type phase comparator.

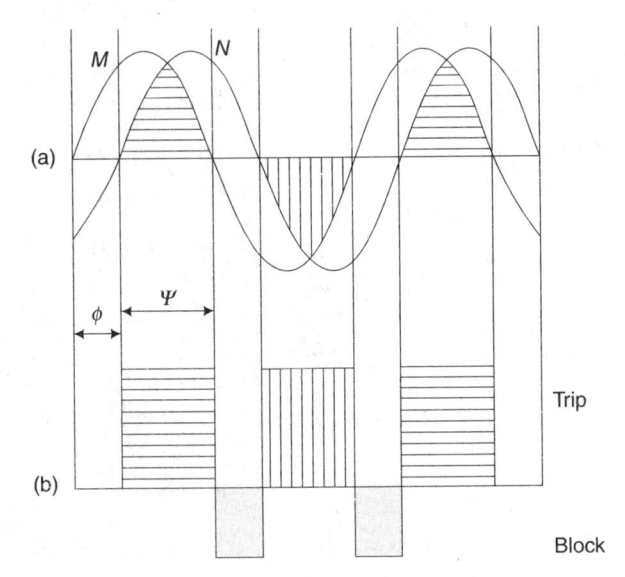

Fig. 2.25 *Period of coincidence of sine wave inputs*

Phase-splitting Type Phase Comparator

In this technique, both inputs are split into two components shifted ± 45° from the original wave, as shown in Fig. 2.26(a). All the four components, which are now available, are fed into an AND gate as shown in Fig. 2.26(b). The tripping occurs when all the four signals become simultaneously positive at any time during the cycle. An AND gate is used as a coincidence detector. The coincidence of all the four signals occurs only when ϕ is less than 90°. The full range of operation is
$$-90° < \phi < 90°$$
It is a technique of direct comparison.

Integrating Type Phase Comparator

In this technique, the coincidence time is measured for each cycle by integrating the output of an AND gate (coincidence detector) to which input signals are applied. Figure 2.25(a) shows two sinusoidal input signals. The hatched area shows the time of overlap (time of coincidence) of the two inputs. During this period, both inputs are positive. This period is represented by ψ. The phase difference between the two inputs is ϕ. The angle $\psi = 180 - \phi$. If ϕ is less than 90°, ψ is greater than 90°. If these two inputs are applied to an AND gate, the output of the gate is a series of square pulses. We get a square wave output during the period of coincidence and no output for the rest of the period of a cycle,. as shown in Fig. 2.25 (b).

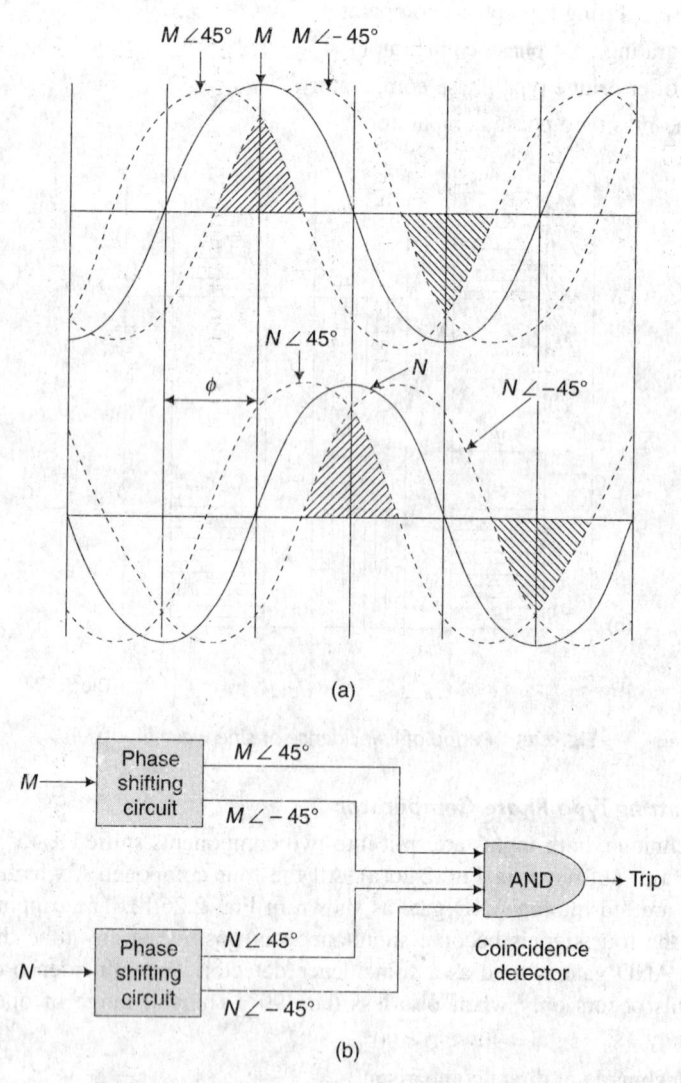

(a)

(b)

Fig. 2.26 *Phase-comparator with phase-split inputs*

Figure 2.27(a) shows the block diagram of a phase comparator. The sinusoidal inputs are first converted into square waves and then are applied to an AND gate. The output of the AND gate is a chain of pulses as shown in Fig. 2.27(b). This is for $\phi < 90°$, i.e. $\psi > 90°$. The relay will provide a trip output. The output of the AND gate is applied to an integrator. The output of the integrator is shown in Fig. 2.27(c). This output is applied to a level detector which finally gives a TRIP signal. The integrating circuit may be employed as shown in Fig. 2.28. The level detector may be a thyristor circuit.

Relay Construction and Operating Principles 59

Figure 2.27(d) and (e) show the outputs of the AND gate and the integrator, respectively. This situation is for $\psi = 90°$ and is the limiting condition. The relay may be set to operate at $\psi = 90°$.

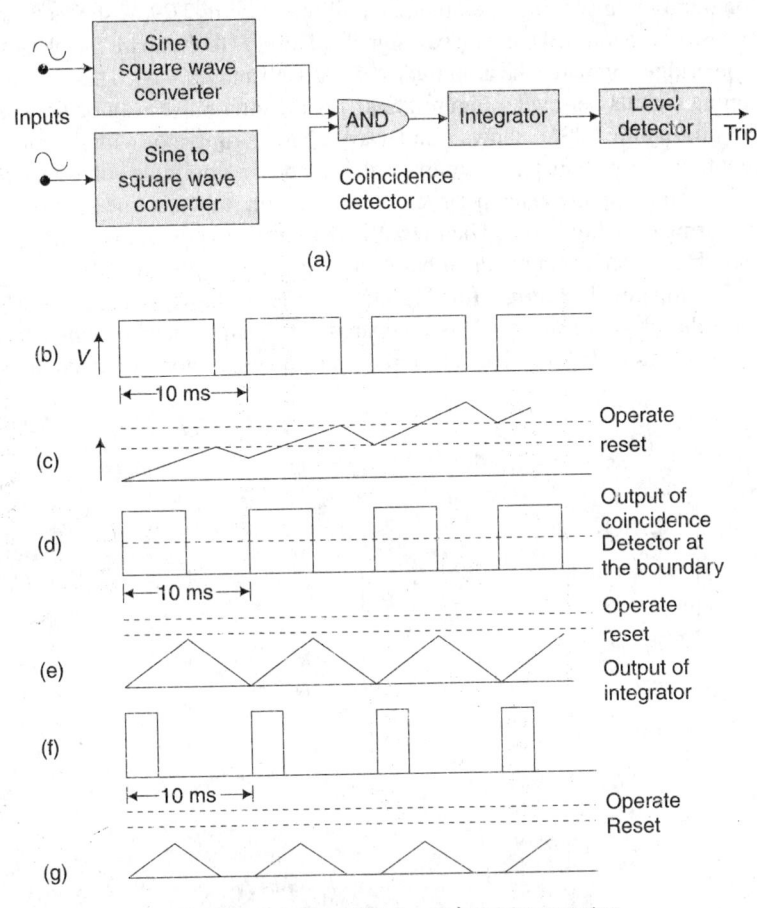

Fig. 2.27 Integrating type phase comparator

Figure 2.27(f) and (g) show the outputs of the AND gate and the integrator, respectively, for $\psi < 90°$. For this condition, the relay does not operate.

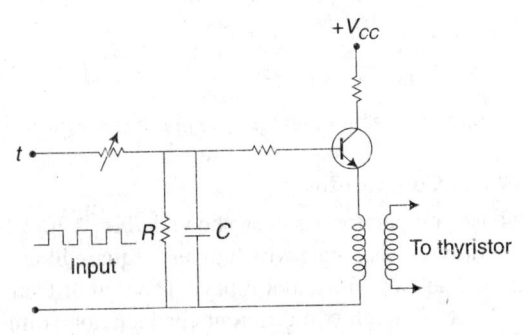

Fig. 2.28 Integrating circuit

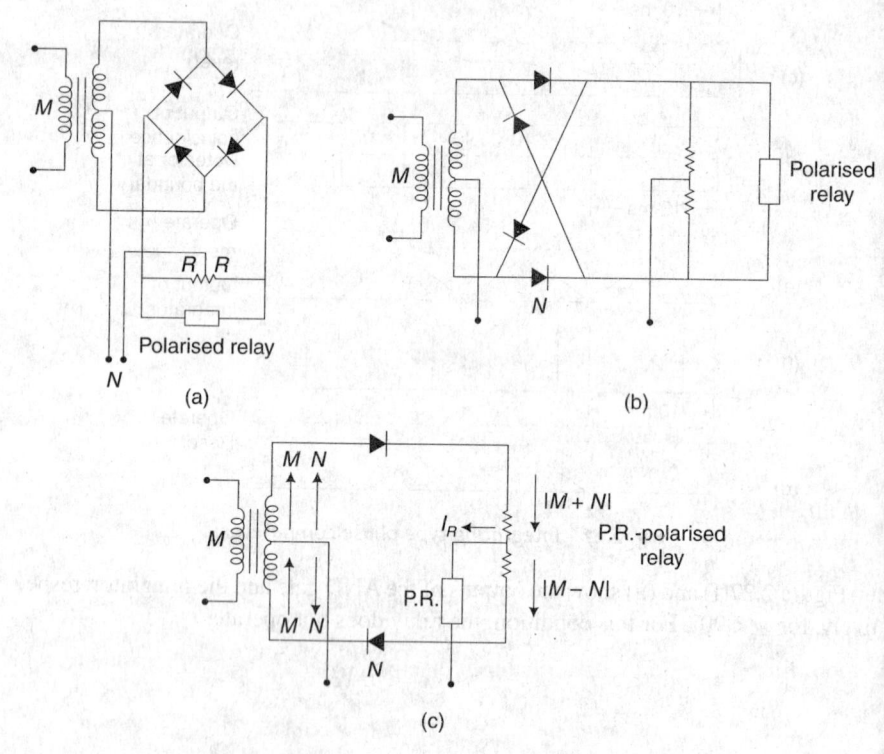

Rectifier Bridge Type Phase Comparator

Rectifier bridge type comparators are widely used for the realisation of distance relay characteristics. For more accurate results the bridge rectifier can be replaced by a precision rectifier employing operational amplifiers. Figure 2.29(a) shows a rectifier bridge phase comparator. There are two signals M and N. To compare the phases of M and N, the bridge compares the amplitudes of $(M + N)$ and $(M - N)$. This circuit gives two tripping signals per cycle. Figure 2.29(b) is the alternative way of drawing the same circuit. Figure 2.29(c) shows a half-wave phase comparator with the directions of current to illustrate how phase comparison is made by amplitude comparison. This circuit gives one tripping signal per cycle. The direction shown is true for a particular moment during the whole cycle when $M > N$ and both have a positive polarity. At other moments, the direction may change but every time the amplitudes of $(M + N)$ and $(M-N)$ are compared. The current flowing in the polarised relay is $I_R = [|M+N|-|M-N|]$. Therefore, the phase of M and N is compared. The output device shown in these figures is a polarised dc relay. It can be replaced by an integrator circuit and thyristor.

Fig. 2.29 Rectifier bridge type phase comparator

Time-bias Type Phase Comparator

A time-bias type phase comparator has been shown in Fig. 2.30(a). In this technique, the inputs are applied to an AND gate which gives a square block output during the coincidence period of the two sinusoidal inputs. The output from the AND gate is fed to another AND gate through two different channels: one directly and the other through a delay circuit. The delay circuit gives a delayed output. The output is delayed by an angle δ from the staring point of the block as shown in Fig. 2.30(b) and (c). The delay δ is kept 90°. If the block and pulse (output of the delay circuit) still coincide. the second AND gate will give an output, as shown in Fig. 2.30(b). If the block and pulse do not coincide, the second AND gate does not give any output as shown in Fig. 2.30(c). This means that the output of the first gate has to persist for a period δ so that the second gate may operate and send a tripping signal. This technique is more suitable for multi-input comparators. However, it is subject to false tripping by a false transient signal whereas phase comparators discussed earlier are not.

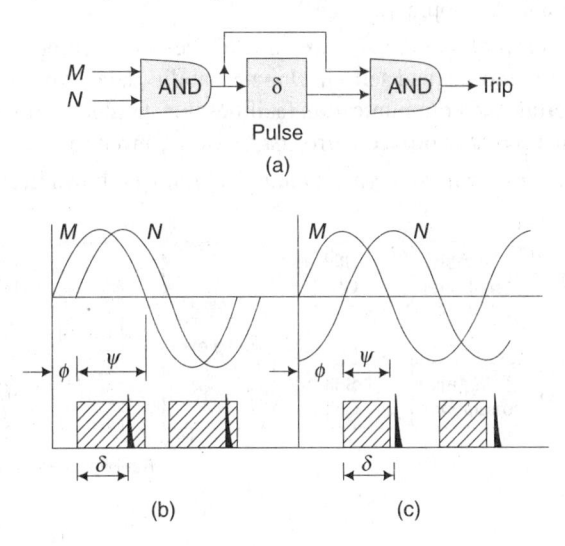

Fig. 2.30 *Time bias type phase comparator*

2.4 NUMERICAL RELAYS

With the tremendous developments in VLSI and computer hardware technology, microprocessors that appeared in the seventies have evolved and made remarkable progress in recent years. Fast and sophisticated microprocessors, microcontrollers, and digital signal processors (DSPs) are available today at low prices. Their application to power system protection have resulted in the availability of compact, faster, more accurate, flexible and reliable protective relays, as compared to the conventional ones.

Numerical relays which are based on numerical (digital) devices e.g. microprocessors, microcontrollers. DSPs etc. are the latest development in the area of power system protection. In these relays, the analog current and voltage signals monitored through primary transducers (CTs and VTs) are conditioned, sampled at specified instants of time and converted to digital form for numerical manipulation, display and recording. Thus, numerical relays, having monitored the current and voltage signals through transducers, acquire the sequential samples of these ac quantities in numeric (digital) data form through the data acquisition system, and process the data numerically using an algorithm to calculate the fault discriminants and make trip decisions. With the continuous reduction in digital circuit costs and increases in their functionality, considerable cost-benefit improvement ensues. At present microprocessor/microcontroller-based numerical relays are widely used. There is a growing trend to develop and use numerical relays for the protection of various components of the modern complex power system. Numerical relaying has become a viable alternative to the traditional relaying systems employing electromechanical and static relays. Intelligent numerical relays using artificial intelligence techniques such as artificial neural networks (ANNs) and Fuzzy Logic Systems are presently under active research and development stage.

The main features of numerical relays are their economy, compactness, flexibility, reliability, self-monitoring and self-checking capability, adaptive capability, multiple functions, metering and communication facilities, low burden on transducers (instrument transformers) and improved performance over conventional relays.

The schematic diagram of a typical numerical relay is shown in Fig. 2.31

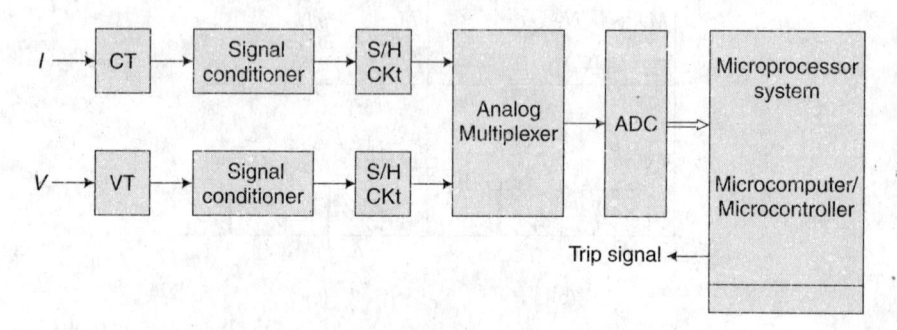

Fig. 2.31 *The schematic diagram of a typical numerical relay*

The levels of voltage and current signals of the power system are reduced by voltage and current transformers (VT and CT). The outputs of the CT and VT (transducers are applied to the signal conditioner which brings real-world signals into digitizer. The signal conditioner electrically isolates the relay from the power system, reduces the level of the input voltage, converts current to equivalent voltage and remove high frequency components from the signals using analog filters. The output of the signal conditioner are applied to the analog interface, which includes sample and hold (S/H) circuits, analog multiplexer and analog-to-digital (A/D) converters. These components sample the reduced level signals and convert their analog levels to equivalent numbers the are stored in memory for processing.

The signal conditioner, and the analog interface (i.e., S/H CKt, analog multiplexer and A/D converter) constitute the data acquisition system (DAS).

The acquired signals in the form of discrete numbers are processed by a numerical relaying algorithm to calculate the fault discriminants and make trip decisions. If there is a fault within the defined protective zone, a trip signal is issued to the circuit breaker.

2.5 COMPARISON BETWEEN ELECTROMECHANICAL RELAYS AND NUMERICAL RELAYS

Sr. No.	Feature	Electromechanical Relay	Numerical Relay
1.	Size	Bigger	Compact
2.	Characteristics	Fixed	Selectable
3.	Flexibility	No flexibility	Flexibility due to programmability.
4.	Communication feature	Not available	Available
5.	Blocking feature	Not available	Available
6.	Self-supervision	Not available	Available
7.	Adaptability	Not adaptable	Adaptable to changing system condition.
8.	Multiple-functions	Not possible	Possible
9.	Accuracy	± 5% or more	± 2%
10.	Speed of operation	Slow	Fast
11.	Burden on Transducers (CTs and VTs)	Very high	Extremely low
12.	Consistancy of calibration	Deteriorate with time	No effect on calibration even after use of 20-25 years
13.	Setting	Through plug setting in fix steps.	Software based settings.
14.	Memory feature	No memory of any type is available.	Several memory features are available.
15.	Maintenance	Cumbersome and frequent maintenance required.	Maintenance free relays
16.	Output relay programming	Not available	Available
17.	Accessibility of relay from remote place	Not possible	Remote accessibility is available
18.	Status of service values	Not available	Available
19.	Safety of personnel	Not adequate due to non-accessibility at remote location.	Adequate safety is provided.
20.	Spares requirement	There is need to stock several items as spare.	Universal designs minimise the spares requirement.
21.	Upward connectivity for a present or future system such as SCADA, EMS etc.	Not possible	Possible

EXERCISES

1. Draw a neat sketch of an induction disc relay and discuss its operating principle.
2. For what type of protective relay will you recommend (i) an induction disc type (ii) induction cup type construction? What measures are taken to minimise the overrun of the disc?
3. What are the merits of induction cup construction over the induction disc construction?
4. Derive an expression for torque produced by an induction relay.
5. Discuss the working principle of a permanent magnet moving coil relay with a neat sketch. State the area of its applications.
6. Describe the operating principle of a moving iron type dc polarised relay. Suggest some suitable area of its applications.
7. What are the different types of electromagnetic relays? Discuss their field of applications.
8. Explain why attracted armature type relays are noisy. What measures are taken to minimize the noise?
9. Discuss why the ratio of reset to pick up should be high.
10. Discuss the working principle, types and applications of thermal relays.
11. What are the advantages of static relays over electromechanical relays?
12. Explain what are amplitude and phase comparators.
13. Discuss how an amplitude comparator can be converted to a phase comparator, and vice versa.
14. Discuss the operating principle of a rectifier bridge phase comparator.
15. Discuss the principle of a coincidence circuit for phase comparator.
16. With a neat sketch, describe the principle of a reed relay. Where is it used?
17. What is a numerical relay? What are its advantages over conventional type relays?
18. Draw the schematic diagram of a numerical relay and briefly describe the functions of its various components.
19. Compare a numerical relay with an electromechanical relay.

Current and Voltage Transformers

3

3.1 INTRODUCTION

Current and voltage transformers (CTs and VTs) are collectively known as *transducers* or *instrument transformers*. They are used to transform the power system currents and voltages to lower magnitudes and to provide isolation between the high-voltage power system and the relays and other measuring instruments (meters) connected to the secondary windings of the transducers. In order to achieve a degree of interchangeability among different manufacturers of relays and meters, the ratings of the secondary windings of the transducers are standardized. The standard current ratings of the secondary windings of the current transformers (CTs) are 5 or 1 ampere. The secondary windings of the voltage transformers (VTs) are rated at 110 V line to line. The current and voltage ratings of the protective relays and meters are same as the current and voltage ratings of the secondary windings of the CTs an VTs respectively. The transducers should be able to provide current and voltage signals to the relays and meters which are faithful reproductions of the corresponding primary quantities. Although in most of the cases the modern transducers are expected to do so, but they can't be ideal and free from the errors of transformation. Hence the errors of transformation introduced by the transducers must be taken into account, so that the performance of the relays can be assessed in the presence of such errors. As the operating time of modern protective relays has reduced to the order of few milliseconds, the transient behavior of current transformers and voltage transformers require more attention.

3.2 CURRENT TRANSFORMERS (CTs)

Current transformers are used to perform two tasks. Firstly, they step down the heavy power system currents to low values that are suitable for the operation of the relays and other measuring instruments (meters) connected to their secondary windings. Secondly, they isolate the relays and meters circuits from the high voltages of the power system. The standard current ratings of the secondary windings of the CTs used in practice are 5A or 1A. Since the current ratings of CT secondary windings are standardized, current ratings for relays and meters are also standardized, so that a degree of interchangeability among different manufacturers of relays and meters can be achieved. A conventional CT of electromagnetic type is similar to a power transformer to some extent since both depend on the same fundamental principle of electromagnetic induction but there are considerable differences in their design and operation. A power transformer is a shunt-operated device while a CT is a series operated device. Current transformers are connected with their primaries in series with the power system (protected circuit) and, because the primary currents are so large, the primary winding has very few turns.

The VA rating of current transformers is small as compared with that of a power transformer. Though the nominal (Continuous) current ratings of the secondary windings of the CTs are 5A or 1A but they must be designed to tolerate higher values for short time of few seconds under abnormal system conditions, e.g., fault conditions. Since the fault currents may be as high as 50 times full-load current, current transformers are designed to withstand these high currents for a few seconds. Protective relays require reasonably accurate reproduction of the normal and abnormal conditions in the power system for correct sensing and operation. Hence, the current transformers should be able to provide current signals to the relays and meters which are faithful reproductions of the primary currents. The measure of a current transformer performance is its ability to accurately reproduce the primary current in secondary amperes. Ideally, the current transformer should faithfully transform the current without any error. But, in practice, there is always some error. The error is both in magnitude and in phase angle. These errors are known as ratio error and phase angle error. The exciting current is the main source of these errors of a CT.

Depending on application, CTs are broadly classified into two categories (1) measuring CTs, and (2) protective CTs. CTs used in conjunction with measuring instruments (meters) are popularly termed as measuring (metering) CTs' and those used in conjunction with protective devices are termed as protective CTs.

3.2.1 Difference Between Measuring and Protective CTs

CTs which are used to step down the primary currents to low values suitable for the operation of measuring instruments (meters) are called measuring or metering CTs. Secondary of the measuring CTs are connected to the current coils of ammeters, wattmeters, energy meters, etc. Since the measurements of electrical quantities are performed under normal conditions and not under fault conditions, the performance of measuring CTs is of interest during normal loading conditions. Measuring CTs are required to give high accuracy for all load currents upto 125% of the rated current. These CTs may have very significant errors during fault conditions, when the currents may be several times their normal value for a short time. This is not significant because metering functions are not required during faults. The measuring CTs should get saturated at about 1.25 times the full-load current so as not to reproduce the fault current on the secondary side, to avoid damage to the measuring instruments.

CTs used in association with protective devices i.e. relays, trip coils, pilot wires etc. are called protective CTs. Protective CTs are designed to have small errors during fault conditions so that they can correctly reproduce the fault currents for satisfactory operation of the protective relays. The performance of protective relays during normal conditions, when the relays are not required to operate, may not be as accurate. When a fault occurs on a power system, the current tends to increase and voltage tends to collapse. The fault current is abnormal and may be 20 to 50 times the full-load current. It may have dc offset in addition to ac component. The fault current for a CT secondary of 5A rating could be 100 to 250 A. Therefore, the CT secondary having a continuous current rating of 5A should have short-time current rating of 100 to 250 A, so that the same is not damaged. Since the ac component in the fault current is of paramount importance for the relays, the protective CT should correctly reproduce it on the secondary side in spite of the dc offset in the primary winding. Hence the dc offset should also be considered while designing the protective CT. The protective CT should not saturate upto 20 to 50 times full-load current.

3.2.2 Core Material of CTs

Figure 3.1 shows the magnetisation Characteristics of (a) cold-rolled grain-oriented silicon steel (3%), (b) hot-rolled silicon steel (4%) and (c) nickel-iron (77% Ni, 14% Fe). It is seen that the nickel-iron core has the qualities of highest permeability, low exciting current, low errors and saturation at a relatively low flux density. Measuring CTs are required to give a high accuracy for all load currents up to 125% of the rated current. Nickel-iron gives a good accuracy up to 5 times the rated current and hence, it is quite a suitable core material for CTs used for meters and instruments. The excessive currents being fed to instruments and meters are prevented during faults on power system due to almost absolute saturation at relatively low flux density.

Cold-rolled grain-oriented silicon steel (3%), which has a high permeability, high saturation level, reasonably small exciting current and low errors is used for the core of the CTs used for protective relays. Such core material has reasonably good accuracy up to 10-15 times the rated current, but when we consider currents that are five times under the rated current, the core material made from nickel-iron alloy fares better.

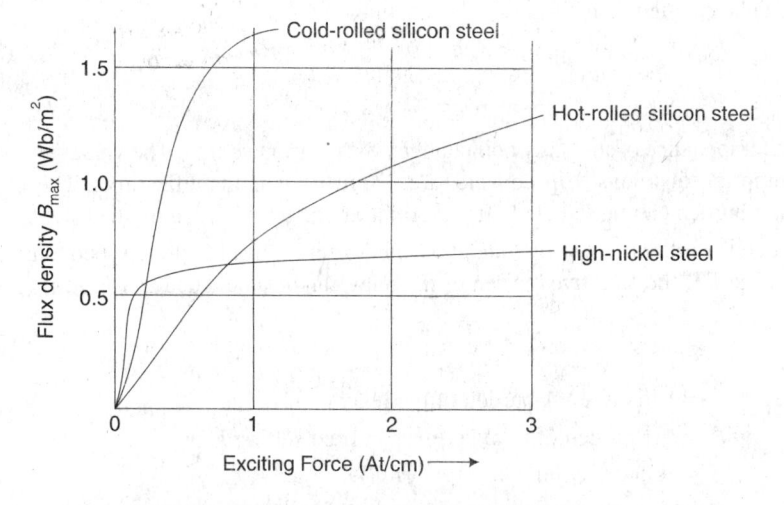

Fig. 3.1 *Magnetisation characteristics of CT cores*

Hot-rolled silicon steel has the lowest permeability. So it is not suitable for CTs. In order to achieve the desired characteristics, composite cores made of laminations of two or more materials are also used in CTs.

3.2.3 CT Burden

The CT burden is defined as the load connected across its secondary, which is usually expressed in volt amperes (VA). It can also be expressed in terms of impedance at the rated secondary current at a given power factor, usually 0.7 lagging. From the given impedance at rated secondary current, the burden in VA can be calculated. Suppose the burden is 0.5 Ω at 5 A secondary current. Its volt amperes will be equal to $I^2R = 5^2 \times 0.5 = 1.25$ VA. The total burden on the CT is that of the relays, meters, connecting leads and the burden due to the resistance of the secondary winding of the CT. The relay burden is defined as the power required to operate the relay. The burden of relays and meters is given by the manufactures or it can be calculated from the manufacturer's specifications as the burden depends on their type and design. The burden of leads depends on their resistance and the secondary current. Lead resistance is appreciable if long wires run from the switchyard to the relay panels placed in the control room. Lead burden can also be reduced using low secondary currents. Usually secondary current of 5 A are used, but current of 2 A or even 1 A can be used to reduce the lead burden. Suppose, the lead resistance is 5 Ω. Then lead burden at 5 A will be $5^2 \times 5 = 125$ VA. The burden at 1 A is only $1^2 \times 5 = 5$ VA. The economy in CT cost and space requirement demands shorter lead runs and sensitive relays. The rating of a large CT is 15 VA. For a 5 A secondary current, the corresponding burden is 0.6 Ω, and for a 1 A secondary current it is 15 Ω.

If rated burden be PVA at rated secondary current I_s amperes, the ohmic impedance of the burden Z_b can be calculated as follows:

$$Z_b = \frac{P}{I_s^2} \text{ ohms} \quad (3.1)$$

If burden power factor is cos ϕ, the values of resistance and reactance of the burden can be calculated as follows:

$$R_b = Z_b \cos \phi \quad (3.2)$$

$$X_b = \sqrt{Z_b^2 - R_b^2} \quad (3.3)$$

The impedance of the relay coil changes with current setting. The values of power consumption of relays, trip coil etc. are given by their manufacturers. The CT of suitable burden can be selected after calculating the total burden on the CT.

When the relay is set to operate at current different from the rated secondary current of the CT, the effective burden of the relay can be calculated as follows:

$$P_e = P_r \left(\frac{I_s}{I_r}\right)^2 \quad (3.4)$$

where, P_e = Effective VA burden of the relay on CT
P_r = VA burden of relay at given current setting I_r
I_s = Rated secondary current of CT
I_r = Current setting of the relay

The rated VA output of the CT selected should be the higher standard value nearest to the calculated value. If the VA rating of the CT selected is very much in excess of the burden, it makes the choice uneconomical and the CT becomes unduly large.

Current and Voltage Transformers

Example 3.1 Calculate the VA output required for a CT of 5 A rated secondary current when burden consists of relay requiring 7.5 VA at 5 A and connecting lead resistance of 0.08 ohm. Suggest the choices of the CT.

Solution: The simplified connection diagram is shown in Fig. 3.2.

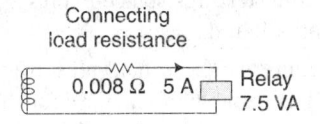

Fig. 3.2 System for Example 3.1

VA required to compensate connection lead resistance
$$= I^2 R = 5^2 \times 0.08 = 25 \times 0.08 = 2.0 \text{ VA}$$
VA required by the relay = 7.5 VA
Hence total VA output required for CT
$$= 7.5 \text{ VA} + 2.0 \text{ VA} = 9.5 \text{ VA}$$
Hence a CT of rating 10 VA and secondary current 5 A may be selected.

Example 3.2 The rated secondary current of a CT is 5 A. The plug setting of a relay is 3.75 A. The power consumption of the relay at this plug setting is 4 VA. Calculate the effective VA burden on the CT.

Solution: Following values are given

P_r = VA burden of the relay at 3.75 A plug setting
= 4 VA
I_s = Rated secondary current of CT = 5 A
I_r = Current setting of the relay = 3.75 A

P_e (The effective VA burden on the CT) is to be calculated.

$$P_e = P_r \left(\frac{I_s}{I_r}\right)^2 = \left(4 \times \frac{5}{3.75}\right)^2 = 7.11 \text{ VA}$$

Hence a CT having a rated VA output of 10 VA can be selected.

3.2.4 Technical Terms of CTs

The following are some of the commonly used terms for current transformers (CTs)

(i) **Rated primary current** The value of the primary current which is marked on the rating plate of the transformer and on which the performance of the CT is specified by the manufacturer.

(ii) **Rated secondary current** The value of the secondary current which is marked on the rating plate of the transformer and on which the performance of the CT is specified by the manufacturer.

(iii) **Rated transformation ratio** The ration of the rated primary current to rated secondary current. It is also called nominal transformation ratio.

(iv) **Actual transformation ratio** The ratio of the actual primary current to the actual secondary current.

(v) **Burden** The value of the load connected across the secondary of CT, expressed in VA or ohms at rated secondary current.

(vi) **Rated burden** The value of the load to be connected across the secondary of CT including connecting lead resistance expressed in VA or ohms on which accuracy requirement is based.

(vii) **Rated short-time current** The r.m.s value of the a.c. component of the current which the CT is capable of carrying for the rated time without being damaged by thermal or dynamic effects.

(viii) **Rated short-time factor** The ratio of rated short-time current to the rated current.

(ix) **Rated accuracy limit primary current** The highest value of primary current assigned by the CT manufacturer, upto which the limits of composit error are complied with.

(x) **Rated accuracy limit factor** The ratio of rated accuracy limit primary current to the rated primary current.

(xi) **Composit error** The r.m.s. value of the difference $(N\, i_s - i_p)$, given by

$$\text{Composit error} = \frac{100}{I_p} \sqrt{\frac{1}{T} \int_0^T (Ni_s - i_p)} \qquad (3.5)$$

Where N = Rated transformation ratio

I_p = r.m.s value of the primary current

i_p = Instantaneous values of the primary current

i_s = Instantaneous values of the secondary current

T = Time period of one cycle in seconds

(xii) **Knee-point voltage** The sinusoidal voltage of rated frequency (system frequency) applied to the secondary terminals of CT, with all other winding being open-circuited, which when increased by 10 per cent, causes the exciting current to increase by 50 per cent. Minimum knee point voltage is specified by the following expression.

$$V_k = KI\,(R_{CT} + Z_s) \qquad (3.6)$$

Where K = A parameter to be specified by the purchaser depending on the system fault level and the characteristic of the relay intended to be used.

I = Rated relay current (1 A or 5 A)

R_{CT} = Resistance of CT secondary winding corrected to 75° C

Z_s = Impedance of the secondary circuit (to be specified by the purchaser)

(xiii) **Rated short-circuit current** The r.m.s. value of primary current which the CT will withstand for a rated time with its secondary winding short-circuited without suffering harmful effects.

(xiv) **Rated primary saturation current** The maximum value of primary current at which the required accuracy is maintained

(xv) **Rated saturation factor** The ratio of rated primary saturation current to rated primary current.

3.2.5 Theory of Current Transformers

Conventional electromagnetic current transformers (CTs) are single primary and single secondary magnetically coupled transformers. Hence, their performance can be analysed from and equivalent circuit commonly used in the analysis of transformers. The equivalent circuit of CT as viewed from secondary side is shown in Fig. 3.3. It is convenient to put the exiting shunt circuit on the secondary side and to refer all quantities to that side, so that I'_p denotes the primary current referred to the secondary side. The exciting current I_0 is deducted from I'_p to excite the core and induce voltage E_s which circulates current I_s.

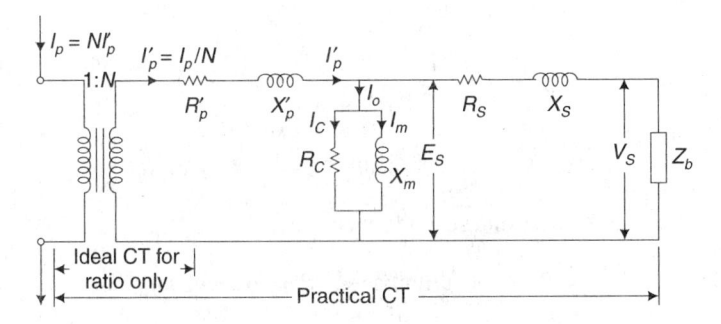

Fig. 3.3 Equivalent circuit of CT as viewed from secondary side

An ideal (perfect) transformer shown in Fig. 3.3 is to provide the necessary ratio change, it has no loss or impedance. All the quantities are referred to the secondary side. In an ideal CT, the primary ampere-turns (AT) is exactly equal in magnitude to the secondary AT and is in precise phase opposition to it. But in practical (actual) CTs errors are introduced both in magnitude and in phase angle. These errors are known as ratio error and phase angle error. The exciting current I_0 is the main source of these errors. Practical CTs do not reproduce the primary currents exactly in magnitude and phase due to these errors.

The errors of practical CTs can best be considered through the study of the phasor diagram shown in Fig. 3.4.

In the equivalent circuit of Fig. 3.3 and phasor diagram of Fig. 3.4,

N = Nominal (rated) transformation ratio or Nominal (rated) CT ratio

$$= \frac{\text{Rated primary current}}{\text{Rated secondary current}} = \frac{\text{Number of secondary turns}}{\text{Number of primary turns}}$$

R_p, X_p = Primary resistance and leakage reactance respectively.

R_s and X_s = secondary resistance and leakage reactance

R'_p and X'_p = Primary resistance and leakage reactance as referred to the secondary side

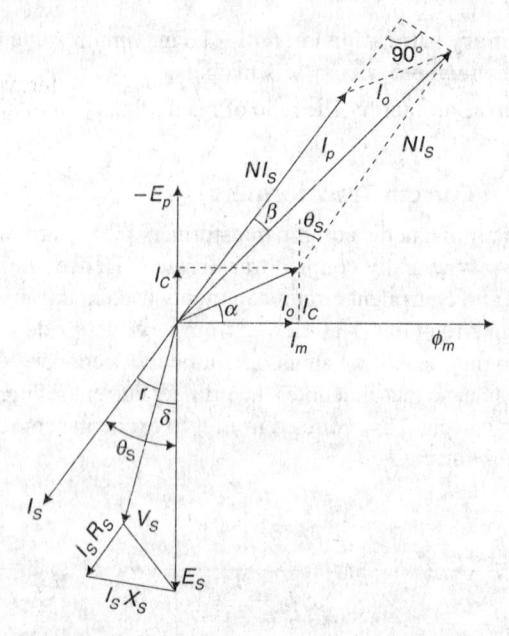

Fig. 3.4 *Phasor diagram of a current transformer*

I_p and I_s = Primary and secondary currents

$I'_p = \dfrac{I_p}{N}$ = Primary current as referred to secondary side

E_s = Secondary induced voltage

V_s = Secondary terminal voltage

E_p = Primary induced voltage

I_0 = CT secondary excitation current

$I_s = I'_p - I_0 = I_p/N - I_0$

   = Secondary current delivered to the burden (load)

I_m = Magnetising component of the exciting current I_0 required to produce flux.

I_c = Iron (core) loss component of the exciting current I_0 supplying core losses (eddy current and hysteresis)

ϕ_m = Main core flux

δ = Phase angle due to secondary winding

γ = Phase angle due to burden

θ_s = Total secondary phase angle ($\theta_s = \delta + \gamma$) phase angle between secondary current and secondary induced voltage.

α = Loss angle due to core excitation

β = Phase angle between primary and reversed secondary current

Z_b = Load impedance (burden on the CT)

Since the core flux ϕ_m is common to both primary and secondary windings, it is taken as reference phase. The induced voltages E_s and E_p lag behind the flux by 90° and their magnitudes are proportional to the secondary and primary turns respectively. The excitation current I_0 is made up of two components I_m and I_c. The secondary current is transferred to the primary side by reversing I_s and multiplying by the turns ratio N. The resultant current flowing in the primary winding 1_p is then the phasor sum of NI_s and I_0.

From the phasor diagram of Fig. 3.4,

$$I_p^2 = (I_c + NI_s \cos\theta_s)^2 + (I_m + NI_s \sin\theta_s)^2$$
$$= (I_0 \sin\alpha + NI_s \cos\theta_s)^2 + (I_0 \cos\alpha + NI_s \sin\theta_s)^2$$

Hence, $I_p = [(I_0 \sin\alpha + NI_s \cos\theta_s)^2 + (I_0 \cos\alpha + NI_s \sin\theta_s)^2]^{1/2}$

$$= (I_0^2 \sin^2\alpha + 2 I_0 \sin\alpha \cdot NI_s \cos\theta_s + N^2 I_s^2 \cos^2\theta_s + I_0^2 \cos^2\alpha$$
$$+ 2 I_0 \cos\alpha \cdot NI_s \sin\theta_s + N^2 I_s^2 \sin^2\theta_s)^{1/2}$$

Neglecting terms containing I_0^2, this becomes

$$I_p = [N^2 I_s^2 (\cos^2\theta_s + \sin^2\theta_s) + 2NI_s I_0 (\sin\alpha \cdot \cos\theta_s + \cos\alpha \sin\theta_s)]^{1/2}$$
$$= [N^2 I_s^2 + 2NI_s \sin(\alpha + \theta_s)]^{1/2}$$

which, to a very close approximation,

$$= NI_s + I_0 \sin(\alpha + \theta_s)$$

Since I_0 is small compared with NI_s.

Hence $I_p = NI_s + I_0 \sin(\alpha + \theta_s)$ \hfill (3.7)

Actual Transformation Ratio

The actual transformation ratio (actual ratio) N_a is given by,

$$N_a = \frac{I_p}{I_s} = \frac{NI_s + I_0 \sin(\alpha + \theta_s)}{I_s}$$

or

$$N_a = N + \frac{I_0}{I_s} \sin(\alpha + \theta_s) \hfill (3.8)$$

Though this expression is only approximate, but it is sufficiently accurate for almost all purposes. The expression can be further expanded as

$$N_a = N + \frac{I_0}{I_s}(\sin\alpha\cos\theta_s + \cos\alpha\sin\theta_s)$$

or

$$N_a = N + \frac{(I_0 \sin\alpha\cos\theta_s + I_0 \cos\alpha\sin\theta_s)}{I_s}$$

or

$$N_a = N + \frac{(I_c \cos\theta_s + I_m \sin\theta_s)}{I_s} \hfill (3.9)$$

Since $I_0 \sin\alpha = I_c$ and $I_0 \cos\alpha = I_m$

If the angle θ_s which is usually fairly small, is assumed zero, then the actual ratio is

$$N_a = N + \frac{I_c}{I_s} \tag{3.10}$$

Phase-angle Error

The angle by which the secondary current phasor, when reversed, differs in phase from the primary current phasor, is the phase angle error of the current transformer.

From the phasor diagram of Fig. 3.4

$$\tan \beta = \frac{I_0 \sin [90° - (\alpha + \theta_s)]}{NI_s + I_0 \cos [90° - (\alpha + \theta_s)]}$$

$$= \frac{I_0 \cos (\alpha + \theta_s)}{NI_s + I_0 \sin (\alpha + \theta_s)}$$

Since the phase angle error β is very small, in practice $\tan \beta = \beta$
Hence, from above expression

$$\beta = \frac{I_0 \cos (\alpha + \theta_s)}{NI_s + I_0 \sin (\alpha + \theta_s)} \tag{3.11}$$

or

$$\beta = \frac{I_0 (\cos \alpha \cdot \cos \theta_s - \sin \alpha \cdot \sin \theta_s)}{NI_s + I_0 \sin (\alpha + \theta_s)}$$

$$\simeq \frac{I_m \cos \theta_s - I_c \sin \theta_s}{NI_s}$$

Since $I_0 \sin (\alpha + \theta_s)$ is small compared with NI_s. Hence phase-angle error β is given by

$$\beta = \frac{I_m \cos \theta_s - I_c \sin \theta_s}{NI_s} \tag{3.12}$$

If the angle θ_s, which is usually fairly small is assumed zero, then the phase angle eror is

$$\beta = \frac{I_m}{NI_s} \tag{3.13}$$

3.2.6 CT Errors

In an ideal (perfect) CT, the secondary current is given by

$$I_s = \frac{I_p}{N}$$

But in a practical (actual) CT, it is

$$I_s = \frac{I_p}{N} - I_0 \tag{3.13a}$$

Thus, the actual CT does not reproduce the primary current exactly in secondary side both in magnitude and phase due to exciting current I_0. The exciting current I_0 is the main source of errors in both measuring and Protective CTs. The error in magnitude is due to error in CT ratio which is called "ratio error" and the error in phase is called "phase-angle error."

Ratio Error (Current Error)

The actual transformation ratio (N_a) is not equal to the rated (nominal) transformation ratio (N) since the primary current is contributed by the exciting current. The error introduced due to this difference in CT ratios is termed as ratio error or current error. The ratio error in percentage is expressed as

$$\text{Per cent ratio error} = \frac{\text{Nominal ratio} - \text{Actual ratio}}{\text{Actual ratio}} \times 100$$

$$= \frac{N - N_a}{N_a} \times 100$$

or

$$\text{Per cent ratio error} = \frac{N - I_p/I_s}{I_p/I_s} \times 100$$

or

$$\text{Per cent ratio error} = \frac{NI_s - I_p}{I_p} \times 100 \qquad (3.14)$$

Where
N = Nominal (rated) ratio

$$= \frac{\text{Rated primary current}}{\text{Rated secondary current}}$$

$$= \frac{\text{Number of secondary turns}}{\text{Number of primary turns}}$$

$$N_a = \text{Actual ratio} = \frac{I_p}{I_s}$$

I_p = Primary Current

I_s = Secondary current

I_p can be calculated by using Equation (3.7) and N_a by using Equation (3.9) or (3.10).

The ratio error is largely dependant upon the value of the iron-loss component I_c of the exciting current.

The ratio error is considered to be positive when the actual ratio of the CT is less than the nominal ratio, i.e., when the secondary current, for a given primary current, is high.

Phase-Angle Error

For a perfect (ideal) current transformer, the phase difference between the primary and reversed secondary phasors is zero. But for an actual transformer, there is always a difference in phase between the two phasors due to the fact that primary current has to supply the components of the exciting current. The phase difference between the primary current phasor and the reversed secondary current phasor is termed as the 'phase-angle error' of the CT. For sinusoidal current, it is said to be positive when the reversed secondary current phasor leads the primary current phasor. The phase angle error β can be calculated by using Equation (3.12) or (3.13).

The phase-angle error is largely dependant upon the value of the magnetizing component I_m of the exciting current.

Ratio Correction Factor

The term 'ratio correction factor' is defined as that factor by which nominal (rated) ratio of a CT is multiplied to obtain the actual (true) ratio. In other words, this factor can be defined as ratio of the actual ratio (N_a) to the nominal ratio (N).

The ratio error in per unit is expressed as

$$\varepsilon = \frac{N - N_a}{N_a} \qquad (3.15)$$

or

$$\varepsilon = \frac{N}{N_a} - 1$$

or

$$1 + \varepsilon = \frac{N}{N_a}$$

or

$$\frac{N_a}{N} = \frac{1}{1 + \varepsilon}$$

Hence,

$$\text{Ratio correction factor } (R_{cf}) = \frac{1}{1 + \varepsilon} \qquad (3.16)$$

3.2.7 Accuracy Class of CTs

The accuracy of any CT is determined essentially by how accurately the CT reproduces the primary current in the secondary. Accuracy class is assigned to the CT with the specified limits of ratio error and phase angle error.

The accuracy of a current transformer is expressed in terms of the departure of its ratio from its true ratio. This is called the ratio error, and is expressed as:

$$\text{Per cent error} = \left[\frac{NI_s - I_p}{I_p}\right] \times 100$$

$$N = \text{Nominal ratio} = \frac{\text{rated primary current}}{\text{rated secondary current}}$$

I_s = Secondary current

I_p = Primary current

The ratio error of a CT depends on its exciting current. When the primary current increases, the CT tries to produce the corresponding secondary current, and this needs a greater secondary *emf*, core flux density and exciting current. A stage comes when any further increase in primary current is almost wholly absorbed in an increased exciting current, and thereby the secondary current hardly increases at all. At this stage, the CT becomes saturated. Thus the ratio error depends on saturation. An accuracy of about 2% to 3% of the CT is desirable for distance and differential relays, whereas for many other relays, a higher percentage can be tolerated.

According to standards followed in U.K., protective CTs are classified as *S*, *T* and *U* type. The errors of these types of CT s are shown in Table 3.1.

Table 3.1 CT Errors

Class	Ratio Error	I_0/I_{sat}	Phase Angle Error in Degrees
S	± 3%	3%	2
T	± 10%	10%	6
U	± 15%	15%	9

I_0 = Exciting current,
I_{sat} = Saturating current

When the primary current increases, at a certain value the core commences to saturate and the error increases. The value of the primary current at which the error reaches a specified limit is known as its *accuracy limit primary current or saturation current*. The maximum value of the primary current for a given accuracy limit is specified by the manufacturer. The CT will maintain the accuracy at the specified maximum primary current at the rated burden. This current is expressed as a multiple of the rated current. The ratio of accuracy limit primary current and rated primary current is known as the rated *accuracy limit factor or saturation factor*, the standard values of which are 5, 10, 15, 20 and 30. The performance of a CT is given at certain multiples of the rated current. According to BSS 3938, rated primary currents of CTs are up to 75 kA and secondary currents 5 A or 1 A.

3.2.8 Transient Behaviour of CTs

For fast relaying (within one or two cycles after the fault inception), it is very essential to know the behaviour of the CT during the first few cycles of a fault, when it carries the transient component in addition to the steady-state component of the fault current. For calculation of the fault current, the power system is considered as a lumped R, L series circuit, and the effect of shunt admittance is neglected.

When a fault occurs on a power system, the fault current is given by

$$i_p = \frac{V_{pm}}{Z_p} \sin(\omega t + \alpha - \phi_p) + e^{-(R_p/L_p)t} \cdot \frac{V_{pm}}{Z_p} \sin(\phi_p - \alpha) \qquad (3.17)$$

where, the subscript p indicates the primary side of the CT.

$\phi_p = \tan^{-1}\left(\dfrac{\omega L_p}{R_p}\right)$ is the phase angle of the primary circuit. The fault is assumed to occur at $t = 0$. The parameter α controls the instant on the voltage wave at which the fault occurs.

In Eq. 3.17 for the fault current, the first term is the sinusoidal steady-state current which is called the symmetrical ac component, while the second term is the unidirectional transient component which starts at a maximum and decays exponentially and is called the dc offset current. The dc offset current causes the total fault current to be unsymmetrical till the transient decays. The waveform of the fault current is shown in Fig. 3.5.

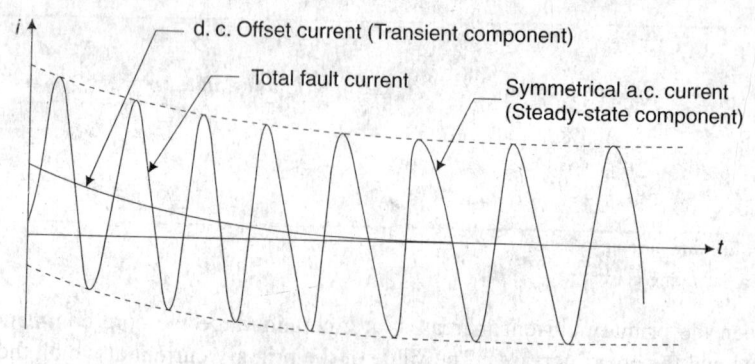

Fig. 3.5 *Waveform of the fault current*

The dc offset current (the transient current) will have a maximum value when $\phi_p - \alpha = \pi/2$ radians and Eq. (3.17) reduces to

$$i_p = I_{pm}[-\cos \omega t + e^{-(R_p/L_p)t}] \qquad (3.18)$$

where
$$I_{pm} = \frac{V_{pm}}{Z_p}, \quad \text{and} \quad Z_p = \sqrt{R_p^2 + (\omega L_p)^2}$$

Normally, as the system rated voltage rises above 100 kV, the circuit becomes highly inductive with negligibly small resistance and the phase angle ϕ_p tending to $\pi/2$ rad (or 90°) and α tending to be zero. In the limit where $\phi_p = \pi/2$ rad, fault at zero voltage (for $\alpha = 0$) gives rise to the maximum fault current asymmetry (current doubling).

The primary fault current referred to the secondary side is

$$i'_p = i_p \left(\frac{N_p}{N_s}\right)$$

and let
$$I'_{pm} = I_{pm} \left(\frac{N_p}{N_s}\right)$$

where N_p and N_s are the number of turns in the primary and secondary, respectively. In order to estimate the transient flux in the core, the magnetizing current will be neglected as this is one of the worst cases and Eq. (3.18) will be used, i.e. the case of maximum transient primary current.

The secondary voltage is

$$v_s = z_s \, i_s = N_s \, (d\phi/dt) \qquad (3.19)$$

where, the subscript s indicates the secondary side of the CT and Z_s is the CT burden.

From Eq. (3.19), the maximum core flux is given by

$$\phi_m = \left(\frac{1}{N_s}\right) \int v_s \, dt \qquad (3.20)$$

Thus, for the steady-state component of i_s, and integrating over a quarter cycle

$$\phi_m = -\left(\frac{1}{N_s}\right) \int_{t=0}^{\pi/2\omega} Z_s \, I'_{pm} \cos \omega t \, dt$$

$$= -\left(\frac{Z_s I'_{pm}}{N_s \omega}\right) \quad (3.21)$$

For the transient component of i_s, and integrating from $t = 0$ to $t = \infty$

$$\phi_m = \left(\frac{1}{N_s}\right) Z_s I'_{pm} \int_{t=0}^{\infty} e^{-\left(\frac{R_p}{L_p}\right)t} dt$$

$$= \left(\frac{Z_s I'_{pm}}{N_s \omega}\right)\left(\frac{\omega L_p}{R_p}\right)$$

$$= \left(\frac{Z_s I'_{pm}}{N_s \omega}\right)\left(\frac{X_p}{R_p}\right) \quad (3.22)$$

Therefore, the component of the core flux due to the transient component of the fault current is X_p/R_p times the component of the core flux due to the steady-state component of the fault current. Considering the worst case of a fault near a large power station, X_p/R_p could be as high as 30, corresponding to a primary-circuit time constant of $L_p/R_p = 0.1$ s, or about 5 cycles.

Assuming that the two fluxes can be added numerically (the worst case), the total core flux is equal to the steady-state component multiplied by the factor

$$1 + \left(\frac{X_p}{R_p}\right) = 1 + \frac{2\pi f L_p}{R_p} = 1 + 2\pi T_p$$

where T_p is the primary circuit time-constant in cycles. The flux waveforms are shown in Fig. 3.6

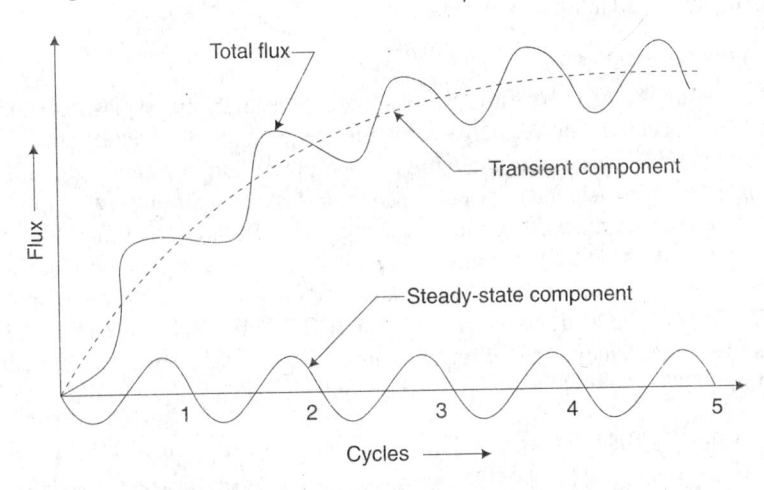

Fig. 3.6 CT core fluxes during transient period

The value of X_p/R_p increases with the system voltage because of the increased spacing of the conductors. The component of the core flux due to transient dc offset current increases X_p/R_p times the flux due to the symmetrical (steady-state) ac current. Therefore, it is clear that there is a large transient flux swing in the magnetic core of the CT. With this large flux-swing during the transient period, the magnetic core of the CT of conventional design will get saturated causing undesirable effects on the performance of the protective relays connected to the secondary of the CT.

In current transformers of conventional design, saturation of the cores due to the transient dc component of the fault current is possible within a few milliseconds, after which their secondary current is fully distorted, resulting in an inaccurate measurement of the fault current by the relay. In order to prevent an adverse effect upon the performance of the relays, the current transformer cores must be greatly enlarged or air-gaps should be provided in the cores.

The choice of the appropriate current transformer design depends upon the protective system requirement. Where the operation of the protective system is not affected by the CT core saturation, as in the case of a plain overcurrent relay, the conventional type of CT may be used. But where a saturation-free current transformation is essential for correct and rapid working of the system protection, the dimension of the CT core must be greatly increased. The increase of core section in such cases leads us to unreasonable core sizes with the use of iron-enclosed cores in high-power systems. Therefore, for such cases, cores having air-gaps, called linear cores have been developed. By providing such air gaps, the time constant of the CT is reduced to a great extent since the current main flux density is diminished. Generally, the flux due to the dc component assumes smaller values, if the CT time constant is reduced. Therefore, CT cores with air gaps are almost free from the problem of saturation and consequent distortion of the secondary current.

The linear cores provide an entirely new solution to a wide range of protective system providing saturation free transformation of transient phenomena with dc offset components of great time constants.

3.2.9 Linear Couplers

The CT core may be of iron or non-ferrous material, usually air or plastic. An iron-cored CT has substantial power output which is adequate for electromagnetic relays. A CT having non-ferrous material as core has low power output which is suitable for static relays but is inadequate for electromagnetic relays. An air or plastic-cored CT has a linear characteristic and is called a linear coupler. Such CTs have no saturation limit and hence show no transient errors, i.e ratio and phase angle errors which arise due to saturation. Problems caused by dc transients are also reduced to a great extent. Such CTs do not have lead resistance problem. A CT with a small air gap in its iron core has also linear characteristic, and has no transient errors. Such CTs are called *transactors*.

3.2.10 Classification of CTs

Current transformers can be classified in various ways depending on the technology used for their construction and operation, their application, their location, etc.

Classification of CTs Based on Technology

CTs can be broadly classified into the following categories, depending on the technology they use for their construction and operation

(i) Electromagnetic CTs
(ii) Opto-electronic CTs
(iii) Rogowski Coil

These CTs are described in subsequent sections.

Classification of CTs Based on Application

Depending on application, CTs are broadly classified into following two categories

(i) Measuring or metering CTs
(ii) Protective CTs

These CTs are described in detail in Section 3.2.1.

Classification of CTs Based on Location

Depending on location for application in the field, CTs are broadly classified into following two categories.

(i) Indoor CTs
(ii) Outdoor CTs

Indoor CTs

CTs designed for mounting inside metal cubicles are known as indoor CTs. Depending upon the method of insulation, these can be further classified as (a) tape insulated, and (b) cast resin (epoxy, polyurethene or polycrete).

In terms of constructional aspects, these can be further classified into three types (a) Bar type CTs (b) slot/window/Ring Type CTs, and (c) wound type CTs.

Outdoor CTs

These CTs are designed for outdoor applications. They use transformer oil or any other suitable liquid for insulation and cooling.

3.2.11 Electromagnetic CTs

Conventional CTs are of electromagnetic type. Electromagnetic CTs work on the fundamental principle of electromagnetic induction. These CTs can be classified into the following two categories:

(i) Bar primary CTs or Toroidal CTs
(ii) Would primary CTs

Bar Primary CTs

Bar primary-type CTs do not contain a primary winding and instead a straight conductor (wire) which is a part of the power system and carries the current acts as the primary. In this type of CTs, the primary does not necessarily form a part of the CT. The primary conductor (wire) that carries the current is encircled by a ring type iron core on which the secondary winding is wound uniformly over the entire periphery to form a toroid. A small gap is left between start and finish leads of the winding for insulation. If the winding requires more than one layer of wire, each layer should be complete. If however, the number of turns of adequate wire gauge would not occupy a whole layer, the turns may be spaced slightly apart so that the winding can be uniformly distributed. Since the secondary winding of this type of CT is toroidally wound, it is also called *toroidal CT*. Figure 3.7 shows a CT of this type.

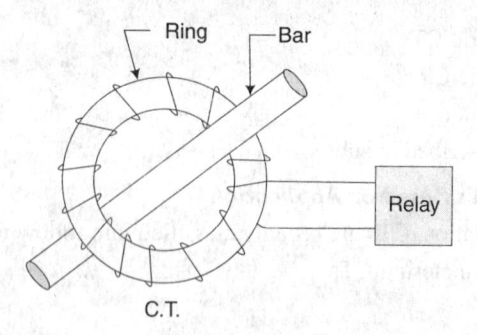

Fig. 3.7 CT Primary-bar and secondary-ring

This type of construction has a negligible leakage flux of both primary and secondary and hence, possesses low reactance. As there is only one primary turn, the primary current should be high enough (about 400 A) to produce sufficient exciting ampere-turns (AT) to give reasonable output. A bushing CT is a sub-class of the bar primary type CT. It is placed over an insulator bushing enclosing a straight conductor. The bushing CT has comparatively more exciting current and a large magnetic path owing to the large diameter of the bushing.

Wound Primary CTs

Wound primary CTs have the primary and secondary windings arranged concentrically, the secondary winding invariably being the inner winding. The core is composed either of hot-rolled silicon steel stampings or, in recent most technique, of cold rolled grain-oriented steel or of nickel-iron alloy. Cores are usually assembled from stampings of E, I, L or C shape depending on whether the core is a simple rectangle or of the three-limb shell-type. Cores of grain-oriented materials should be arranged as far as possible with the flux direction along the grain.

The primary windings of wound primary CTs usually take the form of edge-wound copper strip because this method of winding results in coils best capable of withstanding the electromagnetic forces produced by high values of the primary current. Secondary windings are usually wound from round section enamelled copper wire.

3.2.12 Opto-Electronic CTs

Opto-electronic CTs use light to measure the magnetic field surrounding a current carrying conductor and, based on this measurement, electronics associated with the optics calculate the current flowing in the conductor. Opto-electronic CTs are also known as 'optical current sensors'.

Opto-electronic CTs consist of a light source, photo-detector, optics and electronics coupled to a fibre-sensing head wound around a current carrying conductor. The optical phase modulator is the 'heart' of the optical current sensor technology and it, along with the electronics and optics, provide a highly accurate measurement of current.

Opto-electronic CTs generate two linearly polarized light beams by sending light through a waveguide to a linear polarizer, then to a polarization splitter, and finally to an optical phase modulator. The two linearly polarized light beams are then applied to a polarizing retaining optical fiber that takes them to the "sensing head". A circular polarizer converts one beam to a left circularly polarized beam and the other to a right circularly polarized beam, as shown in Fig 3.8(a) and (b).

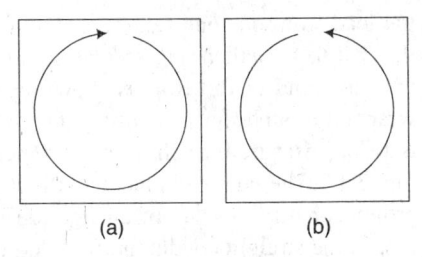

Fig. 3.8 *(a) Left polarized beam (b) Right polarized beam*

The circular polarized beams travel around the phase conductor many times. While encircling the conductor, the magnetic field induced by the current flowing in the conductor creates a differential optical phase shift between the two light beams due to the Faraday effect. As a result of this, one beam is accelerated and the other beam is decelerated. After having gone around the conductor many times, the beams are reflected by a mirror. The reflection changes the polarization, and as a result, the right polarized beam becomes a left polarized beam and the left polarized beam becomes a right polarized beam. The beams travel back around the conductor towards the polarizer. The travel on the return path further accelerates the accelerated beam and further decelerates the decelerated beam. The polarizer changes the circular polarizations to linear polarization.

When there is no current in the conductor, the beams are neither accelerated nor decelerated. In this case, the returning beams are in phase as shown in Fig. 3.9(a). When there is a current in the conductor, the acceleration of one beam and deceleration of the other beam manifests in a phase shift between the two beams as shown in Fig. 3.9(b).

The two light beams are finally routed to the optical detector where the electronics de-modulate them to determine the phase shift. The phase shift between the two light beams is proportional to current in the conductor. An analog or digital signal representing the current is provided by the electronics to the end user.

Opto-electronic CTs provide a reliable method of measuring very high fault currents with significant dc offsets without any type of saturation, as is understood with conventional electromagnetic CTs.

3.2.13 Rogowski Coil

A Rogowski Coil, named after Walter Rogowski, is an 'air-cored' toroidal coil of wire used for measuring alternating current (ac) or high frequency current pulses. It is the modern current sensor. It consists of a helical coil of wire with

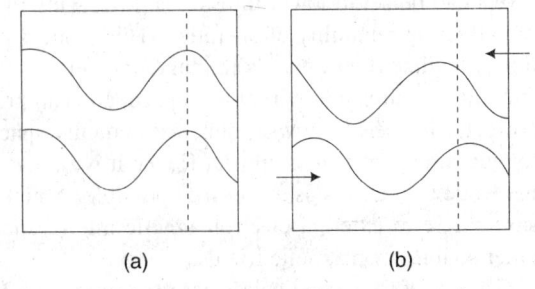

Fig. 3.9 *(a) Beams in phase (b) Acceration of one beam and deceleration of the other beam*

the lead from one end returning through the centre of the coil to the other end, so that both terminals are at the same end of the coil, as shown in Fig. 3.10. It is wrapped around the straight conductor whose current is to be measured. Such an arrangement is shown in Fig. 3.11. The voltage induced in the Rogowski coil is proportional to the rate of change (derivative) of current in the straight conductor. In order to get an output voltage proportional to the current, the output of the Rogowski coil is usually connected to an integrator which may be either active or passive. An active integrator using operational amplifier as shown in Fig. 3.12 is a common solution. Such a simple idea of sensing ac current could not be put into common practice earlier as the electronic integrator design was not so advanced.

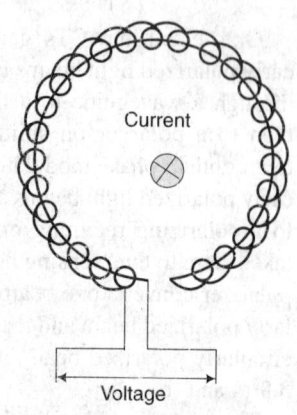

Fig. 3.10 Rogowski coil

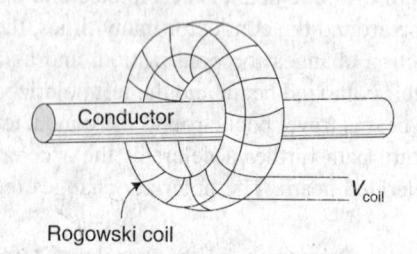

Fig. 3.11 Rogowski coil wrapped around the straight conductor

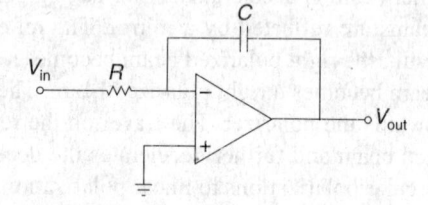

Fig 3.12 An active integrator using op-amp

One advantage of a Rogowski coil over other types of current transformers is that it can be made open-ended and flexible, so that it can be wrapped around a live conductor without disturbing it. Since a Rogowski coil has an air core rather than an iron core, it has a low inductance and does not saturate. Being of low inductance it can respond to very fast changing currents. Since it has no iron core to saturate, it responds linearly to extremely large currents. A conventional current transformer can have its core saturated at very high currents, and the inductance limits its frequency response. Since Rogowski coils are very accurate and do not saturate, protection levels can be set to lower thresholds, increasing the sensitivity of the scheme without affecting reliability of operation. This reduces the stress on protected equipment during faults. The system is immune to external magnetic fields, is simple and user friendly, requires less wiring and space, and can provide metering class accuracy. A correctly formed Rogowski coil, with equally spaced windings, is largely immune to electromagnetic interference. Hence it is important to ensure that the winding of the Rogowski coil is as uniform as possible. A non-uniform winding makes the coil susceptible to external electromagnetic interference due to adjacent conductors or other sources of magnetic frields.

Theory of Rogowski coil is based on Faraday's law of electromagnetic induction which states that the total voltage induced in a closed circuit is proportional to the rate of change of total magnetic flux linking the circuit.

Each turn of the Rogowski coil produces a voltage proportional to the rate of change of the magnetic flux through the turn. Assuming a uniform magnetic flux density throughout the turn, the rate of change of magnetic flux is equal to the rate of change of magnetic flux density (B) times the cross-sectional area (A) of the turn (one of the smallest loops).

The voltage induced in one turn of the Rogowski coil is given by

$$v_{turn} = \frac{d\phi}{dt} = A\frac{dB}{dt} \qquad (3.23)$$

For a coil with N turns, the voltage induced in the coil is

$$v_{coil} = NA\frac{dB}{dt} \qquad (3.24)$$

The flux density B at radial distance r due to a long straight conductor carrying current i is

$$B = \frac{\mu_0 i}{2\pi r} \qquad (3.25)$$

Where r is the perpendicular radial distance from the conductor to the point at which the magnetic flux (field) density is calculated. The direction of the magnetic field being perpendicular to the current i and the radius r, and determined by using the right hand rule.

By substituting the value of B from Eq. (3.25) in Eq. (3.24), the voltage of the whole coil is given by

$$v_{coil} = \frac{NA\mu_0}{2\pi r}\frac{di}{dt} \qquad (3.26)$$

where N = Number of turns on the coil
A = Cross-sectional area of the turn (one of the smallest loops) in m^2
$\mu_0 = 4\pi \times 10^{-7}$ H/m is the permeability of the free space
r = Radius of the coil in m
$\frac{di}{dt}$ = Rate of change of the current

Equation (3.26) can also be written in the following form

$$v_{coil} = \mu_0 nA\frac{di}{dt} \qquad (3.27)$$

where $n = \frac{N}{2\pi r} = \frac{N}{l}$ is the turns per unit length of the winding (coil) in turns/m as $l = 2\pi r$ is the length of the winding (circular coil)

In order to get an output voltage proportional to the current, the output voltage of the Rogowski coil must be integrated.

An active integrator, using an operational amplifier is shown in Fig. 3.12.
The output of the integrator is given by

$$v_{out} = \frac{1}{R_c}\int v_{in}\, dt \qquad (3.28)$$

where $v_{in} = v_{coil}$

After integrating the signal of Eq. (3.27) the total output voltage is

$$v_{out} = \frac{\mu_0 nA}{R_c} i \qquad (3.29)$$

The r.m.s value of the output voltage (v_{out}) is given by

$$V_{out} = \frac{\mu_0 nA}{R_c} I \qquad (3.30)$$

Traditional Rogowski coils consist of wire wound on a nonmagnetic core. Strict design criteria must be followed to obtain a coil immune from nearby conductors and independent of conductor location inside the coil loop. In order to achieve the design criteria to prevent the influence of nearby conductors, the coil is designed with two-wire loops connected in electrically opposite directions. Such design of the coil cancels the electromagnetic fields coming from outside the coil loop. A recent design of the Rogowski coil consists of two wound coils implemented on a pair of printed circuit boards located next to each other as shown in Fig. 3.13(a) and (b).

The Rogowski coil output voltage is in the range of millivolt to several volts and can reliably drive digital devices designed to accept low power signals. Integration of the signals can be performed in the relay by using analog circuitry or digital signal processing techniques or immediately at the coil. Connections to relays can be by wires or through fibreoptic cables.

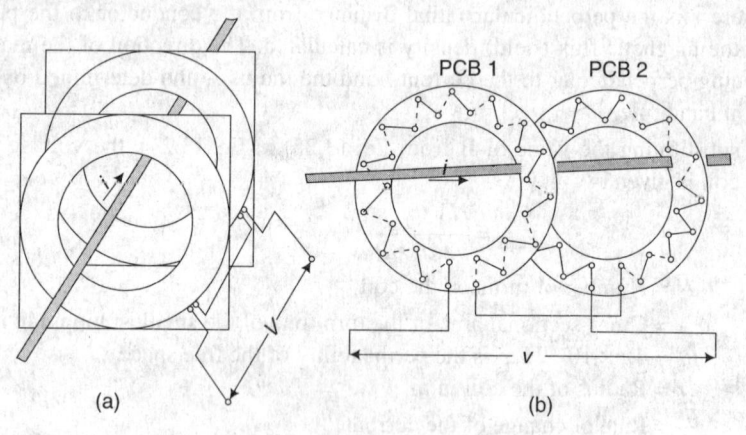

Fig. 3.13 Recent design of Rogowski coil: (a) Front view (b) Side view

The Rogowski coil current sensor has the following characteristics.
  (i) Measurement accuracy reaching 0.1%
 (ii) Wide measurement range (the same coil can measure currents from 1A to over 100 KA)
(iii) Frequency response linear up to 1 MHz
(iv) Short-circuit withstand is unlimited with the window-type design
 (v) Galvanic isolation from the primary conductors (similar to current transformers)
(vi) Can be encapsulated and located around bushings or cables avoiding the need for high insulation
(vii) Can be custom sized for applications
(viii) Can be built as split core style for installation in existing systems

3.2.14 Auxiliary CTs

In many relaying applications, auxiliary CTs are used to provide isolation between the main CT secondary and some other circuit. They are also used for providing an adjustment to the overall current transformation ratio. CTs have standard ratios.

When in any application other than a standard ratio is required, an auxiliary CT is used to provide a convenient method of achieving the desired ratio. The auxiliary CT, however contributes to the overall errors of transformation. The possibility of auxiliary CT's saturation should be taken into account. In order to provide a variable turns ratio, auxiliary CTs with multiple taps are also available. The burden connected into the auxiliary CT secondary is reflected into the main CT secondary, according to the normal rules of transformation.

3.2.15 Flux-summing CT

Zero sequence current in a three-phase system can be obtained by using a single CT, if three phase conductors are passed through the window of a toroidal CT as shown in Fig. 3.14. In this case, the secondary current of the CT is proportional to $I_a + I_b + I_c) = 3I_0$. The CT secondary contains the true zero-sequence current, because this arrangement effectively sums the flux produced by the three phase currents. Such a CT application is possible, if the three phase conductors may be passed through the CT core in close proximity to each other. Hence, such application of CT is possible only in low voltage circuits.

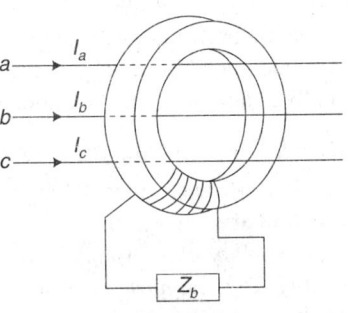

Fig. 3.14 Flux summing CT

3.2.16 High-Voltage CTs

High-voltage CTs of separately mounted post-type are suitable for outdoor service. They are installed in the outdoor switch yard. There are three basic forms of construction of these CTs as shown in Fig. 3.15(a), (b) and (c). In Type A, the cores and secondary windings are contained in an earthed tank at the base of a porcelain insulator and the leads of the fully insulated primary winding are taken up to the top helmet through the porcelain insulator. In Type B, the cores and windings are mounted midway inside the porcelain housing usually with half the major insulation on the primary winding and the other half on the secondary winding and cores. This form of construction is

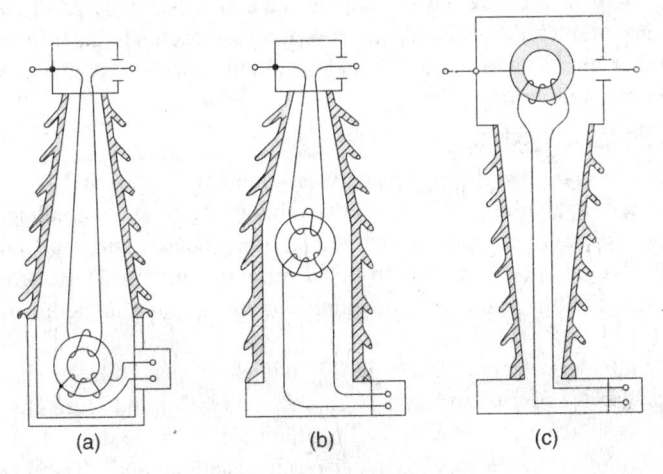

Fig. 3.15 (a) Type A CT (b) Type B CT (c) Type C CT

limited in the volume of cores and windings which can be accomodated without the diameter of the porcelain insulator becoming uneconomically large, but for two or three cores and secondary windings of modest output it offers a compact arrangement.

In Type C, the cores and secondary windings are housed in the helmet or live-tank and the earthed secondary leads brought down the insulator. This form of construction is particularly suitable for use when high primary currents are involved as it permits the use of a short bar primary conductor with consequent easing of the electrodynamics and flux leakage problems. The major insulation may be wholly on the secondary windings and cores or partially on the primary conductor, while the secondary leads require insulating to withstand the system voltage where they pass through the bases of the live-tank and down the porcelain bushing.

The major insulation used in such CTs is usually oil-impregnated paper. An alternative method of insulating for high voltages employs sulphur-hexafluoride (SF_6) gas, usually at a pressure in the region of two to three times the atmospheric pressure.

3.2.17 Open-circuiting of the Secondary Circuit of a CT

If a current transformer has its secondary circuit opened when current is flowing in its primary circuit, there is no secondary mmf (ampere-turns) to oppose that due to the primary current and all the primary mmf acts on the core as a magnetizing quantity. The unopposed primary mmf produces a very high flux density in the core. This high-flux density results in a greatly increased induced voltage in the secondary winding. With rated primary current flowing this induced voltage may be few hundred volts for a small CT but may be many kilovolts for a large high ratio protective CT. With system fault current flowing, the voltage would be raised in nearly direct proportion to the current value.

Such high voltages are dangerous not only to the insulation of the CT and connected apparatus but more important, to the life of the operator. Hence, the secondary circuit of a CT should never be opened while current is flowing in its primary circuit. If the secondary circuit has to be disconnected while primary current is flowing, it is essential first to short-circuit the secondary terminals of the CT. The conductor used for this purpose must be securely connected and of adequate rating to carry the secondary current, including what would flow if a fault occurs in the primary system. Most of the current transformers have a short circuit link or a switch at secondary terminals for short-circuiting purpose.

3.2.18 Modern Trends in CT Design

High-voltage CTs are the oil-filled type. At a system voltage of 400 KV and above there is a severe insulation problem. CTs for this range of system voltage become extremely expensive. Their performance is also limited due to the large dimensional separation of the secondary winding from the primary winding. These problems are overcome using SF_6(gas) and clophen (liquid) as insulation, thus reducing the size and cost of CTs.

A new trend is to use opto-electronic CTs and Rogowski coil current sensors to tackle this problem occuring in Extra High Voltage (EHV) and Ultra High Voltage (UHV) systems. A Rogowski coil and a linear coupler encircle the EHV conductor. A signal proportional to the secondary current is generated and transmitted via the communication channel. Light beam, laser beam and radio frequency are being used to transmit this signal.

It is seen that the secondary voltage supply seldom creates any problem but problems with secondary current supply arise frequently.

3.3 VOLTAGE TRANSFORMERS (VTs)

Voltage transformers (VTs) were previously known as potential transformers (PTs). They are used to reduce the power system voltages to standard lower values and to physically isolate the relays and other instruments (meters) from the high voltages of the power system. The voltage ratings of the secondary windings of the VTs have been standardized, so that a degree of interchangeability among relays and meters of different manufacturers can be achieved. The standard voltage rating of the secondary windings of the VTs used is practice is 110 V line to line or 110/√3 volts line to neutral. Therefore, the voltage ratings of the voltage (pressure) coils of protective relays and measuring instrument (meters) are also 110 V line to line or 110/√3 V line to neutral. The voltage transformers should be able to provide voltage signals to the relays and meters which are faithful reproductions of the primary voltages.

The accuracy of voltage transformers is expressed in terms of the departure of its ratio from its true ration.

3.3.1 Theory of Voltage Transformers

Theory of voltage transformers (VTs) is essentially the same as that of the ordinary power transformer. An ideal (perfect) VT is that in which when rated burden is connected across its secondary, the ratio of voltage applied across the primary to the secondary terminal voltage is equal to the ratio of primary turns to secondary turns and furthermore the two terminal voltages are in precise phase opposition to each other. But in practical (actual) VTs, the above relation doesn't hold good and errors are introduced both in ratio and in phase angle. This can best be explained by the equivalent circuit and phasor diagram for a VT shown in Fig. 3.16 and 3.17 respectively.

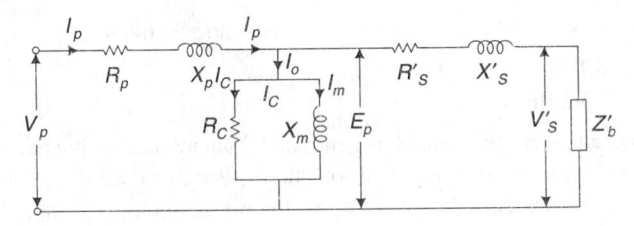

Fig. 3.16 Equivalent circuit of VT as viewed from primary side

The equivalent circuit of Fig. 3.16 is as viewed from primary side and all quantities are referred to that side.

The symbols used in the equivalent circuit and phasor diagram are as follows.

K = Nominal (rated) transformation ratio or Nominal (rated) voltage ratio

$$= \frac{\text{Rated primary voltage}}{\text{Rated secondary voltage}} = \frac{\text{Number of primary turns}}{\text{Number of secondary turns}}$$

v_p = Primary terminal voltage

E_p = Primary induced voltage

v_s = Secondary terminal voltage

E_s = Secondary induced voltage

I_p = Primary current

I_s = Secondary current

R_p, X_p = Primary resistance and leakage reactance respectively

R_s, X_s = Secondary resistance, and leakage reactance.

I_o = VT primary excitation current

I_m = Magnetizing component of the exiting current I_0 required to produce flux

I_c = Iron (core) loss component of the exciting current I_0 supplying core losses.

V'_s = Secondary terminal voltage referred to the primary side = kV_s

I'_s = Secondary current referred to primary side = I_s/K

R'_s = Secondary resistance referred to primary side = $K^2 R_s$

X'_s = Secondary leakage reactance referred to primary side = $K^2 X_s$

Z'_b = Secondary burden referred to primary = $K^2 Z_b$, ϕ = Phase angle of the burden

β = Phase angle error

= Angle between V_p and reversed secondary voltage, V'_s

θ = Angle between V'_s and I_p

ϕ_m = Core flux

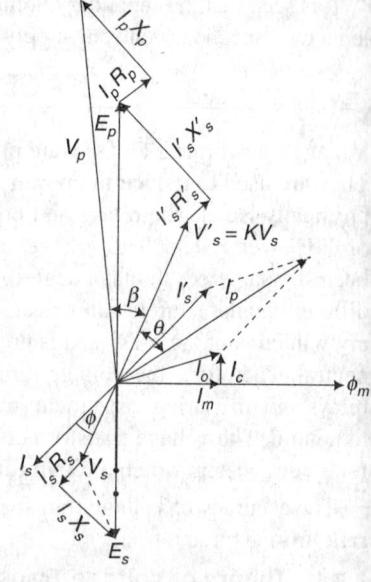

Fig. 3.17 Phasor diagram of a voltage transformer

The secondary terminal voltage is generated from induced voltage in secondary E_s after pahsor subtraction of voltage drops due to secondary winding's resistance and reactance. Secondary current I_s lags V_s by the phase angle of burden. The primary resistive and reactive drops ($I_p R_p$ and $I_p X_p$) are supplied by applied voltage V_p and are subtracted from V_p to derive primary induced voltage E_p. E_p is in opposition to E_s. The angle between V_p and reversed secondary voltage V'_s is termed as the phase angle of transformer denoted by β.

3.3.2 VT Errors

The errors introduced by the use of voltage transformers are, in general, less serious than those introduced by current transformers. It is seen from the phasor diagram that, like current transformers, voltage transformers introduce an error, both in magnitude and in phase, in the measured value of the voltage.

The voltage applied to the primary circuit of the VT cannot be obtained correctly simply by multiplying the voltage across the secondary by the turns ratio K of the transformer.

The divergence of the actual (true) ratio V_p/V_s from nominal (rated) ratio K depends upon the resistance and reactance of the transformer windings as well as upon the value of the exciting current of the transformer.

Ratio Error (Voltage Error)

The ratio error for VTs is defined as the error due to a difference in the actual transformation ratio and the nominal (rated) transformation ratio.

In percentage, it is expressed as

$$\text{Percent ratio error} = \frac{\text{Norminal ratio} - \text{Actual ratio}}{\text{Actual ratio}} \times 100$$

$$= \frac{K - K_a}{K_a} \times 100 = \frac{K - V_p/V_s}{V_p/V_s} \times 100$$

or

$$\text{Percent ratio error} = \frac{KV_s - V_p}{V_p} \times 100 \quad (3.31)$$

where

K = Nominal voltage ratio = $\dfrac{\text{Number of primary turns}}{\text{Number of secondary turns}}$

$K_a = \dfrac{V_p}{V_s}$ = Actual transformation ratio

V_s = Secondary voltage, and

V_p = Primary voltage

Phase Angle Error

The phase difference between the primary voltage and the reversed secondary phasors is the phase angle error of the VT. In order to keep the overall error within the specified limits of accuracy, the winding must be designed to have:

(i) the internal resistance and reactance to an appropriate magnitude, and
(ii) mininum magnetizing and loss components of the exciting current required by the core.

Limits of VT Errors for Protection

The accuracy of VTs used for meters and instruments is only important at normal system voltages, where as VTs used for protection require errors to be limited over a wide range of voltages under fault conditions. This may be about 5% to 150% of nominal voltage. The ratio error and phase angle error for VTs required for protection according to ISS : 3156 (Part III) 1966 are given in Table 3.2.

Table 3.2 Limits of Voltages and Phase Angle Errors for VTs

Class	Ratio Error	Phase Angle Error (in Degrees)
	0.05 to 0.9 times rated primary voltage	
	0.25 to 1.0 times rated output at unity p.f.	
3.0	± 3%	2
5.0	± 5%	5

3.3.3 Type of VTs

Following are three types of voltage transformers.

(i) Electromagnetic type VTs

(ii) Coupling Capacitor Voltage Transformers (CCVTs)
(iii) Opto-electronic VTs

Electromagnetic Type VTs

This type of a VTs is conveniently used up to 132 kV. It is similar to a conventional wound type transformer with additional features to minimise errors. As its output is low, it differs from power transformers in physical size and cooling techniques. In the UK, a 3-phase construction with 5 limbs is used. While in the USA single phase construction is more common. The voltage rating of a VT governs its construction. For lower voltages, up to 3.3 kV, dry type transformers with varnish impregnated and taped windings are quite satisfactory. For higher voltages, oil immersed VTs are used. Recently VTs with windings impregnated and encapsulated in synthetic resins have been developed for higher voltages. This technique has made it possible to use dry type VTs for system voltages up to 66 kV. For voltages above 132 kV, if electromagnetic type VTs to be used, several VTs are connected in cascade. In cascade connection, the primary windings of CTs are connected in series, though each primary is on a separate core. Coupling coils are provided alongwith each primary to keep the effective leakage inductance to a low value. They also distribute the voltage equally. Such an arrangement is conveniently placed in a porcelain enclosure.

Electromagnetic type VTs are used at all power system voltages and are usually connected to the bus. However coupling capacitor voltage transformers (CCVTs) are more economical at higher system voltages.

As the voltage decreases, the accuracy of electromagnetic type VTs decreases but is acceptable down to 1% of normal voltage.

Coupling Capacitor Voltage Transformers (CCVTs)

At higher voltages, electromagnetic type VTs become very expensive and hence it is a common practice to use a capacitance voltage divider as shown in Fig. 3.18. V_2 may be only about 10% or less of the system voltage. This arrangement is called a coupling capacitor voltage transformer (CCVT) or a capacitor type VT and is used at 132 KV and above. CCVT is one of the most common voltage sources for relaying at higher voltages. The reactor L is included to tune the capacitor VT to reduce the ratio and phase angle errors with the variation of VA burden, frequency, etc.

The reactor is adjusted to such a value that at system frequency it resonates with the capacitors. Capacitor VTs are more economical than electromagnetic type in this range of system voltage, particularly where high voltage capacitors are used for carrier-current coupling. The transient performance of a capacitor type VT is inferior to that of an electromagnetic type. A capacitor type VT has the tendency of introducing harmonics in the secondary voltage. High voltage capacitors are enclosed in a porcelain housing. The performance of the voltage divider type capacitor VT is not as good as that of the electromagnetic type.

The performance of high speed distance relays is less reliable with capacitor type VTs. Hence, the decision regarding the choice of a VT will depend whether economy in VT cost or relay performance is more important for a particular power line. Errors

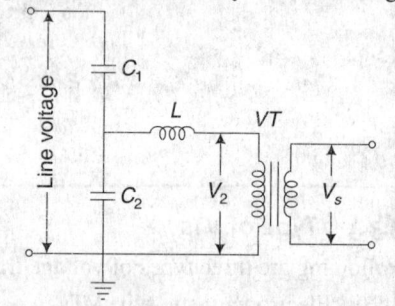

Fig. 3.18 Capacitance voltage divider

of capacitor type VTs can be reduced by reducing its burden. It is due to the fact that the series connected capacitors perform the function of a potential divider if the current drawn by the burden is negligible compared to the current flowing through the capacitors connected in series.

An electronic amplifier having high input impedance and VA output high enough to supply the VA burden can be included in the capacitor type VT arrangement. Such an arrangement gives a good transient response.

Finally, it can be concluded that the secondary voltage supply seldom creates any problem but problems with secondary current supply arise frequently.

Opto-Electronic VTs

The operation of an opto-electronic VT is based on the fact that the voltage difference between the conductor and the ground manifests in an electric field between the two electrodes. The opto-electronics send a circular polarized light beam that travels through an optical fiber up the column. The light beam passes through three strategically placed Pockels cells on the return path. The circular polarization changes to elliptical polarization as the light passes through a cell. The elliptical polarized lights from each cell are sent back to the opto-electronics at the ground level. The weighted measurements of the change of polarization in the three cells is used to determine the voltage difference between the conductor and the ground.

3.4 SUMMATION TRANSFORMER

On many occasions the need to derive a single-phase quantity from three-phase quantities may arise. A summation transformer and sequence filters, etc. are used for the purpose. Figure 3.19(a) shows a schematic diagram of a summation transformer where the primary windings are connected to the output terminals of the line CTs Fig. 3.19(b) and (c) show corresponding phase diagrams. The number of turns between R and Y phases is equal to those between Y and B. But more turns are provided between B and neutral. Table 3.3 shows the output current in terms of the CT rated current for a given fault current in each type of fault.

The output of a summation CT is given by

$$I_{output} = (N + 2) I_R + (N + 1) I_Y + N I_B \qquad (3.32)$$

This can be converted to their symmetrical components. Taking R phase as reference we get,

$$I_{output} = (N + 2) (I_1 + I_2 + I_0) + (N + 1) (a^2 I_1 + a I_2 + I_0) + N (a I_1 + a^2 I_2 + I_0)$$

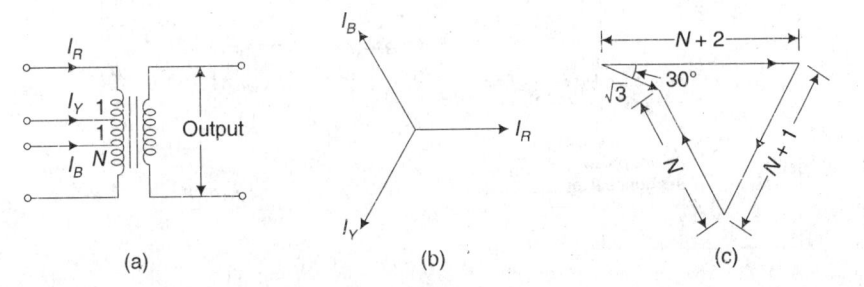

Fig. 3.19 (a) Summation transformer (b) Phasor diagram of 3-phase input current (c) Phasor diagram of summated output for 3-phase balanced input current

$$= 3I_0(N+1) + I_1(2 + a^2 + a^2N + aN + N) + I_2(2 + a + a^2N + aN + N)$$
$$= 3I_0(N+1) + I_1(2 + a^2) + I_2(2 + a)$$
$$= K_0 I_0 + K_1 I_1 + K_2 I_2 \tag{3.33}$$

Table 3.3 Output of Summation Transformer

Type of Fault	R-G	Y-G	B-G	R-Y	Y-B	B-R	R-Y-B
Summation CT turns	N + 2	N + 1	N	1	1	2	3
Output current	14%	16.5%	20%	90%	90%	45%	52%

Using the derived equation, I_{output} can be calculated for different types of faults. Table 3.4 shows the constants K_1, K_2 and K_0 for various types of faults for a summation transformer.

Table 3.4 Constants for different types of faults for summation transformer

Type of Fault	K_0	K_1	K_2
R-G, Y-B, Y-B-G	3(N + 1)	$2 + a^2$	$2 + a$
Y-G, B-R, B-R-G	3(N + 1)	$a(2 + a^2)$	$a^2(2 + a)$
B-G, R-Y, R-Y-G	3(N + 1)	$a^2(2 + a^2)$	$a(2 + a)$

Under certain fault conditions, I_{output} of a summation transformer is negligibly small or is zero. This is the serious drawback of a summation transformer. To overcome this difficulty, a special kind of sequence filter which gives an output of the form $5I_2 - I_1$, is used now-a-days.

3.5 PHASE-SEQUENCE CURRENT-SEGREGATING NETWORK

A general type of phase-sequence filter network can be developed as shown in Fig. 3.20(a). The output of the network or any other kind of summation device can be written in the form given below.

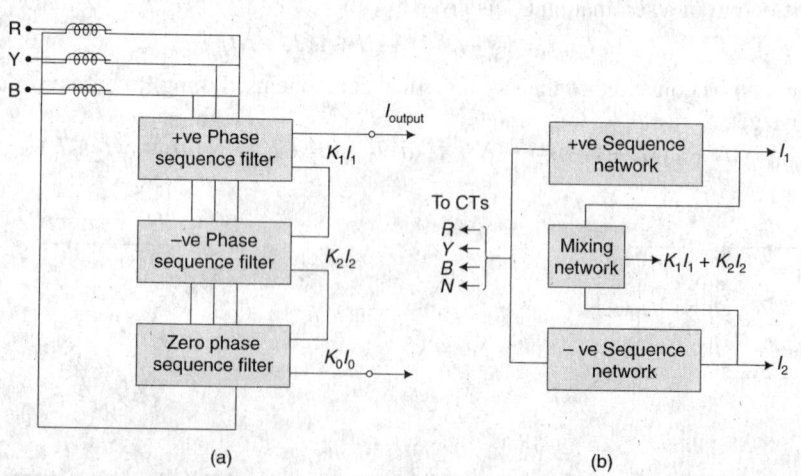

Fig. 3.20 (a) Phase sequence filter network (b) $(K_1 I_1 + K_2 I_2)$ type phase sequence filter network

$$I_{\text{output}} = K_0 I_0 + K_1 I_1 + K_2 I_2 \tag{3.34}$$

The constants K_0, K_1 and K_2 depend on the device which is used to derive a single-phase quantity from the 3-phase quantities.

The phase-sequence filter giving an output in the form of $I_1 - KI_2$ gives the most uniform response for any type of fault. The value of K may be 5 or 6. Fig. 3.20(b) shows a phase sequence filter of this type. Table 3.5 shows the values of constants for general type phase sequence network and $I_1 - KI_2$ type phase-sequence filter.

Table 3.5 Value of Constants K_0, K_1 and K_2, output of phase-sequence filter

Type of Faults	Any Summation Device			$(K_1 - KI_2)$ Type Device
	K_0	K_1	K_2	Output
R-G, Y-B, Y-B-G	K_Z	K_P	K_N	$I_1 - KI_2$
Y-G, B-R, B-R-G	K_Z	aK_P	$a^2 K_N$	$aI_1 - a^2 KI_2$
B-G, R-Y, R-Y-G	K_Z	$a^2 K_P$	aK_N	$a^2 I_1 - aKI_2$

$K = 5$ or 6
K_z = constant for zero-sequence component
K_P = Constant for +ve sequence component
K_N = Constant for –ve sequence component

EXERCISES

1. What are transducers? Why are they required in protection systems?
2. Explain the difference between a CT used for instrumentation and CT used for protection.
3. Explain CT burden. How is it specified?
4. Discuss how saturation affects the accuracy of CTs. Explain the accuracy limit factor or saturation factor.
5. Define the terms: (i) Rated short-time current (ii) Short-time factor (iii) Composit error (iv) Knee point voltage (v) Rated short-circuit current
6. Discuss the theory of CT with the help of equivalent circuit and phasor diagram and derive the expressions for actual transformation ratio and phase angle error.
7. Discuss the causes of ratio and phase angle errors in CTs. How can these errors be minimized?
8. What do you mean by ratio correction factor in CTs? Derive an expression for the same.
9. Discuss the various classification of CTs. Describe the construction of electromagnetic CTs.
10. What is an opto-electronic CT? Discuss its operation.
11. What is Rogowski coil current sensor? Describe its construction and operation and derive the expression for the voltage induced. How can the voltage proportion to the current be obtained?
12. Describe the characteristics of the Rogowski coil current sensor.
13. What is an auxiliary CT? Where is it used?
14. What is a linear coupler? Where is it used?
15. Discuss the different types of VTs with their areas of application.
16. What is a summation transformer? Where is it used?

4 Fault Analysis

4.1 INTRODUCTION

The operation of a power system under normal operating conditions is balanced steady-state three-phase. This condition can be temporarily disturbed due to sudden external or internal changes in the system. A short-circuit or fault occurs when the insulation of the system fails at any location or a conducting object comes in contact with the lines. The insulation of the system fails due to system overvoltages caused by lightning or switching surges, insulation contamination (salt spray or pollution) wind damage etc. Short-circuits (faults) are also caused by external conducting objects coming in contact with the lines, e.g., trees falling across lines, birds shorting the lines. etc. Short-circuit also occurs when the conductor breaks due to wind and ice loading and the broken line conductor falls to the ground. If the broken line conductor does not fall to the ground, there is a failure of the conducting path and the conductor becomes open-circuited. The occurrence of a short-circuit anywhere in the power system is generally termed as a *fault* at that point.

When a fault occurs, excessive current, i.e., current several times that of the normal operating current, flows in the system depending on the nature and location of the fault. If this excessive fault current is allowed to persist for a long period, it may cause thermal damage to the equipment and instability of the interconnected system by pulling synchronous machines out of synchronism due to slow electromechanical transients. Hence, the faulty section of the power system must be isolated as soon as possible after the occurrence of a fault. Relays sense the fault currents within a cycle after fault inception. With proper coordination among relays, faulty sections of the power system are isolated quickly by actuation of the circuit breakers in the next three to four cycles.

The severity of a fault is indicated by the magnitude of the fault current. The magnitude of the fault currents depends on the internal impedance of the generators plus the impedance of the intervening circuit. Normally, faults involve arcs due to flashover of insulators and the fault currents are carried through the arcs. The impedance of the arc which is in path of the fault current constitutes a fault impedance. In general, any impedance other than system impedances which is in the path of the fault current is called 'fault impedance'. A solid fault is said to occur when the fault impedance is zero.

Fault analysis forms an important part of power-system analysis. Fault calculations involve finding the voltage and current distribution throughout the system during the fault. It is essential to determine the values of system voltages and currents during fault conditions, in order to know the current settings of the protective devices to be used for detecting the fault and the ratings of the circuit breakers. Based on fault studies the short-circuit MVA at various points in the system can be calculated.

Faults on power systems are broadly classified into the following two main types:

(i) Symmetrical (Balanced) Faults A fault involving all the three phases (i.e., a three-phase fault) is called a symmetrical (balanced) type of fault. In this type of fault, all the three phases are simultaneously short-circuited. There may be two situations—all the three phases may be short circuited to the ground or they may be short-circuited without involving the ground. Since the network remains electrically balanced during this type of fault, it is also known as balanced fault. Because the network is balanced, it is solved on a per-phase basis. The other two phases carry identical currents except for the phase shift. This type of fault occurs very rarely, but it is the most severe type of fault encountered.

(ii) Unsymmetrical (Unbalanced) Faults An unsymmetrical (unbalanced) fault is any fault other than a 3-phase symmetrical fault: e.g. single line-to-ground (L-G) fault, line-to-line (L-L) fault, and double line-to-ground (2L-G) fault. The path of the fault current from line to ground or line to line may or may not contain impedance. Since any unsymmetrical fault causes unbalanced currents to flow in the system, this type of fault is also known as unbalanced fault. The method of symmetrical components is very useful in an analysis to determine the currents and voltages in all parts of the system after the occurrence of this type of fault.

When a fault occurs at a point in a power system, the corresponding fault MVA is referred to as the fault level at that point, and, unless otherwise stated, it will be taken to refer to a 3-phase symmetrical fault. The fault levels provide the basis for specifying interrupting capacities of circuit breakers. The MVA rating required for a circuit breaker is usually estimated on the assumption that it must clear a 3-phase symmetrical fault because, as that is generally the most severe fault and the worst case, it is reasonable to assume that the circuit breaker can clear any other fault. Hence, the circuit breaker's rated breaking capacity in MVA must be equal to or greater than, the 3-phase fault level MVA.

The main objects of fault analysis are as follows:

(i) Determination of maximum and minimum values of symmetrical fault currents
(ii) Determination of the unsymmetrical fault currents for single line-to-ground (L-G) faults, line-to-line (L-L) faults, double line-to-ground (2L-G) faults, and sometimes for open-circuit faults
(iii) Determination of fault-current distribution and bus bar-voltage levels throughout the system during fault conditions
(iv) Determination of current settings of the protective relays and investigation of their operation during faults
(v) Determination of rated breaking capacity of circuit breakers

The analysis of power system involving generators, transformers and lines of different voltage levels requires transformation of all voltages, and impedances to any selected voltage level. The voltages of generators are transformed in the ratio of transformation and all impedances by the square of ratio of transformation. The transformation of all voltages and impedances to a single voltage level is a cumbersome procedure for a large power system network with several voltage levels. The per-unit system discussed in the next section is the most convenient for power system analysis and will be used throughout this chapter.

4.2 PER-UNIT SYSTEM

An interconnected power system consists of generators, transformers, and lines of different voltage levels. The analysis of such a power system having several different voltage levels requires cumbersome transformation of all voltages and impedances to a single voltage level. The per-unit system has been devised as simpler and the most convenient system for the analysis of power system. In per-unit system, the various physical quantities such as power, voltage, current and impedance are expressed as a decimal fraction or multiples of base quantities. In this system, the different voltage levels disappear, and a power system network involving generators, transformers, and lines reduces to a system of simple impedances.

The per-unit value of any quantity is defined as the ratio of the actual value of that quantity to an arbitrarily chosen base value having the same dimensions. Thus, a per-unit value is dimensionless. Base values are generally indicated by a subscript b, and per-unit values by a subscript pu. The per-unit system is a completely consistent system of units which obeys all the network laws.

A minimum of four base quantities are required to completely define a per-unit system; these are voltage, current, power and impedance. There is inherent correlation between these quantities. Thus, any two base quantities are arbitrarily chosen as the other two base quantities can be derived from the two specified. Since the original quantities voltage (V), current (I), apparent power (S), impedance (Z) can all be represented by complex numbers having phase angles; it is essential that the per-unit system should preserve these phase-angles. Hence, all base values in the denominators are always scalars (real numbers) having no phase-angles, so that the phase angle of a per-unit quantity is the phase angle of the original quantity in its numerator. The per-unit system obeys the rules of complex algebra.

1.0 per-unit value of any quantity is equal to base value of that quantity. When the per-unit value of any quantity is multiplied by 100, it gives the percentage (%) value of that quantity. The per-unit values are relative to 1, whereas the percentage values are relative to 100. The per-unit system has an advantage over the percentage system because the product of two quantities expressed in per unit is expressed in per unit itself, but the product of two quantities expressed in percentage must be divided by 100 to obtain the result in percentage. It can be shown that Ohm's law is not true in the percentage system. Hence any power system data given in the percentage system must be changed to the per-unit system before being used in any calculation.

In power system, usually base kilovoltamperes (kVA) or megavoltamperes (MVA) and base voltage in kilovolts (kV) are the quantities selected to specify the base. For a power system, data are usually given as total three-phase kilovoltamperes (kVA) or megavoltamperes (MVA) and line-to-line voltage in kilovolts (kV). Hence, in three-phase systems, the base values usually chosen are three phase kVA or MVA and line-to-line voltage in kV. For power networks containing transformers, it is convenient to select the same power (kVA or MVA) base for both the sides but the ratio of base voltages on the two sides is kept same as the transformation ratio. Such a selection gives the same per-unit impedance on both sides of the transformer.

Therefore, for per-unit calculations involving transformers in three-phase circuits, the base voltages on the two sides of the transformer have the same ratio as the rated line-to-line voltages on the two sides of the transformer and the kilovoltampere (kVA) or megavoltampere (MVA) base remains the same on both the sides.

In a power system, voltage, current, apparent power (VA, kVA or MVA) and impedance are so related that selection of base values for any two of them determines the base values of the remaining two quantities.

4.2.1 Base and Per-Unit Values in a Single-Phase System

Let the selected base quantities be as follows:

$$\text{Base } VA = (VA)_b$$
$$\text{Base voltage} = V_b \text{ Volts}$$

Then,

$$\text{Base current } I_b = \frac{(VA)_b}{V_b} \text{ A} \tag{4.1}$$

$$\text{Base impedance } Z_b = \frac{V_b}{I_b} = \frac{V_b^2}{(VA)_b} \text{ ohms} \tag{4.2}$$

If the actual impedance is Z (ohms), its per-unit value is given by

$$Z_{pu} = \frac{Z}{Z_b} = \frac{Z \times (VA)_b}{V_b^2} \tag{4.3}$$

For a power system, the practical choice of base values are:

$$\text{Base kVA} = (kVA)_b$$

or

$$\text{Base MVA} = (MVA)_b$$
$$\text{Base Voltage in kV} = (kV)_b$$

Then, base values for current and impedance, and per-unit values can be calculated as follows:

$$\text{Base current } I_b = \frac{(kVA)_b}{(kV)_b} = \frac{(MVA)_b \times 10^3}{(kV)_b} \text{ A} \tag{4.4}$$

$$\text{Base impedance } Z_b = \frac{(kV)_b \times 10^3}{I_b}$$

$$= \frac{(kV)_b^2 \times 10^3}{(kVA)_b} = \frac{(kV)_b^2}{(MVA)_b} \text{ ohms} \tag{4.5}$$

$$\text{Per-unit impedance } Z_{pu} = \frac{Z \text{ (ohms)}}{Z_b \text{ (ohms)}}$$

$$= \frac{Z\text{ (ohms)} \times (kVA)_b}{(kV)_b^2 \times 10^3}$$

$$= \frac{Z\text{ (ohms)} \times (MVA)_b}{(kV)_b^2} \tag{4.6}$$

4.2.2 Base and Per-unit Values in a Three-phase System

For a three-phase system, three-phase kVA or MVA and line-to-line voltage in kV are usually chosen as base values. The base values for current and impedance and per-unit quantities can be calculated directly by using three-phase base quantities. When dealing with balanced three-phase circuits, it is always assumed that the circuit is star-connected, unless otherwise stated i.e., the impedances take line current at line-to-neutral voltage which is equal to line-to-line voltage divided by $\sqrt{3}$. Let the chosen three-phase base quantities be as follows:

Three-phase base kVA = $(kVA)_b$

or

Three-phase base MVA = $(MVA)_b$

Line-to-line base voltage in kV = $(kV)_b$

Assuming the three-phase circuit to be balanced and star-connected.

$$\text{Base current } I_b = \frac{(kVA)_b}{\sqrt{3}\,(kV)_b} = \frac{(MVA)_b \times 10^3}{\sqrt{3}\,(kV)_b}\text{ A} \tag{4.7}$$

$$\text{Base impedance } Z_b = \frac{(kV)_b \times 10^3}{\sqrt{3}\,I_b}$$

$$= \frac{(kV)_b^2 \times 10^3}{(kVA)_b} = \frac{(kV)_b^2}{(MVA)_b}\text{ ohms} \tag{4.8}$$

$$\text{Per-unit impedance } Z_{pu} = \frac{Z\text{ (ohms)}}{Z_b\text{ (ohms)}}$$

$$= \frac{Z\text{ (ohms)} \times (kVA)_b}{(kV)_b^2 \times 10^3}$$

$$= \frac{Z\text{ (ohms)} \times (MVA)_b}{(kV)_b^2} \tag{4.9}$$

4.2.3 Change of Base of Per-unit Quantities

The impedance of individual generators and transformers having different power and voltage ratings are generally expressed in terms of per-unit or per cent quantities based on their own ratings. The impedance of transmission lines are usually expressed by their ohmic values. For power-system analysis, all impedances must be expressed in per-unit on a common system base. In order to accomplish this, common base values for apparent power (kVA or MVA) and voltage in kV are arbitrarily selected, then the per-unit values of impedances on new common base are calculated.

If $Z_{pu,\,old}$ be the per unit impedance on old MVA base $(MVA)_{b,\,old}$ and kV base $(kV)_{b,\,old}$ and $Z_{pu,\,new}$ be the per unit impedance on new MVA base $(MVA)_{b,\,new}$ and kV base $(kV)_{b,\,new}$; then from Eqn. (4.9) $Z_{pu,\,old}$ and $Z_{pu,\,new}$ are given by

$$Z_{pu,\,old} = \frac{Z\,(\text{ohms}) \times (MVA)_{b,\,old}}{(kV)^2_{b,\,old}} \tag{4.10}$$

$$Z_{pu,\,new} = \frac{Z\,(\text{ohms}) \times (MVA)_{b,\,new}}{(kV)^2_{b,\,new}} \tag{4.11}$$

Dividing Equation (4.11) by (4.10), the value of $Z_{pu,\,new}$ is given by

$$Z_{pu,\,new} = Z_{pu,\,old} \times \frac{(MVA)_{b,\,new}}{(MVA)_{b,\,old}} \times \frac{(kV)^2_{b,\,old}}{(kV)^2_{b,\,new}} \tag{4.12}$$

If the old and new kV bases are the same, Eqn. (4.12) reduces to

$$Z_{pu,\,new} = Z_{pu,\,old} \times \frac{(MVA)_{b,\,new}}{(MVA)_{b,\,old}} \tag{4.13}$$

Equations (4.12) and (4.13) can be used to convert per-unit impedances from old bases to new bases. In these equations, $(MVA)_{b,\,old}$ and $(MVA)_{b,\,new}$ are three-phase base MVA, and $(kV)_{b,\,old}$ and $(kV)_{b,\,new}$ are line-to-line base voltage in kV.

4.2.4 Advantages of a Per-Unit System

Following are the advantages of the per-unit system for analysis of power systems.

(i) The impedance of the equipment (generators, transformers, etc), as specified by the manufacturers are in per cent or per unit on the base of their own (nameplate) ratings.

(ii) The per-unit impedances of equipment of the same general type and widely different ratings fall within a narrow range, although their ohmic values differ greatly for equipment of different ratings. For this reason, when the impedance of any equipment is not known definitely, its value can be selected from tabulated average values of per-unit impedance which will be reasonably correct.

(iii) The per-unit system gives us a clear idea about the relative magnitudes of various quantities, such as voltage, current, power and impedance.

(iv) The per-unit values of impedance, voltage and current are the same on both sides of the transformer, provided that the base values on both sides of the transformer are properly related by transformer action. Whereas the analysis using ohmic impedances involves transferring impedances through the transformers, using the square of the turns-ratio, to refer them to a common voltage level. The current calculated at that voltage level needs to be transferred through the transformers, using the inverse turns-ratio, to find the current at any other voltage level.

(v) In power-system analysis using the per-unit system, a transformer can be regarded as equivalent to a series impedance and the transformation of current and voltage can be disregarded.

(vi) The transformer connections in three-phase circuits do not affect the per-unit value of impedances of the equivalent circuit, although the transformer connections do determine the relation between the base voltages on the two sides of the transformer.

(vii) All the circuit laws are valid in the per-unit system, and the power and voltage equations are simplified since the factors of $\sqrt{3}$ and 3 are eliminated in the per-unit system.

(viii) Per-unit values are independent of the type of the power system, i.e., whether the power system is single-phase or three-phase.

(ix) The per-unit system considerably simplifies the analysis of power systems.

(x) The per-unit system is very ideal for the computerized analysis and simulation of complex power system.

4.3 ONE-LINE DIAGRAM

A power system contains a number of generators, transformers, transmission lines, and loads connected together. All power systems are balanced three-phase systems. Since a balanced three-phase system is always solved as a single-phase circuit composed of one of the three lines and a neutral return, it is represented by using only one phase and the neutral. Usually the diagram is simplified further by omitting the neutral and by indicating the components (generators, transformers, lines, etc.) by their standard symbols rather than by their equivalent circuits. Such a simplified diagram of a power system is called a one-line diagram or a single-line diagram. Thus, a one-line diagram represents a power system network by using standard symbols for generators, transformers, lines and loads. The one-line diagram is a convenient practical way of network representation. It supplies the significant information about the system in concise form. The actual three-phase diagram of the system would be quite cumbersome and confusing for a practical size power network. In one-line diagram, generator and transformer connections (star, delta and neutral grounding) are indicated by symbols drawn by the side of the representation of these components and circuit breakers are represented as rectangular blocks. The amount of information included on the one-line diagram depends on the purpose for which the diagram is meant. If the primary function of the diagram is to provide information for load-flow study, relays and circuit breakers are not shown on the diagram. Whereas circuit breakers must be shown on the diagram meant for protection study for maintaining system stability under transient conditions resulting from a fault. Sometimes one-line diagrams include information about the CTs and VTs which connect the relays to the system. Thus, the information to be included on a one-line diagram varies according to the problem to be solved or study to be performed. The one-line diagram of a simple power system is shown in Fig. 4.1.

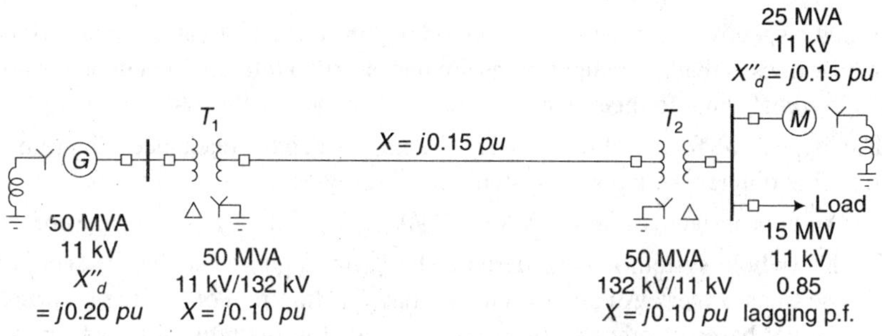

```
                                                      25 MVA
                                                       11 kV
                                                   X"_d = j0.15 pu
              T₁                          T₂        ┌──(M)
    ─(G)─┤ ├─      X = j0.15 pu       ─┤ ├─
      50 MVA                                                   → Load
       11 kV          50 MVA              50 MVA      11 kV    15 MW
       X"_d         11 kV/132 kV         132 kV/11 kV  0.85
    = j0.20 pu      X = j0.10 pu         X = j0.10 pu  lagging p.f.
```

Fig. 4.1 *One-line diagram of a simple power system.*

4.4 IMPEDANCE AND REACTANCE DIAGRAMS

In order to analyze a power system under load conditions or upon the occurrence of a fault, it is essential to draw the per-phase equivalent circuit of the system by using its one-line diagram. The equivalent circuit of the system drawn by combining the equivalent circuits for the various components shown in the one-line diagram is known as the impedance diagram of the system. A generator is represented by an emf in series with an impedance. Transformers and motors are represented by their equivalent circuits. A short line is represented by its series impedance. Medium and long lines are represented by nominal Π circuits, whereas very long lines are represented by equivalent Π circuits. The shunt admittance is usually omitted in the equivalent circuit of the transformer because the magnetizing current of a transformer is usually insignificant compared with the full load current. Since the resistance of a system is very small as compared to its inductive reactance, the resistance is often omitted when making fault calculations. Static loads (i.e., the loads which do not involve rotating machines) have little effect on the total line current during a fault and are usually omitted. Since the generated emfs of synchronous motor loads contribute to the fault current, they are always included in making fault calculations. If the impedance diagram is to be used to determine the current immediately after the occurrence of a fault, the induction motors are taken into account by a generated emf in series with an inductive reactance. Induction motors are omitted in computing the current a few cycles after the occurrence of the fault because the current contributed by an induction motor dies out very quickly after the induction motor gets short circuited.

Static loads are neglected during fault, as voltages dip very low so that currents drawn by them are negligible in comparison to fault currents. If all static loads, all resistances, the shunt admittance of each transformer, and the capacitance of the transmission line are neglected in order to simplify the calculation of the fault current, the impedance diagram reduces to the reactance diagram. These simplifications are applicable to fault calculations only and not to load-flow studies. Since the impedance and reactance diagrams show impedances to balanced currents in a symmetrical three-phase system, they are sometimes called the positive-sequence diagrams. The values of impedances and reactances in the impedance and reactance diagrams respectively are usually expressed in per-unit. The great advantage of using per-unit values is that no computations are necessary to refer an impedance from one side of a transformer to the other.

The steps to be followed for drawing the impedance (or reactance) diagram from the one-line diagram of a power system are as follows:

(i) An appropriate common kVA or MVA base for the system is chosen.
(ii) The whole system is considered to be divided into a number of sections by the transformers. An appropriate kV base is chosen in one of the sections and the kV bases of other sections are calculated in the ratio of transformation.
(iii) Per-unit values of voltages and impedances in all sections are calculated and they are connected as per the topology of the one-line diagram. The resulting diagram is the per-unit impedance diagram.

The following examples illustrate the steps of drawing the impedance diagram from the one-line diagram of a power system.

Example 4.1 Obtain the per-unit impedance (reactance) diagram of the simple power system whose one-line diagram is shown in Fig. 4.1.

Solution: The following simplifying assumptions are made to reduce the impedance diagram to reactance diagram.

(i) All resistances and shunt admittances of the transformers are neglected.
(ii) The capacitance and resistance of the line are neglected so that it is represented as a series reactance only.
(iii) The impedance diagram is meant for fault calculation (short circuit studies). Since the currents drawn by static loads during a fault are negligible as compared to fault currents, static loads are neglected. Therefore, static load has been neglected for drawing the impedance (reactance) diagram.

Common three-phase base MVA of 50 and line-to-line base voltage of 132 kV on the transmission line are selected. Then the voltage base in the generator (G) motor (M) circuits is 11 kV line-to-line. The static load to 15 MW is neglected.

The per-unit reactances of various components of the power system on common base MVA of 50 are calculated below:

Generator (G) : 0.20 pu

Transformer (T_1) : 0.10 pu

Transformer (T_2) : 0.10 pu

Transmission Line : 0.15 pu

Motor (M) : $0.15 \times \dfrac{30}{25}$

$= 0.18 \ pu$

The reactance diagram of the system is shown in Fig. 4.2.

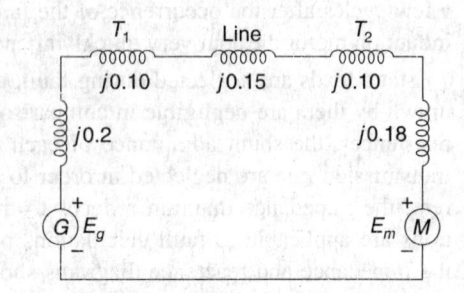

Fig. 4.2 *Reactance diagram of the system of Fig. 4.1.*

Fault Analysis 105

Example 4.2 | The one-line diagram of a simple power system is shown in Fig. 4.3. The data for each device is given below the diagram.

Fig. 4.3 *One-line diagram of the power system*

Generator G_1 : 25 MVA, 6.6 kV, $X'' = 20\%$
Generator G_2 : 25 MVA, 6.6 kV, $X'' = 20\%$
Generator G_3 : 50 MVA, 11 kV, $X'' = 18\%$
Transformer T_1 : 30 MVA, 6.6/33 kV, X = 22%
Transformer T_2 : 30 MVA, 11/33 kV, X = 20%
Load A : 20 MW, 6.6 kV, 0.85 lagging p.f.
Load B : 15 MW, 11 kV, 0.9 lagging p.f.
Synch. Motor M : 25 MVA, 11 kV, $X'' = 25\%$
Transmission Line : 20 ohms/phase. .

Selecting the rating of generator G_3 as the base, draw the per-unit impedance diagram of the system.

Solution: Since the three-phase MVA rating of the generator G_3 is 50 MVA, the common three-phase base MVA is 50 MVA. The line-to-line base voltage in the circuit of generator G_3 is 11 kV. Therefore the line to line base voltages in the transmission line is 33 kV, that in circuits of generators 1 and 2 is 6.6 kV and that in circuit of motor is 11 kV.

For fault calculations, static loads A and B are neglected.

The per-unit reactances of various components on the common MVA base of 50 MVA are calculated below:

$$\text{Generator } G_1 : 0.20 \times \frac{50}{25} = 0.40 \; pu$$

$$\text{Generator } G_2 : 0.20 \times \frac{50}{25} = 0.40 \; pu$$

$$\text{Generator } G_3 : 0.18 \times \frac{50}{50} = 0.18 \; pu$$

$$\text{Transformer } T_1 : 0.22 \times \frac{50}{30} = 0.366 \; pu$$

$$\text{Transformer } T_2 : 0.20 \times \frac{50}{30} = 0.333 \; pu$$

Motor $M : 0.25 \times \dfrac{50}{25} = 0.50\ pu$

Transmission Line : $\dfrac{20 \times 50}{(33)^2} = 0.918\ pu$

The per-unit impedance diagram of the system is shown in Fig. 4.4.

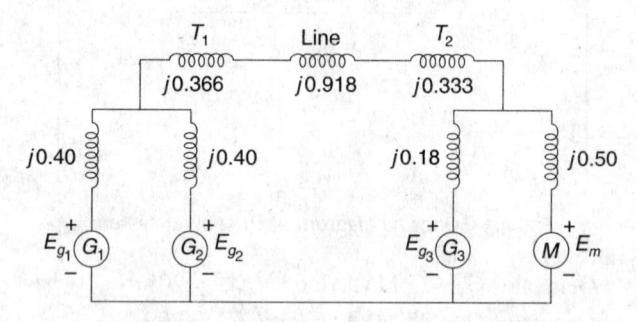

Fig. 4.4 *Per-unit impedance (reactance) diagram of the system of Fig. 4.3*

Example 4.3 | The one-line diagram of a three-phase power system is shown in Fig. 4.5. Select a common three-phase base MVA of 200 and line-to-line base voltage of 22 kV on the generator side. Draw an impedance diagram with all impedances marked in per-unit. The manufacturer's data for each device is given as follows:

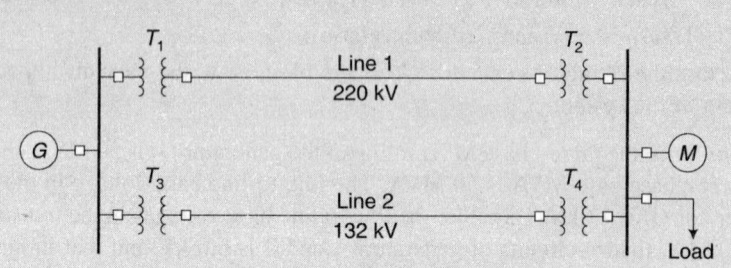

Fig. 4.5 *One-line diagram for Example 4.3*

Generator G : 200 MVA, 22 kV, $X'' = 20\%$
Transformer T_1 : 100 MVA, 22/220 kV, $X = 12\%$
Transformer T_2 : 100 MVA, 220/11 kV, $X = 10\%$
Transformer T_3 : 100 MVA, 22/132 kV, $X = 10\%$
Transformer T_4 : 100 MVA, 132/11 kV, $X = 8\%$
Synchronous Motor M : 60 MVA, 10.5 kV, $X'' = 18\%$
Load : 40 MW, 11 kV, 0.85 Lagging power factor.
Reactances of Line 1 and Line 2 are 40 and 55 ohms respectively.

Solution: Common three-phase base power is 200 MVA. Since the line-to line base voltage on the generator side is 22 kV, line to line base voltage of Line 1 and Line 2 are

220 kV and 132 kV respectively and the line to line base voltage of the synchronous motor circuit is 11 kV.

For fault calculations, static load of 40 MW is neglected in the impedance diagram.

The per-unit reactances of various components on the common MVA base of 200 MVA are calculated below:

$$\text{Generator } G : 0.20 \times \frac{200}{200} = 0.20 \, pu$$

$$\text{Transformer } T_1 : 0.12 \times \frac{200}{100} = 0.24 \, pu$$

$$\text{Transformer } T_2 : 0.10 \times \frac{200}{100} = 0.20 \, pu$$

$$\text{Transformer } T_3 : 0.10 \times \frac{200}{100} = 0.20 \, pu$$

$$\text{Transformer } T_4 : 0.08 \times \frac{200}{100} = 0.16 \, pu$$

$$\text{Synchronous Motor } M : 0.18 \times \left(\frac{200}{60}\right) \times \left(\frac{10.5}{11}\right)^2 = 0.546 \, pu$$

$$\text{Line 1} : \frac{40 \times 200}{(220)^2} = 0.165 \, pu$$

$$\text{Line 2} : \frac{55 \times 200}{(132)^2} = 0.631 \, pu$$

The per-unit impedance diagram of the system is shown in Fig. 4.6

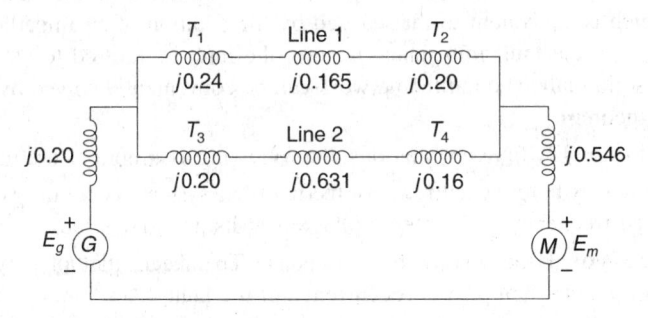

Fig. 4.6 Per-unit impedance (reactance) diagram of the system of Fig. 4.5

4.5 SYMMETRICAL FAULT ANALYSIS

Symmetrical fault is defined as the simultaneous short circuit across all the three phases. This type of fault may involve ground or may not involve ground. Though this fault is very infrequent in occurrence, but it is the most severe type of fault which gives rise to the maximum fault current. Under symmetrical fault condition, the three-phase system behaves like a balanced circuit and can be analyzed on a single-phase basis. Since the network remains electrically balanced during the type of fault, it is also known as balanced fault.

The fault level at any point in MVA is usually taken to refer to a 3-phase symmetrical fault. Since 3-phase symmetrical fault is generally the most severe fault and the worst case, the circuit breaker's rated MVA breaking capacity is based on this fault MVA. The circuit breaker's rated breaking capacity in MVA must be equal to, or greater than, the 3-phase fault level MVA. Since circuit breakers are manufactured in preferred standard sizes, e.g., 250, 500, 750, 1000 MVA, high precision is not necessary when calculating the 3-phase fault level at a point in a power system.

The three-phase symmetrical (balanced) fault on a line causes a collapse of the system voltage accompanied by an immediate reduction of power transmission capability to naught. On the other hand, unsymmetrical (unbalanced) faults partially cripple the line.

Since the flux per pole in the synchronous generator during short-circuit undergoes dynamic change with associated transients in damper and field windings, the reactance of the synchronous generator under short-circuit conditions is a time-varying quantity, and for network analysis three reactances are defined: the sub-transient reactance X_d'', for the first few cycles of the short-circuit current, transient reactance X_d', for the next relatively longer cycles, and the synchronous reactance X_d, thereafter. Since the duration of the short circuit current depends on the time of operation of the protective system, it is not always easy to decide which reactance to use. Generally, the subtransient reactance (X_d'') is used for determining the interrupting capacity of the circuit breakers. In fault studies required for relay setting and coordination, transient reactance (X_d') is used. The values of X_d'', X_d' and X_d are such that $X_d'' < X_d' < X_d$.

A fault at any point represents a change in the structure of the power system network which is equivalent to that caused by the addition of an impedance at the point of fault. If the fault impedance is zero, the fault is referred to as the bolted fault or the solid fault. The faulted network can be conveniently solved by using the Thevenin's theorem.

In order to simplify fault calculations, the following assumptions are made.
(i) Immediately before occurrence of the fault, the system is operating on no-load at rated frequency and system voltage is at its nominal value.
(ii) The emfs of all the generators are in phase. This means that all the generators are in synchronism prior to occurrence of the fault.
(iii) System resistance is neglected and only the inductive reactance of the system is taken into account.
(iv) Shunt admittances of the transformer are neglected.
(v) Shunt capacitances of the transmission line are neglected.

The analysis of a symmetrical fault is easy because the circuit remains completely balanced and calculations can be made only for one phase. The steps in the calculations for a symmetrical fault are as follows:
(i) A one-line diagram of the system is drawn.
(ii) A common base is selected and the per-unit reactances of all the components of the power system as referred to the common base are calculated.

(iii) From the one-line diagram, the single-phase reactance diagram of the system is drawn.

(iv) Thevenin's theorem is applied at the fault location. Keeping the identity of the fault point intact, the reactance diagram is reduced and the Thevenin's equivalent reactance X_{eq} of the system as viewed from the fault point is found. The Thevenin's voltage V_f, the pre-fault voltage at the fault point is taken as $1.0 < 0°$ pu, because the system is unloaded and hence, system voltage at all points is the nominal voltage.

(v) The per-unit value of the fault current is calculated as the ratio of the Thevenin's voltage $V_{f.pu}$ to the Thevenin's equivalent reactance $X_{eq.pu}$, i.e., $V_{f.pu}/X_{eq.pu}$. The fault current is purely reactive as resistance is neglected. The per-unit value of the fault MVA is calculated as $1/X_{eq.pu}$. Using base values, the per-unit values are converted to actual values.

If the normal (rated) system voltage is taken as the base voltage, the per-unit pre-fault voltage at the fault point, $V_{f.pu} = 1.0$ pu. Hence, per-unit fault (short-circuit) current is given by

$$I_{f.pu} = \frac{V_{f.pu}}{X_{eq.pu}} = \frac{1}{X_{eq.pu}} \qquad (4.14)$$

If $(MVA)_b$ be three-phase base MVA and $(kV)_b$ the line-to-line voltage in kV, the base current I_b is given by Eq. (4.7).

Actual fault current, $I_f = pu$ fault current. $I_{f.pu} \times$ base current, I_b

Substituting the values of $I_{f.pu}$ and I_b from Eqs. (4.14) and (4.7) respectively, the actual fault current, I_f is given by

$$I_f = \frac{1}{X_{eq.pu}} \times \frac{(MVA)_b \times 10^3}{\sqrt{3}\,(kV)_b} A \qquad (4.15)$$

The per-unit fault level or fault MVA or short-circuit MVA or short-circuit capacity is given by

$$(MVA)_{f.pu} = V_{f.pu} \times I_{f.pu} = \frac{1}{X_{eq.pu}} \qquad (4.16)$$

The actual value of the fault MVA is given by.

$$(MVA)_f = (MVA)_{f.pu} \times (MVA)_b = \frac{(MVA)_b}{X_{eq.pu}} \qquad (4.17)$$

4.5.1 Sudden Short-circuit of an Unloaded Synchronous Generator

A sudden short-circuit at the terminals of an unloaded synchronous generator is an example of three-phase symmetrical fault. The demagnetizing effect produced by the armature reaction of the synchronous generator under steady state three-phase short circuit condition is taken into account by a reactance X_a in series with induced emf. The series combination of reactance X_a and leakage reactance X_l of the machine is called synchronous reactance X_d (direct axis synchronous reactance in the case of salient pole machines). The steady-state short circuit model of a synchronous machine is shown in Fig. 4.7(c) on a per-phase basis.

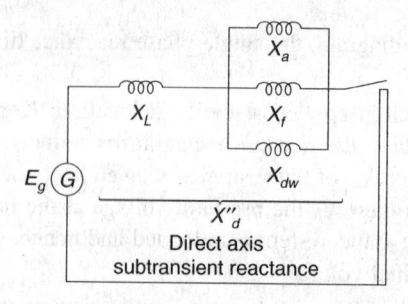

(a) Approximate circuit model during subtransient period of short circuit

X''_d — Direct axis subtransient reactance

(b) Approximate circuit model during transient period of short circuit

X'_d — Direct axis transient reactance

(c) Steady state short circuit model

X_d — Direct axis synchronous reactance

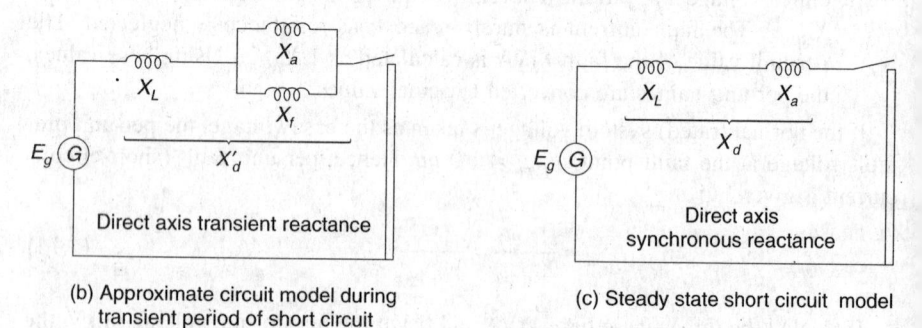

Fig. 4.7 *Circuit models of synchronous generator during short-circuit*

When a sudden three-phase short circuit occurs at the terminals of an unloaded or open-circuited synchronous generator, it undergoes a transient in all the three-phases finally ending up in steady-state condition as described above. The circuit breakers must interrupt the short circuit (fault) current much before steady state conditions are reached. Immediately after occurrence of short-circuit, the dc offset currents appear in all the three-phases, each with a different magnitude since the point on the voltage wave at which short circuit occurs is different for each phase. These dc offset currents are taken into account separately on an empirical basis. Therefore, for short circuit studies, attention is concentrated on symmetrical short-circuit current only. Immediately after occurrence of a short-circuit, the symmetrical short circuit current is limited only by the leakage reactance X_l. Since the air-gap flux cannot change instantaneously (theorem of constant flux linkages), currents appear in the field and damper windings in order to counter the demagnetizing effect of the armature short-circuit current and help the main flux. These currents decay in accordance with the time constants at the windings. The time constant of the field winding which has high leakage inductance is larger than that of the damper winding, which has low leakage inductance. Thus during the initial part of the short circuit, the field and damper windings have transformer currents induced in them so that in the circuit model, their reactances (X_f of field winding and X_{dw} of damper winding) appear in parallel with X_a as shown in Fig. 4.7(a). As the damper winding currents are first to die out, X_{dw} effectively becomes open circuited and at a later stage X_f becomes open circuited. The machine reactance thus changes from the parallel combination of X_a, X_f, and X_{dw} during the initial period of the short circuit to parallel combination of X_a and X_f

[Fig. 4.7(b)] in the middle period of the short circuit, and finally to X_a in steady state as shown in Fig. 4.7(c).

During the initial period of the short circuit, the reactance presented by the machine is given by:

$$X_l + \frac{1}{(1/X_a + 1/X_f + 1/X_{dw})} = X_d'' \quad (4.18)$$

X_d'' is called subtransient reactance (direct-axis subtransient reactance in the case of salient pole machines).

The effective reactance after the damper winding currents have died out is given by:

$$X_l + \frac{1}{(1/X_a + 1/X_f)} = X_d' \quad (4.19)$$

X_d' is called the transient reactance (direct-axis transient reactance in the case of salient pole machines) under steady-state conditions, the reactance is the synchronous reactance X_d of the machine. The values of X_d'', X_d' and X_d are such that $X_d'' < X_d' < X_d$. Thus, under short circuit conditions, the machine offers a time-varying reactance which changes from X_d'' to X_d' and finally to X_d.

The symmetrical trace of a short-circuited stator-current wave is shown in Fig. 4.8. This wave may be divided into three-periods or time regimes namely the subtransient period, lasting only for the first few cycles, during which the current-decrement is very rapid; the transient period, covering a relatively longer time during which the current decrement is more moderate; and finally the steady-state period. During subtransient period, the current is large as the machine offers subtransient reactance, during the middle transient period the machine offers transient reactance, and finally the steady-state period when the machine offers synchronous reactance.

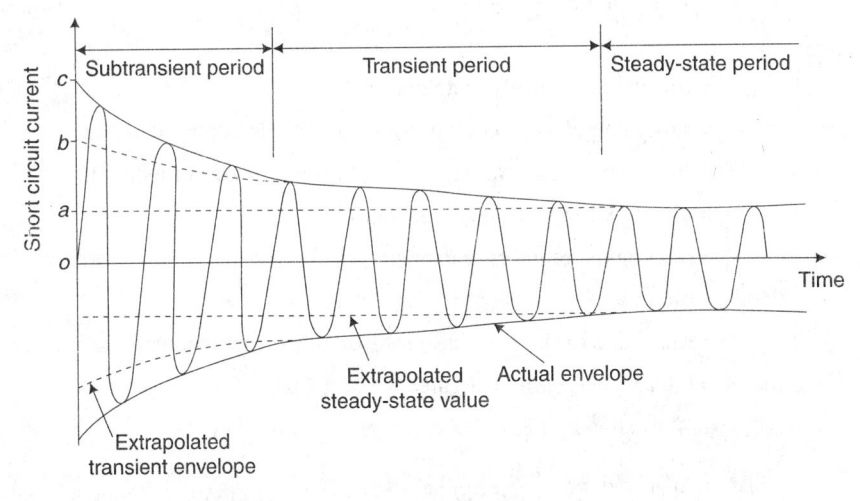

Fig. 4.8 *Symmetrical short-circuit stator current in synchronous machine.*

The envelope of the current wave shape is shown in Fig. 4.9. when the transient envelope is extrapolated backwards in time, the difference $\Delta i''$ between the subtransient and extrapolated transient envelopes is the current corresponding to the damper winding current which decays fast according to the time constant of the damper winding. Similarly, the difference $\Delta i'$ between the transient and steady state-envelopes decays in accordance with the field time constant.

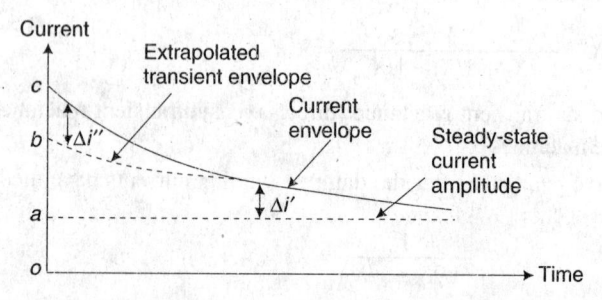

Fig. 4.9 Envelope of symmetrical short circuit current in synchronous machine

The currents during three regimes of subtransient, transient and steady-state are limited primarily by various reactances of the synchronous machine (the armature resistance which is relatively small is neglected). The currents and reactances of synchronous machine during three-phase symmetrical short circuit are defined by the following equations.

$$|I| = \frac{oa}{\sqrt{2}} = \frac{|E_g|}{X_d} \qquad (4.20)$$

$$|I'| = \frac{ob}{\sqrt{2}} = \frac{|E_g|}{X'_d} \qquad (4.21)$$

$$|I''| = \frac{oc}{\sqrt{2}} = \frac{|E_g|}{X''_d} \qquad (4.22)$$

where

$|I|$ = rms value of steady-state current

$|I'|$ = rms value of transient current excluding dc component

$|I''|$ = rms vlaue of subtransient current excluding dc component.

X_d = direct axis synchronous reactance

X'_d = direct axis transient reactance

X''_d = direct axis subtrasient reactance

$|E_g|$ = rms value of the per phase no-load voltage of the generator.

oa, ob, oc = intercepts shown in Figures 4.8 and 4.9

Both $\Delta i''$ and $\Delta i'$ as shown in Fig. 4.9 decay exponentially as

$$\Delta i'' = \Delta i''_o \, e^{-t/\tau_{dw}} = E_g \left(\frac{1}{X''_d} - \frac{1}{X'_d} \right) e^{-t/\tau_{dw}} \qquad (4.23)$$

$$\Delta i' = \Delta i'_o \, e^{-t/\tau_f} = E_g \left(\frac{1}{X'_d} - \frac{1}{X_d} \right) e^{-t/\tau_f} \qquad (4.24)$$

where τ_{dw} and τ_f are time constants of the damper winding and the field respectively.

$\tau_{dw} \ll \tau_f$. At time $t \gg \tau_{dw}$, $\Delta i''$ practically dies out and similarly at time $t \gg \tau_f$, $\Delta i'$ practically dies out. The total rms value of the ac (fundamental frequency) component of the fault current is given by

$$I_{ac} = I + \Delta i' + \Delta i'' \tag{4.25}$$

Substituting the values of I, $\Delta i'$ and $\Delta i''$ from Equation (4.20), (4.24) and (4.23) respectively in Equation 4.25, I_{ac} will be given by

$$I_{ac} = E_g \left[\frac{1}{X_d} + \left(\frac{1}{X_d'} - \frac{1}{X_d} \right) e^{-t/\tau_f} + \left(\frac{1}{X_d''} - \frac{1}{X_d'} \right) e^{-t/\tau_{dw}} \right] \tag{4.26}$$

The maximum value of the dc component of the armature current is

$$I_{dc}(\max) = \sqrt{2} \frac{E_g}{X_d''} = \sqrt{2}\, I'' \tag{4.27}$$

The rms value of the ac component of the fault current will be maximum immediately after the short circuit, i.e., at $t = 0$. Substituting $t = 0$, in Equation (4.26), the maximum rms value of the ac component of the fault current is given by

$$I_{ac}(\max) = \frac{E_g}{X_d''} = I'' \tag{4.28}$$

so, the maximum rms (effective) value of the asymmetrical current (ac + dc) at the beginning of a short circuit is given by

$$I_{\max}(\text{rms}) = \sqrt{I_{dc(\max)}^2 + I_{ac(\max)}^2} = \sqrt{3}\, I'' = \sqrt{3}\, \frac{E_g}{X_d''} \tag{4.29}$$

4.5.2 Short-Circuit Capacity (SCC)

The three-phase symmetrical fault is very rare in occurrence, but it is the most severe type of fault which gives rise to the maximum fault current. When a fault (short-circuit) occurs at a point in a power system, the corresponding fault (short-circuit) MVA is referred to as the fault level or short-circuit capacity (SCC) at that point, and, unless otherwise stated, it is taken to refer to a 3-phase symmetrical short-circuit. High short-circuit capacity (fault MVA) at a point signifies high strength of power system at that point and low equivalent reactance up to that point. Therefore, large loads can be connected at that point. Low-fault level signifies weak system.

The short-circuit capacity or the short circuit MVA at any point indicates the strength of the power system at that point and is used for determining the dimension of a bus bar, and the interrupting capacity of a circuit breaker.

The short-circuit capacity (SCC) or the short-circuit MVA (SC MVA) or the fault MVA or the fault level is defined as

$$(\text{MVA})_f = \sqrt{3} \times (\text{pre-fault line voltage in kV}) \times \text{fault current } I_f \times 10^{-3}$$

Since the system is assumed to be operating at no-load immediately before occurrence of the fault, the pre-fault system voltage at all points is the nominal system voltage.

Hence,
$$(MVA)_f = \sqrt{3} \times \text{(nominal system voltage in kV)} \times I_f \times 10^{-3} \quad (4.30)$$
$$= \sqrt{3} \times V_f \times I_f \times 10^{-3}$$

where I_f in amperes is the rms value of the fault current in a three-phase symmetrical fault at the fault point, and V_f is the nominal system voltage in kV.

Three-phase base megavoltamperes $(MVA)_b$ are related to line-to-line base kilovolts $(kV)_b$ and base current I_b by

$$(MVA)_b = \sqrt{3} \times (kV)_b \times I_b \times 10^{-3} \quad (4.31)$$

Dividing Eq. (4.30) by Eq. (4.31) converts $(MVA)_f$ to per-unit, and we obtain

$$(MVA)_{f.\,pu} = V_{f.\,pu} \times I_{f.\,pu} \quad (4.32)$$

where $V_{f.\,pu} = V_f/(kV)_b$ is the nominal system voltage in per-unit, and

$$I_{f.\,pu} = I_f/I_b \text{ is the per-unit fault current.}$$

If the base kilovolts $(kV)_b$ is equal to the nominal (rated) system voltage V_f in kV, i.e., $(kV)_b = V_f$ in kV, then $V_{f.\,pu} = 1.0$ and from Eq. (4.32), $(MVA)_{f.\,pu}$ is given by

$$(MVA)_{f.\,pu} = I_{f.\,pu} = \frac{V_{f.\,pu}}{X_{eq.\,pu}} = \frac{1}{X_{eq.\,pu}} \quad (4.33)$$

where, $X_{eq.\,pu}$ = Thevenin's equivalent reactance in per-unit

At nominal system voltage the Thevenin equivalent circuit looking back into the system from the fault point is a voltage source (emf) of $1.0 < 0°$ per unit in series with the per-unit reactance $X_{eq.\,pu}$. Therefore, under fault (short-circuit) conditions,

$$X_{eq.\,pu} = \frac{1.0}{(MVA)_{f.\,pu}} \text{ per-unit} \quad (4.34)$$

The actual value of the short-circuit MVA of short circuit capacity at any point in the system is given by Eq. (4.17).

4.5.3 Selection of Circuit Breakers

The selection of a circuit breaker for a particular location depends on the maximum possible fault MVA (short-circuit MVA) to be interrupted with respect to the type and location of the fault, and, generating capacity and synchronous motor load connected to the system. A three-phase symmetrical fault which is very rare in occurrence generally gives the maximum fault MVA and a circuit breaker must be capable of interrupting it. An exception is an L-G (line-to-ground) fault close to a synchronous generator.

Under normal operating conditions, the circuit-breaker contacts are at line voltage and these are to carry the maximum possible fault current immediately after the occurrence of the fault. The maximum value of the fault MVA to be interrupted will be $\sqrt{3} \times$ normal line voltage in kV $\times$ maximum fault current in KA. Therefore, rated MVA interrupting (breaking) capacity of a circuit breaker must be more than or equal to the fault MVA required to be interrupted.

From the current viewpoint, the following two factors are to be considered in selecting the circuit breaker.

(i) The maximum instantaneous current which the circuit breaker must carry (withstand) and
(ii) The total current when the circuit-breaker contacts part to interrupt the circuit.

Two of the circuit-breaker ratings which require the computation of the fault current are: rated momentary current and rated interrupting current. Symmetrical momentary fault current is obtained by using subtransient reactances for synchronous machines. The current which a circuit-breaker must interrupt is usually asymmetrical since it contains the decaying dc component (dc offset current). The asymmetrical momentary fault current is calculated by multiplying the symmetrical momentary fault current by a factor of 1.6 to account for the presence of dc offset current.

The interrupting MVA rating of a circuit-breaker equals $\sqrt{3}$ × (the kilovolts of the bus to which the breaker is connected) × (the current in kA which the breaker must be capable of interrupting when its contacts part). This interrupting current is, of course, lower than the momentary current and depends on the speed of the breaker, such as, 8, 5, 3, or 2 cycles, which is a measure of the time from the occurrence of the fault to the extinction of the arc. Breakers of various speeds are classified by their rated interrupting times. The rated interrupting time of a circuit-breaker is the period between the instant of energizing the trip circuit and the arc extinction on an opening operation, as shown in Fig. 4.10. The tripping delay time which is usually assumed to be ½ cycle for relays to pick up precedes the rated interrupting time.

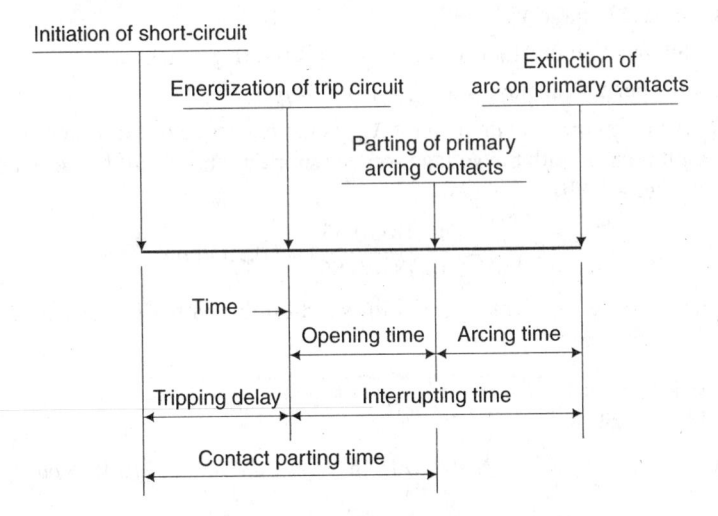

Fig. 4.10 *Definition of interrupting time*

Example 4.4 | The one-line diagram of a simple power system is shown in Fig. 4.1. A symmetrical three-phase fault occurs at the motor terminals. Determine the (i) subtransient fault current; (ii) subtransient fault current in the generator and the motor. Assume the system to be operating at no-load and at the rated voltage immediately before occurrence of the fault.

Solution: The per-unit reactances of various components of the power system on a common MVA base of 50 are calculated in Example 4.1 and the per-unit reactance diagram is shown in Fig. 4.2. The circuit model of the system for fault calculations is shown in Fig. 4.11(a).

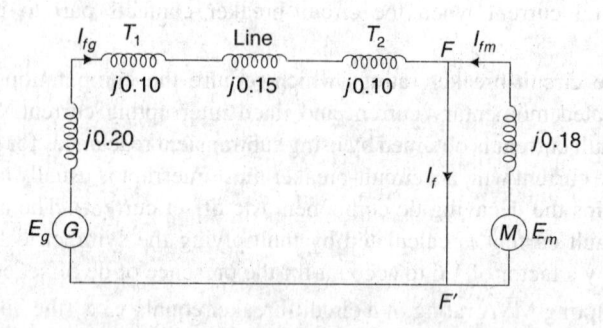

Fig. 4.11(a) Circuit model for Example 4.4

For applying Thevenin's theorem at the fault location, the shorted line across FF' is open-circuited (removed) and the Thevenin's equivalent circuit is drawn with respect to the fault point. The Thevenin's equivalent circuit consists of a voltage source called the Thevenin's voltage in series with the Thevenin's equivalent reactance X_{eq}. The Thevenin's voltage is the open-circuit voltage across FF', which is the pre-fault voltage V_f.

The pre-fault voltage V_f at the motor terminal is the rated voltage of the motor which is the base voltage V_b.

Hence, the per-unit pre-fault voltage $V_{f.\,pu} = 1.0 < 0°$ pu, and

Thevenin's per-unit voltage $= V_{f.\,pu} = 1.0 < 0°$ pu

The Thevenin's equivalent reactnace X_{eq}, is the reactance of the system as viewed from the fault point F with the voltage sources shorted. The circuit for computing X_{eq} is shown in Fig. 4.11(b).

Hence, $$X_{eq.\,pu} = j\frac{(0.18)(0.55)}{(0.18 + 0.55)} = j\,0.1356\ pu$$

The Thevenin's equivalent circuit with short-circuit path FF' showing the fault current I_f is shown in Fig. 4.11(c).

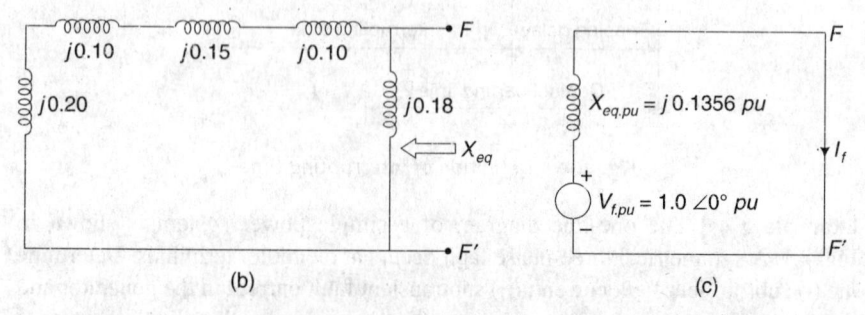

Fig. 4.11 (b) Circuit for computing X_{eq} (c) Thevenin's equivalent circuit

(i) The per-unit subtransient fault current $I_{f.\,pu}$ is then

$$I_{f.\,pu} = \frac{V_{f.\,pu}}{X_{eq.\,pu}} = \frac{1.0}{j\,0.1356} = -j\,7.3746\ pu$$

The base MVA is 50 and the base kV in the motor circuit is 11 kV, i.e., $(MVA)_b$ = 50 MVA and $(kV)_b = 11$ kV
Hence base current in the motor circuit is

$$I_b = \frac{(MVA)_b \times 10^3}{\sqrt{3}\,(kV)_b}\,A$$

$$= \frac{50 \times 10^3}{\sqrt{3} \times 11} = 2627.4\ A$$

The actual fault current $I_f = I_{f.\,pu} \times I_b = 7.3746 \times 2627.4 = 19{,}376$ A

(ii) Referring to Fig. 4.11(a) the per-unit subtransient fault currents in the generator and motor are given by
Per-unit fault current from generator

$$I_{fg.\,pu} = -j\,7.3746 \times \frac{j\,0.18}{j\,0.73}\ pu$$

$$= -j\,1.8184\ pu$$

Per-unit fault current from motor

$$I_{fm.\,pu} = -j\,7.3746 \times \frac{j\,0.55}{j\,0.73}\ pu$$

$$= -j\,5.5562\ pu$$

The base value of the currents in generator and motor circuits, $I_b = 2627.4$ A,
Hence,
The actual value of fault current from generator is

$$I_{fg} = I_{fg.\,pu} \times I_b = 1.8184 \times 2627.4 = 4{,}778\ A$$

and the actual value of the fault current from motor is

$$I_{fm} = I_{fm.\,pu} \times I_b = 5.5562 \times 2627.4 = 14{,}598\ A$$

Example 4.5 Two generators rated 15 MVA, 11 kV, having 15% subtransient reactance are interconnected through transformers and a 120 km long line as shown in Fig. 4.12(a). The reactance of the line is 0.12 ohms/km. The transformers near the generators are rated 30 MVA, 11/66 kV with leakage reactance of 10% each. A symmetrical three-phase fault occurs at a distance of 30 km from one end of the line when the system is on no-load but at rated voltage. Determine the fault current and the fault MVA.

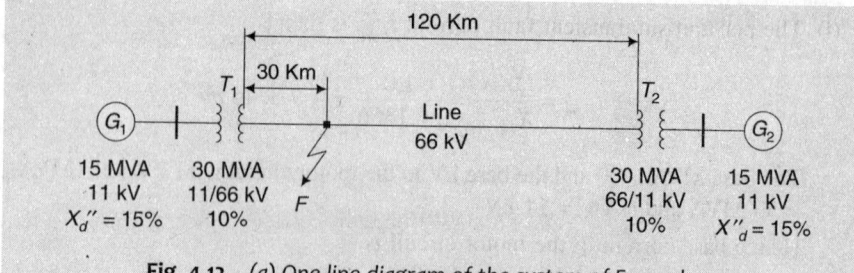

Fig. 4.12 (a) One-line diagram of the system of Example 4.5

Solution: Base MVA for the complete system = 30 MVA

Base kV for generator side = 11 kV

Base kV for the line circuit = 66 kV

The per-unit reactances of various components on the common MVA base of 30 MVA are calculated below:

Generators G_1 and G_2 : $0.15 \times \dfrac{30}{15} = 0.30 \; pu$

Transformers T_1 and T_2 : $0.10 \times \dfrac{30}{30} = 0.10 \; pu$

Line of 30-km length : $\dfrac{(30 \times 0.12) \times 30}{(66)^2} = 0.0248 \; pu$

Line of 90-km length : $\dfrac{(90 \times 0.12) \times 30}{(66)^2} = 0.0744 \; pu$

From these values, the per-unit reactance diagram [Fig. 4.12(b)] is drawn.

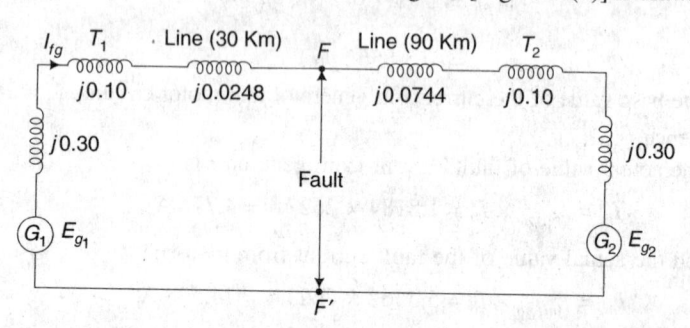

Fig. 4.12 (b) Per-unit reactance diagram of the system of Fig. 4.12(a)

For fault calculation the Thevenin's equivalent circuit is drawn with respect to the fault point F. The pre-fault voltage V_f is the open-circuit voltage across FF'.

Hence, Thevenin's per-unit voltage = $V_{f.pu} = 1.0 \; pu$

The Thevenin's equivalent reactance $X_{eq.pu}$ as viewed from the fault point F with voltage sources short-circuited is the equivalent reactance of two parallel branches whose reactances are

$j\,0.30 + j\,0.10 + j\,0.0248 = j\,0.4248 \; pu$

and
$$j0.30 + j0.10 + j0.0744 = j0.4744 \text{ pu}$$

Hence, the Thevenin's equivalent reactance $X_{eq.pu}$ is given by

$$X_{eq.pu} = \frac{j0.4248 \times j0.4744}{j0.4248 + j0.4744}$$

$$= j0.224 \text{ pu}$$

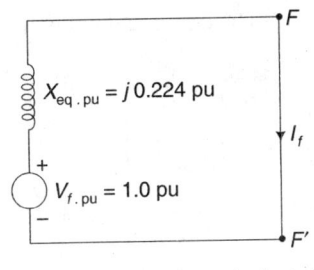

Fig. 4.12(c) Thevenin's equivalent circuit

The Thevenin's equivalent circuit with fault path FF' showing the fault current I_f is shown in Fig. 4.12(c).

Per-unit fault current

$$I_{f.pu} = \frac{V_{f.pu}}{X_{eq.pu}} = \frac{1.0}{j0.224} = -j\,4.46 \text{ pu}$$

Base current in the line circuit is given by

$$I_b = \frac{(MVA)_b \times 10^3}{\sqrt{3}\,(kV)_b}A = \frac{30 \times 10^3}{\sqrt{3} \times 66} = 262.7 \text{ A}$$

The actual fault current

$$I_f = I_{f.pu} \times I_b = 4.46 \times 262.7 = 1172 \text{ A}$$

Fault MVA $= \sqrt{3} \times V_f \times I_f \times 10^{-3} = \sqrt{3} \times 66 \times 1172 \times 10^{-3} = 133.9$ MVA

Alternative Method:

$$\text{Fault MVA, } (MVA)_f = \frac{(MVA)_b}{X_{eq.pu}} = \frac{30}{0.224} = 133.9 \text{ MVA}$$

$$\text{Fault current } I_f = \frac{(MVA)_f \times 10^3}{\sqrt{3} \times (kV)_b} = \frac{133.9 \times 10^3}{\sqrt{3} \times 66} = 1172 \text{ A}$$

Example 4.6 Two 11-kV generators G_1 and G_2 are connected in parallel to a bus-bar. A 66 kV feeder is connected to the bus-bar through a 11/66 kV transformer.

Calculate fault MVA for three-phase symmetrical fault: (a) at the high-voltage terminals of the transformer, and (b) at the load end of the feeder. Find the fault current shared by G_1 and G_2 in each case. The ratings of the equipment are as follows:

Generator G_1 : 11 kV, 50 MVA, $X''_{G1} = 15\%$

Generator G_2 : 11 kV, 25 MVA, $X''_{G2} = 12\%$

Transformer T : 11/66 kV, 75 MVA, $X_T = 10\%$

Feeder : 66 kV, $X_F = 25\,\Omega$

Assume the system to be operating on no load before the fault and neglect the losses.

Solution: The one-line diagram of the system is shown in Fig. 4.13(a).

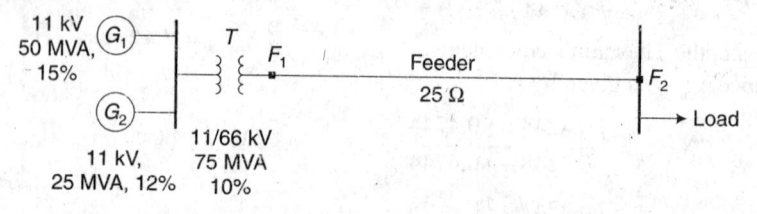

Fig. 4.13(a) One line-diagram for the system of Example 4.6

Let us choose a common base MVA of 75 MVA and base voltage of 11 kV on LT side and 66 kV on the HT side.

The per-unit reactances of the various equipment on the chosen base values are:

$$\text{Generator } G_1 : X''_{G1.pu} = 0.15 \times \frac{75}{50} = 0.225 \, pu$$

$$\text{Generator } G_2 : X''_{G1.pu} = 0.12 \times \frac{75}{25} = 0.36 \, pu$$

$$\text{Transformer } T : X_{T.pu} = 0.10 \, pu$$

$$\text{Feeder} : X_{F.pu} = X_F \times \frac{(MVA)_b}{(kV)_b^2} = 25 \times \frac{75}{(66)^2} = 0.43 \, pu$$

The per-unit reactance diagram of the system is shown in Fig. 4.13(b).

(a) The circuit model of the system for fault at the high voltage terminals of the transformer is shown in Fig. 4.13(c). F_1 is the fault point and $F_1 F'_1$ is the fault path.

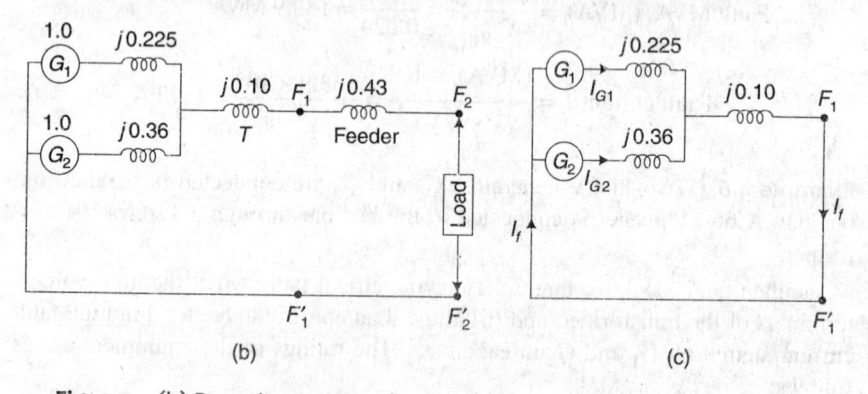

Fig. 4.13 (b) Per unit reactance diagram (c) Circuit model of the system for the fault at high-voltage terminals

The pre-fault voltage at the fault point F_1 is the rated voltage of the HT side of the transformer, (i.e., 66 kV), which is the base voltage of that side.

Hence, the per-unit pre-fault voltage = $V_{f.pu} = 1.0 \angle 0°$ pu, and Thevenin's per-unit voltage = $V_{f.pu} = 1.0 \angle 0°$ pu.

The Thevenin's equivalent reactance $X_{eq.pu}$ of the system as viewed from the fault point F_1 with the voltage sources shorted is computed as follows:

$$X_{eq.pu} = j\,0.10 + \frac{j\,0.225 \times j\,0.36}{(j\,0.225 + j\,0.36)}$$

$$= j\,0.238 \text{ pu.}$$

The Thevenin's equivalent circuit with the fault (short-circuit) path $F_1 F_1'$ showing the fault current I_f is shown in Fig. 4.13(d)

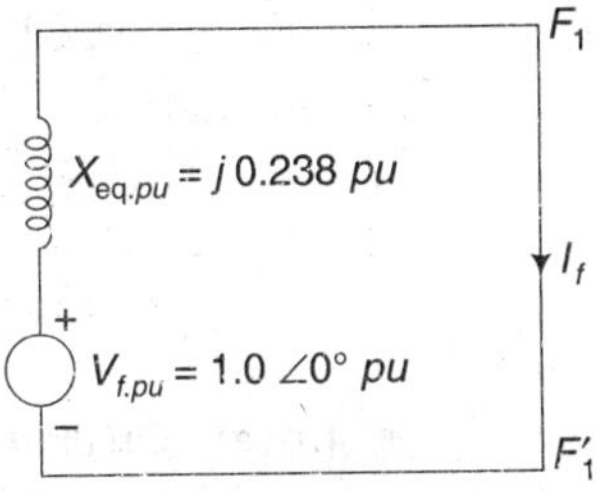

Fig. 4.13(d) Thevenin's equivalent circuit

The per-unit fault current $I_{f.pu}$ is given by.

$$I_{f.pu} = \frac{V_{f.pu}}{X_{eq.pu}} = \frac{1.0}{j\,0.238} = -j\,4.20 \text{ pu.}$$

The base current I_b is given by

$$I_b = \frac{(MVA)_b \times 10^3}{\sqrt{3} \times (kV)_b} = \frac{75 \times 10^3}{\sqrt{3} \times 66} = 656.857 \text{ A}$$

The actual fault current, $I_f = I_{f.pu} \times I_b = 4.20 \times 656.857 = 2758.8$ A

Fault MVA, $(MVA)_f = \dfrac{(MVA)_b}{X_{eq.pu}} = \dfrac{75}{0.238} = 315.126$ MVA

Fault current on LT side = $2758.8 \times \dfrac{66}{11} = 16552.8$ A

The fault current shared by the generators can be found from the circuit shown in Fig. 4.13(c).

$$I_{G1.pu} = I_{f.pu} \times \frac{X_{G2.pu}}{X_{G1.pu} + X_{G2.pu}} = 4.20 \times \frac{0.36}{0.225 + 0.36} = 2.585 \text{ pu}$$

$$I_{G2.pu} = I_{f.pu} \times \frac{X_{G1.pu}}{X_{G1.pu} + X_{G2.pu}} = 4.20 \times \frac{0.225}{0.225 + 0.36} = 1.615 \text{ pu}$$

Base current on LT side, $I_b = \dfrac{(MVA)_b \times 10^3}{\sqrt{3} \times (kV)_b} = \dfrac{75 \times 10^3}{\sqrt{3} \times 11} = 3941.145$ A

Hence, I_{G1}, and I_{G2} in amperes are given by

$I_{G1} = I_{G1.pu} \times I_b = 2.585 \times 3941.145 = 10187.86$ A

$I_{G2} = I_{G2.pu} \times I_b = 1.615 \times 3941.145 = 6364.94$ A

(b) The circuit model of the system for fault at the load end of the feeder is shown in Fig. 4.13(e)

The per-unit open circuit voltage across $F_2 F_2' = V_{f.pu} = 1.0 < 0° \; pu$

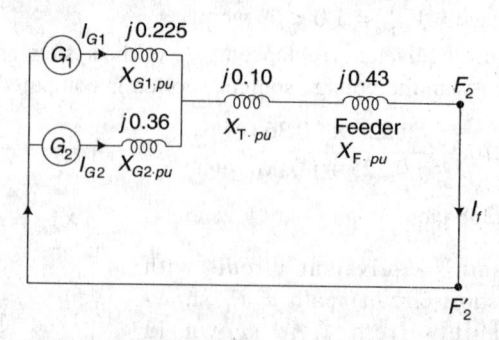

Fig. 4.13(e) Circuit model of the system for fault at load end of feeder

The Thevenin's equivalent reactance $X_{eq.pu}$ is computed from the following circuit, Fig. 4.13(f)

$$X_{eq.pu} = X_{F.pu} + X_{T.pu} + \frac{X_{G1.pu} X_{G2.pu}}{X_{G1.pu} + X_{G2.pu}}$$

$$= j\,0.43 + j\,0.10 + \frac{j\,0.225 \times j\,0.36}{j\,0.225 + j\,0.36}$$

$$= j\,0.668 \; pu.$$

The Thevenin's equivalent circuit with the fault path $F_2 F_2'$ showing the fault current is shown in Fig. 4.13(g).

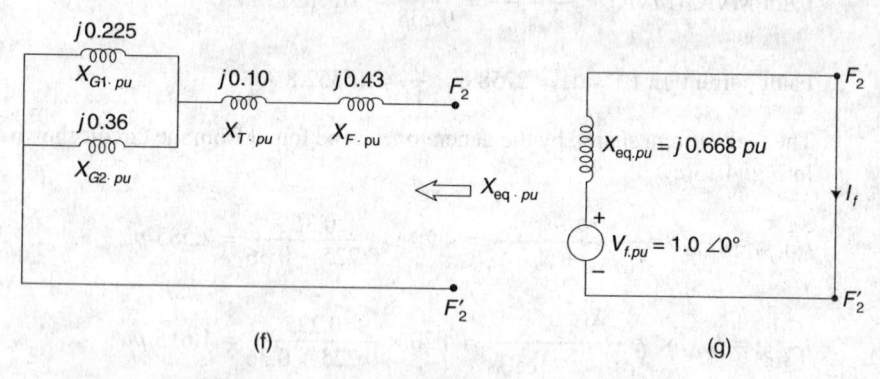

Fig. 4.13 (f) Thevenin's equivalent reactance (g) Thevenin's equivalent circuit

The per-unit fault current, $I_{f.pu} = \dfrac{1.0}{j\,0.668} = -j\,1.497 \; pu.$

Fault current,

$$I_f \text{(on HT side)} = I_{f.pu} \times I_b = 1.497 \times 656.857$$

$$= 983.315 \; A$$

Fault MVA, $(MVA)_f = \dfrac{(MVA)_b}{X_{eq.pu}} = \dfrac{75}{0.668} = 112.275 \; MVA$

The fault current shared by generators is given by [From Fig. 4.13(e)]

$$I_{G1.pu} = 1.497 \times \frac{0.36}{0.225 + 0.36} = 0.9212 \, pu$$

$$I_{G2.pu} = 1.497 \times \frac{0.225}{0.225 + 0.36} = 0.5758 \, pu$$

Base current I_b on LT side = 3941.145 A.

Hence I_{G1} and I_{G2} in amperes are given by

$$I_{G1} = I_{G1.pu} \times I_b = 0.9212 \times 3941.145 = 3630.58 \, A$$

$$I_{G2} = I_{G2.pu} \times I_b = 0.5758 \times 3941.145 = 2269.31 \, A$$

Total fault current I_f on LT side = $(I_{G1} + I_{G2})$

$$= I_{f.pu} \times I_b = 1.497 \times 3941.145$$

$$= 5899.89 \, A$$

Note: From the calculated values of the fault current and fault MVA, it is clear that these values reduce drastically as the fault point moves further and further away from the source. This reduction is mainly due to more impedance coming in the path of the fault current. The MVA rating of the circuit breaker to be placed at any location should be more than the fault MVA (fault level) at that location.

Example 4.7 | A generating station consisting of two generators, one of 20 MVA, 11 kV, 0.18 pu reactance with the unit transformer of 20 MVA, 11/132 kV, 0.08 pu reactance, another of 30 MVA, 11 kV, 0.18 pu reactance with the transformer of 30 MVA, 11/132 kV, 0.12 pu reactance, transmits power over double circuit 132-kV transmission line. Each line is having reactance of 120 ohms per phase. Determine the fault current supplied by the generators to a three-phase solid fault on the 132-kV bus-bar at the receiving end. Neglect pre-fault current and losses. Assume pre-fault generated voltages at rated value.

Solution: The one-line diagram of the system is shown in Fig. 4.14(a).

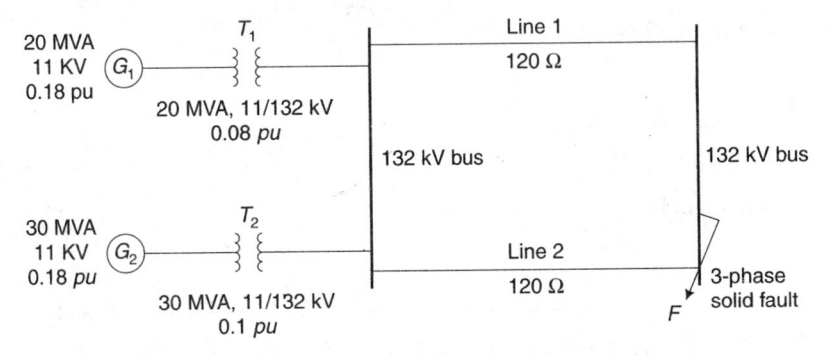

Fig. 4.14(a) *One-line diagram of the system of Example 4.7*

Let us choose the base MVA of 30 and base voltage of 11 kV on LT side. The base voltage on the HT side is 132 kV.

Base current I_b on LT side $= \dfrac{(MVA)_b \times 10^3}{\sqrt{3} \times (kV)_b}$

$= \dfrac{30 \times 10^3}{\sqrt{3} \times 11} = 1575$ A

The per-unit values of the reactances of the various equipment of the system on the chosen base are

$$\text{Generator } G_1 : X''_{G1.pu} = 0.18 \times \dfrac{30}{20} = 0.27 \; pu.$$

$$\text{Generator } G_2 : X''_{G2.pu} = 0.18 \times \dfrac{30}{30} = 018 \; pu.$$

$$\text{Transformer } T_1 : X_{T1.pu} = 0.08 \times \dfrac{30}{20} = 0.12 \; pu$$

$$\text{Transformer } T_2 : X_{T2.pu} = 0.12 \times \dfrac{30}{30} = 0.12 \; pu$$

$$\text{Transmission lines}: X_{L1.pu} = X_{L2.pu} = 120 \times \dfrac{(MVA)_b}{(kV)_b^2}$$

$$= 120 \times \dfrac{30}{(132)^2} = 0.206 \; pu$$

As the system is on no-load with the generated voltage at 1.0 pu, prefault voltage $V_{f.pu}$ at the fault point $F = 1.0 < 0°$ pu.

The per-unit reactance diagram of the system with faulted path FF' is shown in Fig. 4.14(b).

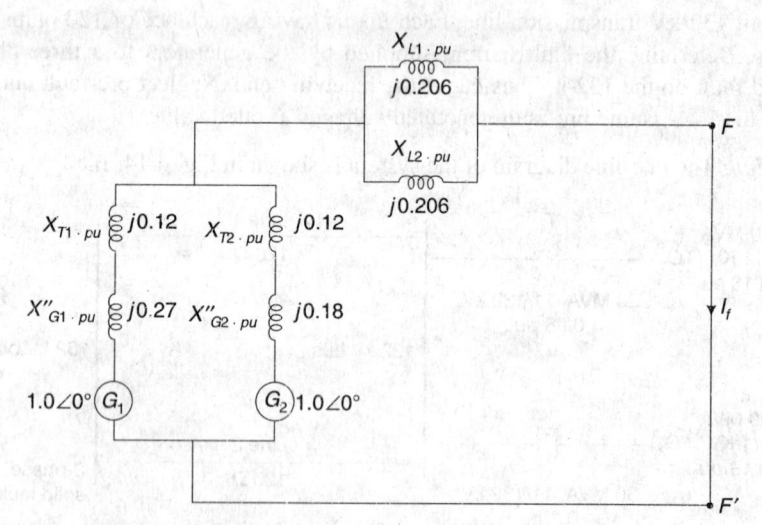

Fig. 4.14(b) *Per-unit reactance diagram with faulted path*

The Thevenin's equivalent reactance X_{eq} as viewed from the fault point F is computed by open-circuiting the faulted path FF' and short-circuiting the voltage sources. The circuit for computing X_{eq} is shown in Fig. 4.14(c).

From the circuit of Fig. 4.14(c), $X_{eq.\,pu}$ is given by.

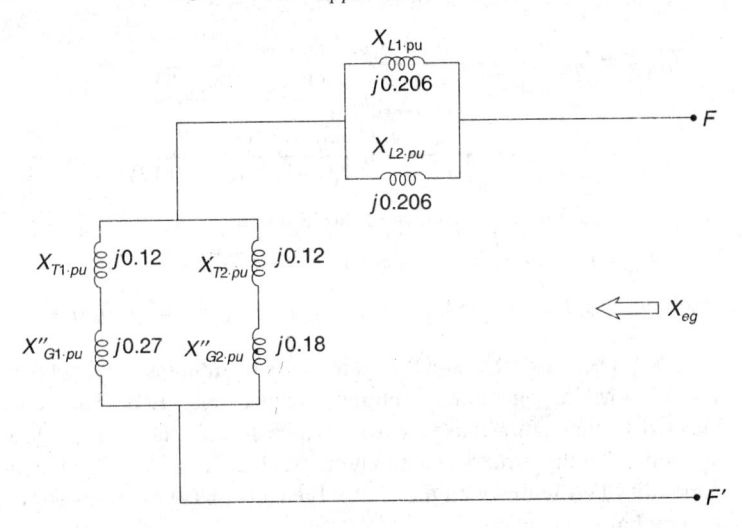

Fig. 4.14(c) Circuit for computing X_{eq}.

$$X_{eq.\,pu} = \frac{X_{L1.\,pu} X_{L2.\,pu}}{(X_{L1.\,pu} + X_{L2.\,pu})} + \frac{(X''_{G1.\,pu} + T_{1.\,pu})(X''_{G2.\,pu} + X_{T2.\,pu})}{(X''_{G1.\,pu} + X''_{G2.\,pu} + X_{T1.\,pu} + X_{T2.\,pu})}$$

$$= \frac{j\,0.206 \times j\,206)}{(j\,0.206 + j\,0.206)} + \frac{(j\,0.27 + j\,0.12)(j\,0.18 + j\,0.12)}{(j\,0.27 + j\,0.18 + j\,0.12 + j\,0.12)}$$

$$= j\,0.103 + j\,0.169 = j\,0.272\ pu.$$

The Thevenin's voltage is the open-circuit voltage across FF' which is the pre-fault voltage V_f.

Hence,

Thevenin's per-unit voltage = $V_{f.\,pu} = 1.0 < 0°\ pu$.

The Thevenin's equivalent circuit with the faulted path FF' is shown in Fig. 4.14(d)

Fig. 4.14(d) Thevenin's equivalent circuit

so, the per-unit fault current,

$$I_{f.\,pu} = \frac{V_{f.\,pu}}{X_{eq.\,pu}} = \frac{1.0}{j\,0.272} = -j\,3.676\ pu$$

The fault current shared by generators G_1 and G_2 are given by

$$I_{G1.\,pu} = I_{f.\,pu} \times \frac{X_{G2.\,pu} + X_{T2.\,pu}}{(X''_{G1.\,pu} + X_{T1.\,pu}) + (X''_{G2.\,pu} + X_{T2.\,pu})}$$

$$= -j\,3.676 \times \frac{j\,0.18 \times j\,0.12}{(j\,0.27 + j\,0.12) + (j\,0.18 + j\,0.12)}$$

$$= -j\,3.676 \times 0.4347 = -j\,1.598\ pu$$

$$I_{G1} = I_{G1.pu} \times I_b = -j\,1.598 \times 1575 = -j\,2517\text{ A}$$

$$I_{G2.pu} = I_{f.pu} \times \frac{X_{G1.pu} + X_{T1.pu}}{(X''_{G1.pu} + X_{T1.pu}) + (X''_{G2.pu} + X_{T2.pu})}$$

$$= -j\,3.676 \times \frac{j\,0.27 + j\,0.12}{(j\,0.27 + j\,0.12) + (j\,0.18 + j\,0.12)}$$

$$= -j\,3.676 \times 0.5652 = -j\,2.078\ pu$$

$$I_{G2} = I_{G2.pu} \times I_b = -j\,2.078 \times 1575 = -j\,3273\text{ A}$$

Total fault current, I_f (on LT side) = $(I_{G1} + I_{G2}) = I_{f.pu} \times I_b = -j\,5790$ A

Example 4.8 | There are two 10-MVA generators of subtransient reactance 15% each and one 5-MVA generator of subtransient reactance 12%. The generators are connected to the station busbars from which load is taken through 3 Nos. 5-MVA, step-up transformers each having reactance of 8%. Calculate the maximum fault MVA with which the circuit breakers on (a) l.v. side, and (b) The h.v. side may have to deal.

Solution: The one-line diagram of the system is shown in Fig. 4.15(a).

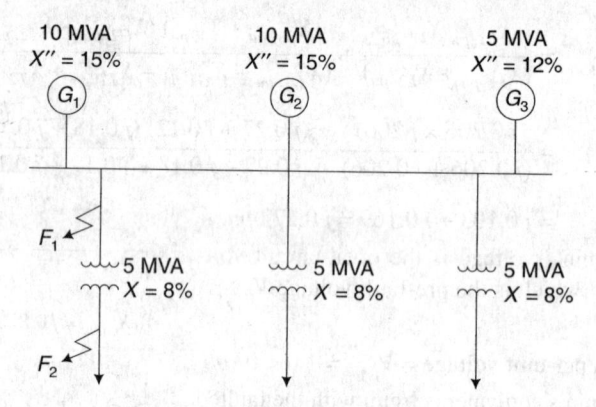

Fig. 4.15(a) One-line diagram of the system of Example 4.8

Let the base MVA be 10

Per-unit reactances of the various equipment on the chosen base are:

$$\text{Generator } G_1 : X''_{G1.pu} = 0.15 \times \frac{10}{10} = 0.15\ pu$$

$$\text{Generator } G_2 : X''_{G2.pu} = 0.15 \times \frac{10}{10} = 0.15\ pu$$

$$\text{Generator } G_3 : X''_{G3.pu} = 0.12 \times \frac{10}{5} = 0.24\ pu$$

$$\text{Each transformer} : X_T = 0.08 \times \frac{10}{5} = 0.16\ pu$$

(a) Fault on l.v. side of the transformer.

The circuit for calculating the Thevenin's equivalent reactance X_{eq} as viewed from the fault point F_1 (with voltage sources short circuited) is shown in Fig. 4.15(b).

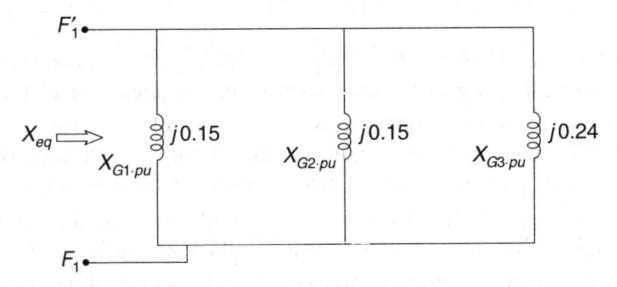

Fig. 4.15(b) Circuit for calculating X_{eq} for fault at F_1

$X_{eq.\,pu}$ for fault at F_1 is given by

$$X_{eq.\,pu} = \frac{\left(j\dfrac{0.15}{2} \times j\,0.24\right)}{\left(j\dfrac{0.15}{2} + j\,0.24\right)} = j\,0.057\ pu$$

Per-unit fault MVA is

$$(MVA)_{f.\,pu} = \frac{1}{X_{eq.\,pu}}$$

and fault MVA is

$$(MVA)_f = \frac{(MVA)_b}{X_{eq.\,pu}} = \frac{10}{0.057} = 175.4\ MVA$$

(b) Fault on h.v. side of the transformer.

The circuit for calculating X_{eq} for fault at F_2 is shown in Fig. 4.15(c). Per-unit value of X_{eq} is given by

$$X_{eq.\,pu} = j\,0.16 + \frac{\left(j\dfrac{0.15}{2} \times j\,0.24\right)}{\left(j\dfrac{0.15}{2} + j\,0.24\right)}$$

Fig. 4.15(c) Circuit for calculating X_{eq} for fault at F_2

$$= j\,0.16 + j\,0.057 = j\,0.217\ pu$$

$$\text{Fault MVA, } (MVA)_f = \frac{(MVA)_b}{X_{eq.\,pu}} = \frac{10}{0.217} = 46\ MVA$$

Example 4.9 A 25-MVA, 11-kV generator with $X''_d = 20\%$ is connected through a transformer, line and a transformer to a bus that supplies four identical motors as shown in Fig. 4.16. Each motor has $X''_d = 25\%$ and $X'_d = 30\%$ on a base of 5 MVA, 6.6 kV. The three-phase rating of the step-up transformer is 25 MVA, 11/66 kV, with a leakage reactance of 10% and that of the step-down transformer is 25 MVA, 66/6.6 kV, with a leakage reactance of 10%. The bus voltage at the motors is 6.6 kV when a three-phase fault occurs at the point F. The reactance of the line is 30 ohms. Assume that the system is operating on no load when the fault occurs. For the specified fault, calculate the following:

(a) The subtransient current in the fault.
(b) The subtransient current in the breaker A.
(c) The momentary current in the breaker A, and
(d) The current to be interrupted by the breaker A, breaker time is 5 cycles.

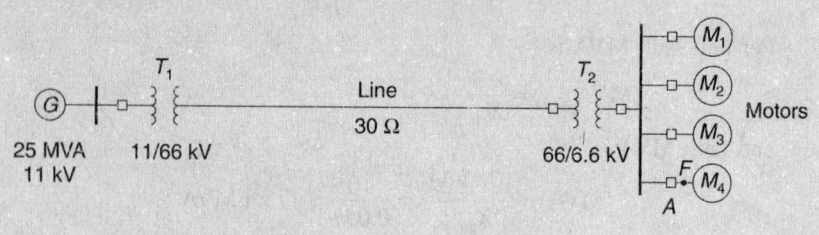

Fig. 4.16 System for Example 4.9

Solution: Let us choose the base MVA of 25. For a generator voltage base of 11 kV, line voltage base is 66 kV and motor voltage base is 6.6 kV.

The per-unit reactances of various components on the chosen base are the following:

$$\text{Generator } G : X''_{dg.\,pu} = j\,0.20\ pu$$

$$\text{Each motor : } X''_{dm.\,pu} = j\,0.25 \times \frac{25}{5}$$

$$= j\,1.25\ pu,$$

$$X'_{dm.\,pu} = j\,0.30 \times \frac{25}{5} = j\,1.5\ pu$$

$$\text{Each Transformer : } X_{T.\,pu} = j\,0.10\ pu.$$

$$\text{Line : } X_{L.\,pu} = j\,30 \times \frac{25}{(66)^2} = j\,0.17\ pu.$$

(a) The reactance diagram of the system is shown in Fig. 4.17(a). The reactance diagram of the system for fault at point F is shown in Fig. 4.17(b). Since the system is initially on no-load, the generator and motor induced emfs are identical. The circuit shown in Fig. 4.17(b) can therefore be reduced to that of Fig. 4.17(c), Fig. 4.17(d) and then to Fig. 4.17(e).

Fault Analysis 129

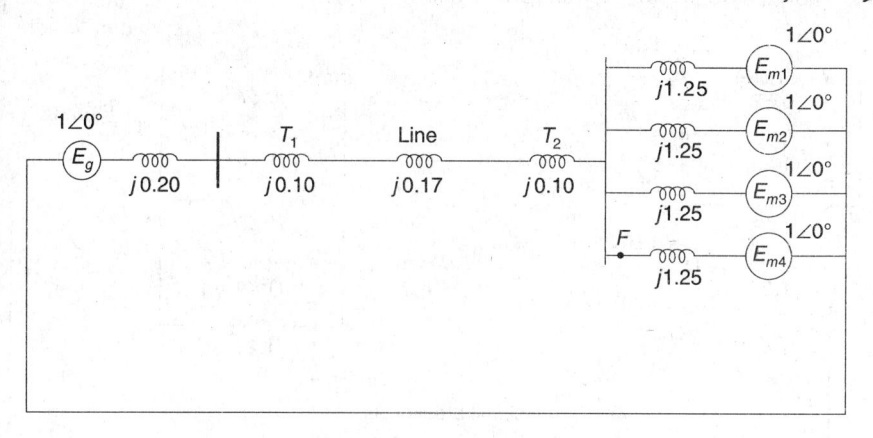

Fig. 4.17(a) The reactance diagram for Example 4.9

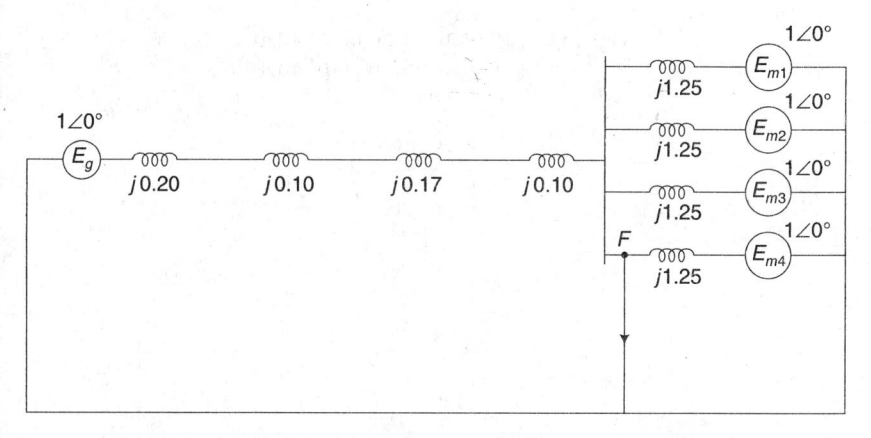

Fig. 4.17(b) Reactance diagram for fault at F

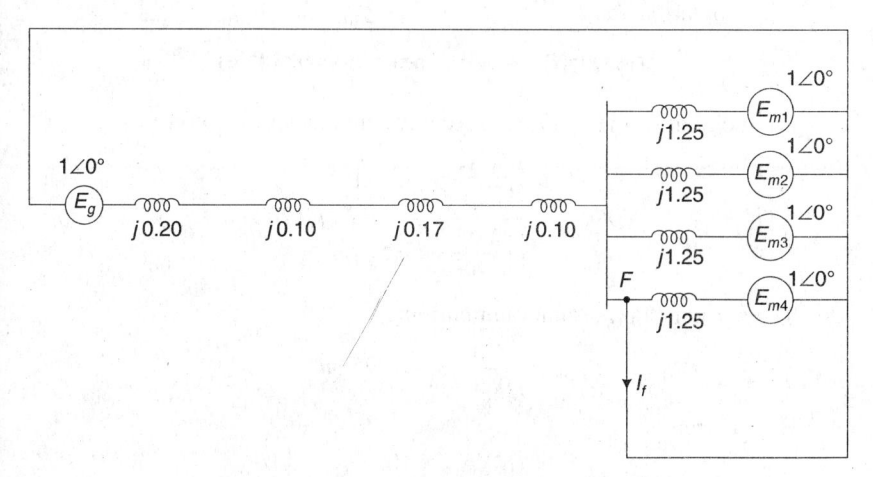

Fig. 4.17(c) Reduced circuit of 4.17(b)

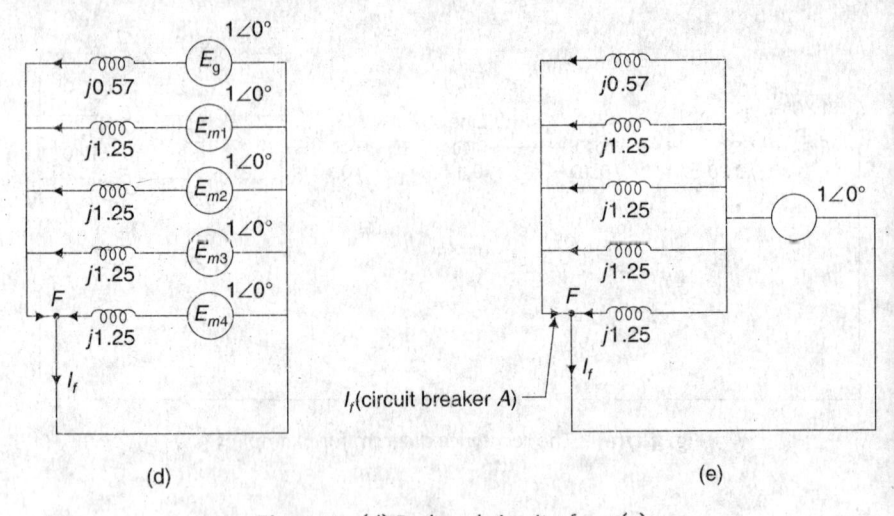

Fig. 4.17 (d) Reduced circuit of 4.17(c)
(e) Reduced circuit of 4.17(d)

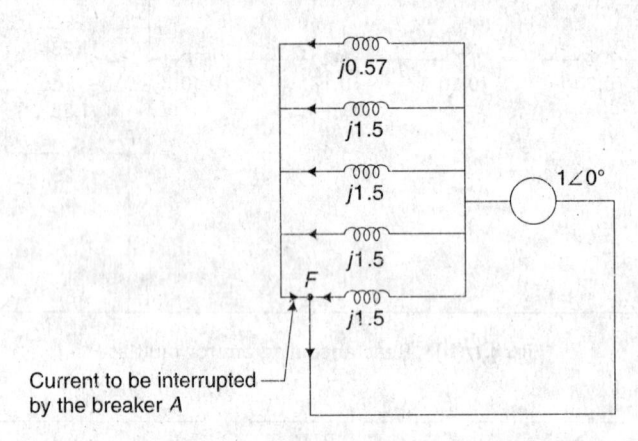

Fig. 4.17(f) Modified reactances of 4.17(e)

From circuit of Fig. 4.17(e), the equivalent reactance is given by

$$X_{eq.\,pu} = \frac{\left(j\dfrac{1.25}{4} \times j\,0.57\right)}{\left(j\dfrac{1.25}{4} + j\,0.57\right)} = j\,0.202\ pu.$$

The per-unit subtransient fault current,

$$I_{f.\,pu} = \frac{1.0}{j\,0.202} = -j\,4.95\ pu$$

Base current,

$$I_b \text{ in 6.6 kV circuit} = \frac{(MVA)_b \times 10^3}{\sqrt{3} \times (kV)_b}$$

$$= \frac{25 \times 10^3}{\sqrt{3} \times 6.6} = 2187 \text{ A}$$

Therefore, $I_f = I_{f.\,pu} \times I_b = 4.95 \times 2187 = 10{,}826 \text{ A}$

Alternative method

From Fig. 4.17(e), $I_{f.\,pu} = 4 \times \dfrac{1}{j\,1.25} + \dfrac{1}{j\,0.57}$

$$= -j\,3.20 - j\,1.75 = -j\,4.95 \text{ pu}.$$

$$I_f = I_{f.\,pu} \times I_b = 4.95 \times 2187 = 10{,}826 \text{ A}$$

(b) From Fig. 4.17(e), current through circuit breaker A is

$$I_{f.\,pu}(A) = 3 \times \frac{1}{j\,1.25} + \frac{1}{j\,0.57} = -j\,2.40 - j\,1.75 = -j\,4.15 \text{ pu}$$

and $I_f(A) = 4.15 \times 2187 = 9076 \text{ A}$

(c) For finding the momentary current through the breaker, the presence of dc offset current is taken into account by multiplying the symmetrical subtransient current by a factor of 1.6.

Hence, momentary current through the breaker A.

$= 1.6 \times I_f(A) = 1.6 \times 9076 = 14{,}521 \text{ A}$

(d) To calculate the current to be interrupted by the breaker, motor subtransient reactance ($X''_{dm} = j\,0.25$) is replaced by transient reactance ($X'_{dm} = j\,0.30$).

Hence, the reactances of the circuit of Fig. 4.17(e) modify to that of Fig. 4.17(f).

Current (symmetrical) to be interrupted by the breaker (as shown by arrow)

$$= 3 \times \frac{1}{j\,1.5} + \frac{1}{j\,0.57} = -j\,3.75 \text{ pu}$$

Allowance for the dc offset value is made by multiplying the symmetrical current by a factor of 1.1.

Therefore, the current to be interrupted by breaker A is

$1.1 \times 3.75 \times 2187 = 9021 \text{ A}$

Example 4.10 A 100-MVA, 13.2 kV genrator is connected to a 100 MVA, 13.2/132 kV transformer. The generator's reactances are $X''_d = 0.15$ *pu*, $X'_d = 0.25$ *pu*, $X_d = 1.25$ *pu* on a 100 MVA base, while the transformer reactance is 0.1 *pu* on the same base. The system is operating on no load, at rated voltage when a three-phase symmetrical fault occurs at the HT terminals of the transformer. Determine

(a) The subtransient, transient and steady state symmetrical fault currents in *pu* and in amperes
(b) The maximum possible dc component
(c) Maximum value of instantaneous current
(d) Maximum rms value of the asymmetrical fault current.

Solution: Let the base MVA be 100. All the reactances are given on this base. For a base voltage of 13.2 kV on L.T. side, the per-unit voltage of the generator, $E_g = 1.0$.

Base current on L.T. side of transformer $= \dfrac{100 \times 10^3}{\sqrt{3} \times 13.2} = 4374$ A

Base current on H.T. side of transformer $= \dfrac{100 \times 10^3}{\sqrt{3} \times 132} = 437.4$ A

(a) (i) Subtransient symmetrical fault current

$$I''_{f.\,pu} = \dfrac{E_g}{X''_{d.\,pu} + X_{t.\,pu}}$$

$$= \dfrac{1.0}{j\,0.15 + j\,0.1} = -j\,4.0\ pu$$

Actual fault current on h.v. side = $4 \times 437.4 = 1749.6$ A

Actual fault current on l.v. side = $4 \times 4374 = 17496$ A

(ii) Transient symmetrical fault current,

$$I'_{f.\,pu} = \dfrac{E_g}{X'_{d.\,pu} + X_{t.\,pu}}$$

$$= \dfrac{1.0}{j\,0.25 + j\,0.1} = -j\,2.857\ pu$$

Actual fault current on h.v. side = $2.857 \times 437.4 = 1249.6$ A

Actual fault current on l.v. side = $2.857 \times 4374 = 12496$ A

(iii) Steady state fault current,

$$I_{f.\,pu} = \dfrac{E_g}{X_{d.\,pu} + X_{t.\,pu}}$$

$$= \dfrac{1.0}{j\,1.25 + j\,0.1} = -j\,0.74\ pu$$

Actual fault current on h.v. side = $0.74 \times 437.4 = 323.67$ A

Actual fault current on l.v. side = $0.74 \times 4374 = 3236.7$ A

(b) Maximum possible dc component,

$$I_{dc(max)} = \sqrt{2}\ I''_{f.\,pu}$$

$$= \sqrt{2} \times 4 = 5.65\ pu$$

(c) Maximum value of instantaneous current

= (Maximum dc component) + (Maximum instantaneous subtransient current)

$= \sqrt{2}\ I''_{f.\,pu} + \sqrt{2}\ I''_{f.\,pu} = 2\sqrt{2}\ I''_{f.\,pu} = 2 \times 1.414 \times 4.0 = 11.31\ pu$

(d) Maximum value of asymmetrical fault current

$$I_{max}\ (rms) = \sqrt{I^2_{dc(max)} + I^2_{ac(max)}} = \sqrt{3}\ I''_{f.\,pu}$$

$$= \sqrt{3} \times 4.0 = 6.928\ pu.$$

4.5.4 Fault Under Loaded Condition of Synchronous Machine

In all the preceding discussion on symmetrical fault, it was assumed that the synchronous machine was operating at no load prior to the occurrence of fault. The load currents are generally neglected during fault calculations because fault currents are much larger than load currents. However, in some situations it becomes necessary to consider the effect of pre-fault load current. The analysis of fault under loaded condition of synchronous machine is very complicated. This section presents the methods of computing fault current when fault occurs under loaded condition.

When a fault occurs under loaded condition of the system, the voltages at different points in the power system deviate from nominal (1 pu) value due to the flow of load current. Therefore, it is first necessary to calculate the pre-fault voltage V_f at the fault point. The fault current is then calculated using this value of V_f. The total current is the phasor sum of the load current and the fault current. If the load is a synchronous motor, the fault would cause the motor to supply power from its stored energy into the fault.

Let us consider that a generator is loaded when the fault occurs. Figure 4.18(a) shows the equivalent circuit of a generator that has a balanced three-phase load. External impedance Z_{ext} is shown between the generator terminals and the point F where the fault occurs. The current flowing before the occurrence of fault at point F is I_L, the pre-fault voltage at the fault point F is V_f, and the terminal voltage of the generator is V_t. When a three-phase fault occurs at point F in the system, the circuit shown in Fig. 4.18(b) becomes the appropriate equivalent circuit (with switch S closed). Here a voltage E_g'' in series with X_d'' supplies the steady-state current I_L when switch S is open, and supplies the current to the fault through X_d'' and Z_{ext} when switch S is closed. If E_g'' can be determined, this current through X_d'' will be I''. With switch S open, we have

$$E_g'' = V_t + j I_L X_d'' \qquad (4.35)$$

and this equation defines E_g'', which is called subtransient internal voltage. Similarly, for the transient internal voltage, we have

$$E_g' = V_t + j I_L X_d' \qquad (4.36)$$

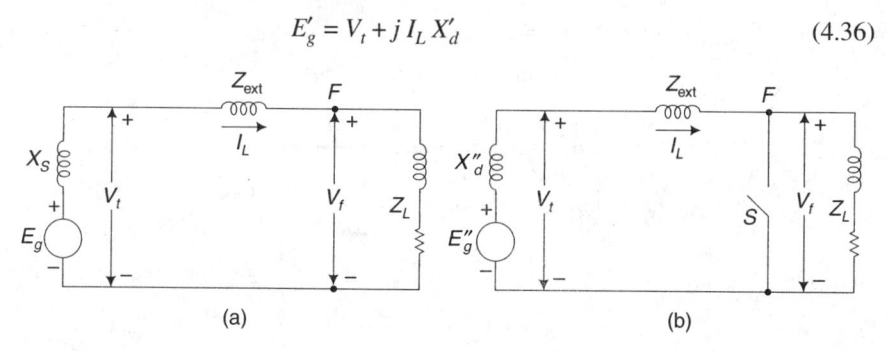

(a)   (b)

Fig. 4.18 *Equivalent circuits for a generator supplying a balanced three-phase load: (a) Steady-state equivalent circuit (b) Circuit for calculation of I″*

Clearly E_g'' and E_g' depend on the value of the load current I_L before the occurrence of the fault. In fact, if I_L is zero (no load case), $E_g'' = E_g' = E_g$, the no load voltage.

Synchronous motors have internal voltages and reactances similar to that of a generator except that the current direction is reversed. During fault conditions these can be replaced by similar equivalent circuits except that the subtransient and transient internal voltages are given by

$$E''_m = V_t - j I_L X''_d \tag{4.37}$$

$$E'_m = V_t - j I_L X'_d \tag{4.38}$$

Systems which contain generators and motors under load conditions may be solved either by Thevenin's Theorem or by the use of transient or subtransient internal voltages. Consideration of pre-fault load current is illustrated in Example 4.11.

Example 4.11 | A synchronous generator and a sychronous motor each rated 50 MVA, 11 kV having 20% subtransient reactance are connected through transformers and a line as shown in Fig. 4.19(a). The transformers are rated 50 MVA, 11/66 kV and 66/11 kV with leakage reactance of 10% each. The line has a reactance of 10% on a base of 50 MVA, 66 kV. The motor is drawing 40 MW at 0.8 pf leading at a terminal voltage of 10.5 kV when a symmetrical three-phase fault occurs at the motor terminals. Find the subtransient current in the generator, motor and fault.

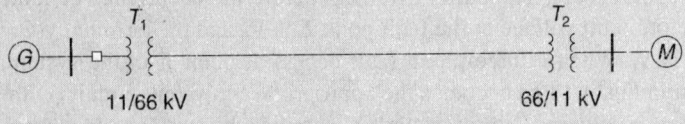

(a) One-line diagram for the system of Example 4.11

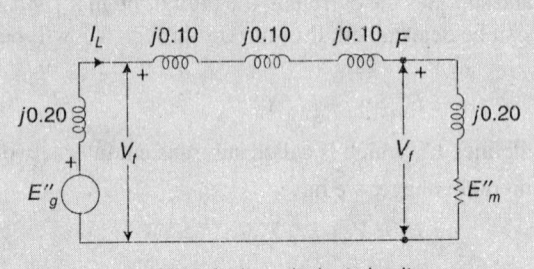

(b) Pre-fault equivalent circuit

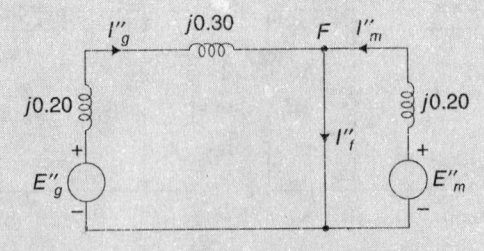

(c) Equivalent circuit during fault

Fig. 4.19 System for Example 4.11

Fault Analysis 135

Solution: All reactances are given on MVA base of 50 and appropriate voltage bases. Hence, the chosen base MVA is 50 and base voltages are 11 kV and 66 kV in LT and HT circuits respectively.

The pre-fault equivalent circuit is shown in Fig. 4.19(b) and the equivalent circuit under faulted condition is shown in Fig. 4.19(c).

Taking the prefault voltage V_f as the reference phasor,

$$V_{f.pu} = \frac{10.5}{11} = 0.954 < 0° \ pu$$

Base current
$$I_b = \frac{50 \times 10^3}{\sqrt{3} \times 11} = 2624 \ A$$

Load current
$$I_L = \frac{40 \times 10^3}{\sqrt{3} \times 10.5 \times 0.8} = 2749 \ A < 36.9° \ A$$

$$I_{L.pu} = \frac{2749}{2624} = 1.047 < 36.9°$$

$$= 1.047 \ (0.8 + j \ 0.6) = 0.837 + j \ 0.628 \ pu$$

From Fig. 4.19(b)
For the generator,

$$V_{t.pu} = V_{f.pu} + I_{L.pu} \ (j \ 0.1 + j \ 0.1 + j \ 0.1)$$
$$= 0.954 + j \ 0.3 \ (0.837 + j \ 0.628)$$
$$= 0.766 + j \ 0.251 \ pu$$
$$E''_{g.pu} = V_{t.pu} + I_{L.pu} \ (j \ 0.20)$$
$$= 0.766 + j \ 0.251 + j \ 0.20 \ (0.837 + j \ 0.628)$$
$$= 0.640 + j \ 0.4184 \ pu$$

For the motor

$$V_{t.pu} = V_{f.pu} = 0.954 < 0°.$$
$$E''_{m.pu} = V_{t.pu} - I_{L.pu} \ (j \ 0.20)$$
$$= 0.954 - j \ 0.20 \ (0.837 + j \ 0.628)$$
$$= 1.079 - j \ 0.1674 \ pu$$

From Fig. 4.19(c)

$$I''_{g.pu} = \frac{E''_{g.pu}}{(j \ 0.50)} = \frac{0.640 + j \ 0.4184}{j \ 0.50} = 0.837 - j \ 1.280 \ pu$$

$$I''_{m.pu} = \frac{E''_{m.pu}}{j \ 0.2} = \frac{1.079 - j \ 0.1674}{j \ 0.2} = -0.837 - j \ 5.395 \ pu$$

Current in the fault,

$$I''_{f.pu} = I''_{g.pu} + I''_{m.pu}$$
$$= 0.837 - j \ 1.280 - 0.837 - j \ 5.395$$
$$= -j \ 6.675 \ pu$$

Base current, $I_b = 2624$ A.

Hence, the actual currents in ampere are

$I''_g = I''_{g.pu} \times I_b = (0.837 - j\,1.28) \times 2624 = 2196.2 - j\,3{,}358.7$ A

$I''_m = I''_{m.pu} \times I_b = (-0.837 - j\,5.395) \times 2624 = -2196.2 - j\,14{,}156.4$ A

$I''_f = I''_g + I''_m = I_{f.pu} \times I_b = -j\,6.675 \times 2624 = -j\,17{,}515$ A

4.5.5 Current Limiting Reactors

When a symmetrical three-phase fault occurs at a point in a power system, the fault current is very large which may damage the equipment of the power system. The fault currents depend upon the generating capacity, pre-fault voltage at the fault point and the total reactance between the generators and the fault point. If X is the total reactance upto the fault point, and V_f is the prefault voltage at the fault point, then neglecting the resistance, the fault current I_f is given by V_f/X. Therefore, by increasing series reactance X of the system, the fault current can be decreased. Current limiting reactors which are large inductive coils wound for high self-inductance and very low resistnace, are used to limit the fault current to a safe value which can be interrupted by circuit breakers. The breaking current capacity of the circuit breakers should be such that the fault currents are less than the breaking current capacity. Sometimes, the fault currents can be so high that the circuit breakers of suitable breaking capacity may not be available. The fault current, then, should be limited by some means so that available or existing circuit breakers can be safely used. When the existing power system is extended by adding more generating stations or more generator units in parallel and installation of additional feeders, the fault current to be interrupted by the same circuit breaker will be greater than earlier. In such a case, the circuit breaker should be either replaced by another one of higher breaking current capacity or the fault current can be limited by means of current limiting reactors. By connecting current limiting reactors at strategic locations, the fault current at several points can be reduced. Hence, current limiting reactors are useful in limiting fault currents to such values so that the circuit breakers can interrupt them. Current limiting reactors may be connected in series with each generator or in series with each feeder or between the busbar section of a Tie-bar system or Ring system.

The reactance of current-limiting reactors should be high enough to limit the fault current, but not so high as to cause excessive voltage drop due to load current.

Terms and Definitions Used with Reactors

Some important terms which are used in connection with current limiting reactors are defined as follows:

(i) **Rated current** It is the rms value of the current which the reactor can carry continuously without exceeding a specified temperature rise dependent on the type of insulation used.

(ii) **Rated short-time current** It is the rms value of the symmetrical fault current which the reactor can carry for specified short time [e.g., 50 KA for 1 Sec].

(iii) **Over-current factor** It is the ratio of rms symmetrical through-fault current to rated current.

(iv) **Over-current time** It is the time in seconds that the reactor can carry the above fault current without suffering damage.

(v) **Rated voltage** It is the line to line system voltage for which the reactor is designed.

(vi) **Short-circuit rating** The reactors should be capable of withstanding the mechanical and thermal stresses during short circuit at its terminals for a specified period of time.

Design Features and Construction of Reactors

The design of current limiting reactors should be such that their reactance should not decrease due to saturation under fault conditions, if fault current is to be limited to a correct predetermined value. Therefore, it is essential that reactors be built with non-magnetic cores which do not saurate due to large fault currents or alternatively with an iron core but with an air gap included in the magnetic circuit. The cross-section of iron core reactor will depend upon the fault current. If the fault current is more than about three times of the rated full load current, an iron-core reactor having essentially constant permeability would require a very large cross-section of the core. Hence, air-core reactors are sometimes used for limiting the fault current. There are two types of air-core reactors, i.e., (i) Unshielded type or Dry type or Cast concrete type (ii) Magnetically shielded type or oil-immersed type.

Unshielded type or dry-type reactors are generally cooled by natural or forced air cooling. These reactors are used only up to 33 kV. For higher voltages oil immersed type reactors are used. Dry type air insulated reactors occupy larger space and they need a large clearance from adjacent constructional work. The reactance of dry-type air-core reactors is almost constant because of the absence of iron.

In the magnetically shielded type or oil-immersed type design, of reactors, laminations of iron shields are provided around the outside conductors so as to avoid the entering of magnetic flux in the surrounding iron parts. The reactance of this type of reactors drops by about 10% during faults due to iron shield. Oil-immersed reactors can be applied to a circuit of any voltage level, for either outdoor or indoor installation.

Types of Reactors

Current limiting reactors are of following two main types.

(i) Unshielded type or dry type

(ii) Magnetically shielded type or Oil-immersed type

Unshielded type In this type of reactor, the winding is usually of bare stranded copper and is embedded in a number of specially shaped concrete slabs. This arrangement of winding provides a very rigid mechanical support against the mechanical forces developed when fault occurs. These forces tend to compress the coil axially and expand it radially. A circular coil is used because all coils tend to assume this shape under fault condition. The individual turns of the winding are inclined with respect to the horizontal plane. A concrete base and porcelain post insulators which support the reactor provide the necessary insulation to earth. This type of reactor is used for system voltage up to 33 kV. This type of reactor is also called concrete type or dry-type reactor.

The disadvantages of this type of reactor are summarized below:
(i) These reactors are not suitable for outside service since they do not conform to the usual metal clad switchgear principle.
(ii) These reactors require large space because the magnetic field due to load current is practically unrestricted.
(iii) It is difficult to cool large coils of the reactors by fans.
(iv) The use of these reactors is limited to system voltage of 33 kV.

Magnetically shielded type The coil of magnetically shielded type reactor is placed in a standard transformer tank and is oil insulated. The magnetic shielding of the reactor prevents the magnetic flux from entering the transformer tank walls and hence avoids considerable losses and heating under normal load conditions which would be prohibitive under fault conditions. One method of shielding makes use of iron laminations, which provide path for the stray magnetic field outside the windings. Unfortunately, this method results in a decrease of reactance under abnormal current conditions. In this arrangement, there is no iron inside the coil. In another method the shields are in the form of short-circuited rings, at earth potential and are supported from non-magnetic end plates. With this arrangement, the reactance does not decrease under fault conditions. In large reactors these shielding rings may be in the form of cooling tubes which perform both functions of magnetic shielding and cooling. This type of reactor is also known as oil-immersed type and can be used up to any voltage level for either outdoor or indoor installation.

The advantages of this type of reactors are
(i) High factor of safety against flashover
(ii) Smaller size
(iii) High thermal capacity

Location of Reactors

The location of a reactor should be such that there is no large voltage drop across it due to load currents under normal conditions and it limits the fault current being fed by the generators into the fault to the safe value which can be lower than the breaking currents of the circuit breakers.

The reactors can be located in the following positions:

1. **Generator reactors** When the reactors are connected in series with each generator as shown in Fig. 4.20, they are called generator reactors. In this case, the reactor is considered as a part of the generator reactance and hence its effect is to protect the generator in the case of any fault beyond the reactors.

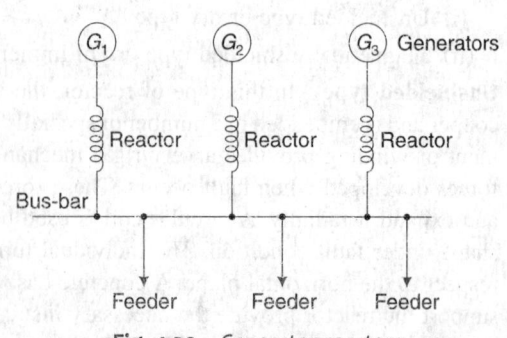

Fig. 4.20 *Generator reactors*

Following are the disadvantages of this method:

(i) Since full-load current is always flowing through the reactor, there is a constant voltage drop and power loss in the reactor even during normal operating conditions.

(ii) When a fault occurs on a feeder, the voltage on the common busbar drops to a low value with the result that the generators and the remaining healthy feeders may also fall out of step.

Due to above disadvantages and also since modern generators have sufficiently large transient reactance to protect them against fault (short-circuit), it is not a common practice to use separate reactors in series with the generators.

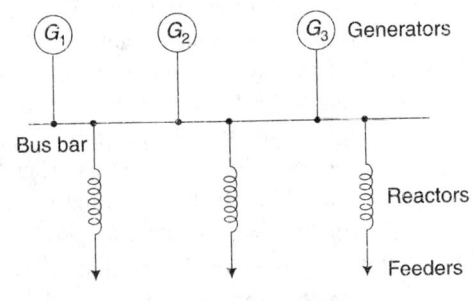

Fig. 4.21 *Feeder reactors*

2. **Feeder reactors** In this arrangement the reactors are connected in series with each feeder as show in Fig. 4.21. Since most of the faults occur on feeders, a large number of reactors are used for such circuits. This method avoids the main disadvantages of the first method since reactors are connected in series with feeders instead of generators. Two main advantages of feeder reactors are as follows:

(i) In the case of a fault on any feeder, the main voltage drop is in its reactor only so that the bus-bar voltage is not affected and there is a little tendency for the generator to lose synchronism.

(ii) Since the fault on a feeder will not affect other feeders, the effects of faults are localised.

Disadvantages are as follows.

(i) This method provides no protection to the generators if a fault occurs on the busbars. But, since the busbar faults are rare, this drawback is not of much importance.

(ii) The disadvantage of constant voltage drop and power loss in the reactors even during normal operating conditions is equally present in this method also.

3. **Busbar reactors** The above two methods of locating reactors suffer from the disadvantage that there is considerable voltage drop and power loss in the reactors due to flow of full-load current through them during normal operating conditions. This disadvantage can be overcome by connecting the reactors in the busbars. Two methods used for this purpose are ring system and tie-bar system.

(a) *Ring System* In this method, the bus-bar is divided into sections and these sections are interconnected through reactors as shown in Fig. 4.22. Here each feeder is generally fed by one generator only. Since each generator supplies its own section of the load and very small amount of power flows through the reactors from other generators, there is low power loss and voltage drop in the reactors under normal operating conditions. The main advantage

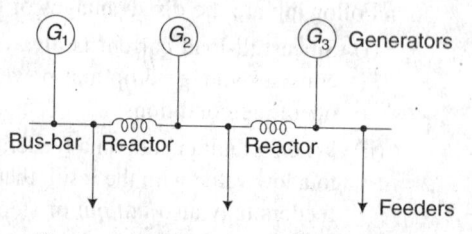

Fig. 4.22 *Bus-bar reactors (ring system)*

of this system is that in the event of fault on any one feeder only one generator (to which the particular feeder is connected) mainly feeds the fault current while the current fed from other generators is small because of being through reactors. Therefore, only that section of bus-bar is affected to which the faulted feeder is connected and the other sections of the bus-bar continue to operate under normal condition.

(b) *Tie-bar System* The tie-bar system is shown in Fig. 4.23. As compared to the ring system, it is clear that the tie-bar system has effectively two reactors in series between sections so that the reactors must have approximately half the reactance of those used in comparable ring system. The tie-bar system makes it possible to connect additional generators to the system without requiring changes in the existing reactors. Thus no modification of the switchgear is required when extension of the system takes place. The main disadvantage of this system is that it requires an additional busbar, i.e., the tie-bar.

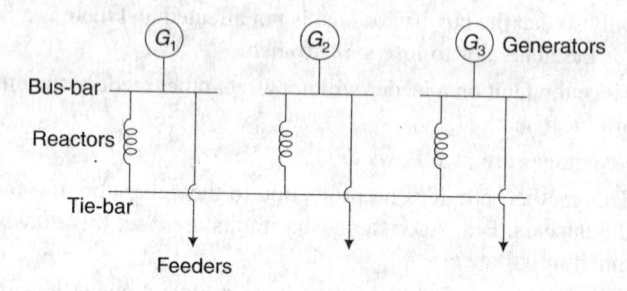

Fig. 4.23 *Busbar reactors (tie-bar system)*

MVA Rating of Reactors

If X be the reactance of the reactor in ohm/phase, I the rated (load) current in amperes, and V the rated line-to-line voltage of the system in kilovolts (kV), then at rated current the voltage between terminals of one phase = $I \times$ volts.

Choosing rated line to line voltage of the system as base kV, and rated current as base current, $(kV)_b = V$ and $I_b = I$

Hence base reactance, $X_b = \dfrac{\dfrac{V}{\sqrt{3}} \times 10^3}{I} = \dfrac{V \times 10^3}{\sqrt{3}\, I}$

Per-unit reactance of the reactor, $X_{pu} = \dfrac{X}{X_b} = \dfrac{\sqrt{3}\ IX}{V \times 10^3}$

$= \dfrac{\sqrt{3}\ IX}{V} \times 10^{-3}$

Base MVA, $(MVA)_b = \sqrt{3}\ VI \times 10^{-3}$

Total 3-phase reactive MVA of the reactor $= 3\,(IX)\,I \times 10^{-6}$
$= 3I^2 X \times 10^{-6}$ MVAr

Per-unit reactive MVA of the reactor, $(MVAr)_{pu}$

$= \dfrac{MVAr}{(MVA)b} = \dfrac{3I^2 X \times 10^{-6}}{\sqrt{3}\ VI \times 10^{-3}}$

$= \dfrac{\sqrt{3}\ IX}{V} \times 10^{-3} = X_{pu}$ \hfill (4.39)

Example 4.12 The 11-kV busbars of a power station are in two sections interconnected through a current limiting reactor X. Each section of the busbars is fed from two 11 kV, 25 MVA generators each having subtransient reactance of 18% on its rating. The MVA rating of the circuit breaker is 500. Find the reactance of the reactor to prevent the circuit breaker from being overloaded if a symmetrical 3-phase fault occurs at one of the outgoing feeders.

Solution: Figure 4.24(a) shows the one-line diagram of the network. Suppose that the fault occurs at point F at one of the outgoing feeders. As per given condition, the fault MVA should not exceed the MVA rating of the circuit breaker, i.e. 500 MVA.

Let the base voltage be 11 kV and base MVA be 25 MVA.

per-unit reactance of each generator $= 0.18\ pu$.

per-unit reactance of the reactor $= X_{pu}$

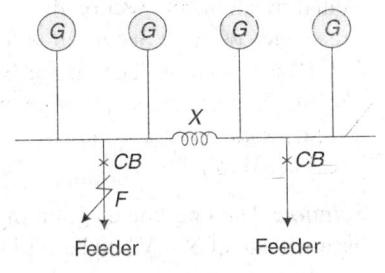

Fig. 4.24(a) System for Example 4.12

The circuit for calculating the equivalent reactance of the network as viewed from the fault point is shown in Fig. 4.24(b).

The circuit of Fig. 4.24(b) reduces to that shown in Fig. 4.24(c)

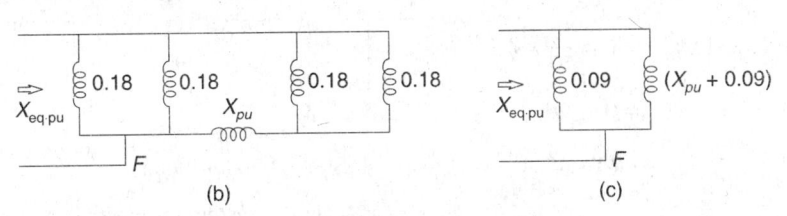

(b) $X_{eq\cdot pu}$, 0.18, 0.18, X_{pu}, 0.18, 0.18, F

(c) $X_{eq\cdot pu}$, 0.09, $(X_{pu} + 0.09)$, F

Fig. 4.24 (b) Equivalent reactance of the network of 4.24(a)
(c) Reduced circuit of 4.24(b)

From Fig. 4.24(c), the per-unit equivalent reactance, $X_{eq.\,pu}$ is given by

$$X_{eq.\,pu} = \frac{0.09\,(X_{pu} + 0.09)}{[0.09 + (X_{pu} + 0.09)]} = \frac{0.09\,(X_{pu} + 0.09)}{(X_{pu} + 0.18)}\,pu$$

The fault MVA is

$$(MVA)_{f.\,pu} = \frac{1}{X_{eq.\,pu}} = \frac{(X_{pu} + 0.18)}{0.09\,(X_{pu} + 0.09)}$$

Per-unit value of the required fault MVA is

$$(MVA)_{f.\,pu} = \frac{500}{25} = 20\,pu$$

Hence

$$20 = \frac{(X_{pu} + 0.18)}{0.09\,(X_{pu} + 0.09)}$$

or

$$X_{pu} = 0.0225\,pu.$$

The ohmic value of the reactance can be calculated as follows:

Base reactance, $\quad X_b = \dfrac{(kV)_b^2}{(MVA)_b} = \dfrac{(11)^2}{25} = 4.84$ ohms.

Hence $\quad X = X_{pu} \times X_b = 0.0225 \times 4.84 = 0.109\,\Omega$

Example 4.13 | The 33 kV busbars of a station are in two sections A and B separated by a reactor. Section A is fed from four 10 MVA generators each having a reactance of 20%. Section B is fed from the grid through a 50 MVA transformer of 10% reactance. The circuit breakers have each a rupturing capacity of 500 MVA. Find the reactance of the reactor to prevent the circuit breakers from being overloaded, if a symmetrical three-phase fault occurs on an outgoing feeder connected to it.

Solution: The one-line diagram of the system is shown in Fig. 4.25(a). Selecting a base voltage of 33 kV and base MVA of 50, the per-unit reactance of generators and transformer are

Generators : $X_{g.\,pu}$

$= \dfrac{50}{10} \times 0.2$

$= 1.0\,pu$

Transformer : $X_t = \dfrac{50}{50} \times 0.1$

$= 0.1\,pu$

Reactor : $X_{pu} = \dfrac{X}{X_b}$

Where X is the actual reactance of the reactor in ohm and X_b is the base reactance.

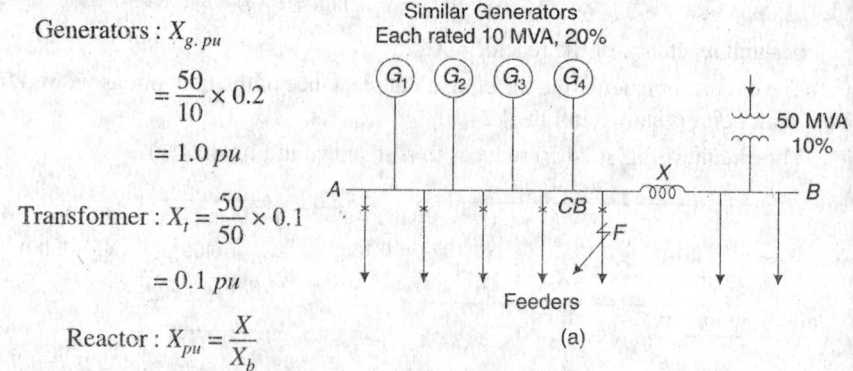

Fig. 4.25(a) System for Example 4.13

For a symmetrical three-phase fault at point F on an outgoing feeder connected to A, the reactance diagram with voltage sources short circuited will be as shown in Fig. 4.25(b). The circuit of Fig. 4.25(b) reduces to that shown in Fig. 4.25(c).

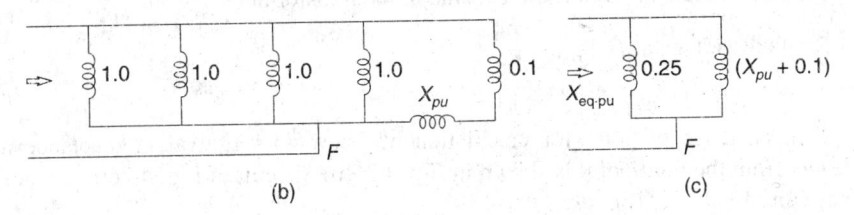

Fig. 4.25 (b) Reactance diagram (c) Reduced circuit for 4.25(b)

From Fig. 4.25(c), the equivalent reactance of the network as viewed from the fault point F is given by

$$X_{eq.\,pu} = \frac{0.25\,(X_{pu}+0.1)}{[0.25+(X_{pu}+0.1)]} = \frac{0.25\,(X_{pu}+0.1)}{(X_{pu}+0.35)}$$

Per-unit fault MVA is

$$(MVA)_{f.\,pu} = \frac{1}{X_{eq.\,pu}} = \frac{(X_{pu}+0.35)}{0.25(X_{pu}+0.1)}$$

Per-unit value of the required fault MVA

$$(MVA)_{f.\,pu} = \frac{500}{50} = 10$$

Hence,

$$10 = \frac{(X_{pu}+0.35)}{0.25(X_{pu}+0.1)}$$

or

$$X_{pu} = 0.066\ pu$$

Actual value,

$$X = X_{pu} \times X_b = 0.066 \times \frac{(33)^2}{50} = 1.437\ \Omega$$

Example 4.14 A generating station has four identical generators G_1, G_2, G_3 and G_4 each of 20 MVA, 11 kV having 20% reactance. They are connected to a busbar which has a busbar reactor of 25% reactance on 20 MVA base, inserted between G_2 and G_3. A 66-kV feeder is taken off from the bus-bars through a 15-MVA transformer having 8% reactance. A symmetrical 3-phase fault occurs at the high voltage terminals of the transformer. Calculate the current fed into the fault.

Solution: The one-line diagram of the system is shown in Fig. 4.26(a).

Let the base MVA chosen be 20 and base kV be 11 kV on generator side and 66 kV on feeder side.

Per-unit reactance of each generator

$$X_{g.\,pu} = 0.20\ pu$$

Per-unit reactance of transformer on new base,

$$X_{t.\,pu} = 0.08 \times \frac{20}{15} = 0.10\ pu$$

Per unit reactance of the reactor
$$X_{pu} = 0.25 \; pu$$
The fault occurs at point F on h.v. side of the transformer.

Pre-fault voltage at F, $V_{f.pu} = \dfrac{66}{66}$
$$= 1.0 < 0°$$

The reactance diagram for calculation of Thevenin's equivalent reactance as viewed from the fault point is shown in Fig. 4.26(b). Circuit of Fig. 4.26(b) reduces to circuit of Fig. 4.26(c).

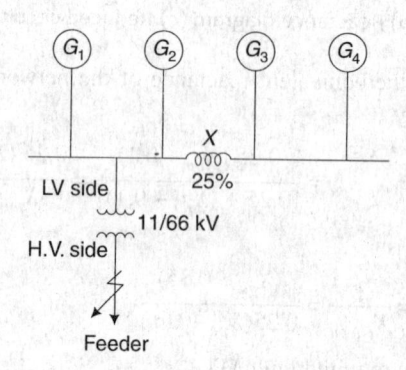

Fig. 4.26(a) System for Example 4.14

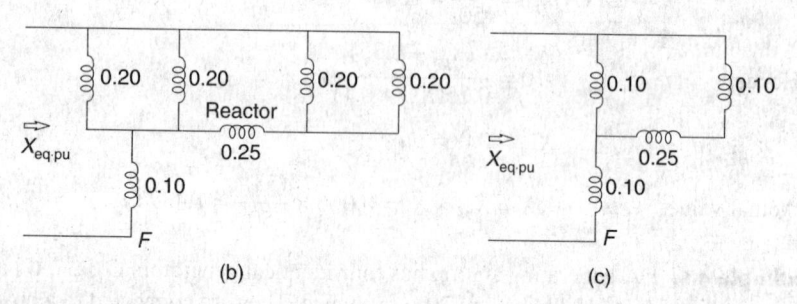

Fig. 4.26 (b) Thevenin's equivalent reactance (c) Reduced circuit of 4.26(b)

From Fig. 4.26, the equivalent reactance is given by
$$X_{eq.\,pu} = 0.10 + \dfrac{0.10 \, (0.25 + 0.10)}{[0.10 + (0.25 + 0.10)]}$$
$$= 0.177 \; pu$$
Thevenin's voltage = $V_{f.pu} = 1.0$
From Thevenin's equivalent circuit of Fig. 4.26(d)
The per-unit fault current,
$$I_{f.pu} = \dfrac{V_{f.pu}}{X_{eq.\,pu}} = \dfrac{1.0}{0.177} = 5.65 \; pu$$

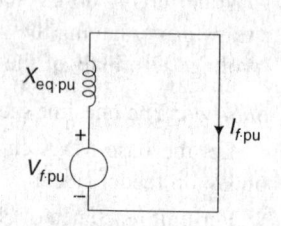

Fig. 4.26(d) Thevenin's equivalent circuit

Base current on h.v. side,
$$I_b = \frac{(MVA)_b \times 10^3}{\sqrt{3} \times (kV)_b} = \frac{20 \times 10^3}{\sqrt{3} \times 66} = 175 \text{ A}.$$

Actual current,
$$I_f = I_{f.pu} \times I_b = 5.65 \times 175 = 988 \text{ A}$$

Per-unit fault MVA,
$$(MVA)_{f.pu} = \frac{1}{X_{eq.pu}} = 5.65 \ pu$$

Fault MVA,
$$(MVA)_f = (MVA)_{f.pu} \times (MVA)_b$$
$$= 5.65 \times 20 \text{ MVA}$$
$$= 113 \text{ MVA}$$

Example 4.15 | The main busbars in a generating station are divided into four sections, each section being connected to a tie-bar by a similar reactor. One 20-MVA, 11-kV generator having subtransient reactance of 20% is connected to each section busbar. When a symmetrical fault takes place between the phases of one of the section busbars, the voltage on the remaining sections fall to 65% of the normal value. Calculate the reactance of each reactor in ohms.

Solution: Let the selected base MVA be 20 and base kV be 11 kV. Referring to Fig. 4.27 let reactors A, B, C, D have a per-unit reactance X_{pu} based on 20 MVA, and their ohmic value be X. The per-unit reactance of each generator on base of 20 MVA is 0.20.

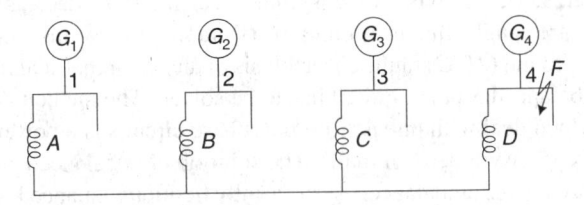

Fig. 4.27 System for Example 4.15

Let the fault occur on the busbars of generator G_4,

Per-unit reactance of G_1 and A in series = $0.20 + X_{pu}$

Per-unit reactance of G_2 and B in series = $0.20 + X_{pu}$

Per-unit reactance of G_3 and C in series = $0.20 + X_{pu}$

Equivalent reactance of the above three groups of reactances in parallel
$$= \frac{0.20 + X_{pu}}{3} = 0.066 + 0.33 \ X_{pu}$$

Equivalent reactance of these three groups to fault via tie-bar
$$= 0.066 + 0.33 \ X_{pu} + X_{pu}$$
$$= 0.066 + 1.33 \ X_{pu}$$

The voltage on sections 1, 2 and 3 drops to 65% of the normal value, i.e., 0.65 pu of the normal value. Therefore, 0.35 pu of per-unit normal voltage is dropped in the reactance of G_1, G_2 and G_3, i.e., the reactance of G_1, G_2 and G_3 which is 0.066 pu, is 0.35 of the total reactance up to fault.

i.e., $\quad\quad\quad\quad 0.066 = 0.35\,(0.066 + 1.33\,X_{pu})$

or $\quad\quad\quad\quad 0.066 = 0.023 + 0.465\,X_{pu}$

or $\quad\quad\quad\quad X_{pu} = \dfrac{0.043}{0.465} = 0.092\ pu$

Therefore the ohmic value of reactance is

$$X = X_{pu} \times X_b = X_{pu} \times \dfrac{(kV)_b^2}{(MVA)_b}$$

$$= 0.092 \times \dfrac{(11)^2}{20} = 0.556\ \text{ohm}$$

4.6 SYMMETRICAL COMPONENTS

The symmetrical fault is very rare in occurrence but it is the most severe type of fault in the power system. The analysis of symmetrical fault is simple because under this fault condition, the three-phase system behaves like a balanced circuit which can be analyzed on a single-phase basis.

The majority of the faults occuring in a power system are of unsymmetrical nature involving one or two phases and not all the three phases. Such faults make the three-phase system unbalanced. The system also becomes unbalanced when loads are unbalanced. Different types of unsymmetrical faults in decreasing frequency of occurrence are single line-to-ground (L-G) fault, line-to-line (L-L) fault, and double line-to-ground ($2L$-G) fault. The analysis of an unsymmetrical fault is difficult because an unbalanced 3-phase circuit has to be solved. The method of symmetircal components which deal with unbalanced polyphase circuits is a powerful technique for the analysis of unsymmetrical faults. The solution of unbalanced circuits is made easier by resolving the unbalanced system into fictitious balanced systems using symmetrical components. The unbalanced three-phase network is resolved into three uncoupled balanced sequence networks by the method of symmetrical components. The sequence networks are connected only at the points of unbalance. Solution of the sequence networks for the fault conditions gives symmetrical current and voltage components which can be superimposed to reflect the effects of the original unbalanced fault currents on the overall system. Analysis by symmetrical components is a powerful tool which makes the calculation of unsymmetrical faults almost as easy and simple as the calculation of symmetrical faults. The method of symmetrical components, proposed by C L Fortecue in 1918 is applicable to any polyphase system, but the three-phase system is of main interest. This method is basically a modelling technique which permits systematic analysis and design of three-phase systems. Decoupling a detailed three-phase network into three simpler sequence networks reveals complicated phenomenon in more simplistic terms.

4.6.1 Fundamentals of Symmetrical Components

The method of symmetrical components was introduced by C L Fortescue, an American Scientist, in 1918. His work proved that an unbalanced system of n related phasors can be resolved into n systems of balanced phasors called the symmetrical components of the original phasors. The n phasors of each set of components are equal in magnitude, and the angles between adjacent phasors of the set are equal. Although the method of symmetrical components is applicable to any unbalanced polyphase system, but its application to a more practical three-phase system is of main interest of discussion here.

According to Fortescue's theorem, three unbalanced phasors (voltages or currents) of a three-phase system can be resolved into three balanced systems of phasors called positive, negative and zero sequence phasors (components). The positive, negative and zero sequence components of the original unbalanced phasors of the three-phase system are called the "Symmetrical Components" which are defined as follows:

1. Positive-sequence components consist of three phasors equal in magnitude, displaced from each other by 120° in phase, and having the same phase sequence as the original phasors.
2. Negative-sequence components consist of three phasors equal in magnitude, displaced from each other by 120° in phase, and having the phase sequence opposite to that of the original phasors.
3. Zero-sequence components consists of three phasors equal in magnitude and in phase with each other (i.e. with zero phase displacement from each other).

The phase sequence in a three-phase system is defined as the order in which the three phases pass through a positive maximum. While solving a problem by symmetircal components, the three phases of the system are customarily designated as a, b, and c in such a manner that the phase sequence of the voltages and currents in the system is abc. Thus the phase sequence of the positive-sequence components of the unbalanced phasors is abc, and the phase-sequence of the negative-sequence components is acb. The original phasors of voltages are designated as V_a, V_b, and V_c; and original phasors of current as I_a, I_b and I_c. The three sets of symmetrical components are designated by the additional subscript 1 for the positive-sequence components, 2 for the negative sequence components, and 0 for the zero-sequence components. The positive-sequence components of V_a, V_b, and V_c are designated by V_{a1}, V_{b1}, and V_{c1}. Similarly, the negative-sequence components are V_{a2}, V_{b2}, and V_{c2}, and the zero-sequence components are V_{a0}, V_{b0}, and V_{c0}. Similarly, the positive, negative and zero-sequence components of I_a, I_b and I_c are designated by I_{a1}, I_{b1}, and I_{c1}, I_{a2}, I_{b2}, and I_{c2}, and I_{a0}, I_{b0}, and I_{c0} respectively.

Since each of the original unbalanced phasors is the sum of its components, the original phasors of voltage can be expressed in terms of their components as

$$V_a = V_{a1} + V_{a2} + V_{a0} \tag{4.40}$$

$$V_b = V_{b1} + V_{b2} + V_{b0} \tag{4.41}$$

$$V_c = V_{c1} + V_{c2} + V_{c0} \tag{4.42}$$

Three sets of symmetrical components of three unbalanced phasors of voltage (V_a, V_b and V_c) are shown in Fig. 4.28. The graphical addition of symmetrical components as per Eqs. (4.40) to (4.42) to obtain V_a, V_b and V_c is indicated by the phasor diagram of Fig. 4.29.

Operator a

By convention, the direction of rotation of the phasors is taken to be counterclockwise. The operator a is defined as an operator which when multiplied to a phasor causes its rotation by 120° in the counterclockwise (positive) direction without changing the magnitude. Such an operator is a complex number of unit magnitude with an angle of 120° and is defined as

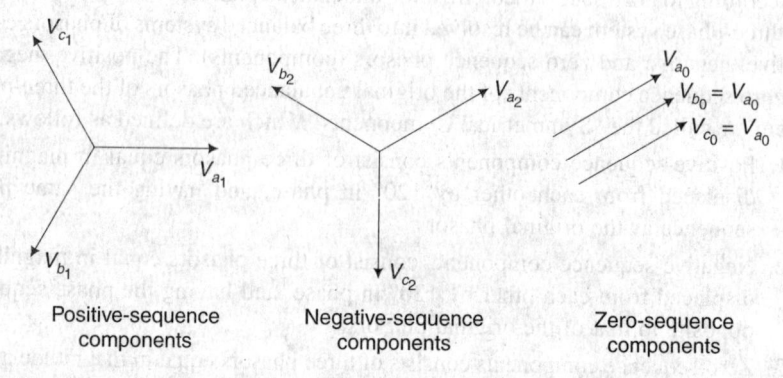

$$a = 1 < 120° = 1\, e^{j2\pi/3} = -0.5 + j\, 0.866$$

Fig. 4.28 Three sets of symmetrical components of three unbalanced phasors.

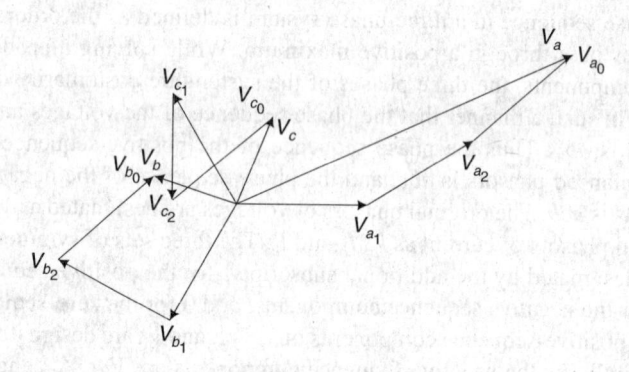

Fig. 4.29 Graphical addition of the symmetrical components to obtain the set of unbalanced phasors V_a, V_b and V_c.

If the operator a is applied to a phasor twice in succession, the phasor is rotated by 240°. Three successive applications of a to a phasor rotate the phasor through 360°.

Thus,
$$a^2 = 1 < 240° = 1 < -120° = 1 e^{-j2\pi/3}$$
$$= -0.5 - j\, 0.866 = a^*$$
$$a^3 = 1 < 360° = 1$$

and
$$1 + a + a^2 = 0$$

The components of phasors b and c can be expressed in terms of the components of phasor a taken as reference, by using the operator a.

Referring to Fig. 4.28, the set of positive, negative and zero sequence phasors are written as

$$V_{a1}, V_{b1} = a^2 V_{a1}, \quad V_{c1} = a V_{a1} \tag{4.43}$$

$$V_{a2}, V_{b2} = a V_{a2}, \quad V_{c2} = a^2 V_{a2} \tag{4.44}$$

$$V_{a0}, V_{b0} = V_{a0}, \quad V_{c0} = V_{a0} \tag{4.45}$$

Equations (4.40) to (4.42) can be expressed in terms of reference phasors V_{a1}, V_{a2} and V_{a0}. Thus

$$V_a = V_{a1} + V_{a2} + V_{a0} \tag{4.46}$$

$$V_b = a^2 V_{a1} + a V_{a2} + V_{a0} \tag{4.47}$$

$$V_c = a V_{a1} + a^2 V_{a2} + V_{a0} \tag{4.48}$$

Equations (4.46) to (4.48) can be expressed in the matrix form as

$$\begin{bmatrix} V_a \\ V_b \\ V_c \end{bmatrix} = \begin{bmatrix} 1 & 1 & 1 \\ a^2 & a & 1 \\ a & a^2 & 1 \end{bmatrix} \begin{bmatrix} V_{a1} \\ V_{a2} \\ V_{a0} \end{bmatrix} \tag{4.49}$$

Equation (4.49) can be written in compact form as

$$[V_p] = [A][V_s] \tag{4.50}$$

Where

$$[V_p] = \begin{bmatrix} V_a \\ V_b \\ V_c \end{bmatrix} = \text{Vectors of the original phasors}, \tag{4.51}$$

$$[V_s] = \begin{bmatrix} V_{a1} \\ V_{a2} \\ V_{a0} \end{bmatrix} = \text{Vector of symmetrical components} \tag{4.52}$$

$$[A] = \begin{bmatrix} 1 & 1 & 1 \\ a^2 & a & 1 \\ a & a^2 & 1 \end{bmatrix} \tag{4.53}$$

Equation (4.50) can be written as

$$[V_s] = [A]^{-1}[V_p] \tag{4.54}$$

Computing $[A]^{-1}$ and utilizing relations of operator a, we get

$$[A]^{-1} = \frac{1}{3} \begin{bmatrix} 1 & a & a^2 \\ 1 & a^2 & a \\ 1 & 1 & 1 \end{bmatrix} \tag{4.55}$$

Equation (4.54) can be written in expanded form as

$$\begin{bmatrix} V_{a1} \\ V_{a2} \\ V_{a0} \end{bmatrix} = \frac{1}{3} \begin{bmatrix} 1 & a & a^2 \\ 1 & a^2 & a \\ 1 & 1 & 1 \end{bmatrix} \begin{bmatrix} V_a \\ V_b \\ V_c \end{bmatrix} \tag{4.56}$$

From Eq. (4.56),

$$V_{a1} = \frac{1}{3}(V_a + a V_b + a^2 V_c) \tag{4.57}$$

$$V_{a2} = \frac{1}{3}(V_a + a^2 V_b + aV_c) \qquad (4.58)$$

$$V_{a0} = \frac{1}{3}(V_a + V_b + V_c) \qquad (4.59)$$

Equations (4.57) to (4.59) give the necessary relationship for obtaining symmetrical components of the original phasors, while Eqs. (4.40) to (4.42) give the relationships for obtaining the original phasors from the symmetrical components.

All the above equations of symmetrical component transformations are valid for set of currents also.

Thus,

$$[I_p] = [A][I_s] \qquad (4.60)$$

and

$$[I_s] = [A]^{-1}[I_p] \qquad (4.61)$$

where

$$[I_p] = \begin{bmatrix} I_a \\ I_b \\ I_c \end{bmatrix} \quad \text{and} \quad [I_s] = \begin{bmatrix} I_{a1} \\ I_{a2} \\ I_{a0} \end{bmatrix}$$

of course $[A]$ and $[A]^{-1}$ are the same as given earlier.

In expanded form the equations (4.60) and (4.61) can be expressed as follows:

(i) For obtaining current phasors form the symmetrical components

$$I_a = I_{a1} + I_{a2} + I_{a0} \qquad (4.62)$$

$$I_b = a^2 I_{a1} + a I_{a2} + I_{a0} \qquad (4.63)$$

$$I_c = a I_{a1} + a^2 I_{a2} + I_{a0} \qquad (4.64)$$

(ii) For obtaining symmetrical components of the original current phsors

$$I_{a1} = \frac{1}{3}(I_a + a I_b + a^2 I_c) \qquad (4.65)$$

$$I_{a2} = \frac{1}{3}(I_a + a^2 I_b + a I_c) \qquad (4.66)$$

$$I_{a0} = \frac{1}{3}(I_a + I_b + I_c) \qquad (4.67)$$

Following observations can be made regarding a three-phase system with neutral return as shown in Fig. 4.30.

(i) The sum of the three line voltages will always be zero. Therefore, the zero sequence component of line voltages is always zero, i.e.,

$$V_{ab0} = \frac{1}{3}(V_{ab} + V_{bc} + V_{ca}) = 0 \qquad (4.68)$$

on the other hand, the sum of phase voltages (line to neutral) may not be zero so that their zero sequence component V_{a0} may exist.

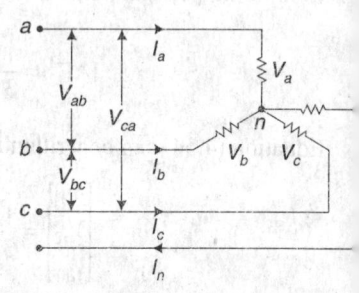

Fig. 4.30 Three-phase system with neutral return

(ii) Since the sum of the three line currents equals the current in the neutral wire, we get

$$I_{a0} = \frac{1}{3}(I_a + I_b + I_c) = \frac{1}{3}I_n \quad (4.69)$$

or
$$I_n = 3I_{a0} \quad (4.70)$$

Hence, the current in the neutral is three times the zero sequence line current.

(iii) If the neutral connection is absent,

$$I_{a0} = \frac{1}{3}I_n = 0 \quad (4.71)$$

i.e., in the absence of neutral connection the zero sequence line corrent is always zero.

4.6.2 Power in Terms of Symmetrical Components

The power in a three-phase circuit can be computed directly from the symmetrical components of voltage and current. The symmetrical component transformation is power invarient, which means that the sum of powers of the three symmetrical components equals the three-phase power.

The total complex power in a three-phase circuit is given by

$$S = P + jQ = V_a I_a^* + V_b I_b^* + V_c I_c^* \quad (4.72)$$

In matrix notation
$$S = [V_p]^T [I_p]^*$$

or
$$S = [AV_s]^T [AI_s]^*$$
$$= [V_s]^T [A]^T [A]^* [I_s]^* \quad (4.73)$$

Now
$$[A]^T [A]^* = \begin{bmatrix} 1 & a^2 & a \\ 1 & a & a^2 \\ 1 & 1 & 1 \end{bmatrix} \begin{bmatrix} 1 & 1 & 1 \\ a & a^2 & 1 \\ a^2 & a & 1 \end{bmatrix}$$

$$= 3 \begin{bmatrix} 1 & 0 & 0 \\ 0 & 1 & 0 \\ 0 & 0 & 1 \end{bmatrix} = 3[U] \quad (4.74)$$

Therefore,
$$S = 3[V_s]^T [U] [I_s]^*$$
$$= 3[V_s]^T [I_s]^*$$
$$= 3[V_{a1} \ V_{a2} \ V_{a0}] \begin{bmatrix} I_{a1}^* \\ I_{a2}^* \\ I_{a0}^* \end{bmatrix}$$

$$= 3V_{a1}I_{a1}^* + 3V_{a2}I_{a2}^* + 3V_{a0}I_{a0}^* \quad (4.75)$$
$$= \text{Sum of symmetrical component powers}$$

4.6.3 Sequence Impedances of Passive Circuit Elements

Power systems which are normally balanced become unbalanced only upon the occurrence of an unsymmetrical fault. Let us discuss the equations of a three-phase circuit when the series impedances are unequal. The conclusion of discussion is important in analysis by symmetrical components. Figure 4.31 shown the unsymmetrical part of a power system with three unequal series impedances Z_a, Z_b and Z_c carrying currents I_a, I_b and I_c respectively. The currents I_a, I_b and I_c return through the neutral (ground) impedance Z_n. The phase voltages at the two ends of the impedances are V_a, V_b, V_c and $V_{a'}$, $V_{b'}$, $V_{c'}$, respectively. This circuit model can represent a three-phase transmission line with active sources (synchronous machines) at each end and with a ground return circuit. If $V_{a'}$, $V_{b'}$, $V_{c'}$ are regarded as zero, the circuit represents three star-connected impedances with neutral return through impedance Z_n. Assuming no mutual inductance (no coupling) between the three impedances, the voltage drop across the impedances Z_a, Z_b, Z_c and Z_n are given by

$$V_a - V_{a'} = V_{aa'}, \text{ the voltage drop in } Z_a \text{ and } Z_n$$

$$V_b - V_{b'} = V_{bb'}, \text{ the voltage drop in } Z_b \text{ and } Z_n$$

$$V_c - V_{c'} = V_{cc'}, \text{ the voltage drop in } Z_c \text{ and } Z_n$$

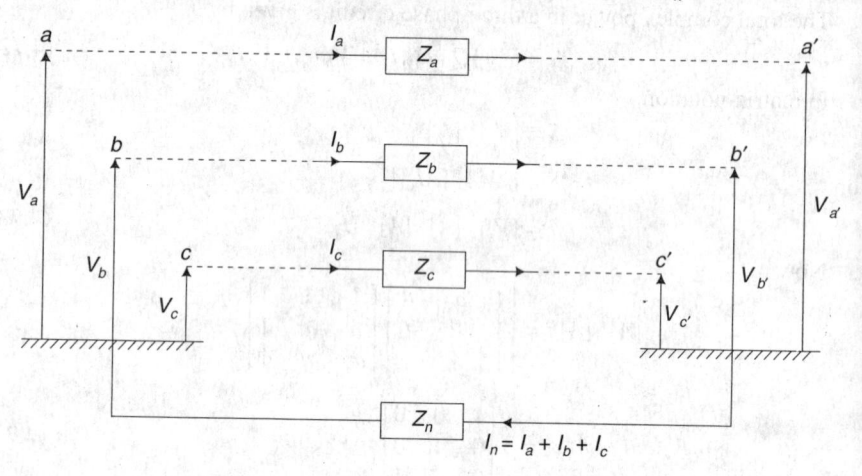

Fig. 4.31 Unbalanced passive circuit elements

We can therefore write

$$V_{aa'} = Z_a I_a + Z_n (I_a + I_b + I_c)$$
$$V_{bb'} = Z_b I_b + Z_n (I_a + I_b + I_c)$$
$$V_{cc'} = Z_c I_c + Z_n (I_a + I_b + I_c)$$

The above equations can be written in the following form

$$V_{aa'} = (Z_a + Z_n) I_a + Z_n I_b + Z_n I_c$$
$$V_{bb'} = Z_n I_a + (Z_b + Z_n) I_b + Z_n I_c$$
$$V_{cc'} = Z_n I_a + Z_n I_b + (Z_c + Z_n) I_c$$

In matrix form, above equations can be written as

$$\begin{bmatrix} V_{aa'} \\ V_{bb'} \\ V_{cc'} \end{bmatrix} = \begin{bmatrix} Z_a + Z_n & Z_n & Z_n \\ Z_n & Z_b + Z_n & Z_n \\ Z_n & Z_n & Z_c + Z_n \end{bmatrix} \begin{bmatrix} I_a \\ I_b \\ I_c \end{bmatrix} \quad (4.76)$$

or
$$[V_p] = [Z][I_p] \quad (4.77)$$

Applying symmetrical component transformation, we get

$$[A][V_s] = [Z][A][I_s]$$

or
$$[V_s] = [A]^{-1}[Z][A][I_s] = [Z_s][I_s] \quad (4.78)$$

$[Z_s] = [A]^{-1}[Z][A] =$ Symmetrical component impedance matrix.

$[Z_s]$ is computed as follows:

$$[Z_s] = \frac{1}{3}\begin{bmatrix} 1 & a & a^2 \\ 1 & a^2 & a \\ 1 & 1 & 1 \end{bmatrix}\begin{bmatrix} Z_a + Z_n & Z_n & Z_n \\ Z_n & Z_b + Z_n & Z_n \\ Z_n & Z_n & Z_c + Z_n \end{bmatrix}\begin{bmatrix} 1 & 1 & 1 \\ a^2 & a & 1 \\ a & a^2 & 1 \end{bmatrix}$$

or

$$[Z_s] = \begin{bmatrix} \frac{1}{3}(Z_a + Z_b + Z_c) & \frac{1}{3}(Z_a + a^2 Z_b + aZ_c) & \frac{1}{3}(Z_a + aZ_b + a^2 Z_c) \\ \frac{1}{3}(Z_a + aZ_b + a^2 Z_c) & \frac{1}{3}(Z_a + Z_b + Z_c) & \frac{1}{3}(Z_a + a^2 Z_b + aZ_c) \\ \frac{1}{3}(Z_a + a^2 Z_b + aZ_c) & \frac{1}{3}(Z_a + aZ_b + a^2 Z_c) & \frac{1}{3}(Z_a + Z_b + Z_c) + 3Z_n \end{bmatrix} \quad (4.79)$$

Substituting $[Z_s]$ from Eq. (4.79) in Eq. (4.78), we can write

$$\begin{bmatrix} V_{aa'1} \\ V_{aa'2} \\ V_{aa'0} \end{bmatrix} = \begin{bmatrix} Z_{11} & Z_{12} & Z_{10} \\ Z_{21} & Z_{22} & Z_{20} \\ Z_{01} & Z_{02} & Z_{00} \end{bmatrix}\begin{bmatrix} I_{a1} \\ I_{a2} \\ I_{a0} \end{bmatrix}$$

Where the symmetrical component self and mutual impedances are

$$Z_{11} = Z_{22} = \frac{1}{3}(Z_a + Z_b + Z_c)$$

$$Z_{00} = \frac{1}{3}(Z_a + Z_b + Z_c) + 3Z_n$$

$$Z_{12} = Z_{20} = Z_{01} = \frac{1}{3}(Z_a + a^2 Z_b + aZ_c)$$

$$Z_{10} = Z_{21} = Z_{02} = \frac{1}{3}(Z_a + aZ_b + a^2 Z_c)$$

From the above equations, it is seen that the voltage drop of any particular sequence will be produced by currents of all sequences (positive, negative and zero). Therefore, the method of symmetrical component does not appear to be simpler than three-phase circuit analysis. However, there is a redeeming feature due to the fact that the impedances of the components of a power system are balanced and an unsymmetrical fault imbalances an otherwise balanced system.

For a balanced system
$$Z_a = Z_b = Z_c = Z_l$$
The Symmetrical component impedance matrix $[Z_s]$ given by Eq. (4.79) now simplifies to

$$[Z_s] = \begin{bmatrix} Z_l & 0 & 0 \\ 0 & Z_l & 0 \\ 0 & 0 & Z_l + 3Z_n \end{bmatrix} = \text{a diagonal matrix} \quad (4.80)$$

Equation (4.78) can now be written in expanded form as
$$V_{aa'1} = Z_l I_{a1} = Z_1 I_{a1}$$
$$V_{aa'2} = Z_l I_{a2} = Z_2 I_{a2}$$
$$V_{aa'0} = (Z_l + 3Z_n) I_{a0} = Z_0 I_{a0} \quad (4.81)$$

Since these equations are decoupled, a voltage drop of any particular sequence is dependent upon the current of that sequence only. The impedances involved in Eq. (4.81) are defined as sequence impedances.

Thus
$$Z_1 = Z_l = \text{positive sequence impedance} \quad (4.82)$$
$$Z_2 = Z_l = \text{negative sequence impedance} \quad (4.83)$$
$$Z_0 = Z_l + 3Z_n = \text{zero sequence impedance} \quad (4.84)$$

The decoupling between sequence currents leads to considerable simplification in the use of the method of symmetrical components in the analysis of unsymmetrical faults. In unsymmetrical fault analysis using the method of symmetrical components, each component (element) of power system will have three values of impedance—one corresponding to each sequence current, i.e. positive sequence impedance Z_1, negative sequence impedance Z_2 and zero sequence impedance Z_0.

4.6.4 Sequence Impedances and Sequence Networks

The voltage drop in any part of a circuit due to current of a particular sequence depends on the impedance of that part of the circuit to current of that sequence. The impedance of any section of a balanced network to current of one sequence may differ from impedance to current of another sequence. For example, the impedance offered by any element of power system to positive-sequence current may not necessarily be the same as offered to negative or zero-sequence current. Hence, there are three sequence impedances corresponding to three sequence (positive, negative and zero sequence) currents.

The impedance offered by a circuit to positive-sequence current is called positive sequence impedance and is represented by Z_1. Similarly, impedances offered by any circuit to negative and zero-sequence currents are respectively called negative seuqence-impedance denoted by Z_2 and zero-sequence impedance denoted by Z_0.

The analysis of unsymmetrical faults in power systems is performed by finding the symmetrical components of the unbalanced currents. Since the component currents of each phase sequence cause voltage drops of that sequence only and are independent of currents of other sequences, currents of each sequence can be considered to flow in an independent network composed of the impedances to the current of that

Fault Analysis 155

quence only. The single-phase equivalent circuit composed of the impedances to the current of any one sequence only is called the sequence network for that particular sequence. The sequence network contains any generated emfs of like sequence. Thus, three-sequence networks can be formed for every power system. Sequence networks carrying the currents I_{a1}, I_{a2}, and I_{a0} are interconnected to represent the different unbalanced fault conditions. Therefore, in order to calculate the effect of a fault by the method of symmetrical components, it is essential to determine the sequence impedances and to combine them to form the sequence networks.

6.5 Sequence Impedances of Power System Elements

A power system consists of synchronous generators, transformers, transmission lines and loads connected together. The impedances offered by various elements of the power system to positive, negative and zero sequence currents is of considerable importance in determining the fault currents in a 3-phase unbalanced system. The impedance of a particular sequence can be found by applying a voltage of that sequence only at the three terminals of the power system element (component) and measuring the current. All elements of the power system are considered to be symmetrical.

The positive and negative sequence impedances of linear, symmetrical and static circuits (e.g., transmission lines, cables, transformers and static loads) are identical and are the same as those used in the analysis of balanced conditions. This is because of the fact that the impedance of such circuits is independent of the phase sequence, provided the applied voltages are balanced. The positive and negative sequence impedances of rotating machines are generally different. The zero sequence impedance depends upon the path taken by the zero sequence current. As the path of zero sequence current is generally different from the path of positive and negative sequence currents, the zero sequence impedance is usually different from positive or negative sequence impedance.

Sequence Impedances of Transmission Lines

A transmission line is a linear and static component of the power system. Considering the line to be fully transposed three-phase line, it is completely symmetrical also. Therefore, the transmission line offers the same impedance to positive and negative sequence currents. In other words, the positive and negative sequence impedances of the line are equal and are the same as the normal impedance of the line. For fault analysis, a transmission line is represented by the series impedance, the shunt branch being neglected. Therefore, for the line $Z_1 = Z_2 =$ series impedance of the line.

The character of zero-sequence impedance of a line is entirely different from either positive or negative sequence impedance. Zero-sequence currents can flow through the line if these currents get a return path. The return path is provided partly by the ground and partly by overhead ground wires. When only zero-sequence currents flow in the line, the currents in all the phases are identical in both magnitude and phase angle. These currents return partly through the ground and partly through the overhead ground wires. The ground wires being grounded at several towers, the return currents through the ground wires are not necessarily uniform along the entire length. The magnetic field created due to the flow of zero-sequence currents through the transmission line, ground and ground wires is very different from that due to the flow of positive or negative sequence currents. The zero sequence impedance of transmission lines usually ranges from 2 to 4 times the positive sequence impedance.

Sequence Impedances of Synchronous Machines

The impedances offered by rotating machines to currents of the three sequences (positive, negative and zero) are generally different for each sequence. The positive sequence impedance of a synchronous machine is equal to the synchronous impedance of the machine. The sequence impedances of the synchronous machine depend upon the phase order of the sequence current relative to the direction of rotation of the rotor. The magnetic field produced by negative sequence armature current rotates in the direction opposite to that of the rotor having the dc field winding. That is, the net flux rotates at twice synchronous speed relative to the rotor. The field winding has no influence because the field voltage is associated with the positive-sequence variables. The flux produced by positive-sequence current is stationary with respect to the rotor, whereas the flux produced by the negative-sequence current is sweeping rapidly over the face of the rotor. The currents produced in the field and damper windings due to the rotating armature flux keep the flux from penetrating the rotor. This condition is similar to the rapidly changing flux immediately upon the occurrence of a fault at the terminals of a machine. The path of the flux is same as that encountered in evaluating the subtransient reactance. Therefore, the negative sequence and subtransient reactances of the machine are equal, i.e.

$$X_2 = X_d'' \tag{4.85}$$

Zero-sequence is the impedance offered by the machine to the flow of the zero-sequence current. Since, a set of zero sequence currents are all identical, the resultant air-gap flux would be zero and there is no reactance due to armature reaction. The machine offers a very small reactance due to the leakage flux. Therefore, the zero-sequence reactance (X_0) of the machine is approximately equal to the leakage reactance (X_l), i.e.,

$$X_0 \approx X_l \tag{4.86}$$

Sequence Impedances of Transformers

For fault analysis, the transformer is modeled as the equivalent series leakage impedance. The positive sequence impedance of a transformer equals its leakage impedance. Since the transformer is a static device, its leakage impedance will not change if the phase sequence of the applied balanced voltages is reversed. Therefore, the negative-sequence impedance of a transformer is equal to the positive-sequence impedance which is equal to the leakage impedance. i.e.,

$$Z_1 = Z_2 = Z_l \tag{4.87}$$

where, Z_l is the leakage impedance.

The zero sequence impedance of a transformer depends on two factors: (i) winding connection and (ii) neutral connection to the ground. In case of Δ connected windings, zero-sequence currents cannot be present in the line but may circulate in the delta. In case of star connected windings with isolated neutral, zero-sequence currents will be absent due to non-availability of return path. On the other hand, if the neutral of the star connected winding is grounded, zero-sequence currents may flow. The zero-sequence impedance is either the same as positive and negative sequence impedances, or is infinite depending upon the winding connections. i.e.

$$Z_0 = Z_1 = Z_2, \quad \text{if there is return path through ground for the zero-sequence current} \tag{4.88}$$

and $\quad Z_0 = \text{Infinite}, \quad$ if there is no return path for the
zero-sequence current. (4.89)

Most of the transformers used in power systems are either Y-Δ or Δ-Y. In a Y-Δ or a Δ-Y transformer, the positive sequence line voltage on HV side leads the corresponding line voltage on the LV side by 30°. The corresponding phase-shift for the negative sequence voltage is –30°, i.e. the negative sequence line voltage on for that particular sequence. The sequence network the LV side by 30°. For either Y-Δ or Δ-Y connections, phases are labelled in such a way that positive sequence quantities on the HV side lead their corresponding quantities on the LV side by 30°.

The equivalent circuit for the zero-sequence impedance depends on the winding connections and also upon whether the neutrals are grounded or not. The neutral impedance plays an important role in the equivalent circuit. When the neutral is grounded through an impedance Z_n, in the equivalent circuit the neutral impedance appears as $3Z_n$ in the path of I_0, because $I_n = 3I_0$.

Sequence Impedances of Static Loads

The positive, negative and zero-sequence impedances of balanced Y-and Δ-connected static loads are equal, i.e.,

$$Z_1 = Z_2 = Z_0 = Z \tag{4.90}$$

where Z is the impedance per-phase of the load

4.6.6 Sequence Networks of Unloaded Synchronous Machines

An unloaded synchronous machine (generator or motor) grounded through an impedance Z_n is shown in Fig. 4.32. E_a, E_b and E_c are the induced emfs of the three phases. When a fault (not shown in the figure) occurs at the terminals of the machine, currents I_a, I_b and I_c flow in the lines. If the fault involves ground, the current flowing into the neutral from ground through Z_n is $I_n = I_a + I_b + I_c$. Unbalanced line currents due to fault can be resolved into their symmetrical components I_{a1}, I_{a2} and I_{a0}.

The knowledge of the equivalent circuits presented by the machine to the flow of positive, negative and zero sequence currents, respectively is essential for the fault analysis. Since a synchronous machine is designed with symmetrical windings to generate balanced three phase voltages, it generates voltages of positive sequences only. Drawing the sequence networks of the synchronous machine is simple.

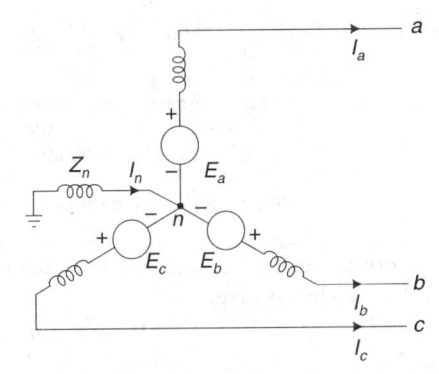

Fig. 4.32 Three-phase synchronous generator with grounded neutral

Positive Sequence Network

Since the synchronous machine generates balanced three-phase voltages of positive sequence only, the positive-sequence network consists of an emf (voltage source) in series with the positive-sequence impedance of the machine. Since the machine is not loaded, the generated emf in the positive-sequence network is the no-load terminal voltage to neutral. Under fault conditions, the machine offers a time-varying reactance which changes from subtransient reactance (X_d'') to transient reactance (X_d') and finally to steady state synchronous reactance (X_d). If armature resistance of the machine is assumed negligible, the positive-sequence impedance is equal to reactance of the machine. Therefore, the reactance in the positive-sequence network is X_d''', X_d' or X_d, depending on whether subtransient, transient, or steady-state conditions are to be studied. The positive-sequence impedance (Z_1) of the machine in the positive-sequence network is

$$Z_1 = j X_d'' \quad \text{(if subtransient condition is of interest)} \quad (4.91)$$
$$= j X_d' \quad \text{(if transient condition is of interest)} \quad (4.92)$$
$$= j X_d \quad \text{(if steady-state condition is of interest)} \quad (4.93)$$

Figure 4.33(a) shows the three-phase positive sequence network model of the synchronous machine through which the positive-sequence components (I_{a1}, I_{b1} and I_{c1}) of the unbalanced currents (I_a, I_b and I_c) are considered to flow. Z_n does not appear in this model as $I_n = (I_{a1} + I_{b1} + I_{c1}) = 0$. Since this network is balanced, it can be represented by the single-phase network model of Fig. 4.33(b) for analysis purposes. The reference bus for a positive-sequence network is at neutral potential. Further, since no current flows from ground to neutral, the neutral is at ground potential.

(a) Three-phase model

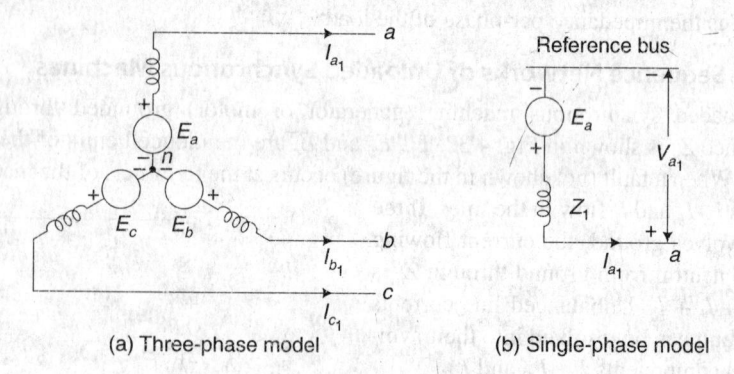

(b) Single-phase model

Fig. 4.33 *Positive-sequence network of synchronous machine*

From Fig. 4.33(b), the positive sequence voltage of terminal a with respect to the reference bus is given by

$$V_{a1} = E_a - I_{a1} Z_1 \quad (4.94)$$

Negative Sequence Network

Since the synchronous machine has zero negative sequence generated voltages, the negative-sequence network of the machine contains no emfs but includes the impedance offered by the machine to negative-sequence current. The flow of negative sequence currents in the stator produces a rotating magnetic field which rotates in the direction opposite to that of the positive sequence field, i.e., at double synchronous speed with respect to the rotor. Therefore, the currents induced in the rotor field and damper windings are at double the rotor frequency. While sweeping over the rotor surface, the reluctances of direct and quadrature axes are alternately presented to the negative sequence mmf. The negative sequence impedance offered by the machine with consideration given to the damper windings, is often given by

$$Z_2 = \frac{jX_d'' + X_q''}{2} \tag{4.95}$$

Figures 4.34(a) and (b) show negative sequence models of synchronous machine, on a three-phase and single-phase basis respectively. In this case also, the reference bus is at neutral potential which is the same as the ground potential. Z_n does not appear in the model as $I_n = I_{a2} + I_{b2} + I_{c2} = 0$

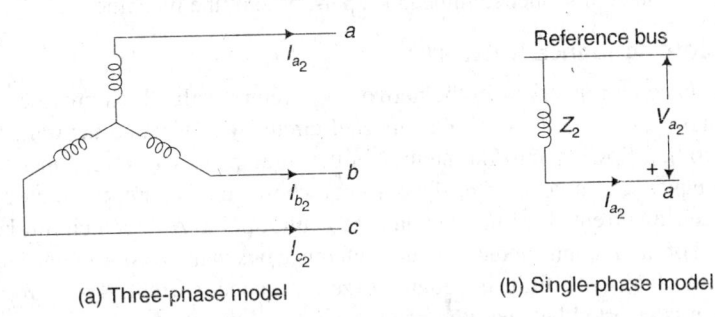

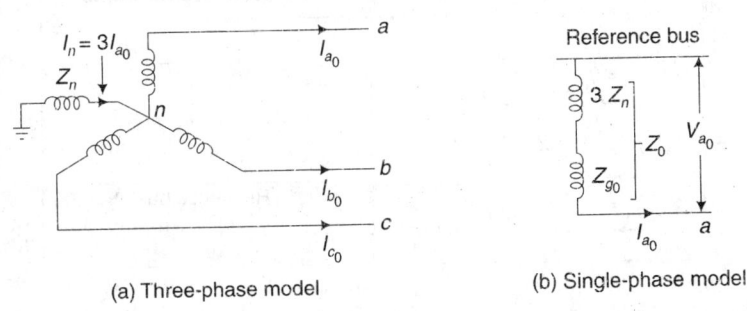

Fig. 4.34 Negative sequence network of synchronous machine

From Fig. 4.34(b), the negative sequence voltage of terminal a with respect to reference bus is given by

$$V_{a2} = -I_{a2} Z_2 \tag{4.96}$$

Zero-Sequence Network

Since no zero-sequence voltages are induced in a synchronous machine, the zero-sequence network contains no emfs but includes the impedance offered to zero-sequence current. Three magnetic fields produced by the zero-sequence currents are in time-phase but are distributed in space phase by 120°. Therefore, the resultant air gap field due to zero-sequence currents is zero. Hence, the reactance offered by the rotor winding to the flow of zero-sequence currents is leakage reactance only.

Three-phase and single-phase zero-sequence network models are shown in Fig. 4.35(a) and (b) respectively. In Fig. 4.35(a), the current flowing in the impedance

Fig. 4.35 Zero-sequence network of synchronous machine

Z_n between neutral and ground is $I_n = 3I_{a0}$. The reference bus is at neutral potential which is same as the ground potential.

From Fig. 4.35(b), the zero-sequence voltage of terminal a with respect to reference bus is

$$V_{a0} = -3Z_n I_{a0} - Z_{g0} I_{a0}$$
$$= -I_{a0}(3Z_n + Z_{g0}) = -I_{a0} Z_0 \qquad (4.97)$$

where Z_0 is the total zero-sequence impedance through which the per-phase zero sequence current I_{a0} flows. Z_0 is given by

$$Z_0 = 3Z_n + Z_{g0} \qquad (4.98)$$

where Z_{g0} is the zero sequence impedance per-phase of the machine.

4.6.7 Zero-sequence Networks

Zero-sequence currents flow in the network if a return path which provides a completed circuit exists. In case of Y connected circuit with isolated neutral, the sum of the currents flowing into the neutral in the three-phases is zero. Since there is no zero-sequence components of the currents whose sum is zero, the impedance to zero-sequence current is infinite beyond the neutral point. An open circuit between the neutral of the Y-connected circuit and reference bus in the zero-sequence network indicates this fact, i.e., infinite impedance to zero-sequence current. A zero-impedance connection is inserted between the neutral point and the reference bus of the zero-sequence network, if the neutral of the Y-connected circuit is grounded through zero-impedance. If the neutral of the Y-connected circuit is grounded through the impedance Z_n, an impedance of $3Z_n$ must be inserted between the neutral and reference bus of the zero-sequence network.

Since a Δ-connected circuit provides no return path for zero-sequence currents, it offers infinite impedance to zero-sequence line currents.

Hence the zero-sequence network is open at the Δ connected circuit. However, zero-sequence currents may flow in the legs of a delta since it is a closed series circuit for circulating single-phase currents.

Zero-sequence Networks of 3-phase Loads

The zero-sequence networks for Y and Δ-connected 3-phase loads are shown in Fig. 4.36.

Fig. contd...

Fig. contd...

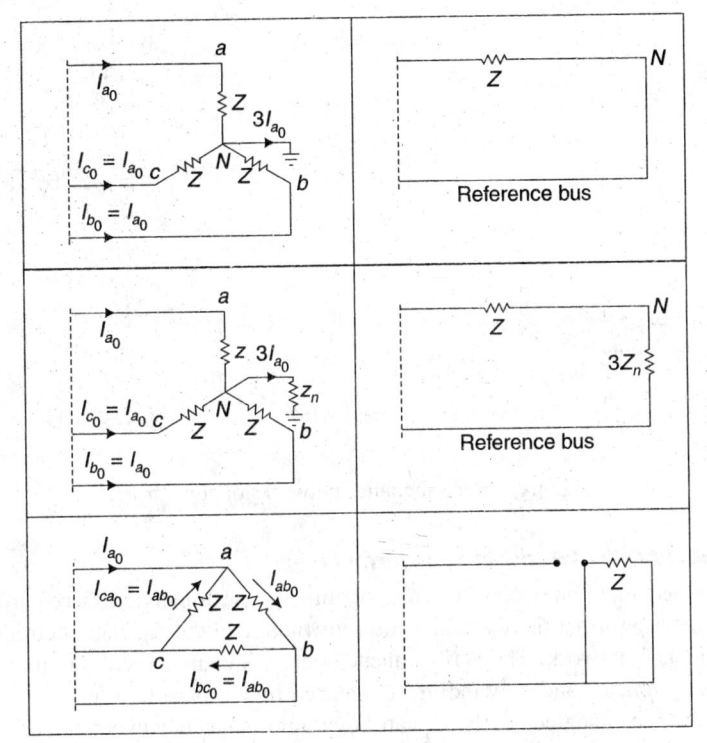

Fig. 4.36 Zero-sequence networks for Y and Δ connected loads.

Zero-sequence Diagrams of Generators

Zero-sequence diagrams of Y and Δ-connected generators are shown in Fig. 4.37.

Fig. contd...

Fig. contd...

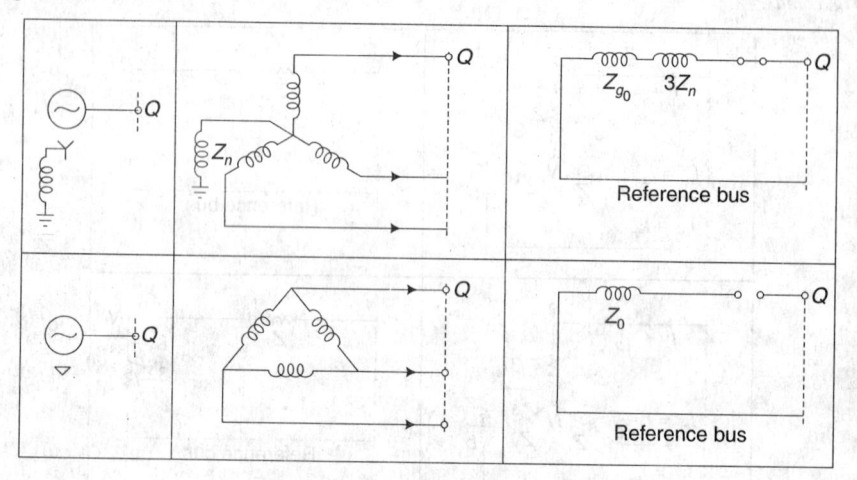

Fig. 4.37 *Zero-sequence networks of generators.*

Zero-sequence Networks of Transformers

The zero-sequence equivalent circuits of three-phase transformers require special attention. Transformer theory guides the construction of the equivalent circuit for the zero-sequence network. The zero-sequence network changes with combinations of the primary and secondary windings connected in Y and Δ. In a transformer, if the core reluctance is neglected, there is an exact mmf balance between the primary and secondary. This means that if relatively small magnetizing current is neglected, the current flows in the primary of a transformer only if there is a current in the secondary. The primary current is determined by the secondary current and the turns ratio of the windings, again with magnetizing current neglected. Five possible connections of two-winding transformers have been discussed here. The zero-sequence equivalent circuits of all the possible connections of transformers are shown in Fig. 4.38. The possible paths for the flow of zero-sequence current are shown by arrows on the connection diagram. Absence of an arrow on the connection diagram indicates that the transformer connection is such that zero-sequence current cannot flow. The corresponding points on the connection diagram and equivalent circuit are identified by letters P and Q. The justification of the equivalent circuit for each case of connection is as follows

Case 1: *Y-Y* transformer bank with any one neutral grounded

If any one of the two neutrals of a Y-Y transformer is ungrounded, zero-sequence current cannot flow in either winding because the absence of a path through one winding prevents flow of current in the other.

Hence, an open circuit exists for zero-sequence current between the two parts of the system connected by the transformer.

Case 2: *Y-Y* transformer bank with both neutrals grounded

In this case, a path through the transformer exists for the flow of zero-sequence currents in both windings via the two grounded neutrals. Hence, in the zero-sequence network, points P and Q on the two sides of the transformer are connected by the zero-sequence impedance of the transformer.

Case 3: Y-Δ transformer bank with grounded Y neutral

When the neutral of star side is grounded, zero-sequence currents can flow in star because of availability of path to ground through the star and the corresponding induced currents can circulate in the delta. The induced zero-sequence currents which circulate within the delta only to balance the zero-sequence currents in the star cannot flow in the lines connected to the delta. Hence, the equivalent circuit must have a path from the line P on the star side through the zero-sequence impedance of the transformer to the reference bus and an open circuit must exist between the line Q on the delta side and the reference bus. If the neutral of the star is grounded through an impedance Z_n, its effect is indicated by including an impedance of $3Z_n$ in series with the zero-sequence impedance of the transformer in the zero-sequence network.

Case 4: Y-Δ transformer bank with ungrounded star

This is the special case of case 3 where the impedance Z_n between neutral and ground is infinite. The impedance $3Z_n$ in the equivalent circuit of case 3 also becomes infinite. Therefore, zero-sequence current cannot flow in the transformer windings.

Case 5: Δ-Δ transformer bank

Since a Δ circuit provides no return path for the zero-sequence current, it cannot flow in or out of Δ-Δ transformer, although it can circulate within the Δ windings.

The zero-sequence networks of transformers for the above five cases are shown in Fig. 4.38.

Fig. contd...

Fig. contd...

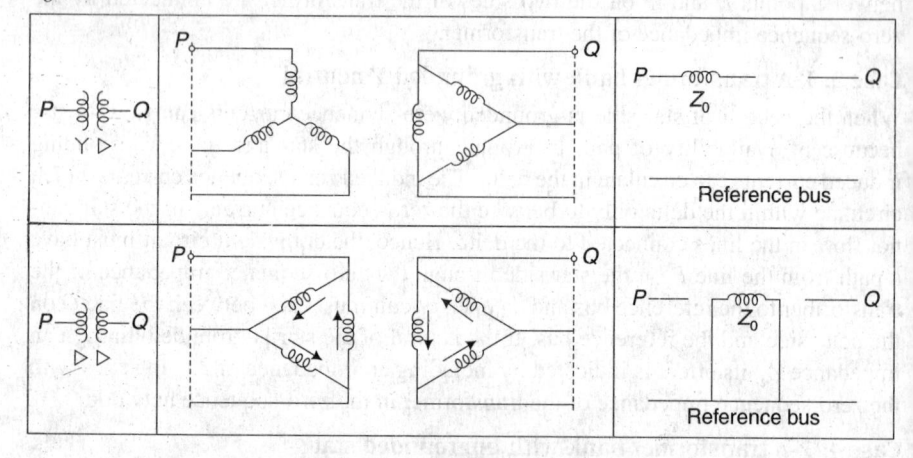

Fig. 4.38 *Zero-sequence networks of three-phase transformer banks.*

4.6.8 Formation of Sequence Networks of a Power System

A power system network consists of various components such as synchronous machines, transformer and lines. The sequence networks of all these components have been discussed in the preceding sections. The complete sequence network of a power system can be drawn by combining the sequence networks of the various components. The positive sequence network is drawn by examination of the one-line diagram of the system. This network contains the positive sequence voltages of synchronous machines (generators and motors) and positive sequence impedances of various components. The negative-sequence network can be easily drawn from positive-sequence network. The negative and positive-sequence impedances for static components (transformers and lines) are identical and each machine is represented by its negative-sequence impedance. Since synchronous machines don't generate voltages of negative-sequence, the negative-sequence network contains no voltage source, i.e. negative-sequence voltage is zero. The reference bus for both positive and negative-sequence networks is the system neutral. Any impedance connected between a neutral and ground is not included in these networks as neither of these sequence currents can flow through such an impedance.

Complete zero-sequence network of the system can be easily drawn by combining zero sequence subnetworks for various parts of the system. No voltage sources are present in the zero-sequence network. Any impedance included between generator or transformer neutral and ground becomes three times its value in a zero sequence network.

Example 4.16 Figure 4.39 shows the one-line diagram of a power system. Draw positive, negative, and zero sequence networks of the system. The system data is given below:

Generator G_1 : 30 MVA, 11 kV, $X_1 = 0.20$ pu, $X_2 = 0.20$ pu, $X_0 = 0.05$ pu
Transformer T_1 : 30 MVA, 11/220 kV, $X_1 = X_2 = X_0 = 0.07$ pu
Transformer T_2 : 50 MVA, 11/220 kV, $X_1 = X_2 = X_0 = 0.08$ pu

Generator G_2 : 50 MVA, 11 kV, $X_1 = X_2 = 0.25\ pu$, $X_0 = 0.06\ pu$
Line : $X_1 = X_2 = 150\ \Omega$, $X_0 = 500\ \Omega$.

Fig. 4.39 One-line diagram of the system

Solution: Let the selected base MVA be 50 and the base kV in the generator ckt and LT side of the transformer be 11. Then base voltage for HT side of the transformer and the line will be 220 kV.

All reactances are calculated on the common base MVA of 50 and the concerned base voltage.

Generator G_1 : $X_1 = X_2 = 0.20 \times \dfrac{50}{30} = 0.33\ pu$,

$$X_0 = 0.05 \times \dfrac{50}{30} = 0.083\ pu$$

Transformer T_1 : $X_1 = X_2 = X_0 = 0.07 \times \dfrac{50}{30} = 0.117\ pu$

Generator G_2 : $X_1 = X_2 = 0.25\ pu$, $X_0 = 0.06\ pu$.

Transformer T_2 : $X_1 = X_2 = X_0 = 0.08\ pu$.

Line : $X_1 = X_2 = \dfrac{150 \times 50}{(220)^2} = 0.155\ pu$,

$$X_0 = \dfrac{500 \times 50}{(220)^2} = 0.516\ pu$$

Pu value of neutral reactor, $X_n = \dfrac{2 \times 50}{(11)^2} = 0.826\ pu$

The positive, negative and zero sequence networks of the system are shown in Fig. 4.40(a), (b) and (c) respectively.

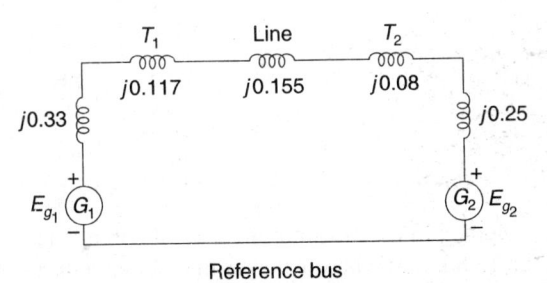

Fig. 40(a) Positive-sequence network

Fig. contd...

Fig. contd...

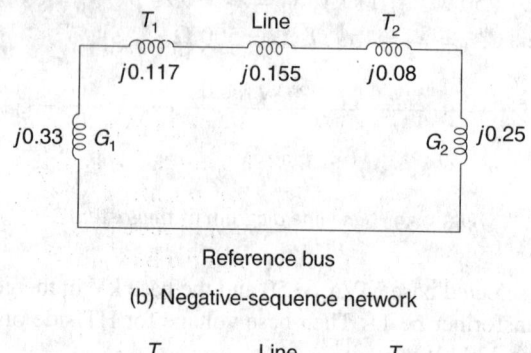

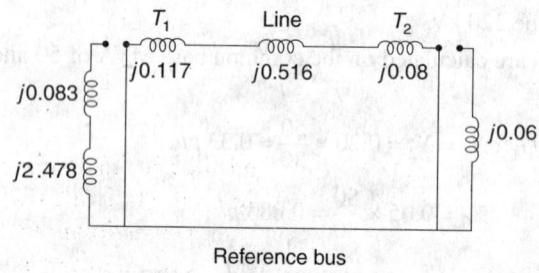

```
        T₁        Line       T₂
       j0.117    j0.155     j0.08
j0.33  G₁                    G₂  j0.25
```

Reference bus

(b) Negative-sequence network

```
         T₁        Line       T₂
        j0.117    j0.516     j0.08
j0.083
                                   j0.06
j2.478
```

Reference bus

(c) Zero-sequence network

Fig. 4.40 *Sequence networks for the system of Fig. 4.39*

Example 4.17 A 30 MVA, 13.8 kV, three-phase generator has a subtransient reactance of 15%. It supplies two motors over a transmission line having transformers at both ends as shown in the one-line diagram of Fig. 4.41. The motors are rated 15 MVA and 10 MVA, both 12.5 kV with a subtransient reactance of 20%. The transformers are both rated 50 MVA, 13.2/132 kV, with a reactance of 10%. The series reactance of the line is 90 ohms. The negative sequence reactance of each machine is equal to subtransient reactance. The zero-sequence reactance of machines is 6%. The zero-sequence reactance of line is 300 Ω. Current limiting reactor connected in neutrals of generator and motor M_1 are 2 Ω. Draw the sequence networks of the system.

```
       T₁                              T₂        e
  G  a  b         Line          c  d        M₁
 2Ω                                              2Ω
                                              f  M₂
```

Fig. 4.41 *One-line diagram of the system*

Solution: Let the common MVA base chosen be 30 MVA and the voltage base in the generator circuit be 13.8 kV. The base voltages in other circuits are as follows:

$$\text{Base voltage in transmission line} = 13.8 \times \frac{132}{13.2} = 138 \text{ kV}$$

Base voltage in motor circuit = $138 \times \dfrac{13.2}{132} = 13.8$ kV

The reactances of generator, transformers, line and motors are converted to *pu* values on appropriate bases as follows:

For Generator G_1: Base MVA = 30, Base voltage = 13.8 kV

$$X_1 = X_2 = X''_d = 0.15 \, pu, \, X_0 = 0.06 \, pu$$

For transformers T_1 and T_2: Base MVA = 30, Base voltage on LT side = 13.8 kV, Base voltage on HT side = 138 kV

The reactances of transformers on appropriate bases are

$$X_1 = X_2 = X_0 = 0.1 \times \dfrac{30}{50} \times \left(\dfrac{13.2}{13.8}\right)^2 = 0.0548 \, pu$$

For transmission line: Base MVA = 30, Base voltage = 138 kV
The reactances are

$$X_1 = X_2 = \dfrac{90 \times 30}{(138)^2} = 0.142 \, pu, \, X_0 = \dfrac{300 \times 30}{(138)^2} = 0.472 \, pu$$

For motors M_1 and M_2: Base MVA = 30, Base voltage = 13.8 kV
The reactances of motor M_1 are

$$X_1 = X_2 = X''_d = 0.2 \times \dfrac{30}{15} \times \left(\dfrac{12.5}{13.8}\right)^2 = 0.328 \, pu$$

$$X_0 = 0.06 \times \dfrac{30}{15} \times \left(\dfrac{12.5}{13.8}\right)^2 = 0.098 \, pu$$

The reactances of motor M_2 are

$$X_1 = X_2 = X''_d = 0.2 \times \dfrac{30}{10} \times \left(\dfrac{12.5}{13.8}\right)^2 = 0.492 \, pu$$

$$X_0 = 0.06 \times \dfrac{30}{10} \times \left(\dfrac{12.5}{13.8}\right)^2 = 0.147 \, pu$$

Reactance of current limiting reactor of generator

$$Z_n = X_n = 2 \times \dfrac{30}{(13.8)^2} = 0.315 \, pu$$

$$3Z_n = 3X_n = 3 \times 0.315 = 0.945 \, pu$$

Reactance of current limiting reactor of motor M_1

$$Z_n = X_n = 2 \times \dfrac{30}{(13.8)^2} = 0.315 \, pu$$

$$3Z_n = 3X_n = 3 \times 0.315 = 0.945 \, pu$$

Using the values of the above reactances, the positive, negative and zero-sequence networks are as shown in Fig. 4.42(a), (b) and (c) respectively.

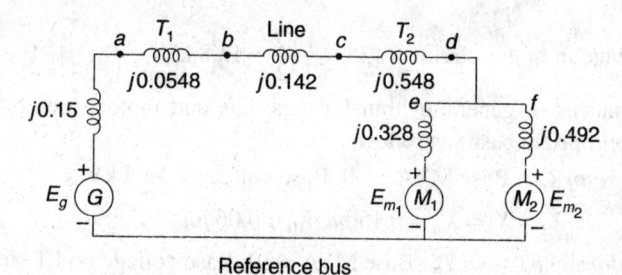

(a) Positive-sequence network

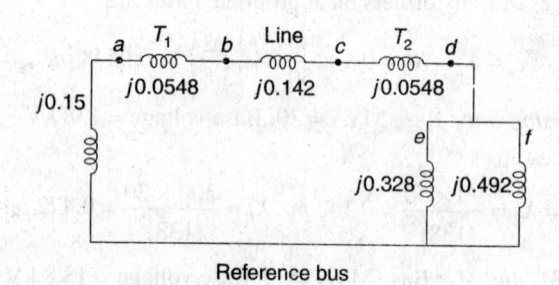

(b) Negative-sequence network

(c)

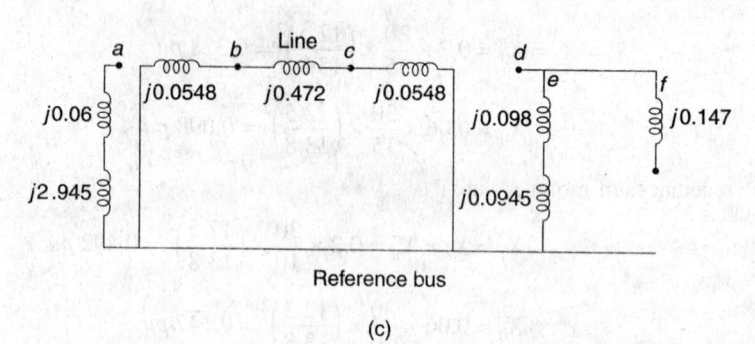

Fig. 4.42 *Sequence networks for the system of Fig. 4.41*

4.7 UNSYMMETRICAL FAULT ANALYSIS

Most of the faults which occur on a power system are of unsymmetrical nature involving one or two phases and not all the three phases. Unsymmetrical faults make the three-phase system unbalanced. The path of the fault current from line to ground or line to line may or may not contain impedance. Since any unsymmetrical fault causes unbalanced currents to flow in the system, this type of fault is also known as unbalanced fault. Various types of unsymmetrical faults that occur on power systems are broadly classified into the following two main types:

1. Shunt-type Faults This type of unsymmetrical faults in decreasing frequency of occurrence are as follows:
   (i) Single line-to-ground (L-G) fault
   (ii) Line-to-line (L-L) fault
   (iii) Double line-to-ground (2L-G) fault

2. Series-type Faults One or two open conductors also result in unsymmetrical faults. The conductors may open either through the breaking of one or two conductors or through the action of fuses and other devices that may not open the three phases simultaneously. Hence, series type of unsymmetrical faults are as follows:
   (i) One conductor open
   (ii) Two conductors open

Unsymmetrical fault analysis involves determination of the currents and voltages in all parts of the system after the occurrence of the unsymmetrical fault.

The method of symmetrical components which deal with unbalanced polyphase circuits is a powerful tool for the analysis of unsymmetrical faults. The balanced three-phase network is resolved into three uncoupled balanced sequence networks by the method of symmetrical components. The sequence networks are connected only at the points of unbalance. Solution of the sequence networks for the fault conditions gives symmetrical current and voltage components which can be superimposed to reflect the effects of the original unbalanced fault currents on the overall system. Hence, the method of symmetrical components discussed in Section 4.6 is very useful in analysis of unsymmetrical faults and has been fully exploited in this section.

Let us consider a general power network shown in Fig. 4.43. If V_a, V_b and V_c be the voltages of lines a, b and c respectively with respect to the ground, the occurrence of fault at point F in the system results in flow of currents I_a, I_b and I_c from the system.

It is assumed that the system is operating at no load before the occurrence of a fault. Hence, the positive sequence voltages of all the synchronous machines will be identical and equal to the prefault voltage at F. Let this voltage be denoted as E_a. The positive, negative and zero sequence networks presented by the power system as seen from the fault point F are schematically represented by Figs. 4.44(a), (b) and (c). On each sequence network, the reference bus is indicated by a thick line and the fault point is identified as F. Thevenin's equivalents of the three sequence networks are shown in Figs. 4.45 (a), (b) and (c) respectively. Since only the positive sequence network contains the voltage E_a and there is no coupling between the sequence networks, the sequence voltages at F can be expressed in terms of sequence currents and Thevenin sequence impedances in the same way as Eqs. (4.94), (4.96) and (4.97). In matrix form the equations for sequence voltages at F can be expressed as

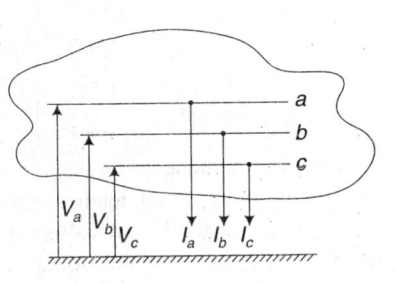

Fig. 4.43 *A general power network*

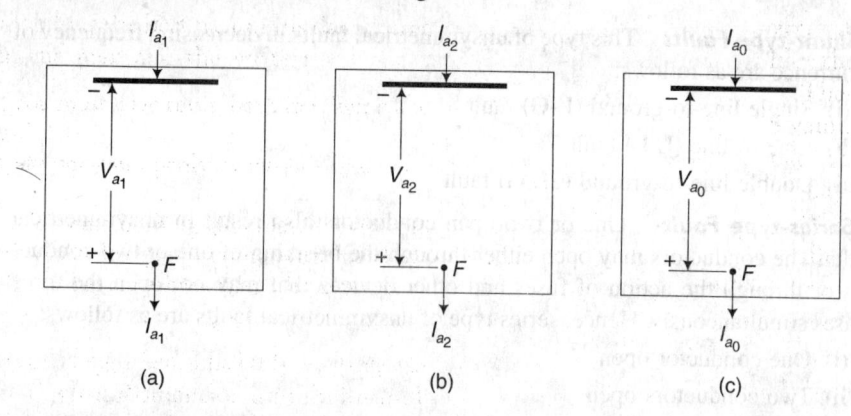

Fig. 4.44 Sequence networks as seen from fault point F

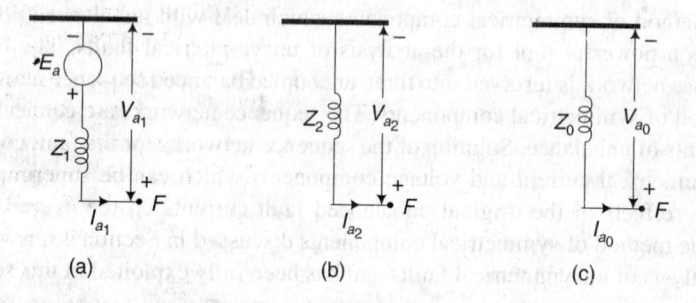

Fig. 4.45 Thevenin equivalents of the sequence networks as seen form fault point F

$$\begin{bmatrix} V_{a1} \\ V_{a2} \\ V_{a0} \end{bmatrix} = \begin{bmatrix} E_a \\ 0 \\ 0 \end{bmatrix} - \begin{bmatrix} Z_1 & 0 & 0 \\ 0 & Z_2 & 0 \\ 0 & 0 & Z_0 \end{bmatrix} \begin{bmatrix} I_{a1} \\ I_{a2} \\ I_{a0} \end{bmatrix} \quad (4.99)$$

Since the sequence currents and voltages are constrained by the type of the fault, a particular type of fault leads to a particular connection of sequence networks. After drawing the connection of sequence networks depending upon the type of the fault, the sequence currents and voltages and fault currents and voltages can be easily computed. The steps to be followed to compute unsymmetrical fault current are as follows:

(i) The one-line diagram of the system is drawn.
(ii) The positive, negative and zero-sequence networks are assembled individually and the fault point location F is marked on each diagram.
(iii) The sequence networks are replaced by their Thevenin equivalents at the fault point F. The Thevenin voltage source in the positive sequence network is the pre-fault voltage at F.
(iv) Thevenin equivalent sequence networks are connected at the fault point F according to the fault condition.
(v) Fault currents and voltages are computed from the resulting network.

4.7.1 Single Line-to-Ground (L-G) Fault

Figure 4.46 shows a line-to-ground fault at point F on phase a in a power system through a fault impedance Z_f.

At the fault point F, the currents from the power system and the line-to-ground voltages are expressed as follows:

I_b = Current from phase b = 0 (4.100)
I_c = Current from phase c = 0 (4.101)
V_a = Voltage of phase a at the fault point $F = I_a Z_f$ (4.102)
I_a = Current fed to the fault by phase a

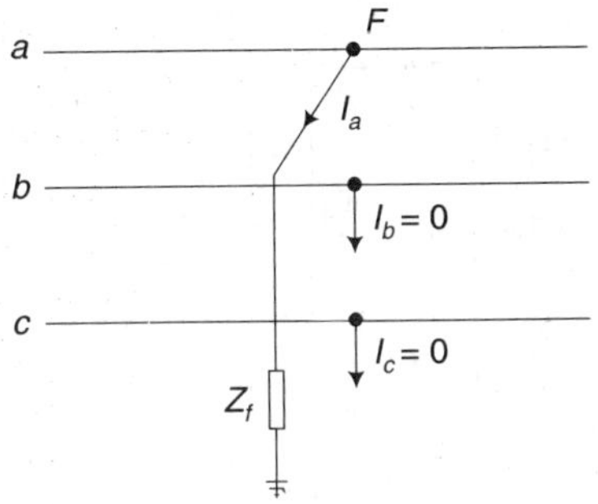

Fig. 4.46 Single line-to-ground (L-G) fault at F

The symmetrical components of the fault currents, according to Eq. (4.61) are

$$\begin{bmatrix} I_{a1} \\ I_{a2} \\ I_{a0} \end{bmatrix} = \frac{1}{3} \begin{bmatrix} 1 & a & a^2 \\ 1 & a^2 & a \\ 1 & 1 & 1 \end{bmatrix} \begin{bmatrix} I_a \\ 0 \\ 0 \end{bmatrix} \quad (4.103)$$

Thus,
$$I_{a1} = I_{a2} = I_{a0} = \frac{1}{3} I_a \quad (4.104)$$

Equation (4.102) can also be expressed in terms of symmetrical components as

$$V_{a1} + V_{a2} + V_{a0} = I_a Z_f = 3 I_{a1} Z_f \quad (4.105)$$

From Eqs. (4.104) and (4.105) it is clear that all sequence currents are equal and the sum of sequence voltages equals $3I_{a1}Z_f$. Therefore, these equations suggest a series connection of the three sequence networks through an impedance $3Z_f$ as shown in Figs. 4.47(a) and (b).

From the series connected sequence networks shown in Fig. 4.47(b), we get

$$I_{a1} = I_{a2} = I_{a0} = \frac{E_a}{(Z_1 + Z_2 + Z_0) + 3Z_f} \quad (4.106)$$

The fault current I_a is then given by

$$I_a = 3I_{a1} = \frac{3E_a}{(Z_1 + Z_2 + Z_0) + 3Z_f} \quad (4.107)$$

The above results can also be obtained directly from Eqs. (4.104) and (4.105) by using V_{a1}, V_{a2} and V_{a0} from Eq. (4.99).

Thus,
$$(E_a - I_{a1}Z_1) + (-I_{a2}Z_2) + (-I_{a0}Z_0) = 3I_{a1}Z_f$$

or $$[(Z_1 + Z_2 + Z_0) + 3Z_f] I_{a1} = E_a$$

or
$$I_{a1} = \frac{E_a}{(Z_1 + Z_2 + Z_0) + 3Z_f} \quad (4.108)$$

172 Power System Protection and Switchgear

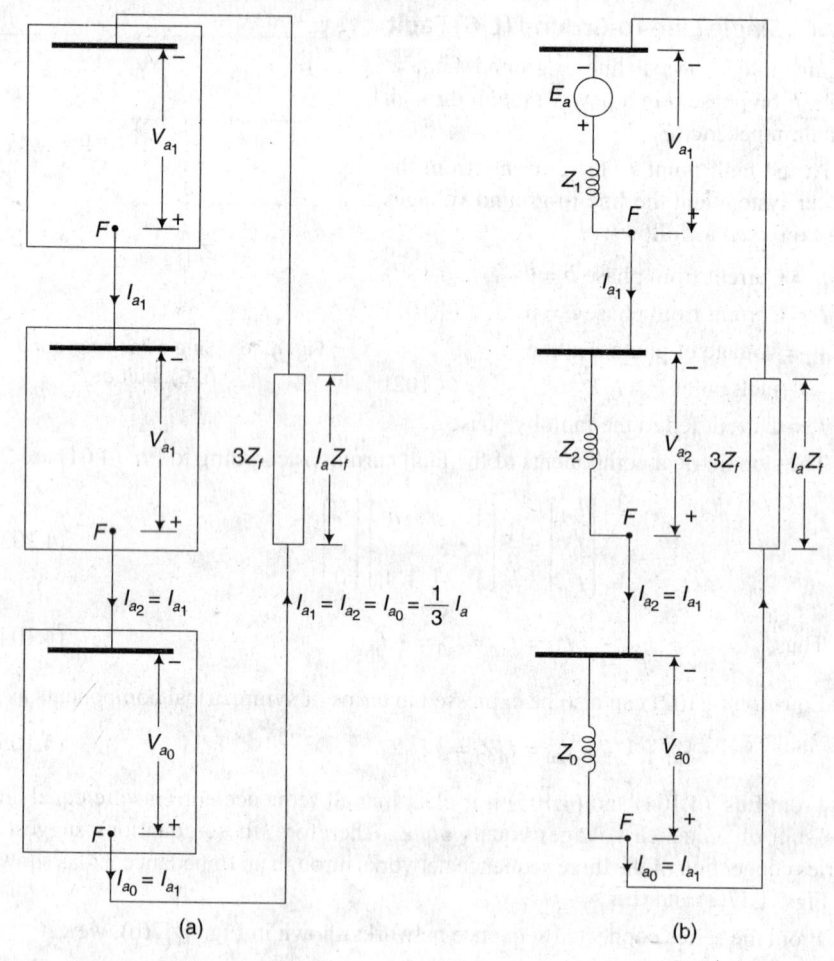

Fig. 4.47 Connection of sequence networks for a single line-to-ground (L-G) fault

The voltages of the healthy phases (lines) a and b to ground under fault conditions are computed as follows.

From Eq. (4.47), V_b is given by

$$V_b = a^2 V_{a1} + a V_{a2} + V_{a0}$$
$$= a^2 (E_a - I_{a1} Z_1) + a (-I_{a2} Z_2) + (-I_{a0} Z_0)$$
$$= a^2 (E_a - I_{a1} Z_1) - a I_{a1} Z_2 - I_{a1} Z_0$$

Substituting for I_{a1} from (4.106) and reorganizing, we get

$$V_b = E_a \frac{3a^2 Z_f + (a^2 - a) Z_2 + (a^2 - 1) Z_0}{(Z_1 + Z_2 + Z_0) + 3Z_f} \qquad (4.109)$$

The expression for V_c can be similarly obtained from Eq. (4.48).

$$V_c = aV_{a1} + a^2V_{a2} + V_{a0}$$
$$= a(E_a - I_{a1}Z_1) + a^2(-I_{a2}Z_2) + (-I_{a0}Z_0)$$
$$= a(E_a - I_{a1}Z_1) - a^2 I_{a1}Z_2 - I_{a1}Z_0$$

Substituting for I_{a1} and reorganizing, we get.

$$V_c = E_a \frac{3aZ_f + (a - a^2)Z_2 + (a-1)Z_0}{(Z_1 + Z_2 + Z_0) + 3Z_f} \quad (4.110)$$

Fault Occuring Under Loaded Conditions

If the power system supplies balanced load during pre-fault condition, only positive sequence load currents flow in the system. Since a generator does not generate negative and zero-sequence voltages, the negative and zero-sequence components of the load current during pre-fault balanced load condition are zero. Therefore, negative and zero-sequence networks during fault under loaded conditions are the same as without load. Under the fault condition, the positive-sequence component of the fault current is to be added to the pre-fault load current. Hence, the positive-sequence network under fault condition must of course carry the load current. In order to account for load current, the synchronous machines in the positive-sequence network are replaced by subtransient (X_d''), transient (X_d') or synchronous reactances depending upon time at which the fault currents are to be determined after the occurrence of the fault. The positive-sequence network which takes the pre-fault load current into account is used in the connection of sequence networks shown in Fig. 4.47(a) for computing sequence currents under fault condition.

If the positive-sequence network is replaced by its Thevenin equivalent network as shown in Fig. 4.47(b), the Thevenin voltage is equal to the pre-fault voltage V_f at the fault point F under loaded conditions. The Thevenin equivalent impedance is the impedance between fault point F and the reference bus of the positive sequence network (with voltage sources short circuited).

Comparison of Fault Currents Under Single-line-to-Ground and 3-Phase Symmetrical Faults

Though a 3-phase symmetrical fault is the most severe types of fault which gives rise to the maximum fault current, but under certain special situations an L-G fault can be more severe and can cause greater fault current than a three-phase symmetrical fault. A single line-to-ground fault located close to large generating units can be more severe than a 3-phase symmetrical fault at the same location.

If it is assumed that the sequence impedances Z_1, Z_2 and Z_0 are pure reactances (X_1, X_2 and X_0) and $Z_f = 0$, the fault currents are

Under three-phase symmetrical fault:

Fault current, $\quad I_a = \dfrac{E_a}{jX_1} \quad (4.111)$

Under line-to-ground (L-G) fault:

Fault current, $$I_a = \frac{3E_a}{j(X_1 + X_2 + X_3)} \qquad (4.112)$$

From practical importance point of view, three cases of occurrence of fault are as follows:

(i) Fault at the terminals of the generator with solidly grounded neutral For a synchronous machine (generator or motor) $X_0 \ll X_1$ and it can be assumed that $X_1 = X_2$ (for subtransient conditions X_1 and X_2 are exactly equal). When these values are substituted in Eqs. (4.111) and (4.112), it is seen that the current for a single line-to-ground fault is more than the current for a three-phase symmetrical fault. Hence, it can be concluded that a single line-to-ground fault at the terminals of a generator with solidly grounded neutral is more severe than a 3-phase symmetrical fault at the same location.

(ii) Fault at the terminals of a generator with neutral grounded through reactance X_n Three-phase symmetrical fault current is not affected by the method of grounding. However, the fault current for a single line-to-ground fault is given by

$$I_a = \frac{3E_a}{j(X_1 + X_2 + X_0 + 3X_n)} \qquad (4.113)$$

The relative severity of 3-phase symmetrical fault and the single line-to-ground fault depends on the value of X_n. If X_n is large, the fault current under single line-to-ground fault is less than that under 3-phase symmetrical fault. If X_n is very small, the fault current under single line-to-ground fault may be more than the fault current under 3-phase symmetrical fault. Since a term $3X_n$ appears in the denominator of Eq. (4.113), even a small value of X_n is sufficient to make the value of the single line-to-ground fault current less than 3-phase fault current. If $X_n = \frac{1}{3}(X_1 - X_0)$, the line currents for the two cases are equal. If $X_n < \frac{1}{3}(X_1 - X_0)$, a single line-to-ground fault is more severe than a 3-phase symmetrical fault. If $X_n > \frac{1}{3}(X_1 - X_0)$, a 3-phase fault is more severe than a single-line to ground (L-G) fault.

(iii) Fault on transmission line In case of a transmission line $X_0 \gg X_1$ so that when a fault occurs on the line sufficiently away from the generator terminals, 3-phase symmetrical fault current is more than the fault current for a single line-to-ground fault.

4.7.2 Line-to-Line (L-L) Fault

A line-to-line fault at point F in a power system between phases b and c through a fault impedance Z_f is shown in Fig. 4.48.

The conditions at the fault can be expressed by the following equations:

$$I_a = 0 \qquad (4.114)$$
$$I_b = -I_c \qquad (4.115)$$

$$V_b - V_c = I_b Z_f \qquad (4.116)$$

and, hence

$$[I_p] = \begin{bmatrix} I_a \\ I_b \\ I_c \end{bmatrix} = \begin{bmatrix} 0 \\ I_b \\ -I_b \end{bmatrix} \qquad (4.117)$$

The symmetrical components of the fault current are given by

$$\begin{bmatrix} I_{a1} \\ I_{a2} \\ I_{a0} \end{bmatrix} = \frac{1}{3} \begin{bmatrix} 1 & a & a^2 \\ 1 & a^2 & a \\ 1 & 1 & 1 \end{bmatrix} \begin{bmatrix} 0 \\ I_b \\ -I_b \end{bmatrix} \qquad (4.118)$$

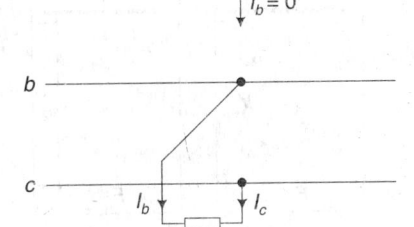

Fig. 4.48 Line-to-line (L-L) fault through impedance Z_f

from which we get

$$I_{a2} = -I_{a1} \qquad (4.119)$$
$$I_{a0} = 0 \qquad (4.120)$$

The symmetrical components of voltages at F under fault conditions are

$$\begin{bmatrix} V_{a1} \\ V_{a2} \\ V_{a0} \end{bmatrix} = \frac{1}{3} \begin{bmatrix} 1 & a & a^2 \\ 1 & a^2 & a \\ 1 & 1 & 1 \end{bmatrix} \begin{bmatrix} V_a \\ V_b \\ V_b - I_b Z_f \end{bmatrix} \qquad (4.121)$$

Writing the first two equations from the above equation, we get

$$3V_{a1} = V_a + (a + a^2) V_b - a^2 I_b Z_f$$
$$3V_{a2} = V_a + (a + a^2) V_b - a I_b Z_f$$

and therefore

$$3(V_{a1} - V_{a2}) = (a - a^2) I_b Z_f = j\sqrt{3}\, I_b Z_f \qquad (4.122)$$

I_b is given by

$$I_b = a^2 I_{a1} + a I_{a2} + I_{a0}$$
$$= (a^2 - a) I_{a1} \qquad \text{(since, } I_{a2} = -I_{a1};\ I_{a0} = 0\text{)}$$
$$= -j\sqrt{3}\, I_{a1} \qquad (4.123)$$

substituting I_b from Eq. (4.123) in Eq. (4.122), we get

$$V_{a1} - V_{a2} = I_{a1} Z_f \qquad (4.124)$$

Equations (4.119) and (4.124) provide the basis for interconnection of positive and negative sequence networks in parallel through a series impedance Z_f as shown in Figs. 4.49(a) and(b). Since $I_{a0} = 0$ as per Eq. (4.120), the zero sequence network is on open circuit and hence remains unconnected.

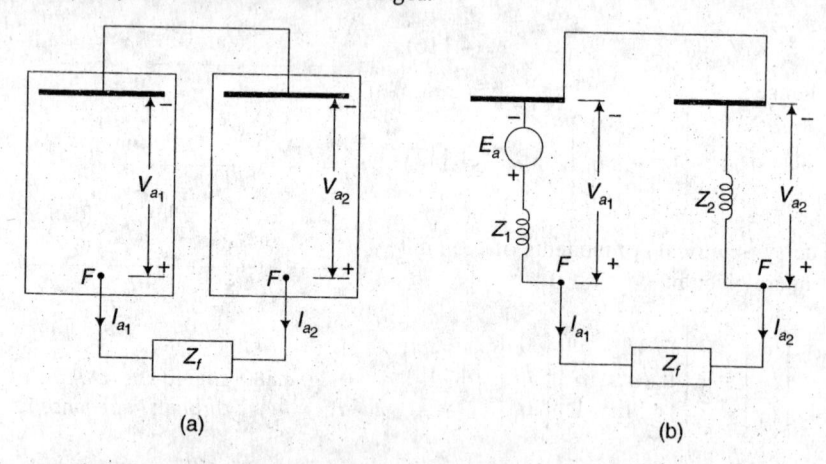

Fig. 4.49 *Connection of sequence networks for a line-to-line (L-L) fault*

In terms of Thevenin's equivalent network, we get from Fig. 4.49(b),

$$I_{a1} = \frac{E_a}{Z_1 + Z_2 + Z_f} \tag{4.125}$$

From Eq. (4.123), we get

$$I_b = -I_c = \frac{-j\sqrt{3}\, E_a}{Z_1 + Z_2 + Z_f} \tag{4.126}$$

Knowing I_{a1}, the voltages at the fault can be found as follows:

$$V_{a1} = E_a - I_{a1} Z_1 \tag{4.127}$$

$$V_{a2} = -I_{a2} Z_2 = I_{a1} Z_2 \tag{4.128}$$

$$V_{a0} = -I_{a0} Z_0 = 0 \tag{4.129}$$

Hence,
$$V_a = V_{a1} + V_{a2}$$
$$= (E_a - I_{a1} Z_1) + I_{a1} Z_2$$
$$= E_a - I_{a1} (Z_1 - Z_2)$$

Substituting the value of I_{a1} from Eq. (4.125), we get

$$V_a = E_a - \frac{E_a (Z_1 - Z_2)}{Z_1 + Z_2 + Z_f}$$

$$= E_a \left[\frac{(Z_1 + Z_2 + Z_f) - (Z_1 - Z_2)}{(Z_1 + Z_2 + Z_f)} \right]$$

or
$$V_a = E_a \left(\frac{Z_f + 2Z_2}{Z_1 + Z_2 + Z_f} \right) \tag{4.130}$$

$$V_b = a^2 V_{a1} + a V_{a2}$$
$$= a^2 (E_a - I_{a1} Z_1) + a I_{a1} Z_2$$
$$= a^2 E_a - I_{a1} (a^2 Z_1 - a Z_2)$$
$$= a^2 E_a - \frac{E_a (a^2 Z_1 - a Z_2)}{(Z_1 + Z_2 + Z_f)}$$
$$= E_a \frac{a^2 (Z_1 + Z_2 + Z_f) - (a^2 Z_1 - a Z_2)}{(Z_1 + Z_2 + Z_f)}$$

or
$$V_b = E_a \left(\frac{-Z_2 + a^2 Z_f}{Z_1 + Z_2 + Z_f} \right) \tag{4.131}$$

Similarly,
$$V_c = E_a \left(\frac{-Z_2 + a Z_f}{Z_1 + Z_2 + Z_f} \right) \tag{4.132}$$

Check:
$$V_b - V_c = \frac{E_a (a^2 - a) Z_f}{Z_1 + Z_2 + Z_f}$$
$$= (a^2 - a) I_{a1} Z_f = I_b Z_f$$

Note: Current caused by any fault that does not involve ground, cannot contain zero sequence component.

4.7.3 Double Line-to-Ground (2L-G) Fault

A double line-to-ground fault at F in a power system is shown in Fig. 4.50. The faulted phases are b and c which are shorted to the ground through the fault impedance Z_f.

The conditions at the fault are expressed as

$$I_a = I_{a1} + I_{a2} + I_{a0} = 0 \tag{4.133}$$
$$V_b = V_c = (I_b + I_c) Z_f \tag{4.134}$$

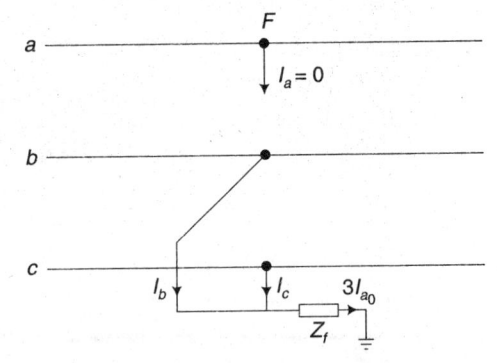

Fig. 4.50 Double line-to-ground fault

Since,
$$\begin{bmatrix} I_{a1} \\ I_{a2} \\ I_{a0} \end{bmatrix} = \frac{1}{3} \begin{bmatrix} 1 & a & a^2 \\ 1 & a^2 & a \\ 1 & 1 & 1 \end{bmatrix} \begin{bmatrix} 0 \\ I_b \\ I_c \end{bmatrix} \quad \text{(As } I_a = 0 \text{ in this case)}$$

Hence,
$$I_{a0} = \frac{1}{3} (I_b + I_c)$$

or
$$(I_b + I_c) = 3 I_{a0}$$

Therefore, Eq. (4.134) can be written as

$$V_b = V_c = (I_b + I_c) Z_f = 3I_{a0} Z_f \qquad (4.135)$$

The symmetrical components of voltages are given by

$$\begin{bmatrix} V_{a1} \\ V_{a2} \\ V_{a0} \end{bmatrix} = \frac{1}{3} \begin{bmatrix} 1 & a & a^2 \\ 1 & a^2 & a \\ 1 & 1 & 1 \end{bmatrix} \begin{bmatrix} V_a \\ V_b \\ V_c \end{bmatrix} \qquad (4.136)$$

from which, we get

$$V_{a1} = V_{a2} = \frac{1}{3} [V_a + (a + a^2) V_b] \qquad \text{(As } V_b = V_c\text{)} \qquad (4.137)$$

$$V_{a0} = \frac{1}{3} (V_a + 2V_b) \qquad (4.138)$$

Subtracting Eq. (4.137) from Eq. (4.138),

$$V_{a0} - V_{a1} = \frac{1}{3} (2 - a - a^2) V_b = V_b = 3I_{a0} Z_f$$

or

$$V_{a0} = V_{a1} + 3I_{a0} Z_f \qquad (4.139)$$

By observing Eqs. (4.133), (4.137) and (4.139), the connection of sequence networks is drawn in Figs. 4.51(a) and (b).

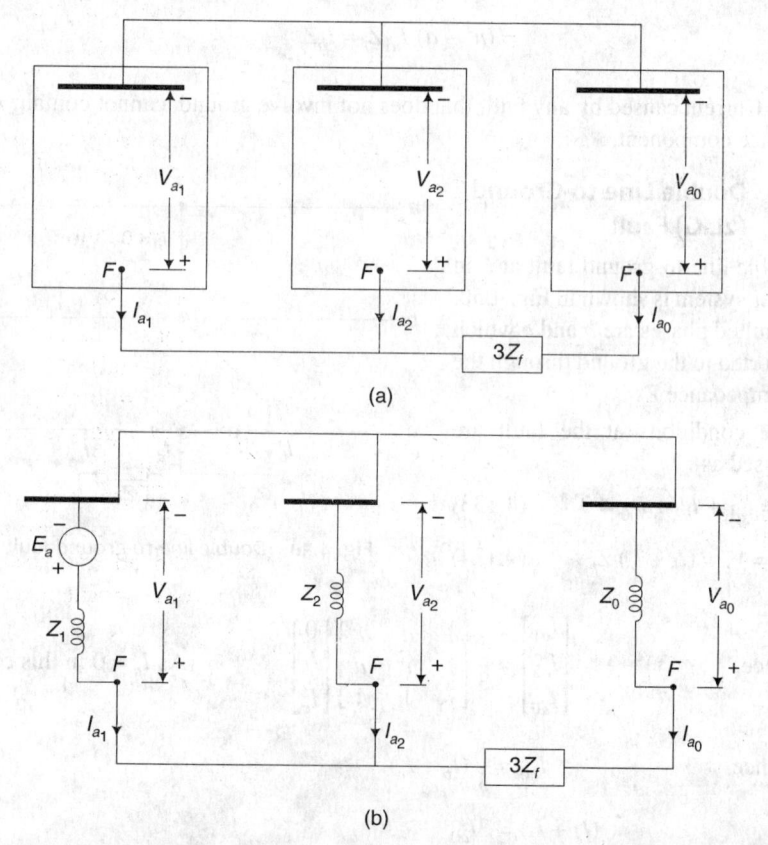

Fig. 4.51 Connection of sequence networks for a double line-to-ground (2L-G) fault at F

Fault Analysis **179**

In terms of the Thevenin's equivalents with respect to fault point F, we can write from Fig. 4.51(b).

$$I_{a1} = \frac{E_a}{Z_1 + Z_2 \| (Z_0 + 3Z_f)}$$

or

$$I_{a1} = \frac{E_a}{Z_1 + Z_2 (Z_0 + 3Z_f)/(Z_2 + Z_0 + 3Z_f)} \quad (4.140)$$

Values of I_{a2} and I_{a0} from Fig. 4.51(b) are

$$I_{a2} = -I_{a1} \frac{(Z_0 + 3Z_f)}{(Z_2 + Z_0 + 3Z_f)} \quad (4.141)$$

$$I_{a0} = -I_{a1} \frac{Z_2}{(Z_2 + Z_0 + 3Z_f)} \quad (4.142)$$

In case of a solid/bolted fault, $Z_f = 0$ in all the above derivations.

Example 4.18 A 10 MVA, 6.6 kV generator has percentage reactances to positive, negative and zero sequence currents of 15%, 10% and 5% respectively and its neutral is solidly grounded. The generator is unloaded and excited to its rated voltage. Calculate the sequence components of currents and voltages, the fault current, phase currents and phase voltages when a single line to ground fault develops on phase a of the generator. Phase sequence is abc.

Solution: On base MVA of 10 and base voltage of 6.6 kV

$$Z_1 = j\,0.15\ pu$$
$$Z_2 = j\,0.10\ pu$$
$$Z_0 = j\,0.05\ pu$$

Referring to Fig. 4.52,

$$Z_n = 0,\ Z_f = 0$$

For L-G fault on phase 'a'

$$I_b = I_c = 0 \text{ and } V_a = 0$$

Symmetrical components of currents

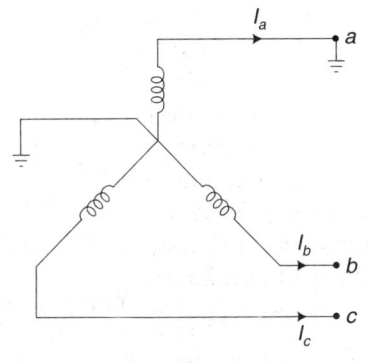

Fig. 4.52 System for Example 4.18

$$I_{a1} = I_{a2} = I_{a0} = \frac{E_a}{(Z_1 + Z_2 + Z_0)}$$

$$= \frac{1.0}{(j\,0.15 + j\,0.10 + j\,0.05)}$$

$$= \frac{1.0}{j\,0.3} = -j\,3.33\ pu.$$

Line Currents

Fault current

$$I_f = I_a = 3I_{a1} = 3(-j\,3.33) = -j\,10\ pu.$$

$$I_b = I_c = 0$$

Base current

$$= \frac{10 \times 10^3}{\sqrt{3} \times 6.6} = 875\ A$$

Line currents in amps are
$$I_a = 10 \times 875 = 8750 \text{ A}$$
$$I_b = I_c = 0$$

Symmetrical components of voltages
$$V_{a1} = E_a - I_{a1} Z_1 = 1.0 - (j\,3.33)(j\,0.15)$$
$$= 1.0 + j^2 (0.5)$$
$$= 1.0 - 0.5 = 0.5 \; pu$$
$$V_{a2} = - I_{a2} Z_2 = -I_{a1} Z_2 = - (-j\,3.33)(j\,0.10)$$
$$= - 0.33 \; pu$$
$$V_{a2} = - I_{a0} Z_0 = -I_{a1} Z_0 = - (-j\,3.33)(j\,0.05)$$
$$= - 0.17 \; pu.$$

Line to ground voltages at fault are
$$V_a = V_{a1} + V_{a2} + V_{a0}$$
$$= 0.5 - 0.33 - 0.17 = 0$$
$$V_b = a^2 V_{a1} + a V_{a1} + V_{a0}$$
$$= (-0.5 - j\,0.866)\,0.5 + (-0.5 + j\,0.866)(-0.33) - 0.17$$
$$= (-0.25 + 0.165 - 0.17) + (-j\,0.433 - j\,0.286)$$
$$= (-0.255 - j\,0.719)$$
$$= 0.763 < 258° \; pu$$
$$V_c = a V_{a1} + a^2 V_{a2} + V_{a0}$$
$$= (-0.5 + j\,0.866)(0.5) + (-0.5 - j\,0.866)(-0.33) - 0.17$$
$$= (-0.25 + 0.165 - 0.17) + (j\,0.433 + j\,0.286)$$
$$= (-0.255 + j\,0.719) = 0.763 < 102°$$

Since phase value of the base voltage $= \dfrac{6.6}{\sqrt{3}} = 3.81$ kV
$$V_a = 0 \text{ kV}$$
$$V_b = 0.763 \times 3.81 < 258° = 2.90 < 258°$$
$$V_c = 0.763 \times 3.81 < 102° = 2.90 < 102°$$

Example 4.19 A 3-phase, 30 MVA, 6.6 kV generator having 20% reactance is connected through a 30 MVA 6.6/33 kV delta-star connected transformer of 10% reactance to a 33 kV transmission line having a negligible resistance and a reactance of 5 ohms. At the receiving end of the line there is a 30 MVA 33/6.6 kV delta-star connected transformer of 10% reactance stepping down the voltage to 6.6 kV. The neutrals of both the transformers are solidly grounded.

Draw the one-line diagram, and the positive, negative and zero sequence networks of this system and determine the fault currents for the following types of fault at the receiving station L.V. busbars

(a) Three-phase symmetrical fault
(b) A single line-to-ground fault
(c) Line-to-line fault

For a generator, assume –ve sequence reactance to be 75% that of +ve sequence.

Solution: One-line diagram of the system is shown in Fig. 4.53. The fault occurs at point F. Let us choose a common base MVA of 30 MVA and base voltage of 6.6 kV on LV side. Then base voltage on HV side, i.e., of transmission line is 33 kV.

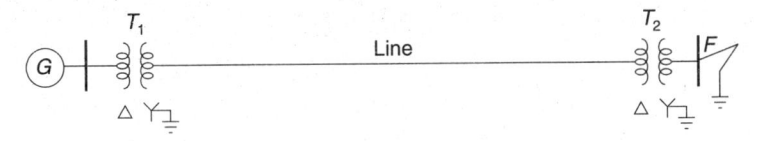

Fig. 4.53 *One-line diagram of the system*

The sequence impedances on the common base MVA and the concerned base voltages are

Generator G : $Z_{g1} = j\,0.20\ pu$, $Z_{g2} = j\,0.20 \times 0.75$
$= j\,0.15\ pu$, Z_{g0} Z_{g0} is not necessary to know.

Transformers : $Z_{T1} = Z_{T2} = Z_{T0} = j\,0.10\ pu$

Transmission line : $Z_{L1} = Z_{L2} = j\,5 \times \dfrac{30}{(33)^2}$

$= j\,0.138\ pu$ Z_{L0} is not necessary to know.

Positive, negative and zero sequence networks are then drawn as shown in Figs. 4.54(a), (b) and (c).

From sequence networks of Fig. 4.54 it is clear that

$$Z_1 = Z_{g1} + Z_{T1} + Z_{L1} + Z_{T1}$$
$$= j\,(0.20 + j\,0.10 + j\,0.138 + j\,0.10)$$
$$= j\,0.538\ pu$$
$$Z_2 = Z_{g2} + Z_{T2} + Z_{L2} + Z_{T2}$$
$$= j\,(0.15 + j\,0.10 + j\,0.138 + j\,0.10)$$
$$= j\,0.488\ pu$$
$$Z_0 = Z_{T0} = j\,0.10\ pu$$

Pre-fault voltage at $F = E_a = 1.0\ pu$.

(a) When 3-phase symmetrical fault occurs at F. In this case only the positive-sequence network need be considered

$$I_f = \dfrac{E_a}{Z_1} = \dfrac{1.0}{j\,0.538} = -j\,1.86\ pu.$$

(a) Positive-sequence network

Reference bus

E_a, Z_{g_1}, Z_{T_1}, Z_{L_1}, Z_{T_1}, F

(b) Negative-sequence network

Reference bus

Z_{g_2}, Z_{T_2}, Z_{L_2}, Z_{T_2}, F

(c) Zero-sequence network

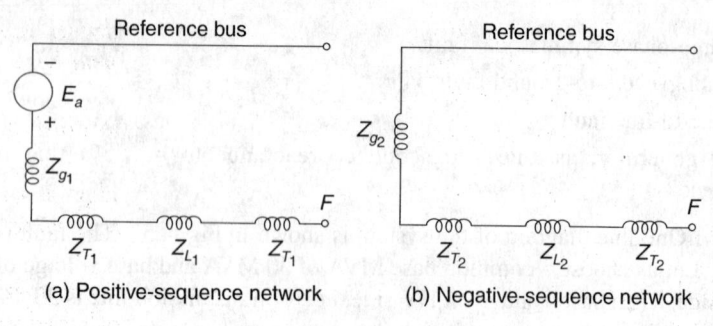

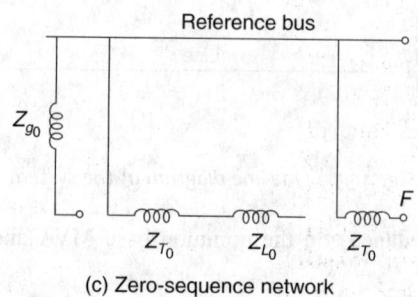

Reference bus

Z_{g_0}, Z_{T_0}, Z_{L_0}, Z_{T_0}, F

Fig. 4.54 *Sequence networks*

$$\text{Base current} = \frac{30 \times 10^3}{\sqrt{3} \times 6.6} = 2624 \text{ A}$$

Actual fault current in amperes = $1.86 \times 2624 = 4880$ A

(b) For single line-to-ground (L-G) fault

(Assuming 'a' phase to be faulted)

$$I_f = I_a = \frac{3E_a}{Z_1 + Z_2 + Z_0} = \frac{3 \times 1.0}{j(0.538 + 0.488 + 0.10)}$$

$$= \frac{3.0}{j\,1.126} = -j\,2.66\ pu$$

Fault current in amperes

$$I_f(A) = 2.66 \times 2624 = 6980 \text{ A}$$

(c) For line-to-line fault (Assuming fault between 'b' & 'c')

Fault currents are

$$I_b = \frac{-j\sqrt{3}\,E_a}{Z_1 + Z_2} = \frac{-j\sqrt{3} \times 1.0}{j(0.538 + 0.488)} = \frac{-j\,1.732}{j\,1.026} = -1.688\ pu$$

$$I_c = -I_b = \frac{j\sqrt{3}\,E_a}{Z_1 + Z_2} = \frac{j\,1.732}{j\,1.026} = 1.688\ pu.$$

Fault current in amperes

$$I_c = -I_b = 1.688 \times 2624 = 4429 \text{ A}$$

Fault Analysis

Example 4.20 | A 3-phase power system is represented by one-line diagram as shown in Fig. 4.55.

Fig. 4.55 System for Example 4.20

The rating of the equipment are

Generator G : 15 MVA, 6.6 kV, $X_1 = 15\%$, $X_2 = 10\%$
Transformers : 15 MVA, 6.6 kV delta/33 kV, $X_1 = X_2 = X_0 = 6\%$
Line reactances : $X_1 = X_2 = 2\,\Omega$ and $X_0 = 6\,\Omega$

Find the fault current for a ground fault on one of the busbars at B.

Solution: Selecting a common base MVA of 15 MVA and base voltage of 6.6 kV in the LV circuit, base voltage in transmission line is 33 kV.

The positive, negative and zero-sequence reactances of the line on selected base are

$$X_1 = X_2 = 2 \times \frac{15}{(33)^2} = 0.0275 \; pu$$

$$X_0 = 6 \times \frac{15}{(33)^2} = 0.0826 \; pu$$

The sequence reactances of various equipment are tabulated below:

	Positive-Seq. (X_1)	Negative-Seq. (X_2)	Zero-Seq. (X_0)
Generator G	0.15 pu	0.10 pu	–
Transformer T_1	0.06 pu	0.06 pu	0.06 pu
Transformer T_2	0.06 pu	0.06 pu	0.06 pu
Line	0.0275 pu	0.0275 pu	0.0826 pu

The sequence diagrams for the fault on one of the busbars at B are drawn in Fig. 4.56.

(a) Positive-sequence network

(b) Negative-sequence network

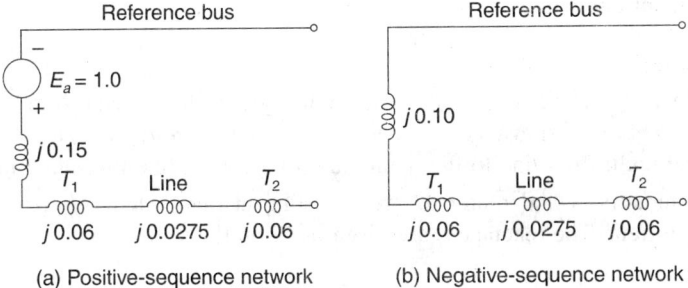

Fig. contd...

Fig. contd...

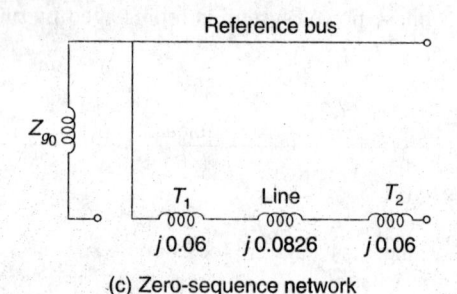

```
                    Reference bus
         ┌──────────────┬──────────────┬──────────────┐
         │              │              │              │
        Z₉₀            T₁    Line     T₂             
         │          ─000─  ─000─   ─000─
         └──────────  j0.06  j0.0826  j0.06
```

(c) Zero-sequence network

Fig. 4.56 Sequence diagrams

Form these, we get

$$Z_1 = j\,0.15 + j\,0.06 + j\,0.0275 + j\,0.06$$
$$= j\,0.2975\ pu$$
$$Z_2 = j\,0.10 + j\,0.06 + j\,0.0275 + j\,0.06$$
$$= j\,0.2475\ pu$$
$$Z_0 = j\,0.06 + j\,0.0826 + j\,0.06$$
$$= j\,0.2026$$

For line to ground fault on phase 'a' at busbars:

$$\text{Fault current } I_f = I_a = \frac{3E_a}{(Z_1 + Z_2 + Z_0)}$$

$$= \frac{3 \times 1.0}{j\,0.2975 + j\,0.2475 + j\,0.2026}$$

$$= \frac{3.0}{j\,0.7476}$$

$$= -j\,4.01\ pu$$

$$\text{Base current} = \frac{15 \times 10^3}{\sqrt{3} \times 6.6}$$

$$= 1312\ A$$

Fault current in amperes = 4.01×1312

$$= 5261\ A$$

Example 4.21 | The power system shown in Fig. 4.57 has a dead short circuit at the mid point of the transmission line. Find the fault current for (a) a single line-to-ground fault, (b) a line-to-line fault, and (c) a double line-to-ground fault.

Both generators G and motor M are operating at their rated voltage. Neglect prefault current. The reactances are given in *pu* on the same base.

Fault Analysis

```
X₁ = 0.30                                                      X₁ = 0.25
X₂ = 0.20                        X₁ = X₂ = 0.15                X₂ = 0.15
X₀ = 0.05    T₁        Line    F  X₀ = 0.40         T₂         X₀ = 0.05
    G──────3E────────────────────•───────────────────3E──────M
    Y       △ Y                                      Y △      Y
         x = 0.10                Fault                       x = 0.10
```

Fig. 4.57 System for Example 4.21

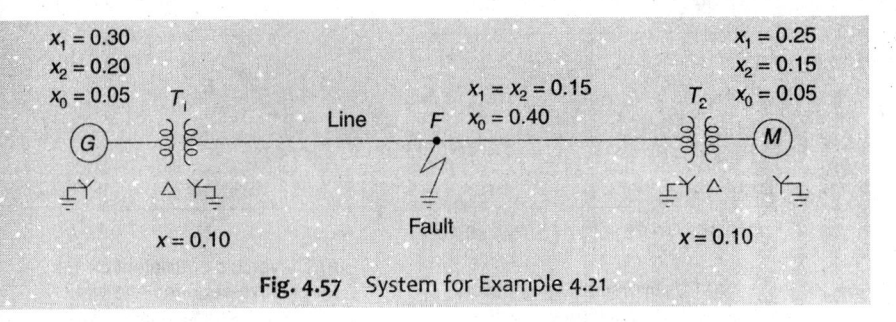

Solution: The sequence reactances of various equipment are tabulated below:

	Positive-Seq. (X_1)	Negative-Seq. (X_2)	Zero-Seq. (X_0)
Generator G	0.30 pu	0.20 pu	0.05 pu
Motor M	0.25 pu	0.15 pu	0.05 pu
Transformer T_1	0.10 pu	0.10 pu	0.10 pu
Transformer T_2	0.10 pu	0.10 pu	0.10 pu
Line	0.15 pu	0.15 pu	0.40 pu

For a fault at middle point of the line, i.e., at F, the sequence networks of the system are shown in Fig. 4.58(a), (b) and (c) and their corresponding Thevenin's equivalent circuits in Fig. 4.59(a), (b) and (c), In the positive sequence network $V_f = E_a$ is shown as the pre-fault voltage at point F.

Thevenin equivalent impedances of each of the sequence network as viewed from the fault point F are calculated from the sequence diagrams:

$$Z_1 = \frac{(j\,0.30 + j\,0.10 + j\,0.075)\,(j\,0.075 + j\,0.10 + j\,0.25)}{(j\,0.475 + j\,0.425)}$$

$$= \frac{j\,0.475 \times j\,0.425}{j\,0.90} = j\,0.224\ pu$$

$$Z_2 = \frac{(j\,0.20 + j\,0.10 + j\,0.075)\,(j\,0.075 + j\,0.10 + j\,0.15)}{(j\,0.375 + j\,0.325)}$$

$$= \frac{j\,0.375 \times j\,0.325}{j\,0.70} = j\,0.174$$

$$Z_0 = \frac{(j\,0.10 + j\,0.20)\,(j\,0.20 + j\,0.10)}{(j\,0.30 + j\,0.30)}$$

$$= \frac{j\,0.30 \times j\,0.30}{j\,0.60} = j\,0.15\ pu$$

Since pre-fault current is to be neglected, the pre-fault voltage at the fault point F, $V_f = E_a = 1.0\ pu$

Fig. 4.58 Sequence networks for fault at middle point

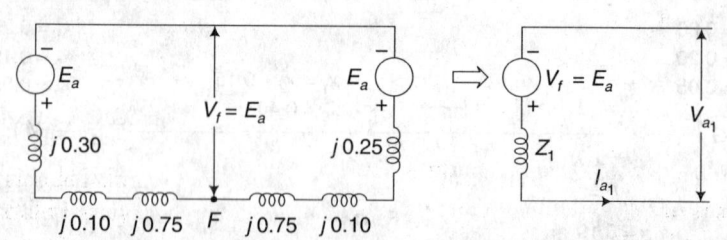

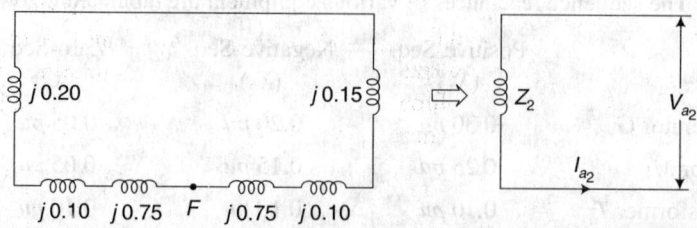

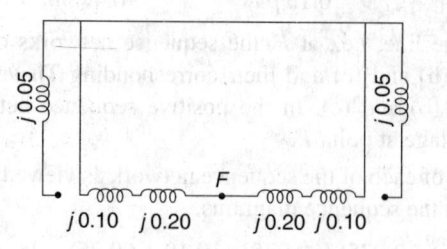

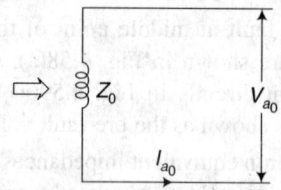

Fig. 4.59 Corresponding equivalent circuits

(a) Positive-sequence network
(a) Thevenin's equivalent circuit for positive-sequence network
(b) Negative-sequence network
(b) Thevenin's equivalent of negative-sequence network
(c) Zero-sequence network
(c) Thevenin's equivalent of zero-sequence network

(a) **Single line to ground fault**

$$I_f = I_a = \frac{3E_a}{Z_1 + Z_2 + Z_0}$$

$$= \frac{3 \times 1.0}{(j\,0.224 + j\,0.174 + j\,0.15)}$$

$$= \frac{3.0}{j\,0.548} = -j\,5.47\ pu$$

(b) **Line-to-line fault**

Assuming line-to-line fault between phases 'b' and 'c'

$$I_f = I_c = -I_b = \frac{j\sqrt{3}\,E_a}{Z_1 + Z_2}$$

$$= \frac{j\sqrt{3} \times 1.0}{(j\,0.224 + j\,0.174)}$$

$$= \frac{j\,1.732}{j\,0.398} = 4.35\ pu$$

(c) **Double line to ground fault**

Assuming the fault between lines b and c and ground.

$$I_{a1} = \frac{E_a}{Z_1 + Z_2 Z_0/(Z_2 + Z_0)}$$

$$= \frac{1.0}{\left[j\,0.224 + \dfrac{j\,0.174 \times j\,0.15}{(j\,0.174 + j\,0.15)}\right]}$$

$$= \frac{1.0}{\left[j\,0.224 + \dfrac{j\,0.174 \times j\,0.15}{j\,0.324}\right]}$$

$$= \frac{1.0}{(j\,0.224 + j\,0.080)}$$

$$= \frac{1.0}{j\,0.304} = -j\,3.29\ pu$$

$$I_{a2} = -I_{a1} \frac{Z_0}{(Z_2 + Z_0)}$$

$$= j\,3.29 \times \frac{j\,0.15}{j\,0.324} = j\,1.52$$

$$I_{a0} = -I_{a1} \frac{Z_2}{(Z_2 + Z_0)}$$

$$= j\,3.29 \times \frac{j\,0.174}{(j\,0.174 + j\,0.15)}$$

$$= j\,3.29 \times \frac{j\,0.174}{j\,0.324} = j\,1.76\ pu$$

Fault current in phase b

$$I_b = a^2 I_{a1} + a I_{a2} + I_{a0}$$

$$= (-0.5 - j\,0.866)(-j\,3.29) + (-0.5 + j\,0.866)(j\,1.52) + j\,1.76$$

$$= j\,1.64 - 2.85 - j\,0.76 - 1.32 + j\,1.76$$

$$= (-4.17 + 2.64)\ pu$$

Fault current in phase c

$$I_c = a I_{a1} + a^2 I_{a2} + I_{a0}$$

$$= (-0.5 + j\,0.866)(-j\,3.29) + (-0.5 - j\,0.866)(j\,1.52) + j\,1.76$$

$= j\,1.64 + 2.85 - j\,0.76 + 1.32 + j\,1.76$

$= (4.17 + j\,2.64)\,pu$

Fault current in ground

$= (I_b + I_c) = 3Ia_0$

$= 3 \times j\,1.76 = j\,5.28\,pu$

Example 4.22 | A 15 MVA, 6.6 kV star connected generator has positive, negative and zero sequence reactance of 20%, 20% and 10% respectively. The neutral of the generator is grounded through a reactor with 5% reactance based on generator rating. A line-to-line fault occurs at the terminals of the generator when it is operating at rated voltage. Find currents in the line and also in the generator reactor

(a) When the fault does not involve the ground

(b) When the fault is solidly grounded

Solution: Let us choose the rating of the generator as base, i.e. base MVA of 15 MVA and base voltage of 6.6 kV. The generator with fault between lines 'b' and 'c' is shown in Fig. 4.60.

(a) When the fault does not involve ground per unit impedances of the generator are

$Z_1 = j\,0.20\,pu$

$Z_2 = j\,0.20\,pu$

$Z_0 = Z_{g0} + 3\,Z_n$

$= j\,0.10 + 3 \times j\,0.05$

$= j\,0.25\,pu$

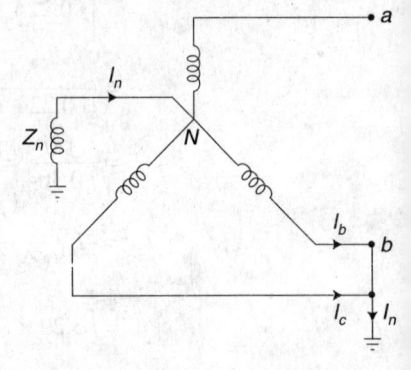

Fig. 4.60 System for Example 4.22

Assuming line to line fault between phases 'b' and 'c', the currents are

$I_a = 0$

$I_b + I_c = 0$

The symmetrical components of the current are

$I_{a1} = \dfrac{E_a}{Z_1 + Z_2} = \dfrac{1.0}{j\,0.20 + j\,0.20} = \dfrac{1.0}{j\,0.40} = -j\,2.5\,pu$

$I_{a2} = -I_{a1} = j\,2.5\,pu$

$I_{a0} = 0$

Line currents are

$I_b = a^2 I_{a1} + a I_{a2} + I_{a0}$

$= a^2 I_{a1} - a I_{a1} = (a^2 - a)\,I_{a1}$

$= -j\,\sqrt{3}\,I_{a1} = (-j\,1.732) \times (-j\,2.5)$

$$= -4.33\, pu = 4.33 < 180° \, pu$$
$$I_c = -I_b = 4.33\, pu = 4.33 < 0° \, pu.$$

Base current $= \dfrac{15 \times 10^3}{\sqrt{3} \times 6.6} = 1312$ A

Line currents in amperes are
$$I_b = 4.33 \times 1312 < 180° = 5680 < 180°\ \text{A}$$
$$I_c = 4.33 \times 1312 < 0° = 5680 < 0°\ \text{A}$$

(b) When the fault is solidly grounded (Fig. 4.60)

For fault between phases 'b' and 'c' involving ground, we have

$$I_{a1} = \dfrac{E_a}{[Z_1 + Z_2 Z_0/(Z_2 + Z_0)]}$$

$$= \dfrac{1.0}{\left[j\,0.20 + \dfrac{j\,0.20 \times j\,0.25}{(j\,0.20 + j\,0.25)}\right]} = \dfrac{1.0}{j\,0.20 + j\,0.11}$$

$$= \dfrac{1.0}{j\,0.31} = -j\,3.22\ pu.$$

$$I_{a2} = -I_{a1}\dfrac{Z_0}{Z_2 + Z_0} = j\,3.22 \times \dfrac{j\,0.25}{j\,0.45} = j\,1.78\ pu$$

$$I_{a0} = -I_{a1}\dfrac{Z_2}{Z_2 + Z_0} = j\,3.22\,\dfrac{j\,0.20}{j\,0.45} = j\,1.43\ pu.$$

$$I_b = a^2 I_{a1} + a I_{a2} + I_{a0}$$
$$= (-0.5 - j\,0.866)(-j\,3.22) + (-0.5 + j\,0.866)(j\,1.78) + j\,1.43$$
$$= j\,1.61 - 2.79 - j\,0.89 - 1.54 + j\,1.43$$
$$= (-4.33 + j\,2.15)\ pu = 4.83 < 151°\ pu$$

Actual value in amperes
$$I_b = 4.83 \times 1312 = 6337\ \text{A}$$

$$I_c = a I_{a1} + a^2 I_{a2} + I_{a0}$$
$$= (-0.5 + j\,0.866)(-j\,3.22) + (-0.5 - j\,0.866)(j\,1.78) + j\,1.43$$
$$= j\,1.61 + 2.79 - j\,0.89 + 1.54 + j\,1.43$$
$$= 4.33 + j\,2.15 = 4.83 < 29°\ pu$$

Actual value in amperes
$$I_c = 4.83 \times 1312 = 6337\ \text{A}$$
$$I_n = (I_b + I_c) = 3 I_{a0} = 3 \times j\,1.43 = j\,4.29\ pu$$

Actual value in amperes
$$I_n = 4.29 \times 1312 = 5628\ \text{A}$$

Therefore line current in phases b and c will be 6337 A and neutral current 5628 A.

Example 4.23 | A 30 MVA, 13.8 kV generator with neutral grounded through a 1-ohm resistance, has a three-phase fault MVA of 200 MVA. Calculate the fault current and the terminal voltages for a single line-to-ground fault at one of the terminals of the generator. The negative and zero sequence reactances of the machine are 0.10 pu and 0.05 pu respectively. Neglect pre-fault current, and losses. Assume the pre-fault generated voltage at the rated value. The fault is of dead short-circuit type.

Solution: Let us choose base quantities to be the rating of the generator. So, base MVA = 30 and base voltage (line to line) = 13.8 kV.

$$\text{Base current} = \frac{30 \times 10^3}{\sqrt{3} \times 13.8} = 1255 \text{ A}$$

$$\text{Fault MVA} = \frac{1}{X''} \times \text{Base MVA}$$

or

$$200 = \frac{1}{X''} \times 30$$

(where, X'' = Sub-transient reactance in pu)

or

$$X'' = \frac{30}{200} = 0.15 \ pu$$

X_1 = positive sequence reactance under Sub-transient condition in pu
$\quad = X'' = 0.15 \ pu$
$X_2 = 0.10 \ pu, X_0 = 0.05 \ pu$

Let the single line-to ground (L-G) fault occurs on phase 'a' of the generator, as shown in Fig. 4.61.

The pu value of the grounding resistance

$$= 1 \times \frac{30}{(13.8)^2} = 0.157 \ pu$$

$Z_n = (0.157 + j\,0) \ pu$

The sequence impedances of the generator are

$Z_1 = jX_1 = j\,0.15 \ pu$
$Z_2 = jX_2 = j\,0.10 \ pu$
$Z_0 = Z_{g0} + 3Z_n$
$\quad = j\,0.05 + 3(0.157 + j\,0)$
$\quad = 0.471 + j\,0.05$

As the machine is unloaded before occurrence of the fault, the prefault voltage, $V_f = E_a = 1.0 < 0° \ pu$.

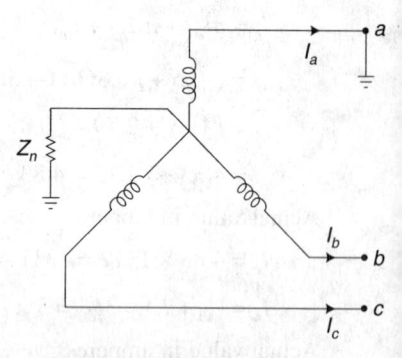

Fig. 4.61 System for Example 4.23

Since the fault is of dead short-circuit type, $Z_f = 0$

$$I_{a1} = I_{a2} = I_{a0} = \frac{E_a}{Z_1 + Z_2 + Z_0} = \frac{1.0 < 0°}{j(0.15 + 0.10) + (0.471 + j0.05)}$$

$$= \frac{1.0 < 0°}{0.471 + j0.30} = \frac{1.0 < 0°}{0.558 < 36°} = 1.792 < -36° \; pu$$

Fault current at phase $a = I_a = 3I_{a1} = 3 \times 1.792 = 5.376 \; pu$

Actual value of fault current in amperes

$$I_a = 5.376 \times 1255 = 6747 \; A$$

$$I_b = I_c = 0$$

The terminal voltages can be calculated in the following manner:

The sequence voltages are

$$V_{a1} = E_a - I_{a1}Z_1$$
$$V_{a2} = -I_{a2}Z_2 = -I_{a1}Z_2 \quad (\text{As } I_{a2} = I_{a1})$$
$$V_{a0} = -I_{a0}Z_0 = -I_{a1}Z_0 \quad (\text{As } I_{a0} = I_{a1})$$

So, line-to-neutral voltages at the terminals during fault are

$$V_a = V_{a1} + V_{a2} + V_{a0} = 0$$
$$V_b = a^2 V_{a1} + a V_{a2} + V_{a0}$$
$$V_c = a V_{a1} + a^2 V_{a2} + V_{a0}$$

So, the line voltages are

$$V_{ab} = V_a - V_b$$
$$V_{bc} = V_b - V_c$$
$$V_{ca} = V_c - V_a$$

Example 4.24 Three 6.6 kV, 10 MVA, three-phase, star connected generators are connected to a common set of busbars. Each generator has a reactance to positive sequence current of 18%. The reactance to negative and zero-sequence currents are 75% and 30% of the positive sequence value. If a ground fault occurs on one busbar, determine the fault current (a) if all the generator neutrals are solidly grounded, (b) if only one of the generator's neutral is solidly grounded and others are isolated, and (c) if one of the generator's neutral is grounded through a resistance of 1.5 ohm and the others are isolated.

Solution: Let us assume base quantities to be the rating of the generator. So, base MVA = 10 and base voltage = 6.6 kV (line to line).

$$\text{Base current} = \frac{(\text{MVA})_b \times 10^3}{\sqrt{3} \; (\text{kV})_b} = \frac{10 \times 10^3}{\sqrt{3} \times 6.6} = 875 \; A$$

(a) All the generator neutrals are solidly grounded and a ground fault occurs on one busbar i.e. on one phase (say phase 'a') as shown in Fig. 4.62(a).

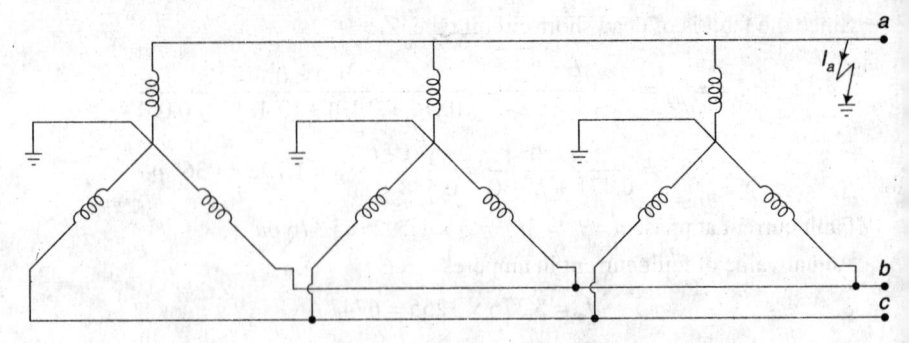

Fig. 4.62(a) Ground fault on one busbar

The sequence impedances of the individual generators are

$$Z_{g1} = j\,0.18\ pu$$
$$Z_{g2} = j\,0.18 \times 0.75 = j\,0.135\ pu$$
$$Z_{g0} = j\,0.18 \times 0.30 = j\,0.054\ pu.$$

When the three generators are operating in parallel, the resultant impedances will be $\frac{1}{3}$ rd, i.e.,

$$Z_1 = j\,\frac{0.18}{3} = j\,0.06\ pu$$
$$Z_2 = j\,\frac{0.135}{3} = j\,0.045\ pu$$
$$Z_0 = j\,\frac{0.054}{3} = j\,0.018\ pu$$

Therefore for an L-G fault, we get

Fault current,
$$I_f = I_a = 3I_{a1} = \frac{3E_a}{Z_1 + Z_2 + Z_0}$$
$$= \frac{3 \times 1}{j\,(0.06 + 0.045 + 0.018)}$$
$$= \frac{3}{j\,0.123} = -j\,24.39\ pu$$

Actual fault current in amperes
$$= -j\,24.39 \times 875 = -j\,21,3$$
$$= -j\,21,341\ A$$

(b) Refer Fig. 4.62(b). In this case the neutral of only one generator is solidly grounded and that of others are isolated. Since the neutrals of the second and third generators are isolated, their zero sequence impedances do not come into picture. The positive and negative sequence impedances will be the same as in case (a).

Therefore, $Z_1 = j\,0.06\ pu$, $Z_2 = j\,0.045\ pu$ and

$$Z_0 = Z_{g0} = j\,0.054\ pu$$

Therefore, fault current $I_f = I_a = \dfrac{3E_a}{Z_1 + Z_2 + Z_0}$

$$= \dfrac{3 \times 1}{j\,0.06 + j\,0.045 + j\,0.054}$$

$$= \dfrac{3}{j\,0.159} = -j\,18.87\ pu$$

Actual value of fault current in amperes $= -j\,18.87 \times 875$

$$= -j\,16{,}511\ A$$

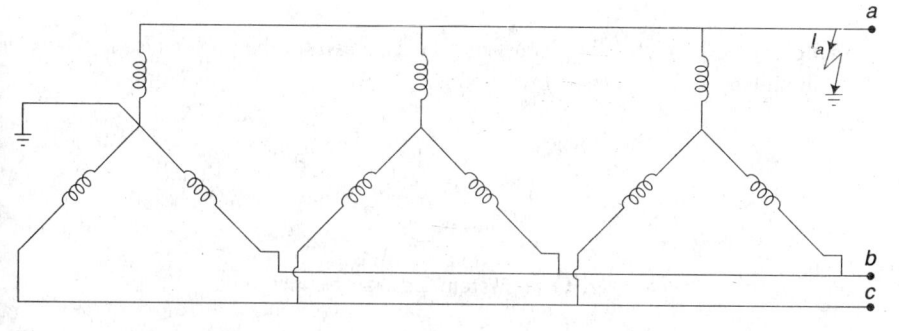

Fig. 4.62(b) One generator's neutral grounded, others isolated

(c) Refer Fig. 4.62(c). In this case

The pu value of the grounding resistance

$$= 1.5 \times \dfrac{10}{(6.6)^2} = 0.344\ pu$$

Hence,

$$Z_n = (0.344 + j\,0)\ pu$$

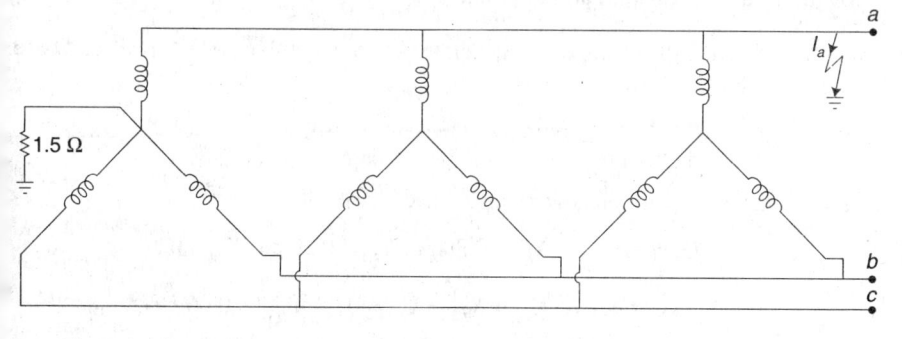

Fig. 4.62(c) One generator's neutral grounded through 1.5 Ω, others isolated

$$Z_0 = Z_{g0} + 3Z_n = j\,0.054 + 3\,(0.344 + j\,0)\ pu$$

$$= (1.032 + j\,0.054)$$

Fault current, $I_f = I_a = \dfrac{3 \times 1}{j\,0.06 + j\,0.045 + (1.032 + j\,0.054)}$

$= \dfrac{3}{1.032 + j\,0.159}$

$= (2.84 - j\,0.437)\ pu$

Actual fault current in amperes $= (2.84 - j\,0.437) \times 875$ A

$= (2485 - j\,382)$ A

Magnitude of the fault current $= \sqrt{(2485)^2 + (382)^2}$

$= 2514$ A

Example 4.25 | The following data may be assumed for the power system whose one-line diagram is shown in Fig. 4.63.

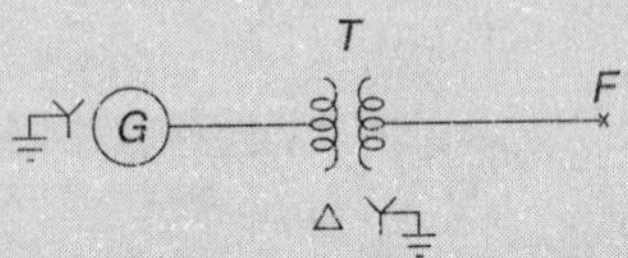

Fig. 4.63 System for Example 4.25

Device	MVA capacity	Line voltage (kV)	% Reactance Positive sequence	% Reactance Negative sequence	% Reactance Zero sequence
Generator G	30	11	30	20	10
Transformer T	40	11/110	8	8	8

The given values of percentage reactances are based on the rating of each particular device. If a double line to ground fault occurs at F, find the current flowing in the conductors at that point.

Solution: Selecting the rating of the generator as base, i.e., base MVA = 30, and base voltage = 11 kV

Base current $= \dfrac{30 \times 10^3}{\sqrt{3} \times 11} = 1575$ A

Pu reactances of various devices on selected base

Generator $G : X_{g1} = 0.30\ pu,\ X_{g2} = 0.20\ pu,\ X_{g0} = 0.10\ pu$

Transformer $T : X_{T1} = X_{T2} = X_{T0} = 0.08 \times \dfrac{30}{40} = 0.06\ pu$

Let the fault be between phases 'b' and 'c' involving the ground, then the sequence diagrams will be as shown in Fig. 4.64.

Fault Analysis

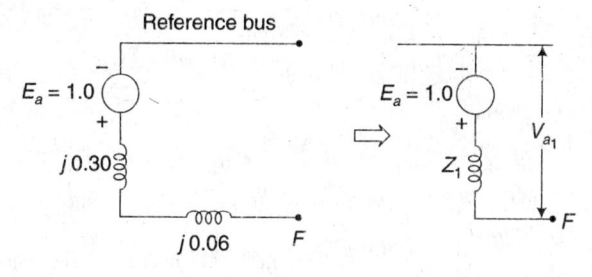

(a) Positive-sequence network

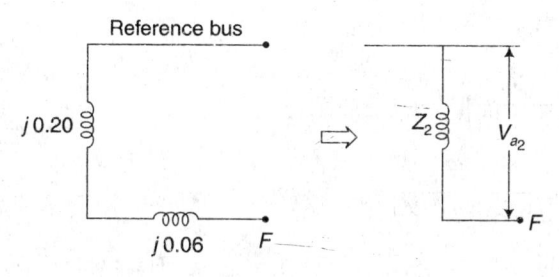

(b) Negative-sequence network

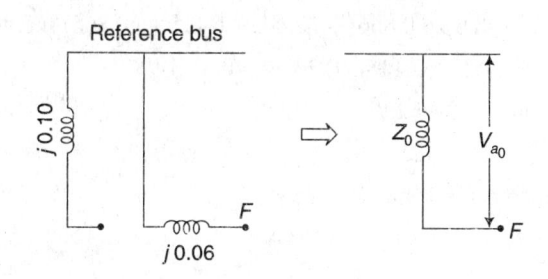

(c) Zero-sequence network

Fig. 4.64 *Sequence networks*

From which, $Z_1 = jX_{g1} + jX_{T1} = j\,0.30 + j\,0.06 = j\,0.36\ pu$

$Z_2 = jX_{g2} + jX_{T2} = j\,0.20 + j\,0.06$

$\qquad = j\,0.26\ pu$

$Z_0 = jX_{T0} = j\,0.06\ pu$

Connection of sequence networks for a double line to ground fault at F is shown in Fig. 4.65

From Fig. 4.65, symmetrical components of line currents are

$$I_{a1} = \dfrac{E_a}{\left[Z_1 + \dfrac{Z_2 Z_0}{(Z_2 + Z_0)}\right]} = \dfrac{1.0}{\left[j\,0.36 + \dfrac{j\,0.26 \times j\,0.06}{(j\,0.26 + j\,0.06)}\right]}$$

$$= \frac{1.0}{(j\,0.36 + j\,0.049)}$$

$$= \frac{1.0}{j\,0.409} = -j\,2.44\ pu.$$

$$I_{a2} = -I_{a1}\left(\frac{Z_0}{Z_2 + Z_0}\right)$$

$$= j\,2.44 \times \frac{j\,0.06}{(j\,0.26 + j\,0.06)}$$

$$= j\,0.46\ pu$$

$$I_{a0} = -I_{a1}\left(\frac{Z_2}{Z_2 + Z_0}\right)$$

$$= j\,2.44 \times \frac{j\,0.26}{(j\,0.26 + j\,0.06)} = j\,1.98\ pu$$

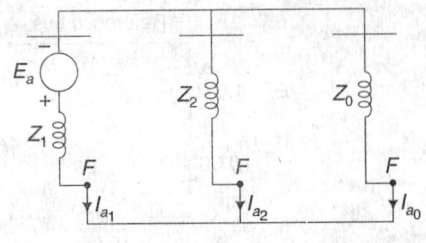

Fig. 4.65 Connection of sequence networks for a double line to ground fault at F

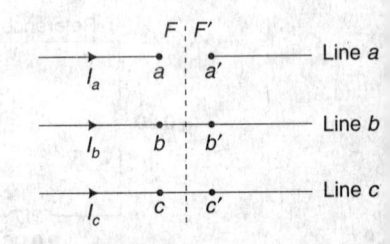

Fig. 4.66(a) One conductor open

Line currents are

$$I_b = a^2 I_{a1} + a I_{a2} + I_{a0}$$

$$= (-0.5 - j\,0.866)(-j\,2.44) + (-0.5 + j\,0.866)(j\,0.46) + j\,1.98$$

$$= j\,1.22 - 2.11 - j\,0.23 - 0.40 + j\,1.98$$

$$= -2.51 + j\,2.97$$

$$= 3.89 < 125°\ pu$$

Current in amperes $= 3.89 \times 1575 < 125°$

$$= 6127 < 125°\ A$$

$$I_c = a I_{a1} + a^2 I_{a2} + I_{a0}$$

$$= (-0.5 + j\,0.866)(-j\,2.44) + (-0.5 - j\,0.866)(j\,0.46) + j\,1.98$$

$$= j\,1.22 + 2.11 - j\,0.23 + 0.40 + j\,1.98$$

$$= 2.51 + j\,2.97$$

$$= 3.89 < 55°$$

Current in amperes $= 3.89 \times 1575 < 55°$

$$= 6127 < 55°\ A$$

4.7.4 Series Faults

Faults discussed in previous sections are shunt type faults which result from short-circuits. A break in one or two conductors in a 3-phase circuit leads to an open conductor fault which is in series with the line. When fuses are used in 3-phase circuits, blowing off of fuses in one or two phases may lead to an open conductor fault. Open-conductor faults are called series type of faults. The unbalanced conditions created by open conductor faults in the system can also be analyzed by the method of symmetrical components.

One Conductor Open

Let the conductor a be broken between points $a - a'$ leading to one conductor open condition as shown in Fig. 4.66(a). The system ends on the sides of the fault are identified as F, F', whereas the conductor ends are identified as aa', bb' and cc'.

The current and voltages at the fault point are given by

$$I_a = 0 \qquad (4.143)$$
$$V_{bb'} = V_{cc'} = 0 \qquad (4.144)$$

The current and voltages in terms of symmetrical components are expressed as

$$I_{a1} + I_{a2} + I_{a0} = 0 \qquad (4.145)$$

and

$$\begin{bmatrix} V_{aa'1} \\ V_{aa'2} \\ V_{aa'0} \end{bmatrix} = \frac{1}{3}\begin{bmatrix} 1 & a & a^2 \\ 1 & a^2 & a \\ 1 & 1 & 1 \end{bmatrix}\begin{bmatrix} V_{aa'} \\ 0 \\ 0 \end{bmatrix}$$

or

$$V_{aa'1} = V_{aa'2} = V_{aa'0} = \frac{1}{3} V_{aa'} \qquad (4.146)$$

Fig. 4.66(b) Connection of sequence networks for one conductor open

Equations (4.145) and (4.146) provide the basis for interconnection of sequence networks in parallel as shown in Fig. 4.67.

Two Conductors Open

Figure 4.67 shows the fault at FF' with conductors b and c open. The currents and voltages under this fault condition are expressed as

$$V_{aa'} = 0 \qquad (4.147)$$
$$I_b = I_c = 0 \qquad (4.148)$$

Currents and voltages in terms of symmetrical components can be written as

$$V_{aa'1} + V_{aa'2} + V_{aa'0} = 0 \qquad (4.149)$$

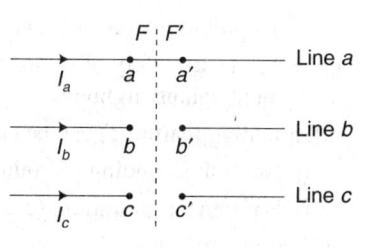

Fig. 4.67 Two conductors open

and

$$\begin{bmatrix} I_{a1} \\ I_{a2} \\ I_{a0} \end{bmatrix} = \frac{1}{3}\begin{bmatrix} 1 & a & a^2 \\ 1 & a^2 & a \\ 1 & 1 & 1 \end{bmatrix}\begin{bmatrix} I_a \\ 0 \\ 0 \end{bmatrix}$$

or

$$I_{a1} = I_{a2} = I_{a0} = \frac{1}{3} I_a \qquad (4.150)$$

Equations (4.159) and (4.150) suggest a series connection of positive, negative and zero sequence networks as shown in Fig. 4.68. Sequence currents and voltage can now be computed.

4.8 GROUNDING (EARTHING)

In a power system, the term 'grounding' or 'earthing' means electrically connecting the neutrals of the system (i.e., the neutral points of star-connected 3-phase windings of power transformers, generators, motors, earthing transformers, etc.) or non-current carrying metallic parts of the electrical equipment to the general mass of earth (ground). Since the large majority of faults in the power system involve ground, power system grounding has a significant effect on the protection of all the components of the power system. Hence power system grounding is very important. A number of points from generating stations to consumer's installations are grounded (earthed).

The principal purposes of grounding are as follows:

(i) To minimize transient over voltages due to lightning and ground faults
(ii) To allow sufficient fault current to flow safely for proper operation of protective relays during ground faults
(iii) To eliminate persistent arcing grounds
(iv) To provide improved service reliability
(v) To ensure safety of personnel against electric shocks and protection of equipment against lightning

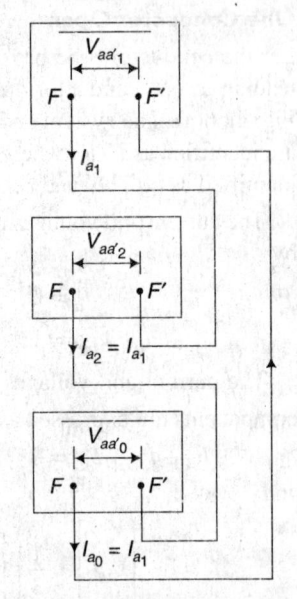

Fig. 4.68 Connection of sequence networks for two conductors open

Grounding (earthing) can be classified into the following two categories:

(i) Neutral grounding (earthing) or system grounding (earthing)
(ii) Equipment grounding (earthing) or safety grounding (earthing)

Neutral grounding deals with the grounding of the system neutrals to ensure system security and protection. Grounding of neutrals in power stations and substations belongs to this category. Neutral grounding is also called, 'system grounding'. Equipment grounding deals with grounding of non-current carrying metallic parts of the equipment to ensure safety of personnel and protection against lightning. Equipment grounding is also called 'safety grounding'.

4.8.1 Terms and Definitions

(i) Ground Electrode (Earth Electrode) Any wire, rod, pipe, plate or an array of conductors, embedded in ground (earth) horizontally or vertically is known as the ground electrode or earth electrode. For distribution systems, the ground electrode may consist of a rod, about 1m long, driven vertically into the ground. In power stations and substations, elaborate grounding systems known as ground mats or grounding grids are used rather than individual rods.

(ii) Resistance of Ground Electrode The resistance offered by the ground electrode to the flow of current into the ground is called resistance of ground electrode. This resistance is not the ohmic resistance of the electrode but represents the resistance of the mass of ground (earth) surrounding the ground electrode. Its numerical value is equal to the ratio of the voltage of ground electrode with respect to a remote point to the current dissipated by it.

(iii) Ground Current (Earth Current) The current dissipated by the ground electrode into the ground is called the ground current or earth current.

(iv) Ground Mat (Ground Grid or Ground Mesh) It is formed by placing mild steel bars placed in X and Y directions in mesh formation in the soil at a depth of about 0.5 m below the surface of substation floor in the entire area of the substation except the foundations. The crossings of the horizontal bars in X and Y directions are welded. In order to form ground electrodes, several vertical galvanised steel pipes are inserted in the ground and their heads are solidly connected to the ground mat by means of horizontal grounding rods/grounding strips. Ground mat (ground grid or ground mesh) is also called earth mat (earth grid or earth mesh).

(v) Grounding Risers (Earthing Risers) These are generally mild steel rods bent in vertical and horizontal shapes and welded to the ground mat at one end and brought directly upto equipment/structure foundation.

(vi) Grounding Connections These are in the form of galvanised steel strips or electrolytic copper flats or strips/stranded wires and are used for final connection (bolted/clamped/welded) between the grounding riser and the points to be grounded.

(vii) Step Potential (Step Voltage) It is the voltage between the feet of a person standing on the floor of the substation, with 0.5 m spacing between two feet (one step), during the flow of ground fault current through the grounding system.

(viii) Touch Potential (Touch Voltage) It is the voltage between the fingers of raised hand touching the faulted structure and the feet of the person standing on a substation floor. The value of touch potential should be such a low value that the person should not get a shock even if the grounded structure is carrying a fault current.

(ix) Mesh Potential It is the maximum touch potential within a mesh of the grid.

(x) Transferred Potential It is a special case of touch potential where a potential is transferred into or out of the substation.

(xi) Arcing Ground It is intermittent, repeating line-to-ground arc through air insulation on overhead line/exposed conductor due to charging and discharging of line to ground capacitance.

(xii) Coefficient of Grounding (C_g) The coefficient of grounding (C_g) is defined as the ratio of the highest line-to-ground voltage of the healthy lines during a single line-to-ground (L-G) fault to the normal line to line system voltage. The coefficient of grounding is also known as 'coefficient of earthing'.

(xiii) Effectively Grounded System A system or a portion thereof is said to be effectively grounded when, for all points on the system or specified portion thereof,

the ratio of zero-sequence reactance to positive sequence reactance is not greater than three (i.e., $X_0/X_1 \leq 3$) and the ratio of zero-sequence resistance to positive sequence reactance is not greater than one (i.e., $R_0/X_1 \leq 1$) for any condition of operation. The coefficient of grounding, in this case, is less than 80% (i.e., $c_g < 80\%$). Thus, the maximum line to ground voltage for effectively grounded system does not exceed 80% of the line voltage. When the neutral is directly connected to the ground without any intentional resistance or reactance (i.e., the neutral is solidly grounded), the condition for effective grounding is satisfied since $Z_n = 0$. Hence, the term 'effectively grounded' is now used in place of the older term solidly grounded' for the reason of definition.

(xiv) Non-effective Grounding When an intentional resistance or reactance is connected between neutral and ground, it is called non-effective grounding. The coefficient of grounding for non-effective grounding is greater than 80% (i.e., $c_g > 80\%$). Hence in this case, the maximum line to ground voltage of healthy lines in the event of single line-to-ground fault is more than 80% of line to line voltage.

(xv) Grounding Transformer A transformer intended primarily to provide a neutral point for grounding purposes is called 'grounding transformer'. Grounding transformer is also called 'earthing transformer'.

4.8.2 Neutral Grounding (Neutral Earthing)

Neutral grounding or neutral earthing means electrically connecting the neutrals of the system (i.e., the neutral points of star-connected 3-phase windings of power transformers, generators, motors, grounding transformers etc.) to low resistance ground (ground electrode/ground mat) either directly or through some circuit element (resistance or reactance). The neutral grounding is an important aspect of the power system design since the performance of the system in terms of the faults, stability, protection, etc., is greatly affected by the state of the neutral. Neutral grounding does not have any effect on the operation of a 3-phase system under balanced steady-state conditions.

However, the currents and voltages during ground fault conditions are greatly influenced by the type of grounding of neutral. There are three classes of neutral grounding: ungrounded (isolated neutral), impedance-grounded (resistance or reactance), and effectively grounded (neutral solidly grounded). An ungrounded neutral system is connected to ground through the natural shunt capacitance. Neutral grounding (neutral earthing) is also called 'system grounding (system earthing)'.

In earlier years, power systems were operated without neutral grounding. Such systems were called ungrounded systems or isolated neutral systems. The operation of the ungrounded system was thought to be natural at that time because the ground connection was not useful for the actual transfer of power. Many difficulties were being encountered in ungrounded systems as the systems were growing in terms of power transmitted, voltage level and distance of transmission. Ungrounded systems experience repeated arcing grounds and overvoltages of the healthy lines during single line to ground fault. The overvoltage causes stress on the insulation of the lines and equipment and may result in insulation breakdown. The ground fault protection of ungrounded systems is difficult because the capacitive fault current is so small in magnitude that the same cannot operate the ground fault relays and hence the ground fault cannot be easily sensed. Due to these disadvantages, ungrounded systems are not used these days.

The modern power systems operate with neutral grounding at every voltage level. In present-day power systems, there are various voltage levels between generation and distribution. It is desirable to provide at least one grounding at each voltage level.

Neutral grounding can be broadly classified into the following two categories:
(i) Effective grounding
(ii) Non-effective grounding

In effective grounding, the neutral is directly connected to ground without inserting any intentional impedance (resistance or reactance). Effective grounding was previously called 'solid grounding'. The coefficient of grounding of effectively grounded system is less than 80%. In non-effective grounding, the neutral is connected to ground through impedance (resistance or reactance). The coefficient of grounding of non-effectively grounded system is greater than 80%.

Ungrounded System

This is power system without an intentional connection to ground. However, it is connected to ground through the natural shunt capacitance of the system to ground. The ungrounded system is also called *isolated neutral system*.

In an isolated neutral system, the voltage of the neutral is not fixed and may float freely. Under balanced system conditions, the voltage of the neutral is held at ground due to the presence of the natural shunt capacitance of the system. The three-phase conductors have, then, the phase voltages with respect to ground. Thus, as shown in Fig. 4.69(a), in the normal balanced system, neutral (N) equals ground (G) and hence, $V_{aN} = V_{aG}$, $V_{bN} = V_{bG}$, and $V_{cN} = V_{cG}$. When a ground fault occurs on any conductor, the phase-to-neutral voltages and the phase-to-ground voltages are quite different.

With a ground fault on any line conductor, the faulted line conductor assumes the voltage of ground and so, the voltage of the neutral with respect to ground attains the voltage of the conductor. This is illustrated in Fig. 4.69(b). The neutral is thus, shifted from ground and the two healthy lines b and c will experience the line voltage (i.e., $\sqrt{3}$ times the phase voltage) with respect to ground.

From Fig. 4.69(b), the voltage drop around the right-hand triangle bGN is given by.

$$V_{bG} - V_{bN} - V_{NG} = 0 \qquad (4.151)$$

and around the left-hand triangle CGN

$$V_{CG} - V_{CN} - V_{NG} = 0 \qquad (4.152)$$

Also,

$$V_{aN} + V_{NG} = 0 \qquad (4.153)$$

From the basic equations,

$$V_{aG} + V_{bG} + V_{cG} = 3 V_0 \quad \text{(Where, } V_0 = \text{zero-sequence voltage)} \quad (4.154)$$

$$V_{aN} + V_{bN} + V_{cN} = 0 \qquad (4.155)$$

202 Power System Protection and Switchgear

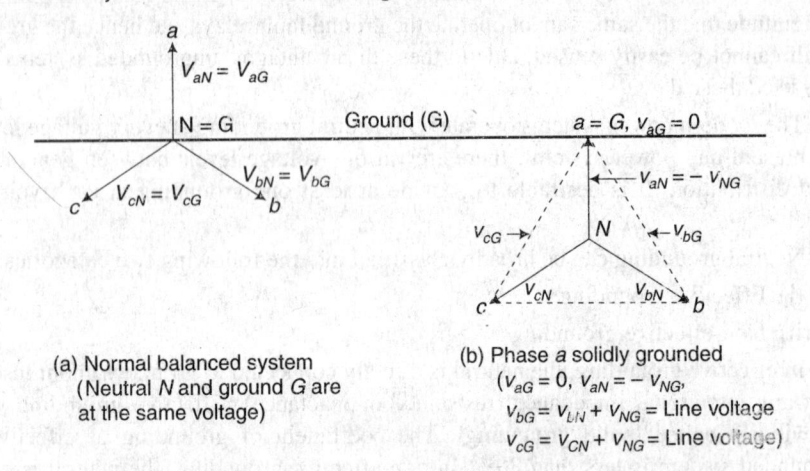

Fig. 4.69 Voltage shift for a single line-to-ground fault on phase 'a' of an ungrounded system

(a) Normal balanced system (Neutral N and ground G are at the same voltage)

(b) Phase a solidly grounded
($V_{aG} = 0$, $V_{aN} = -V_{NG}$,
$V_{bG} = V_{LN} + V_{NG}$ = Line voltage
$V_{cG} = V_{cN} + V_{NG}$ = Line voltage)

Subtracting Eq. (4.155) from (4.154), substituting Eqs. (4.151) through (4.153), and with $V_{aG} = 0$, we get

$$V_{aG} - V_{aN} + V_{bG} - V_{bN} + V_{cG} - V_{cN} = 3V_0$$

or
$$V_{NG} + V_{NG} + V_{NG} = 3V_0$$

or
$$3V_{NG} = 3V_0$$

or
$$V_{NG} = V_0 \tag{4.156}$$

Thus the neutral shift is the zero-sequence voltage (V_0). In the balanced system of Fig. 4.69(a), $N = G$ and V_0 is zero and there is no neutral shift.

A simple ungrounded neutral system is shown in Fig. 4.70(a). The line conductors have capacitances between one another and to ground, the former being delta-connected while the latter are star-connected. The delta-connected capacitances have little effect on the grounding characteristics of the system and, therefore, can be disregarded. The circuit then reduces to the one shown in Fig. 4.70(b).

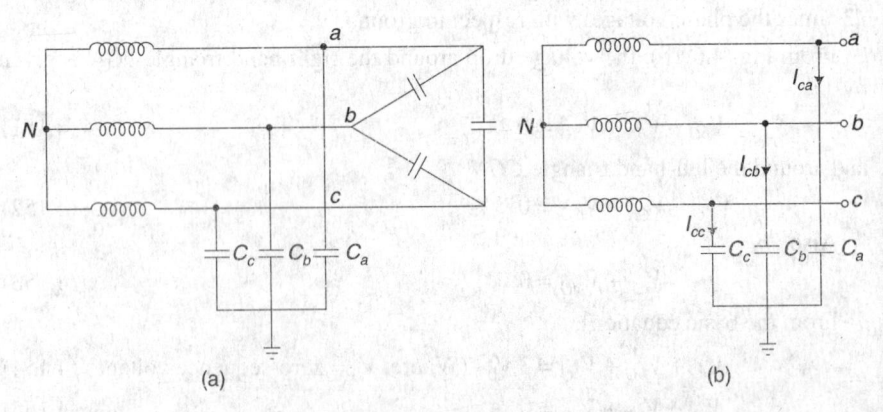

Fig. 4.70 Ungrounded neutral system

In the system having perfectly transposed line, each conductor will have the same capacitance to ground, i.e., $c_a = c_b = c_c = c$. Therefore, under normal conditions (i.e., steady state and balanced conditions), the line to neutral charging currents I_{ca}, I_{cb}, I_{cc} will form a balanced set of currents as shown in Fig. 4.71. The phasors V_{aN}, V_{bN} and V_{cN} which represent the phase to neutral voltage of each phase have the same magnitude and are displaced from one another by 120°. The charging currents I_{ca}, I_{cb} and I_{cc} lead their respective phase voltages by 90° and are equal.

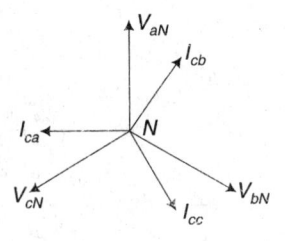

Fig. 4.71 Phasor diagram of Fig. 4.70(b)

Let, $\qquad V_{aN} = V_{bN} = V_{cN} = V_p \qquad$ (Phase Voltage)

In magnitude each of the charging current is

$$I_{ca} = I_{cb} = I_{cc} = I_c = \frac{V_p}{X_c} = \frac{V_p}{1/w_c} = V_p w c \qquad (4.157)$$

Where, $\quad V_p$ = Phase voltage $\qquad$ (i.e., line to neutral voltage)

$X_c = \dfrac{1}{w_c}$ = capacitive reactance of the line to ground.

The resultant of balanced phasor currents is zero and so no current flows to ground.

In case of a line-to-ground fault in line 'a' at point F, the circuit becomes as shown in Fig. 4.72(a). Under this fault condition, the faulty line a takes up the ground voltage while the voltages of the remaining two healthy lines b and c rise from phase values to line value.

When a line-to-ground fault occurs on any line, the capacitance to ground of that line gets shorted and does not come into picture for analysis. That is why the capacitance to ground of line a is not shown in Fig. 4.72(a). The capacitive currents becomes unbalanced and the components of the capacitive fault current I_f flow through the healthy lines b and c, and their capacitances and returns to the system via ground, the fault and the faulty line a. Thus the capacitive fault current (i.e., the current in line a) has two components: one, I_{ba}, through line b, capacitance c_b and the fault to line a; second, I_{ca}, through line c, capacitance c_c and the fault to line a. The voltages driving the currents I_{ba} and I_{ca} are V_{ba} and V_{ca} respectively. These currents lead their respective voltages by 90° and their phasor sum is equal to the fault current I_f as shown in the phasor diagram of Fig. 4.72(b).

$$I_{ba} = \frac{V_{ba}}{X_c} = \frac{\sqrt{3} V_p}{X_c} = \frac{\sqrt{3} V_p}{1/w_c} = \sqrt{3} V_p wc = \sqrt{3} I_c \qquad (4.158)$$

Similarly $\quad I_{ca} = \dfrac{V_{ca}}{X_c} = \dfrac{\sqrt{3} V_p}{X_c} = \dfrac{\sqrt{3} V_p}{1/wc} = \sqrt{3} V_p wc = \sqrt{3} I_c \qquad (4.159)$

Where $I_c = V_p wc$, is the per phase capacitive (charging) current under normal conditions.

The capacitive fault current I_f in line a is the phasor sum of I_{ba} and I_{ca}.

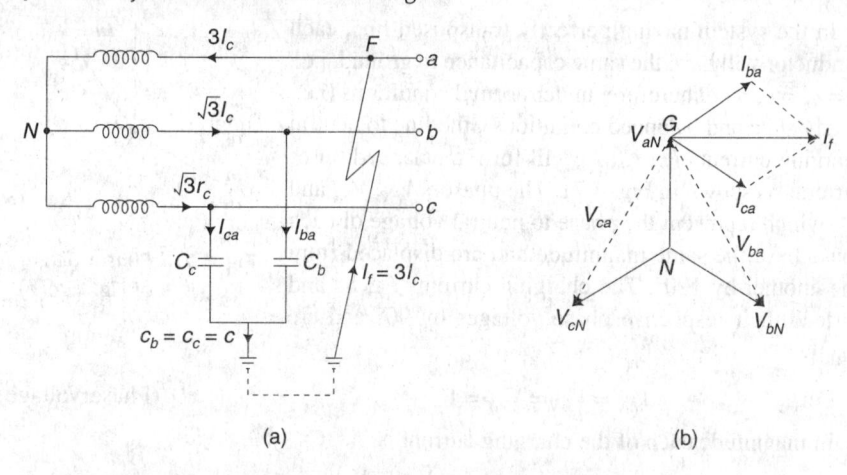

Fig. 4.72 (a) A ground fault on one of the lines of the ungrounded system.
(b) Phasor diagram for fault on line 'a'

From the phasor diagram of 4.72(b),

$$I_f = \sqrt{3}I_{ba} = \sqrt{3}I_{ca} = \sqrt{3} \times \sqrt{3}I_c = 3I_c \qquad (4.160)$$

$$= 3 \times \text{Per phase capacitive current under normal conditions.}$$

Thus the capacitive fault current I_f which is three times the normal per phase capacitive current I_c, flows into the ground and in the faulty line. After occurrence of the ground fault, when the gap between the fault point F and the ground breaks down and the fault path becomes ionized, the capacitance discharges through the fault and capacitive current flows. If the capacitive current exceeds about 4 amperes, it is sufficient to maintain an arc in the ionized path of the fault, even though the medium causing the fault has cleared itself. The repeated charging and discharging of the line to ground capacitance results in the flow of capacitive current and repeated arcs between line and ground. The persistence of the arc due to the flow of capacitive current gives rise to a phenomenon known as "arcing grounds".

The cyclic charging and discharging of the system capacitance through the fault under arcing ground conditions results in severe voltage oscillations reaching 5 to 6 times normal voltage. The build-up of high voltages may result in insulation breakdown. Thus, a temporary fault may grow into a permanent fault due to arcing grounds.

The following conclusions can be drawn from the performance of an ungrounded system during a line to ground fault.

(i) Under a single line-to-ground fault, the voltage of the faulty phase (line) becomes equal to ground voltage and the voltages of the two remaining healthy phases (lines) with respect to ground rise from their normal phase to neutral voltages to full line value (i.e., $\sqrt{3}$ times the normal phase value). This causes stress on the insulation.

(ii) The voltage of the neutral is not fixed and may float freely. Hence neutral is not stable.

(iii) The capacitive current in the two healthy phases increases to $\sqrt{3}$ times the normal value.
(iv) The capacitive fault current (I_f) in the faulty phase is 3 times the normal per phase capacitive current.
(v) Persistent arcing ground occurs. Due to arcing ground, the system capacitance is charged and discharged in a cyclic order. This results in high-frequency oscillations being superimposed on the whole system and the phase-voltage of healthy phases may rise to 5 to 6 times its normal value. The overvoltages may result in insulation breakdown.
(vi) The ground fault protection of ungrounded systems becomes difficult because the capacitive fault current is so small in magnitude that it may not be sufficient to operate-protective devices (i.e., ground fault relays).
(vii) Because of the capacitive fault current being small, there will be very little effect on the neighbouring communication circuit. However, the long duration of the arc which may be as long as 30 minutes, may offset this advantage.

In spite of many disadvantages as discussed above, the following are the few advantages of ungrounded (isolated) neutral system.
(i) In case of a single line-to-ground fault, the fault current is very small. Hence, it is possible to maintain the supply with a fault on one line.
(ii) Interference with communication lines is reduced because of the absence of zero sequence currents.

The advantages of ungrounded neutral system are of negligible importance as compared to their disadvantages. With the growth of power systems in terms of power transmitted, voltage level and distance of transmission, many difficulties were being encountered in ungrounded systems. Therefore, ungrounded systems are no more used. The modern power systems operate with neutral grounding at every voltage level.

Grounded System

In order to avoid dangerous and inconvenient situations arising in ungrounded systems during ground faults, the neutrals of the modern power systems are grounded. The systems having their neutrals grounded are called 'grounded systems'. The neutral is grounded either directly or through resistance or reactance. The neutral grounding provides a return path to the zero-seuqence current, and hence the ground fault current in a grounded system may be too large compared to the small capacitive fault current in an ungrounded system. The large ground fault current cannot be allowed to be in the system for a long time and needs to be interrupted by the circuit breaker. In a grounded system, the ground fault current is sufficient to operate the ground fault relays. The advantages of neutral grounding are discussed in the following section.

Advantages of Neutral Grounding

The modern power systems operate with their neutrals grounded. The advantages of the neutral grounding are as follows:
(i) Persistent arcing grounds are eliminated and hence the system is not subjected to overvoltages due to arcing grounds.

(ii) The neutral point remains stable and is not shifted.

(iii) The voltages of healthy phases (lines) with respect to ground remain at normal value. They do not increase to $\sqrt{3}$ times normal value as in the case of an ungrounded (isolated neutral) system.

(iv) The ground-fault relaying is relatively simple because sufficient amount of ground-fault current is available to operate ground fault relays.

(v) The life of insulation is increased due to prevention of voltage surges caused by arcing grounds. Hence, maintenance, repairs and breakdowns are reduced and service continuity is improved.

(vi) By employing resistance or reactance in ground connection, the ground-fault current can be controlled.

(vii) The overvoltages due to lightning are discharged to ground.

(viii) Improved service reliability is provided because of limitation of arcing grounds and prevention of unnecessary tripping of circuit-breakers.

(ix) Greater safety is provided to personnel and equipment.

Types of Neutral Grounding

The following are the various methods of neutral grounding.
  (i) Effective grounding (solid grounding)
  (ii) Resistance grounding
  (iii) Reactance grounding
  (iv) Resonant grounding (arc suppression coil grounding)
  (v) Voltage transformer grounding
  (vi) Grounding through grounding transformer

(i) Effective grounding (solid grounding) The term 'effective grounding' has replaced the older term 'solid grounding'. In this case, the neutral is directly connected to ground without any intentional impedance between neutral and ground. Here, $Z_n = 0$ and the condition for effective grounding is satisfied. The coefficient of grounding, in this case, is less than 80%.

When the single line to ground (L-G) fault occurs on any line (phase) of the effectively grounded system, the capacitive current is completely neutralized by the fault current preventing arcing ground to occur. Let us consider a L-G fault at point F on line c of the effectively grounded system as shown in Fig. 4.73(a). The neutral and the faulty line c are at ground voltage. The reversed phasor of V_{cN} is shown by dotted line. The voltages of the healthy phases a and b remain unchanged (i.e., phase-to-ground voltages). The neutralization of the capacitive current by the fault current I_f is shown in the phasor diagram of Fig. 4.73(b).

The neutral current I_N is given by:

$$I_N = I_f + (I_{ca} + I_{cb}) = I_f + I_c \quad \text{(phasor sum)}$$

An analysis of the fault by symmetrical components gives

$$I_f = \frac{3V_p}{Z_1 + Z_2 + Z_0}$$

(a) L-G fault on line c

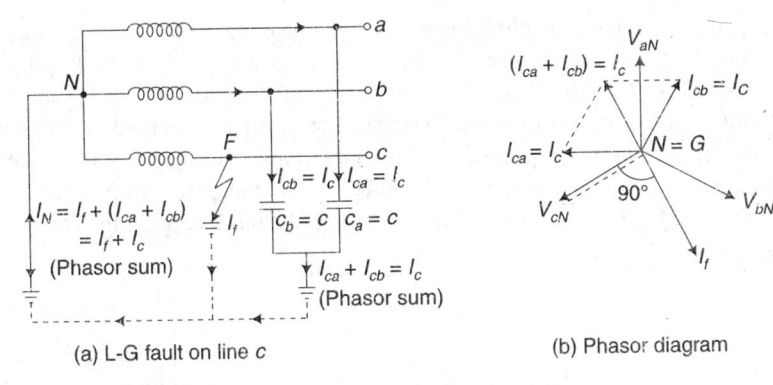

(b) Phasor diagram

Fig. 4.73 *Effectively (solidly) grounded system.*

Since $Z_1 + Z_2 + Z_0$ is predominantly inductive, I_f lags behind the phase to neutral voltage of the faulted phase by nearly 90°. This is shown in Fig. 4.73(b). As the fault current I_f is inductive and much larger than the capacitive current, the neutral current I_N is essentially inductive in nature.

The complete cancellation of the capacitive current by the fault current can be explained through the phasor diagram shown in Fig. 4.73(b). The capacitive currents flowing in the healthy lines a and b are I_{ca} and I_{cb} respectively. Both I_{ca} and I_{cb} are equal to I_c. The resultant capacitive currents I_c is the phasor sum of I_{ca} and I_{cb}. In addition to these capacitive currents, the power source supplies the fault current I_f. This fault current will go from fault point F to ground, then to neutral point N and back to the fault point through the faulty line. The path of resultant capacitive current I_c is capacitive and that of I_f is inductive. The two currents are in phase opposition and hence I_c is completely cancelled by I_f. Therefore, the arcing ground phenomenon and overvoltage conditions do not occur.

The use of 'effective grounding' is limited only to systems where the normal circuit impedance is sufficient to keep the ground fault current within safe limits.

(ii) Resistance grounding In this method of neutral grounding, the neutral of the system is connected to ground through a resistor. The value of the grounding resistance R inserted between the neutral and ground should be so chosen that the ground fault current is less than the symmetrical 3-phase fault current and the power loss in the resistance is not excessive. The value of R should be neither very low nor very high. If the value of grounding resistance R is very low, the ground fault current will be large and the system becomes similar to the effectively (solidly) grounded system. On the other hand, if the grounding resistance R is very high, the system conditions become similar to ungrounded system. The resistor used for neutral grounding is generally a metallic resistance mounted on insulators in a metallic frame. For voltages below 6.6 KV liquid resistors are also used. Metallic resistors do not alter with time and require little or no maintenance. The Neutral Grounding Resistor (NGR) is normally designed to carry its rated current for a short period, usually 30 seconds.

Figure 4.74(a) shows a resistance grounded system with ground fault at point F on line c. When a line-to-ground fault occurs on any line, the neutral is displaced and the maximum voltage across healthy lines may become equal to line-to-line value. The phasor diagram showing currents for single line to ground (L-G) fault on phase c is illustrated in Fig. 4.74(b). Capacitive currents I_{ca} and I_{cb} lead voltages V_{ac} and V_{bc} by 90°. Fault current I_f lags behind the phase voltage of the faulted phase c by an angle ϕ, which depends upon the grounding resistance R and impedance of the system up to the fault point F. The fault current I_f can be resolved into two components; one in phase with the faulty phase voltage ($I_f \cos \phi$) and the other lagging the faulty phase voltage by 90° ($I_f \sin \phi$).

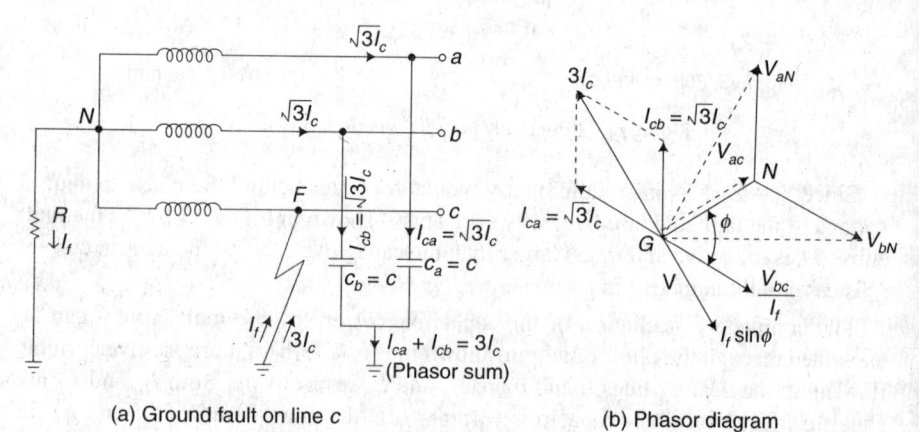

(a) Ground fault on line c (b) Phasor diagram

Fig. 4.74 *Resistance grounded system*

The lagging component of the fault current ($I_f \sin \phi$) is in phase opposition to the resultant capacitive current (phasor sum of I_{ca} and I_{cb}). If the value of grounding resistance R is so adjusted that $I_f \sin \phi$ is equal to resultant capacitive current, the arcing ground is completely eliminated and the operation of the system becomes that of effectively (solidly) grounded system. If the value of grounding resistance R is made sufficiently large so that the lagging component of the fault current ($I_f \sin \phi$) is less than resultant capacitive current, the operation of the system approach that of ungrounded system with the risk of transient overvoltages due to arcing grounds.

Resistance grounding is usually employed for the systems operating at voltages between 3.3 kV and 33 kV.

The following are the main features of resistance grounded systems.

1. The ground fault current is materially lower than that in effectively grounded system but higher than capacitive ground fault current.
2. The ground fault relaying is simple and satisfactory.
3. The power dissipation in the grounding resistance has damping effect. This eliminates arcing ground and improves the stability of the system.
4. The inductive interference to the neighbouring communication circuit is also lesser than that in effectively grounded system because of lower ground fault current.
5. Transient ground faults are converted into controlled current faults.

6. Since the system neutral is almost invariably displaced during ground faults, the equipment require insulation for higher voltages.
7. The system becomes costlier than the effectively grounded system.

(iii) Reactance grounding In this case, the neutral is grounded through impedance, the principal element of which is reactance (i.e., the impedance is having predominant reactance and negligible resistance) as hown in Fig. 4.75. In fact, whether a system is reactance grounded or effectively grounded depends upon the ratio of X_0/X_1. For reactance grounded system, X_0/X_1 ratio is more than 3.0 whereas it is less than 3.0 for effectively grounded system. Reactance grounding lies between effective grounding and resonant grounding. The value of reactance is such that the ground fault current is within safe limits.

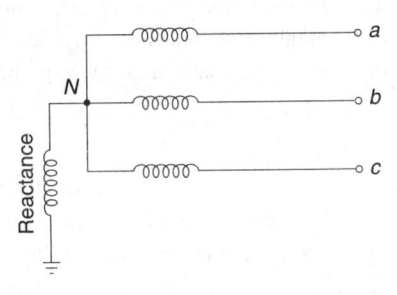

Fig. 4.75 Reactance grounded system.

Since reactance grounding provides additional reactance, the capacitive currents are neutralized. Hence this method of neutral grounding may be used for grounding the neutral of circuits and equipment where high charging currents are involved such as transmission lines, underground cables, synchronous motors, synchronous capacitors, etc. For networks where capacitance is relatively low, resistance grounding is preferred.

(iv) Resonant grounding (arc suppressing coil grounding) Resonant grounding is a special case of reactance grounding. In this case an iron-cored reactor connected between the neutral and ground is capable of being tuned to resonate with the capacitance of the system when a line-to-ground (L-G) fault occurs [Fig. 4.76(a)]. The iron-cored reactor is known as arc suppression coil or Peterson coil or ground fault neutralizer. Arc suppression coil (Peterson coil or ground fault neutralizer) is provided with tappings which permit selection of reactance of the coil depending upon the capacitance to be neutralized. The capacitance to be neutralized depends upon the length of the transmission line. The phasor diagram for fault on phase c is shown in Fig. 4.76(b).

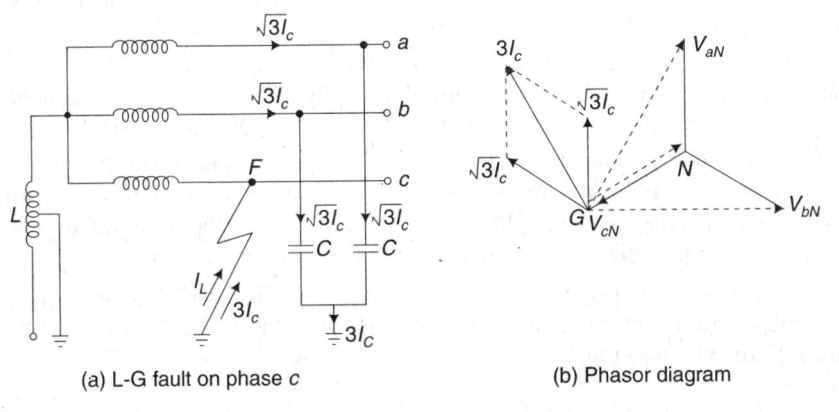

(a) L-G fault on phase c

(b) Phasor diagram

Fig. 4.76 Resonant grounded system.

In case of resonant grounding, the value of the inductive reactance of the arc suppression coil is such that the fault current I_L exactly balances the charging current. In an ungrounded system, when a ground fault occurs on any one line, the voltages of the healthy phases is increased by $\sqrt{3}$ times (i.e., $\sqrt{3}\,V_p$). Hence the charging currents become $\sqrt{3}I_c$ per phase, were I_c is the charging current of line to ground of one phase. The resultant charging current I_{cr} which is the phasor sum of the charging currents of the healthy phases (i.e., $\sqrt{3}I_c$) becomes 3 times the normal line to neutral charging current of one phase, as shown in phasor diagram of Fig. 4.76(b).

Hence,

$$I_{cr} = 3I_c = 3\frac{V_p}{1/wc} = 3V_p wc \qquad (4.161)$$

If L is the inductance of the arc suppression coil (Peterson coil or ground-fault neutralizer) connected between the neutral and the ground, then

$$I_L = \frac{V_p}{wL} \qquad (4.162)$$

In order to obtain satisfactory cancellation (neutralization) of arcing grounds, the fault current I_L flowing through the arc suppression coil should be equal to resultant charging current I_{cr}.

Therefore, for balance condition

$$I_L = I_{cr}$$

or $\qquad\qquad\dfrac{V_p}{wL} = 3V_p wc \quad$ [From Eqs. (4.161) and (4.162)]

or

$$L = \frac{1}{3w^2 c} \qquad (4.163)$$

The inductance of the arc suppression coil is calculated from Eq. (4.163).

The method of resonant grounding is usually confined to medium-voltage overhead transmission lines which are connected to the system generating source through intervening power transformers. The use of this method of grounding reduces the line interruption due to transient line-to-ground faults which will not be possible with other methods of grounding. In this method of grounding, the tendency of a single phase to ground fault developing into two or three-phase fault is decreased.

Normal time-rating of arc suppression coils used on systems on which permanent ground faults can be located and removed promptly is ten minutes whereas continuous time-rated coils are used on all other systems. In such cases, it is also usual to provide automatic means as shown in Fig. 4.77 to bypass the arc suppression coil after some time. If for any reason more current flows through the arc suppression coil, a circuit breaker closes after a certain time-lag and the ground-fault current flows through the parallel circuit by-passing the arc-suppression coil.

The circuit breaker (C.B.) which is normally open is closed by the trip coil when the relay operates after a predetermined time. Thus the fault current is by-passed through the resistor branch.

(v) Voltage transformer grounding In this method of neutral grounding, the primary of a single-phase voltage transformer is connected between the neutral and the ground as shown in Fig. 4.78. A low resistor in series with a relay is connected across the secondary of the voltage transformer. The voltage transformer acts as a very high reactance grounding device and does not assist in mitigating the overvoltage conditions which are associated with ungrounded (isolated) neutral system. A system grounded through voltage transformer operates virtually as an ungrounded system. A ground fault on any phase produces a voltage across the relay which causes operation of the protective device.

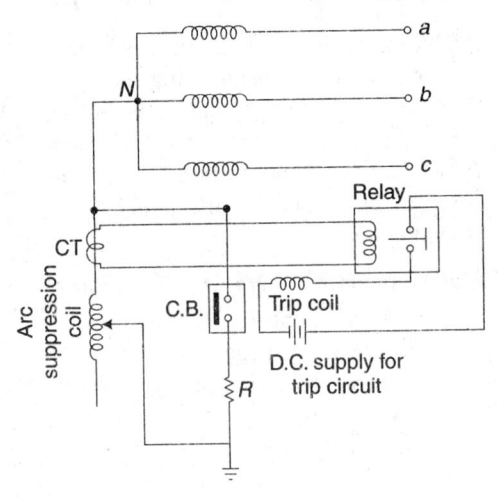

Fig. 4.77 *Connection of arc suppression coil.*

(vi) Grounding through grounding transformer In cases where the neutral of a power system is not available for grounding or where the transformer or generator is delta connected, an artificial neutral grounding point is created by using a zig-zag transformer called 'grounding transformer'. This transformer does not have secondary winding and it is a core-type transformer having three limbs built in the same fashion as that of the power transformer. Each limb of the transformer has two identical windings wound differentially (i.e., directions of current in the two windings on each limb are opposite to each other) as shown in Fig. 4.79(a). Since the two identical windings on each limb are wound differentially, under normal conditions, the total flux in each limb is negligibly small and, therefore, the transformer draws very little magnetising current. The grounding transformer is of short-time rating, usually 10 seconds to 1 minute. Therefore, the size of such transformer is small as compared to the power transformer of the same rating.

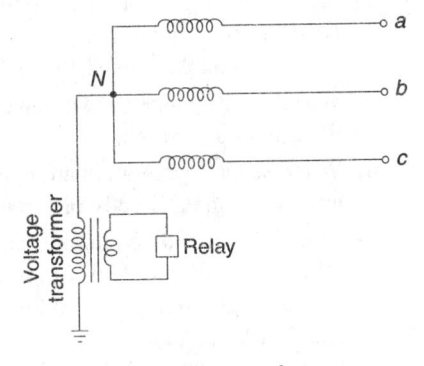

Fig. 4.78 *Voltage transformer grounding of neutral.*

If a zig-zag transformer is not available for grounding, a special small size star-delta transformer can be used without loading the delta side as shown in Fig. 4.79(b). This transformer is also known as grounding transformer and it is a step-down transformer. The star connected primary is connected to the system and its neutral is grounded. The secondary is in delta and generally does not supply any load but provides a closed path for triple harmonic currents to circulate in them. Under normal conditions, the current in the transformer is only its own magnetizing current. However, large current may flow in the event of a single line to ground fault condition. Hence, it should be of sufficient rating to withstand the effect of line to ground faults.

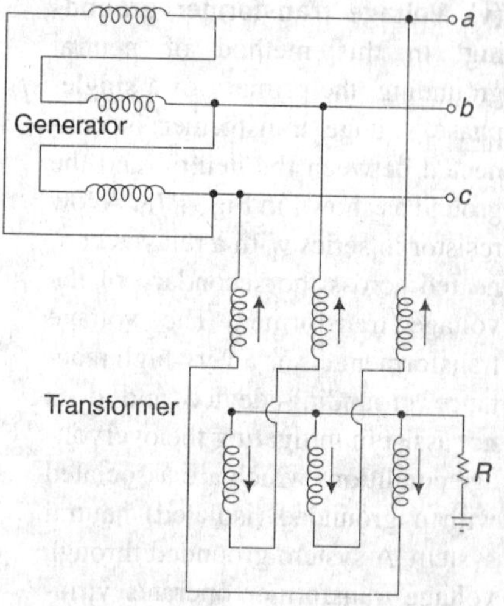

Present Practice in Neutral Grounding

(i) Generally, one grounding is provided at each voltage level. In power system, there are various voltage levels between generation and distribution. At least one grounding is normally provided at each voltage level.

Fig. 4.79(a) *Zig-zag transformer for neutral grounding*

(ii) Grounding is provided at the power source end and not at the load end.
If power source is delta connected, grounding is provided by means of grounding transformer rather than grounding at load end.

(iii) Grounding is provided at each major source bus section.

(iv) The generators are normally provided with resistance grounding whereas synchronous motors and synchronous capacitors are provided with reactance grounding.

(v) When several generators are operating in parallel, only one generator neutral is grounded. If more neutral are grounded, disturbance is created by the zero sequence components of the circulating currents.

(vi) When there are one or two power sources, no switching equipment is used in the grounding circuit.

(vii) When several generators are connected to a common neutral bus, the bus is grounded either directly or through reactance.

(viii) Effectively (solid) grounding is used for the systems of low voltages up to 600 volts and for systems of high voltages above 33 kV whereas resistance or reactance grounding is used for systems of medium voltages between 3.3 kV and 33 kV.

4.8.3 Equipment Grounding (Safety Grounding)

Equipment grounding means electrically connecting non-current carrying metallic parts (i.e., metallic frame, metallic enclosure, etc.) of the equipment to ground. The object of equipment grounding is to ensure safety against electrical shocks to operating personnel and other human or animal bodies by discharging electrical energy to ground. Equipment grounding also ensures protection against lightning. If insulation fails, there will be a direct contact of the live conductor with the metallic parts (i.e., frame) of the equipment. Any person in contact with the metallic part of this equipment will be subjected to electrical shock which can be fatal. Thus equipment grounding eals with grounding of non-current carrying metallic parts of the equipment to ensure safety of personnel and protection against lightning. Equipment grounding is also called 'safety grounding'.

Under fault conditions, the non-current carrying metallic parts of an electrical installation such as frames, enclosures, supports, fencing, etc., may attain high voltages with respect to the ground so that any person or stray animal touching these or approaching these will be subjected to voltage which may result in the flow of a current through the body of the person or the animal of such a value as may prove fatal. The magnitude of the tolerable current flowing through the body is related to duration.

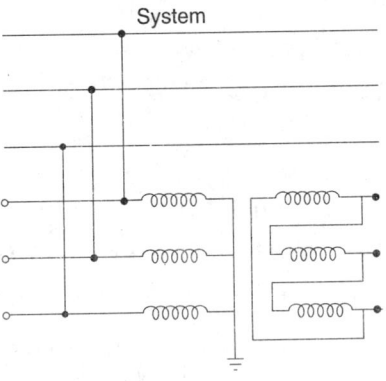

Fig. 4.79(b) *Star-delta transformer for grounding.*

It has been found experimentally that the safe value of the current which a human body can tolerate is given by

$$I_b = \frac{0.165}{\sqrt{t}} \quad \text{for } t < 3 \text{ secs} \quad (4.164)$$

and

$$I_b = 9 \text{ mA} \quad \text{for } t > 3 \text{ secs} \quad (4.165)$$

where I_b is the rms value of the body current in amps and t is the time in seconds

Tolerable Step and Touch Voltages

When a fault occurs, the flow of fault current to ground results in voltage gradient on the surface of the ground in the vicinity of the grounding system. This voltage gradient may affect a person in two ways, viz., step or foot to foot contact and hand to both feet or touch contact.

Step voltage

It is the voltage between the feet of a person standing on the floor of the substation, with 0.5 m spacing between two feet (i.e., one step), during the flow of ground fault current through the grounding system. Fig. 4.80 shows the circuit for step contact. R_f is the grounding resistance of one foot (in ohms), and R_b is the resistance of body (in ohms).

The grounding resistance of one foot (R_f) may be assumed to be $3p_s$ where p_s is the resistivity of the soil near the surface of ground. R_b is assumed to be 1000 ohms.

Therefore, tolerable value of the step voltage is given by [from Fig. 4.80(b)].

$$V_{step} = (R_b + 2R_f) I_b \quad (4.166)$$

Substituting the values of R_b, R_f and I_b [from Eqs. (164) and (165)], we get

$$V_{step} = (1000 + 6p_s) 0.165/\sqrt{t} \quad \text{for } t < 3 \text{ sec} \quad (4.167)$$

$$= (1000 + 6p_s) 0.009 \quad \text{for } t > 3 \text{ sec}$$

(a) A person standing on the floor of the substation

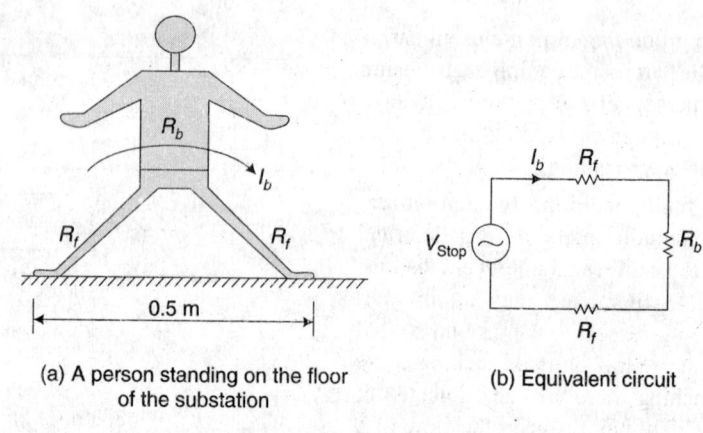

(b) Equivalent circuit

Fig. 4.80 Step contact

Touch voltage

It is the voltage between the fingers of raised hand touching the faulted structure and the feet of the person standing on substation floor. Figure 4.81 shows the circuit for touch contact.

(a) Person touching faulted structure

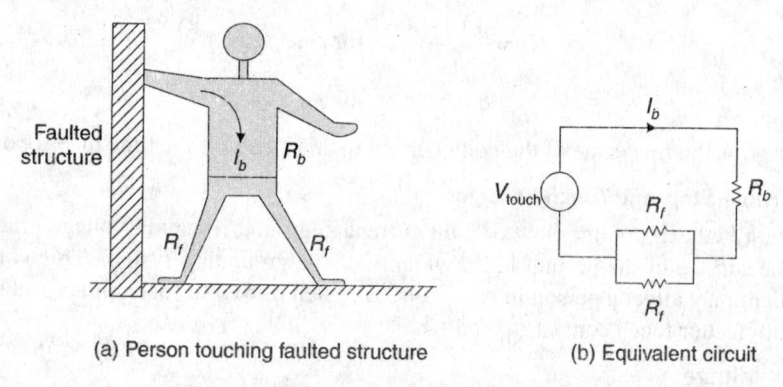

(b) Equivalent circuit

Fig. 4.81 Touch contact

From Fig. 4.81(b), the tolerable value of touch voltage is given by

$$V_{touch} = (R_b + R_f/2) I_b \qquad (4.168)$$
$$= (1000 + 1.5\, p_s)\, 0.165/\sqrt{t} \qquad \text{for } t < 3 \text{ sec}$$
$$= (1000 + 1.5\, p_s)\, 0.009 \qquad \text{for } t > 3 \text{ sec}$$

The actual values of step voltage and touch voltage should be less than the tolerable values given by Eqs. (4.167) and (4.168) respectively.

Functions of Substation Grounding System

(i) To ensure safety against electrical shocks to operating and maintenance personnel
(ii) To discharge electrical charges to ground during ground faults and lightning
(iii) To provide grounding of overhead shielding wires

(iv) To ensure freedom from electromagnetic interference in communication and data processing equipment in the substation

EXERCISES

1. What do you mean by a fault? Discuss the causes of fault on a power system. What are the types of faults that occur in any power system?
2. What is the necessity of fault analysis? What do you mean by fault level at any point on the power system?
3. What is a per-unit system? What is its significance in power system analysis? What are the advantages of per-unit system? How can the per unit value of impedance on one base be changed to per unit value on a different base?
4. What is an impedance diagram of a power system? What are the steps to be followed for drawing the impedance (reactance) diagram from the one-line diagram of a power system?
5. Distinguish between symmetrical and unsymmetrical faults. What assumptions are made to simplify symmetrical fault calculations?
6. What do you mean by symmetrical short circuit capacity of a power system? How can it be calculated?
7. Discuss the steps to be followed in the calculations for a symmetrical fault.
8. Discuss the criterion for the selection of circuit breaker for a particular location in a power system.
9. A generating station consisting of two generators, one of 20 MVA, 11 kV, 0.20 pu reactance with the unit transformer of 11/132 kV, 20 MVA, 0.08 pu reactance, and another of 30 MVA, 11 kV, 0.20 pu reactance with the transformer of 11/132 kV, 30 MVA, 0.10 pu reactance, transmits power over two 132 kV, 3 phase transmission lines each with reactance of 300 ohm per phase.

   The 132 kV lines are bussed at both ends. Determine the current in the windings of generators when a 3-phase symmetrical fault takes place on the 132 kV bus bars at the receiving end station (**Ans.** 1510 A, 2110 A)

10. The power system shown in Fig. 4.82 is initially at no load. Calculate the subtransient fault current that results when a three-phase symmetrical fault occurs at F, given that the transformer voltage on high voltage side is 66 kV.
    (**Ans.** $-j\ 2.735\ pu$)

50 MVA, 13.8 kV
$x''_d = 0.25\ pu$
G_1

F

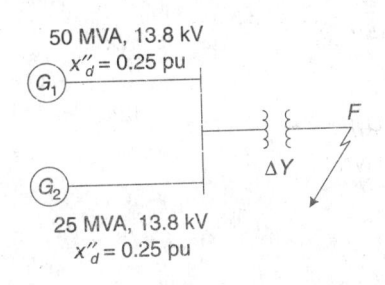

ΔY

G_2
25 MVA, 13.8 kV
$x''_d = 0.25\ pu$

Fig. 4.82

11. Figure 4.83 shows a power system where an infinite bus feeds power to two identical synchronous motors through a step down transformer. The ratings are given on the diagram. Neglecting cable impedances, find the subtransient fault current and subtransient current at breaker B, when 3-phase fault occurs at F.

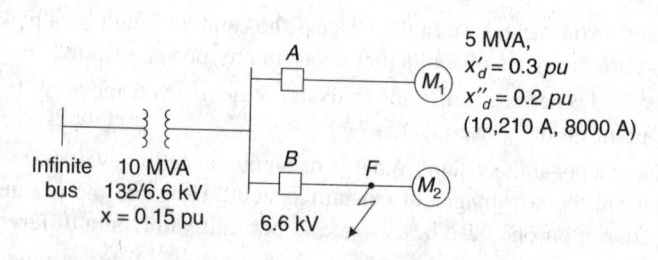

```
                              A                    5 MVA,
                              □                    x'_d = 0.3 pu
                                           (M_1)   x"_d = 0.2 pu
                                                   (10,210 A, 8000 A)
         ┤┠                   B      F
  Infinite   10 MVA           □      ↙
    bus      132/6.6 kV                    (M_2)
             x = 0.15 pu     6.6 kV
```

Fig. 4.83

12. Three 25 MVA generators each having a reactance of 0.2 pu are operating in parallel. They feed a transmission line through a 75 MVA transformer having a per unit reactance of 0.05 pu. Find the fault MVA for a fault at the sending end of the line.
(**Ans.** 300 MVA)

13. A 20 MVA, 11 kV, 3-phase generator having a transient reactance of 0.2 pu feeds a 20 MVA, 11 kV, 3-phase synchronous motor having a transient reactance of 0.5 pu through a short line. The line has a reactance of 0.30 ohm. A three-phase fault occurs at the motor terminals when the motor is taking an input of 15 MW at unity p.f. The line voltage across motor terminals at the time of fault is 10.5 kV. Find the total current supplied by the generator and motor to the fault. (**Ans.** $783.5 - j\,3990$ A, $783.5 - j\,1995$ A)

14. What are the types of reactors used for limiting the symmetrical fault current? Discuss the various locations of the current limiting reactors.

15. There are two generators connected to a common 6.6 kV bus bar and a feeder is taken out through a step-up transformer as shown in Fig. 4.84. Find the ohmic value of the current limiting feeder reactor in order that fault current is limited to four times the full load current, if there is a 3-phase fault just near the reactor.
(**Ans.** 0.128 ohm)

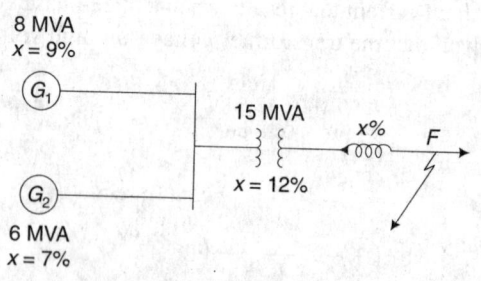

```
         8 MVA
         x = 9%
          (G_1)
                         15 MVA
                         ┤┠         x%    F
                                   ─000─  ↙
          (G_2)         x = 12%
         6 MVA
         x = 7%
```

Fig. 4.84

Fault Analysis 217

16. The section bus bars A and B in Fig. 4.85 are linked by a bus bar reactor rated at 5 MVA with 10% reactance drop. On bus bar A there are two generators each of 10 MVA with 10% reactance and on B two generators, each of 8 MVA with 12% reactance. Find the steady MVA load fed into a dead short circuit between all phases on B with bus bar reactor in circuit. **(Ans. 174 MVA)**

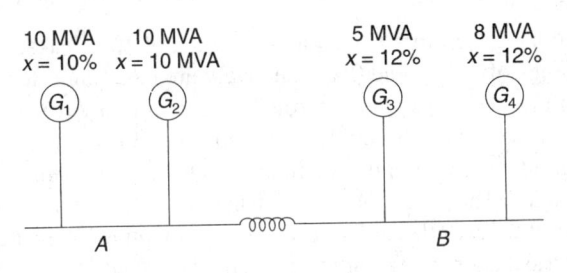

```
10 MVA      10 MVA              5 MVA       8 MVA
X = 10%    X = 10 MVA           X = 12%    X = 12%
  (G₁)      (G₂)                 (G₃)       (G₄)
   |         |                    |          |
   |         |      0000          |          |
   |_____|_____----_____|_____|
         A                              B
```

Fig. 4.85

17. Each of the three generators in a generating station has a short circuit reactance of 10% based on the respective ratings of 75 MVA, 90 MVA and 110 MVA. Each machine is connected to its own sectional bus bar and each bus bar is connected to a tiebar through a reactor of 5% reactance based on the rating of the generator connected to it. Calculate the short circuit current in amperes and the MVA fed into the short circuit occurring between the bus bars of the section to which the 110 MVA generator is connected. The generator voltage is 33 kV. **(Ans. 32,200 A; 1830 MVA)**

18. A hydro station has five 80 MVA, 11 kV generators each of 10% reactance connected in parallel. The generated voltage is stepped up to 220 kV by three 160 MVA transformers, each of 15% reactance, also connected in parallel. The 220 kV system supplies power to two sub-stations over two feeders having reactances of 12.1 ohms and 24.2 ohms per phase. Calculate for each sub-station separately, the rms value of the symmetrical fault current when a 3-phase short circuit (fault) occurs at the sub-station. Assume that the generators are not loaded prior to the fault. **(Ans. 3220 A, 2480 A)**

19. For the system shown in Fig. 4.86 determine the fault current and rupturing capacity of the breaker on feeder C.

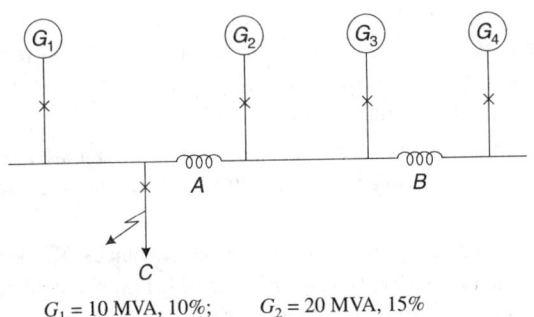

```
        (G₁)        (G₂)        (G₃)        (G₄)
         |           |           |           |
         X           X           X           X
         |           |           |           |
         |____000____|           |____000____|
         X     A                       B
         ↓
         C
```

G_1 = 10 MVA, 10%; G_2 = 20 MVA, 15%
G_3 = 20 MVA, 15%; G_4 = 12.5 MVA, 12.5%
A = 8 MVA, 5%;   B = 10 MVA, 5%

Fig. 4.86

20. A small generating station has two generators of 2.5 MVA and 5 MVA and percentage reactances 8 and 6 respectively. The circuit breakers are rated 150 MVA. Due to increase in system load it is intended to extend the system by a supply from the grid via a transformer of 10 MVA rating and 7.5% reactance. If the system voltage is 3.3 kV, find the reactance necessary to protect the switchgear. (**Ans. 0.226 ohm**)

21. Three 6.6 kV generators G_1, G_2 and G_3, each of 10% leakage reactance and MVA ratings of 40, 50 and 25 respectively are interconnected electrically, as shown in Fig. 4.87, by a tie bar through current limiting reactors, each of 12% reactance based upon the rating of the machine to which it is connected. A three-phase feeder is supplied from the bus bar of generator G_1 at a line voltage of 6.6 kV. The feeder has a resistance of 0.06 Ω/phase and an inductive reactance of 0.12 Ω/phase. Estimate the maximum MVA that can be fed into a symmetrical short circuit at the far end of the feeder. (**Ans. 212 MVA**)

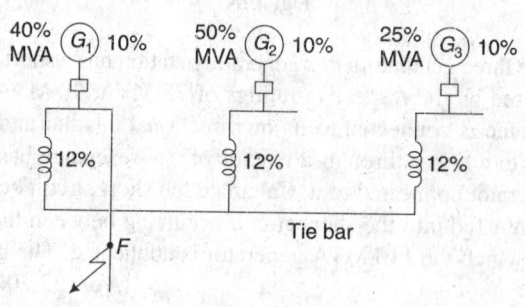

Fig. 4.87

22. Discuss the principle of symmetrical components. Derive the necessary equations to convert phase quantities into symmetrical components, and vice versa

23. Derive the necessary equation to determine the fault current for a single line to ground fault. Discuss the interconnection of sequence networks for this type of fault.

24. Discuss the situation under which a single line-to-ground (L-G) fault can be more severe than a 3-phase symmetrical fault.

25. Show that positive and negative sequence currents are equal in magnitude but out of phase by 180° in case of a line-to-line fault. Draw a diagram showing the interconnection of positive and negative sequence networks. Why zero-sequence network remains unconnected?

26. Derive equations for sequence currents in case of a double line to ground fault. Draw a diagram showing interconnection of sequence networks for this type of fault.

27. The positive, negative, and zero-sequence reactances of a 20-MVA, 13.2 kV synchronous generator are 0.3 pu, 0.2 pu and 0.1 pu respectively. The generator is solidly grounded and is not loaded. A line-to-ground fault occurs on phase a. Neglecting all the resistance, determine the fault current. (**Ans. 4374 A**)

28. The one-line diagram of a system is shown in Fig. 4.88. The ratings of the generator and transformers are given on the diagram. Fault occurs at point F. Determine the fault current for
   (i) 3-phase symmetrical fault
   (ii) L-G fault
   (iii) L-L fault
   (iv) 2L-G fault.
   Assume system to be initially on no load.

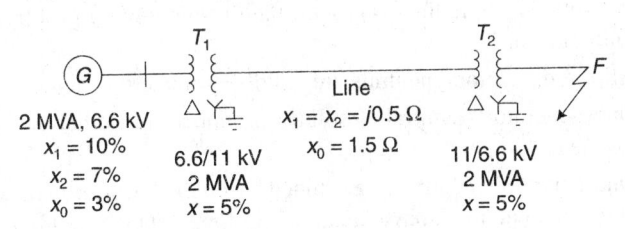

Fig. 4.88

2 MVA, 6.6 kV
$X_1 = 10\%$
$X_2 = 7\%$
$X_0 = 3\%$

6.6/11 kV
2 MVA
$X = 5\%$

Line
$X_1 = X_2 = j0.5\ \Omega$
$X_0 = 1.5\ \Omega$

11/6.6 kV
2 MVA
$X = 5\%$

(Ans. $j\,4.8$ pu; 6.87 pu; 4.48 pu; 9.48 pu)

29. Six 6.6 kV, 3-phase generators are connected to a common set of bus bars. Each has positive, negative and zero sequence reactances of 0.90 Ω, 0.72 Ω and 0.30 Ω respectively. A ground fault occurs on one bus bars. Determine the value of the fault current if
   (i) All generator neutrals are solidly grounded.
   (ii) If one generator neutral is grounded through a resistance of 0.2 ohm, the others being isolated.
   (iii) If only one generator neutral is solidly grounded, the others being isolated. (Ans. 35,700 A; 13,940 A; 20054.74 A)

30. Two generators rated 11 kV, 5 MVA having $X_1 = 0.15$, pu $X_2 = 0.12$ pu, $X_0 = 0.10$ pu are operating in parallel. A single line-to-ground fault occurs on the bus bar. Calculate the fault current if
   (i) Both generator neutrals are solidly grounded,
   (ii) Only one generator neutral is solidly grounded,
   (iii) Both neutrals are isolated period

31. A double line to ground fault occurs at F in the system shown in Fig. 4.89. Draw the sequence networks for the system, and calculate the line current I_b. [Ans. $4469 < 123.8°$ A]

35 MVA, 11 kV
$X_1 = 0.6$ pu,
$X_2 = 0.4$ pu
$X_0 = 0.2$ pu

35 MVA, 11/110 kV
$X_1 = X_2 = X_0 = 0.05$ pu

Fig. 4.89

32. For a generator the ratio of fault current for line-to-line fault and three-phase faults is 0.866. The positive sequence reactance is 0.15 pu. Calculate negative sequence reactance.

$$\text{Hint:} \quad \frac{\text{L-L fault}}{\text{3-phase fault}} = \frac{1.732 \, X_1}{X_1 + X_2}$$

33. Three 6.6 kV, 3-phase, 10 MVA generators are connected to a grid. The positive sequence reactance of each generator is 0.15 pu while the negative and zero-sequence reactances are 75% and 30% of positive sequence reactances respectively. A single line-to-ground fault occurs on the grid bus. Determine the fault current if
   (i) All the generators' neutrals are solidly grounded
   (ii) One generator's neutral is solidly grounded and other two neutrals are isolated.
   (iii) One generator neutral is grounded through 0.3 ohm resistance and the other two neutrals are isolated. **(Ans. 25600 A, 19810 A, 10680 A)**

34. What are the principal purposes of grounding? Differentiate between neutral grounding and equipment grounding?

35. Define the following terms:
   (a) Step voltage
   (b) Touch voltage
   (c) Coefficient of grounding
   (d) Arcing ground
   (e) Effectively grounded system

36. Discuss the performance of an ungrounded system during a line-to-ground fault.

37. Show that when a single line-to-ground fault occurs in an ungrounded system, the neutral shift is equal to the zero-sequence voltage, and voltage of the healthy lines with respect to the ground is equal to the line-voltage.

38. Discuss the advantages of neutral grounding what are the various types of neutral grounding?

39. Explain the advantages and disadvantages of various types of grounding and indicate where each type is commonly employed.

40. Discuss the present practice in neutral grounding.

41. What is the object of equipment grounding? Briefly describe the functions of substation grounding system.

5 Overcurrent Protection

5.1 INTRODUCTION

A protective relay which operates when the load current exceeds a preset value, is called an *overcurrent relay*. The value of the preset current above which the relay operates is known as its pick-up value. Overcurrent relays offer the cheapest and simplest form of protection. These relays are used for the protection of distribution lines, large motors, power equipment, industrial systems, etc. Overcurrent relays are also used on some subtransmission lines which cannot justify more expensive protection such as distance or pilot relays. A scheme which incorporates overcurrent relays for the protection of an element of a power system, is known as an overcurrent protection scheme or overcurrent protection. An overcurrent protection scheme may include one or more overcurrent relays.

At present, electromechanical relays are widely used for overcurrent protection. The induction disc type construction, as shown in Fig. 2.9(b) is commonly used. With the development of numerical relays based on microprocessors or micorcontrollers, there is a growing trend to use numerical overcurrent relays for overcurrent protection.

5.2 TIME-CURRENT CHARACTERISTICS

A wide variety of time-current characteristics is available for overcurrent relays. The name assigned to an overcurrent relay indicates its time-current characteristic as describe below.

5.2.1 Definite-time Overcurrent Relay

A definite-time overcurrent relay operates after a predetermined time when the current exceeds its pick-up value. Curve (a) of Fig. 5.1 shows the time-current characteristic for this type of relay. The operating time is constant, irrespective of the magnitude of the current above the pick-up value. The desired definite operating time can be set with the help of an intentional time-delay mechanism provided in the relaying unit.

5.2.2 Instantaneous Overcurrent Relay

An instantaneous relay operates in a definite time when the current exceeds its pick-up value. The operating time is constant, irrespective of the magnitude of the current, as shown by the curve (a) of Fig. 5.1. There is no intentional time-delay. It operates in 0.1s or less. Sometimes the term like "high set" or "high speed" is used for very fast relays having operating times less than 0.1s.

5.2.3 Inverse-time Over-current Relay

An inverse-time overcurrent relay operates when the current exceeds its pick-up value. The operating time depends on the magnitude of the operating current. The operating time decreases as the current increases. Curve (b) of Fig. 5.1 shows the inverse time-current characteristic of this types of relays.

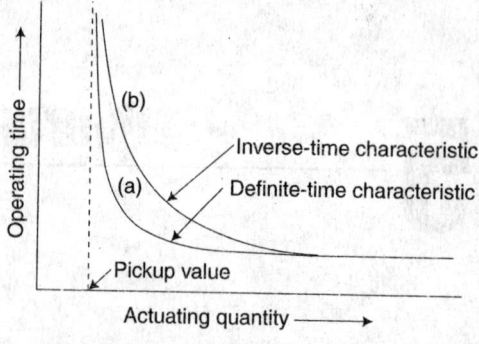

Fig. 5.1 *Definite-time and inverse-time characteristics of overcurrent relays*

5.2.4 Inverse Definite Minimum Time Overcurrent (I.D.M.T) Relay

This type of a relay gives an inverse-time current characteristic at lower values of the fault current and definite-time characteristic at higher values of the fault current. Generally, an inverse-time characteristic is obtained if the value of the plug setting multiplier is below 10. For values of plug setting multiplier between 10 and 20, the characteristic tends to become a straight line, i.e. towards the definite time characteristic. Figure 5.2 shows the characteristic of an I.D.M.T. relay along with other characteristics. I.D.M.T. relays are widely used for the protection of distribution lines. Such relays have a provision for current and time settings which will be discussed later on.

5.2.5 Very Inverse-time Overcurrent Relay

A very inverse-time overcurrent realy gives more inverse characteristic than that of a plain inverse relay or the I.D.M.T. relay. Its time-current characteristic lies between an I.D.M.T. characteristic and extremely inverse characteristic, as shown in Fig. 5.2. The very inverse characteristic gives better selectivity than the I.D.M.T. characteristic. Hence, it can be used where an I.D.M.T. relay fails to achieve good selectivity. Its recommended standard time-current characteristic is given by

$$t = \frac{13.5}{I-1}$$

The general expression for time-current characteristic of overcurrent relays is given by

$$t = \frac{K}{I^n - 1}$$

The value of n for very inverse characteristic may lie between 1.02 and 2.

Very inverse time-current relays are recommended for the cases where there is a substantial reduction of fault current as the distance from the power source increases. They are particularly effective with ground faults because of their steep characteristic.

5.2.6 Extremely Inverse-time Overcurrent Relay

An extremely inverse time overcurrent relay gives a time-current characteristic more inverse than that of the very inverse and I.D.M.T. relays, as shown in Fig. 5.2. When I.D.M.T. and very inverse relays fail in selectivity, extremely inverse relays are employed. I.D.M.T. relays are not suitable to be graded with fuses. Enclosed fuses have time-current characteristics according to the law

$$I^{3.5} t = K$$

The electromechanical relay which gives the steepest time-current characteristic is an extremely inverse relay. The time-current characteristic of an extremely inverse relay is $I^2 t = K$. Its characteristic is not good enough to be graded with fuses. But the best that can be done with electromechanical relay is to use extremely inverse relays to grade with fuses.

An extremely inverse relay is very suitable for the protection of machines against overheating. The heating characteristics of machines and other apparatus is also governed by the law $I^2 t = K$. Hence, this type of relays are used for the protection of alternators, power transformers, earthing transformers, expensive cables, railways trolley wires, etc. The rotors of large alternators may be overheated if an unbalanced load or fault remains for a longer period on the system. In such a case, an extremely inverse relay, in conjunction with a negative sequence network is used. By adjusting the time and current settings, a suitable characteristic of the relay is obtained for a particular machine to be protected.

A relay should not operate on momentary overloads. But it must operate on sustained short circuit current. For such a situation, it is difficult to set I.D.M.T. relays. An extremely inverse relay is quite suitable for such a situation. This relay is used for the protection of alternators against overloads and internal faults. It is also used for reclosing distribution circuits after a long outage. After long outages, when the circuit breaker is reclosed there is a heavy inrush current which is comparable to a fault current. An I.D.M.T. relay is not able to distinguish between the rapidly decaying inrush current of the load and the persistent high current of a fault. Hence, an I.D.M.T. relay trips again after reclosing. But an extremely inverse relay is able to distinguish between a fault current and inrush current due to its steep time-current characteristic. Therefore an extremely inverse relay is quite suitable for the load restoration purpose.

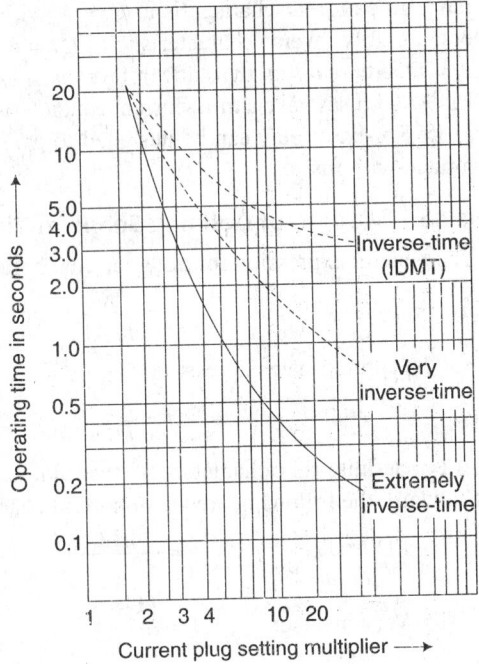

Fig. 5.2 I.D.M.T., very inverse-time and extremely inverse-time characteristics

5.2.7 Special Characteristics

Overcurrent relays, having their time-current characteristics steeper than those of extremely inverse relays are required for certain industrial applications. These relays have time-current characteristic $I^n = K$ with $n = 2$. To protect rectifier transformers, a highly inverse characteristic of $I^8 t = K$ is required. The characteristics having $n = 2$ are realised by static relays or microprocessor-based overcurrent relays. Enclosed fuses have a time-current characteristic of $I^{3.5} t = K$. A static relay or microprocessor-base relay can be designed to give $I^{3.5} t = K$ characteristic, suitable to be graded with fuses.

5.2.8 Method of Defining Shape of Time-current Characteristics

The general expression for time-current charactgeristics is given by

$$t = \frac{K}{I^n - 1}$$

The approximate expression is

$$t = \frac{K}{I^n}$$

For definite-time characteristic, the value of n is equal to 0. According to the British Standard, the following are the important characteristics of overcurrent relays.

(i) I.D.M.T.: $\quad t = \dfrac{0.14}{I^{0.02} - 1}$

(ii) Very inverse: $\quad t = \dfrac{13.5}{I - 1}$

(iii) Extremely inverse: $t = \dfrac{80}{I^2 - 1}$

The inverse time-current characteristics obtained from the above expressions are not straight line characteristics. A microprocessor-based relay can easily give straight line characteristics of the form $t = K/I^n$ with any value of n. These characteristics are straight line characteristics on log t/log I graph. The advantage of such simplified time-current curves is the saving in time in calculating relay time settings.

5.2.9 Technique to Realise Various Time-Current Characteristics using Electromechanical Relays

The magnetic circuit of an overcurrent relay can be designed to saturate above a certain value of the actuating current. Below this value of the actuating current, the relay gives an inverse characteristic. Above the saturation value of the current, the relay gives a straight line characteristic, parallel to the current-axis. It means that whatever may be the value of the current above saturation value the operating time remains constant.

If the core is designed to saturate at the pick-up value of the current, the relay gives a definite time-characteristic. If the core is designed to saturate at a later stage, an I.D.M.T. characteristic is obtained. If the core saturates at a still later stage, a very inverse characteristic is obtained. If the saturation occurs at a very late stage, the relay give an extremely inverse characteristic.

5.3 CURRENT SETTING

The current above which an overcurrent relay should operate can be set. Suppose that a relay is set at 5 A. It will then operate if the current exceeds 5 A. Below 5 A, the relay will not operate. There are a number of tappings on the current coil, available for current setting, as shown in Fig. 2.9 and Fig. 2.10. The operation of the relay requires a certain flux and ampere turns. The current settings of the relay are chosen by altering the number of turns of the current coil by means of a plug PS in Fig. 2.9 and 2.10. The plug-setting (current-setting) can either be given directly in amperes or indirectly as percentages of the rated current. An overcurrent relay which is used for phase-to-phase fault protection, can be set at 50% to 200% of the rated current in steps of 25%. The usual current rating of this relay is 5 A. So it can be set at 2.5 A, 3.75 A, 5 A, ..., 10 A. When a relay is set at 2.5 A, it will operate when current exceeds 2.5 A. When the relay is set at 10 A, it will operate when current exceeds 10 A. The relay which is used for protection against ground faults (earth-fault relay) has settings 20% to 80% of the rated current in steps of 10%. The current rating of an earth-fault relay is usually 1 A.

If time-current curves are drawn, taking current in amperes on the X-axis, there will be one graph for each setting of the relay. To avoid this complex situation, the plug setting multipliers are taken on the X-axis. The actual r.m.s. current flowing in the relay expressed as a multiple of the setting current (pickup current) is known as the plug setting multiplier (PSM). Suppose, the rating of a relay is 5 A and it is set at 200%, i.e. at 10 A. If the current flowing through the relay is 100 A, then the plug setting multiplier will be 10. The PSM = 4 means 40 A of current is flowing, PSM = 6 means 60 A of current is flowing and so on.

If the same relay is set at 50%, i.e. at 2.5 A, the PSM = 4 means 10 A; PSM = 6 means 15 A; PSM = 10 means 25 A and so on.

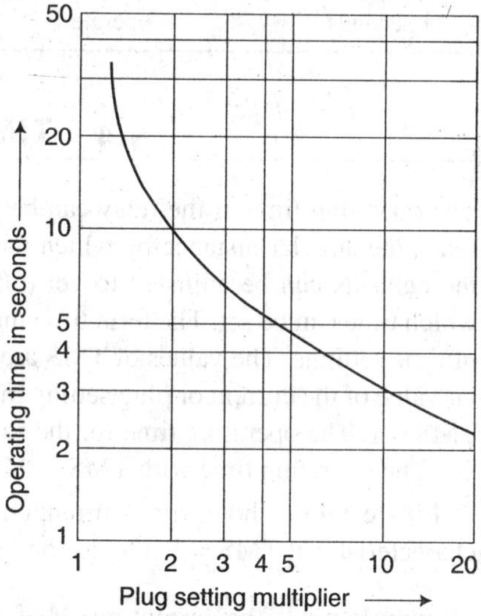

Hence, PSM can be expressed as

$$\text{PSM} = \frac{\text{Secondary current}}{\text{Relay current setting}}$$

$$= \frac{\text{Primary current during fault, i.e. fault current}}{\text{Relay current setting} \times \text{CT ratio}}$$

Fig. 5.3 *Standard I.D.M.T. characteristic*

While plotting the time-current characteristic, if PSM is taken on the X-axis, there will be only one curve for all the settings of the relay. Figure 5.3 shows a time-current characteristic with PSM on the X-axis. The curve is generally plotted on log/log graph. Only this curve will give the operating time for different settings of the relay. Suppose the relay is set at 5 A. The operating times for different currents are shown in Table 5.1.

If the same relay is set at 10 A, the corresponding operating times for different currents are shown in Table 5.2, using the same curve of Fig. 5.3

Table 5.1

Current in Amperes	5	10	20	50
PSM	1	2	4	10
Operating time in seconds	No operation	10	5	3

Table 5.2

Current in Amperes	5	10	20	40	100
PSM	less than 1	1	2	4	10
Operating time in seconds	Relay will not operate	No operation	10	5	3

5.4 TIME SETTING

The operating time of the relay can be set at a desired value. In induction disc type relay, the angular distance by which the moving part of the relay travels for closing the contacts can be adjusted to get different operating time. There are 10 steps in which time can be set. The term time multiplier setting (TMS) is used for these steps of time settings. The values of TMS are 0.1, 0.2, ..., 0.9, 1. Suppose that at a particular value of the current or plug setting multiplier (PSM), the operating time is 4 s with TMS = 1. The operating time for the same current with TMS = 0.5 will be 4 × 0.5 = 2 s. The operating time with TMS = 0.2 will be 4 × 0.2 = 0.8 s.

Figure 5.4 (a) shows time-current characteristics for different values of TMS. The characteristic at TMS = 1 can also be presented in the form shown in Fig. 5.4 (b).

Example 5.1 | The current rating of an overcurrent relay is 5 A. The relay has a plug setting of 150% and time multiplier setting (TMS) of 0.4. The CT ratio is 400/5. Determine the operating time of the relay for a fault current of 6000 A. At TMS = 1, operating time at various PSM are given in the Table 5.3

Table 5.3

PSM	2	4	5	8	10	20
Operating time in seconds	10	5	4	3	2.8	2.4

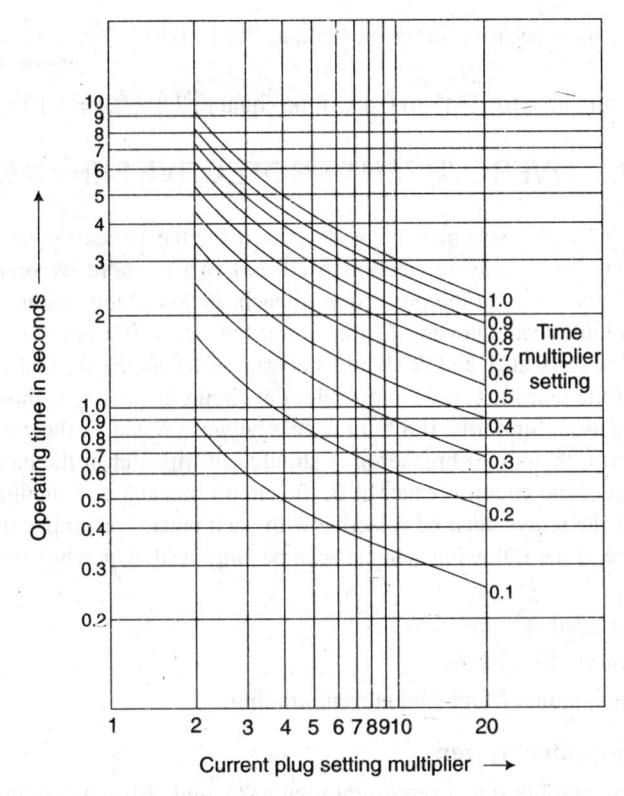

Fig. 5.4 (a) *Time-current characteristics for different values of TMS*

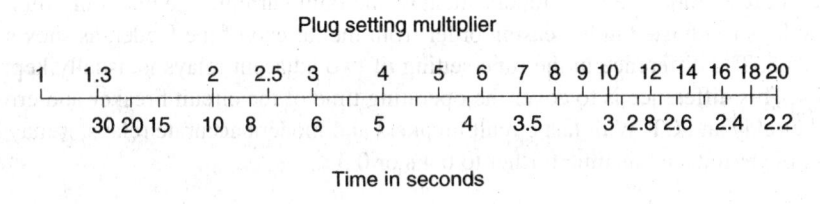

Fig. 5.4 (b) *Logarithmic scale for I.D.M.T. relay at TMS = 1*

Solution: CT ratio = 400/5 = 80

Relay current setting = 150% of 5 A = 1.5 × 5 A = 7.5 A

$$\text{PSM} = \frac{\text{Secondary current}}{\text{Relay current setting}}$$

$$= \frac{\text{Primary current (fault current)}}{\text{Relay current setting} \times \text{CT ratio}}$$

$$= \frac{6000}{7.5 \times 80} = 10$$

The operating time from the given table at PSM of 10 is 2.8 s. This time is for TMS = 1.

The operating time for TMS of 0.4 will be equal to $2.8 \times 0.4 = 1.12$ s.

5.5 OVERCURRENT PROTECTIVE SCHEMES

Overcurrent protective schemes are widely used for the protection of distribution lines. A radial feeder may be sectionalised and two or more overcurrent relays may be used, one relay for the protection of each section of the feeder, as shown in Fig. 5.5. If a fault occurs beyond C, the circuit breaker at substation C should trip. The circuit breakers at A and B should not trip as far as the normal operation is concerned. If the relay at C fails to operate, the circuit breaker at B should trip as a back-up protection. Similarly, if a fault occurs between B and C, the circuit breaker at B should trip; the circuit breaker at A should not trip. But in the case of failure of a relay and/or the circuit breaker at B, the circuit breaker at A should trip. Thus, it is seen that the relays must be selective with each other. For proper selectivity of the relays, one of the following schemes can be employed, depending on the system conditions.

  (i) Time-graded system
  (ii) Current-graded system
  (iii) A combination of time and current grading.

5.5.1 Time-graded System

In this scheme, definite-time overcurrent relays are used. When a definite-time relay operates for a fault current, it starts a timing unit which trips the circuit breaker after a preset time, which is independent of the fault current. The operating time of the relays is adjusted in increasing order from the far end of the feeder, as shown in Fig. 5.5. The difference in the time setting of two adjacent relays is usually kept at 0.5 s. This difference is to cover the operating time of the circuit breaker and errors in the relay and CT. With fast circuit breakers and modern accurate relays, it may be possible to reduce this time further to 0.4 s or 0.3 s.

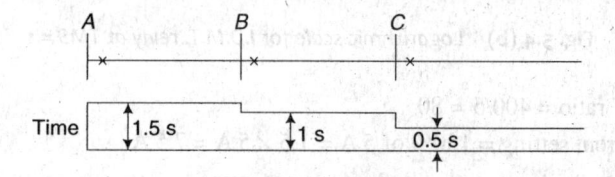

Fig. 5.5 *Time-graded overcurrent protection of a feeder*

When a fault occurs beyond C, all relays come into action as the fault current flows through all of them. The least time setting is for the relay placed at C. So it operates after 0.5 s and the fault is cleared. Now the relays at A and B are reset. If the relay or circuit breaker at C fails, the fault remains uncleared. In this situation, after 1 s, the relay at B will operate and the circuit breaker at B will trip. If the circuit breaker at B also fails to operate, after 1.5 s, circuit breaker at A will trip.

The drawback of this scheme is that for faults near the power source, the operating time is more. If a fault occurs near the power source, it involves a large current and hence it should be cleared quickly. But this scheme takes the longest time in clearing the heaviest fault, which is undesirable because the heaviest fault is the most destructive.

This scheme is suitable for a system where the impedance (distance) between substations is low. It means that the fault current is practically the same if a fault occurs on any section of the feeder. This is true for a system in which the source impedance Z_s is more than the impedance of the protected section, Z_1. If the neutral of the system is grounded through a resistance or an impedance, Z_s is high and $Z_s/(Z_s + Z_1)$ is not sufficiently lower than unity. In this situation, the advantage of inverse-time characteristic cannot be obtained. So definite relays can be employed, which are cheaper than I.D.M.T. relays. Definite-time relays are popular in Central Europe.

5.5.2 Current-graded System

In a current-graded scheme, the relays are set to pick-up at progressively higher values of current towards the source. The relays employed in this scheme are high set (high speed) instantaneous overcurrent relays. The operating time is kept the same for all relays used to protect different sections of the feeder, as shown in Fig. 5.6. The current setting for a relay corresponds to the fault current level for the feeder section to be protected.

Ideally, the relay at B should trip for faults any where between B and C. But it should not operate for faults beyond C. Similarly, the relay at A should trip for faults between A and B. The relay at C should trip for faults beyond C. This ideal operation is not achieved due to the following reasons.

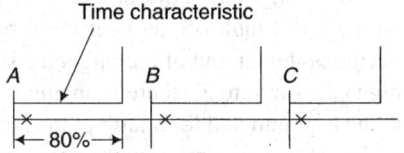

Fig. 5.6 *Instantaneous overcurrent protection of a feeder*

(i) The relay at A is not able to differentiate between faults very close to B which may be on either side of B. If a fault in the section BC is very close to the station B, the relay at A 'understands' that it is in section AB. This happens due to the fact that there is very little difference in fault currents if a fault occurs at the end of the section AB or in the beginning of the section BC.

(ii) The magnitude of the fault current cannot be accurately determined as all the circuit parameters may not be known.

(iii) During a fault, there is a transient conditions and the performance of the relays is not accurate.

Consequently, to obtain proper discrimination, relays are set to protect only a part of the feeder, usually about 80%. Since this scheme cannot protect the entire feeder, this system is not used alone. It may be used in conjunction with I.D.M.T. relays, as shown in Fig. 5.7.

The performance of instantaneous relays is affected by the dc component of transients. The

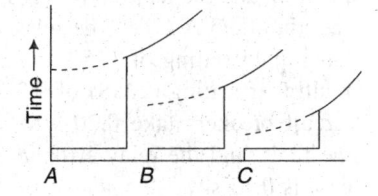

Fig. 5.7 *Combined instantaneous and I.D.M.T. protection*

error introduced by the dc offset component causes the relay to overreach. Higher the X/R ratio of the system, greater is the problem. A dc filter is used to overcome this problem. In the USA an instantaneous relay, employing induction cup type construction is used for this purpose as it is less sensitive to the d.c. offset component. A less expensive solution is to employ a relay as shown in Fig. 2.2(b). This arrangement also provides a high reset to pick-up ratio, more than 90%.

The current-graded scheme is used where the impedance between substations is sufficient to create a margin of difference in fault currents. For such a system Z_s is smaller compared to Z_1. The advantage of this system as compared to the time-graded scheme is that the operating time is less near the power source.

5.5.3 Combination of Current and Time-grading

This scheme is widely used for the protection of distribution lines. I.D.M.T. relays are employed in this scheme. They have the combined features of current and time-grading. I.D.M.T. relays have current as well as time setting arrangements. The current setting of the relay is made according to the fault current level of the particular section to be protected. The relays are set to pickup progressively at higher current levels, towards the source. Time setting is also done in a progressively increasing order towards the source. The difference in operating times of two adjacent relays is kept 0.5 s.

An inverse time-current characteristic is desirable where Z_s is small compared with Z_1. If a fault occurs near the substation, the fault current is $I = E/Z_s$. If a fault occurs at the far end of the protected section, the fault current $I = E/(Z_s + Z_1)$. If Z_1 is high compared to Z_s, there is an appreciable difference in the fault current for a fault at the near end and for a fault at the far end of the protected section of the feeder. For such a situation, a relay with inverse-time characteristic would trip faster for a fault near the substation, which is a very desirable feature. Inverse time relays on solidly grounded systems have an advantage. Definite-time characteristic is desirable where Z_s is large compared to Z_1. An I.D.M.T. characteristic is a compromise. At lower values of fault current, its characteristic is an inverse-time characteristic. At higher values of fault current, it gives a definite-time characteristic.

Though I.D.M.T. relays are widely used for the protection of distribution systems and some other applications, in certain situations very inverse and extremely inverse relays are used instead of I.D.M.T. relays. This has already been discussed in section 5.2.5 and 5.2.6.

Example 5.2 | An earth fault develops at point F on the feeder shown in the Fig. 5.8, and the fault current is 16000 A. The IDMT relays at points A and B are fed via 800/5 A CTs: The relay at B has a plug setting of 125% and time multiplier setting (TMS) of 0.2. The circuit breakers take 0.20 s to clear the fault, and the relay error in each case is 0.15 s.

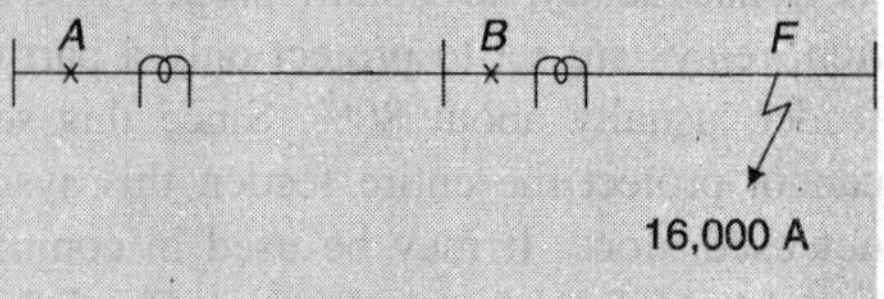

Fig. 5.8 System for Example 5.2

For a plug setting of 200% on the relay A, determine the minimum TMS on that relay for it not to operate before the circuit breaker at B has cleared the fault. A relay operating time curve is same as shown in Fig. 5.3.

Solution: The primary current in both relays is 16,000 A

$$\text{CT ratio} = 800/5 = 160$$

Thus secondary current

$$= \frac{\text{Primary current}}{\text{CT ratio}}$$

$$= \frac{16{,}000}{160} = 100 \text{ A}$$

For relay at B,

$$\text{Current setting} = 125\% \text{ of } 5 \text{ A} = 1.25 \times 5 \text{ A}$$
$$= 6.25 \text{ A}$$

$$\text{PSM} = \frac{\text{Secondary current}}{\text{Relay current setting}}$$

$$= \frac{100}{6.25} = 16$$

From the curve in Fig. 5.3, the operating time at PSM of 16 for a TMS of 1 = 2.5 s

Since TMS of relay at B = 0.2,

Operating time of B = 0.2 × 2.5 s = 0.50 s

Discrimination time = time for breaker at B + twice relay error

$$= 0.20 + 2 \times 0.15 = 0.50 \text{ s}$$

This is because one relay may run rapidly while the second runs slowly. moreover, the relay at A does not reset until the breaker at B has interrupted the fault current. Any overshoot of the relay A has been neglected.

Hence time for relay at A = operating time for B + discrimination time

$$= 0.50 \text{ s} + 0.50 \text{ s} = 1.00 \text{ s}$$

Secondary current in A = 100 A

For relay at A, current setting = 200% of 5 A

$$= 2 \times 5 \text{ A} = 10 \text{ A}$$

Thus, $\quad \text{PSM} = \dfrac{100}{10} = 10$

From the curve in Fig. 5.3, the operating time at PSM of 10 for a TMS of 1 = 3.0 s.

But actual time required = 1.00 s

Hence required TMS for relay at A = 1.00/3.0

$$= 0.33$$

i.e., the minimum value of TMS of relay at A must be 0.33.

Example 5.3 | A 20 MVA transformer, which may be called upon to operate at 25% overload, feeds 11-kV busbars through a circuit breaker; other circuit breakers supply outgoing feeders. The transformer circuit breaker is equipped with 1000/5 A CTs and the feeder circuit breakers with 500/5 A CTs and all sets of CTs feed induction-type overcurrent relays. The relays on the feeder circuit breakers have a 125% plug setting and a 0.4 time setting. If a three phase fault current of 7500 A flows from the transformer to one of the feeders, find the operating time of the feeder relay, the minimum plug setting of the transformer relay, and its time setting assuming a discriminative time margin of 0.5 second. The time-current characteristic of the relays is same as shown in Fig. 5.3.

Solution:
Feeder

$$\text{Secondary current} = 7500 \times \frac{5}{500} = 75 \text{ A}$$

Relay current setting = 125% of 5 A = 1.25 × 5 = 6.25 A

$$\text{PSM} = \frac{\text{Secondary current}}{\text{Relay current setting}} = \frac{75}{6.25} = 12$$

From the curve in Fig. 5.3, the operating time at PSM of 12 for a TMS of 1 = 2.8 s

Since TMS of the relay = 0.4,

Operating time of the relay = 0.4 × 2.8 = 1.12 s

Transformer

$$\text{Overload current} = \frac{(1.25 \times 20) \times 10^3}{\sqrt{3} \times 11} = 1312 \text{ A}$$

$$\text{Secondary current} = 1312 \times \frac{5}{1000} = 6.56 \text{ A}$$

$$\text{Plug Setting Multiplier (PSM)} = \frac{6.56}{\text{PS} \times 5}$$

Where PS means plug setting of the relay.

Since the transformer relay must not operate to overload current, its plug setting multiplier (PSM) must be less than 1, i.e., PS × 5 > 6.56. Thus plug setting (PS) > 6.56/5 > 1.31% or 131%.

The plug settings are restricted to standard values (See section 5.3) in intervals of 25%, so the nearest value is 150%.

$$\text{Secondary fault current} = 7500 \times \frac{5}{1000} = 37.5 \text{ A}$$

Relay current setting = 150% of 5 A = 1.5 × 5 A = 7.5 A

$$\text{PSM} = \frac{\text{Secondary fault current}}{\text{Relay current setting}} = \frac{37.5}{7.5} = 5$$

The operating time from the curve in Fig. 5.3 at PSM of 5 and TMS of 1 = 4.7 seconds

But, actual operating time required

$\quad$ = Operating time of feeder relay + discriminative time margin
$\quad$ = 1.12 sec + 0.50 sec
$\quad$ = 1.62 sec

Hence required time multiplier setting (TMS) = 1.62/4.7 = 0.345

Example 5.4 | Two relays R_1 and R_2 are connected in two sections of a feeder as shown in Fig. 5.9. CTs are of ratio 1000/5 A. The plug setting of relay R_1 is 100% and R_2 is 125%. The operating time characteristics of the relays is same as given in Table 5.3 of Example 5.1.

The time multiplier setting of the relay R_1 is 0.3. The time grading scheme has a discriminative time margin of 0.5 s between the relays. A three-phase short circuit at F results in a fault current of 5000 A. Find the actual operating times of R_1 and R_2. What is the time multiplier setting (TMS) of R_2.

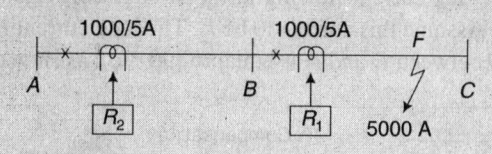

Fig. 5.9 System for Example 5.4

Solution: CT secondary current = $5000 \times \dfrac{5}{1000} = 25$ A

Relay R_1

$\quad$ Plug setting = 100%
$\quad$ Current setting = 5 A

$$\text{PSM of } R_1 = \frac{\text{Secondary current}}{\text{Relay current setting}} = \frac{25}{5} = 5$$

Operating time of the relay at PSM of 5 and TMS of 1 from the table of Example 5.1 = 4 seconds.

Since TMS of the relay R_1 is 0.3, the actual operating time of the relay = 0.3×4 = $0.3 \times 4 = 1.2$ seconds

Relay R_2

$\quad$ Plug setting = 125%

Relay current setting = 125% of 5 A = $1.25 \times 5 = 6.25$ A

$$\text{PSM} = \frac{\text{Secondary current}}{\text{Relay current setting}} = \frac{25}{6.25} = 4$$

Operating time at PSM of 4 and TMS of 1 from the table of Example 5.1 = 5 seconds

$\quad$ Actual operating time of R_2 = Operating time of R_1 + time grading margin
$\quad\quad\quad$ = 1.2 + 0.5
$\quad\quad\quad$ = 1.7 seconds

Hence, $\quad\quad$ TMS = 1.7/5 = 0.34

5.6 REVERSE POWER OR DIRECTIONAL RELAY

Figure 5.10(a) shows an electromechanical directional relay. A directional relay is energised by two quantities, namely voltage and current. Fluxes ϕ_1 and ϕ_2 are set up by voltage and current, respectively. Eddy currents induced in the disc by ϕ_1 interact with ϕ_2 and produce a torque. Similarly, ϕ_2 also induces eddy currents in the disc, which interact with ϕ_1 and produce a torque. The resultant torque rotates the disc. The torque is proportional to $VI \cos \phi$, where ϕ is the phase angle between V and I. The torque is maximum when voltage and current are in phase. To produce maximum torque during the fault condition, when the power factor is very poor, a compensating winding and shading are provided, as shown in Fig. 5.10(a).

Earlier it has been mentioned that the torque produced by an induction relay is given by $T = \phi_1 \phi_2 \sin \theta \propto I_1 I_2 \sin \theta$, where ϕ_1 and ϕ_2 are fluxes produced by I_1 and I_2, respectively. The angle between ϕ_1 and ϕ_2 or I_1 and I_2 is θ. If one of the actuating quantities is voltage, the current flowing in the voltage coil lags behind voltage by approximately 90°. Assume this current to be I_2. The load current I (say I_1) lags V by ϕ. Then the angle θ between I_1 and I_2 is equal to $(90 - \phi)$, as shown in Fig. 5.10(b).

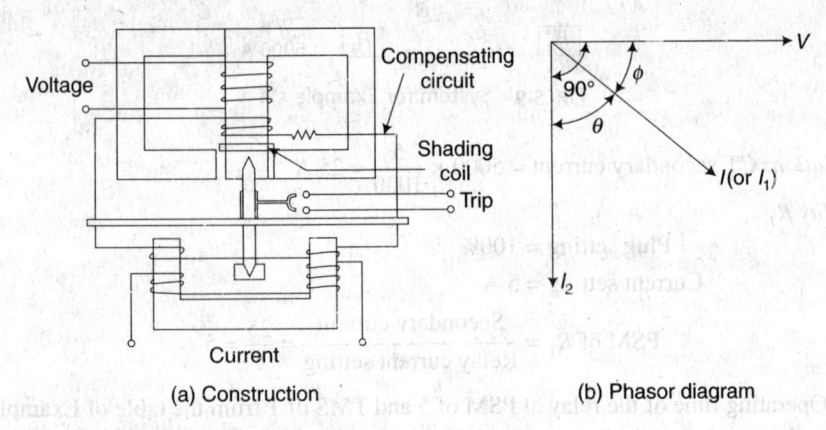

Fig. 5.10 *Induction disc type directional relay*

(a) Construction (b) Phasor diagram

$T = I_1 I_2 \sin (90 - \phi) \propto I_1 I_2 \cos \phi \propto VI \cos \phi$

An induction cup construction can also be used to produce a torque proportional to $VI \cos \phi$. The arrangement is shown in Fig. 5.11. Two opposite poles are energised by voltage and the other two poles by current. Here voltage is a polarising quantity. The polarising quantity is one which produces one of the two fluxes. The polarising quantity is taken as a reference with respect to the other quantity which is current in this case.

Torque produced is positive when $\cos \phi$ is positive, i.e. ϕ is less than 90°. When ϕ is more than 90° (between 90° and 180°), the torque is

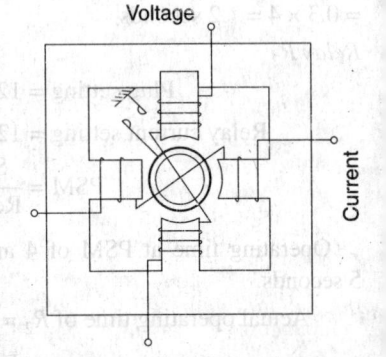

Fig. 5.11 *Induction cup type directional relay*

negative. At a particular relay location, when power flows in the normal direction, the relay is connected to produce negative torque. The angle between the actuating quantities supplied to the relays is kept $(180° - \phi)$ to produce negative torque. If due to any reason, the power flows in the reverse direction, the relay produces a positive torque and it operates. In this condition, the angle between the actuating quantities ϕ is kept less than $90°$ to produce a positive torque. This is shown in Fig. 5.12(a). For normal flow of power, the relay is supplied with V and $-I$. For reverse flow, the actuating quantities become V and I. Torque becomes $VI \cos \phi$, i.e. positive. This can be achieved easily by reversing the current coil, as shown in Fig. 5.12(b).

Relaying units supplied with single actuating quantity discussed earlier are non-directional overcurrent relays. Non-directional relays are simple and less expensive than directional relays.

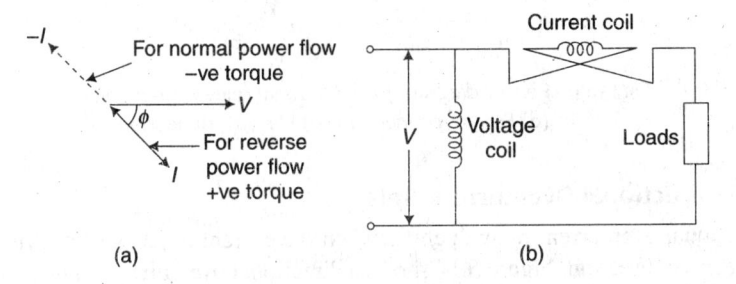

Fig. 5.12 (a) Phasor diagram for directional relay
(b) Connection of current coil for reverse power relay

5.6.1 Directional Relay Connections

When a close-up fault occurs, the voltage becomes low and the directional relay may not develop sufficient torque for its operation. Under certain fault conditions the power factor may be very low due to which insufficient torque is developed. If the relay is connected in the normal way to develop a torque proportional to $VI \cos \phi$, these types of problems cannot be overcome. To get sufficient torque during all types of faults, irrespective of their locations with respect to the relays, the relay connections are to be modified. Each relay is energised by current from its respective phase and voltage from the other two phases.

There are two methods of connections, one of them is known as the $30°$ connection and the other the $90°$ connection. In the $30°$ connection the current coil of the relay of phase A is energised by phase current I_A and line voltage $V_{A\text{-}C}$. Similarly, the relay in phase B is energised by I_B and $V_{B\text{-}A}$, the relay in phase C with I_C and $V_{C\text{-}B}$, as shown by the phasor diagram, Fig. 5.13(a). The relay is designed to develop maximum torque when its current and voltage are in phase. This condition with present connection is satisfied when the system power factor is 0.866 lagging. See Ref. 2 for details.

The $90°$ connection gives better performance under most circumstances. In this connection, the relay in phase A is energised by I_A and $V_{B\text{-}C}$, B phase relay by I_B and $V_{C\text{-}A}$ and C phase relay by I_C and $V_{A\text{-}B}$, as shown in Fig. 5.13(b). The relays are designed to develop maximum torque when the relay current leads voltage by

45° and have internal compensation. For all types of faults, L-L, L-G, 2L-G, 3-ϕ, the phase angle seen by the relay is well below 90°. This connection also ensures adequate voltage polarisation, except for a three-phase close-up fault when the voltages on all phases become very small. For three-phase symmetrical faults the 90° connection is better than the 30° connection (see Ref. 2 for more details).

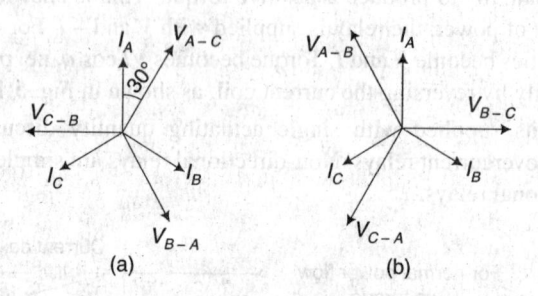

Fig. 5.13 *Phasor diagram for directional relay connections:*
(a) For 30° connection (b) For 90° connection

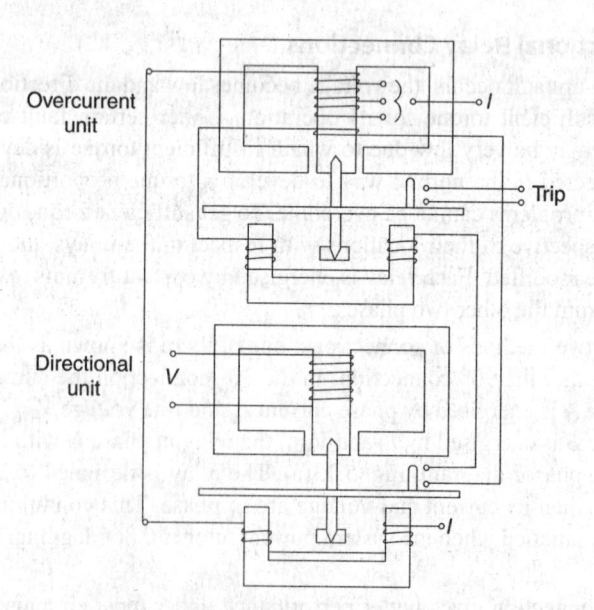

5.6.2 Directional Overcurrent Relay

A directional overcurrent relay operates when the current exceeds a specified value in a specified direction. Figure 5.14 shows a directional overcurrent relay. It contains two relaying units, one overcurrent unit and the other a directional unit. For directional control, the secondary winding of the overcurrent unit is kept open. When the

Fig. 5.14 *Directional overcurrent relay*

directional unit operates, it closes the open contacts of the secondary winding of the overcurrent unit. Thus, a directional feature is attributed to the overcurrent relay. The overcurrent unit may be of either a wattmeter or shaded pole type. In shaded pole type, the opening is made in the shading coil which is in this case a wound coil instead of an ordinary copper strip.

5.7 PROTECTION OF PARALLEL FEEDERS

Figure 5.15 shows an overcurrent protective scheme for parallel feeders. At the sending end of the feeders (at A and B), non-directional relays are required. The symbol $\leftrightarrow$ indicates a non-directional relay. At the other end of feeders (at C and D),

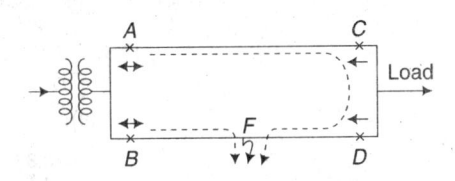

Fig. 5.15 Protective scheme for parallel feeder

directional overcurrent relays are required. The arrow mark for directional relays placed at C and D indicate that the relay will operate if current flows in the direction shown by the arrow. If a fault occurs at F, the directional relay at D trips, as the direction of the current is reversed. The relay at C does not trip, as the current flows in the normal direction. The relay at B trips for a fault at F. Thus, the faulty feeder is isolated and the supply of the healthy feeder is maintained.

If non-directional relays are used at C and D, both relays placed at C and D will trip for a fault at F. This is not desired as the healthy feeder is also tripped. Due to this very reason relays at C and D are directional overcurrent relays. For faults at feeders, the direction of current at A and B does not change and hence relays used at A and B are non-directional.

5.8 PROTECTION OF RING MAINS

Figure 5.16(a) shows an overcurrent scheme for the protection of a ring feeder. Figure 5.16(b) is another way of drawing the same scheme. Compared with radial feeders, the protection of ring feeders is costly and complex. Each feeder requires two relays. A non-directional relay is required at one end and a directional relay at the other end. The operating times for relays are determined by considering the grading, first in one direction and then in the other direction, as shown in Fig. 5.16.

If a fault occurs at F_1 as shown in Fig. 5.16(a), the relays at C' and D' will trip to isolate the faulty feeder. The relay at C will not trip as the fault current is not flowing in its tripping direction though its operating time is the same as that of C'. Similarly, the relays at B and D will not trip as the fault currents are not in their tripping direction, though their operating time is less than the operating time of B' and D' respectively. Figure 5.16(b) is an alternative way of drawing the same scheme. In this figure, loads, though present are not shown on buses A, B and D so as to make the figure simple to understand. If a fault occurs at F_2, the relays at A' and D will trip.

Fig. 5.16(c) shows a scheme involving even greater number of feeders.

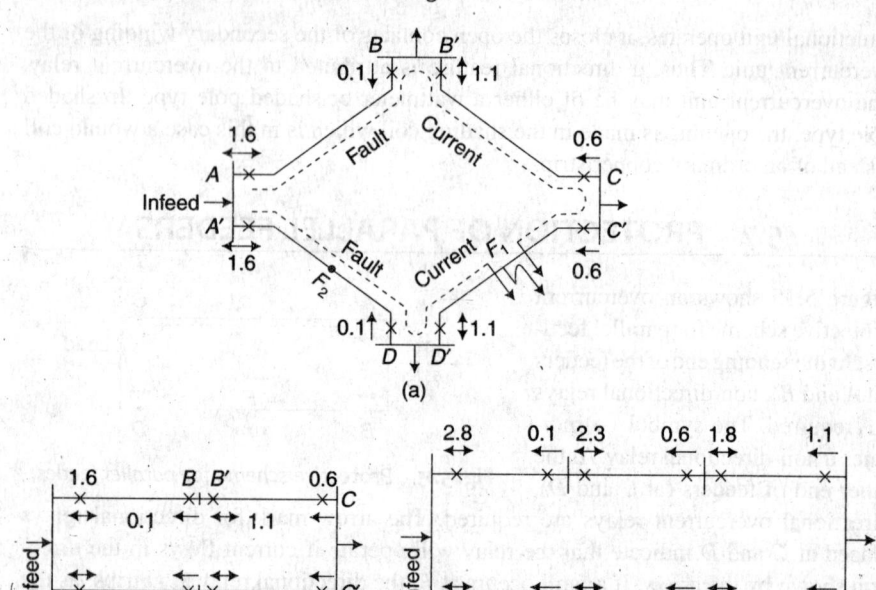

Fig. 5.16 *Protection of ring feeder*

5.9 EARTH FAULT AND PHASE FAULT PROTECTION

A fault which involves ground is called an *earth fault*. Examples are—single line to ground (L-G) fault and double line to ground (2L-G) fault. Faults which do not involve ground are called *phase faults*. The protective scheme used for the protection of an element of a power system against earth faults is known as earth fault protection. Similarly, the scheme used for the protection against phase faults is known as phase fault protection.

5.9.1 Earth Fault Relay and Overcurrent Relay

Relays which are used for the protection of a section (or an element) of the power system against earth faults are called earth fault relays. Similarly, relays used for the protection of a section of the power system against phase faults are called phase fault relays or overcurrent relays. The operating principles and constructional features of earth fault relays and phase fault relays are the same. They differ only in the current levels of their operation. The plug setting for earth fault relays varies from 20% to 80% of the CT secondary rating in steps of 10%. Earth fault relays are more sensitive than the relays used for phase faults. The plug setting for phase fault relays varies from 50% to 200% of the CT secondary rating in steps of 25%. The name phase fault relay or phase relays is not common. The common name for such relays is overcurrent relay. One should not confuse this term with the general meaning of overcurrent

Overcurrent Protection 239

relay. In a general sense, a relay which operates when the current exceeds its pick-up value is called an overcurrent relay. But in the context under consideration, i.e. phase fault protection and earth fault protection, the relays which are used for the protection of the system against phase faults are called overcurrent relays.

5.9.2 Earth Fault Protective Schemes

An earth fault relay may be energised by a residual current. As shown in Fig. 5.17(a), i_a, i_b and i_c are currents in the secondary of CTs of different phases. The sum ($i_a + i_b + i_c$) is called residual current. Under normal conditions the residual current is zero. When an earth fault occurs, the residual current is non-zero. When it exceeds pick-up value, the earth fault relay operates. In this scheme, the relay operates only for earth faults. During balanced load conditions, the earth fault relay carries no current; hence theoretically its current setting may be any value greater than zero. But in practice, it is not true as ideal conditions do not exist in the system. Usually, the minimum plug setting is made at 20% or 30%. The manufacturer provides a range of plug settings for earth fault relay from 20% to 80% of the CT secondary rating in steps of 10%.

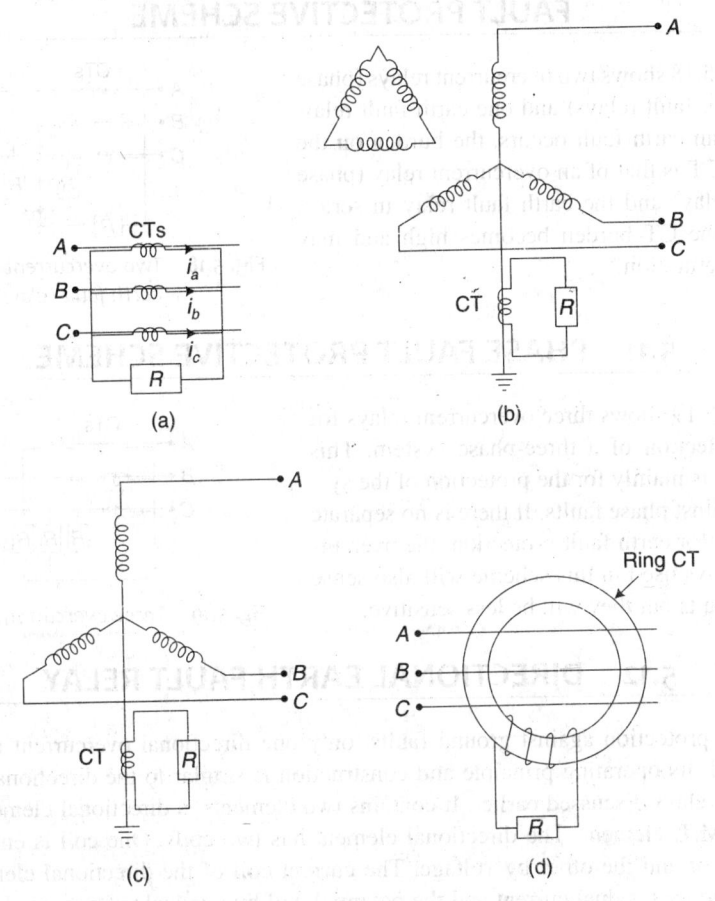

Fig. 5.17 *Various earth fault protective schemes*

The magnitude of the earth fault current depends on the fault impedance. In case of an earth fault, the fault impedance depends on the system parameter and also on the type of neutral earthing. The neutral may be solidly grounded, grounded through resistance or reactance. The fault impedance for earth faults is much higher than that for phase faults. Hence, the earth fault current is low compared to the phase fault currents. An earth fault relay is set independent of load current. Its setting is below normal load current. When an earth fault relay is set at lower values, its ohmic impedance is high, resulting in a high CT burden.

Figure 5.17(b) and 5.17(c) show an earth fault relay used for the protection of transformer and an alternator, respectively. When an earth fault occurs, zero-sequence current flows through the neutral. It actuates earth fault relay.

Figure 5.17(d) shows the connection of an earth fault relay using a special type of CT known as a core-balance CT, which encircles the three-phase conductors.

5.10 COMBINED EARTH FAULT AND PHASE FAULT PROTECTIVE SCHEME

Figure 5.18 shows two overcurrent relays (phase to phase fault relays) and one earth fault relay. When an earth fault occurs, the burden on the active CT is that of an overcurrent relay (phase fault relay) and the earth fault relay in series. Thus, the CT burden becomes high and may cause saturation.

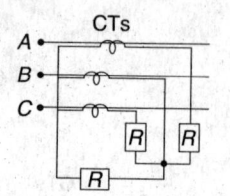

Fig. 5.18 *Two overcurrent and one earth fault relays*

5.11 PHASE FAULT PROTECTIVE SCHEME

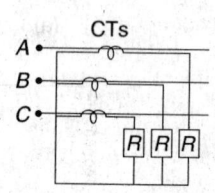

Figure 5.19 shows three overcurrent relays for the protection of a three-phase system. This scheme is mainly for the protection of the system against phase faults. If there is no separate scheme for earth fault protection, the overcurrent relays used in this scheme will also sense earth faults but they will be less sensitive.

Fig. 5.19 *Three overcurrent relays*

5.12 DIRECTIONAL EARTH FAULT RELAY

For the protection against ground faults, only one directional overcurrent relay is required. Its operating principle and construction is similar to the directional overcurrent relays discussed earlier. It contains two elements, a directional element and an I.D.M.T. element. The directional element has two coils. One coil is energised by current and the other by voltage. The current coil of the directional element is energised by residual current and the potential coil by residual voltage, as shown in Fig. 5.20(a). This connection is suitable for a place where the neutral point is not available. If the neutral of an alternator or transformer is grounded, connections are

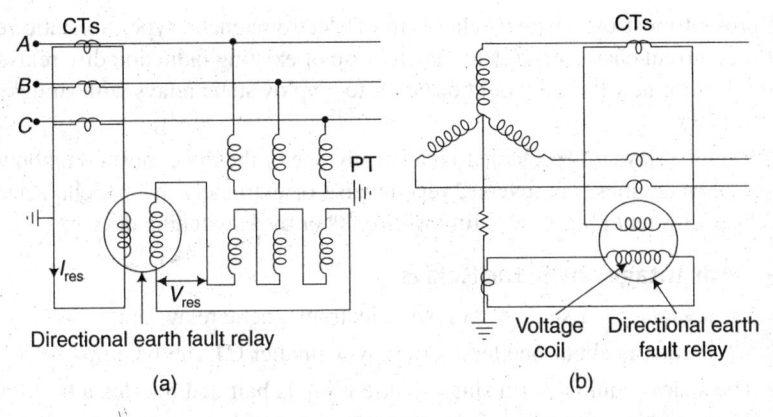

Fig. 5.20 Connection of a directional earth fault relay

made as shown in Fig. 5.20(b). If the neutral point is grounded through a VT, the voltage coil of the directional earth fault relay may be connected to the secondary of the VT. The I.D.M.T. element has a plug setting of 20% to 80%.

A special five limbs VT which can energise both the earth fault relay as well as the phase fault relays, as shown in Fig. 5.21, may be used.

5.13 STATIC OVERCURRENT RELAYS

The general expression for the operating time of a time-current relay is

$$t = \frac{K}{I^n - 1}$$

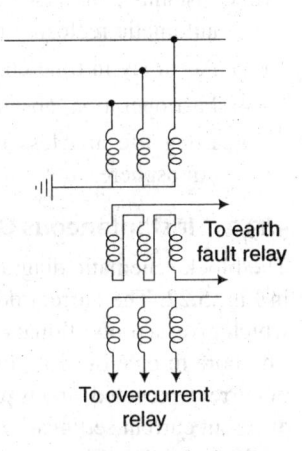

Fig. 5.21 Five limb VT

Inverse time electromagnetic relays produce time-current curves according to this law only up to a few times the CT rating because of magnetic saturation. The time-current characteristic does not follow a simple mathematical equation and it is very difficult to obtain consistency between the characteristics of individual relays. The time-current characteristic of static relays depends on the R-C circuit which can be precisely controlled. In the static relays circuit, components are linear, thereby it becomes easier to produce characteristics according to the above law. With an electromagnetic unit, the maximum value of n which is the index of I may be only up to 2. With static relay time-current characteristic with higher values of n can easily be realised. The characteristic of the form $t = K/I^n$ can be realised which will give straight line characteristic on log t/log I graph. Since the time-current characteristics, given by $t = K/I^n$ are not asymptotic to the pick-up value of the current, a separate device to control pick-up is required. Similarly, for an I.D.M.T. relay, a separate unit will provide the definite time portion of the characteristic. With straight line curves there is a great saving in computing relay time settings.

At present most overcurrent relays are of electromagnetic type. So static relays with time-current characteristics to match those of existing induction disc relays can be used. But for new lines it would be better to employ static relays with straight line characteristics.

With static relays it is possible to realise any one of the three most common time-current characteristics, i.e. inverse, very inverse or extremely inverse characteristic using three different plug-in R-C timing circuits or by a switching device.

5.13.1 Advantages of Static Relays

The main advantages of static relays over electromagnetic relays are:
  (i) CT burden is about one tenth, thereby a smaller CT can be employed.
  (ii) The space required for a single-phase relay is half and that for a three-phase relay is about one third. Consequently, the panel space and overall cost of installation are reduced. This helps in miniaturization of control equipment.
  (iii) Instantaneous reset can easily be achieved. This allows the application of automatic reclosing of circuit breaker.
  (iv) Accuracy in time-current characteristics.
  (v) Fast operation, absence of mechanical inertia and bouncing of contacts.
  (vi) Long life and less maintenance, immunity to vibration, dust and polluted atmosphere.

5.13.2 Instantaneous Overcurrent Relay

The block schematic diagram of the static instantaneous overcurrent relay is shown in Fig. 5.22. The current derived from the main CT is fed to the input transformer which gives a proportional output voltage. The input transformer has an air gap in the iron core to give linearity in the current/voltage relationship up to the highest value of current expected, and is provided with tappings on its secondary winding to obtain different current settings. The output voltage of the transformer is rectified through a rectifier and then filtered at a single stage to avoid undesirable time delay in filtering so as to ensure high speed of operation. A limiter made of a zener diode is also incorporated in the circuit to limit the rectified voltage to safe values even when the input current is very high under fault conditions. A fixed portion of the rectified and filtered voltage (through a potential divider) is compared against a preset pick-up value by a level detector and if it exceeds the pick-up value, a signal through an amplifier is given to the output device which issues the trip signal. The output device may either be a static thyristor circuit or an electromagnetic slave relay.

Input current → Input transformer → Rectifier, filter → Level detector → Amplifier → Output device → Trip signal

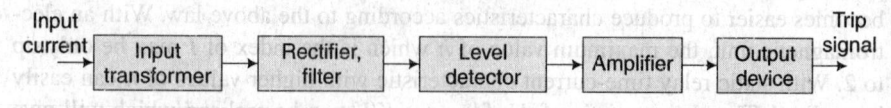

Fig. 5.22 *Block diagram of static instantaneous overcurrent relay*

5.13.3 Definite Time Overcurrent Relay

The operating time of a definite time overcurrent relay is constant, irrespective of the level of the fault current. In this case, an intentional time delay is introduced through a timing circuit. Figure 5.23 shows the simplified block diagram of a definite time overcurrent relay. The input current signal derived from the main CT is converted to a proportional voltage signal by the input transformer and then rectified, filtered and compared with the preset threshold value of the level detector (1). If the voltage exceeds the preset threshold value, the level detector gives an output voltage, thereby the charging of the capacitor C of the RC timing circuit starts. As soon as the voltage across the capacitor exceeds the preset threshold value (V_T) of level detector (2), a signal through the amplifier is given to the output device which issues the trip signal. Potentiometers P_1 and P_2 are used for current setting and time setting, respectively.

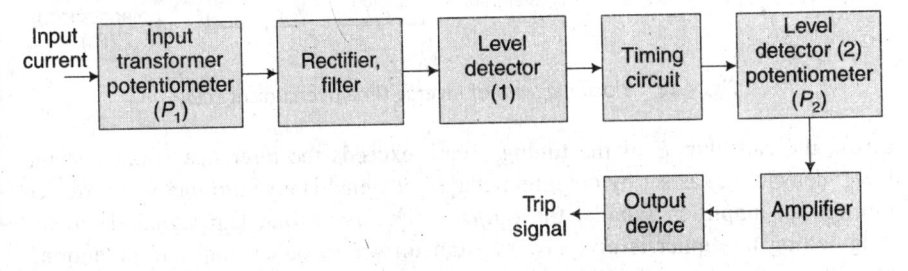

Fig. 5.23 Block diagram of definite time overcurrent relay

If V_T is the threshold value of the level detector, the time T_C required to reach this voltage depends upon the charging time of the capacitor C of the RC timing circuit, given by,

$$T_C = RC \log_e \left[\frac{V}{V - V_T} \right]$$

where V is the voltage applied to the capacitor. If V, R and C are constant, the charging time for a given value of V_T will be constant. The time T_C can be varied by varying R-C combinations and V_T. In this case since the capacitor charging is done from a fixed d.c. output voltage of level detector (1), the operating time of the relay for particular values of R and C of the timing circuit and V_T of the level detector (2) will be constant for different values of the fault current.

5.13.4 Inverse-time Overcurrent Relay

The operating time of the inverse-time overcurrent relay decreases with increasing fault current. For this relay with inverse-time characteristic, the charging of the capacitor of timing circuit takes place from a voltage proportional to current.

The block diagram of the inverse-time overcurrent relay is shown in Fig. 5.24. The current signal is converted to a proportional voltage signal by the input transformer and then rectified, filtered and compared with a reference voltage of the level detector (1) set by the potentiometer P_1. Under normal conditions, i.e. when the input current is low, switch S_1 is ON, shortcircuiting the capacitor C of the RC timing circuit and switch S_2 is OFF. As soon as the input voltage exceeds the preset reference voltage of the level detector (1), switch S_1 is switched OFF and switch S_2 is switched ON and the charging of capacitor C of the timing circuit starts from a voltage proportional to the current. Switches S_1 and S_2 are made of static components. When the voltage

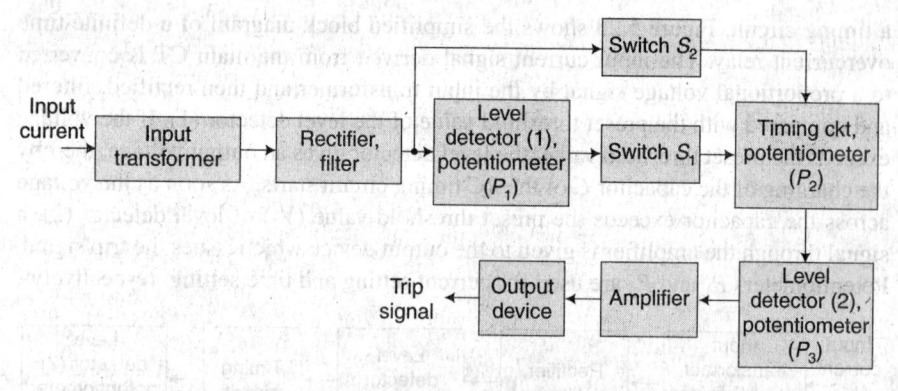

Fig. 5.24 Block diagram of inverse-time overcurrent relay

across the capacitor C of the timing circuit exceeds the reference voltage of the level detector (2) as set by potentiometer P_3, a signal is given to the output device through an amplifier. Finally, the output device issues the trip signal. Here the plug setting multiplier is given by the transformer secondary tap and potentiometer P_1 and the time multiplier setting is determined by potentiometers P_2 and P_3.

5.13.5 Directional Overcurrent Relay

The directional overcurrent relay incorporates a directional unit which responds to power flow in a specified direction. The directional relay senses the direction of power flow by means of a phase difference (ϕ) between voltage (V) and current (I). When ϕ exceeds a certain predetermined value and the current is above the pick-up value, the directional overcurrent relay operates. The directional relay is a double actuating quantity relay with one input as current I from CT and the other input as voltage V from VT.

In case of electromagnetic directional overcurrent relays, discrimination is affected when voltage drops down to very low values under fault conditions. In static directional relays, this problem is less serious because the static comparators used in these relays are inherently very sensitive and they can give reliable performance up to 1% of system voltage which is well within the minimum fault voltage.

Figure 5.25 shows the simplified block diagram of the directional overcurrent relay. The inputs V and I are applied to phase comparator. A phase shifter is added in voltage input circuit before applying it to the phase comparator to achieve the maximum output of the phase comparator under fault conditions. The output of the phase comparator is given to the level detector and then to the output device through an amplifier. If the output of the phase comparator exceeds the preset reference voltage of the level detector, the output device issues the trip signal.

There are two main types of phase comparators used for the purpose. One of these is the Hall effect types comparator which has been used in USSR, whereas all other countries have preferred the rectifier bridge type of comparator due to its lower cost and the higher outputs obtainable as compared to the Hall elements.

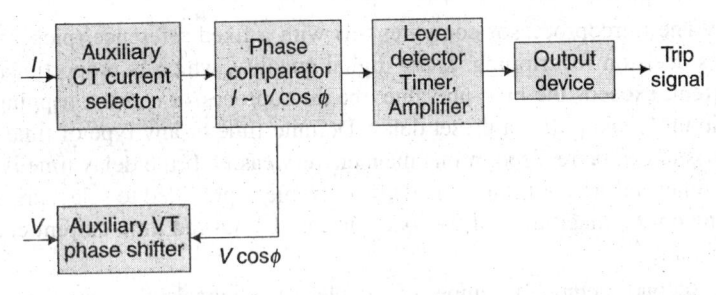

Fig. 5.25 Simplified block diagram of static directional overcurrent relay

5.14 NUMERICAL OVERCURRENT RELAYS

Numerical overcurrent relays are the latest development in the area of protection. These relays have been developed because of tremendous advancement in VLSI and computer hardware technology. A numerical overcurrent relay acquires sequential samples of the current signal (i.e., proportional voltage signal) in numeric (digital) data form through the data acquisition system (DAS) and processes the data numerically using a numerical filtering algorithm to extract the fundamental frequency component of the current and make trip decisions. Depending on the processor and processing tool used for processing of the current signal, numerical overcurrent relays are of the following types:

(i) Microprocessor-based overcurrent relays
(ii) Microcontroller-based overcurrent relays
(iii) Digital Signal Processor (DSP)-based overcurrent relays
(iv) Field Programmable Gate Arrays (FPGA)-based overcurrent relays.
(v) Artificial Neural Network (ANN) based numerical relays.

A protection scheme which incorporates numerical overcurrent relays for the protection of an element of a power system, is known as a numerical overcurrent protection scheme or numerical overcurrent protection.

5.14.1 Microprocessor-based Overcurrent Relays

Modern power networks require faster, more accurate and reliable protective schemes. Microprocessor-based protective schemes are capable of fulfilling these requirements. They are superior to electromechanical and static relays. These schemes have more flexibility due to their programmable approach when compared with static relays which have hardwired circuitry. With the same interfacing circuitry, a number of characteristics can be realised using different programs. Microprocessor-based schemes are more compact, accurate, reliable and fast.

There are two methods to realise overcurrent characteristics. One method employs a precision rectifier to convert ac signals to dc signals. Since the microprocessor-based system cannot process current signals, a voltage signal proportional to load current is obtained. The ac voltage signal is rectified and then converted to a digital quantity. The mircoprocessor compares this with a fixed reference (pick-up value) and takes decision for tripping. If the digital quantity, which is proportional to the load current, exceeds the pick-up value, the microprocessor sends a tripping signal to the circuit breaker after a preset delay. Definite-time or any type of time-current characteristic can be realised in this manner, very easily. If the delay time is a fixed one, the relay acts as a definite-time relay. If a table is provided to select preset time depending on the magnitude of the load current, the desired time-current characteristic is obtained.

In the second method, a number of samples of ac signals over one cycle or half cycle can be taken. From these samples rms values can be computed. The computed rms values of the load current are individually compared with the pick-up value and the desired definite-time or time-current characteristic is realised.

More details have been provided in Chapter 12.

5.14.2 Microcontroller-based Overcurrent Relays

Microcontrollers are single chip microcomputers in which the microprocessor (CPU), memory and I/O ports are all fabricated on a single chip. In a microcontroller based overcurrent relay, a microcontroller is used to perform all the functions of the relay. measures the current by acquiring them in numeric (digital) data form through a data acquisition system, processes the data numerically and makes trip decisions.

5.14.3 DSP-based Overcurrent Relay

In this relay, a dedicated digital signal processor (DSP) is used to perform all the functions of the relay, e.g. data acquisition, data processing, calculation of fault discriminants and making trip decisions.

5.14.4 FPGA-based Overcurrent Relay

In this relay a Field Programmable Gate Array (FPGA) is used to perform the functions of the overcurrent relay.

5.14.5 ANN-based Overcurrent Relay

In this relay, a properly trained and tested artificial neural network (ANN) is used for processing of the acquired digital signal and making the trip decisions.

EXERCISES

1. What are the various types of overcurrent relays? Discuss their area of applications.
2. Explain current setting and time setting.
   The current rating of an overcurrent relay is 5 A. PSM = 2, TMS = 0.3 CT ratio = 400/5, Fault current = 4000 A. Determine the operating time of the relay. At TMS = 1, operating time at various PSM are:

PSM	2	4	5	8	10	20
Operating time in seconds	10	5	4	3	2.8	2.4

3. An overcurrent relay of current rating 5 A and setting 150% is connected to the secondary of CT of ratio 400/5. Calculate the current in lines for which the relay picks up. **(Ans. 600 A)**

4. An earth-fault starting relay has a setting of 30%, and a current rating of 5 A. It is connected to a CT of ratio 500/5. Calculate pick-up current in primary for which the earth fault relay operates **(Ans. 150 A)**

5. The time-current (PSM) characteristic of an overcurrent relay for TMS of 1 is given in the Table 5.4.

Table 5.4

PSM	2	3	5	7	10	13	15	18	20
Operating time in seconds	10	6.8	4.4	3.4	2.8	2.5	2.4	2.3	2.2

If the current plug setting is adjusted to 50% and the time multiplier is adjusted to, 0.75, calculate the time of operation of the relay when the fault current is 3000 A and the relay is connected to a CT ratio 400/5. **(Ans. 1.8 sec.)**

6. Under what circumstances are overcurrent relays having very inverse and extremely inverse characteristics used?

7. Compare the time-current characteristics of inverse, very inverse and extremely inverse overcurrent relays. Discuss their area of applications.

8. Describe the techniques used to realise various time-current characteristics using electromechanical relays.

9. Can a relay, having a time-current characteristic steeper than extermely inverse relay, be realised using electromechanical construction?

10. What are the various overcurrent protective schemes? Discuss their merits, demerits and field of applications.

11. Two relays R_1 and R_2 are connected in two sections fo a feeder as shown in Fig. 5.26.
    Relay R_1 : CT ratio = 300/5, plug setting = 50%, TMS = 0.3
    Relay R_2 : CT ratio = 500/5, plug setting = 75%.
    A fault at F results in a fault current of 3000 A. Find TMS of R_2 to give time-grading margin of 0.5 sec between the relays.

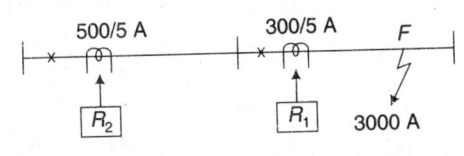

Fig. 5.26

The operating time characteristic of relays is given in the Table 5.3 of Example 5.1.

(Ans. 0.4

12. Explain what is meant by transient over-reach as applied to high set instantaneous overcurrent relays. What measures are taken to overcome this difficulty?
13. Why IDMT relays are widely used for overcurrent protection.
14. Describe the operating principle, constructional features and area of applications of reverse power or directional relay. What is a directional overcurrent relay?
15. Discuss a protective scheme for parallel feeders.
16. Distinguish between an earth fault relay and an overcurret relay. Discuss various methods to energise an earth fault relay.
17. How is a directional earth fault relay energise?
18. What are the advantages of static relays over electromechanical relays.
19. What are the advantages of numerical overcurrent relays over converntional Overcurrent relays?
20. What are the various types of numerical overcurrent relay? How can numerical overcurreont relay be realised?
21. What do you mean by numerical overcurrent protection scheme? How can a numerical protection scheme be developed using an ANN?

6 Distance Protection

6.1 INTRODUCTION

Distance protection is a widely used protective scheme for the protection of high and extra high voltage (EHV) transmission and sub-transmission lines. This scheme employs a number of distance relays which measure the impedance or some components of the line impedance at the relay location. The measured quantity is proportional to the line-length between the location of the relay and the point where the fault has occurred. As the measured quantity is proportional to the distance along the line, the measuring relay is called a distance relay. Overcurrent relays have been found unsuitable for the protection of transmission lines because of their inherent drawbacks of variable reach and variable operating time due to changes in source impedance and fault type. Distance relays have been developed to overcome the problems associated with the use of overcurrent relays for the protection of transmission lines.

Modern distance relays provide high speed fault clearance. They are used where overcurrent relays become slow, and there is difficulty in grading time-overcurrent relays for complicated networks. They are used for the protection of transmission and subtransmission lines at 220 kV, 132 kV, 66 kV, and 33 kV. Sometimes, they are also used at 11 kV. For 132 kV and 220 kV systems, the recent trend is to use carrier current protection. The relaying units used in carrier current protection are distance relays. They operate under the control of carrier signals. In case of the failure of carrier signal, they act as back-up protection.

A distance protection scheme is a non-unit system of protection. A single scheme provides both primary and back-up protection.

The most important and versatile family of relays is the distance-relay group. It includes the following types:

(i) Impedance relays
(ii) Reactance relays
(iii) MHO relays
(iv) Angle impedance relays
(v) Quadrilateral relays
(vi) Elliptical and other conic section relays

6.2 IMPEDANCE RELAY

An impedance relay measures the importance of the line at the relay location. When fault occurs on the protected line section, the measured impedance is the impedance of the line section between the relay location and the point of fault. It is proportional to the length of the line and hence, to the distance along the line. In general, the term impedance can be applied to a resistance alone, a reactance alone or a combination of the two. But in distance relaying terminology the term impedance includes both resistance as well as reactance.

6.2.1 Operating Principle of an Impedance Relay

To realise the characteristics of an impedance relay, current is compared with voltage at the relay location. The current produces a positive torque (operating torque) and the voltage produces a negative torque (restraining torque). The equation for the operating torque of an electromagnetic relay can be written as

$$T = K_1 I^2 - K_2 V^2 - K_3$$

where K_1, K_2 and K_3 are constants, K_3 being the torque due to the control-spring effect.

Neglecting the effect of the spring used, which is very small, the torque equation can be written as

$$T = K_1 I^2 - K_2 V^2$$

For the operation of the relay, the following condition should be satisfied.

$$K_1 I^2 > K_2 V^2 \quad \text{or} \quad K_2 V^2 < K_1 I^2$$

or
$$\frac{V^2}{I^2} < \frac{K_1}{K_2}$$

or
$$\frac{V}{I} < K \quad \text{where } K \text{ is a constant}$$

or
$$Z < K$$

For static and microprocessor-based relays, I is compared with V. For the operation of the relay, the following condition should be satisfied.

$$K_1 I > K_2 V \quad \text{or} \quad K_2 V < K_1 I$$

or
$$\frac{V}{I} < \frac{K_1}{K_2} \quad \text{or} \quad Z < K$$

The above expression explains that the relay is on the verge of operation when the ratio of V to I, i.e. the measured value of line impedance is equal to a given constant. The relay operates if the measured impedance Z is less than the given constant.

6.2.2 Impedance Relay Characteristic

Figure 6.1 shows the operating characteristic of an impedance relay in terms of voltage and current. In case of an electromagnetic relay, the characteristic is slightly bent near the origin due to the effect of the control spring. In case of microprocessor based or static relay, the characteristic will be a straight line.

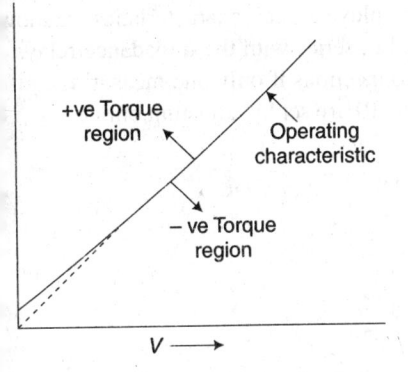

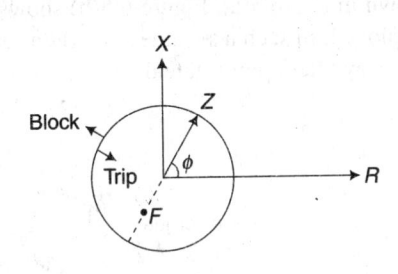

Fig. 6.1 Operating characteristic of an impedance relay

Fig. 6.2 Operating characteristic of an impedance relay on the R-X diagram

A more useful way is to draw a distance relay characteristic on the $R\text{-}X$ diagram. Figure 6.2 shows an impedance relay characteristic on the $R\text{-}X$ diagram, where $Z = $ represents a circle and $Z < K$ indicates the area within the circle. Thus, it is seen that the zone within the circle is the operating zone of the relay. Its radius is $Z = K$, which is the setting of the relay. K is equal to the impedance of the line which is to be protected. ϕ is the phase angle between V and I. As the operating characteristic is a circle, the relay operation is independent of the phase angle ϕ. The operation depends on the magnitude of Z. If a fault point is on the protected section of the line, it will be within the circle. For this condition, the relay will operate and send a tripping signal to the circuit breaker. The region outside the circle is the blocking zone. If a fault point lies in this zone, i.e. it is beyond the protected section of the line, the relay will not respond. In such a situation, the fault point may lie in the protection zone of some other relay.

The operating time of the relay is constant, irrespective of the fault location within the protected section, as shown in Fig. 6.3.

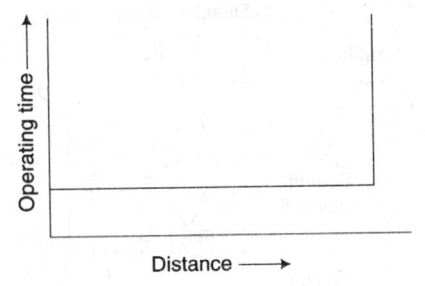

Fig. 6.3 Operating time characteristic of an impedance relay

6.2.3 Directional Units Used with Impedance Relays

It is evident from the impedance relay characteristic on the $R\text{-}X$ diagram that an impedance relay is a non-directional relay. As its characteristic is a circle, the relay will trip for a fault point lying within the circle, irrespective of the fact that the fault point lies either in the forward direction or in the reverse direction. For example, the relay will trip for a fault point F which is behind the relay location, i.e. in the reverse direction as shown in Fig. 6.2. It is always desired that a relay should operate for faults lying only in the forward direction. It should never operate for fault points lying in the reverse direction. To restrict the tripping zone in the forward direction only, a directional unit is included in the protective scheme. At any location, three impedance relays and a directional unit are employed. Their characteristics are shown in Fig. 6.4. The directional unit is connected in series with the impedance relays shown in Fig. 6.5(a). Figure 6.5(b) shows connections if only one measuring unit employed. In such a scheme zone II and zone III are set by adjusting ohmic reach of the relay after appropriate delay.

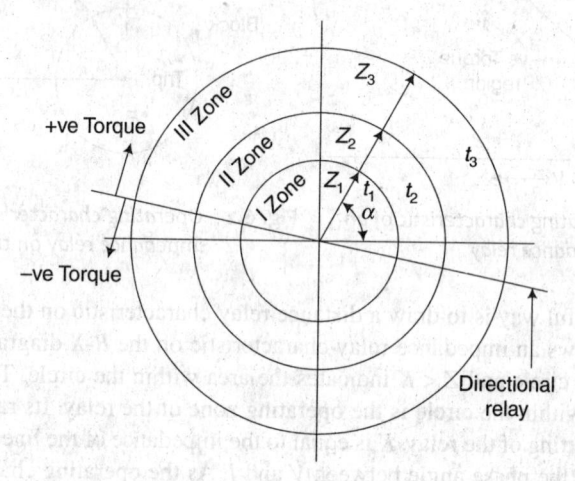

Fig. 6.4 Characteristics of three-zone impedance relays with directional unit

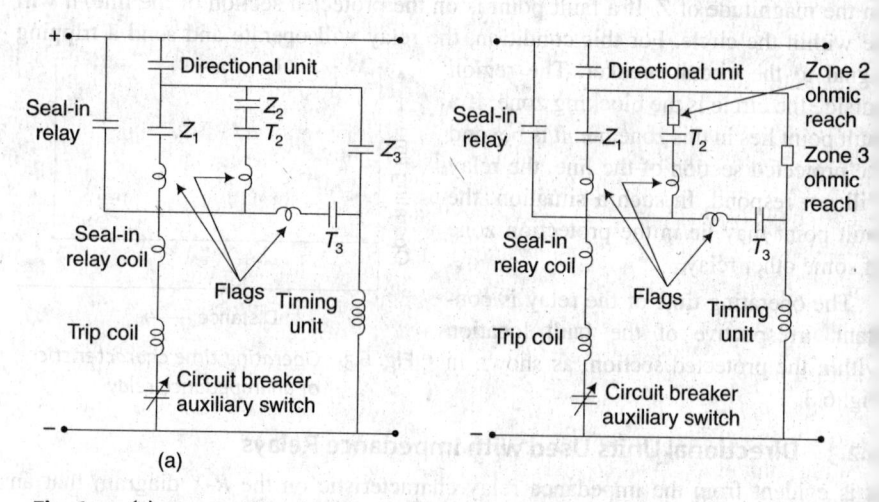

Fig. 6.5 (a) Connections of impedance relays (b) Connections of one measuring unit

The directional unit has a straight line characteristic as shown in the figure. It allows impedance relays to see only in the forward direction. The torque equation of the directional unit is given by $T = KVI \cos(\phi - \alpha)$, neglecting spring-effect. Here, ϕ is the angle between V and I, and α is the angle of maximum torque. For the operation of the relay, T should be positive and hence,

$$KVI \cos(\phi - \alpha) > 0$$

or

$$\cos(\phi - \alpha) > 0 \quad \text{or} \quad (\phi - \alpha) < \pm 90°$$

The directional relay employs two pairs of contacts, one pair is placed in series with the contacts of the impedance relay. The other pair is connected to energise an auxiliary relay. The contacts of the auxiliary relay, when deenergised, short circuit the current coil of the impedance relay. The contacts of the auxiliary relay are opened when a fault occurs in the forward direction. This type of a control is essential to avoid a contact race between the impedance relay and the directional relay in interconnected or double circuit lines. See details in Ref. 5.

6.2.4 Protective Scheme Using Impedance Relays

Three units of impedance relays are required at a particular location for three zones of protection. It is normal practice to adjust the first unit to protect only up to 80%-90% of the protected line. The protected zone of the first unit is called the first zone of protection. It is a high speed unit and is used for the primary protection of the protected line. Its operation is instantaneous; about 1 to 2 cycles. This unit is not set to protect the entire line to avoid undesired tripping due to overreach. If the relay operates for a fault beyond the protected line, this phenomenon is called overreach. Overreach may occur due to transients during the fault condition.

The main purpose of the second unit is to protect the rest of the protected line, which is beyond the reach of the first unit. The setting of the second unit is so adjusted that it operates the relay even for arcing faults at the end of the line. To achieve this, the unit must reach beyond the end of the line. In other words, its setting must take care of underreach caused by arc resistance. Underreach is also caused by intermediate current sources, errors in data available for initial setting of the relay and errors in CT, VT and measurements performed by the relay. To take into account the underreaching tendency caused by these factors, the normal practice is to set the second zone reach up to 50% of the shortest adjoining line section. The protective zone of the second unit is known as the second zone of protection. The second zone unit operates after a certain time delay. Its operating time is usually 0.2 s to 0.5 s.

The third zone of protection is provided for back-up protection of the adjoining line. Its reach should extend beyond the end of the adjoining line under the maximum underreach which may be caused by arcs, intermediate current sources and errors in CT, VT and measuring units. The protective zone of the third stage is known as the third zone of protection. The setting of the third zone covers the first line, i.e. the protected line plus the longest second line plus 25% of the third line. The time-delay for the third unit is usually 0.4 s to 1 s. Figure 6.6 shows the operating time of impedance relays and is known as the stepped time-distance characteristic. A_1, A_2 and A_3 are operating times for the I, II, and III zone relays (placed at A) respectively. Similarly, B_1, B_2, B_3 are operating times for the I, II, and III zone relays, placed at B, respectively. Because of the cost factor and the panel space, it is not

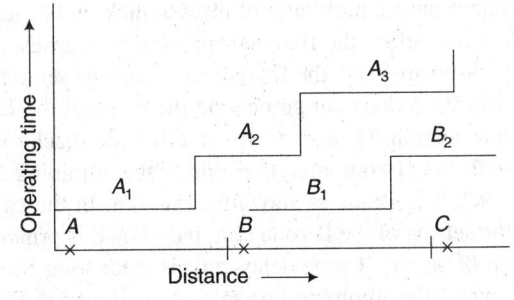

Fig. 6.6 *Stepped time-distance characteristics of impedance relays*

possible to use three measuring units for the 3 zones of protection. In a modern distance protective system, only one measuring unit is employed for all the three zones of protection. The ohmic reach is progressively increased by the timing unit to obtain the distance settings for the II and III zones. Sometimes two units can be employed, one unit for the I and II zone and a separate unit for the III zone. The distance setting for the II zone is done by the timing unit.

6.2.5 Connections of Impedance Relays

The circuit connections for circuit breaker trip coil, the contacts of directional and impedance relays, flags, timer, etc. are shown in Fig. 6.5(a). Z_1, Z_2 and Z_3 represent impedance relays. T_2 and T_3 are contacts of the timer for the second and the third unit, respectively. Figure 6.4 shows the characteristics of directional and impedance relays, where t_1, t_2 and t_3 are the operating time of the impedance relays. The characteristic circle for Z_1 is the smallest, the circle for Z_3 is the largest and the circle for Z_2 is intermediate. If any fault point lies within the circle Z_1 and it is in the forward direction, the directional relay and all the three impedance relays operate. Due to the operation of the first unit and the directional unit, the circuit breaker trips in a very short time period of t_1. A timer is energised whenever the directional unit and Z_3 operate. After a definite time-delay, the timer closes the contact T_2 first and then after some more delay, the contact T_3 also closes. The delay times for T_2 and T_3 are independently adjustable. Therefore, if a fault point lies in the characteristic circle Z_2 but outside the circle Z_1, the circuit breaker trips after the closure of the contact T_2, in time t_2. If a fault point lies within the characteristic circle Z_3 but outside the circles Z_1 and Z_2, the circuit breaker trips after the closure of the contact T_3, in time t_3. Whenever a relay operates, its flag indicates its operation. A seal-in relay is used to bypass the contacts of the main relays to save their costly and delicate contacts. Once the contacts of the main relay are closed and the current passes through the trip coil, the coil of the seal-in relay is energised and its contacts are closed. The circuit breaker auxiliary switch is a normally closed switch. When the circuit breaker trips, the auxiliary switch is opened to prevent unnecessary drainage of the battery. If only one measuring unit is employed, the circuit connection can be modified, as shown in Fig. 6.5(b).

6.2.6 Special Cases of Zone II and Zone III Settings

With the II zone relaying units, transient overreach need not be considered if the relays have a high ratio of reset to pick-up because any transients causing overreach die out before the II zone tripping time elapses. If relays with a low ratio of reset to pick-up are used, the II zone relay must be set to have a reach short enough so that its overreach does not go beyond the reach of the I zone relaying unit of the adjoining line section. In other words, the II zone reach must be short enough to be selective with the II zone relaying unit of the adjoining line section under maximum overreach conditions as shown in Fig. 6.6. In the case of short adjoining line sections, the setting of the II zone unit, based on this principle becomes difficult. To tackle the problem, the II zone delay-time is made long enough to be selective with the II zone time of the adjoining line section, as shown in Fig. 6.7.

Distance Protection

The third zone unit is provided for the back-up protection of the adjoining line section. It should be set to reach beyond the end of the longest adjoining line section under the conditions of maximum underreach which may arise due to arcs, intermediate current sources and errors in CT, VT and measuring units. This is achieved with additional time delay, as shown in Fig. 6.8. Compare Fig. 6.8 with Fig. 6.6 which is for normal selectivity of the III zone unit. The reach of any unit should not be so long as to cause the relay to operate under any load condition or fail to reset if operated previously due to any reason.

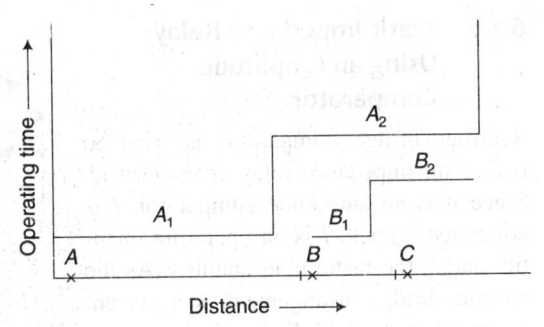

Fig. 6.7 Second zone adjustment with additional time for selectivity with relay of a very short adjoining line

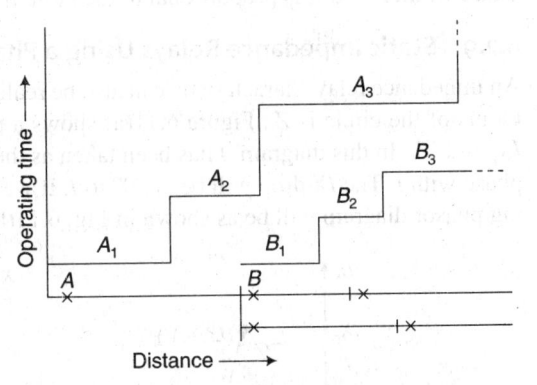

Fig. 6.8 Third zone adjustment with additional time to provide back-up protection for adjoining line

6.2.7 Electromechanical Impedance Relay

Induction cup type construction is used to realise an impedance relay characteristic. This construction is compact and robust. It produces nonvibrating torque. It is less affected by dc transients and possesses high speed and sensitivity. It gives a perfect circular characteristic. For such an impedance relay, $\frac{V}{I} \leq K$. IZ_r is to be compared with V if an amplitude comparator is used. But the induction cup construction is a phase comparator and hence, $(I + V)$ and $(I - V)$ are used as the actuating quantities, as shown in Fig. 6.9. With the introduction of Z_r in the voltage circuit, the current flowing in this circuit is VZ_r.

Therefore, $\left(I + \dfrac{V}{Z_r}\right)$ is the polarising quantity and $(I - V/Z_r)$ is the operating quantity. The polarising quantity produces one of the fluxes. In case of phase comparators it provides reference for phase angle measurement.

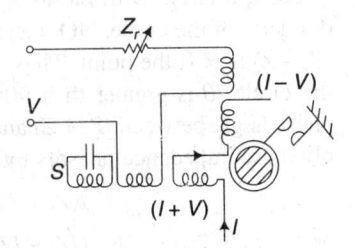

Fig. 6.9 Induction cup type impedence relay

6.2.8 Static Impedance Relay Using an Amplitude Comparator

Rectifier bridge comparator is used to realise an impedance relay characteristic. Since it is an amplitude comparator, I is compared with V. I is an operating quantity and V the restraining quantity. As the rectifier bridge arrangement is a current comparator, it is supplied with the operating current I_0 and restraining current I_r, as shown in Fig. 6.10. I_0 is proportional to the load current I, and I_r is proportional to the system voltage V.

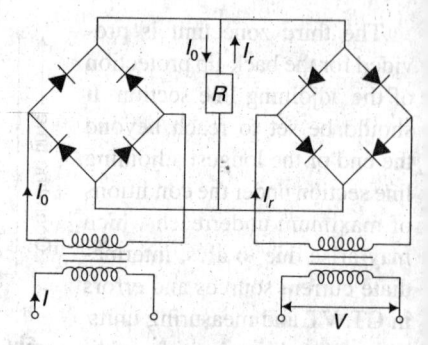

Fig. 6.10 Static impedance relay unit using amplitude comparator

6.2.9 Static Impedance Relays Using a Phase Comparator

An impedance relay characteristic can also be realised using a phase comparator. The radius of the circle is Z_r. Figure 6.11(a) shows a phasor diagram showing V, I, IR_r, IX_r and IZ_r. In this diagram, I has been taken as the reference. The IR drop will be in phase with I. The IX drop will be at 90° to I. If we divide all phasors by I, the resulting phasor diagram will be as shown in Fig. 6.11(b).

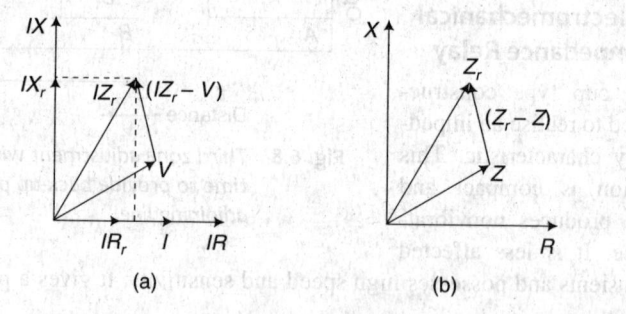

Fig. 6.11 (a) Phasor diagram showing V, I and voltage drops (b) Impedance diagram

Now, a circle with radius Z_r is drawn, as shown in Fig. 6.12. Since NM is the diameter of the circle, NO = Z_r and phasor NP = Z_r + Z. The angle between (Z_r + Z) (Z_r − Z) is θ. If the point P lies within the circle, θ is less than 90°. If P falls outside the circle, θ is greater than 90°. Thus, to realise an impedance characteristic, the phase angle between (Z_r + Z) and (Z_r − Z) is to be compared with ± 90°. If we multiply these impedance phasors by I the resulting phasors are

$$I(Z_r + Z) \quad \text{and} \quad I(Z_r - Z)$$
or
$$(IZ_r + IZ) \quad \text{and} \quad (IZ_r - IZ)$$
or
$$(IZ_r + V) \quad \text{and} \quad (IZ_r - V)$$

Therefore, to realise an impedance relay characteristic using a phase comparator, the phase angle θ between ($IZ_r + V$) and ($IZr − V$) is compared with ± 90°. If θ is less than ± 90°, the point lies within the characteristic circle. ($IZ_r + V$) is the polarising input and ($IZ_r − V$) is the operating input.

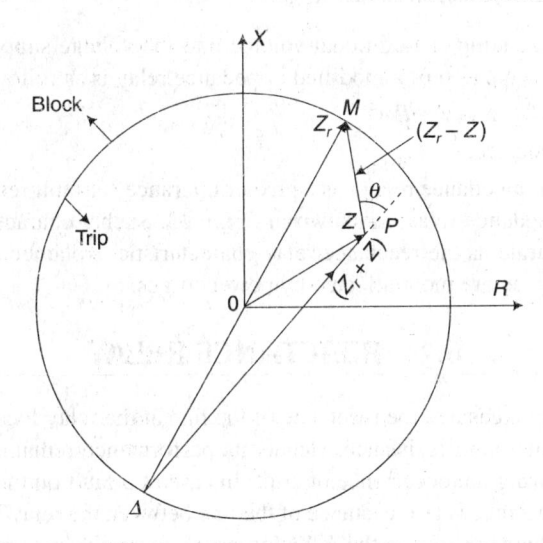

Fig. 6.12 Impedance relay

6.2.10 Microprocessor-based Impedance Relay

The microprocessor computes line impedance at relay locations using I_{rms} and V_{rms}. There are a number of techniques which can be used for this type of computation. One of the techniques uses I_{dc} and V_{dc} for computation as these are proportional to I_{rms} and V_{rms}. V_{ac} and I_{ac} are rectified using rectifiers to obtain V_{dc} and I_{dc}. These rectifiers employ ICs and diodes. Now line impedance is computed and the microprocessor issues a trip signal to the circuit breaker if the fault point lies within its protected section.

In another method, the microprocessor takes samples of the voltage and current signals over half a cycle or one full cycle to compute V_{rms} and I_{rms}. Then the line impedance is computed from these quantities. If the fault point lies within the protected section, a trip signal is issued.

After taking samples, the microprocessor may use digital filter techniques to compute R and X at the relay location and take a decision to trip the circuit breaker if the fault point lies within the protected section. See details of microprocessor impedance relay in Chap. 12.

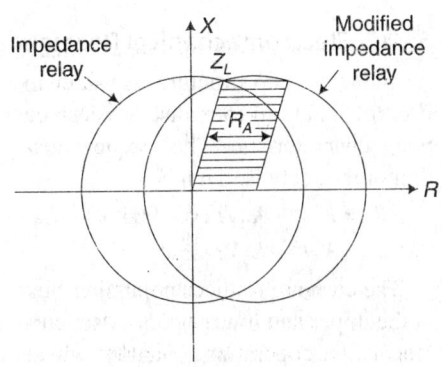

6.2.11 Modified Impedance Relay

Figure 6.13 shows the characteristics of a modified impedance relay. Its characteristic is a circle on the R-X diagram. It is similar to that of an impedance relay but has a shifted characteristic. To realise a modified impedance characteristic, the standard impedance characteristic is shifted outwards along the R-axis by a current bias. This is

Fig. 6.13 Modified impedance relay

achieved by introducing an additional voltage into the voltage supply circuit of the relay. The torque equation of a modified impedance relay is as follows.

$$T = K_1 I^2 - K_2 (V + IR)^2$$

where R is a resistance.

The modified impedance relay has a greater tolerance for fault resistance as compared to the impedance relay, as shown in Fig. 6.13. Such a characteristic is not as effective or accurate as the reactance relay characteristic explained in the next section. Moreover, they are more affected by power surges.

6.3 REACTANCE RELAY

A reactance relay measures the reactance of the line at the relay location, and is not affected by variations in resistance. Hence, its performance remains unaffected by arc resistance during the occurrence of fault. In case of a fault on the protected line, the measured reactance is the reactance of the line between the relay location and the fault point. Its characteristic on the R-X diagram is a straight line, parallel to R-axis as shown in Fig. 6.14(a).

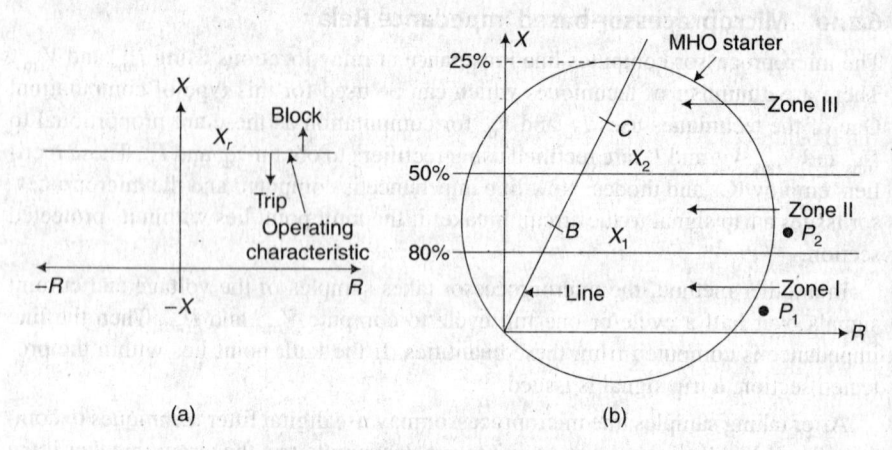

Fig. 6.14 (a) Operating characteristic of a reactance relay
(b) Reactance relay with starting unit

6.3.1 Electromechanical Reactance Relay

An induction cup structure, as shown in Fig. 6.15, is used to realise a reactance relay characteristic. The torque equation of the relay is given by

$$T = K_1 I^2 - K_2 VI \cos(90 - \phi) - K_3$$
$$= K_1 I^2 - K_2 VI \sin \phi - K_3$$

The current produces polarising flux in the upper and lower poles. Also, current is the operating quantity which produces flux in the right-hand side

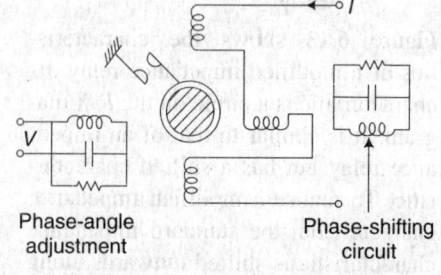

Fig. 6.15 Induction cup type reactance relay

pole. The flux in the right-hand side pole is out of phase with that in the upper and lower poles because of the secondary winding which is closed through a phase shifting circuit and is placed on the right-hand side pole. The interaction of the polarising flux and the flux in the right-hand side pole produces an operating torque $K_1 I^2$. The winding placed on the left-hand side pole produces a flux which interacts with the polarising flux to produce a restraining torque. There is a phase-angle adjustment circuit connected in series with the voltage coil. The restraining torque is proportional to $VI \cos(90 - \phi)$. The angle between the actuating quantities which are proportional to V and I can be changed to realise the desired characteristic. In this case, the angle between the actuating quantities is kept $(90 - \phi)$. The relay operates when $K_1 I^2 > K_2 VI \sin \phi$, neglecting K_3 which is a constant for the spring's torque. Thus, we have

$$\frac{V}{I} \sin \phi < \frac{K_1}{K_2}$$

or $\quad Z \sin \phi < K \quad$ or $\quad X < K$

The characteristic of the reactance relay on the R-X diagram is shown in Fig. 6.14(a). It will operate when the measured value of the reactance is less than the predetermined value K. It is a non directional relay as it will also operate for the negative values of X. The negative value of X means that the fault is behind the relay location, i.e. in the reverse direction. A directional unit, having a circular characteristic is used in conjunction with reactance relays. The directional unit also acts as the III unit of the distance scheme. The I and II units are reactance units as shown in Fig. 6.14(b). The I unit is a high speed unit to protect 80% to 90% of the protected line. The II unit protects up to 50% of the adjacent line. The III unit is a back-up unit to protect the whole of the adjacent line. The time-distance characteristic is a stepped characteristic, as shown in Fig 6.6

Why the directional unit used with reactance relays should have a circular characteristic needs further explanation. Under normal conditions, with a load of high power factor, the reactance measured by the reactance relay may be less than its setting. Such points have been shown in Fig. 6.14(b) by P_1 lying in the I zone of protection and P_2 in the II zone of protection. To prevent false trippings under such conditions, the reactance relay should be supervised by a fault-detecting unit (starting unit) which limits its area on the R-X diagram. Hence, its characteristic should be a circular one. A directional unit with a straight line characteristic, as used with an impedance relay cannot be used in this case. With this type of a directional unit, the reactance relay will not trip under conditions of a high power factor load.

The starting unit detects faults and also serves the function of the III zone unit. Its connection is shown in Fig. 6.16.

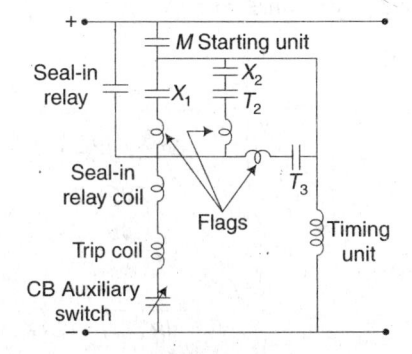

Fig. 6.16 *Connections of reactance relay*

6.3.2 Static Reactance Relay Using an Amplitude Comparator

Figure 6.17 shows a rectifier bridge type amplitude comparator to realise a reactance relay. The actuating quantities to be compared are $(I - V/2X_r)$ and $V/2X_r$. The relay operates when

$$\left| I - \frac{V}{2X_r} \right| > \left| \frac{V}{2X_r} \right|$$

Multiplying both sides by $2X_r$, we get,

$$|2IX_r - V| > |V|$$

Dividing both sides by I, we get

$$\left| 2X_r - \frac{V}{I} \right| > \left| \frac{V}{I} \right| \text{ or } |2X_r - Z| > |Z|$$

where X_r is the reactance of the line to be protected.

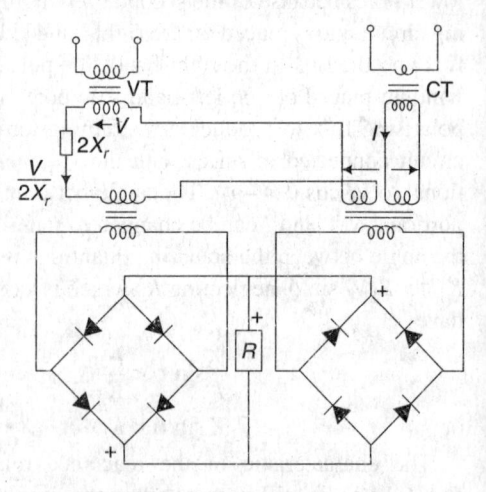

Fig. 6.17 Static reactance relay

When the above condition is satisfied, the characteristic realised is a reactance relay characteristic. Proof of this will be given later on while discussing the angle impedance relay as the reactance relay is a special case of an angle impedance relay.

6.3.3 Static Reactance Relay Using a Phase Comparator

Figure 6.18(a) shows a phasor diagram showing voltage, current and voltage drops. If we divide all vectors by I, the vectors of Fig. 6.18(b) are obtained. A perpendicular line MK is drawn from the point M. A horizontal line LN is drawn through the point M. As MK is parallel to IX_r the phase angle between IX_r and $(IZ_r - V)$ is equal to the angle between MK and $(IZ_r - V)$, i.e. θ. If the point P is below the horizontal line LN, θ is less than $\pm 90°$. If P is above LN, θ is greater than $\pm 90°$. Therefore, a reactance relay characteristic can be realised by comparing the phase angle between IX_r and $(IZ_r - V)$ with $\pm 90°$.

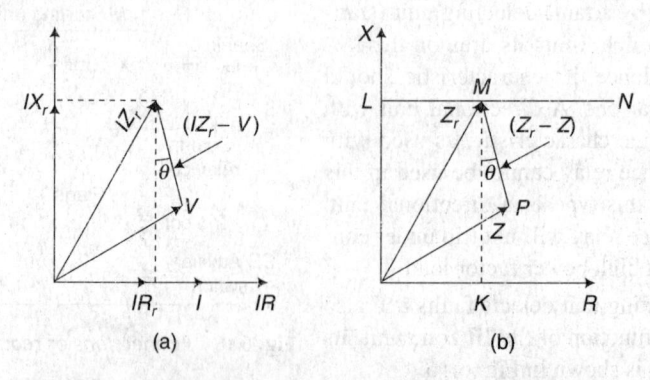

Fig. 6.18 (a) Phasor diagrams showing V, I and voltage drop
(b) Phasor diagram for reactance relay

The reactance relay characteristic can also be realised if the phase angle between IX_r and $(IX_r - V)$ is compared with $\pm 90°$. The vector diagram for this condition has been shown in Fig 6.19

6.3.4 Microprocessor-based Reactance Relay

A reactance relay can be realised using a microprocessor by comparing I_{dc} with $V \sin \phi$. Alternatively, X can be measured at the relay location using differential equations, Fast Fourier transforms, walsh functions or any other digital technique and it may be compared with the preset value of X. More details are given in Chap. 12.

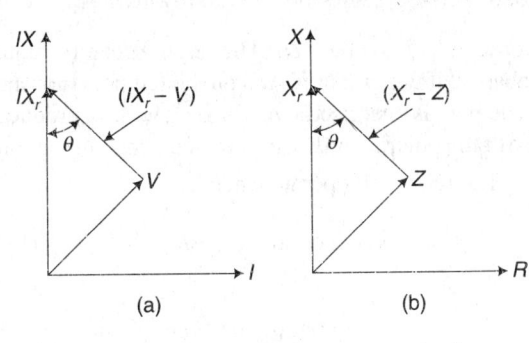

Fig. 6.19 *Realisation of reactance relay by comparing IX_r and $(IX_r - V)$*

6.4 MHO (ADMITTANCE OR ANGLE ADMITTANCE) RELAY

A MHO relay measures a component of admittance $|Y| \angle \theta$. But its characteristic, when plotted on the impedance diagram (R-X diagram) is a circle, passing through the origin. It is inherently a directional relays as it detects the fault only in the forward direction. This is obvious from its circular characteristic passing through the origin, as shown in Fig. 6.20. It is also called an admittance or angle admittance relay. It is called a MHO relay because its characteristic is a straight line when plotted on an admittance diagram (G-B axes).

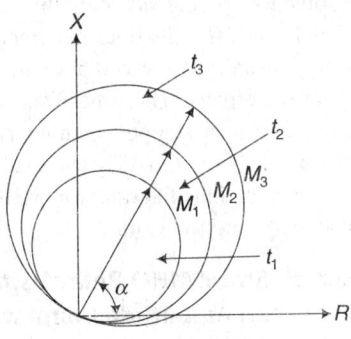

Fig. 6.20 *Characteristics of MHO relay*

6.4.1 Electromechanical MHO Relay

An induction cup structure, as shown in Fig. 6.21 is used to realise a MHO characteristic. The torque equation is given by

$$T = K_1 VI \cos(\phi - \alpha) - K_2 V^2 - K_3$$

The upper and lower poles are energised by a voltage V to produce a polarising flux. The series capacitor provides memory action which will be explained later on. The left pole is energised by a current which is the operating quantity. The flux produced by I interacts with the polarising flux to give an operating

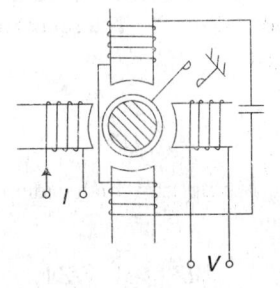

Fig. 6.21 *Induction cup type MHO relay*

torque $K_1 VI \cos(\phi - \alpha)$. The angle α can be adjusted by varying resistance in the phase shifting circuit placed on the left pole (not shown in the figure). The right-hand side pole is energised by voltage. The flux produced by the right side pole interacts with the polarising flux to produce a restraining torque $K_2 V^2$.

The relay will operate when

$$K_1 VI \cos(\phi - \alpha) > K_2 V^2 \quad \text{or} \quad \frac{1}{V} \cos(\phi - \alpha) > \frac{K_2}{K_1}$$

or $$Y \cos(\phi - \alpha) > \frac{K_2}{K_1} \quad \text{or} \quad \frac{1}{Y \cos(\phi - \alpha)} < K$$

or $$\frac{Z}{\cos(\phi - \alpha)} < K \quad \text{or} \quad M < K$$

Three units of MHO relays are used for the protection of a section of the line. The I unit is a high speed unit to protect 80%–90% of the line section. The II unit protects the rest of the line section, and its reach extends up to 50% of the adjacent line section. The III unit is meant for back-up protection of the adjacent line section. The II and III units operate after a preset delay, usually 0.2 s to 0.5 s and 0.4 s to 1 s respectively. The time-distance characteristic is a stepped characteristic, as shown in Fig. 6.6. Figure 6.22 shows the connection diagram for MHO units placed at one location.

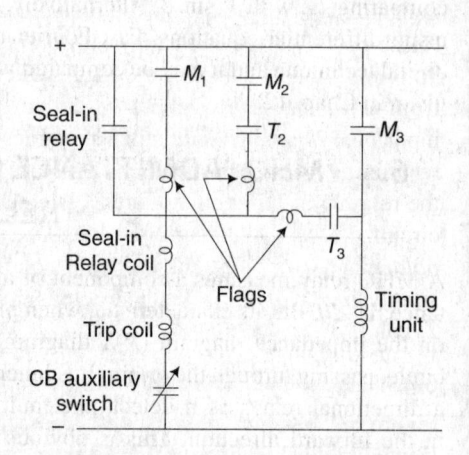

Fig. 6.22 Connections of MHO relays

6.4.2 Static MHO Relay Using an Amplitude Comparator

Figure 6.23 shows a rectifier bridge type amplitude comparator to realise a MHO characteristic. The actuating quantities to be compared are I and $(V/Z_r - I)$. The relay will operate, when

$$I > \left| \frac{V}{Z_r} - I \right|$$

Multiplying both sides by Z_r, we get

$$|IZ_r| > |V - IZ_r|$$

Dividing both sides by I, we get

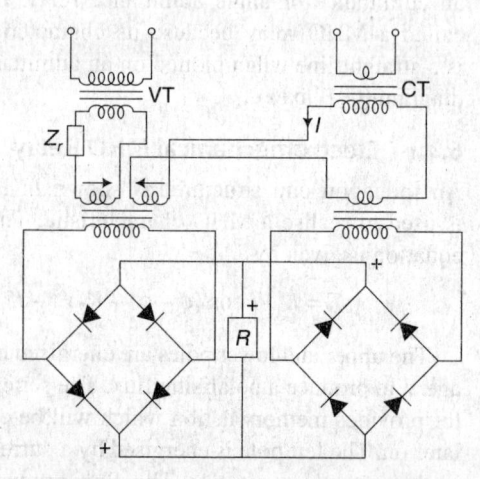

Fig. 6.23 Schematic diagram of a static MHO relay

$$|Z_r| > \left|\frac{V}{I} - Z_r\right| \quad \text{or} \quad |Z_r| > |Z - Z_r|$$

When the above condition is satisfied, the characteristic obtained will be a MHO characteristic, as shown in Fig. 6.24. Z_r is the radius of the MHO circle, which is equal to the impedance of the voltage circuit. If a fault point Z lies within the circle, $|Z_r| > |Z - Z_r|$. If a fault point lies on the circumference of the circle, $|Z_r| = |Z - Z_r|$. If the fault point is outside the circle, $|Z_r| < |Z - Z_r|$. The above conditions are also true if the point P is anywhere on AB. When the fault point is very close to the relay location (close-up fault), the relay may fail to operate.

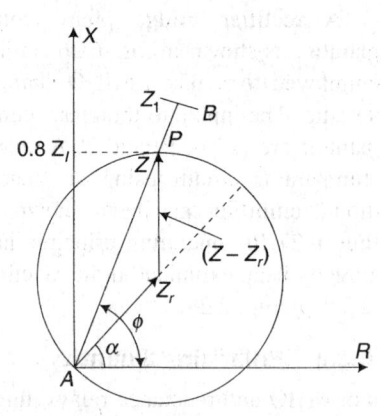

Fig. 6.24 MHO characteristic

To overcome this difficulty, a voltage called the polarising voltage, which is obtained from a pair of healthy phases is added to the actuating quantities. The operating input of the modified actuating quantities corresponds to $V_p/Z_p - I + V_r/Z_r$ and the restraining input corresponds $V_p/Z_p + I - V_r/Z_r$, where Z_r is equal to the impedance of the relay restraining circuit and Z_p is equal to the impedance of the relay polarising circuit. V_p is the polarising voltage. A relay using polarising voltage is known as a *polarised MHO relay*. But the word polarised is frequently omitted and the relay is simply called a MHO relay. The details of a polarised MHO relay can be seen in Ref. 2, vol. I, page 385.

6.4.3 Static MHO Relay Using a Phase Comparator

Figure 6.25(a) shows a phasor diagram showing voltage, current and voltage drops. If we divide all phasors of this diagram by I, the resulting phasor diagram will be as shown in Fig. 6.25(b). The phase angle between V and $(IZ_r - V)$ is θ. Now draw a circle with Z_r as diameter, as shown in Fig. 6.25(c). If the point P lies within the circle, θ is less than $\pm 90°$. If P lies outside the circle, θ is greater than $\pm 90°$. Therefore, to realise a MHO characteristic, the phase angle θ between $(IZ_r - V)$ and V is compared with $\pm 90°$.

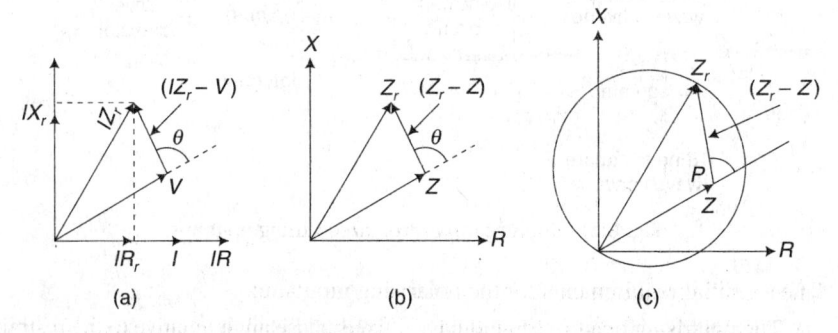

Fig. 6.25 (a) Phasor diagram showing V, I and voltage drop
(b) Impedance diagram (c) MHO characteristic

A rectifier bridge phase comparator, as shown in Fig. 6.26 can be employed to realise a MHO characteristic. The inputs to the phase comparator are $(IZ_r - V)$ and V. A phase comparator circuit using an operational amplifier has been shown in Fig. 6.27. Its operating principle has already been explained in the Section 2.3.5(c), Fig. 2.29.

6.4.4 Polarising Quantity

Fig. 6.26 *Rectifier bridge phase comparator*

For MHO and reactance relays, three inputs are used, though the comparator employed is a two-input comparator. These are (i) operating input (current I), (ii) restraining input (voltage V) and (iii) polarising input. The polarising input is current in the case of a reactance relay, and voltage in the case of a MHO relay. The relay uses the first two quantities for impedance (or a component of the impedance) measurement, which are derived from the current and voltage associated with the fault. The third quantity, i.e. the polarising quantity is a reference for determining the phase-sense of the operating current. If a terminal fault occurs, the voltage at the relay location becomes zero. In case of a reactance relay, the polarising input is current and hence, the relay operates even though the terminal voltage is zero. The restraining quantity V is zero. This will not prevent the operation of the relay. On the other hand, in case of a MHO relay, the polarising input is V. If it is derived from the faulty phase, it will become zero in case of a terminal fault. Therefore, the MHO relay will fail to operate for terminal faults. To overcome this difficulty, the polarising input may be derived from the healthy phases, as discussed in the next section.

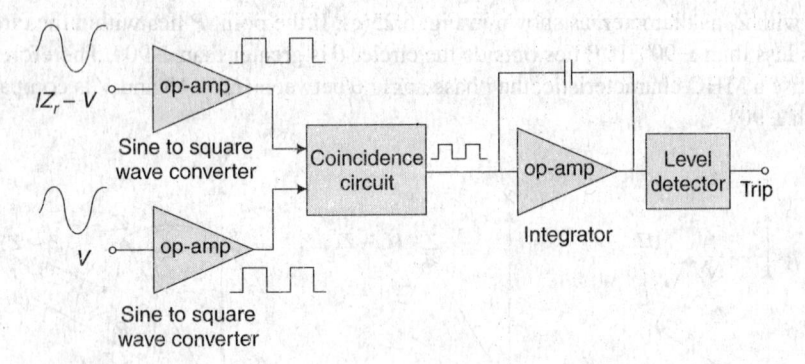

Fig. 6.27 *Phase comparator circuit using op-amps*

The essential requirements for the polarising input are:
(i) The polarising quantity should have a fixed phase angle relative to the restraint voltage.

(ii) The magnitude of the polarising quantity is of no importance. But in no case should it be zero. In case of terminal faults, when the restraint voltage is zero, the polarising input should not be zero.

The polarising voltage V_p can be related to the restraint voltage V_r by an angle θ, such that

$$\frac{V_p}{V_r} = C \angle \theta, \quad \text{where } C \text{ can have any value}$$

The polarising quantities are also used with directional relays. While discussing the directional relays, methods to derive a polarising voltage have been discussed.

6.4.5 Polarised MHO Relay

In a polarised MHO relay, the inputs to the phase comparator are $(IZ_r - V)$ and V_p, where

V = voltage at the relay point. During a fault, it becomes the fault voltage.

I = Current at the relay point. During a fault, it is the fault current.

Z_r = MHO relay setting

V_p = polarising voltage.

IZ_r is the operating quantity and V is the restraint voltage. V_p is the polarising input which exists even if V is zero as in the case of terminal faults. Therefore, a polarised MHO relay operates when terminal faults occur, as the phase comparison is made between $(IZ_r - V)$ and V_p, though $V = 0$ or is negligibly small. For a self-polarised MHO relay in which memory has not been used, the input quantities are $(IZ_r - V)$ and V. For such a simple MHO relay, when V becomes zero, phase comparison is not possible and the relay fails to operate.

If V_p and V, applied to the measuring unit are in phase, the diameter of the MHO circle will be equal to Z_r. If the polarising and restraint currents are displaced by an angle β, the characteristic will remain a circle but Z_r becomes a chord of the circle, as shown in Fig. 6.28.

In case of an amplitude comparator, if the polarising voltage is to be used, the inputs are $(V_p/Z_p - I + V/Z_r)$ and $(V_p/Z_p - I - V/Z_r)$.

The following methods are used in practice to obtain a polarising voltage for a polarised MHO relay.

(i) V_p can be derived from the fault voltage through a resonant circuit tuned to the system frequency (i.e. faulty phase voltage with memory).

(ii) It may be derived from the healthy phases through a suitable phase-shifting circuit.

(iii) It may be a combination of the faulty phase voltage and the healthy phase voltage.

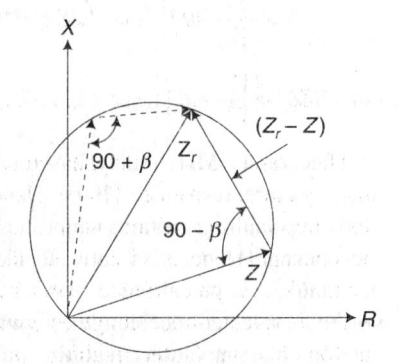

Fig. 6.28 General case of polarised MHO relay

In the last two methods, when V_p is derived either fully or partly from the healthy phases, the relay fails to operate if a three-phase terminal fault occurs. In such a situation, the fault is cleared by an offset MHO relay which is used as a back-up relay. An offset MHO relay, which has current bias in the voltage circuit, operates even in case of three-phase terminal faults, (see Section 6.4.6). In a number of cases, high-set overcurrent relays have been used to clear three-phase terminal faults at a high speed.

If V_p is derived from the faulty phase, the relay is said to be a self-polarised relay. If it is derived from the healthy phases, it is called a cross-polarised relay. Fully cross-polarised means that V_p is fully derived from the healthy phases.

In the first method in which V_p is derived from the faulty phase, it is possible, by using memory, to maintain the polarising input for a short time even after the occurrence of the fault. Therefore, when a terminal fault occurs, the polarising input is maintained sufficiently long to cause the operation of the relay. In this method, the polarising current does not maintain the constant phase relation with respect to the faulty phase voltage. The phase angle of the faulty phase voltage changes when a fault occurs but the phase angle of the current of the memory circuit is maintained at the original value. The change in the phase-shift is negligible. The phase-angle shift also occurs due to the variation in the supply frequency. The supply frequency may vary from 47 c/s to 51 c/s but the resonant circuit resonates at a fixed frequency. To minimise the error, the memory is restricted to about three cycles at the most. Therefore, the relay must be very fast. The most serious drawback of this method is that this method is not effective when the line is energised. This drawback can be removed by the supply using voltage from the busbar instead of the line.

6.4.6 Offset MHO Relay

Figure 6.29 shows an offset MHO characteristic. A rectifier bridge type amplitude comparator, as shown in Fig. 6.23 can be used to realise the offset MHO characteristic. The actuating quantities to be compared are I and $(V/Z_r - nI)$. Only a fraction of the CT output current is injected into the restraint circuit. Thus n is fraction, i.e. $n < 1$. The relay operates when

$$|I| > \left|\frac{V}{Z_r} - nI\right| \quad \text{or} \quad |IZ_r| > |V - nIZ_r|$$

$$\text{or} \quad |Z_r| > \left|\frac{V}{I} - nZ_r\right| \quad \text{or} \quad |Z_r| > |Z - nZ_r|$$

The offset MHO relay has more tolerance to arc resistance. It can also see a close-up fault and a fault which lies behind the busbar. Hence, it is able to clear busbar faults. A typical value of offset is 10% of the protected line length. It will operate for close-up faults resulting in $V = 0$. When $V = 0$, the relay operates because $|I| > |0 - nI|$ condition is satisfied, n being less than 1.

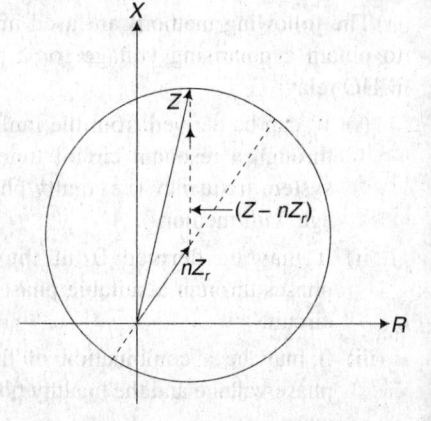

Fig. 6.29 Offset MHO characteristic

In a distance protective scheme employing MHO relays, the third unit may be an offset MHO, as shown in Fig. 6.30. The III zone unit provides busbar zone back-up protection in such a scheme. The main applications of offset MHO relays are:
  (i) busbar zone back-up
  (ii) carrier starting unit in distance/carrier blocking schemes
  (iii) power swing blocking.

The second and third applications will be discussed later. When a fault occurs, the voltage, current and phase angle change instantaneously, whereas in case of power swings, they change slowly. This property is utilised for the out of step blocking relay. The III zone offset unit operates with some time-delay. When a fault occurs in the zone of the II unit, it operates first and its tripping is not blocked. In case of power swings, the III zone unit operates first and blocks the tripping of the II zone unit. The offset characteristic gives a sufficient time-delay for the III zone unit for this purpose.

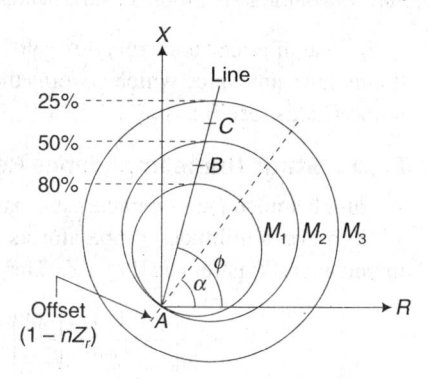

Fig. 6.30 *MHO relays scheme with III unit an offset MHO*

6.5 ANGLE IMPEDANCE (OHM) RELAY

An angle impedance relay measures a component of the impedance of the line at the relay location. It is also called an ohm relay. Its characteristic on the R-X diagram is a straight line and it is inclined to the R-axis at any angle, as shown in Fig. 6.31. The reactance relay is a particular case of an angle impedance relay. The angle impedance relay is used in conjunction with other relays, for example it is used to limit the area of the MHO relay on the R-X diagram to make it less sensitive to power surges. In this particular application, the angle impedance relay is called a blinder. This concept will be discussed later on.

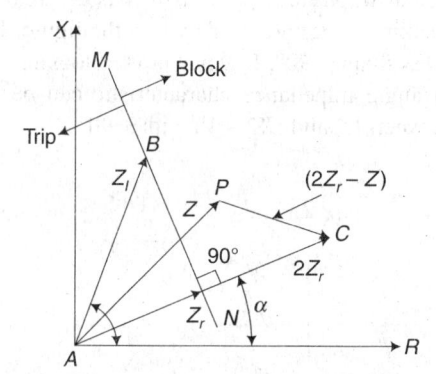

Fig. 6.31 *Characteristic of angle-impedance relay*

6.5.1 Electromechanical Angle Impedance Relay

For this kind of a relay, an induction cup construction, as shown in Fig. 6.15 is used. The torque equation of the relay is given by

$$T = K_1 I^2 - K_2 VI \cos(\phi - \alpha) - K_3$$

In case of a reactance relay, $\alpha = 90°$. But in the case of an angle impedance relay, it may have any value which governs the inclination of the characteristic with respect to the R-axis (see Fig. 6.31).

6.5.2 Static Angle Impedance Relay Using an Amplitude Comparator

An angle impedance characteristic can be realised by comparing $(I - V/2Z_r)$ and $V/2Z_r$ by an amplitude comparator as shown in Fig. 6.17, except that the restraint impedance $2X_r$ is replaced by $2Z_r$. The relay operates when

$$\left| I - \frac{V}{2Z_r} \right| > \left| \frac{V}{2Z_r} \right| \quad \text{or} \quad |2IZ_r - V| > |V|$$

or

$$\left| 2Z_r - \frac{V}{I} \right| > \left| \frac{V}{I} \right| \quad \text{or} \quad |2Z_r - Z| > |Z|$$

Figure 6.31 shows the characteristic of an angle impedance relay. P represents any point and the phasor $AP = Z$. The angle impedance characteristic is represented by MN, which is the perpendicular bisector of $2Z_r$. If the point P is on the left hand side of MN, then $PC > AP$, i.e. $|2Z_r - Z| > |Z|$. If the point P lies on MN, $PC = AP$, i.e. $|2Z_r - Z| = |Z|$. If the point P lies on the right hand side of MN, $PC < AP$, i.e. $|2Z_r - Z| < |Z|$.

6.5.3 Static Angle Impedance Relay Using a Phase Comparator

Figure 6.32(a) shows a phasor diagram showing voltage, current and voltage drops. If we divide the phasors by I, the phasors of Fig. 6.32(b) are obtained. A line perpendicular to phasor Z_r is drawn. θ is the angle between Z_r and $(Z_r - Z)$. Tripping and blocking zones are as shown in the figure. If the point P lies in the tripping zone, θ is less than $\pm 90°$. If θ lies in the blocking zone, it is greater than $\pm 90°$. Therefore, an angle impedance characteristic can be relaised by comparing the phase angle between IZ_r and $(IZ_r - V)$ with $\pm 90°$.

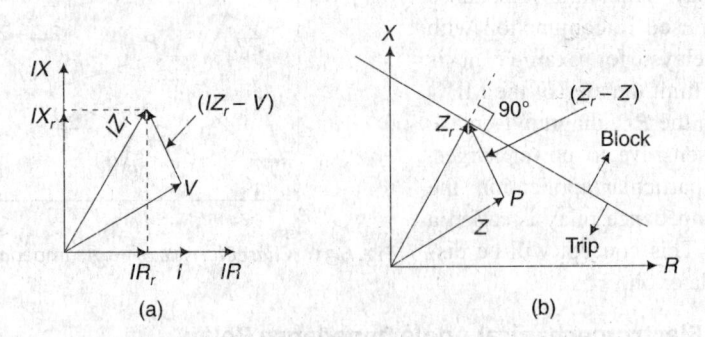

Fig. 6.32 (a) Phasor diagram showing V, I and voltage drops
(b) Angle-impedance characteristic using phase comparator

6.5.4 Microprocessor-based Angle Impedance Relay

A microprocessor can compare the amplitude of $(2IZ_r - V)$ and V very conveniently to realise an angle impedance relay.

6.6 INPUT QUANTITIES FOR VARIOUS TYPES OF DISTANCE RELAYS

Static relays employ either voltage comparator or current comparator. Table 6.1 shows voltage inputs for different types of distance relays. Table 6.2 shows current inputs to realise various distance relay characteristics. The directional relay is not a distance relay. It has been included in these tables as it is used in conjunction with impedance relays. Its characteristic is a straight line passing through the origin.

Table 6.1 Voltage inputs for different types of distance relays

Types of Relays	Amplitude Comparator		Phase Comparator (90°)	
	Operating quantity	Restraining quantity	Operation quantity	Polarising quantity
Impedance	IZ_r	V	$IZ_r - V$	$IZ_r + V$
Directional	$IZ_r + V$	$V - IZ_r$	IZ_r	V
Reactance	$2IX_r - V$	V	$(IZ_r - V)$ or $(IX_r - V)$	IX_r
MHO	IZ_r	$2V - IZ_r$	$IZ_r - V$	V
Offset MHO	$I(Z_r - Z_0)$	$2V - I(Z_r + Z_0)$	$IZ_r - V$	$V - IZ_0$
Angle Impedance	$2IZ_r - V$	V	$IZ_r - V$	IZ_r

Table 6.2 Current inputs for different types of distance relays

Types of Relays	Amplitude Comparator		Phase Comparator (90°)	
	Operating quantity	Restraining quantity	Operating quantity	Polarising quantity
Impedance	I	$\dfrac{V}{Z_r}$	$I - \dfrac{V}{Z_r}$	$I + \dfrac{V}{Z_r}$
Directional	$I + \dfrac{V}{Z_r}$	$\dfrac{V}{Z_r} - I$	I	$\dfrac{V}{Z_r}$
Reactance	$I - \dfrac{V}{2X_r}$	$\dfrac{V}{2X_r}$	I	$I - \dfrac{V}{X_r}$
MHO	I	$\dfrac{2V}{Z_r} - I$	$I - \dfrac{V}{Z_r}$	$\dfrac{V}{Z_r}$
Offset MHO	$2I - \left(\dfrac{V}{Z_r} + \dfrac{V}{Z_0}\right)$	$\dfrac{V}{Z_r} - \dfrac{V}{Z_0}$	$I - \dfrac{V}{Z_r}$	$\dfrac{V}{Z_o} - I$
Angle Impedance	$2I - \dfrac{V}{Z_r}$	$\dfrac{V}{Z_r}$	I	$I - \dfrac{V}{Z_r}$

6.7 SAMPLING COMPARATOR

A sampling comparator compares the amplitudes of input quantities. So it is a kind of an amplitude comparator. One or both input signals are sampled and compared by the comparator. When one signal is sampled, it is compared with the average value of the other signal. Sampling of input signals can be carried out once every cycle or once every half cycle. If sampling is carried out once every cycle, the scheme is a slower one but its circuit is simpler and less expensive. If sampling is to be carried out every half cycle, the scheme is a faster one but its circuit is complex and more expensive.

6.7.1 Realisation of Reactance Relay Using Sampling Comparator

A reactance relay characteristic is realised by comparing the instantaneous value of the voltage at the moment when current is zero with the rectified current. Figure 6.33 shows the instantaneous value of the voltage at the moment when current is zero. It

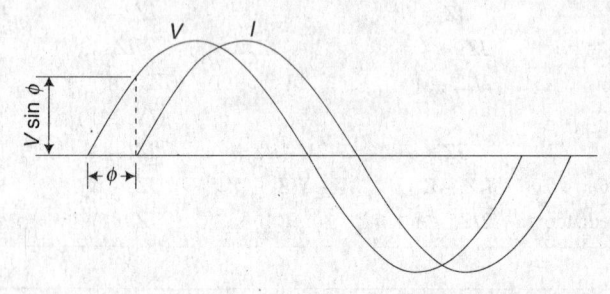

Fig. 6.33 *Instantaneous value of voltage at the moment of current zero*

is equal to $V \sin \phi$. For the operation of the relay, the condition to be satisfied is as follows.

$$K_1 I_{dc} > V \sin \phi \quad \text{or} \quad \frac{V \sin \phi}{I_{dc}} < K_1$$

or $\quad \dfrac{V \sin \phi}{I_{rms}} < K,\quad$ as I_{rms} is proportional to I_{dc}

or $\quad Z \sin \phi < K \quad$ or $\quad X < K$

6.7.2 Realisation of MHO Relay Using a Sampling Comparator

A MHO relay characteristic can be realised by comparing the instantaneous values of the current at the moment of voltage peak with the rectified voltage. Figure 6.34 shows the instantaneous value of the current at the moment of voltage maximum. It is equal to $I \cos \phi$. For the operation of the relay, the condition to be satisfied is as follows.

$$I \cos \phi > K_1 V_{dc} \quad \text{or} \quad \frac{I}{V_{dc}} \cos \phi > K_1$$

or $\quad \dfrac{I}{V_{rms}} \cos \phi > K_2,\quad$ as V_{dc} is proportional to V_{rms}

or $\quad Y \cos \phi > K_2 \quad$ or $\quad \dfrac{1}{Y \cos \phi} < \dfrac{1}{K_2}$

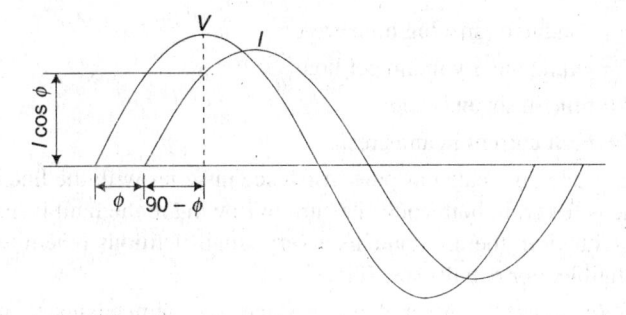

Fig. 6.34 *Instantaneous value of current at the moment of voltage peak*

or

$$M < K$$

If a design angle α is introduced while feeding the voltage and current signals to the relay, the above expression is modified and is given by

$$\frac{1}{Y \cos (\phi - \alpha)} < K$$

By changing α, a MHO characteristic can be shifted towards the R-axis to make it more tolerant to arc resistance

Other relaying characteristics can also be realised using a sampling comparator. See details in Ref. 8.

6.8 EFFECT OF ARC RESISTANCE ON THE PERFORMANCE OF DISTANCE RELAYS

If a flashover from phase to phase or phase to ground occurs, an arc resistance is introduced into the fault path. The arc resistance is appreciable at higher voltages. The arc resistance is added to the impedance of the line and hence, the resultant impedance which is seen by distance relays is increased. In case of ground faults, the resistance of the earth is also introduced into the fault path. The earth resistance includes the resistance of the tower, tower footing resistance and earth return path. Earth resistance and arc resistance combined together are known as fault resistance. In case of phase to phase faults, the fault resistance consists of only arc resistance as there is no earth resistance in this case.

The arc resistance is given by the Warrington formula:

$$R_{arc} = \frac{29 \times 10^3 \cdot l}{I^{1.4}} \; \Omega$$

where l = length of arc in metres in still air and I = fault current in amperes.

Initially, l will be equal to the conductor spacing for phase faults, and the distance from phase conductor to the tower for ground faults. The arc length is increased by the cross winds which usually accompany a lightning storm.

The arc resistance, taking into account the wind velocity and time is given by

$$R_{arc} = \frac{16300 \, (1.75 \, S + vt)}{I^{1.4}} \; \Omega$$

where S = conductor spacing in metres,
         V = wind velocity in km per hour,
         t = time in seconds and
         I = fault current in amperes.

The arc resistance is treated as pure resistance in series with the line impedance. When the line is fed from both ends, the current flowing in the fault is fed from both sides. In this situation, the arc contains a very small fictitious reactive component which is negligible. For details, see Ref. 1.

Figure 6.35(a) shows the effect of arc resistance on an impedance relay. The relay has been set to protect a line of impedance Z_l. If a fault occurs at the point F and an arc resistance R is introduced, the relay will measure $(Z_F + R)$. Z_F is the impedance of the line up to the point F. If the value of the arc resistance is greater than R, the impedance measured by the relay will be greater than the radius of the circle, and the relay will fail to operate. Thus, with arc resistance R, the relay just operates. The maximum length of the line which can be protected is OF when arc resistance is R. It is seen that the arc resistance causes the relay to underreach.

Figure 6.35(b) shows the effect of arc resistance on a MHO relay. The characteristic angle of the relay is the same as the characteristic angle ϕ of the line. For a fault at the point F, the actual line impedance is Z_F but the impedance measured by the relay is $(Z_F + R)$ which is less than Z_l. This shows that arc resistance causes underreach.

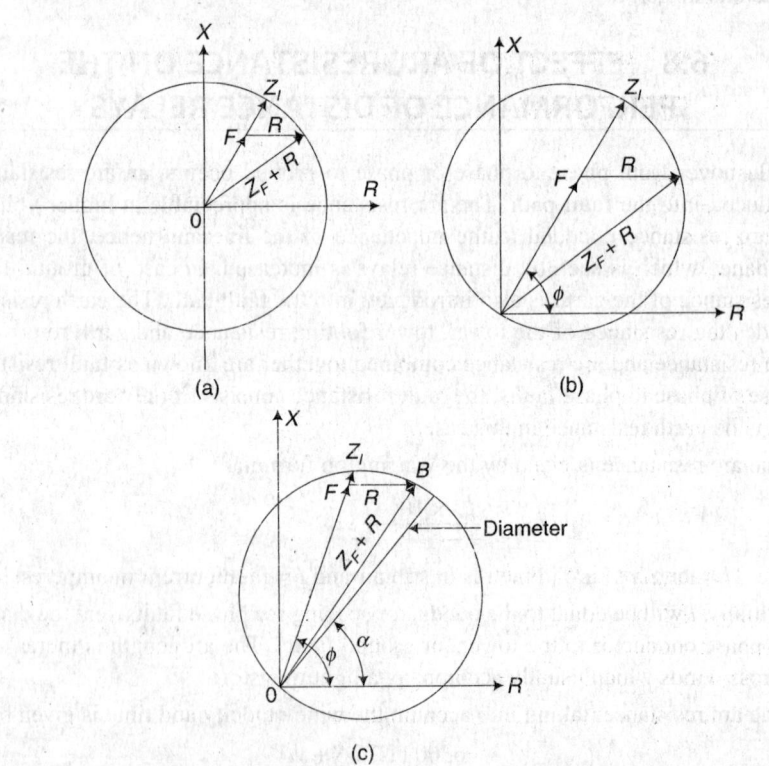

Fig. 6.35 *(a) Effect of arc resistance on impedance relay (b) Effect of arc resistance on MHO relay (c) MHO circle shifted towards R-axis*

If the MHO circle is shifted towards the R-axis by making the characteristic angle of MHO circle α less than the characteristic angle of the line ϕ, the resulting characteristic will tolerate a greater value of arc resistance. Figure 6.35(c) shows such a characteristic. In this case, $(Z_F + R)$ may be even greater than Z_l, but it is less than the diameter of the circle. In such a case, the relay will operate so long as the point $(Z_F + R)$ remains within the characteristic circle. In this case, the relay setting is equal to the diameter of the MHO circle. The maximum length of the line which can be protected is given by $Z_l = OB \cos(\phi - \alpha)$ where OB is the diameter of the circle. The values of α which are used for different system voltages are as follows:

System voltage in kV	400	275	132	66	33	11
	75	75	60	60	45	45

The inclination of the relay characteristic towards R-axis for lower voltage lines is more and such characteristics have a greater tolerance for arc resistance. When line impedance angle is more than 60°, the inclination towards the R-axis is also reduced, otherwise the accuracy of the relay is affected. In other words, with lower values of α arc tolerance is greater but relay accuracy is reduced. Thus, with longer values of α, arc tolerance is less but relay accuracy is higher.

The arc resistance affects the performance of different types of distance relays to different extents. Figure 6.36 shows the characteristics of a MHO, reactance and impedance relays on the R-X diagram to protect the same line. If a fault occurs at the point F with arc resistance R_1, the MHO relay fails to operate but the impedance and reactance relays will operate. If the values of the arc resistance is R_2, the MHO and impedance relays fail to operate but the reactance relay will operate. This shows that the MHO relay is most affected, the impedance relay is moderately affected and the reactance relay is least affected by arc resistance. As the reactance relay measures only reactance it is not at all affected by arc resistance.

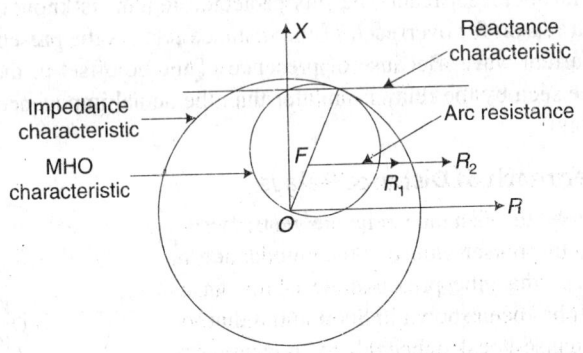

Fig. 6.36 *Effect of arc resistance on distance relays*

6.8.1 Fault Area on Impedance Diagram

Figure 6.37 shows the effect of fault resistance on the impedance diagram vectorially. If a fault occurs at B, without resistance, the relay located at A will measure the impedance as AB. If there is an arc resistance R_F, it will be added to AB, vectorially. Now the relay at A will measure the impedance AC. The horizontal lines show the values of fault resistance for faults at different points of the line AB. The area $ABCD$ is called the fault area.

If the line is fed at only one end, the horizontal lines representing arc resistance are of equal length. If the line is fed at both end BC will be greater than AD. The fault resistance is more when a fault occurs at the remote end. It is due to the fact that only a part of the total fault current flows through the relay. The other part of the fault current is fed from the other end of the line.

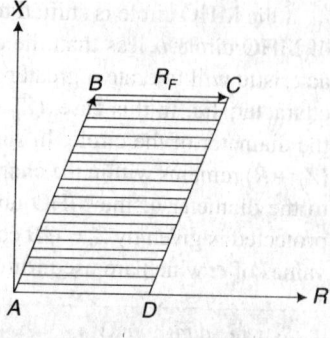

Fig. 6.37 *Fault area on an impedance diagram*

6.9 REACH OF DISTANCE RELAYS

A distance relay operates when the impedance (or a component of the impedance) as seen by the relay is less than a preset value. This preset impedance (or a component of the impedance) or corresponding distance is called the reach of the relay. In other words, it is the maximum length of the line up to which the relay can protect. Distance relays have underreaching and over-reaching tendencies depending on the fault conditions. When a distance relay fails to operate even when the fault point is within its reach, but it is at the far end of the protected line; it is called under-reach. The main reason for under-reach is the presence of arc resistance in the fault. Due to presence of arc resistance, the impedance seen by the relay is more than the actual impedance of the line up to the fault point. Hence are resistance causes underreach of the distance relay. The tendency of a distance relay to operate even when a fault point is beyond its preset reach (i.e., its protected length) is known as over-reach. The important reason for overreach of the distance relay is the presence of dc offset in the fault current wave. Because of presence of the dc offset in the fault current, the impedance seen by the relay is smaller than the actual impedance of the line up to the fault point.

6.9.1 Under-reach of Distance Relays

The tendency of the distance relay to restrain (not to operate) at the preset value of the impedance or impedances less than the preset value is known as under-reach. It has been shown in Section 6.8 that the arc resistance causes the distance relay to under-reach. Due to presence of arc resistance, the impedance seen by the relay appears to be more than the actual value of the impedance up to the fault point and the relay tends to underreach. Figure 6.38 shows the effect of arc resistance on reach of the distance relay.

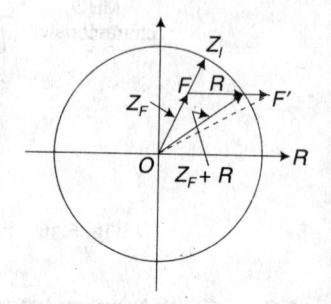

Fig. 6.38 *Under-reach of distance relay*

The relay at O has been set to protect a line of impedance Z_1. If a fault occurs at the point F and an arc resistance R is introduced, the impedance seen by the relay will be $(Z_F + R)$. Z_F is the impedance of the line upto the fault point F. In this case the impedance seen by the relay, i.e., $(Z_F + R)$ is equal to the radius of the circle, i.e., Z_1 which is more than the actual value of the impedance upto the fault point, i.e., Z_F. If the arc resistance as shown by FF' is greater than R the impedance OF' as seen by the relay will be greater than the radius of the circle such that F' lies outside the operating region of the relay and the relay will fail to operate. Though the actual impedance of the line upto the fault point (i.e., Z_F) is less than Z_1, but the relay fails to operate as the impedance seen by the relay appears to be more than Z_1 due to presence of arc resistance. This shows that the arc resistance causes the distance relay to underreach.

The arc resistance causes the underreach of distance relays to different extents. The extents of underreach of different relays are summarized below:

Relay	Under-reach due to arc resistance
Impedance relay	Moderate
Reactance relay	None
MHO relay	Maximum
Quadrilateral relay	None

6.9.2 Overreach of Distance Relays

The tendency of a distance relay to operate at impedances larger than its preset value, i.e., when the fault point is beyond its preset reach, is known as overreach. The main reason for overreach is the presence of dc offset in the fault current wave. The distance relay is prone to overreach on a transient fault consisting of a dc offset. All high-speed distance relays tend to see more current due to the presence of dc offset. The rms value of the fault current with and without dc offset is as follows:

$$I = \sqrt{I_{ac}^2 + I_{dc}^2} \quad \text{with dc offset}$$

$$I = I_{ac} \quad \text{without dc offset}$$

Since the rms value of fault current with a dc offset is greater than the rms value of a pure alternating current having symmetrical wave, the impedance seen by the relay appears to be less than the actual value, and the relay tends to overreach. The impedance seen by the relay with and without dc offset are given by

$$Z_{off} = \frac{V}{I} = \frac{V}{\sqrt{I_{ac}^2 + I_{dc}^2}} \quad \text{with offset}$$

$$Z_{no\text{-}off} = \frac{V}{I} = \frac{V}{I_{ac}} \quad \text{without dc offset}$$

$$Z_{off} < Z_{no\text{-}off}$$

The transient overreach is defined as

$$\text{Percent transient overreach} = \frac{Z_{\text{off(m)}} - Z_{\text{no-off(m)}}}{Z_{\text{no-off(m)}}} \times 100$$

where

$Z_{\text{off(m)}}$ = the maximum impedance for which the relay will operate with an offset current wave, for a given adjustment.

$Z_{\text{no-off(m)}}$ = the maximum impedance for which the relay will operate for a pure symmetrical ac for the same adjustment.

The transient overreach increases as the system angle $\tan^{-1} X/R$ increases.

Figure 6.39 shows the overreach of a distance relay located at A. The source impedance is neglected. The distance relay has been adjusted to reach point $F = 11.00$ ohms under no-offset fault current as shown in Fig. 6.39(a). The output of the VT is 110 V. The fault current for a fault at F is

$$I_{\text{no-offset}} = \frac{110 \text{ V}}{11 \text{ ohms}} = 10 \text{ A}$$

where $Z_{st} = 11.00$ ohms is the setting or steady-state reach of the relay.

Now, suppose that the fault point F is moved to a new point $F' = 13.75$ ohms as shown in Fig. 6.39(b) and let the fault current has a dc offset of 6.0 A. Thus,

$$I_{ac} = \frac{110 \text{ V}}{13.75 \text{ ohms}} = 8.0 \text{ A}$$

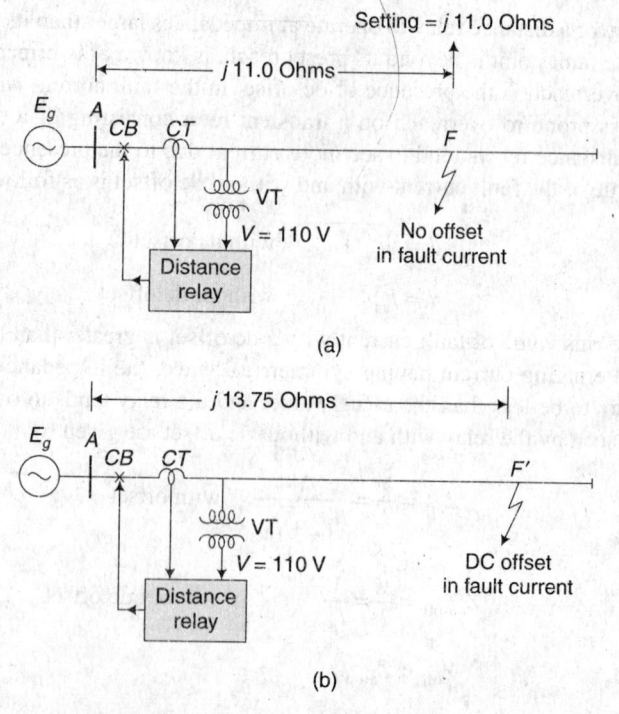

Fig. 6.39 *Transient overreach of distance relay*

Since $I_{dc} = 6.0$ A,

$$I = \sqrt{I_{ac}^2 + I_{dc}^2} = \sqrt{(8.0)^2 + (6.0)^2} = 10.0 \text{ A}$$

Therefore,

$$Z_{fictitious} = \frac{110 \text{ V}}{10.0 \text{ A}} = 11.0 \text{ ohms}$$

Since the fictitious impedance due to the dc offset now equals the setting, the relay operates. Though, the fault point is beyond the steady-state reach of the relay, but the relay operates as the fictitious impedance due to the dc offset equals the setting of the relay. This is the case of overreach of the distance relay. The percentage overreach in this case is

$$\text{Percentage overreach} = \frac{13.75 - 11.00}{11.00} \times 100 = 25\%$$

Therefore, in order to minimise the transient overreach, the distance relay should have better design. Relay designers have not been able to reduce the overreach of the distance relay to less than 10% of its setting (i.e., steady-state reach). In practice, the percentage overreach is taken between 10% and 20%. Therefore, the distance relays cannot be set 100% of the line length to provide high-speed primary protection to the entire line. The high-speed distance relay of zone-1 is adjusted to 80% to 90% of the line length. The margin of 10% to 20% is left for transient overreach, error in calculation of line impedance and errors in transducers (CTs and VTs).

6.10 EFFECT OF POWER SURGES (POWER SWINGS) ON THE PERFORMANCE OF DISTANCE RELAYS

Consider a transmission line which connects two generating stations, as shown in Fig. 6.41. The current flowing through the transmission line depends upon the phase difference between the voltage generated at the two ends of the line. The phase difference is equal to the rotor angle. The phase angle between the generated voltages changes during disturbances which may arise because of the removal of a fault or a sudden change in the load. During disturbances, the rotor of the generator swings around the final steady state value. When the rotor swings, the rotor angle changes and the current flowing through the line also changes. Such currents are heavy and they are known as power surges. So long as the phase angle between the generated voltages goes on changing, the current 'seen' by the relay is also changing. Therefore, the impedance measured by the relay also varies during power swings. Thus, a power surge 'seen' by the relay appears like a fault which is changing its distance from the relay location. The characteristic of some important distance relays and power surge are shown on the R-X diagram, Fig. 6.40. It is evident from the figure that the relay characteristic occupying greater area on the R-X diagram remains under the influence of the power surges. The MHO relay having the least area on the R-X diagram is least affected. The impedance relay characteristic has more area than the MHO relay but lesser area than a reactance relay. Therefore, while it is more affected than the MHO relay, it is affected less than the reactance relay. In other words, it is moderately affected. The reactance relay occupying the largest area is most affected.

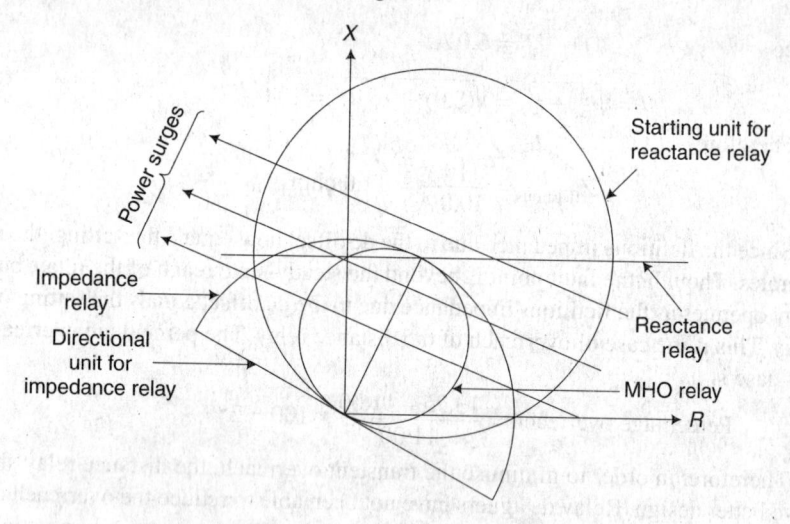

Fig. 6.40 Effect of power surges on distance relays

6.10.1 Power Swing Analysis

Figure 6.41 shows a section of a transmission line with generating stations beyond either end of the line section. The generated voltages are E_A and E_B, respectively. The voltage at the relay location is V. Impedances are as shown in the figure. The current flowing through the line is given by

$$I = \frac{E_A - E_B}{Z_A + Z_L + Z_B}$$

$$= \frac{E_A - E_B}{Z_T} \quad \text{(where } Z_T = Z_A + Z_L + Z_B\text{)}$$

$$V = E_A - IZ_A$$

The impedance 'seen' by the relay is given by

$$Z = \frac{V}{I} = \frac{E_A - IZ_A}{I} = \frac{E_A}{I} - Z_A$$

$$= \frac{E_A Z_T}{E_A - E_B} - Z_A$$

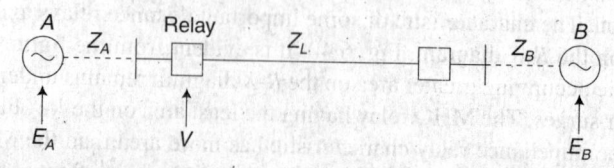

Fig. 6.41 One line diagram of a system to illustrate loss of synchronism

If E_A leads E_B by an angle δ and $E_A/E_B = n$, the above expression is written as

$$Z = \frac{ne^{i\delta}}{ne^{i\delta} - 1} Z_T - Z_A$$

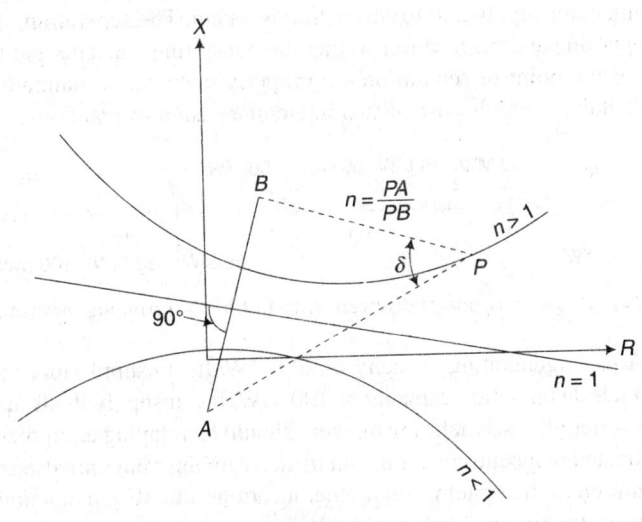

Fig. 6.42 *General loss of synchronism characteristics*

This equation represents a family of circles with n as parameter and δ as variable, as shown in Fig. 6.42. With $n = 1$, the power swing locus is a straight line perpendicular to AB ($AB = Z_T$). With $n > 1$ and $n < 1$, the power swing loci are circles, as shown in the figure.

The centres of the circles lie on the extension of the line Z_T, as shown in Fig. 6.43. For $n > 1$, the distance from B to the centre of circle $= Z_T/(n^2 - 1)$

Radius of circle $= \dfrac{nZ_T}{n^2 - 1}$

These are shown in Fig. 6.43.

For $n < 1$, the circles are symmetrical to those for $n > 1$, but with their centres beyond A, as shown in Fig. 6.42. The same formulae can be used for the radius and the distance of the centre of the circle by putting $1/n$ in place of n. So long as the power swing locus remains in the characteristic zone of a distance relay on the R-X diagram, the relay will see a fault and it will have a tendency to trip. Whether the relay will trip or not depends on the period for which the swing locus influ-

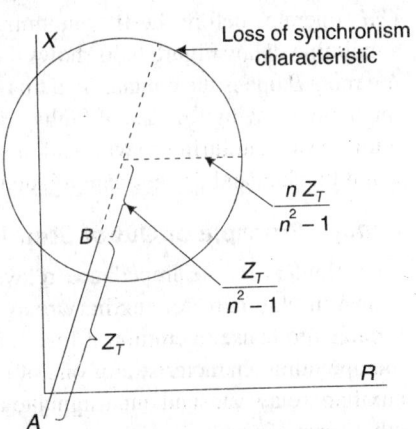

Fig. 6.43 *Graphical construction of loss of synchronism characteristic*

ences the relay characteristic. This has been explained in the previous section (Sec 6.10)

6.10.2 Out of Step Tripping and Blocking Relays

When generators of a power system lose synchronism, all ties between them are opened to maintain supply and to resynchronise them. The separation of the system is made at certain selected location so that the generating capacity and the loads on either side of the point of separation are properly matched to maintain the supply Figure 6.44 shows a simple case of two interconnected power stations.

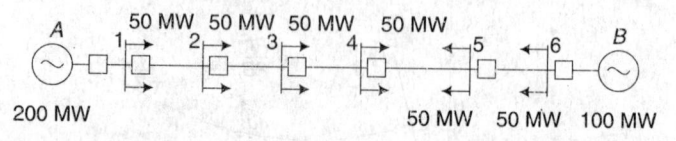

Fig. 6.44 Two interconnected system with out of step tripping relay at point 4

Station A has a generating capacity of 200 MW. It can supply loads up to points 4. Station B has a generating capacity of 100 MW. It can supply loads at point 6 and 5. So, if the system loses synchronism, there should be relaying equipment at point 4 for separation and resynchronisation. Out of step tripping relays are used at this point for the separation of the system. At all other locations, out of step blocking relays are used to prevent the tripping of distance relays due to power swings.

6.10.3 Principle of Out of Step Blocking Relay

When a fault occurs on a system, the voltage, current and the phase angle change instantly, whereas during power swings, these quantities change more slowly. This fact is utilised to cause blocking of the distance relay on power swings without preventing tripping under fault conditions.

In a MHO distance scheme, the I and II units are MHO relays and the III unit is an offset MHO relay, as shown in Fig. 6.45. During a power swing, the III zone unit OM_3 operates before the II zone unit M_2. OM_3 picks up a blocking relay B with a small time delay. Figure 6.46 shows the connection of the blocking relay. The blocking relay B opens the contact B so that the operation of the relay M_2 is prevented during a power swing. In case of faults, M_2 operates first and trips the circuit breaker. A fault may occur during a swing and it must be cleared. From this consideration, OM_3 is not blocked and it takes care of such faults.

6.10.4 Principle of Out of Step Tripping Relay

Two blinders (angle impedance relays) are employed for out of step tripping, as shown in Fig. 6.46. An auxiliary relay is used with each blinder. These blinders are arranged to cause operation of one of the auxiliary relays if the impedance crosses the operating characteristic of one of them before the other, in either direction. The auxiliary relay can send out a signal to an alarm, trip a circuit breaker or initiate some other form of control.

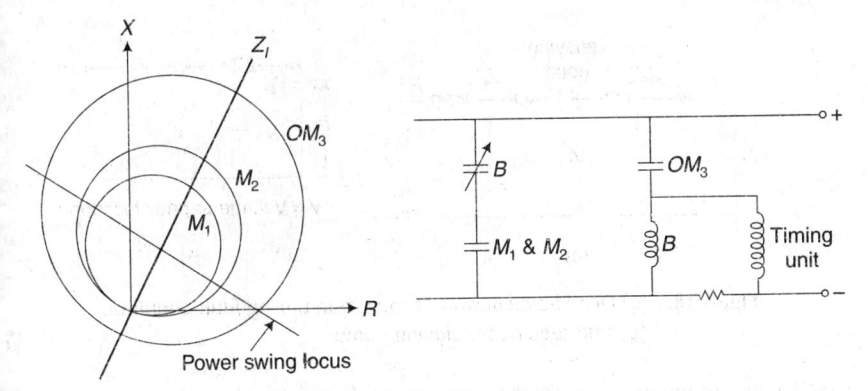

Fig. 6.45 Offset III zone characteristic for power swing blocking

Fig. 6.46 Connection of out of step blocking relay

6.11 EFFECT OF LINE LENGTH AND SOURCE IMPEDANCE ON DISTANCE RELAYS

Figure 6.48(a) shows a one-line diagram of the system during fault condition. Z_s is the source impedance behind the relay. Z_L is the line impedance from the relaying point to the fault point F. The current flowing through the relay is given by

$$I = \frac{E}{Z_S + Z_L}$$

The voltage at the relay location which is applied to the distance relay is

$$V = IZ_L = \frac{EZ_L}{Z_S + Z_L} = \frac{E}{\frac{Z_S}{Z_L} + 1}$$

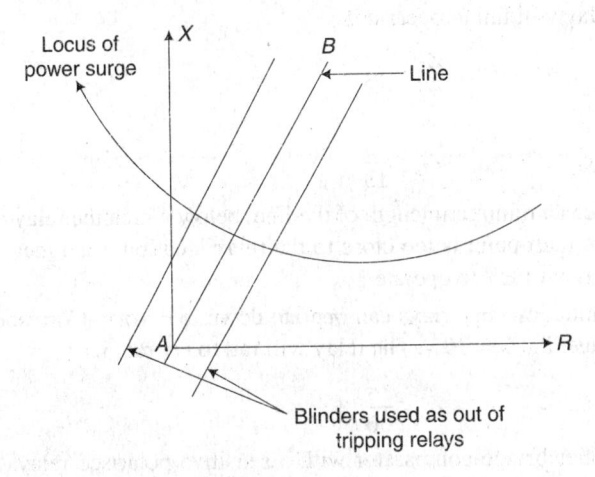

Fig. 6.47 Out of step tripping relays

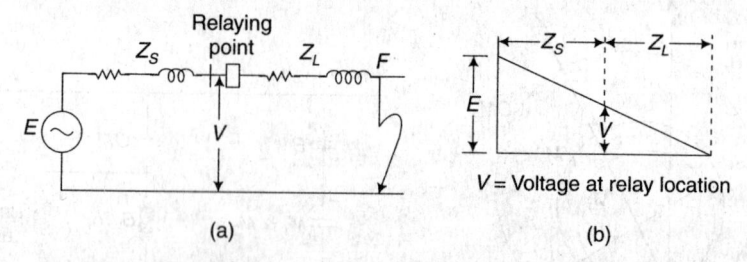

Fig. 6.48 *(a) One-line diagram of the system during fault condition*
(b) Voltage at the relaying point

Relay manufactures specify the minimum voltage for the relay operation. For example an induction cup MHO relay can operate down to 8 volts within 5% accuracy. We can find the limiting value of Z_S/Z_L for $V = 8$ V.

$$8 = \frac{E}{\frac{Z_S}{Z_L} + 1}$$

E can be taken equal to the normal secondary CT voltage, which is 110 V

$$8 = \frac{110}{\frac{Z_S}{Z_L} + 1}$$

$$\frac{Z_S}{Z_L} = \frac{110}{8} - 1 = 13.75 - 1 \cong 13$$

If the value of Z_S/Z_L is less than 13, the voltage at the relay point is more than 8 volts and the relay will operate. If the ratio Z_S/Z_L is more than 13, the voltage at the relay point is less than 8 volts and the relay will fail to operate. Z_S is constant for the system under consideration. The value of Z_L depends on the position of the fault point. The relay will fail to operate if

$$\frac{Z_S}{Z_L} > 13$$

or

$$Z_L < \frac{Z_S}{13}$$

Thus, there is a minimum length of the line, below which the relay cannot protect the line. If the fault point is too close to the relay location, such that Z_L is less than $Z_S/13$, the relay will fail to operate.

Modern induction cup relays can operate down to 3 volts. Corresponding to this value of voltage, $Z_S/Z_L = 36$ V. The relay will fail to operate if

$$Z_L < \frac{Z_S}{36}$$

For a rectifier bridge comparator with a sensitive polarised relay, the minimum operating voltage is 3.5 V, and correspondingly, $Z_S/Z_L \cong 30$.

6.12 SELECTION OF DISTANCE RELAYS

The effect of arc resistance and power surges plays an important role in the selection of distance relays for a particular distance protective scheme. As the reactance relay remains unaffected by arc resistance, it is preferred for ground fault relaying. A reactance relay is also used for phase fault relaying in case of a short line. As the impedance of a short line is small, the value of arc resistance is comparable to the line impedance and it may cause an appreciable error. Therefore, a relaying unit independent of arc resistance, i.e. a reactance relay is suitable for the protection of short lines against phase faults. But the reactance relay is more affected by power surges than the MHO and impedance relays. In case of short lines, power surges remain for a shorter period and hence their effect is unimportant. The predominating factor is therefore, the effect of arc resistance.

A conductance relay is more tolerant to arc resistance than the MHO and impedance relays. It is also more economical than a reactance relay as it does not require a directional unit. It can be used where the problem of arc resistance is moderate. It is applicable to both distribution lines, overhead and cables.

As reactance relays are affected by power surges more than impedance and MHO relays, they are not suitable for longer lines. The effect of power surges stays for a longer period in case of long lines and hence, a relay which is least affected by power surges is preferred for the protection of long lines. The MHO unit is less affected by power surges than the impedance and reactance relays and hence, it is best suited for the protection of long lines against phase faults. But it is most affected by arc resistance. As the impedance of a long line is large, the arc resistance will not cause appreciable error and its effect can be neglected. Thus, the predominating factor for the selection of a distance relay for long lines is the effect of power surges. In case of very long line, elliptical or quadrilateral relays are best suited as they occupy the least area on the R-X diagram and hence are least affected by power surges.

An impedance relay is moderately affected by both power surges as well as arc resistance. So it is better suited for medium lines for phase fault relaying. There is no sharp dividing line which can decide the choice of a distance relay for a particular application. Actually there is a lot of overlapping between the areas of application where one or another type of distance relay is best suited. The above discussions simply describe the basic principles. Practical experience also plays an important role in the selection of distance relays for a particular situation.

6.13 MHO RELAY WITH BLINDERS

Though a MHO relay is quite suitable for the protection of long lines, for the protection of extra long lines (ELD), its area on the R-X diagram may be too large to trip under power swing conditions. Hence, further reduction in the area of a MHO characteristic is necessary to make the relay suitable for ELD lines. In the case of electromagnetic relays, two blinders have been used to reduce the characteristic area of a MHO relay, as shown in Fig. 6.49. These blinders are angle impedance relays which are connected in series with the MHO relay. Each of them allows the MHO relay to see only in a particular direction.

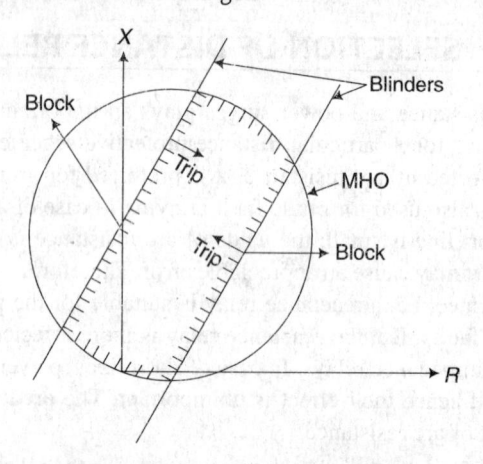

Fig. 6.49 *MHO relay with blinders*

6.14 QUADRILATERAL RELAY

A quadrilateral characteristic is best suited to ELD lines. It possesses an ideal distance relay characteristic. Its characteristic can be designed to just enclose the fault area of the line to be protected, see the explanation of the term fault area in Sec. 6.8.1. Such a characteristic is least affected by power surges, fault resistance and overloads. Thus, it is seen that a quadrilateral characteristic is an ideal characteristic for the protection of ELD lines. It is also quite suitable for short and medium lines. A static relay gives a better quadrilateral characteristic than a combination of electromagnetic relays.

To realise a quadrilateral or any other multilateral characteristic a multi-input phase comparator is employed. The term multi-input comparator is used for a comparator which has more than two inputs.

If more than one two-input comparators are used to realise a complex characteristic, the relay will be faster than a single multi-input comparator. But the drawback of the combination of two-input comparators is that the outputs of each comparator do not go active at the same time. To overcome this difficulty, the output of each comparator has to be prolonged for a short time which sometimes leads to erratic tripping.

6.14.1 Realisation of Quadrilateral Characteristic

To realise a quadrilateral characteristic, a multi-input phase comparator, as shown in Fig. 6.50, is employed. In a multi-input comparator, all the input signals are compared with each other. The resultant characteristic is the area enclosed by the lines and circles resulting from all these comparsions.

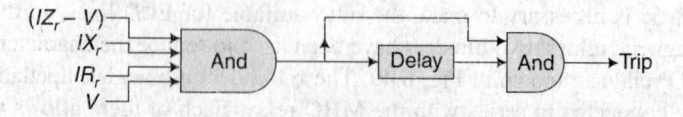

Fig. 6.50 *Multi-input phase comparator for quadrilateral characteristic*

Distance Protection

The phase comparison between IX_r and V gives a straight line characteristic, as shown in Fig. 6.51(a). The characteristics resulting from the phase comparison of other inputs are shown in Fig. 6.51(b), (c), (d) and (e). The proof for the characteristics shown in Fig. 6.51(d) and (e) has already been given in section 6.3.3 and 6.4.3, respectively. The proofs for the other characteristics are given at the end of this chapter in the appendix. The resultant characteristic is shown in Fig. 6.51(f). The MHO circle will not interfere with the rectangular characteristic as the diameter of the circle passes through the corners of the rectangle.

The desired characteristic is shown in Fig. 6.51(g). In this case, the MHO circle is an undesired element, which can be eliminated if at least one of the quantities $(IZ_r - V)$ and V is pulsed. To take care of the resistance fault near the bus, a characteristic shown in Fig. 6.51(h) is obtained giving a 10° shift in the characteristic. This is obtained by shifting IX_r and IR_r by 10°. For this, X_r is replaced by an impedance having a phase angle of 80°. A capacitor is placed in parallel with R_r for 10° shift in IR_r.

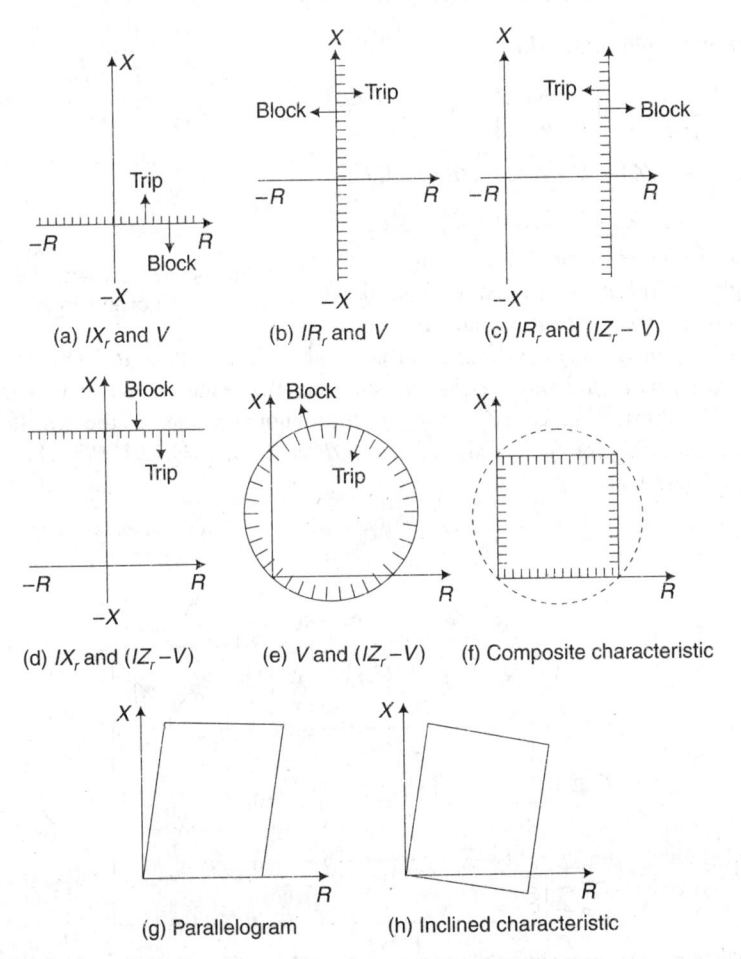

Fig. 6.51 Various characteristics resulting from the phase comparison of inputs indicated in Fig. 6.50

(a) IX_r and V
(b) IR_r and V
(c) IR_r and $(IZ_r - V)$
(d) IX_r and $(IZ_r - V)$
(e) V and $(IZ_r - V)$
(f) Composite characteristic
(g) Parallelogram
(h) Inclined characteristic

6.15 ELLIPTICAL RELAY

An elliptical characteristic is less affected by power surges as it occupies less area on the R-X diagram compared to other types of distance relays, as shown in Fig. 6.52. To realise an elliptical relay, a three input amplitude comparator, as shown in Fig. 6.53 is employed. The inputs are $(V - IZ_r)$, $(V - IKZ_r)$ and MI as shown in figure. The corresponding bridge currents are $\dfrac{V - IZ_r}{R}$, $\dfrac{V - IKZ_r}{R}$ and MI. M is a constant which is made equal to $\dfrac{Z_r + KZ_r}{R}$.

For the operation of the relay, the following condition should be satisfied.

$$\left|\dfrac{V - IZ_r}{R}\right| + \left|\dfrac{V - KIZ_r}{R}\right| < \left|I\dfrac{(Z_r + KZ_r)}{R}\right|$$

or $\quad |V - IZ_r| + |V - KIZ_r| < |I(Z_r + KZ_r)|$

or $\quad |Z - Z_r| + |Z - KZ_r| < |Z_r + KZ_r|$

This is an equation for an ellipse passing through the origin as shown in Fig. 6.54(a). The sum of the distance of any point on the ellipse from the foci is constant and it is equal to the length of the major-axis. The first two terms of the above equation represent the distance of two foci from the curve. The third term, i.e. $(Z_r + KZ_r)$ is the major diameter of the ellipse. In the figure F_1 and F_2 are foci of the ellipse. $OP = Z$, $OF_1 = Z_r$, $F_1P = Z - Z_r$, $F_2P = Z - KZ_r$ and $OF_2 = KZ_r$.

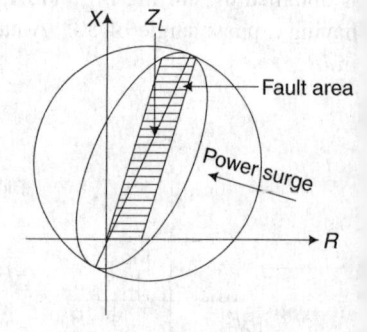

Fig. 6.52 *Ellipse less affected by power surges*

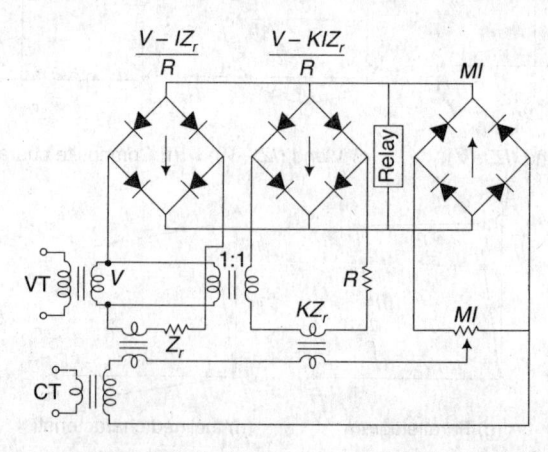

Fig. 6.53 *Three-input amplitude comparator to realise elliptical characteristic*

To take care of the resistance fault close to the bus, an offset elliptical characteristic which overlaps the origin as shown in Fig. 6.54(b) is used. The offset characteristic, having one focus i.e., F_2 at the origin is given by the equation.

$$|Z - Z_r| + |Z| < |Z_r + 2KZ_r|$$

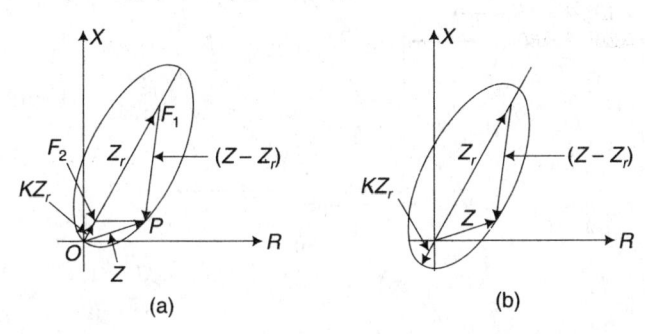

Fig. 6.54 (a) Ellipse passing through origin (b) Offset ellipse with one focus at the origin

Due to its rounded characteristic, it is only suitable for III zone of protection as a back-up relay.

6.16 RESTRICTED MHO RELAY

To realise a MHO characteristic, the phase angle θ between $(IZ_r - V)$ and V is compared with $\pm 90°$. If θ is compared with some other angle less than $\pm 90°$, the resulting characteristics are as shown in Fig. 6.55(a). The equation for relay operation is written as

$$-(90° - \varepsilon) \le \theta \le (90° - \varepsilon)$$

or
$$-\lambda \le \theta \le \lambda$$

where λ is an angle less than 90° i.e., $\lambda = 90° - \varepsilon$.

The restricted MHO characteristic can also be realised using an amplitude comparator. The inputs are

$(2V - IZ'_r)$ and IZ'_r, where,

$Z'_r = Z_r \operatorname{cosec} \psi$, arc $Z'_r = \theta \pm (\pi/2 - \psi)$,

$\lambda = 180° - \psi$,

Z_r = diameter of the MHO circle,

Z'_r = chord of restricted MHO, and

θ = angle of Z_r.

For operating conditions,

$$(2V - IZ_r) < IZ'_r$$

Restricted MHO characteristic is shown in Fig. 6.56(a). Restricted offset MHO characteristic is shown in Fig. 6.56(b).

Fig. 6.55 *(a) MHO and restricted MHO characteristics*
(b) Offset MHO and restricted offset MHO

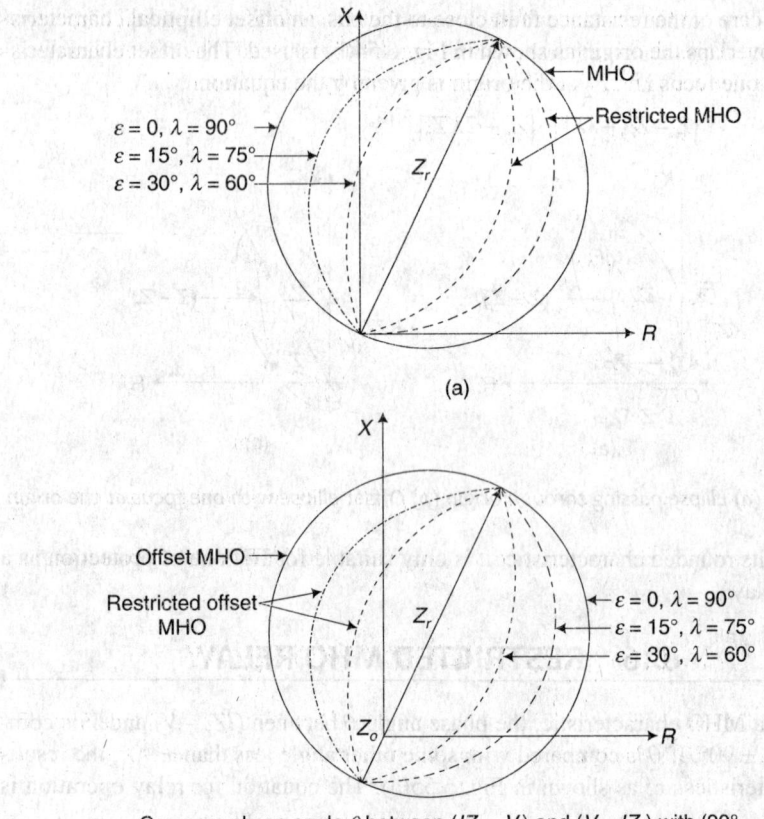

Compare phase angle θ between $(IZ_r - V)$ and $(V - IZ_o)$ with $(90° - \varepsilon)$

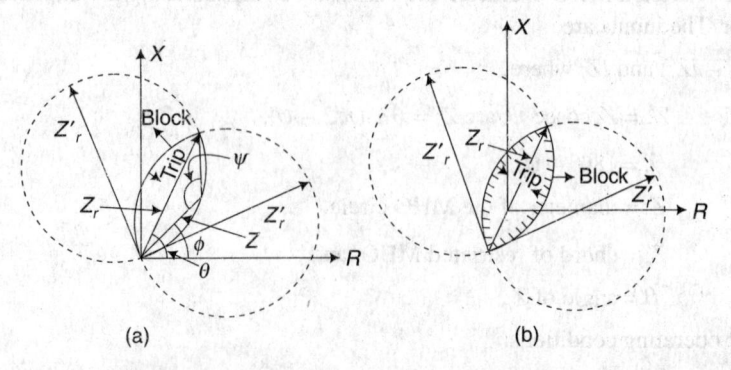

Fig. 6.56 *(a) Restricted MHO characteristics using an amplitude comparator*
(b) Restricted offset MHO relay

6.17 RESTRICTED IMPEDANCE RELAY

To realise an impedance relay characteristic, the phase angle θ between $(IZ_r + V)$ and $(IZ_r - V)$ is compared with $\pm 90°$. If θ is compared with some other angle less than $\pm 90°$, the resulting characteristic is the restricted impedance characteristic, as shown in Fig. 6.57. The operating equation is written as

$$-(90° - \varepsilon) \leq \theta \leq (90° - \varepsilon)$$

or

$$-\lambda \leq \theta \leq \lambda$$

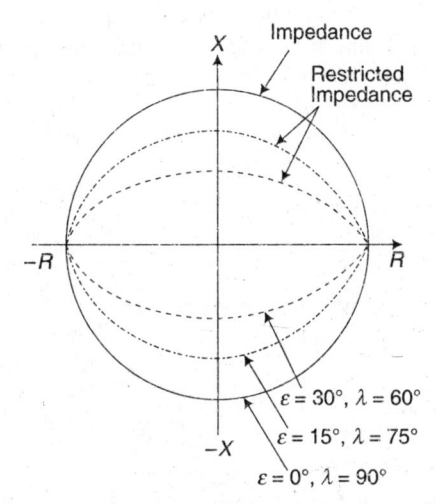

Fig. 6.57 Impedance and restricted impedance characteristics

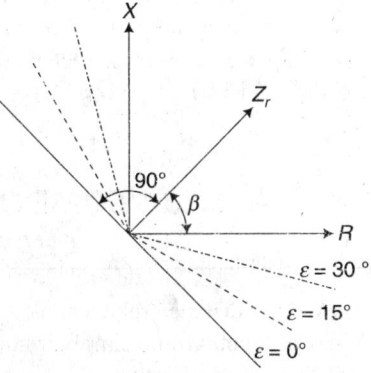

$\varepsilon = 30°, \lambda = 60°$
$\varepsilon = 15°, \lambda = 75°$
$\varepsilon = 0°, \lambda = 90°$

6.18 RESTRICTED DIRECTIONAL RELAY

The directional relay has a straight line characteristic passing through the origin, as shown in Fig. 6.58. It is realised by comparing the phase angle $(\theta - \beta)$ with $\pm 90°$, where θ is the phase angle between IZ_r and V. If $(\theta - \beta)$ is compared with an angle less than 90°, the restricted directional characteristics, as shown in Fig. 6.58 are obtained. The relay trips when the following condition is satisfied.

$$-(90° - \varepsilon) \leq (\theta - \beta) \leq (90° - \varepsilon)$$

$$-\lambda \leq (\theta - \beta) \leq \lambda$$

The restricted directional relay can also be realised using an amplitude comparator. For a simple directional relay, the inputs to the amplitude comparator are $(IZ_r + V)$ and $(IZ_r - V)$. For restricted directional characteristic, the inputs are $K(IZ_r + V)$ and $(IZ_r - V)$. The relay operates when

$$|K(IZ_r + V)| > |(IZ_r - V)|$$

Fig. 6.58 Restricted directional relay

or
$$|K(Z_r + Z)| > |(Z_r - Z)|$$

where,
$$K = \sqrt{\frac{Z_r^2 + Z^2 - 2ZZ_r \cos \lambda}{Z_r^2 + Z^2 + 2ZZ_r \cos \lambda}}$$

6.19 RESTRICTED REACTANCE RELAY

An angle impedance relay is realised by comparing the phase angle θ between IZ_r and $(IZ_r - V)$ with $\pm 90°$. If θ is compared with an angle λ which is less than $90°$ as shown in Fig. 6.59, the restricted reactance relay characteristic is obtained. The relay operates when the following condition is satisfied.

$$-\lambda \leq \theta \leq \lambda$$

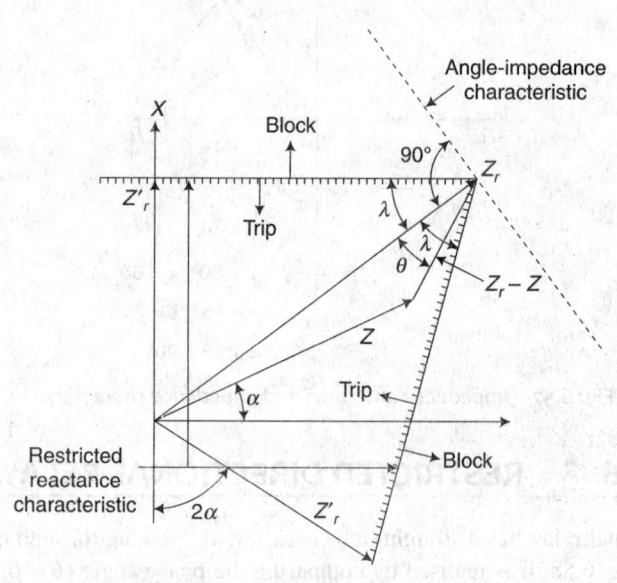

Fig. 6.59 *Restricted reactance relay*

The restricted reactance characteristic can also be realised using an amplitude comparator. In an amplitude comparator, the inputs are V and $(2IZ_r - V)$ to realise an angle impedance relay characteristic. For restricted reactance relay, the inputs are V and $(2IZ'_r - V)$. Tripping occurs when

$$|V| < |(2IZ'_r - V)| \text{ or } |Z| < |(2Z'_r - Z)|$$

6.20 SOME OTHER DISTANCE RELAY CHARACTERISTICS

Many other distance relay characteristics have been realised and discussed in texts. A few of them will be simply introduced in this section to make the reader familiar with these characteristics.

6.20.1 Hyperbola

A hyperbola characteristic as shown in Fig. 6.60 can be realised using a multi-input comparator. For the operation of the relay, the following condition should be satisfied.

$$|Z - Z_r| - |Z - KZ_r| < |Z_r + KZ_r|$$

This expression is similar to that for an elliptical characteristic, except that the second term is negative. Hyperbolic characteristics can be used as blinders.

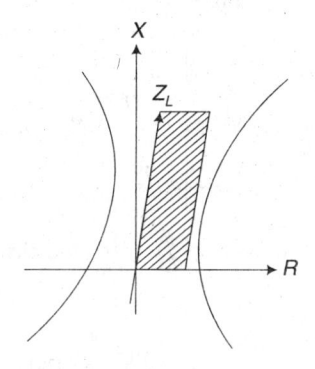

Fig. 6.60 *Hyperbolic characteristic*

6.20.2 Parabola

A parabolic characteristic, as shown in Fig. 6.61, can also be realised. Such a characteristic can be used as a blinder, i.e., to limit the area of another relay on the *R-X* diagram.

6.20.3 Limacons

Limacons are also used as blinders. Their characteristic is shown in Fig. 6.62.

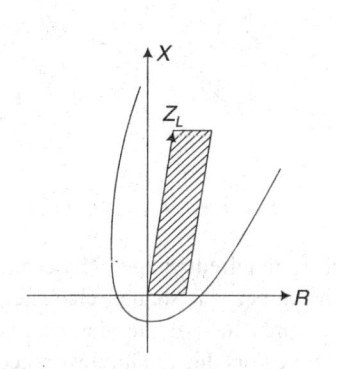

Fig. 6.61 *Parabolic characteristic*

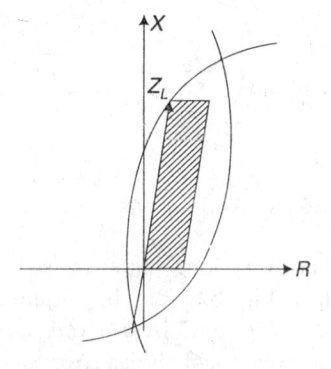

Fig. 6.62 *Limacons*

6.20.4 Specially Shaped Characteristics

From the relay performance point of view, specially shaped characteristics, other than circular ones, have been developed. These include some well-known curves also. A multi-input comparator is employed to realise specially shaped characteristics to provide greater immunity to power swings and heavy system loading. The characteristic curve shown in Fig. 6.63(a) comprises of intersecting circular characteristics. This characteristic has a reduced area on the *R-X* diagram and hence is less susceptible to power surges. It is suitable for the protection of very long line. This can be realised using a scheme shown in Fig. 6.63(b). Figure 6.64 and 6.65 show some typical characteristics.

292 Power System Protection and Switchgear

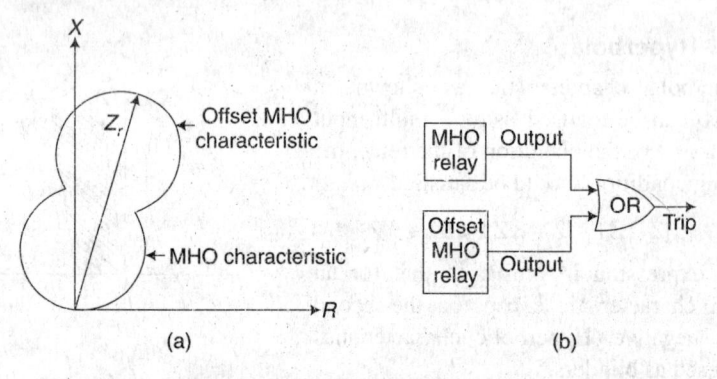

Fig. 6.63 (a) Resultant of MHO and offset MHO characteristics
(b) Scheme to realise characteristic shown in Fig. 6.63(a)

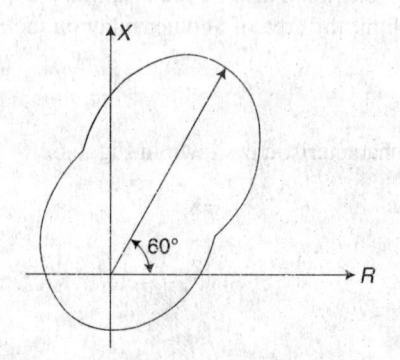

Fig. 6.64 A typical offset characteristic

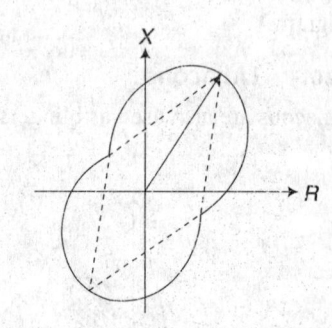

Fig. 6.65 A typical characteristic

Where very long starter or back-up reach is required, shaped characteristics as shown in Fig. 6.66 can be employed. An asymmetrical shaped characteristic, as shown in Fig. 6.67 is also very useful. The upper-half of the characteristic is to provide a reasonably broad coverage to allow for errors due to fault resistance, power

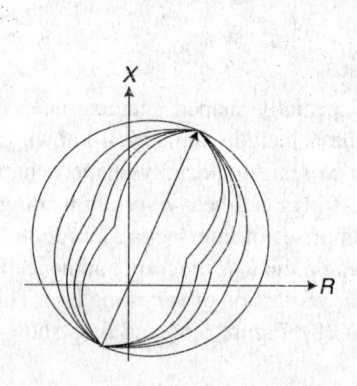

Fig. 6.66 Specially shaped characteristics

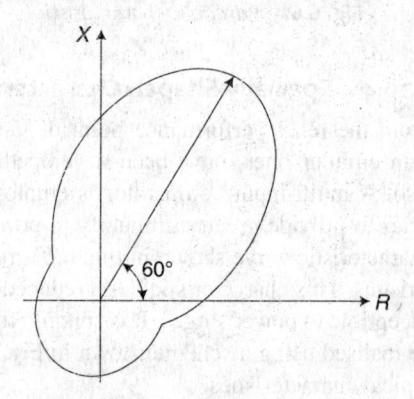

Fig. 6.67 An asymmetrically shaped characteristic

system data, etc. and the lower half of the characteristic is chosen to be a narrow one to give good discrimination with load impedance. This characteristic gives very good performance where only one measuring unit is employed in a switched distance scheme.

Specially shaped characteristics can also be realised using multi-relay schemes. But such schemes become complex and expensive. There is also a time-coordination problem in a multi-relay scheme. A multi-input relay is superior to the multi-relay scheme. It is capable of producing a wide variety of complex and specially shaped characteristics. Multi-input phase comparators are commonly used. Sometimes multi-input amplitude comparator or hybrid (combination of amplitude and phase comparator) comparators are also employed.

6.21 SWIVELLING CHARACTERISTICS

We have already discussed that a reactance relay is suitable for short lines and MHO relay for long lines. It will be better if a relay changes its characteristic when the line impedance Z_L changes. The fault resistance increases when the source impedance Z_S increases. So it will be still better if the relay characteristic changes with changing Z_S/Z_L. A circular relay characteristic is desirable for low Z_S/Z_L and a straight line characteristic for high Z_S/Z_L. For intermediate values of Z_S/Z_L intermediate characteristic is desirable. Figure 6.68 shows such characteristics. To realise such characteristics, a fully cross-polarised MHO relay is used. The polarising voltage is taken from the healthy phases. A phase comparator is employed to compare $(IZ_r - V)$ and V_{pol}, where V_{pol} = polarizing voltage derived from the healthy pahses, V = fault voltage, I = fault current, and Z_r = distance relay setting.

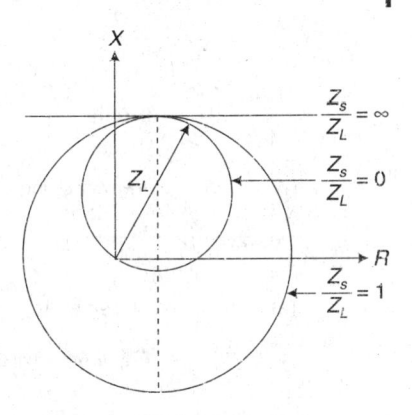

Fig. 6.68 Swivelling characteristic

Such a relay also operates reliably when a close-up fault occurs. It has the ability to operate on a large arc resistance and high resistance faults. But it may overreach in case of high resistance faults for low values of Z_S/Z_L. This problem can be overcome by introducing non-linearity in the impedance of the cross-polarising circuit, i.e. the circuit to realise voltage from the unfaulted phase pair.

With single-line to ground faults, Z_S/Z_L ratio is reduced due to high ground resistance. For such a situation, a reactance characteristic is desirable but the relay under discussion will give a MHO characteristic. Thus, we see that the swivelling is not so effective with single-phase to ground faults. However, its performance for L-G faults will be quite satisfactory on lines with ground wires with metal supports, i.e. with lines with good earthing (i.e. low value of earth resistance).

6.22 CHOICE OF CHARACTERISTICS FOR DIFFERENT ZONES OF PROTECTION

For the protection of lines against line to line faults, MHO relays are used for I zone and II zone. For III zone, offset MHO type of a relay is used. The third zone unit is an offset unit and is placed at station A if the line section to be protected is not too long. It can also be used for initiating power swing blocking. Such an arrangement is shown in Fig. 6.69(a). The offset unit is set with the backward reach of about 10% of the full forward reach. An offset MHO characteristic takes care of close-up faults, including 3-phase faults. In case of long line section, the III zone unit is placed at the remote end of the line so that its characteristic circle is smaller and hence it is less affected by power swings. The arrangement is shown in Fig. 6.69(b). In this position, it has better discrimination between faults and loads.

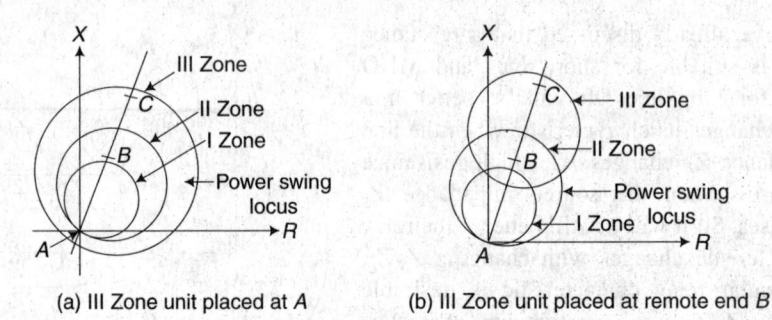

(a) III Zone unit placed at A (b) III Zone unit placed at remote end B

Fig. 6.69 *Three-step MHO characteristic*

For ground faults and short lines a reactance relay is preferable though it is more expensive. A quadrilateral relay is suitable for long as well as short lines. For I and II zones, quadrilateral relays passing through the origin are used, as shown in Fig. 6.70(a). The third unit is an offset unit. For a long line section, the third unit is placed at the remote end of the line, as shown in Fig. 6.70(b).

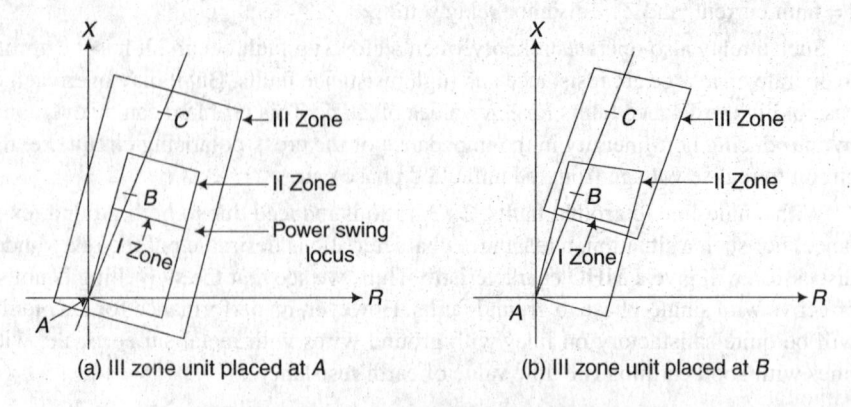

(a) III zone unit placed at A (b) III zone unit placed at B

Fig. 6.70 *Three-step quadrilateral characteristic*

6.23 COMPENSATION FOR CORRECT DISTANCE MEASUREMENT

If only one measuring unit is used for all types of phase and ground faults, a complex switching arrangement is required to switch on the corresponding voltage and current. Detailed analysis is given below. Different types of switching systems will be discussed later on. The impedance measured by a protective relay depends on the type of fault. If a distance relay is energised by line to line voltage and line current, the impedance seen by the relay will be $2Z_1$ for phase to phase fault, and $\sqrt{3}\,Z_1 \angle 30°$ for a three-phase fault. If the relay is fed with phase voltage and phase current, the impedance seen is $(Z_1 + Z_2 + Z_3)/3$ for a line to ground fault. But it is not true for all positions of the fault point. It depends on the number of sources and the number of earthed neutrals available at the time. It is the usual practice to provide separate measuring units for phase to phase faults and phase to ground faults. The distance relay which protects the system against phase to phase faults is also supposed to protect against three-phase faults. To measure the same impedance for phase to phase and three-phase faults, the measuring unit is energised by line to line voltage and the difference between the currents in the corresponding two phases, as given bleow.

Relay	Voltage	Current
A – B phase-pair	V_{AB}	$I_A - I_B$
B – C phase-pair	V_{BC}	$I_B - I_C$
C – A phase-pair	V_{CA}	$I_C - I_A$

If only one measuring unit is used for all the phase-pairs, the corresponding voltage and difference in currents of the involved phases is supplied by a switching system.

For phase to ground faults, the measuring units are energised by phase to neutral voltage and corresponding phase current, plus a fraction of the residual current, as shown on the next page.

Relay	Voltage	Currents
A-phase	V_A	$I_A + \frac{1}{3}(K-1)I_{res}$
B-phase	V_B	$I_B + \frac{1}{3}(K-1)I_{res}$
C-phase	V_C	$I_C + \frac{1}{3}(K-1)I_{res}$

where, $K = Z_0/Z_1$ and $I_{res} = I_A + I_B + I_C = 3I_0$.

If only one unit of distance relay is used for all the three phases, corresponding voltage and current are supplied from a switching system.

6.23.1 Phase-fault Compensation

Although the same relays are employed for both phase to phase and three-phase faults, they do not measure the same impedance between the fault point and the relay location for each type of fault unless proper compensation is provided. Let us first examine a case of a distance relay which is used for phase fault protection and it is supplied with line to line voltage, say V_{BC} and line current I_B. The impedance seen by the relay is given by

$$Z = \frac{V_{BC}}{I_B} = \frac{V_B - V_C}{I_B}$$

$$V_B = V_{B1} + V_{B2} + V_{B0} = a^2 V_{A1} + a V_{A2} + V_{A0}$$

It can also be written as

$$V_B = a^2 V_1 + a V_2 + V_0$$

Similarly

$$V_C = a V_1 + a^2 V_2 + V_0$$

Writing the sequence voltage drops as

$$V_1 = I_1 Z_1,\ V_2 = I_2 Z_2 \text{ and } V_0 = I_0 Z_0 \text{ we get}$$

$$V_B - V_C = (a^2 I_1 Z_1 + a I_2 Z_2 + I_0 Z_0) - (a I_1 Z_1 + a^2 I_2 Z_2 + I_0 Z_0)$$

$$Z = \frac{V_{BC}}{I_B} = \frac{(a^2 - a)(I_1 Z_1 - I_2 Z_2)}{a^2 I_1 + a I_2 + I_0}$$

or

$$Z = \frac{Z_1 (a^2 - a)(I_1 - I_2)}{a^2 I_1 + a I_2 + I_0} \quad (\text{as } Z_2 = Z_1 \text{ for a feeder})$$

For a phase to phase fault, $I_2 = -I_1$ and $I_0 = 0$,

$$\therefore\quad Z = \frac{2 I_1 Z_1 (a^2 - a)}{I_1 (a^2 - a)} = 2 Z_1$$

for a three-phase fault, $I_2 = I_0 = 0$,

$$\therefore\quad Z = \frac{I_1 Z_1 (a^2 - a)}{a^2 I_1} = Z_1 (1 - a^2) = \sqrt{3}\, Z_1 \angle 30°$$

From the above analysis it is clear that the impedance seen by the relay is not same for phase to phase and three-phase faults. Therefore, it is necessary to provide some form of compensation so that the relay measures the same impedance for both types of fault. Now consider the case where the relay is energised by V_{BC} and $(I_B - I_C)$.

$$I_B - I_C = (a^2 I_1 + a I_2 + I_0) - (a I_1 + a^2 I_2 + I_0)$$
$$= (a^2 - a)(I_1 - I_2)$$

The impedance seen by the relay is

$$Z = \frac{V_{BC}}{I_B - I_C}$$

$$= \frac{Z_1 (a^2 - a)(I_1 - I_2)}{(a^2 - a)(I_1 - I_2)} = Z_1$$

We see that the impedance measured by the relay is independent of the sequence components of current. So the relay is fully compensated. It measures the same impedance for both phase to phase

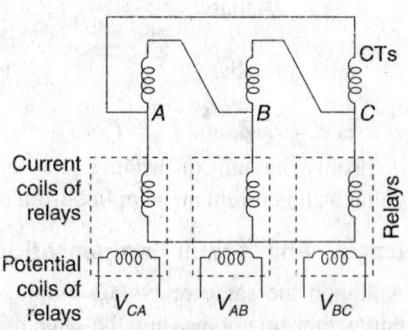

Fig. 6.71 Compensated phase-fault connections

and three-phase faults, including the case where these faults also involve ground. To implement this type of compensation, the CTs are connected in delta and relays in star configuration, as shown in Fig. 6.71.

6.23.2 Earth-fault Compensation

An uncompensated earth-fault relay is energised by V_A and I_A. The impedance seen by such a relay is

$$Z = \frac{V_A}{I_A} = \frac{V_1 + V_2 + V_0}{I_1 + I_2 + I_0}$$

$$= \frac{I_1 Z_1 + I_2 Z_2 + I_0 Z_0}{I_1 + I_2 + I_0}$$

For a line to ground fault, $I_1 = I_2 = I_0$. This relation is true for the phase-sequence component in the fault as well as in the line. Also $Z_1 = Z_2$.

$Z = (2Z_1 + Z_0)/3 = Z_E$, where Z_E is the earth-fault loop impedance.

If the system has a number of sources connected to it or has multiple earthing, or when both of these conditions exist, as in the case of the 132 kV transmission system, there will be a number of paths to the fault. In such a situation, $I_1 = I_2 = I_0$ relation no longer holds for the line currents, although it still holds for the current in the fault. Consequently, the impedance seen by an uncompensated relay depends on a number of sources and earthed neutrals available at the time.

As the sequence components of current in the faulty phase are no longer equal, it shows that currents are present in the healthy phases as well as the faulty phase. For the compensation of the relay, this condition can be utilised,

The voltage applied to the relay is

$$V_A = I_1 Z_1 + I_2 Z_2 + I_0 Z_0$$

By adding and subtracting $I_0 Z_1$, we get

$$V_A = (I_1 + I_2 + I_0) Z_1 + I_0 (Z_0 - Z_1)$$
$$\text{as } Z_1 = Z_2$$
$$= I_A Z_1 + I_0 (Z_0 - Z_1)$$
$$= Z_1 \left[I_A + I_0 \frac{(Z_0 - Z_1)}{Z_1} \right]$$

If a relay has a voltage V_A applied to it and carries the current $I_A + (Z_0 - Z_1) I_0 / Z_1$, the impedance seen by the relay is Z_1. This is known as residual current compensation.

$$I_0 = (I_A + I_B + I_C)/3$$

$(I_A + I_B + I_C)$ is known as residual current. Figure 6.72 shows residual

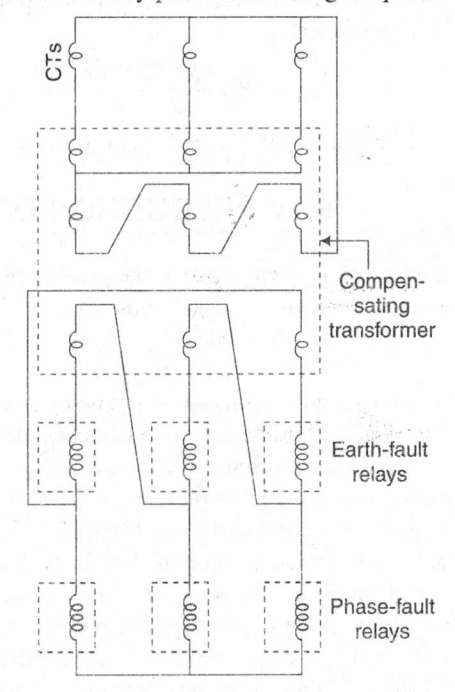

Fig. 6.72 Method of applying compensation for phase-fault relays and residual compensation for earth-fault relays

compensation for earth-fault relays. This figure shows the combined connections for phase-fault and earth-fault relays. Voltage coils of relays are not shown in the figure. The voltage coils of earth fault relays are energised by the phase voltage and those of the phase fault relays by line voltage.

6.23.3 Sound-Phase Compensation

If we put $\dfrac{Z_0}{Z_1} = K$, V_A is given by

$$V_A = Z_1 I_A + \frac{(I_A + I_B + I_C)}{3}(Z_0 - Z_1)$$

$$= Z_1 \left[I_A + (I_A + I_B + I_C)\frac{(K-1)}{3} \right]$$

This equation can be re-written as

$$V_A = Z_1 \left[I_A \left(1 + \frac{K-1}{3}\right) + I_B \frac{(K-1)}{3} + I_C \frac{(K-1)}{3} \right]$$

$$= Z_1 \left(1 + \frac{K-1}{3}\right) \left[I_A + I_B \left(\frac{K-1}{K+2}\right) + I_C \left(\frac{K-1}{K+2}\right) \right]$$

If the relays has a voltage V_A applied to it and carries a current equal to $[I_A + I_B(K-1)/(K+2) + I_C(K_1-1)/(K+2)]$ i.e. faulty phase current together with a fraction $(K-1)/(K+2)$ or $(Z_0 - Z_1)/(2Z_1 + Z_0)$ of the current in each of the healthy phases, it will see the impedance

$$Z = Z_1 \frac{(2+K)}{3} = \frac{2Z_1 + Z_0}{3} = Z_E$$

For details and connections of this type of compensation, see Ref. 2.

6.24 REDUCTION OF MEASURING UNITS

Distance relays are used for the protection of transmission or subtransmission lines against phase to phase and phase to ground faults. For phase to phase faults, one set consisting of 3 distance relays is required for each pair of lines ($A - B$, $B - C$ and $C - A$). Thus, 9 distance relays are required. Each set requires 3 units as there are 3 zones of protection; one relay being necessary for each zone. Similarly, 9 relays are required for phase to ground faults; III units for each phase. Thus, 18 relays are required for providing three time-distance steps for the protection of a feeder against phase to phase and phase to ground faults. Besides 18 distance relays, some auxiliary relays, fault detectors and flag indicators are also required. This type of scheme will be very costly and complex. In practice an effort is made to reduce the number of relays employed. In modern distance relaying schemes, only one measuring unit is provided for all the three protective zones. The ohmic reach is progressively increased by a timing unit, to obtain the distance settings for the II and III zones; the I zone being instantaneous. Sometimes two MHO measuring units are used for III zones of protection. One MHO unit is used for the I and II zone, and a separate unit for the III zone which is an offset MHO unit.

There are different types of distance protection schemes which reduce the number of measuring relays (i.e. distance relays in the schemes under discussion) to reduce the cost of protection. The number of measuring units depends upon the technical requirements but in selecting a particular scheme the economic consideration also must be taken into account. Some important schemes have been described below.

6.24.1 Scheme Using Six Measuring Units

In this scheme, one measuring unit is used for all the III zones through a timer unit. Three units are used for phase to ground faults; one unit for each phase. Three units are used for phase to phase faults; one unit for each pair of phases.

6.24.2 Scheme Using Twelve Measuring Units

In this scheme, only one unit is employed for the I and II zone. A timing unit sets their reach. A separate unit is employed for the III zone. Separate relaying units are provided for phase to phase and phase to ground faults. Thus, 12 relaying units are employed in this scheme.

6.25 SWITCHED SCHEMES

A switched scheme may employ only one, two or three measuring units. Such schemes are less costly and hence they can be used for lines operating at comparatively lower voltages such as 66 kV, 33 kV or 11 kV.

In a switched scheme, an appropriate switching arrangement is necessary to switch on the corresponding voltage and current for the measuring units. The various types of switched schemes have been discussed below.

6.25.1 Delta-Wye Switching

In this scheme, 3 measuring units are employed. The same unit is used for phase as well as ground faults. Normally the units are connected for phase faults. They are enerised by delta voltage and delta current. When ground faults occur they are switched to Y connections. Thus, phase faults are cleared instantaneously but ground faults after a small delay. The tripping time is about one cycle for phase faults and about five cycles for ground faults.

The measuring units may be impedance, MHO or reactance units depending on the technical requirements of the system in a particular situation. Polyphase overcurrent or MHO type fault detectors can be used to start the timing unit on phase fault in the case of impedance or MHO type measuring units. For ground faults residual current or power relays are employed as fault detectors.

When the measuring units are reactance type distance relays, switching to Y connection is done only for single line to ground faults. For double line to ground faults switching to Y connection is prevented by a suitable arrangement in the switching system and the delta connection is retained in this situation. On double line to ground fault, the current in the leading of the two fault phases may be almost in phase with the corresponding Y voltage, and because of this fact, the relay may measure zero or negative reactance, which could cause an undesired tripping of the circuit breaker.

This scheme has been in use in Europe but the large number of contacts required make the system complex and costly. Hence, most manufacturers employ six reactance units where high speed operation (0.03) is required. Only one reactance unit is employed where 0.15 s operating time can be accepted.

6.25.2 Interphase Switching

Only two measuring units are employed in this scheme, one unit for phase faults and the other for ground faults. These units are switched to appropriate phase or phase pair by fault detectors.

For phase to phase faults, a MHO type measuring unit and instantaneous overcurrent fault detectors are employed. On a phase to phase fault the measuring unit is energised by phase to phase voltage such as V_{AB} or V_{BC} or V_{CA}, depending on the phase pair involved in the fault. The current coils are energised by the difference of currents of the two phases involved in the fault, such as $(I_A - I_B)$ or $(I_B - I_C)$ or $(I_C - I_A)$. With these actuating quantities, the relay measures the same Z_1 for both phase to phase and three-phase faults. Figure 6.73 shows a schematic diagram for this scheme.

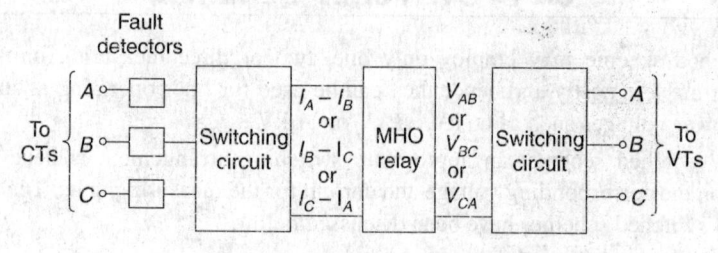

Fig. 6.73 Single MHO unit for phase faults

For ground faults, a reactance type measuring unit and under-voltage relays as fault detectors are employed. A MHO type starting unit is also included for the reactance relay. For correct measurement of Z_1 under all types of fault conditions, residual current compensation is used, (see Sec. 6.23.2). The energising current is phase current plus $(Z_0 - Z_1) I_0/Z_1$. For fault on phase A, the current is $I_A + (Z_0 - Z_1) I_0/Z_1 = I_A + KI_0$. V_A is applied to the voltage coils. A switching circuit is employed to switch on the appropriate phase voltage and phase current to the measuring unit. Figure 6.74 shows a schematic diagram for this scheme.

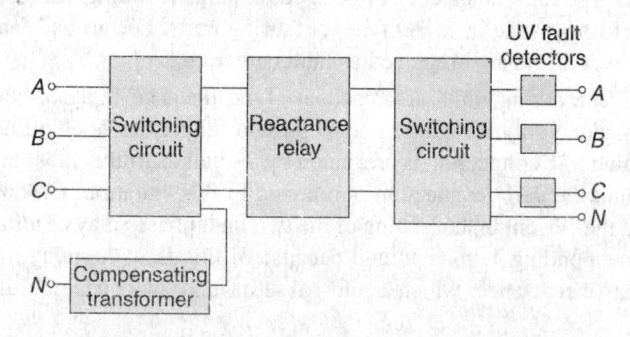

Fig. 6.74 Single reactance relay for ground faults

A cheaper alternative of interphase switching scheme is to employ a switched distance scheme only for phase to phase faults and to employ residual current relays for ground faults.

6.25.3 Switch Scheme Using Only One Measuring Unit

For voltage of 33 kV and below, only one distance relay can be employed for the protection of the system against all types of faults. The measuring unit may be MHO, reactance or impedance type, depending upon the technical requirements of the system. If a reactance relay is used as a measuring unit, a MHO type directional unit is also used. The operating and restraining coils of the directional unit receive the same actuating quantities which is switched for the measuring unit. The polarising coil of the directional unit uses the same voltage as the restraining coil. But to ensure the operation for close up faults, 5% of the voltage from the phase or phase pair leading to the faulty phase is added to the polarising voltage. In England, France and the USA, a reactance relay is preferred in the switched scheme which employs a single measuring unit. In Germany, impedance relays have been used. In some schemes MHO units have also been employed.

Fault detectors are used in each phase to detect faults. In case of phase to phase faults, phase to phase voltage and difference of the current of the faulty phases are applied to the measuring unit. In case of ground faults, phase voltage and phase current are switched on.

6.26 AUTO-RECLOSING

About 80-90% of faults on overhead transmission and distribution lines are transient in nature. These faults disappear if the line circuit breakers are tripped momentarily to isolate the line. The disconnection of the line from the system permits the arc to extinguish. The line is re-energised again by reclosing circuit breakers to restore normal supply after the arc path becomes sufficiently de-ionised. Automatic reclosing of circuit breakers is known as auto-reclosing. The flashover across an insulator due to lightning is the most common cause of transient faults. Other possible causes of such faults are swinging of line conductors and temporary contact with external conducting objects.

The remaining 10-20% of faults are either semi-permanent or permanent. The most common cause of semi-permanent fault is a small tree branch falling on the line. It requires one or more reclosures to burn the tree branch. Permanent faults are caused by broken conductor, broken insulator or a wire falling on tower or ground. A permanent fault must be located and repaired before the supply is restored. Auto-reclosing is not recommended in case of cable breakdowns because the breakdown of the insulation in cables causes a permanent defect.

The radial lines are most benefited by automatic reclosure because they are connected to the power supply only at one end. The quicker the power supply is restored, the better. If a line is fed at both ends, breakers at both ends trip simultaneously on the occurrence of a fault. The separate generators at the two ends of the line drift apart in their phase relationship. Therefore, the automatic reclosure of breakers must be done before the generators drift too apart to maintain synchronism. Thus, we see that a further benefit, particularly to the EHV system is the maintenance of system stability and synchronism.

In the case of interconnected system, loss of a line is not important and automatic reclosure is not very essential. There is more than one power source and hence no area is deprived of supply due to the loss of a line. If there is only one important line connecting two networks and it is to be kept in at all costs, then an automatic reclosure is necessary.

6.26.1 Single-shot Auto-reclosing

Most of the faults on EHV transmission lines are due to flashover across insulators caused by lightning. Because of the height of EHV lines, tree branches are unlikely to cause faults. If some physical conducting objects are dropped on EHV lines by birds, they are vaporised instantly due to large amount of power in the arc. Consequently, there is no need for more than one reclosure in case of EHV transmission lines. In a single-shot auto-reclosing scheme, only one reclosure is made. The reclosure should be made as quickly as possible so that there should not be an appreciable drift in phase angle between the voltages at the two ends of the open line.

If there are circuit breakers at both ends of an EHV line, they should be tripped and reclosed simultaneously. For simultaneous tripping, a carrier channel, microwave, radio link or a temporary extension of zone 1 can be employed. If one circuit breaker trips before the other, there will be an effective reduction in the dead time. It may jeopardize the chances of a successful reclosure. The arc path takes a certain period of time to become sufficiently deionised. Therefore, before reclosing the circuit breakers, the line must remain de-energised for a certain period for de-ionisation of the arc path which depends on the system voltage. The time for de-ionisation is equal to the dead time of the circuit breaker, which is defined as the time between the fault arc being extinguished and the circuit breaker contacts remaking. Table 6.3 shows de-energisation time for different lines.

Table 6.3 De-energisation time required for automatic reclosing				
Line Voltage in kV	66	132	220	400
Minimum de-energisation (time in seconds)	0.1	0.17	0.28	0.5

6.26.2 Multi-shot Auto-reclosing

On lines at a voltage of 33 kV or below, the faults might be caused by external objects, such as tree-branches, vines, etc. falling on the line. Tree-branches fall on the lines because their height may be less than that of nearby trees. Such objects may not be burnt clear at the first reclosure, and hence, subsequent reclosures are required. A multi-shot auto-reclosing scheme provides more than one automatic reclosures. On radial lines, one instantaneous reclosure is provided, followed by 2 or 3 more delayed reclosures if necessary. The statistical fig. show that 80% of the faults are cleared after the first reclosure. The second reclosure is made after a delay of 15 to 45 seconds. About 10% of the remaining faults are cleared after the second reclosure.

The third reclosure is made after 60 to 120 s. Less than 2% of faults require the third reclosure. If a fault is not cleared after three reclosures, there is an automatic lock-out of the reclosing relay. The usual practice is to reclose the circuit breakers three-times. The fourth reclosure, if required, can be made by hand. If the fourth reclosure fails, there is clear indication of a permanent fault. The remaining faults are permanent faults which require detection and repair by the maintenance staff. A multi-shot scheme is often recommended in forest areas. If several trip and reclose operations are performed, the maintenance of circuit breaker increases. The vacuum circuit breakers are almost maintenance-free and hence are more suitable for multi-shot reclosing.

6.26.3 Single-phase (Single-pole) Auto-reclosing

If two systems are interconnected by a single tie line, the tripping of all three phases on a fault causes an immediate drift in phase angle of the voltages of the two systems. On the other hand, if only the faulty phase is tripped during earth fault conditions, the synchronising power still flows through the healthy phases. In a single-phase auto-reclosing (single-pole auto-reclosing) scheme, only the faulty phase pole of the circuit breaker is tripped and reclosed. For any multi-phase fault, all the three phases are simultaneously tripped and reclosed. In a single-phase auto-reclosing, each phase of the circuit breaker is segregated and provided with its own closing and tripping mechanism. The main disadvantage is a longer de-ionsiation time. When a faulty phase is disconnected, a capacitive coupling between the healthy phases and the faulty phase tends to maintain the arc, resulting in a longer de-ionisation time. This may cause interference with communication circuits.

A complex relaying scheme is required to detect and select the faulty phase in case of single phase auto-reclosing scheme. Therefore, this scheme is more complex and expensive as compared to the three-phase auto-reclosing scheme.

6.26.4 Three-phase Auto-reclosing

In a three-phase auto-reclosing scheme, all the three phases are tripped and reclosed when a fault occurs on the system, irrespective of types of the fault. Its relaying scheme is simpler and less expensive than the single phase auto-reclosing scheme. It is faster because of less de-ionising time.

6.26.5 Delayed Auto-reclosing Scheme

When two sections of a power system are connected through a number of transmission systems, there is little chance of drifting them apart in phase and losing synchronism. On such a system, delayed auto-reclosing can be employed. The C.E.G.B. employs delayed auto-reclosing with dead times of the order of 5-6 s. The fault arc de-ionisation times and circuit breaker operating characteristics do not present problems in such a scheme. Before reclosing, power swings are allowed to settle down. All tripping and reclosing schemes may be three-phase only, simplifying the control circuits compared to single-phase schemes. An analysis of their relative performance on the C.E.G.B. EHV system shows that high-speed auto-reclosings are successful in 68% of reclosure attempts, whereas delayed reclosings in 77.5%.

APPENDIX

A 6.1 To Realise Special Straight Line Characteristic, Coincident with R-axis

Figure 6.75(a) is a general phasor diagram showing V, I and voltage drops. If the phase angle θ between IX_r and V is compared with $\pm 90°$, the straight line characteristic as shown in Fig. 6.75(b) is obtained. If the point P moves towards the R-axis, θ increases and becomes equal to $90°$ when P lies on the R-axis. If P moves towards the $-$ R-axis, the value of θ is minus and it is $-90°$ when P lies on the $-$ R-axis. If P lies in the tripping area, θ is less than $\pm 90°$. If P lies in the blocking zone, θ is greater than $\pm 90°$.

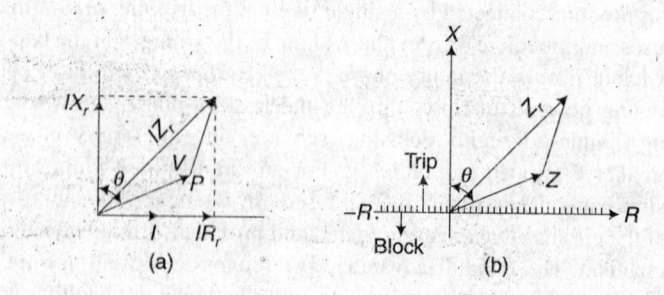

Fig. 6.75 (a) General phasor diagram showing V, I and voltage drops
(b) Straight line characteristic, the line coincident with R-axis. Relay trips when X > 0

A 6.2 To Realise a Straight Line Characteristic, Coincident with X-axis

Figure 6.76(a) is a general phasor diagram showing voltage, current and voltage drops. If the phase angle θ between IR_r and V is compared with $\pm 90°$, the straight line characteristic as shown in Fig. 6.76(b) is obtained. The R-axis has been taken as reference for the angle θ. If the point P moves towards the X-axis, θ increases and it becomes $90°$ when P lies on the X-axis. If the point P moves towards the $-$ X-axis, θ is negative, and it becomes $-90°$ when P lies on the $-$ X-axis. If P lies in the tripping zone, θ is less than $\pm 90°$. If P lies in the blocking zone, θ is greater than $\pm 90°$.

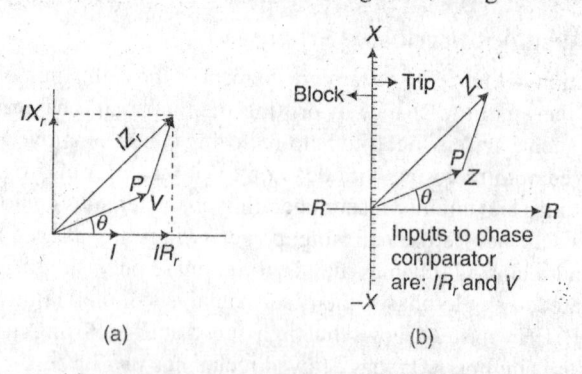

Fig. 6.76 (a) General phasor diagram showing V, I and voltage drops
(b) Straight line characteristic coincident with X-axis

A 6.3 To Realise a Straight Line Characteristic, Parallel to X-axis

Figure 6.77(a) is a general phasor diagram showing voltage, current and voltage drops. A perpendicular MN from the point M, as shown in Fig. 6.77(b), has been drawn. The phase angle θ between $(IZ_r - V)$ and IR_r is compared with $\pm 90°$. If the point P lies on the left hand side of the line MN, θ is less than $90°$. If P lies on the right hand side of MN, θ is greater than $90°$. If the point P lies in the tripping zone, θ is less than $\pm 90°$. If P lies in the blocking zone, θ is greater than $\pm 90°$. If P lies above the R-axis, θ is positive. If P lies below R-axis θ is $-ve$.

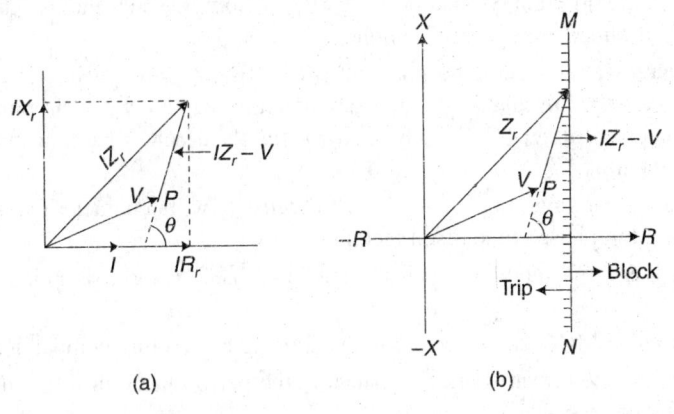

Fig. 6.77 (a) General phasor diagram showing V, I and voltage drops
(b) Straight line characteristic parallel to X-axis

EXERCISES

1. In what way is distance protection superior to overcurrent protection for the protection of transmission lines?
2. What is an impedance relay? Explain its operating principle. Discuss how it is realised using the (i) electromagnetic principle, (ii) amplitude comparator, (iii) phase comparator.
3. Explain impedance relay characteristic on the R-X diagram. Discuss the range setting of three impedance relays placed at a particular location. Discuss why the I zone unit is not set for the protection of 100% of the line.
4. Draw and explain the circuit connection of three impedance relays together with the directional relay, circuit breaker trip coil, CB auxiliary switch, flags, seal-in relay, etc.
5. Explain stepped time-distance characteristics of three distance relaying units used for I, II and III zone of protection.
6. Discuss how (i) an electromechanical and (ii) a static reactance relay is realised. Explain its characteristic on the R-X diagram. Is it a directional relay? If not what type of directional unit is recommended for it? Is a directional relay which has a straight line characteristic suitable for it?

7. Discuss how (i) an electromechanical and (ii) a static MHO relay is realised. Explain its characteristic on the *R-X* diagram.
8. Draw and explain the circuit connections of three MHO units used at a particular location for three zones of protection.
9. What is the difference between a polarised MHO and a simple MHO relay? What are self-polarised and cross-polarised MHO relays?
10. Draw and explain the circuit connections of three reactance units used at a particular location for the III zones of protection.
11. What are different types of distance relays? Compare their merits and demerits. Discuss their field of applications.
12. What is an offset MHO characteristic? Discuss how such a characteristic is realised. Discuss its field of applications. Under what circumstances is a reversed offset MHO relay used for the third zone of a transmission line protection?
13. What is an angle impedance relay? Discuss how its characteristic is realised using the phase comparison technique.
14. How is a reactance relay characteristic realised using a sampling comparator?
15. How is a MHO characteristic realised using a sampling comparator?
16. Discuss the effect of arc resistance on the performance of different types of distance relays.
17. Draw impedance, reactance and MHO characteristics to protect the 100 per cent of the line having $(2.5 + j6)$ ohm impedance. A fault may occur at any point on the line through an arc resistance of 2 ohms. Determine the maximum percentage of line section which can be protected by each type of relay.

(**Ans.** 83%, 100% and 77%).

Hint. Draw relay characteristics as shown in Fig. 6.78. AB is the line to be protected. Make $CD = EF = 2$ ohms.

The line section protected by impedance relay $= \dfrac{AC}{AB} \times 100$

The line section protected MHO relay $= \dfrac{AE}{AB} \times 100$

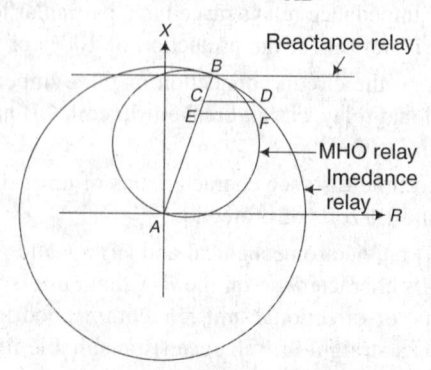

Fig. 6.78

18. Discuss the effect of power surges on the performance of different types of distance relays.
19. Discuss the selection of distance relays for the protection of long, medium and short lines against (i) ground faults and (ii) phase faults.
20. Explain why a MHO characteristic is preferred for the protection of long lines against phase faults, whereas a reactance relay is preferred for ground faults.
21. Explain why a reactance relay is preferred for the protection of short lines against both, phase faults as well as ground faults.
22. What are blinders? In what circumstances are they used in conjunction with a MHO relay?
23. What do you understand by out of step tripping? Discuss the operating principle of an out of step tripping relay.
24. What do you understand by out of step blocking? Discuss the operating principle of an out of step blocking relay.
25. Derive an expression for the minimum voltage at which the relay can operate. What is the minimum voltage for the operation of a relay, if $Z_S/Z_L = 13$?
26. Discuss how an elliptical characteristic is realised using static comparators. Why is an elliptical characteristic used only for back-up protection?
27. What are restricted MHO characteristic? How are such characteristics realised using a phase comparator?
28. Discuss what is swivelling characteristic. How is it realised using a phase comparator? What are its advantages?
29. What is quadrilateral characteristic? In what way is it superior to other characteristics? How is it realised using a phase comparator?
30. Show that a relay used for phase to phase faults and three-phase faults does not measure the same impedance for both types of faults if they are energized by line voltage and line currents. Discuss what measures are taken so that the relay can measure the same impedance for both types of faults.
31. What are switched distance relaying schemes? Describe them in brief.
32. What is delta-wye switching scheme for distance protection?
33. What is interphase switching scheme for distance protection?
34. Discuss a switched scheme which uses only one measuring unit for all types of phase as well as ground faults.
35. What are the advantages of auto-reclosing?

    What are (i) Single-shot auto-reclosing, (ii) Multi-shot auto-reclosing, (iii) Single-phase auto-reclosing, (iv) Three-phase auto-reclosing and (v) Delayed auto-reclosing?

36. Figure 6.79 shows distance protection for a section of a power system. The I zone setting at A and B is 150 ohms.

    (a) What will be impedance seen by the relay at A for a fault at F_1?
    Will the relay at A operate before the circuit breaker at B has tripped?

(b) Will the relay at B trip for a fault at F_1 before the circuit breaker at A has tripped?

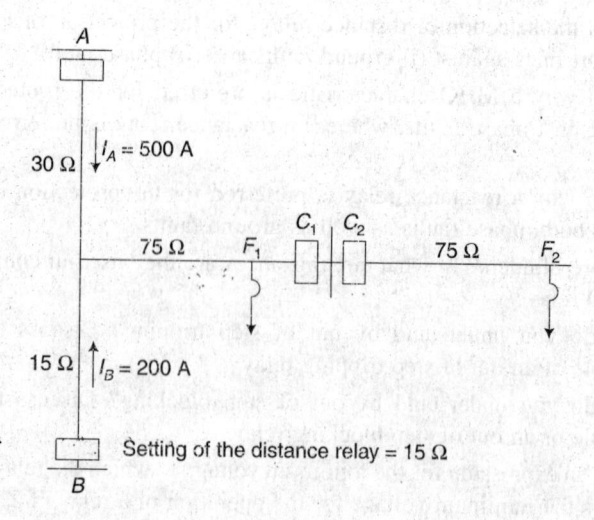

Fig. 6.79

(c) If the circuit breaker C_2 fails for a fault at F_2, will the fault be cleared by relays at A and B?

(d) How will the fault at F_2 be cleared?

Solution:

(a) For a fault at F_1, the voltage drop from A to F_1
= 500 × 30 + (500 + 200) × 75 V.

The impedance seen by the relay at A

$$= \frac{500 \times 30 + (500 + 200) \times 75}{500} = 135 \text{ }\Omega.$$

The setting of the distance relay at A = 150 Ω.

Therefore, the relay at A will see the fault at F_1 and trip before the circuit breaker at B has tripped.

(b) The voltage from B to F_1 for a fault at F_1
= 200 × 15 + (200 + 500) × 75 V.

The impedance measured by the relay at B

$$= \frac{200 \times 15 + (200 + 500) \times 75}{200} = 277.5 \text{ }\Omega$$

Therefore, the relay at B will not trip before the circuit breaker at A has tripped.

When the circuit breaker at A has tripped, the relay at B will measure the impedance = (15 + 75) = 90 Ω and trip.

(c) For a fault at F_2 the impedance measured by the relay at A

$$= \frac{500 \times 30 + (500 + 200) \times (75 + 75)}{500} = 240 \, \Omega$$

The relay at A will not operate.

For a fault at F_2, the impedance measured by the relay at B

$$= \frac{200 \times 15 + (200 + 500) \times (75 + 75)}{200} = 540 \, \Omega$$

The relay at B will not see the fault at F_2.

(d) The fault at F_2 will be cleared by back-up protection.

37. Distance relays are located at the point C of the power system shown in Fig. 6.80. The distance relays are MHO type. Their distance and time settings are as follows:

Zone	Distance Setting	Time Setting
I	80% of line CD	High speed
II	120% of line CD	0.30 second
III	200% of line CD	0.50 second

The X/R ratio is 1.732 and the centres of MHO circles lie at 60° lag on the R-X diagram. The generator A speeds up and loses synchronism with respect to generator B. The system is transmitting synchronising power equal to base kVA at base voltage over line CD towards B at 0.90 p.f. lagging. The rate of slip is constant at 0.10 slip-cycle per second. Will the MHO relay operate?

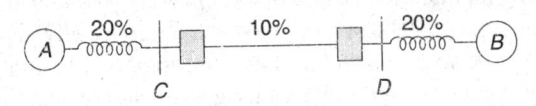

Fig. 6.80

Ans: The I zone relay is unaffected

The II zone unit does not operate

The III zone unit operates.

38. What type of protection will you recommend for a power line oprating at (i) 11 kV, (ii) 33 kV, (iii) 66 kV, (iv) 132 kV, (v) 220 kV, (vi) 400 kV and (vii) 750 kV?

39. Discuss the merits and demerits of single-phase and three-phase auto-reclosing.

7 Pilot Relaying Schemes

7.1 INTRODUCTION

Pilot relaying schemes are used for the protection of transmission line sections. They fall into the category of unit protection. In these schemes, some electrical quantities at the two ends of the transmission line are compared and hence they require some sort of interconnecting channel over which information can be transmitted from one end to the other. Such an interconnecting channel is called a pilot. Three different types of such channels are presently in use, namely wire pilot, carrier-current pilot and microwave pilot. A wire pilot may be buried private cables or alternatively, rented Post Office or private telephone lines. A carrier-current pilot is one in which a low-voltage, high frequency signal (50 kHz–700 kHz) is used to transmit information from one end of the line to the other. In this scheme, the pilot signal is coupled directly to the same high voltage line which is to be protected. This type of pilot is also called a power line carrier. A microwave pilot is a radio channel of very high frequency, 450 to 10,000 MHz. Wire pilot schemes are usually economical for distances up to 30 km. Carrier-current schemes are more economical for longer distances. When the number of services requiring pilot channels exceeds the technical or economical capabilities of carrier-current pilot, the microwave pilot is employed. A distance range up to 150 km is possible in a flat country, otherwise it is limited by hills and building. The link may operate up to 40 to 60 km without repeater station. The system is applicable only where there is a clear line of sight between stations. The power requirement for signal transmission is less than a watt because highly directive antennas are employed.

7.2 WIRE PILOT PROTECTION

In a wire pilot relaying scheme, two wires are used to carry information signals from one end of the protected line to the other. A wire pilot may be buried cable or a pair of overhead auxiliary wires other than the power line conductors. The scheme is a unit protection and operates on the principle of differential protection. The comparison is made between the CT secondary currents at the two ends of the line. As the pilot channels are very expensive, a single phase current is derived from three-phase currents at each end of the line, thereby using only a pair of pilot wires to carry information signal. For short lines, wire-pilot schemes are less expensive than carrier-current schemes because terminal equipment is simpler and cheaper. It is more reliable because of its simplicity. From cost considerations, the break-even point is about 15-30 km, but the distance is usually limited due to the attenuation of the signal caused by distributed capacitance and series resistance, rather than the cost. For short important lines, wire-pilot relaying is recommended. For unimportant lines, slow-speed overcurrent relays are employed. For long lines, carrier-current schemes are cheaper and more reliable than wire-pilot schemes.

The two alternative operating principles which are used for most of the practical schemes are circulating current principle and balanced voltage principle. Most wire pilot schemes use amplitude comparison in circulating current scheme since they are easier to apply to multi-ended lines and are less affected by pilot capacitance.

7.2.1 Circulating Current Scheme

Figure 7.1 shows the schematic arrangement for the circulating current principle. Figure 7.2 shows the schematic diagram of a practical scheme employing circulating current principle. The scheme is suitable for pilot loop resistance up to 1000 Ω and inter-core capacitances up to 2.5 microfarad. Polarities of the secondary voltage of CT have been marked in Fig. 7.1 for normal or external fault condition. Currents flowing in pilot wires and relay coils caused by CT secondary voltages are also

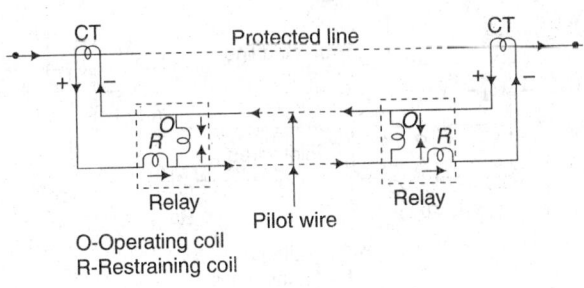

Fig. 7.1 *Schematic diagram of circulating current principle*

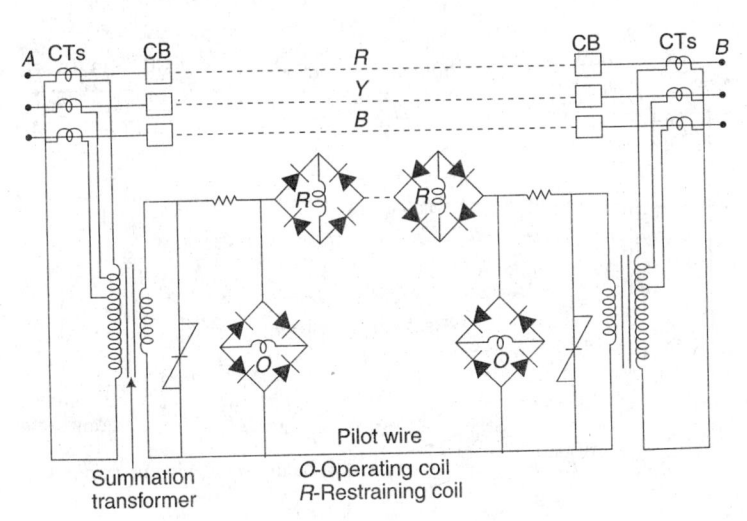

Fig. 7.2 *Practical scheme based on circulating current principle*

shown in the figure. In a circulating current scheme, the current circulates normally through the terminal CT and pilot wires. Under normal conditions and in case of external faults, current does not flow thorough the operating coil. In case of internal faults, the polarity of the remote end CT is reversed and hence current flows through the operating coil of the relay.

7.2.2 Balanced Voltage (or Opposed Voltage) Scheme

Figure 7.3 shows a schematic diagram of the balanced voltage principle. Polarities of CTs and direction of currents shown in the figure are for normal condition or external fault. In this scheme, current does not normally circulate through pilot wires. The operating coil of the relay is placed in series with the pilot wire and hence current does not flow through the pilot wires under normal conditions and in case of external faults. In case of internal faults, the polarity of the remote end CT is reversed, and hence current flows through the pilot wires and operating coils of the relays. Figure 7.4 shows a practical scheme based on the balanced voltage principle. It is called the Solkar system (Reyrolle). The capacitor shown in the figure is used to tune the operating circuit to the fundamental frequency component. The scheme is suitable for 7/0.029 pilot loops up to 400 Ω.

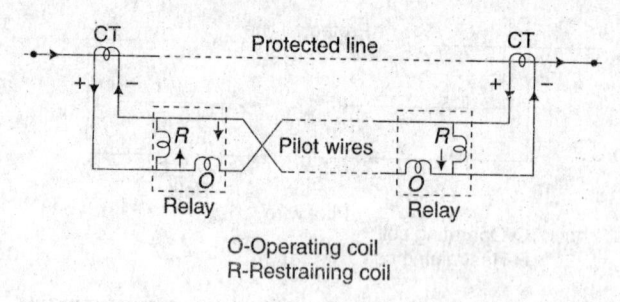

Fig. 7.3 Schematic diagram of balanced voltage principle

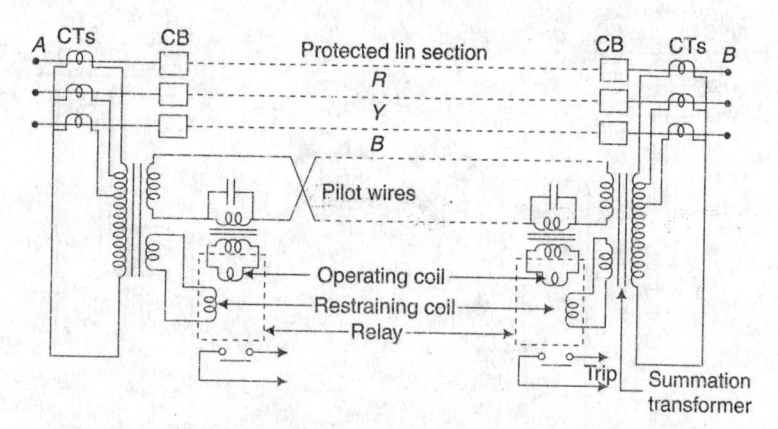

Fig. 7.4 Practical scheme employing balanced voltage principle

7.2.3 Transley Scheme (AEI)

This scheme is a balanced voltage scheme with the addition of a directional feature. Figure 7.5 shows the schematic arrangement of the scheme. An induction disc type relay is used at each end of the protected line section. The secondary windings of the relays are interconnected in opposition as a balanced voltage system by pilot wires. The upper magnet of the relay carries a summation winding to receive the output of current transformers. Under normal conditions and in case of external faults, no current circulates through the pilot wires and hence through the lower magnets of the relays. In these conditions, no operating torque is produced. In case of internal faults, current flows through the pilot wires and the lower electromagnets of the relay. In this condition, the relay torque is produced from the interaction of the two fluxes, one of which is produced directly from the local CT secondary current flowing through the upper magnet of the relay. The second flux is produced by the current flowing through the lower magnet. The current flowing through the lower magnet may be relatively small. Therefore, this scheme is suitable for fairly long pilots having loop resistance up to 1000 Ω. It is worth noting that the scheme is a phase comparison voltage balanced scheme. For details, see vol. I of Ref. 1 and vol. II of Ref. 2.

7.2.4 Transley S Protection

It is the latest wire pilot system which employs solid state technology. It is a very compact system. Its performance is superior to other schemes described earlier. It operates on the circulating current principle in contrast with its electromechanical predecessor described in the Section 7.2.3. Phase comparators are employed for measurement. For details, see Ref. 5, Sec.15.8.4. This scheme can be used up to pilot loop resistance of 1000 Ω. However, when pilot isolation transformers are employed, which is an optional feature to introduce an insulting barrier capable of withstanding 15 kV, the scheme can operate up to pilot loop resistance of 2500 ohms with the help of primary tappings range available on the isolation transformers.

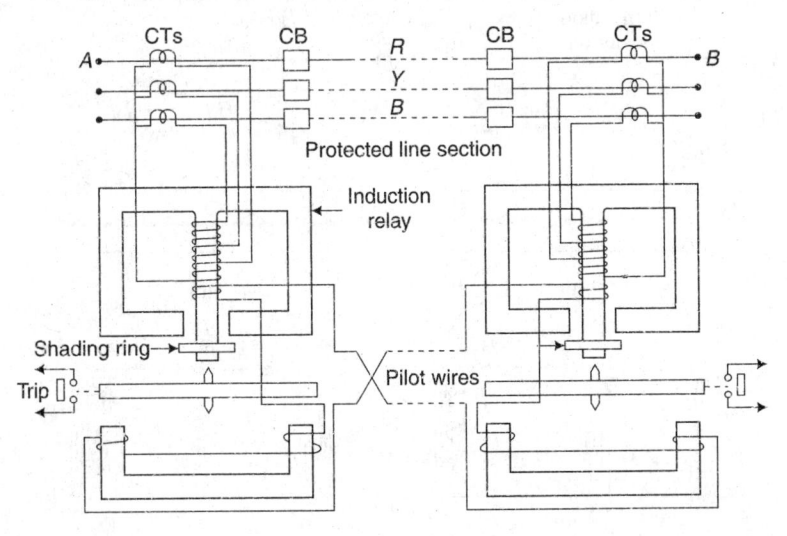

Fig. 7.5 *Transley scheme*

7.2.5 Half-wave Comparison Scheme

The circuit connection of this scheme is similar to that of the circulating current scheme but the operating principle is different. Figure 7.6(a) shows the schematic diagram of the scheme. The relay has only an operating coil, no restraining coil. The rectifiers are connected so as to allow the current through the operating coil only during an internal fault. The resistances R_A and R_B are made slightly greater than the pilot loop resistance R_p. CTs at A and B are connected in such a way that the polarities of the voltage applied to pilot are as shown in Fig. 7.6(b) and (c) in case of external fault or during normal conditions. If the polarity of the voltage applied to pilot at A is positive, as shown in Fig. 7.6(b), the resistance R_B is short circuited by the rectifier connected across it. The voltage applied to the operating coil at B is negative. Theoretically, the voltage applied to the operating coil at A is zero, but practically it becomes slightly negative because R_A is slightly greater than R_P. Thus no relay operates in case of external fault or during normal conditions. If the polarity of the voltage applied to the pilot is positive at B, the resistance R_A is short circuited as shown in Fig. 7.6(c). In this situation, voltage applied to the relay at A is negative. The voltage applied to the relay at B is slightly negative. Thus no relay operates.

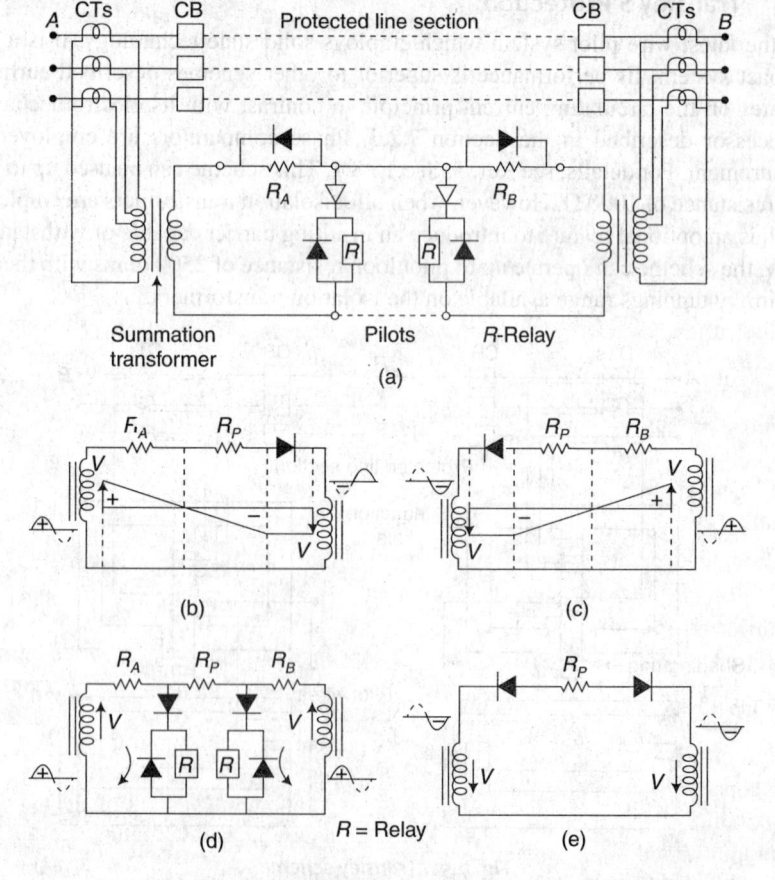

Fig. 7.6 *Half-wave comparison scheme*

In case of internal fault, the voltage applied to both relays is positive during the positive half cycle as shown in Fig. 7.6(d). Both relays operate in this condition. During the negative half cycle, the voltage applied at both ends of the pilot is negative, as shown in Fig. 7.6(e). An additional half-wave rectifier is placed across each relay coil to perpetuate current during the dead half cycle. Non-linear resistors are used to protect the CTs from overvoltages during dead half cycle when the two CTs would otherwise be open circuited.

7.3 CARRIER CURRENT PROTECTION

This is the most widely used scheme for the protection of EHV and UHV power lines. In this scheme a carrier channel at high frequency is employed. The carrier signal is directly coupled to the same high voltage line that is to be protected. The frequency range of the carrier signal is 50 kHz to 700 kHz. Below this range, the size and the cost of coupling equipment becomes high whereas above this range, signal attenuation and transmission loss is considerable. The power level is about 10-20 W. In this scheme, the conductor of the power line to be protected are used for the transmission of carrier signals. So the pilot is termed as a power line carrier.

With the rapid development of power systems and the large amount of interconnection involved, it has become very essential to have high speed protective schemes. Carrier current schemes are quite suitable for EHV and UHV power lines. They are faster and superior to distance schemes. Distance protective schemes are non-unit type schemes. They are fast, simple and economical and provide both primary and back-up protection. The main disadvantage of conventional time-stepped distance protection is that the circuit breakers at both ends of the line do not trip simultaneously when a fault occurs at one of the end zones of the protected line section. This may cause instability in the system. Where high voltage auto-reclosing is employed, non-simultaneous opening of the circuit breakers at both ends of the faulted section does not provide sufficient time for the de-ionisation of gases. The carrier current protection or any other unit protection does not suffer from these disadvantages. In unit protection, circuit breakers trip simultaneously at both ends. It is capable of providing high speed protection for the whole length of the protected line section.

In a carrier current scheme, the carrier signal can be used either to prevent or initiate the tripping of a protective relay. When the carrier signal is used to prevent the operation of the relay, the scheme is known as carrier-blocking scheme. When the carrier signal is employed to initiate tripping, the scheme is called a carrier intertripping or transfer tripping or permissive tripping scheme.

Carrier current schemes are cheaper and more reliable for long lines compared to wire pilot schemes, even though the terminal equipment is more expensive and more complicated. In some cases, the carrier signal may be jointly utilised for telephone communication, supervisory control, telemetering as well as relaying. Thus, the cost of carrier equipment chargeable to relaying work can be reduced. The coupling capacitors required for carrier signal can be used also as potential dividers to supply reduced voltage to instruments, relays etc. This eliminates the use of separate potential transformers.

There are two important operating techniques employed for carrier current protection namely the phase comparison technique and directional comparison technique. In the phase comparison technique, the phase angle of the current entering one end is compared with the phase angle of the current leaving the other end of the protected line section. If the currents at both the ends of the line are in phase, there is no fault on the protected line section. This will be true during normal conditions or in case of external faults. In case of faults on the protected line section, the two currents will be 180° out of phase. In this scheme, the carrier signal is employed as a blocking pilot.

In the directional comparison technique, the direction of power flow at the two ends of the protected line section is compared. During normal conditions and in the case of external faults, the power must flow into the protected line section at one end and out of it at the other end. In case of an internal fault, the power flows inwards from both ends.

7.3.1 Phase Comparison Carrier Current Protection

In this scheme, the phase angle of the current entering one end of the protected line section is compared with the current leaving the other end. Figure 7.7 shows the schematic diagram of the phase comparison scheme. The line trap is a parallel resonant circuit tuned to the carrier frequency connected in series with the line conductor at each end of the protected line section. This keeps carrier signal confined to the protected line section and does not allow the carrier signal to flow into the neighbouring sections. It offers very high impedance to the carrier signal but negligible impedance to the power frequency current. There are carrier transmitter and receivers at both the end of the protected line. The transmitter and receiver are connected to the power line through a coupling capacitor to withstand high voltage and grounded through an inductance.

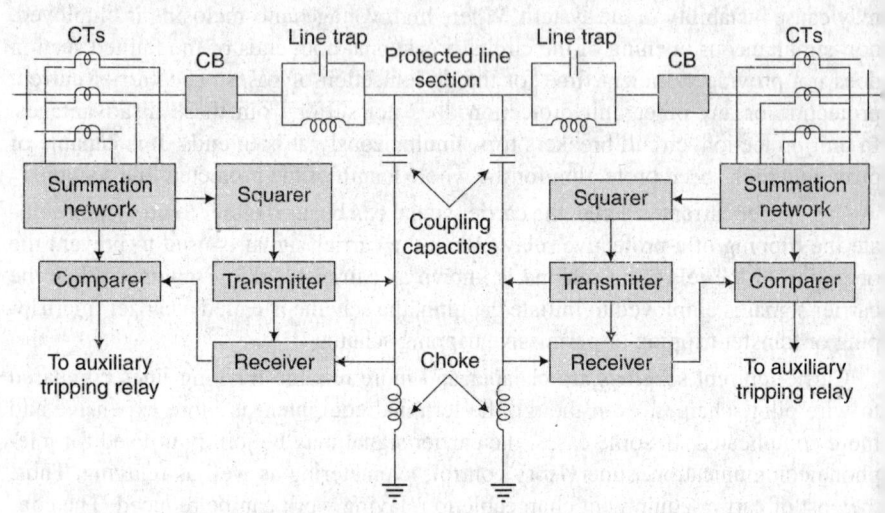

Fig. 7.7 *Schematic diagram of phase comparison carrier current protection*

The coupling capacitor consists of porcelain-clad, oil-filled stack of capacitors connected in series. It offers very high impedance to power frequency current but low impedance to carrier frequency current. On the other hand, the inductance offers a low impedance to power frequency current and high impedance to carrier frequency current. Thus the transmitter and receiver are insulated from the power line and effectively grounded at power frequency current. But at carrier frequency they are connected to the power line and effectively insulated from the ground.

For the transmission of carrier signal either one phase conductor with earth return or two phase conductors can be employed. The former is called phase to earth coupling and the latter is called phase to phase coupling. The phase to earth coupling is less expensive as the number of coupling capacitors and line traps required is half of that needed for phase to phase coupling. However the performance of phase to phase coupling is better compared to phase to earth coupling because of lower attenuation and lower interference levels.

The half-cycle blocks of carrier signals are injected into the transmission line through the coupling capacitor. Fault detectors control the carrier signal so that it is started only during faults.

The voltage outputs of the summation network at stations A and B are 180° out of phase during normal conditions. This is because the CT connections at the two ends are reversed. The carrier signal is transmitted only during positive half cycle of the network output. Figure 7.8 shows the transmission of carrier signal during external fault and internal fault conditions. Wave (a) shows the output of the summation network at A. Wave (b) shows the carrier signal transmitted by the transmitter at A. Wave (c) shows the output of the summation network at B for external fault at C. Wave (d) shows the carrier signal transmitted by the transmitter at B. Thus for an external fault, carrier signals are always present in such a way that during one half-cycle, signals are transmitted by the transmitter at A and during the next half-cycle by the transmitter at B. As the carrier signal is a blocking signal and it is always present the relay does not trip. For an internal fault, the polarity of the network output voltage at B is reversed, as shown by the wave (e). The carrier signal sent by the transmitter at B is shown by wave (f). In case of internal faults, carrier signals are transmitted only during one half-cycle and there is no signal during the other half-cycle. As the carrier signal is not present during the other half-cycle, the relay operates and the circuit breaker trips. The comparator receives carrier signal from the receiver. So long as the carrier is present, it does not give an output to the auxiliary tripping relay. When the comparator does not receive the carrier signal, it gives an output to the auxiliary tripping relay.

The ideal phase difference between carrier blocks is 180° for internal faults and zero degree for external faults. In practice, it is kept 180° ± 30° for internal faults because of

(i) the phase displacement between emfs at the ends of the protected line section.
(ii) through current being added to the fault current at one end and subtracted at the other.
(iii) errors produced by CTs.

The phase comparison scheme provides only primary protection. The back-up protection is provided by supplementary three step distance relays for phase and ground faults. In a phase comparison scheme, the relay does not trip during swings

318 Power System Protection and Switchgear

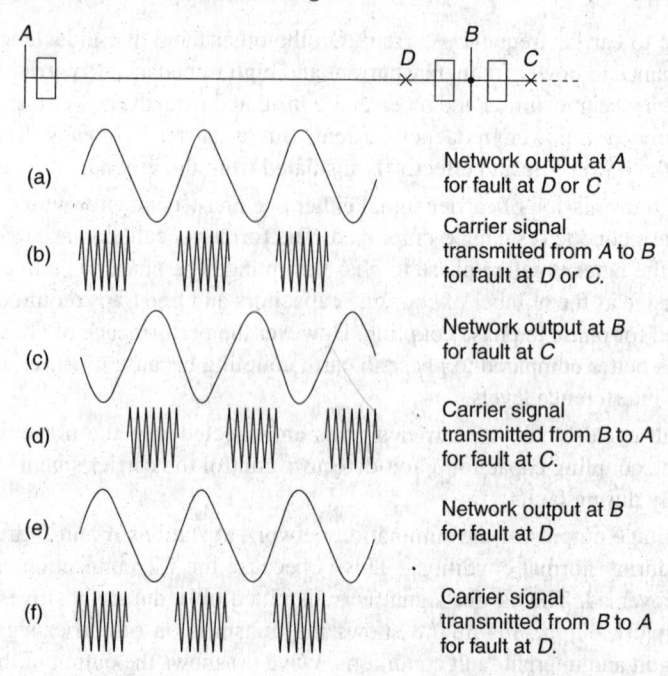

Fig. 7.8 *Transmission of carrier signals during internal and external fault conditions*

(a) Network output at *A* for fault at *D* or *C*
(b) Carrier signal transmitted from *A* to *B* for fault at *D* or *C*.
(c) Network output at *B* for fault at *C*
(d) Carrier signal transmitted from *B* to *A* for fault at *C*.
(e) Network output at *B* for fault at *D*
(f) Carrier signal transmitted from *B* to *A* for fault at *D*.

or out of step conditions or because of zero sequence current induced from a parallel line, if there is no fault on the protected line section. It is used as a primary protection for all long distance overhead EHV and SHV transmission lines.

The length of transmission line which can be protected by phase comparison scheme is limited by phase shifts produced by the following factors.

(i) The propagation time, i.e. the time taken by the carrier signal to travel from one end to other end of the protected line section (up to 0.06° per km).
(ii) The time of response of the band pass filter (about 5°).
(iii) The phase shift caused by the transmission line capacitance (up to 10°).

7.3.2 Carrier Aided Distance Protection

The main disadvantage of unit protection scheme is that they do not provide back-up protection to the adjacent line section. A distance scheme is capable of providing back-up protection but it does not provide high-speed protection for the whole length of the line. The circuit breakers do not trip simultaneously at both ends for end-zone faults. The most desirable scheme will be one which includes the best features of both, unit protection and distance protection. This can be achieved by interconnecting the distance relays at both ends of the protected section by carrier signals. Such schemes provide instantaneous tripping for the whole length of the line as well as back-up protection. The following are the three types of such schemes.

(i) Carrier transfer or intertripping scheme
(ii) Carrier acceleration scheme
(iii) Carrier blocking scheme

Carrier Transfer or Carrier Intertripping Scheme

The following are important types of transfer tripping schemes.
  (a) Direct transfer tripping (Under-reaching scheme)
  (b) Premissive under-reach transfer tripping scheme
  (c) Premissive over-reach transfer tripping scheme

(a) Direct transfer tripping (under-reaching scheme) In this scheme, three-stepped distance relays are placed at each of the protected line. Consider the protective scheme for line AB. The time-distance characteristics of the relays placed at A and B are shown in Fig. 7.9. When a fault occurs at F_3, the I zone high-speed relay operates at B and trips the circuit breaker. But the circuit breaker at A does not trip instantaneously. Therefore, for instantaneous tripping of the circuit breaker at A, a carrier signal is transmitted from B to A. A receive relay RR is included in the trip circuit as shown in Fig. 7.10(a). Thus, the circuit breaker at A also trips instantaneously for any fault at F_3, i.e. an end-zone fault. The disadvantage of this scheme however is that there may be undesirable tripping due to mal-operation or accidental operation of the signalling channel. The operation of the I zone relay at end B initiates tripping at that end as well as a carrier transmission. The scheme in which the I zone relay is used to send carrier signal to the remote end of the protected line section is called "transfer trip under-reaching scheme".

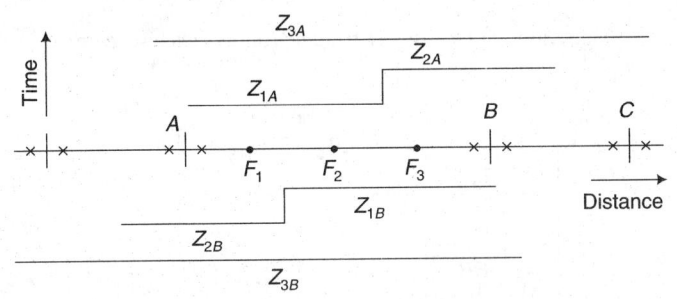

Fig. 7.9 Stepped time-distance characteristics of relays for direct transfer tripping (under-reach scheme)

If the fault occurs at F_1, the zone 1 relay operates and trips the circuit breaker at end A. It also sends a carrier signal to B. The receipt of the carrier signal at end B initiates tripping of the circuit breaker immediately. When a fault occurs at F_2, circuit breakers at both ends trip simultaneously. Distance relays provide back-up protection for adjacent lines which is obvious from Fig. 7.10(a) as the contacts T_2 and T_3 operate after a certain time delay. Figure 7.10(b) shows a signal sending arrangement. Figure 7.10(c) shows a solid state logic for the trip circuit.

In this scheme, the carrier signal is transmitted over the faulty line. Therefore, there is an additional attenuation of the carrier signal.

(b) Permissive under-reach transfer tripping scheme To overcome the possibility of undesired tripping by accidental operation or mal-operation of the signaling channel, the receive relay is supervised by the zone 2 relay. The zone 2 relay contact is placed in series with the receive relay RR as shown in Fig. 7.11(a). For an internal end-zone fault, contact Z_2 is closed. RR is also closed after receipt of the carrier signal from the other end and it trips the circuit. When there is no fault in the end

320 Power System Protection and Switchgear

(a) Trip circuit (electromagnetic)

(b) Arrangement to send carrier signal

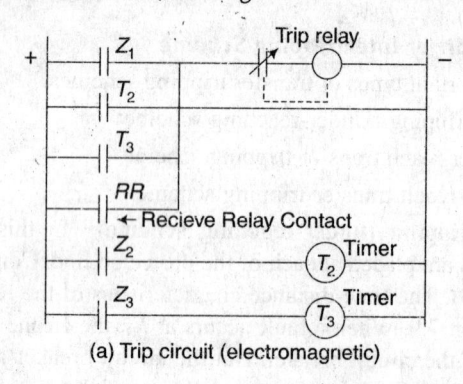

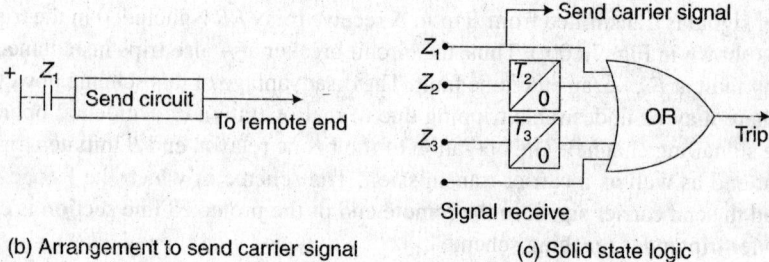

(c) Solid state logic

Fig. 7.10 *Direct transfer tripping (under-reach scheme)*

(a) Trip circuit (electromagnetic)

(b) Arrangement to send carrier signal

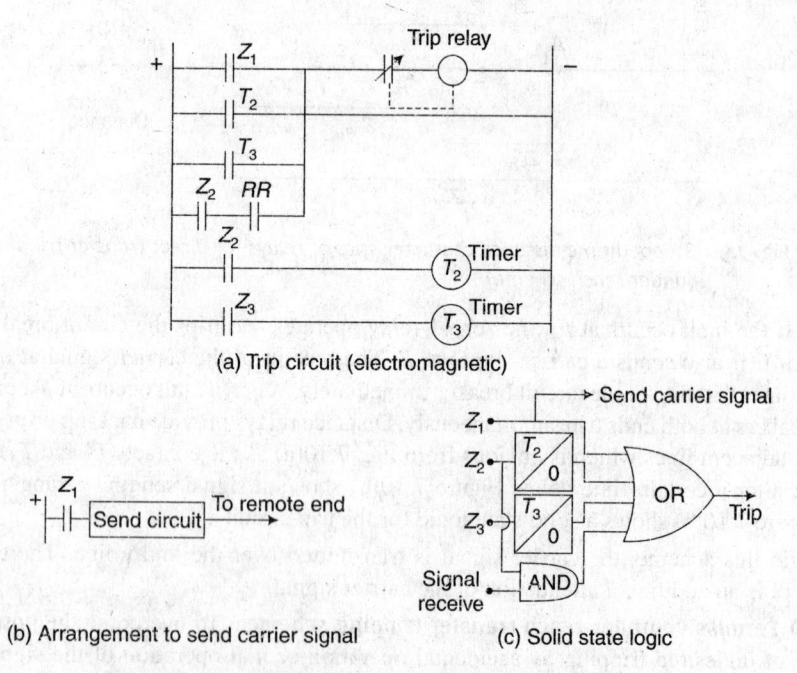

(c) Solid state logic

Fig. 7.11 *Permissive under-reach transfer tripping scheme*

zone, Z_2 will not operate. As the contact Z_2, placed in series with RR is open, the circuit breaker will not trip even if there is a mal-operation of the carrier signal. Thus, it prevents false tripping. Figure 7.11(b) shows the schematic diagram of the signal sending arrangement. Zone 1 unit is arranged to send a carrier signal. Figure 7.11(c) shows the solid state logic for the trip circuit. In this scheme also the carrier signal is transmitted over the faulty line section which causes an additional attenuation of the carrier signal.

(c) Permissive over-reach transfer tripping scheme In this scheme, the zone 2 unit is arranged to send a carrier signal to the remote end of the protected section of the line. In this case, it is essential that the recieve relay contact is supervised by a directional relay. Figure 7.12(a) shows its trip circuit. Zone 2 relay is used to monitor the recieve relay contact RR. The unit at zone 2 must be a directional unit (it may be a MHO unit) to ensure that tripping does not take place unless the fault is within the protected section. This scheme is also know as a directional comparison scheme. In this scheme, direct transfer tripping cannot be employed because a carrier signal is transmitted even for an external fault which lies within the protective zone of the zone 2 relay. Figure 7.12(b) shows a signal sending arrangement. Figure 7.12(c) shows its solid state logic. The scheme in which the second zone relay is used to transmit carrier signal to the remote end of the protected line section is called "Over-reach Transfer Scheme". The second zone relay is set to reach beyond the far end of the line. Its use as a signal transmitter does not make any undesired tripping when fault occurs in the overlapped section of the adjacent line. It is due to the fact that the second zone relay used in this scheme is a directional unit and it also monitors the recieve relay. In this scheme also, the carrier signal is transmitted over the faulty line section which causes additional attenuation of the carrier signal.

(a) Trip circuit

(b) Arrangement to send carrier signal

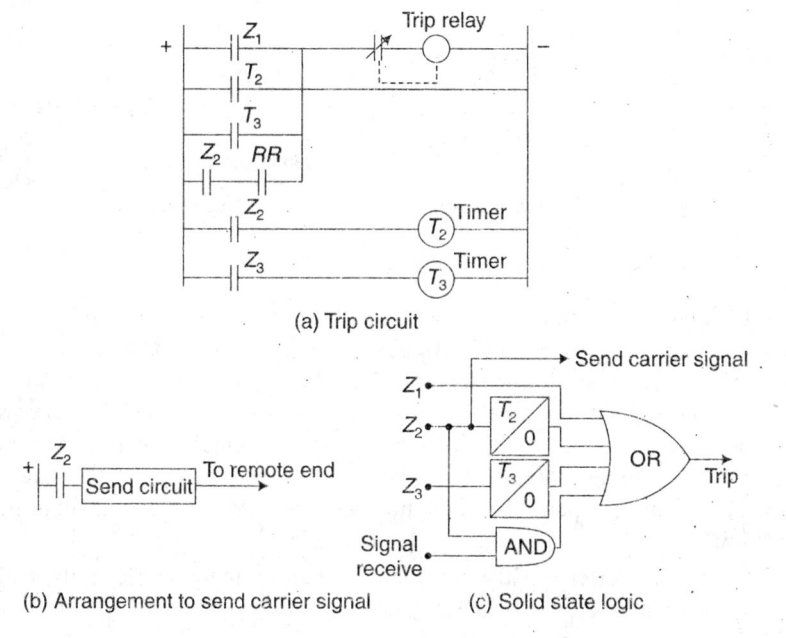

(c) Solid state logic

Fig. 7.12 *Permissive over-reach transfer tripping scheme*

Carrier Acceleration Scheme

In this scheme, the carrier signal is used to extend the reach of the zone 1 unit to zone 2, thereby enabling the measuring unit to see the end-zone faults. When an end-zone fault occurs, the relay trips at that end and sends a carrier signal to the remote end. This scheme employs a single measuring unit for zone 1 and zone 2 unit (MHO unit). The zone 1 unit is arranged to send the carrier signal to the other end. The receive relay contact is arranged to operate a range change relay as shown in Fig. 7.13(a). On receipt of the carrier signal from the other end, the range change relay extends the reach of the mho unit from zone 1 to zone 2 immediately. Thus, the clearance of fault at the remote end is accelerated.

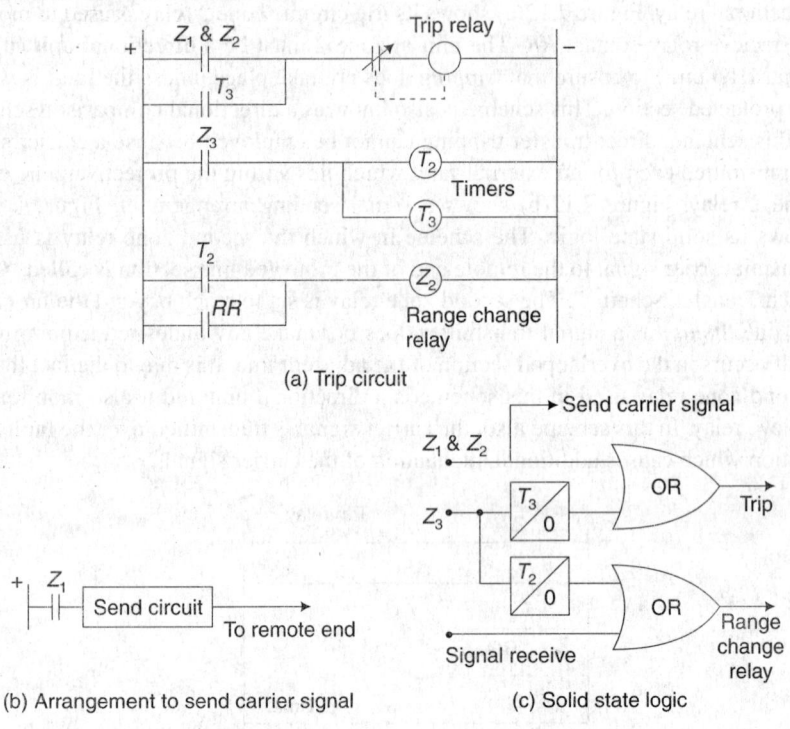

Fig. 7.13 *Carrier acceleration scheme*

If the carrier fails, the fault will be cleared in zone 2 operating time. This scheme is not as fast as permissive transfer tripping schemes as some time is required for the operation of the mho unit after its range has been changed from zone 1 to zone 2. But it is more reliable because the zone 2 relay operates only when it sees a fault in its operating zone. It does not operate due to accidental or mal-operation of the carrier channel. In this scheme, the carrier signal is transmitted over the faulty line. So its effectiveness depends upon the transmission of the carrier signal during such conditions.

In carrier acceleration and intertripping schemes, if the carrier fails, end-zone faults will take a longer time to be cleared.

Carrier Blocking Scheme

In this scheme, the carrier signal is used to block the operation of the relay in case of external faults. When a fault occurs on the protected line section, there is no transmission of the carrier signal. The blocking schemes are particularly suited to the protection of multi-ended lines.

In this scheme the zone 3 unit looks in the reverse direction and it sends a blocking signal to prevent the operation of zone 2 unit at the other end for an external fault. When a fault occurs at F_1 (see Fig. 7.14), it is seen by zone 1 relays at both ends A and B. Consequently, the fault is cleared instantaneously at both ends of the protected line. The carrier signal is not transmitted by the reverse looking zone 3 unit because it does not see the fault at F_1.

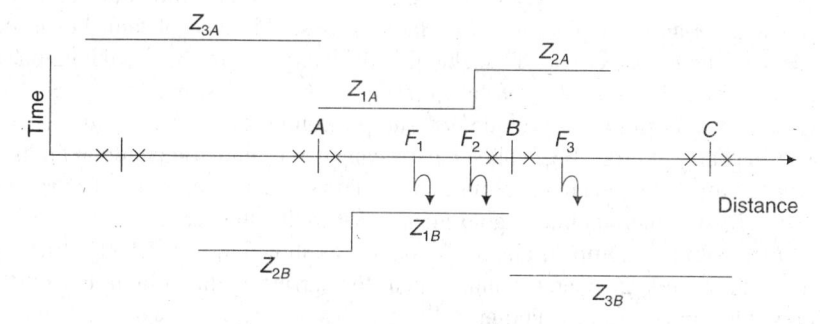

Fig. 7.14 Stepped time-distance characteristics of relays for carrier blocking scheme

When a fault occurs on F_2, which is an end-zone fault, it is seen by zone 2 units at both ends A and B and also by zone 1 unit at B. The fault is cleared by zone 1 unit at B and instantaneously by the zone 2 unit at A. The zone 2 unit has two operating times,

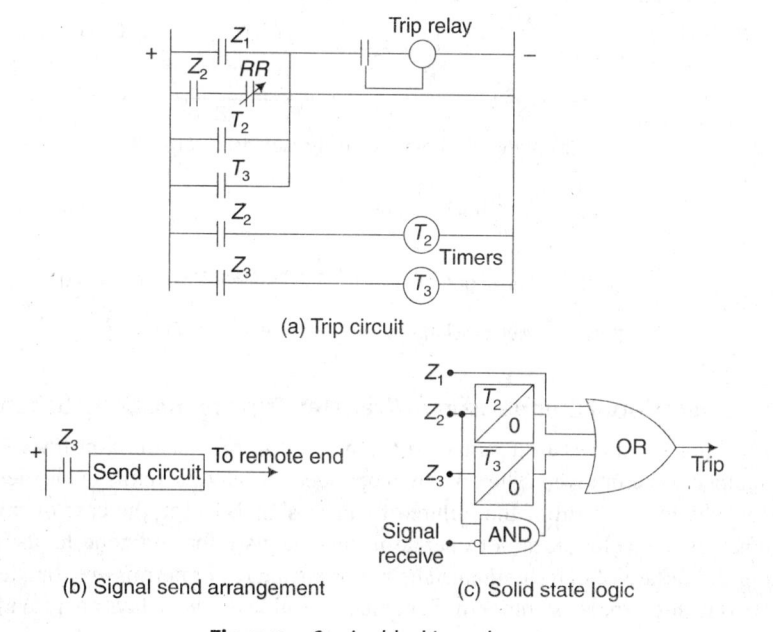

(a) Trip circuit

(b) Signal send arrangement

(c) Solid state logic

Fig. 7.15 Carrier blocking scheme

one instantaneous and other delayed. The instantaneous operation is through Z_2 and RR, see Fig. 7.15(a). The delayed operation is through T_2. As the fault is an internal one, there is no transmission of the carrier signal.

When a fault occurs at F_3, it is seen by the forward looking zone 2 unit A and the reverse looking zone 3 unit at B. It is an external fault. Normally, it has to be cleared by the zone 1 unit associated with line BC. So to prevent the operation of zone 2 unit at A, a carrier signal is transmitted by the reverse looking zone 3 unit at B. If this fault is not cleared instantaneously by the relays of line BC, the zone 2 relay at A will trip after the zone 2 time lapse, as back-up protection.

Reverse Looking Relay with Offset Characteristic

In the blocking scheme, the reverse looking relay becomes inherently slow in operation for a close-up three phase external fault because this type of fault lies at the boundary of the characteristic. To tackle this difficulty, the reverse looking zone 3 relay with offset characteristic, as shown in Fig. 7.16(a) is desired. The offset characteristic is also desirable to provide back-up protection for bus bar faults after the zone 3 time delay. As the offset characteristic can also see end-zone internal faults, it will send a blocking signal to the remote end, unless some measures are taken against the same. To stop the blocking signal for internal faults, the carrier sending circuit is provided with an additional closed Z_2 contact, as shown in Fig. 7.16(b). When an internal fault occurs, Z_2 operates and it opens the carrier sending circuit even if the fault point lies in zone 3 jurisdiction.

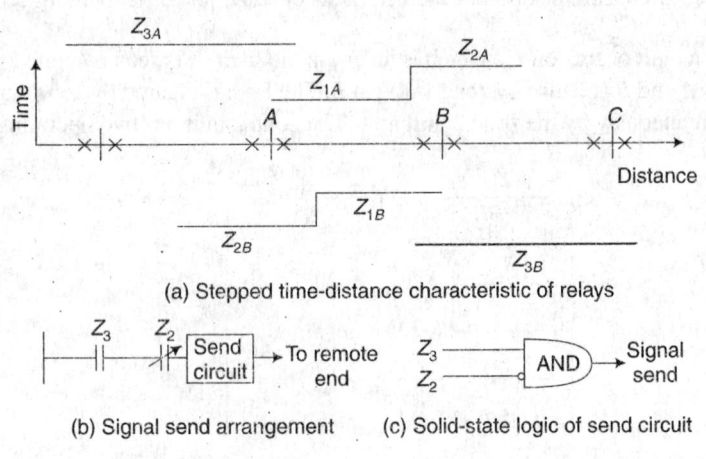

Fig. 7.16 Carrier blocking scheme with offset zone 3 relay

(a) Stepped time-distance characteristic of relays

(b) Signal send arrangement

(c) Solid-state logic of send circuit

7.3.3 Operational Comparison of Transfer Trip and Blocking Schemes

The blocking scheme is best suited to the protection of transmission lines where auto-reclosing is employed. The relay will operate for end-zone faults in the blocking scheme even in the event of the failure of carrier signal. But in the case of transfer trip schemes, the relay does not operate instantaneously for end-zone faults in the event of the failure of carrier signal. The major advantage of permissive transfer trip scheme is its high speed. In blocking schemes, intentional time delays are provided to ensure that the carrier signal has been received. In transfer schemes, such intentional time delays are not necessary. The carrier acceleration scheme is more economical than the carrier intertripping and carrier blocking schemes. It has a speed which is intermediate to these schemes. Even during internal faults, sufficient fault current may not flow through a weak end terminal to operate a protective relay. However, the circuit breaker at the weak end has to be opened because even a low fault current can maintain the fault. To maintain stability and achieve high-speed auto-reclosing, the circuit breakers at both ends must trip simultaneously. In permisive under-reach and acceleration schemes, zone 1 relay at the strong infeed terminal do not see end-zone faults near the weak end terminal. The relay at the weak end terminal does not operate and hence the carrier signal is not transmitted from the weak end. Thus, the strong end terminal does not receive a carrier signal. Therefore, these schemes are not suitable for the protection of such lines.

In the permissive over-reach scheme, zone 2 relay at the strong infeed end can see an end-zone fault near the weak end terminal. Therefore, it operates and sends a carrier signal to the weak end. The relay at the weak end terminal does not operate and there is no transmission of the carrier signal from the weak end to the strong infeed end for end-zone faults. Therefore, the standard over-reach scheme is not suitable for this type of application. By providing additional logic in the distance scheme at the weak end, circuit breakers at both ends can be tripped instantaneously. For details see Ref. 5.

In case of the carrier blocking scheme, relays at the strong infeed end operate for internal faults. For external faults a reverse looking zone 3 unit placed behind the weak end terminal also operates and sends a carrier signal to the strong end to block the operation of the relay. But at the weak end, relays meant for the operation in case of internal faults do not operate. Therefore, tripping at this end is possible only by direct intertripping.

Details for the protection of multi-ended lines and parallel lines can be seen in Ref. 5.

7.3.4 Optical Fibre Channels

Optical fibres are fine strands of glass. They behave as wave guides for light. As they are capable of transmitting light over considerable distance, they can be used to provide an optical communication link with enormous information carrying capacity. The optical link is inherently immune to electromagnetic interference. At the sending end, the transmitter converts the electrical pulses into light pulses. The light pulses are transmitted through the optical fibre and subsequently decoded at the remote terminal. At the receiving end, the light pulses are converted into electrical pulses. The terminal equipment uses modern digital techniques of Pulse Code Modulation. Using A/D converters, the analog relaying quantity can be sampled and transmitted as coded information to the remote terminal. At the remote terminal, the decoded information is compared with the locally derived signal.

The distance over which effective communication is feasible depends on the attenuation and dispersion of the communication link. The quality of fibre and the wavelength of the optical source governs the attenuation and dispersion. Optical fibres are available in cable form. Such cables can be laid in cable trenches. There is a trend to associate optical fibres with power cables themselves. Optical fibres are embedded within the cable conductors, either earth or phase conductor. Optical fibre communication is expected to play an important role in future protection signalling, either as a dedicated protective relaying link or as channels within a multiplexed system which carry all communication traffic such as voice, telecontrol, telemetering and protection signalling.

EXERCISES

1. Explain the term 'pilot' with reference to power line protection. What are the different types of pilots which are presently employed? Discuss their fields of application.
2. Discuss the limitations of wire pilot protection. What type of pilot is used for the protection of EHV and UHV transmission lines?
3. What are the important operating principles which are used in wire pilot schemes? Discuss the Transley scheme of wire pilot protection.
4. What is carrier current protection? For what voltage range is it used for the protection of transmission lines? What are its merits and demerits? With neat sketches, discuss the phase comparison scheme of carrier current protection?
5. What is unit protection? What are its merits and demerits? How does carrier aided distance protection give better performance than carrier current protection?
6. What is carrier aided distance protection? What are its different types? Discuss the permissive under-reach transfer tripping scheme of protection.
7. What is the difference between permissive under-reach and permissive over-reach transfer tripping schemes of protection? What is the advantage of these schemes over direct transfer tripping under-reach scheme?
8. What is a carrier blocking scheme? Discuss its merits and demerits over other types of carrier-aided distance protection.
9. Discuss the merits of optical fibre channel which can be used as a communication link. What is its future prospect?

8 Differential Protection

8.1 INTRODUCTION

Differential protection is a method of protection in which an internal fault is identified by comparing the electrical conditions at the terminals of the electrical equipment to be protected. It is based on the fact that any internal fault in an electrical equipment would cause the current entering it, to be different from that leaving it. Differential protection is one of the most sensitive and effective methods of protection of electrical equipment against internal faults. This principle of protection is capable of detecting very small magnitudes of the differential currents. The only drawback of the differential protection principle is that it requires currents from the extremities of a zone of protection, which restricts its application to the protection of electrical equipment, such as generators, transformers, motors of large size, bus zones, reactors, capacitors, etc. The differential protection is called unit protection because it is confined to protection of a particular unit (equipment) of a power plant or substation.

The main component of a differential protection scheme is the differential relay which operates when the phasor difference of two or more similar electrical quantities exceeds a predetermined value. Most differential relays are of current differential type. CTs are placed on both sides of each winding of a machine. The outputs of their secondaries are applied to the relay coils. Careful attention must be paid to the polarity marks (dot markings) on these CTs while connecting them. The relay compares the current entering a machine winding and leaving the same. Under normal conditions or during any external (through) fault, the current entering the winding is equal to the current leaving the winding. But in the case of an internal fault on the winding, these are not equal. This difference in the current actuates the relay. Thus, the relay operates for internal faults and remains inoperative under normal conditions or during external (through) faults. In case of bus-zone protections, CTs are placed on both sides of the busbar.

Differential protection is universally applicable to all elements of the power system, such as generators, transformers, motor of large size, bus zones, reactors, capacitors, etc. A pair of identical current transformers (CTs) are fitted on both end of the protected element (equipment). The CTs are of such a ratio that their secondary currents are equal during the normal conditions or external (through) faults. The secondary of the CTs are interconnected by a pilot-wire circuit incorporating an instantaneous overcurrent relay. While connecting CTs careful attention must be paid to the dot (polarity) marks placed on the CTs. In order to trace the direction of the currents, following rule can be applied:

"When current enters the dot mark on the primary side of a CT, the current must leave the similar dot mark on the secondary side". The polarity connections of the CTs are such that their secondary currents are in the same direction in pilot wires, during normal conditions or external (through) faults. The operating coil of the overcurrent relay is connected across the CT secondary at the middle of pilot wires.

All differential protection schemes generate a well defined protected zone which is accurately delimited by the locations of CTs on both the ends of the protected element.

Overcurrent and distance protection systems discussed in Chapters 5 and 6 respectively are non-unit systems of protection which discriminate by virtue of time-grading. They are comparatively simple and cheap and require no pilot-wire circuits, but their fault clearance times are too slow for circuits operating at 33 kV and above, and for expensive equipment such as large generators, transformers and motors. Differential protection systems require pilot-wire circuits for interconnecting secondaries of CTs incorporating overcurrent relays. These protective systems can inherently discriminate without the use of time-grading and are fast in fault clearance. Differential protection systems are unit protection systems which are instantaneous and their time setting are independent of other protective systems.

As described earlier, the differential principle compares the secondary currents of the CTs in all the circuits into and out of the protected zone or area. It is easier to apply differential principle for the protection of generators, transformers, buses, motors and so on, because the terminals of each of these devices are at one geographical location where secondaries of the CTs and relays can be directly interconnected through pilot wires. For EHV/UHV transmission lines, where the terminals and CTs are widely separated by considerable distances (often by many kilometres), it is not practically possible to use differential relays as described above. Still, the differential principle provides the best protection and is still widely used for EHV/UHV transmission lines. A communication channel known as pilot such as wire pilot, carrier-current pilot, microwave pilot, or fiber optic cable is used for comparison of information between the various terminals. These systems are described in Chapter 7.

8.2 DIFFERENTIAL RELAYS

A differential relay is a suitably connected overcurrent relay which operates when the phasor difference of currents at the two ends of a protected element exceeds a predetermined value. Most of the differential relays are of current differential type.

The following are the various types of differential relays.

(i) Simple (basic) differential relay
(ii) Percentage (biased) differential relay
(iii) Balanced (opposed) voltage differential relay.

Depending on the type of differential relay employed as the main constituent in differential protection, the following are the various types of differential protection.

(i) Simple (basic) differential protection
(ii) Percentage (biased) differential protection
(iii) Balanced (opposed) voltage differential protection,

The construction and operation of various differential relays are described in concerned sections of differential protection.

8.3 SIMPLE (BASIC) DIFFERENTIAL PROTECTION

The main constituent of a simple differential protection scheme is a simple differential relay. A simple differential relay is also called basic differential relay. A simple differential relay is an overcurrent relay having operating coil only which carries the phasor difference of currents at the two ends of a protected element. It operates when the phasor difference of secondary currents of the CTs at the two ends of the protected element exceeds a predetermined value. The secondary of the CTs at the two ends of the protected element are connected together by a pilot-wire circuit. The operating coil of the overcurrent relay is connected at the middle of pilot wires. The differential protection scheme employing simple differential relay is called Simple differential protection or Basic differential protection. The simple differential protection scheme is also called circulating current differential protection scheme of Merz-Price protection scheme.

8.3.1 Behaviour of Simple Differential Protection during Normal Condition

Figure 8.1 illustrates the principle of simple differential protection employing a simple differential relay. The CTs are of such a ratio that their secondary currents are equal under normal conditions or for external (through) faults. If the protected element (equipment) is either a 1:1 ratio transformer or a generator winding or a busbar, the two currents on the primary side will be equal under normal conditions and external (through) faults. Hence, the ratios of the protective CTs will also be identical.

If n be the CT ratio, the secondary current of CT_1 $(I_{s1}) = I_L/n$, secondary current of CT_2 $(I_{s2}) = I_L/n$, and the secondary load current $(I'_L) = I_L/n$.

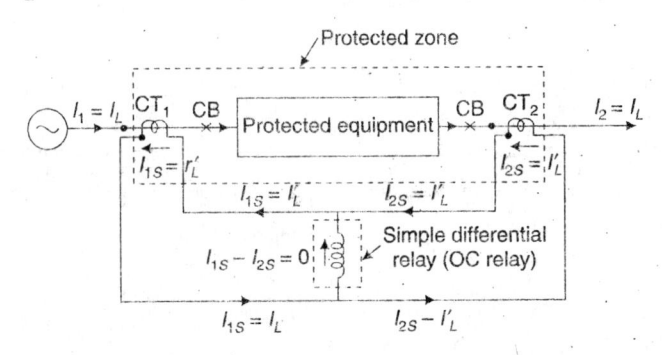

Fig. 8.1 Simple differential protection scheme behaviour under normal condition. ($I_1 = I_2 = I_L$ and $I_{1S} = I_{2S} = I'_L$, hence $I_{1S} - I_{2S} = 0$)

Figure 8.1 shows the combination of the two CTs and an instantaneous overcurrent relay (acting as simple differential relay) in the difference or spill circuit, as the constituents of the simple differential protection scheme. The normally open contacts of this overcurrent (OC) relay are wired to trip the circuit breakers. Under normal conditions the secondary currents I_{1S} and I_{2S} of CT_1 and CT_2 respectively are equal to the secondary load current I'_L. The secondary currents, under normal conditions, simply circulate through the secondary windings of the two CTs and the pilot leads connecting them, and there is no current through the spill or difference circuit, where the instantaneous overcurrent (OC) relay is connected. Hence, the OC relay does not operate to trip the circuit breakers (CBs). Since the currents circulate in the CT secondaries this differential protection scheme is called "circulating current differential protection scheme" or "Merz-Price protection scheme". The boundaries of the protected zone is determined by the locations of the CTs.

8.3.2 Behaviour of Simple Differential Protection during External Fault

Figure 8.2 illustrates the behaviour of the simple differential protection scheme under external (through) fault. An external (through) fault occurs outside the protected zone, i.e., outside the CT locations and not in the protected equipment. As in the case of normal conditions, the current through the OC relay is zero. The secondary currents I_{1S} and I_{2S} of the CTs are equal to the secondary fault current I'_F, and the differential (spill) current $(I_{1S} - I_{2S})$ flowing through the operating coil of the OC relay is zero. Hence the OC relay does not operate under external (through) faults.

In Fig. 8.2,

I_1 = Current entering the protected zone, = I_F

I_2 = Current leaving the protected zone = I_F

Hence, $I_1 = I_2 = I_F$, and through fault current = $\dfrac{I_1 + I_2}{2} = I_F$

$I_{1S} = I_{2S} = I'_F = I_F/n$, where n is the CT ratio

Secondary value of through fault current = $\dfrac{I_{1s} + I_{2s}}{2} = I'_F$

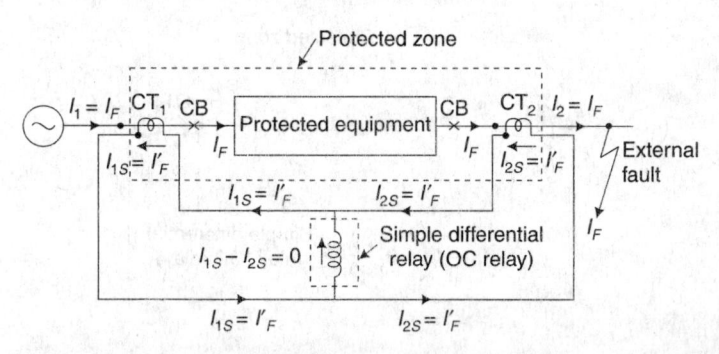

Fig. 8.2 *Behaviour of simple differential protection scheme on external (through) fault.* $(I_{1s} - I_{2s}) = 0$ *(No current in OC relay)*

8.3.3 Behaviour of Simple Differential Protection during Internal Fault

Figure 8.3 illustrates the behaviour of simple differential protection scheme during internal faults in a single-end-fed system. An internal fault occurs inside the protected zone, i.e., within the CT locations. In case of internal fault the current entering the protected zone is I_1 whereas that leaving it is I_2, such that $I_1 = I_2 + I_f$ i.e. $I_1 \neq I_2$. Hence, the two secondary currents (I_{1s} and I_{2s}) through CTs are not equal and the differential or spill current ($I_{1s} - I_{2s}$) flows through OC relay. If the differential current ($I_{1s} - I_{2s}$) is higher than the pick-up value of the overcurrent relay, the relay will operate and both the circuit breakers will be tripped out isolating the protected equipment from the system.

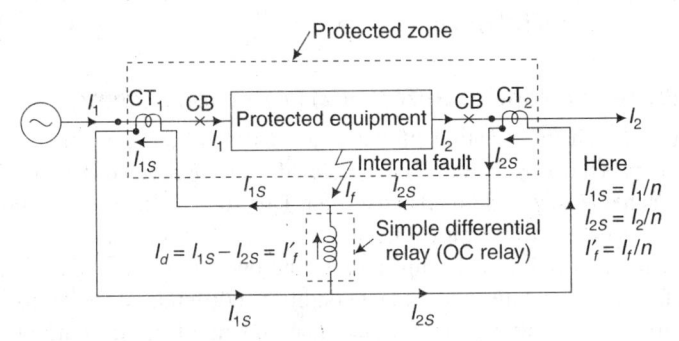

$I_1 = I_2 + I_f$, Differential current (I_d) = $I_{1s} - I_{2s} = I'_f$
If both CTs (CT$_1$ and CT$_2$) have same CT ratio (n), $I_{s1} = I_1/n$, $I_{s2} = I_{s2}/n$ and $I'_f = I_f/n$
If both CTs have different CT ratios (n_1 and n_2) as in the case of transformer,
$I_{s1} = I_1/n_1$ and $I_{s2} = I_2/n_2$

Fig. 8.3 *Behaviour of simple differential protection during internal fault*

If an internal fault occurs in a double-end-fed system as shown in Fig. 8.4, the right-hand current in Fig. 8.3 is reversed, and both pilot currents (I_{s1} and I_{s2}) add and flow through the OC relay to cause tripping. If the protected zone is fed from one end only (conventionally from the left) as shown in Fig. 8.3, the right-hand current (I_2) in case of an internal fault is either very small or zero and the left-hand current (I_1) dominates the differential current ($I_{1s} - I_{2s}$) which flows through the OC relay to cause trippling. If the differential current ($I_{1s} - I_{2s}$) is higher than the pick-up value of the overcurrent relay, the relay will operate. For this reason, when considering the sensitivity of a differential protection system to an internal fault, a single-end-fed system is usually assumed. In case of an internal fault in a single-end-fed system, the left hand CT (CT$_1$) carrying the fault current (I_f) is said to be active while the right hand CT (CT$_2$) is said to be idling. Since, during an internal fault in a single-end-fed system only, a voltage exists across the relay, almost the same voltage exists across idling CT which therefore takes a small exciting current.

Since an internal fault fed from one end only represents the worst case, the usual procedure for the study of a differential (unit) protection system is to assume a maximum external fault to check relay stability, then to subtract or delete the right-hand current to check tripping due to an internal fault in a single-end-fed system only.

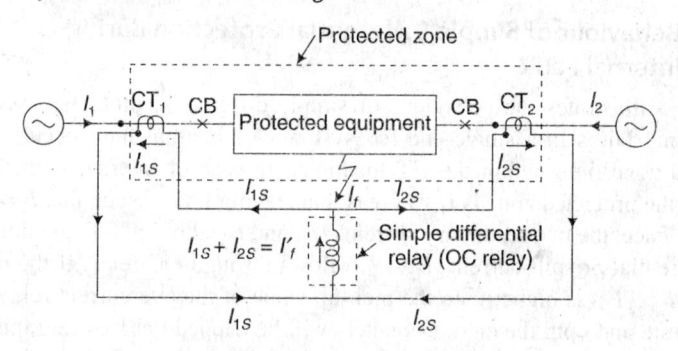

Fig. 8.4 Operation of simple differential protection scheme during internal fault in a double-end-fed system

8.3.4 Protected Zone of Differential Protection Scheme

The zone between the location of the two CTs is called 'protected zone' of the differential protection scheme. The protected zone generated by the differential protection scheme is well-defined and closed. This zone includes everything between the two CTs as shown in Figures 8.1, 8.2, 8.3 and 8.4. Any fault between the two CTs is treated as an 'internal fault' and all the other faults are 'external faults' or 'through faults'. Therefore, under ideal condition a differential protection scheme is expected to respond only to internal faults, and restrain from tripping under normal conditions or during external (through) faults. Thus, an ideal differential protection scheme should be able to respond to the smallest internal fault and restrain from tripping even during the largest external (through) fault. In practice, it is difficult to achieve this ideal case, especially for very heavy external (through) faults because of the non-ideal nature of the various components of the differential protection system. The disadvantages of simple (basic) differential protection system caused by non-ideal nature of the various components and other factors are discussed in the following section.

8.3.5 Disadvantages of Simple Differential Protection Scheme

Simple differential protection schemes discussed so far are based on the assumptions that both the CTs are ideal and matched exactly. Ideal CTs don't have any error and always reproduce the primary currents accurately in their secondaries. Exactly matched CTs have exactly identical saturation characteristics and give identical secondary currents at all values of primary currents up to the highest fault current which can be expected. However, in practice, it is difficult to get ideal, identical, and exactly matched CTs. CTs used in practice will not always give the same secondary current for the same primary current, even if they are commercially identical, i.e. of the same rating and manufactured by the same manufacturer. The difference in secondary current, even under normal conditions, is caused by constructional errors and CT errors (ratio and phase angle errors). CT errors depend upon the saturation characteristics of CTs and CT burdens which in turn depend on the lengths of the pilot wires (cables) and the impedance of the relay coil.

Because of the problems discussed above, simple differential protection scheme suffers from the following disadvantages.

(i) The pilot wires through which CTs are connected are usually made up in the form of paper-insulated, lead-sheathed cable. The impedance of such pilot cables which depends on their lengths generally causes a slight difference between the CT secondary currents at the two ends of the protected section. If the relay is very sensitive, then very small differential current flowing through the relay may cause it to operate even if there is no fault existing.

(ii) There is possibility of maloperation of the relay due to currents circulating via paths other than the main pilot loop. The circuits for these unwanted currents is through the capacitance between the cores of the pilot cable. They are due to heavy through currents, caused by external (through) faults, which induce voltages in the pilot wires, this voltage, in turn, drives a current through the pilot capacitance. The effect increases with length of pilot cable and hence imposes a limit on the length of power circuit which can be protected.

(iii) Accurate matching of CTs cannot be achieved due to constructional errors and pilot cable impedances. If the CTs are not exactly matched, they may not have identical secondary currents, even though their primary currents are equal. Unequal secondary currents of CTs leads to flowing of a differential (spill) current through the relay. This spill current is especially large for a heavy external (through) fault. If the spill current exceeds the current setting of the relay, its maloperation may occur.

(iv) During severe external (through) fault conditions, large primary currents flowing through CTs cause unequal currents in their secondaries due to inherent difference in CT characteristics and unequal dc offset components in the fault currents. A conventional CT will saturate at some value of primary current and under these conditions, a considerable proportion of the primary current is used up in exciting the core and as a result there may be large ratio error and/or phase angle error. If CTs have different saturation levels, their ratio and phase angle errors will be different. Because of difference in CT errors, CT secondary currents have difference in magnitude and/or in phase and as a result, a differential (spill) current flows through the relay as shown in Fig. 8.5. Since both ratio and phase angle errors of CTs aggravate as primary current increases, the differential current (I_d) increases as the through fault current increases, causing the relay to maloperate. Ideally, the secondary currents (I_{1s} and I_{2s}) of both the CTs during through faults and normal conditions would be equal in magnitude and in phase with each other, and thus the differential current (I_d) would be zero. However, in practice CTs have ratio and phase angle errors. As shown in Fig. 8.5, both the CTs have nominal ratio of n, CT_1 has an actual ratio of n_{1a} and phase angle error of β_1 while CT_2 has an actual ratio of n_{2a} and phase angle error of β_2. There will be difference in magnitude of secondary currents of CTs due to ratio error and difference in phase of secondary currents due to phase angle error. The phasor diagram of currents is shown in Fig. 8.6. If the differential current ($I_{1s} - I_{2s}$) will exceed the current setting (pick-up value) of the relay, there will be maloperation of the relay.

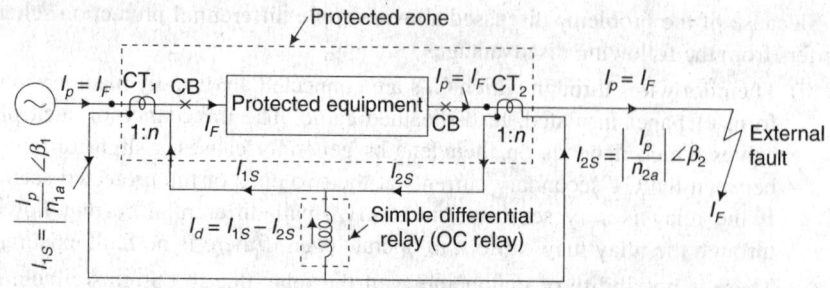

334 Power System Protection and Switchgear

Fig. 8.5 Differential current because of CT errors

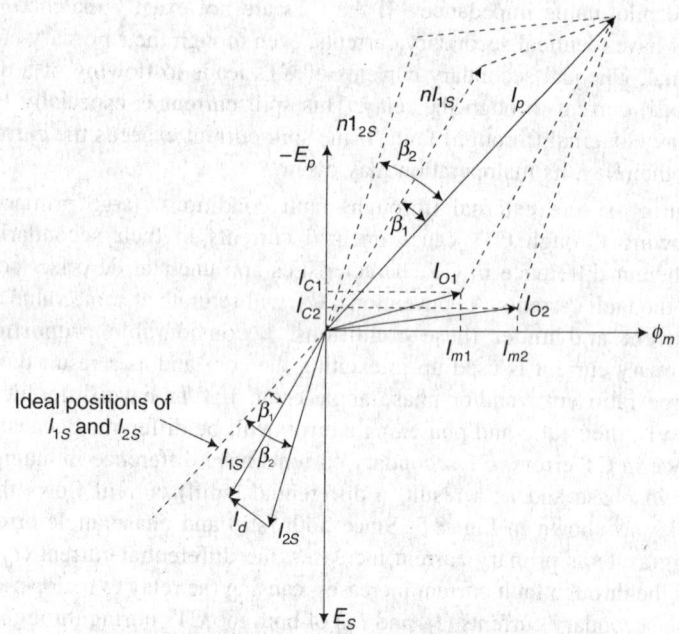

n = Nominal ratio of both CTs (CT$_1$ and CT$_2$)
n_{1a} and n_{2a} = Actual ratios of CT$_1$ and CT$_2$ respectively
β_1 and β_2 = Phase angle errors of CT$_1$ and CT$_2$ respectively.

$I_{1s} = \dfrac{I_p}{n_{1a}} \angle \beta_1$, $I_{2s} = \dfrac{I_p}{n_{2a}} \angle \beta_2$, I_p = primary current of CTs.

Differential current (I_d) = Phasor difference of I_{1s} and I_{2s}

Fig. 8.6 Phasor diagram of currents showing differential current due to phase angle errors of CTs

Simple differential protection scheme with equivalent circuits of CTs is shown in Fig. 8.7. The differential current flowing through the relay is due to difference in exciting currents of CTs during external (through) faults. Since the exciting currents of the two CTs will generally vary widely, there will be a substantial differential current (I_d) during external fault conditions. If the differential current will be higher than the pick up value of the overcurrent relay, the relay will maloperate and the simple differential protection scheme will lose its stability. Therefore, in practice, the simple differential protection scheme cannot be used without further modifications.

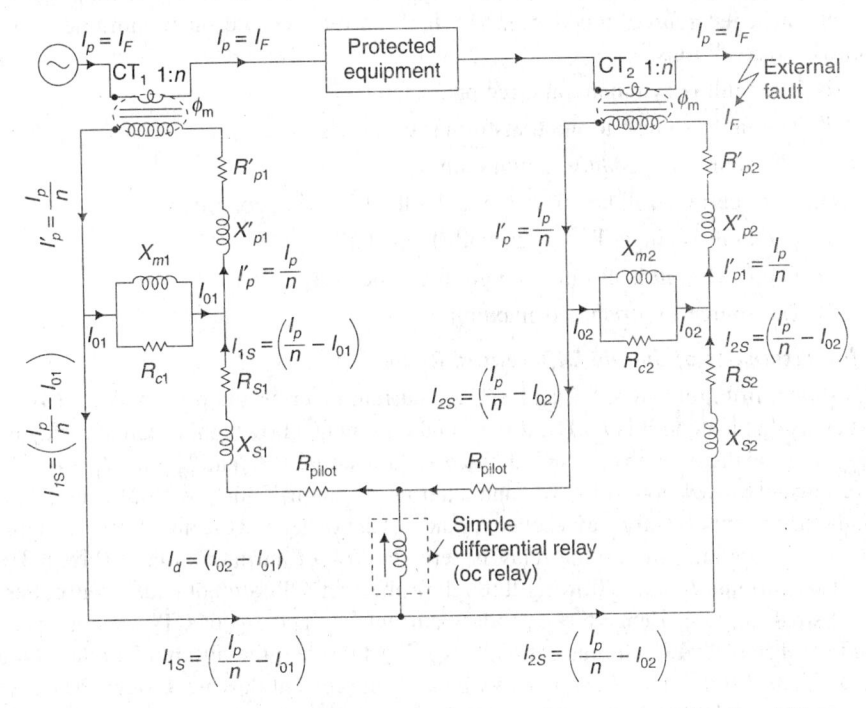

Fig. 8.7 Simple differential protection scheme with equivalent circuits of CTs

Referring to Fig. 8.7, we have

$$I_{1s} = \left(\frac{I_p}{n} - I_{01}\right) \tag{8.1}$$

$$I_{2s} = \left(\frac{I_p}{n} - I_{02}\right) \tag{8.2}$$

where, n is the nominal ratio of both the CTs and I_{01} and I_{02} are exciting currents of CT_1 and CT_2 respectively.

Therefore,

Differential current $(I_d) = (I_{1s} - I_{2s}) = (I_{02} - I_{01})$ \hfill (8.3)

Thus, the dissimilarity between two CTs due to imperfections in their manufacture (constructional errors) and difference in their saturation characteristics would result in the unequal secondary currents even with the same current flowing through the CT primaries and an out-of-balance (differential) current would flow through the relay even under normal operation and/or external fault condition. If the differential current (I_d) flowing through the relay would exceed its pick-up value, there would be maloperation of the relay which is not desirable.

8.3.6 Characteristics of Simple Differential Protection Scheme

The differential relay should operate only on an internal fault and should not operate under normal power flow and external (through) fault conditions. Both internal and external fault currents depend on the type and location of the fault and possible variation in the source impedance. The fault current will be maximum under the following conditions:

(i) The fault is symmetrical three phase fault.
(ii) The fault is close to the transformer terminals.
(iii) The source impedance is minimum.

The fault current will be minimum under the following conditions:

(i) The fault is single line-to-ground (L–G) fault.
(ii) The L–G fault is close to the grounded neutral.
(iii) The source impedance is maximum.

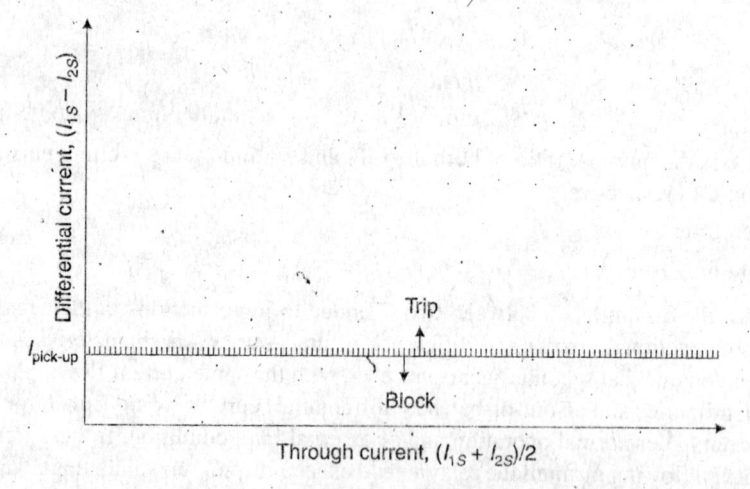

Characteristics of Simple Differential Relay

As shown in Figures 8.1, 8.2 and 8.3, the current entering the protected zone is I_1 whereas that leaving it is I_2. The differential current (I_d) flowing through the relay is ($I_{1s} - I_{2s}$) and the through current is defined as the average of I_{1s} and I_{2s}, i.e., ($I_{1s} + I_{2s}$)/2. For normal power flow (Fig. 8.1) and external (through) fault (Fig. 8.2) conditions both the currents I_1 and I_2 (and hence I_{1s} and I_{2s}) are equal and the differential current ($I_{1s} - I_{2s}$) flowing through the relay is zero. In case of an internal fault (Fig. 8.3), the two currents I_1 and I_2 flowing through primary of CTs are not equal. I_2 is either very small or zero. Hence the secondary currents (I_{1s} and I_{2s}) of CTs are not equal and the differential (spill) current ($I_{1s} - I_{2s}$) flows through the differential relay (OC relay). The differential relay operates whenever the current flowing through the relay exceeds its pick-up value (current setting) $I_{pick-up}$. The characteristic of simple differential relay is shown in Fig. 8.8. The characteristic shows that whatever be the value of the through current ($I_{1s} + I_{2s}$)/2, the relay operates whenever the differential

Fig. 8.8 *Characteristics of simple differential relay*

current $(I_{1s} - I_{2s})$ in the relay exceeds its pick-up value $I_{\text{pick-up}}$. Therefore, the operation of the simple differential relay depends on the value of the differential current $(I_{1s} - I_{2s})$ and the operation is irrespective of the value of through current $(I_{1s} + I_{2s})/2$. As shown in Fig. 8.8, the characteristic of the simple differential relay is a horizontal line displaced vertically by $I_{\text{pick-up}}$.

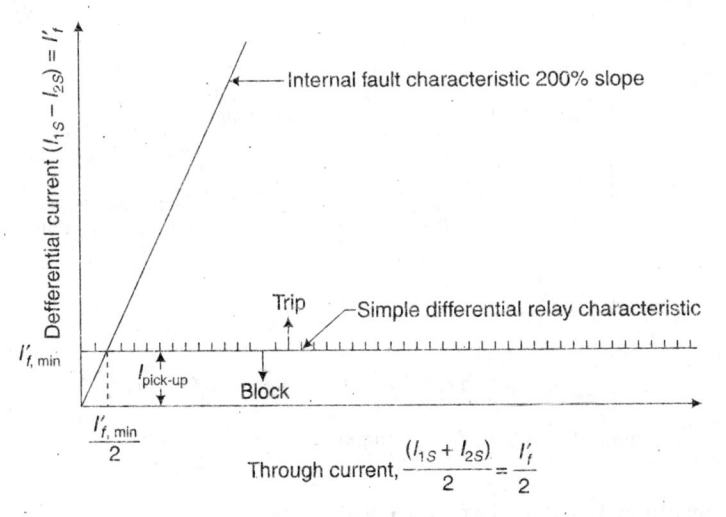

Internal Fault Characteristics

Figure 8.3 Shows an internal fault. In this case, the current leaving the protected zone (I_2) is either very small or zero.

If $I_2 = 0$, $I_{2s} = 0$

Then, the differential current, $(I_{1s} - I_{2s}) = I_{1s} = I'_f$ \hfill (8.4)

and, the through current, $\dfrac{(I_{1s} + I_{2s})}{2} = \dfrac{I_{1s}}{2} = \dfrac{I'_f}{2}$ \hfill (8.5)

Therefore, slope $= \dfrac{(I_{1s} - I_{2s})}{(I_{1s} + I_{2s})/2} = \dfrac{I'_f}{I'_f/2} = 2.0$, i.e., 200% \hfill (8.6)

Thus, the internal fault characteristic on $(I_{1s} + I_{2s})/2$ versus $(I_{1s} - I_{2s})$ diagram has a 200% slope (i.e., a slope of 2.0), as shown in Fig. 8.9. When the relay characteristic, is super imposed upon internal fault characteristic, the point of intersection of the two characteristics gives the minimum internal fault current required for the operation of the relay $(I'_{f,\min})$

$$I'_{f,\min} = I_{f,\min}/n, \quad \text{where } n \text{ is the CT ratio} \hfill (8.7)$$

Fig. 8.9 Internal fault characteristics

External Fault Characteristics

Figure 8.2 shows an external or through fault. Ideally, the secondary currents of both CTs would be equal, and thus the differential (spill) current flowing through the relay would be zero, and the relay, rightly, does not operate. However, in practice, secondary currents of the CTs are not equal due to unequal exciting currents and as a result there is a differential (spill) current in the relay, which increases as the through fault-current increases, causing the relay to maloperate. In this case, the differential (spill) current is usually called out-of-balance (unbalance) current of CTs for through faults. Out-of-balance current of CTs is shown in Fig. 8.7. Referring back to Fig. 8.7, we have

Primary currents of CTs $(I_p) = I_F$

$$I_{1s} = \left(\frac{I_p}{n} - I_{01}\right) = \left(\frac{I_F}{n} - I_{01}\right) = (I'_F - I_{01}) \tag{8.8}$$

$$I_{2s} = \left(\frac{I_p}{n} - I_{01}\right) = \left(\frac{I_F}{n} - I_{02}\right) = (I'_F - I_{02}) \tag{8.9}$$

Therefore,

through fault current, $\dfrac{(I_{1s} + I_{2s})}{2} = I'_F - \dfrac{(I_{01} + I_{02})}{2} \approx I'_F$ (8.10)

Differential (out-of-balance) current, $(I_{1s} - I_{2s}) = (I_{02} - I_{01})$ (8.11)

Figure 8.10 shows the external fault and simple differential relay characteristics. The point of intersection of these two characteristics gives the maximum external fault current, $I'_{F,\,max}$, beyond which the relay maloperates and the differential protection scheme loses stability.

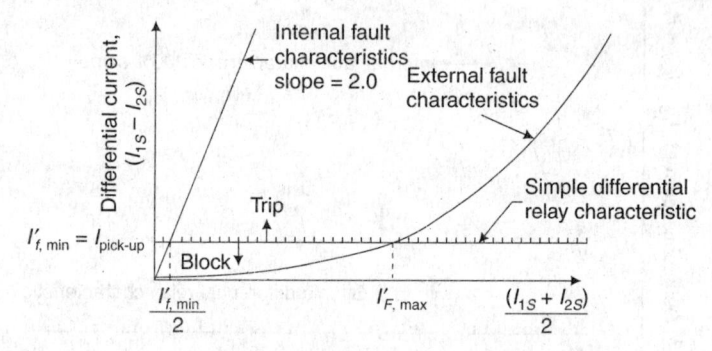

$I'_{f,\,min} = I_{pick\text{-}up}$, and $I'_{F,\,max} = I_{F,\,max}/n$, where $n = CT$ ratio

Fig. 8.10 Simple differential relay, internal fault and external fault characteristics

8.3.7 Stability Limit and Stability Ratio

From the external fault characteristic shown in Fig. 8.10, it is clear that the differential (spill) current $(I_{1s} - I_{2s})$ increases as the through-fault current increases. As the through fault current goes on increasing, there comes a stage when the differential current exceeds the pick-up value of the simple differential relay (overcurrent relay) in the spill path. This causes the relay to operate, disconnecting the protected equipment from rest of the system. This is clearly a case of maloperation of the relay, because the relay has operated on external (through) fault. In such cases, the differential protection scheme is said to have lost stability. In order to signify the ability of the differential protection scheme to maintain stability during external (through) faults, the term 'through fault stability limit' is defined as the maximum external (through) fault current for which the scheme will remain stable. If the external (through) fault current will increase beyond the stability limit current, the relay will maloperate and the differential protection scheme will lose stability. In Fig. 8.10, $I'_{F,\,max}$ is through-fault stability limit current. Having ensured that the relay will be stable to the maximum internal fault current, it is also necessary to check that it will trip to the minimum internal fault current, assuming an infeed from one end only. The minimum internal fault current which will cause the relay to operate is $I'_{f,\,min}$ as shown in Figs. 8.9 and 8.10. The minimum internal fault current required for the correct operation of the relay ($I'_{f,\,min}$) is decided by pick-up value ($I_{pick-up}$) of the overcurrent relay in the spill path. The term 'stability ratio (S)' is used to signify the spread between $I'_{f,\,min}$ and $I'_{f,\,max}$.

The stsability ratio, S, of any differential protection scheme is defined as the ratio of the maximum through-fault current beyond which the relay maloperates ($I'_{F,\,max}$) to the internal fault current for which the relay operates satisfactorily ($I'_{f,\,min}$). $I'_{F,\,max}$ and $I'_{f,\,min}$ are shown in Fig. 8.10.

Therefore,

$$\text{Stability ratio } (S) = \frac{I'_{F,\,max}}{I'_{f,\,min}} \tag{8.12}$$

where, $I'_{F,\,max}$ = Maximum external (through) fault current within which the relay will not maloperate, and beyond which the relay will maloperate.

$I'_{f,\,min}$ = Minimum internal fault current for which the relay will correctly operate, if fault current is less the relay will not operate.

The stability ratio should be as high as the design permits. The desired values vary from 100 to 300. For the simple differential relay the stability ratio is very poor, so a percentage or biased differential relay has been developed.

8.4 PERCENTAGE OR BIASED DIFFERENTIAL RELAY

The disadvantage of moloperation of the simple differential relay due to CT errors during heavy external (through) faults is overcome by using percentage differential relay which is also called biased differential relay. Percentage differential relay provides high sensitivity to light internal faults with high security (high restraint) for external faults and makes differential protection scheme more stable.

The schematic diagram of the percentage (biased) differential relay is shown in Fig. 8.11. This relay has two coils. One coil is known as restraining coil or bias coil which restrains (inhibits) the operation of the relay. The another coil is the operating coil which produces the operating torque for the relay. When the operating torque exceeds the restraining torque, the relay operates. The operating coil is connected to the mid-point of the restraining coil as shown in Fig. 8.11. N_r and N_0 are the total number of turns of the restraining coil and the operating coil respectively. Since the restraining coil is tapped at the centre, it forms two sections with equal number of turns, $N_r/2$. The restraining coil is connected in the circulating current path in such a way that current I_{1s} flows through one section of $N_r/2$ turns and I_{2s} flows through the another section of $N_r/2$, so that the complete restraining coil of N_r turns receives the through fault current of $(I_{1s} + I_{2s})/2$. The operating coil, having N_0 number of turns, is connected in the difference (spill) path, so that it receives the differential (spill) current, $(I_{1s} - I_{2s})$.

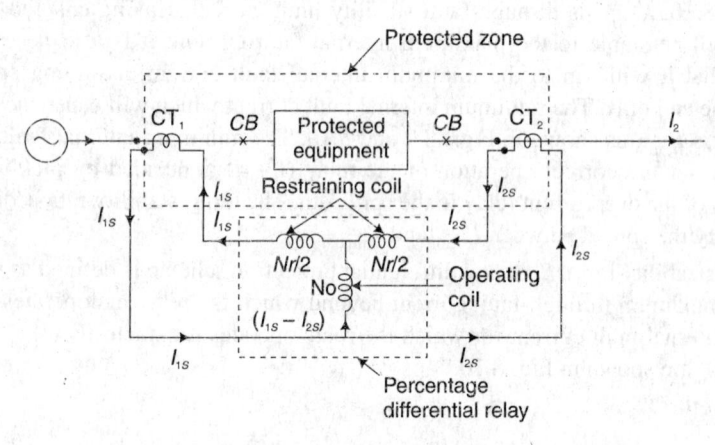

Fig. 8.11 *Percentage (biased) differential relay*

The operating condition of the percentage differential relay can be derived as follows:

The relay operates if the operating torque produced by the operating coil is more than the restraining torque produced by the restraining coil. As the torque is proportional to the ampere-turns (AT), the relay will operate when the ampere-turns of the operating coil $(AT)_0$, will be greater than ampere-turns of the restraining coil, $(AT)_r$.

Ampere-turns of the left-hand section of the restraining coil = $\dfrac{N_r}{2} I_{1s}$

Ampere-turns of the right-hand section of the restraining coil = $\dfrac{N_r}{2} I_{2s}$

Total ampere-turns of the restraining coil, $(AT)_r = \dfrac{N_r}{2}(I_{1s} + I_{2s})$

$$= N_r \dfrac{(I_{1s} + I_{2s})}{2} \qquad (8.13)$$

Thus it can be assumed that the entire N_r turns of the restraining coil carries a current $(I_{1s} + I_{2s})/2$. The current $(I_{1s} + I_{2s})/2$ which is the average of the secondary currents of the two CTs (CT$_1$ and CT$_2$) is known as the 'through current' or restraining current, I_r. Hence

$$I_r = (I_{1s} + I_{2s})/2 \qquad (8.14)$$

The ampere-turns of the operating coil, $(AT)_0 = N_0(I_{1s} - I_{2s})$ (8.15)

Neglecting spring restraint, the relay will operate when,

$$(AT)_0 > (AT)_r$$

or $$N_0 (I_{1s} - I_{2s}) > N_r \frac{(I_{1s} + I_{2s})}{2}$$

or $$(I_{1s} - I_{2s}) > \frac{N_r}{N_0} \frac{(I_{1s} + I_{2s})}{2}$$

or $$I_d > K I_r \quad (8.16)$$

where, $I_d = (I_{1s} - I_{2s})$ is the differential current through the operating coil. Hence it is also called the differential operating current

$I_r = (I_{1s} + I_{2s})/2$ is the restraining current or through current,

and $K = \dfrac{N_r}{N_0} =$ Slope or Bias

K(slope or bias) is generally expressed as a percentage value.

The relay will be on the verge of operation when:

$$(I_{1s} - I_{2s}) = \frac{N_r}{N_0} \frac{(I_{1s} + I_{2s})}{2}$$

or $$I_d = K I_r \quad (8.17)$$

Thus, at the threshold of operation of the relay, the ratio of the differential operating current (I_d) to the restraining current (I_r) is a fixed percentage; and for operation of the relay the differential operating current must be greater than this fixed percentage of the restraining (through fault) current. Hence, this relay is called 'percentage differential relay'. The percentage differential relay is also known as 'biased differential relay'. It is customary to express the slope (or bias) of the relay as a percentage. Thus, a slope (bias) of 0.2 is expressed as 20% slope (bias).

Under normal and external (through) fault conditions, the restraining torque produced by the restraining coil is greater than the operating torque produced by the operating coil, hence the relay is inoperative. When internal fault occurs, the operating torque becomes more than the restraining torque and the relay operates.

If the effect of control spring is taken into account, the equation 8.17 can be written as

$$I_d = K I_r + K_s \quad (8.18)$$

where K_s accounts for the effect of spring.

Thus, the operating characteristic of the percentage (biased) differential relay is a straight line having a slope of K and an intercept K_s on the y-axis. All points above the straight line represent the condition where the operating torque is greater than the restraining torque and hence fall in the trip region of the relay. All points below the straight line belong to the blocking (restraining) region. The operating characteristic of this relay is shown in Fig. 8.12.

The typical values of K (slope or bias) are 10%, 20%, 30% and 40%. Clearly, a relay with a slope of 10% is far more sensitive than a relay with a slope of 40%. The slope of the relay determines the trip zone.

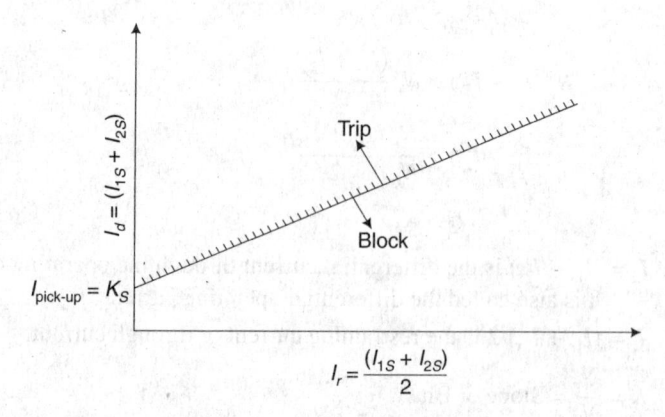

Fig. 8.12 *Operating characteristic of percentage differential relay*

The percentage differential relay does not have a fixed pick-up value. The relay automatically adapts its pick-up value to the restraining (through) current. As the restraining (through fault) current goes on increasing, the restraining torque also increases and the relay is prevented from maloperation. Since the restraining winding biases the relay towards restraint, it is also known as the biasing winding.

8.4.1 Stability Limit and Stability Ratio

The characteristic of the percentage differential relay, superimposed on the external fault and internal fault characteristic is shown in Fig. 8.13. The slope of internal fault characteristic is 2 (200%) as found in Section 8.3.6. The point of intersection of the relay and internal fault characteristics gives the minimum internal fault current required for the correct operation of the relay ($I'_{f,\,min}$). The point of intersection of the relay and external fault characteristics gives the maximum external fault current, $I'_{F,\,max}$ beyond which the relay maloperates and the differential protection scheme loses stability. $I'_{F,\,max}$ is known as 'through fault stability limit'.

It is clear from Fig. 8.13 that the maximum through-fault current ($I'_{F,\,max}$) which the percentage differential relay can accomodate without maloperation increases. $I'_{F,\,max}$ increases with slope of the relay's characteristic. The minimum internal fault current ($I'_{f,\,min}$) for which the relay operates satisfactorily does not deteriorate much. Thus, the through-fault stability and the stability ratio of the percentage differential relay increases and they are substantially better than that of the simple differential relay.

The stability ratio is given by

$$\text{Stability ratio } (S) = \frac{I'_{F,\,max}}{I'_{f,\,min}} \tag{8.19}$$

The immunity of the percentage differential relay to maloperation on external (through) faults can be increased by increasing the slope of the characteristic. As the

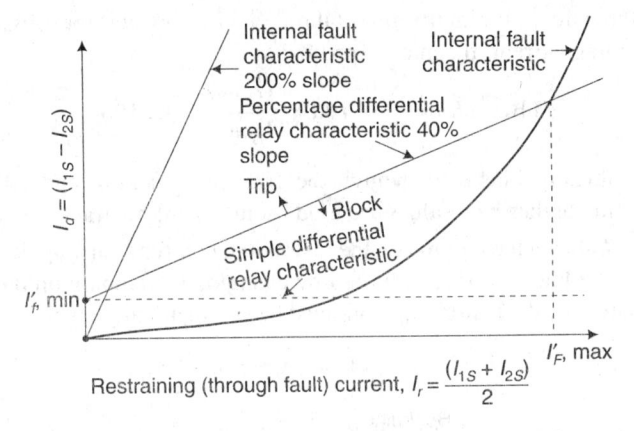

Fig. 8.13 Increased "through fault stability limit" and stability ratio of percentage differential relay

percentage (biased) differential relay is very stable on external (through) faults, it is universally used to protect generators and transformers.

8.4.2 Types of Percentage Differential Relays

Percentage differential relays are of two types: (i) fixed-percentage type, and (ii) variable-percentage type. The fixed-percentage type differential relays have percentage taps on the restraining coil to fix the percentage bias (slope) of the relay. Both the halves of the restraining coil are symmetrically tapped, so that operating coil could be connected to the midpoint of the restraining coil for a particular fixed-percentage bias.

The variable-percentage type differential relays do not have percentage taps on the restraining coil. At low external (through) fault currents the percentage is low since at these levels the CT performance is usually quite good. At high through-fault currents, where the CT performance may not be as good, a high percentage differential characteristic is provided. This gives increased sensitivity with high security.

8.4.3 Settings of Percentage Differential Relay

The percentage differential relay has following two types of settings.

(i) Basic Setting or Sensitivity Setting Basic setting of the relay is the minimum current in the operating coil only (zero bias) which will operate the relay. It is expressed as a percentage of rated current. It is defined as follows:

$$\% \text{Basic Setting} = \frac{\text{Minimum current in operating coil only to cause operation}^*}{\text{Rated current of the operating coil}} \times 100$$

*when the bias is zero, i.e., $N_r = 0$.

Typical values of basic setting might be 10 to 20% for generator and 20% for transformer.

(ii) Bias Setting The bias (K) of the relay is the ratio of the number of turns in the restraining coil (N_r) to the number of turns in the operating coil (N_0). It can also be defined as the ratio of minimum current through operating coil for causing operation to the restraining current. It can be expressed as

$$\%\text{Bias}(K) = \frac{N_r}{N_0} \times 100 = \frac{(I_{1s} - I_{2s})}{(I_{1s} + I_{2s})/2} \times 100 \tag{8.20}$$

Typical values of bias setting might be 10% for generator and 20 to 40% for transformer (the higher bias values are used for tap-changing transformers).

Figure 8.14 shows the tapping on the restraining coil for adjusting the bias (slope) of the relay. The bias (slope) is adjusted by changing the tapping on the restraining coil. Both halves of the restraining coil must be symmetrically tapped.

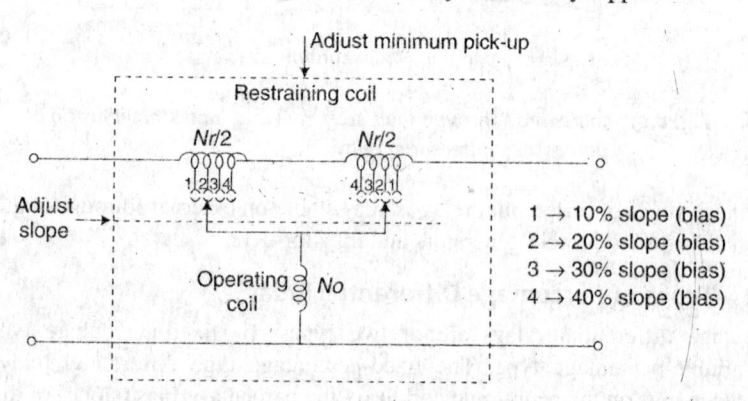

Fig. 8.14 *Bias setting of percentage differential relay*

Example 8.1 | Figure 8.15 shows percentage differential relay applied to the protection of a generator winding. The relay has a 0.1 amp minimum pick-up and 10% slope of its operating characteristic on $(I_{1s} + I_{2s})/2$ versus $(I_{1s} - I_{2s})$ diagram. A high resistance ground fault occured near the grounded neutral end of the generator winding while generator is carrying load. As a consequence, the currents flowing at each end of the winding is shown in the figure. Assuming CT ratios of 400/5 amperes, will the relay operate to trip the breaker.

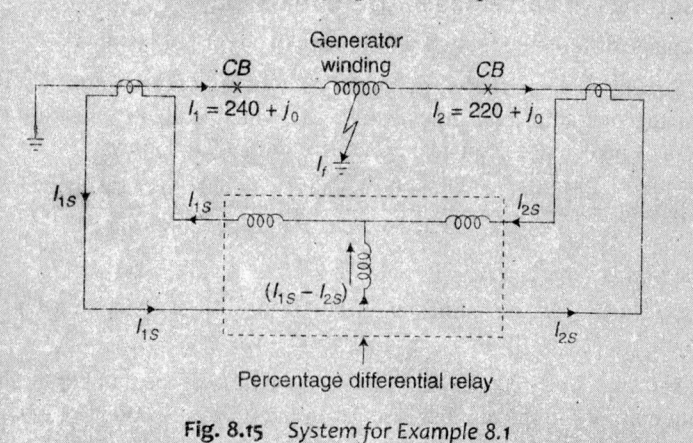

Fig. 8.15 *System for Example 8.1*

Differential Protection

Solution: Given: $I_1 = 240 + j0$, $I_2 = 220 + j0$, and CT ratio = 400/5
Therefore, CT secondary currents will be

$$I_{1s} = \frac{(240 + j0) \times 5}{400} A = 3 + j0 \text{ A}$$

$$I_{2s} = \frac{(220 + j0) \times 5}{400} A = 2.75 + j0 \text{ A}$$

Differential operating current $(I_d) = I_{1s} - I_{2s} = 3 - 2.75 = 0.25$ A

(i.e. current in the operating coil)

$$\text{Restraining current } (I_r) = \frac{(I_{1s} + I_{2s})}{2} = \frac{(3 + 2.75)}{2} = 2.875 \text{ A}$$

(i.e. the current in the restraining coil)

Slope of the characteristic, $K = 10\% = 0.1$

The differential operating current required for the operation of the relay corresponding to current of 2.875 A in the restraining coil $= KI_r = 0.1 \times 2.875 = 0.2875$ A. Since the actual current in the operating coil is 0.25 A, the relay will not operate to trip the circuit breaker.

Example 8.2 A generator winding is protected by using a percentage differential relay whose characteristic is having a slope of 10%. A ground fault occured near the terminal end of the generator winding while generator is carrying load. As a consequence, the currents flowing at each end of the winding are shown in the Fig. 8.16. Assuming CT ratios of 500/5 amperes, the relay operate to trip the circuit breakers.

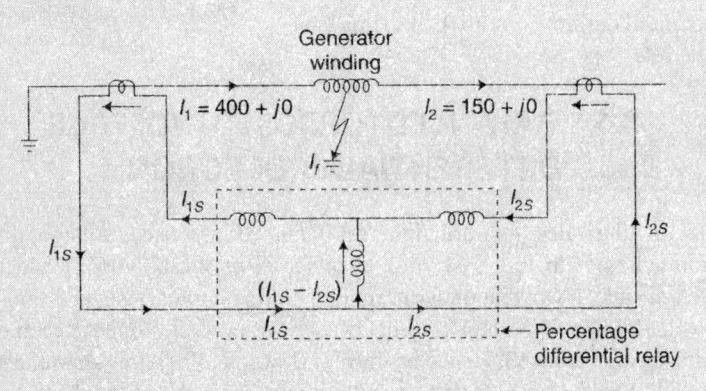

Fig. 8.16 *System for Example 8.2*

Solution: Given: $I_1 = 400 + j0$ A, $I_2 = 150 + j0$, and CT ratios = 500/5 A
Therefore, CT secondary currents will be:

$$I_{1s} = \frac{(400 + j0) \times 5}{500} A = (4 + j0) \text{ A}$$

$$I_{2s} = \frac{(150 + j0) \times 5}{500} \text{ A} = (1.5 + j0) \text{ A}$$

Differential operating current $(I_d) = (I_{1s} - I_{2s}) = (4 - 1.5) \text{ A} = 2.5 \text{ A}$

(i.e., current in the operating coil)

$$\text{Restraining current } (I_r) = \frac{(I_{1s} + I_{2s})}{2} = \frac{(4 + 1.5)}{2} \text{ A} = 2.75 \text{ A}$$

(i.e. current in the restraining coil)

Slope of the characteristic, $K = 10\% = 0.1$

The differential operating current required for the operation of the relay corresponding to restraining current of 2.75 A = KI_r = 0.1 × 2.75 = 0.275 A.

Since, the actual current in the operating coil is 2.5 A, the relay will operate to trip the circuit breaker.

8.5 DIFFERENTIAL PROTECTION OF 3-PHASE CIRCUITS

Figure 8.17 illustrates the differential protection of three-phase circuits. Under normal conditions the three secondary currents of CTs are balanced and their phasor sum is zero. Therefore no current flows through the relay coil during normal conditions.

When fault occurs in the protected zone, the balance is disturbed and differential current flows through the operating coil of the relay. If the differential current is more than the pick-up value, the relay operates.

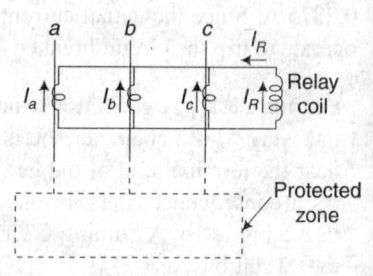

Fig. 8.17 *Differential protection of 3-phase circuit*

8.6 BALANCED (OPPOSED) VOLTAGE DIFFERENTIAL PROTECTION

Figure 8.18 illustrates the principle of balanced (opposed voltage differential protection scheme. In this case, the secondaries of the CTs (CT_1 and CT_2) are connected in such a way that under normal operating conditions and during external (through) faults, the secondary currents of CTs on two sides oppose each other and their voltages are balanced. Hence no current flows in pilot wires and relays. During internal fault, however, a differential (spill) current proportional to $(I_1 - I_2)$ in case of single-end-fed system and proportional to $(I_1 + I_2)$ in case of double-end-fed system flows through the relay coils. If the differential (spill) current flowing through the relay coils is higher than the pick-up value, the relays operate to isolate the protected equipment from the system.

Since no current flows through the secondaries of CTs under normal operating conditions, the whole of the primary ampere-turns are used in exciting the CTs. This creates large flux causing saturation of CTs and inducing high overvoltages which can damage the insulation of CT secondaries. For this reason, the CTs used in such protective scheme are air-core type so that the core does not get saturated and overvoltages are not induced during zero secondary current under normal operating conditions.

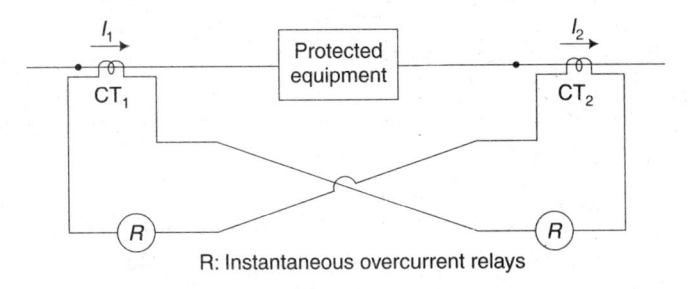

Fig. 8.18 *Balanced (opposed) voltage differential protection*

EXERCISES

1. What do you mean by 'differential protection'? Why is differential protection called 'unit-protection'?
2. Describe the various types and working principle of differential relays.
3. What is a simple differential protection scheme? Describe its behaviour during normal, external fault and internal fault conditions.
4. What are the disadvantages of the simple differential protection scheme?
5. Explain the following terms with respect to the simple differential protection scheme. (i) Spill current (ii) Minimum internal fault current for operation of the relay (iii) Maximum external fault current within which the relay will not maloperate (iv) through fault stability limit (v) Stability ratio.
6. With the help of a phasor diagram, discuss the differential current in the simple differential relay due to phase angle errors of CTs.
7. With the help of equivalent circuits of CTs, discuss the differential current in the simple differential relay due to difference in exciting currents of CTs.
8. Describe the construction and operating principle of the percentage differential relay. How the percentage differential relay overcomes the drawbacks of the simple differential relay?
9. Explain the terms 'through fault stability limit' and 'stability ratio' with respect to the percentage differential relay.
10. Why is percentage differential relay more stable than simple differential relay?
11. How can the 'through fault stability limit' of the percentage differential relay be increased?
12. Prove that the slope of the internal fault characteristics for a single-end-fed system is 200% and for a double-end-fed system it is more than 200%.
13. What are the types of percentage differential relays?

14. Discuss the basic setting and bias setting of percentage differential relay. What are the typical values of these settings for generator and transformer protection.
15. Describe the balanced (opposed) voltage differential protection scheme.

9 Rotating Machines Protection

9.1 INTRODUCTION

Rotating machines include synchronous generators, synchronous motors, synchronous condensers and induction motors. The protection of rotating machines involves the consideration of more possible failures or abnormal operating conditions than any other system equipment. The protection scheme for any machine is influenced by the size of the machine and its importance in the system. The failures involving short circuits are usually detected by some type of overcurrent or differential relay. Electromechanical, static or microprocessor-based relays can be used stand-alone or in combination with one another to achieve the desired degree of security and dependability. The failures of mechanical nature use mechanical devices or depend upon the control circuits for removing the problem.

9.2 PROTECTION OF GENERATORS

A generator is the most important and costly equipment in a power system. As it is accompanied by prime mover, excitation system, voltage regulator, cooling system, etc., its protection becomes very complex and elaborate. It is subjected to more types of troubles than any other equipment. A modern generating set is generally provided with the following protective schemes.

(i) **Stator protection**
   (a) Percentage differential protection
   (b) Protection against stator inter-turn faults
   (c) Stator-overheating protection

(ii) **Rotor protection**
   (a) Field ground-fault protection
   (b) Loss of excitation protection
   (c) Protection against rotor overheating because of unbalanced three-phase stator currents

(iii) **Miscellaneous**
   (a) Overvoltage protection
   (b) Overspeed protection
   (c) Protection against motoring
   (d) Protection against vibration

(e) Bearing-overheating protection
(f) Protection against auxiliary failure
(g) Protection against voltage regulator failure

9.2.1 Stator Protection

(a) Percentage Differential Protection Figure 9.1(a) shows the schematic diagram of percentage differential protection. It is used for the protection of generators above 1 MW. It protects against winding faults, i.e. phase to phase and phase to ground faults. This is also called biased differential protection or longitudinal differential protection. The polarity of the secondary voltage of CTs at a particular moment for an external fault has been shown in the figure. In the operating coil, the current sent by the upper CT is cancelled by the current sent by the lower CT and the relay does not operate. For an internal fault, the polarity of the secondary voltage of the upper CT is reversed, as shown in Fig. 9.1(b). Now the operating coil carries the sum of the currents sent by the upper CT and the lower CT and it operates and trips the circuit breaker.

The percentage differential protection does not respond to external faults and overloads. It provides complete protection against phase to phase faults. It provides protection against ground faults to about 80 to 85 per cent of the generator wind- ings. It does not

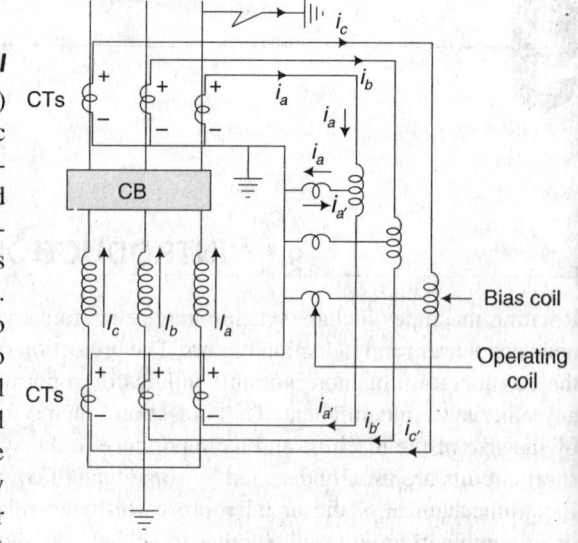

Fig. 9.1(a) Percentage differential protection for external fault condition (instantaneous current directions shown for external fault condition)

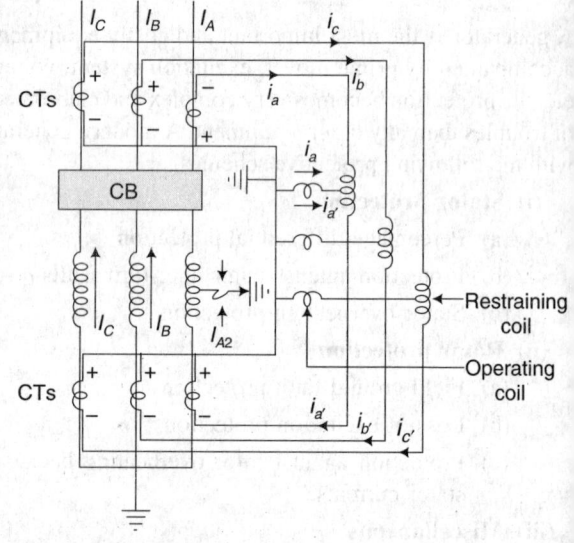

Fig. 9.1(b) Percentage differential protection of generator (instantaneous current directions shown for internal fault condition)

provide pro-tection to 100 per cent of the winding because it is influenced by the magnitude of the earth fault current which depends upon the method of neutral grounding. When the neutral is grounded through an impedance, the differential protection is supplemented by sensitive earth fault relays which will be described later.

Due to the difference in the magnetising currents of the upper and the lower CTs, the current through the operating coil will not be zero even under normal loading conditions or during external fault conditions. Therefore, to provide stability on external faults, bias coils (restraining coils) are provided. The relay is set to operate, not at a definite current but at a certain percentage of the through current. To obtain the required amount of biasing a suitable ratio of the restraining coil turns to operating coil turns is provided. High-speed percentage-differential relays having variable ratio or percentage slope characteristics are preferred. The setting of the bias coils varies from 5% to 50% and that of the relay coil (operating coil) from 10% to 100% of the full load current.

In case of stator faults, the tripping of circuit breaker to isolate the faulty generator is not sufficient to prevent further damage as the generator will still continue to supply power to the fault until its field excitation is suppressed. Therefore, the percentage differential relays initiate an auxiliary relay which in turn trips the main circuit breaker, trips the field circuit breaker, shuts down the prime mover, turns on CO_2 if provided and operates an alarm.

Restricted earth-fault protection by differential system

When the neutral is solidly grounded, it is possible to provide protection to complete winding of the generator against ground faults. However, the neutral is grounded through resistance to limit ground fault currents. With resistance grounding it is not possible to protect the complete winding against ground faults. The percentage of winding protected depends on the value of the neutral grounding resistor and the relay setting. The usual practice is to protect 80 to 85% of the generator winding against ground fault. The remaining 15-20% from neutral end is left unprotected. In Fig. 9.2 for phase to ground fault, it can be seen that the relay setting for the differential protection is determined by the value of the neutral grounding resistor and the percentage of winding to be protected.

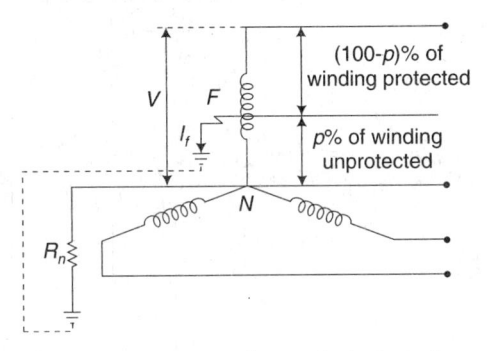

Fig. 9.2 Percentage of unprotected winding against phase to ground fault

If the ground fault occurs at the point F of the generator winding, the voltage V_{FN} is available to drive the ground-fault current I_f through the neutral to ground connection. If the fault point F is nearer to the neutral point N, the forcing voltage V_{FN} will be relatively less. Hence, ground fault current I_f will reduce. It is not practicable to keep the relay setting too sensitive to sense the ground fault currents of small magnitudes. Because, if the relay is made too sensitive, it may respond during through faults or other faults due to inaccuracies of CTs, saturation of CTs etc. Hence, a practice is to protect 80-85% of the generator winding against phase to ground faults and to leave the 15-20% portion of the winding from neutral end unprotected.

In Fig. 9.2, let $p\%$ of the winding from the neutral remains unprotected. Then $(100-p)\%$ of the winding is protected. The ground fault current I_f is given by

$$I_f = \frac{p}{100} \frac{V}{R_n} \qquad (9.1)$$

where V is the line to neutral voltage and R_n is the neutral grounding resistance.

For the operation of the relay, the fault current must be greater than the relay pick-up current.

Percentage differential protection for a Y-connected generator with only four leads brought out

When the neutral connection is made within the generator and only the neutral terminal is brought out, the percentage differential protection can be provided, as shown in Fig. 9.3. This scheme protects the generator winding only against ground

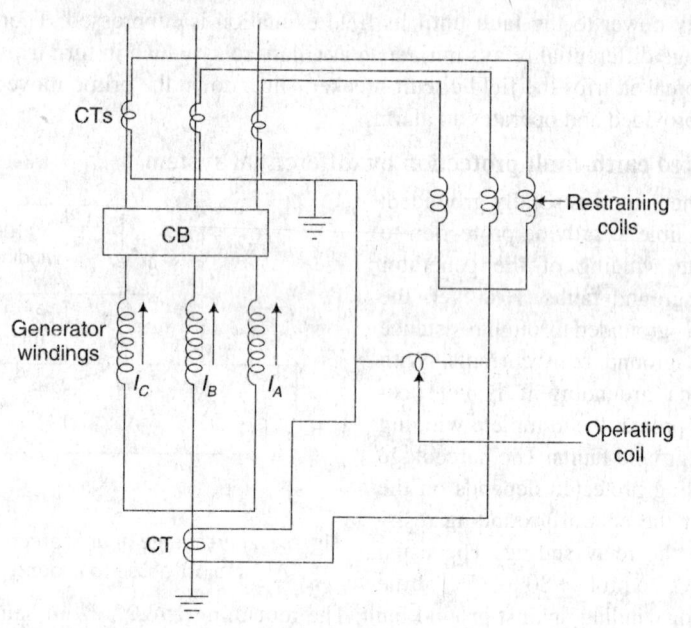

Fig. 9.3 *Percentage differential protection for Y connected generator with only four leads brought out*

faults. It does not protect it against phase faults. This is also called restricted earth fault protection because the relays operate only for ground faults within the protected windings.

Generator-transformer unit protection

In modern power systems, each generator is directly connected to the primary winding of a power transformer. The primary winding of the power transformer is connected in delta configuration and the secondary winding in Y configuration. The secondary winding which is the H.V. winding is connected through a circuit breaker to the H.V. bus. The generator and transformer is considered as a single unit and the percentage differential protection is provided for the combined unit. This type of unit protection has been discussed later while discussing transformer protection.

Example 9.1 | An 11 kV, 100 MVA alternator is grounded through a resistance of 5 Ω. The CTs have a ratio 1000/5. The relay is set to operate when there is an out of balance current of 1 A. What percentage of the generator winding will be protected by the percentage differential scheme of protection?

Solution: Primary earth-fault current at which the relay operates

$$= \frac{1000}{5} \times 1 = 200 \text{ A}$$

Suppose $p\%$ of the winding from the neutral remains unprotected.

$$\text{The fault current} = \frac{p}{100} \times \frac{11 \times 10^3}{\sqrt{3} \times 5}$$

For the operation of the relay, the fault current must be greater than the relay pick-up current

$$\frac{p}{100} \times \frac{11 \times 10^3}{\sqrt{3} \times 5} > \frac{1000}{5} \times 1 \quad \text{or} \quad p > 15.75$$

This means that 15.75% of the winding from the neutral is not protected. In other words, $100 - p = 100 - 15.75 = 84.25\%$ of the winding from the terminal is protected.

Note: Near the neutral point voltage stress is less and therefore, phase to earth faults are not likely to occur.

Example 9.2 | An 11 kV, 100 MVA alternator is provided with differential protection. The percentage of winding to be protected against phase to ground fault is 85%. The relay is set to operate when there is 20% out of balance current. Determine the value of the resistance to be placed in the neutral to ground connection.

Solution: (a) Primary earth-fault current at which the relay operates

$$= \frac{100 \times 10^3}{\sqrt{3} \times 11} \times \frac{20}{100} = 1049.759 \text{ A}$$

Suppose that the percentage of winding which remains unprotected is

$$p = 100 - 85 = 15\%.$$

$$\text{The fault current} = \frac{p}{100} \times \frac{11 \times 10^3}{\sqrt{3} R_n},$$

where R_n is the resistance in the neutral connection

$$\therefore \quad \frac{15}{100} \times \frac{11 \times 10^3}{\sqrt{3} R_n} = 1049.759$$

$$\therefore \quad R_n = \frac{15 \times 11 \times 10^3}{100 \times \sqrt{3} \times 1049.759} = 0.91 \, \Omega$$

(b) Protection against Stator Interturn Faults Longitudinal percentage differential protection does not detect stator interturn faults. A transverse percentage differential protection, as shown in Fig. 9.4 is employed for the protection of the generator against stator interturn faults. This type of protection is used for generators having parallel windings separately brought out to the terminals. The coils of modern large steam turbine driven generators usually have only one turn per phase per slot and hence they do not need interturn fault protection. Hydro generators having parallel windings in each phase employ such protection which thus provides back-up protection and detects interturn faults. This scheme is also known as split-phase protection.

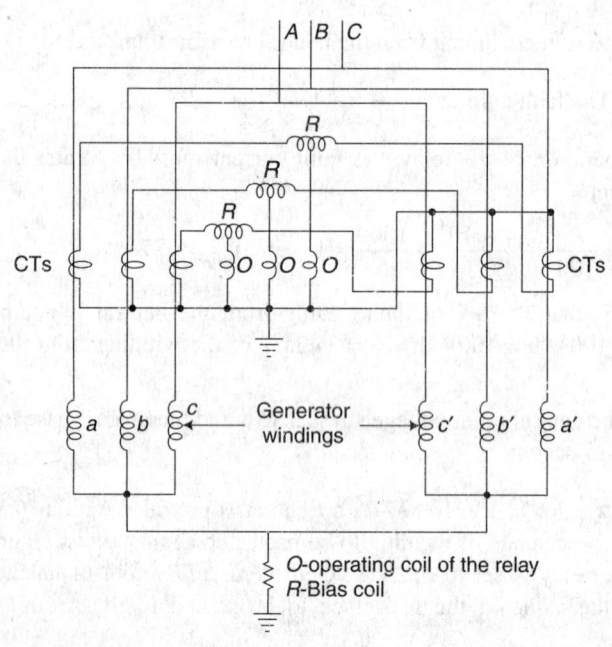

Fig. 9.4 *Transverse percentage differential protection for multi-winding generators*

A faster and more sensitive split-phase protection as shown in Fig. 9.5 can be employed. In this scheme, a single CT having double primary is used. No bias is necessary because a common CT is employed so that errors due to CT differences do not occur.

Interturn protection based on zero-sequence component

If generators do not have access to parallel windings, a method based on zero-sequence voltage measurment can be employed for the protection against stator interturn faults. This type of scheme will also be applicable to single winding generators having multi-turn per phase per slot to protect against interturn faults. Figure 9.6 shows the schematic diagram of interturn protection by zero-sequence voltage measurement across the machine.

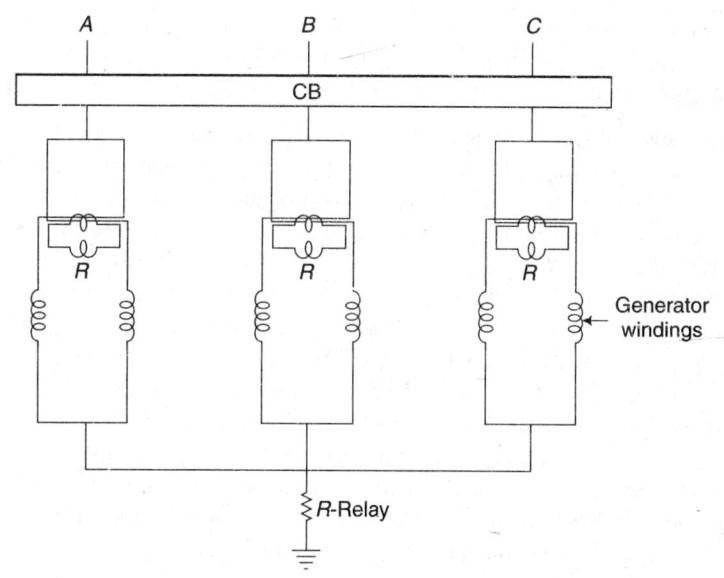

Fig. 9.5 *Split-phase protection of generator using double primary CTs*

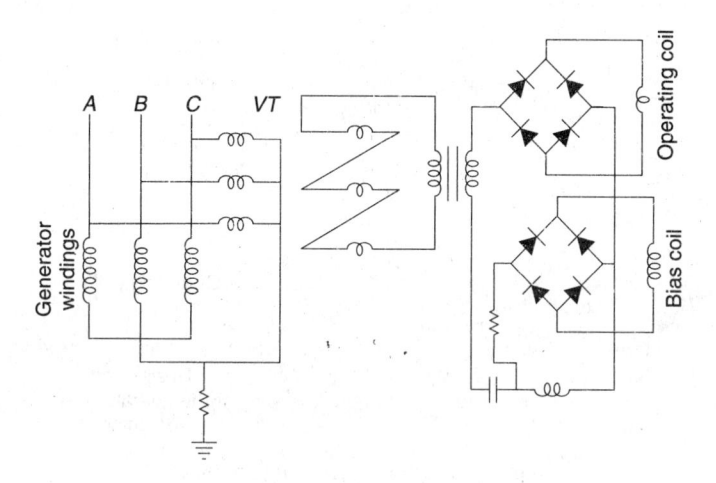

Fig. 9.6 *Interturn protection of generator using zero-sequence voltage*

The zero-sequence voltage does not exist during normal conditions. If one or more turns of a phase are short circuited, the generated emf contains zero-sequence component. A voltage transformer as shown in the figure is employed to extract zero-sequence component. The secondary winding of the voltage transformer is in open-delta connection to provide the zero-sequence component of the voltage to the protective relay. A filter is provided to extract a third harmonic component from the VT output and apply it as the relay bias.

The zero-sequence voltage is also produced in case of an external earth fault. But most of this voltage appears across the earthing resistor. A very small amount, 1 or 2 per cent, appears across the generator. Therefore, the zero-sequence voltage is measured across the generator windings at the line terminals rather than the zero-sequence voltage to the earth as shown in the figure to activate the relay on the occurrence of internal faults.

(c) Stator-overheating Protection Overheating of the stator may be caused by the failure of the cooling system, overloading or core faults like short-circuited laminations and failure of core bolt insulation. Modern generators employ two methods to detect overheating both being used in large generators (above 2 MW). In one method, the inlet and outlet temperatures of the cooling medium which may be hydrogen/water are compared for detecting overheating. In the other method, the temperature sensing elements are embedded in the stator slots to sense the temperature. Figure 9.7 shows a stator overheating relaying scheme. When the temperature exceeds a certain preset maximum temperature limit, the relay sounds an alarm. The scheme employs a temperature detector unit, relay and Wheatstone-bridge for the purpose. The temperature sensing elements may either be thermistors, thermocouples or resistance temperature indicators. They are embedded in the stator slots at different locations. These elements are connected to a multi-way selector switch which checks each one in turn for a period long enough to operate an alarm relay.

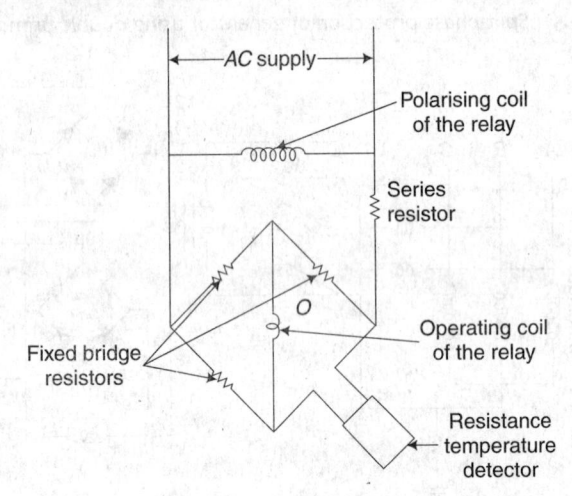

Fig. 9.7 Stator-overheating protection

For small generators, a bimetallic strip heated by the secondary current of the CT is placed in the stator circuit. This relay will not operate for the failure of the cooling system.

Thermocouples are not embedded in the rotor winding as this makes slip ring connections very complicated. Rotor temperature can be determined by measuring the winding resistance. An ohm-meter type instrument, energised by the rotor voltage and current and calibrated in temperature is employed for the purpose.

9.2.2 Rotor Protection

(a) Field Ground-fault Protection As the field circuit is operated ungrounded, a single ground fault does not affect the operation of the generator or cause any damage. However, a single rotor fault to earth increases the stress to the ground in the field when stator transients induce an extra voltage in the field winding. Thus, the probability of the occurrence of the second ground fault is increased. In case a second ground fault occurs, a part of the field winding is bypassed, thereby increasing the current through the remaining portion of the field winding. This causes an unbalance in the air-gap fluxes, thereby creating an unbalance in the magnetic forces on opposite sides of the rotor. The unbalancing in magnetic forces makes the rotor shaft eccentric. This also causes vibrations. Even though the second ground fault may not bypass enough portion of the field winding to cause magnetic unbalance, the arcing at the fault causes local heating which slowly distorts the rotor producing eccentricity and vibration.

Figure 9.8 shows the schematic diagram of rotor earth protection. A dc voltage is impressed between the field circuit and earth through a polarised moving iron relay. It is not necessary to trip the machine when a single field earth fault occurs. Usually an alarm is sounded. Then immediate steps are taken to transfer the load from the faulty generator and to shut it down as quickly as possible to avoid further problems.

In case of brushless machines, the main field circuit is not accessible. If there is a partial field failure due to short-circuiting of turns in the main field winding, it is detected by the increase in level of the field current. A severe fault or short-circuiting of the diode is detected by a relay monitoring the current in the exciter control circuit.

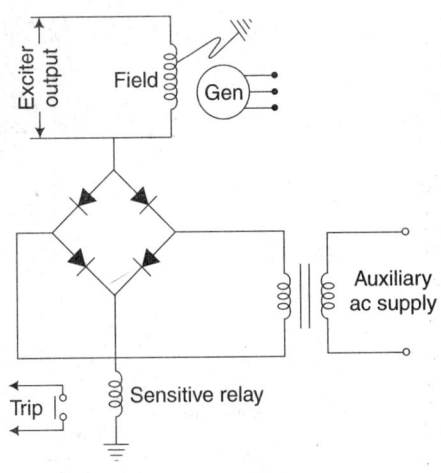

Fig. 9.8 *Earth fault protection*

(b) Loss of Excitation When the excitation of a generator is lost, it speeds up slightly and operates as an induction generator. Round-rotor generators do not have damper windings and hence they are not suitable for such an operation. The rotor is overheated quickly due to heavy induced currents in the rotor iron. The rotors of salient pole generators are not overheated because they have damper windings which carry induced currents. The stators of both salient and non-salient pole generators are overheated due to wattless current drawn by the machines as magnetising current from the system. The stator overheating does not occur as quickly as rotor overheating. A large machine may upset the system stability because it draws reactive power from the system when it runs as an induction generator whereas it supplies reactive power when it runs as a generator. A machine provided with a quick-acting automatic voltage regulator and connected to a very large system may run for several minutes as an induction generator without harm.

Field failure may be caused by the failure of excitation or mal-operation of a faulty field breaker. A protective scheme employing offset mho or directional impedance relay having a characteristic as shown in Fig. 9.9 is recommended for large modern generators. When a generator loses its excitation, the locus of the equivalent generator impedance moves from the first quadrant to the fourth quadrant, irrespective of initial conditions. This type of locus is not traced in any other conditions. The relay trips the field breaker and the generator is disconnected from the system.

(c) Protection against Rotor Overheating because of Unbalanced Three-phase Stator Currents The negative sequence component of unbalanced stator currents cause double frequency current to be induced in the rotor iron. If this component becomes high severe overheating of the rotor may be caused. The unbalanced condition may arise due to the following reasons:
  (i) When a fault occurs in the stator winding
  (ii) An unbalanced external fault which is not cleared quickly

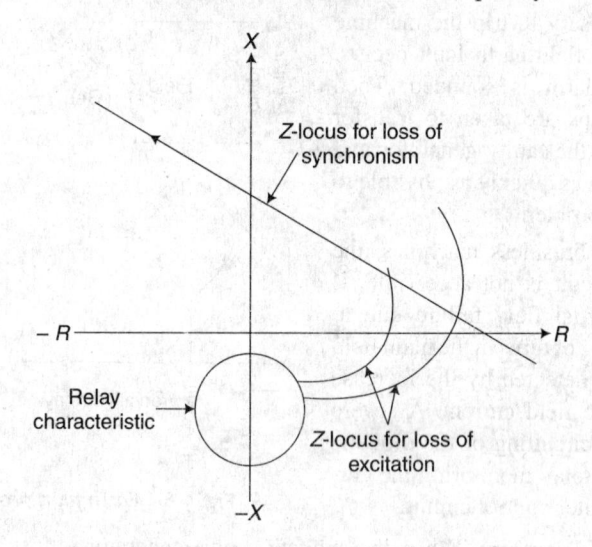

Fig. 9.9 Loss of excitation relay characteristic

  (iii) Open-circuiting of a phase
  (iv) Failure of one contact of the circuit breaker

The time for which the rotor can be allowed to withstand such a condition is related by the expression

$$I_2^2 t = K$$

where I_2 is the negative sequence component of the current, t = time, and K = a constant which depends on the type of generating set and its cooling system,

$K = 7$ for turbo-generator with direct cooling

$ = 60$ for a salient-pole hydro generator.

Figure 9.10 shows a protective scheme using a negative sequence filter and relay. The overcurrent relay used in the negative phase sequence protection has a long operating time with a facility of range setting to permit its characteristic to be matched to $I_2^2 t$ characteristic of the machine. A typical time range of the relay is 0.2 to 2000 s. It has a typical construction with a special electromagnet. It has shaded pole construction with a Mu-metal shunt. The negative sequence filter gives an output proportional to I_2. It actuates an alarm as well as the time-current relay which has a very inverse characteristic. The alarm unit also starts a timer which is adjustable from 8% to 40% of negative sequence component. The timer makes a delay in the alarm to prevent the alarm from sounding unnecessarily on unbalanced loads of short duration.

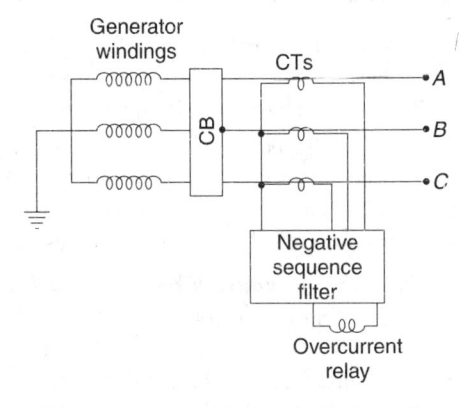

Fig. 9.10 *Protection against unbalanced stator currents*

9.2.3 Miscellaneous

(a) Overvoltage Protection Overvoltage may be caused by a defective voltage regulator or it may occur due to sudden loss of electrical load on generators. When a load is lost, there is an increase in speed and hence the voltage also increases. In case of a steam power station, it is possible to bypass the steam before the speed reaches a limit above which a dangerous overvoltage can be produced. In steam power stations, the automatic voltage regulator controls the overvoltages which is associated with overspeed. In hydro-stations it is not possible to stop or divert water flow so quickly and overspeed may occur. Therefore, overvoltage relays are provided with hydro and gas-turbine sets. But overvoltage relays are not commonly used with turbo-alternators.

(b) Overspeed Protection A turbo-generator is provided with a mechanical overspeed device. The spseed governor normally controls its speed. It is designed to prevent any speed rise even with a 100 per cent load rejection. An emergency centrifugal overspeed device is also incorporated to trip emergency steam valves when the speed exceeds 110 percent.

Large turbosets are sometimes provided with overspeed relays. In USA out of step tripping relays are used which cut off steam when the generator is 180° out of synchronism and has slipped one pole. In UK a sensitive fast power relay is used to determine whether the power output falls below a certain value or reversed.

Severe electrical faults also cause overspeed and hence HV circuit breakers, field circuit breakers and the steam turbine valves are tripped simultaneously. As water flow cannot be stopped quickly, hydrosets are provided with overspeed protection. The setting of overspeed relays for hydrosets is 140%. Overspeed relays are also provided with gas-turbine sets.

(c) Protection against Motoring When the steam supply is cut off, the generator runs as a motor. The steam turbine gets overheated because insufficient steam passes through the turbine to carry away the heat generated by windage loss. Therefore, a protective relay is required for the protection of the steam turbine. Generally, the relay operates when power output falls below 3%. A sensitive reverse power relay is available which has an operating setting of about 0.5% of the generator's output.

Hydrosets sometimes require protection against motoring. Cavitation problems arise in water turbines at low water flow. Protection is provided by reverse power relay having an operating setting of 0.2 to 2% of the rated power. A diesel set and gas turbine require 25% and 50% setting respectively.

(d) Protection against Vibration and Distortion of Rotor Vibration is caused by overheating of the rotor or some mechanical failure or abnormality. The overheating of the rotor is caused due to unbalanced stator currents or rotor ground faults. Overheating of the rotor distorts it, thereby causing eccentricity. Eccentric running produces vibration. Protection provided for unbalanced stator currents and rotor ground faults minimise vibration. A vibration measuring device is also used for steam turbine sets. Such a device detects the vibration which is caused either by electrical or mechanical causes. An alarm is actuated if vibration takes place. The vibration detector is mounted rigidly on one of the bearing pedestals of a horizontal shaft machine or on the upper guide-bearing of vertical shaft bearing. For details see Ref. 5.

(e) Bearing Overheating Protection Temperature of the bearing is detected by inserting a temperature sensing device in a hole in the bearing. For large machines where lubricating oil is circulated through the bearing, an oil flow device is used to detect the failure of oil cooling equipment. An alarm is actuated when the bearing is overheated or when the circulation of the lubricating oil fails.

(f) Protection against Auxiliary Failure The power plant auxiliaries are very important for the running of the generating sets. High grade protective equipment is employed for their reliable operation. For large generating sets, protection against loss of vacuum and loss of boiler pressure are provided. Such failures are due to the failure of the associated auxiliaries. So the protection provided for the loss of vacuum and loss of boiler pressure provides to some extent protection against the auxiliary failure. In the case of such failures generating sets are shut down. Protection against the failure of induced draught fans is also provided.

(g) Protection against Voltage Regulator Failure Modern quick response automatic voltage regulators are very complex. They are subject to component failures. Suitable protective devices are provided against their failure. A definite time dc overcurrent relay is provided which operates when there is overcurrent in the rotor circuit for a period longer than a prescribed limit. In such a situation, the excitation is switched to a predetermined value for manual control.

The supply for the regulator reference voltage is given from a separate voltage transformer. Protection is also required against the failure of the regulator reference voltage. An under voltage relay is used for this purpose. A better approach is to use a voltage balance relay which compares the voltage derived from the instrument transformer with the voltage derived from the voltage regulator transformer. If there is mal-operation of the voltage regulator due to the failure of the reference voltage, the relay operates and switches the excitation to a predetermined value for manual control.

(h) Protection against Pole Slipping In case of system disturbances after the operation of circuit breaker or when heavy load is thrown or switched on, the generator rotor may oscillate. Consequently, variations in current, voltage and power factor may take place. Such oscillations may disappear in a few seconds. Therefore, in such a situation, tripping is not desired. In some cases, angular displacement of the rotor exceeds the stability limit and the rotor slips a pole pitch. If the disturbance is over, the generator may regain synchronism. If it does not, it should be tripped.

An alternative approach is to trip the field switch and allow the machine to run as an asynchronous machine, thereby removing the oscillations from the machine. Then the load is reduced to a low value at which the machine can re-synchronise itself. If the machine does not re-synchronise, the field switch is reclosed at the minimum excitation setting. This will cause the machine to re-synchronise smoothly.

(i) Field Suppression When a fault occurs in the generator winding, the circuit breaker trips and the generator is isolated from the system. However, the generator still continues to feed the fault as long as the excitation is maintained, and the damage increases. Therefore, it is desirable to suppress the field as quickly as possible. The field cannot be destroyed immediately. The energy asssociated with the flux must be dissipated into an external device. To achieve this, the field winding is connected to a discharge resistor to absorb the stored energy. The discharge resistor is connected in parallel with the field winding before opening the field circuit breaker.

(j) Back-up Protection Overcurrent relays are used as back-up protection. As the synchronous impedance of a turbo-generator is more than 100%, the fault current, may fall below the normal load current. Therefore, standard time-overcurrent relays cannot be employed for back-up protection. A voltage controlled overcurrent can be employed for such a purpose. A better alternative is to use reactance or impedance type distance relays.

In addition to overcurrent relays, the stator protection is generally supplemented by sensitive earth fault relays. Relays having inverse time characteristic are used. An earth fault relay is connected to a CT placed in the neutral point earthing lead. This is an unrestricted protection and hence, it is to be graded with the feeder protection.

EXERCISES

1. Enumerate the relaying schemes which are employed for the protection of a modern alternator.
   Describe with a neat sketch, the percentage differential protection of a modern alternator.
2. What is transverse or split phase protection of an alternator? For what type of fault is this scheme of protection employed? With a neat sketch discuss the working principle of this scheme.
3. What type of protective device is used for the protection of an alternator against overheating of its (a) stator, (b) rotor? Discuss them in brief.
4. What type of a protective scheme is employed for the protection of the field winding of the alternator against ground faults?

5. Discuss the protection employed against loss of excitation of an alternator.
6. Are the protective devices employed for the protection of an alternator against (a) overvoltage, (b) overspeed, (c) motoring? Discuss them in brief.
7. Are the protective devices employed for the protection of an alternator against (a) vibration and distortion of the rotor, (b) bearing overheating, (c) auxiliary failure, (d) voltage regulator failure? Briefly describe them.
8. What is pole slipping phenomenon in case of an alternator? What measures are taken if pole slipping occurs?
9. What do you understand by field suppression of an alternator? How is it achieved?
10. Is any back-up protection employed for the protection of an alternator? If yes, discuss the scheme which is used for this purpose.
11. An 11 kV, 100 MVA generator is grounded through a resistance of 6 Ω. The CTs have a ratio of 1000/5. The relay is set to operate when there is an out of balance current of 1 A. What percentage of the generator winding will be protected by the percentage differential scheme of protection **(Ans.** 18.9%)
12. An 11 kV, 100 MVA generator is provided with differential scheme of protection. The percentage of the generator winding to be protected against phase to ground fault is 80%. The relay is set to operate when there is 15% out of balance current. Determine the value of the resistance to be placed in the neutral to ground connection. **(Ans.** 1.61 Ω)
13. A 5 MVA, 6.6 kv Y (star) connected generator has resistance per phase of 0.5 Ω and synchronous reactance per phase of 2 Ω. It is protected by a differential relay which operates when the out of balance current exceeds 30% of the load current. Determine what proportion of the generator winding is unprotected if the star point is grounded through a resistor of 6.5 Ω.

(Ans. 22.8%)

[**Hint.** If p % of the winding remains unprotected, the impedance of this portion of the winding $= \dfrac{p}{100}(0.5 + j\, 2.0)$

$= p\,(0.005 + j\, 0.02)$

Grounding resistance $= 6.5$ Ω

Total impedance in the fault circuit $= 6.5 + 0.005p + j\, 0.02p$

$= (6.5 + 0.005p)$ very approx

neglecting reactance portion which is small.

14. The neutral point of a 50 MVA, 11 kv generator is grounded through a resistance of 5 Ω, the relay is set to operate when, there is an out of balance current of 1.5 A. The CTs have a ratio of 1000/5. What percentage of the winding is protected against a ground fault and what should be the minimum value of the grounding resistance to protect 90% of the winding.

(Ans. 76.4%, 2.12 Ω)

15. Discuss the faults and various abnormal operating conditions of induction motors and protection provided against each.

16. Describe the protection of low voltage induction motor using conductor starter.
17. Explain the various methods of short circuit and ground fault protection of motors.
18. Discuss the causes of motor failure of both electrical and mechanical origin.
19. Explain the terms, 'single phasing' and 'phase reversal'. In what form protection is provided against single phasing and phase reversal.
20. Discuss the differential protection scheme for large motors.
21. What additional protection is required for synchronous motors?

10 Transformer and Buszone Protection

10.1 INTRODUCTION

The power transformer is a major and very important equipment in a power system. It requires highly reliable protective devices. The protective scheme depends on the size of the transformer. The rating of transformers used in transmission and distribution systems range from a few kVA to several hundred MVA. For small transformers, simple protective device such as fuses are employed. For transformers of medium size overcurrent relays are used. For large transformers differential protection is recommended.

Bus zone protection includes, besides the busbar itself, all the apparatus connected to the busbars such as circuit breakers, disconnecting switches, instrument transformers and bus sectionalizing reactors etc. Though the occurrence of bus zone faults are rare, but experience shows that bus zone protection is highly desirable in large and important stations. The clearing of a bus fault requires the opening of all the circuits branching from the faulted bus.

10.2 TRANSFORMER PROTECTION

10.2.1 Types of Faults Encountered in Transformers

The faults encountered in transformer can be placed in two main groups.
 (a) External faults (or through faults)
 (b) Internal faults

External Faults

In case of external faults, the transformer must be disconnected if other protective devices meant to operate for such faults, fail to operate within a predetermined time. For external faults, time graded overcurrent relays are employed as back-up protection. Also, in case of sustained overload conditions, the transformer should not be allowed to operate for long duration. Thermal relays are used to detect overload conditions and give an alarm.

Internal Faults

The primary protection of transformers is meant for internal faults. Internal faults are classified into two groups.

(i) *Short circuits in the transformer winding and connections* These are electrical faults of serious nature and are likely to cause immediate damage. Such faults are detectable at the winding terminals by unbalances in voltage or current. This type of faults include line to ground or line to line and interturn faults on H.V. and L.V. windings.

(ii) *Incipient faults* Initially, such faults are of minor nature but slowly might develop into major faults. Such faults are not detectable at the winding terminals by unbalance in voltage or current and hence, the protective devices meant to operate under short circuit conditions are not capable of detecting this type of faults. Such faults include poor electrical connections, core faults, failure of the coolant, regulator faults and bad load sharing between transformers.

10.2.2 Percentage Differential Protection

Percentage differential protection is used for the protection of large power transformers having ratings of 5 MVA and above. This scheme is employed for the protection of transformers against internal short circuits. It is not capable of detecting incipient faults. Figure 10.1 shows the schematic diagram of percentage differential protection for a $Y - \Delta$ transformer. The direction of current and the polarity of the CT voltage shown in the figure are for a particular instant. The convention for marking the polarity for upper and lower CTs is the same. The current entering end has been marked as positive. The end at which current is leaving has been marked negative.

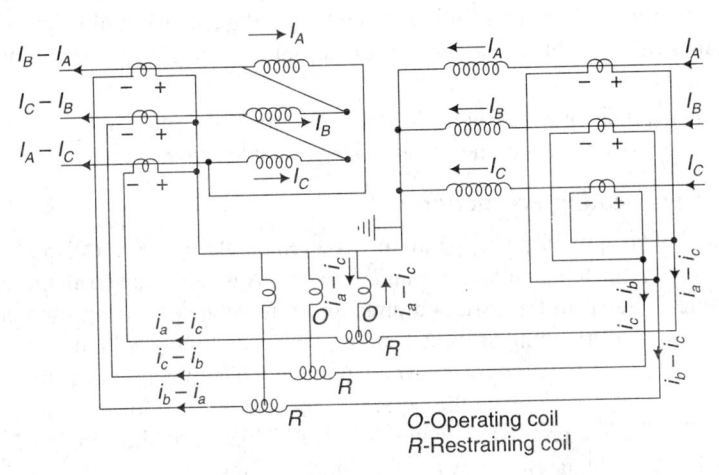

Fig. 10.1 Percentage differential protection for $Y - \Delta$ connected transformer

O and R are the operating and restraining coils of the relay, respectively. The connections are made in such a way that under normal conditions or in case of external faults the current flowing in the operating coil of the relay due to CTs of the primary side is in opposition to the current flowing due to the CTs of the secondary side. Consequently, the relay does not operate under such conditions. If a fault occurs on the winding, the polarity of the induced voltage of the CT of the secondary side is reversed. Now the currents in the operating coil from CTs of both primary and secondary side are in the same direction and cause the operation of the relay. To supply the matching current in the operating winding of the relay, the CT which are on the star side of the transformer are connected in delta. The CTs which are on the delta side of the transformer are connected in star. In case of $Y - \Delta$ connected transformer there is a phase shift of 30° in line currents. Also the above mentioned CTs connections also correct this phase shift. Moreover, zero sequence current flowing on the star side of the transformers does not produce current outside the delta on the other side. Therefore, the zero sequence current should be eliminated from the star side. This condition is also fulfilled by CTs connection in delta on the star side of the transformer.

In case of star/star connected transformer CTs on both sides should be connected in delta. In case of star/star connected transformer, if star point is not earthed, CTs may be connected in star on both sides. If the star point is earthed and CTs are connected in star, the relay will also operated for external faults. Therefore, it is better to follow the rule that CTs associated with star-connected transformer windings should be connected in delta and those associated with delta windings in star.

The relay settings for transformer protection are kept higher than those for alternators. The typical value of alternator is 10% for operating coil and 5% for bias. The corresponding values for transformer may be 40% and 10% respectively. The reasons for a higher setting in the case of transformer protection are.

(i) A transformer is provided with on-load tap changing gear. The CT ratio cannot be changed with varying transformation ratio of the power transformer. The CT ratio is fixed and it is kept to suit the nominal ratio of the power transformer. Therefore, for taps other than nominal, an out of balance current flows through the operating coil of the relay during load and external fault conditions.

(ii) When a transformer is on no-load, there is no-load current in the relay. Therefore, its setting should be greater than no-load current.

10.2.3 Overheating Protection

The rating of a transformer depends on the temperature rise above an assumed maximum ambient temperature. Sustained overload is not allowed if the ambient temperature is equal to the assumed ambient temperature. At lower ambient temperature, some overloading is permissible. The overloading will depend on the ambient temperature prevailing at the time of operation. The maximum safe overloading is that which does not overheat the winding. The maximum allowed temperature is about 95°C. Thus the protection against overload depends on the winding temperature which is usually measured by thermal image technique.

In the thermal image technique, a temperature sensing device is placed in the transformer oil near the top of the transformer tank. A CT is employed on LV side to supply current to a small heater. Both the temperature sensing device and the heater are placed in a small pocket. The heater produces a local temperature rise similar to that of the main winding. The temperature of the sensing element is similar to that of the winding under all conditions. In a typical modern system the heat sensitive element is a silicon resistor or silistor. It is incorporated with the heating element and kept in a thermal moulded material. The whole unit forms a thermal replica of the transformer winding. It is in the shape of a small cylinder and it is placed in the pocket in the transformer tank about 25 cm below the tank top, which is supposed to be the hottest layer in the oil. The silistor is used as an arm of a resistance bridge supplied from a stabilised dc source. An indicating instrument is energized from the out of balance voltage of the bridge. Also the voltage across the silistor is applied to a static control circuit which controls cooling pumps and fans, gives warning of overheating, and ultimately trips the transformer circuit breakers.

10.2.4 Protection against Magnetising Inrush Current

When an unloaded transformer is switched on, it draws a large initial magnetising current which may be several times the rated current of the transformer. This initial magnetising current is called the magnetising inrush current. As the inrush current flows only in the primary winding, the differential protection will see this inrush current as an internal fault. The harmonic contents in the inrush current are different than those in usual fault current. The dc component varies from 40 to 60%, the second harmonic 30 to 70% and the third harmonic 10 to 30%. The other harmonics are progressively less. the third harmonic and its multiples do not appear in CT leads as these harmonics circulate in the delta winding of the transformer and the delta connected CTs on the Y side of the transformer. As the second harmonic is more in the inrush current than in the fault current, this feature can be utilised to distinguish between a fault and magnetising inrush current.

Figure 10.2 shows a high speed biased differential scheme incorporating a harmonic restraint feature. The relay of this scheme is made insensitive to magnetic inrush current. The operating principle is to filter out the harmonics from the differential current, rectify them and add them to the percentage restraint. The tuned circuit

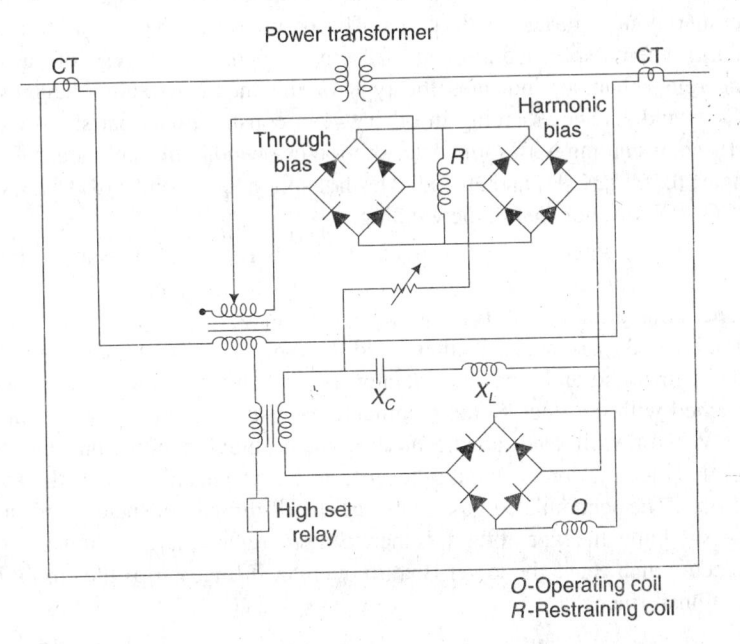

Fig. 10.2 *Harmonic restraint relay*

$X_C X_L$ allows only current of fundamental frequency to flow through the operating coil. The dc and harmonics, mostly second harmonics in case of magnetic inrush current, are diverted into the restraining coil. The relay is adjusted so as not to operate when the second harmonic (restraining) exceeds 15% of the fundamental current (operating). The minimum operating time is about 2 cycles.

The dc offset and harmonics are also present in the fault current, particularly if CT saturates. The harmonic restraint relay will fail to operate on the occurrence of an internal fault which contains considerable harmonics due to an arc or saturation of the CT. To overcome this difficulty, an instantaneous overcurrent relay (the high set unit) is also incorporated in the harmonic restraint scheme. This relay is set above the maximum inrush current. It will operate on heavy internal faults in less than one cycle.

In an alternative scheme, known as harmonic blocking scheme, a separate blocking relay whose contacts are in series with those of a biased differential relay, is employed. The blocking relay is set to operate when the second harmonic is less than 15% of the fundamental.

10.2.5 Buchholz Relay

It is a gas actuated relay. It is used to detect incipient faults which are initially minor faults but may cause major faults in due course of time. The Buchholz relay is used to supplement biased differential protection of the transformer because the Buchholz relay cannot necessarily detect short circuits within the transformer or at the terminals.

When a fault develops slowly, it produces heat, thereby decomposing solid or liquid insulating material in the transform. The decomposition of the insulating material produces inflammable gases. The operation of the Buchholz relay gives an alarm when a specified amount of gas is formed. The analysis of gas collected in the relay chamber indicates the type of the incipient fault. The presence of: (a) C_2H_2 and H_2 shows arcing in oil between constructional parts; (b) C_2H_2, CH_4 and H_2 shows arcing with some deterioration of phenolic insulation, e.g. fault in tap changer; (c) CH_4, C_2H_4 and H_2 indicates hot spot in core joints; (d) C_2H_4, C_3H_6, H_2 and CO_2 shows a hot spot in the winding.

There is a chamber to accommodate Buchholz relay, in between the transformer tank and the conservator as shown in Fig. 10.3 (a). A simple diagram to explain the operating principle of Buchholz relay is shown in Fig. 10.3 (b). When gas accumulates, the oil level falls down and thus the float also comes down. It causes an alarm to sound and alert the operator. For reliable operation, a mercury switch is attached with the float. Some manufacturers use open-topped bucket in place of a bob. When the oil level falls because of gas accumulation, the bucket is filled up with oil. Thus, the force available to operate the contacts is greater than with hollow floats. The accumulated gas can be drawn off through the petcock via a pipe for analysis to know the type of fault. If there is a severe fault, large volumes of gases are produced which cause the lower float to operate. It finally trips the circuit breakers of the transformer.

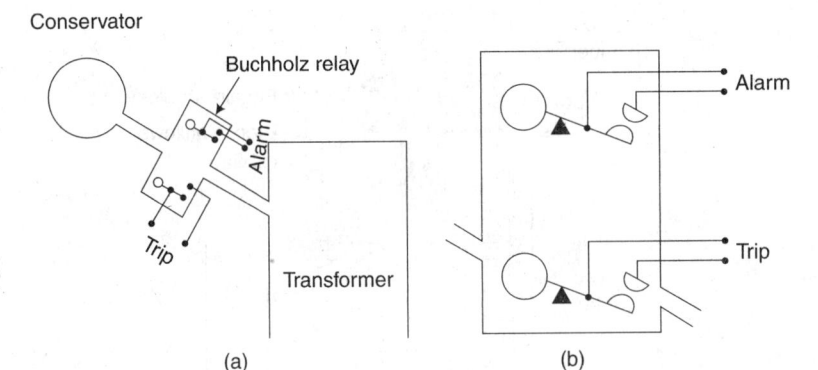

Fig. 10.3 *(a) Transformer tank, Buchholz relay and conservator (b) Buchholz relay*

The buchholz relay is a slow acting device, the minimum operating time is 0.1 s, the average time 0.2 s. Too sensitive settings of the mercury contacts are not desirable because they are subjected to false operation on shock and vibration caused by conditions like earthquakes, mechanical shock to the pipe, tap changer operation and heavy external faults. This can be reduced by improved design of the mercury contact tubes.

10.2.6 Oil Pressure Relief Devices

An oil pressure relief device is fitted at the top of the transformer tank. In its simplest form, it is a frangible disc located at the end of an oil relief pipe protruding from the top of the transformer tank. In case of a serious fault, a surge in the oil is developed, which bursts the disc, thereby allowing the oil to discharge rapidly. This avoids the explosive rupture of the tank and the risk of fire.

The drawback of the frangible disc is that the oil which remains in the tank after rupture is left exposed to the atmosphere. This drawback can be overcome by employing a more effective device: a spring controlled pressure relief valve. It operates when the pressure exceeds 10 psi but closes automatically when the pressure falls below the critical level. The discharged oil can be ducted to a catchment pit where random discharge of oil is to be avoided. The device is commonly employed for large power transformers of the rating 2 MVA and above but it can also be used for distribution transformers of 200 kVA and above.

10.2.7 Rate of Rise of Pressure Relay

This device is capable of detecting a rapid rise of pressure, rather than absolute pressure. Its operation is quicker than the pressure relief valve.

It is employed in transformers which are provided with gas cushions instead of conservators. Figure 10.4 shows a modern sudden pressure relay which contains a metallic bellows full of silicone oil. The bellows is placed in the transformer oil. The relay is placed at the bottom of the tank where maintenance jobs can be performed conveniently. It operates on the principle of rate or increase of pressure. It is usually designed to trip the transformer.

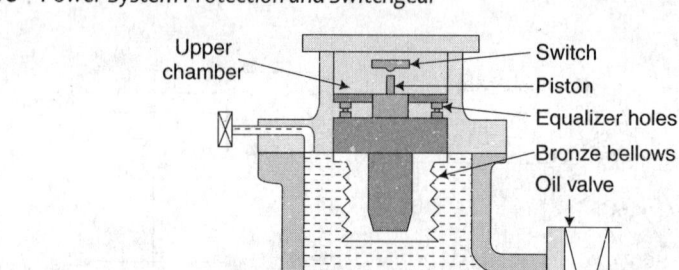

Fig. 10.4 Sudden pressure relay

10.2.8 Overcurrent Relays

Overcurrent relays are used for the protection of transformers of rating 100 kVA and below 5 MVA. An earth fault tripping element is also provided in addition to the overcurrent feature. Such relays are used as primary protection for transformers which are not provided with differential protection. Overcurrent relays are also used as back-up protection where differential protection is used as primary protection.

For small transformers, overcurrent relays are used for both overload and fault protection. An extremely inverse relay is desirable for overload and light faults, with instantaneous overcurrent relay for heavy faults. A very inverse residual current relay with instantaneous relay is suitable for ground faults.

10.2.9 Earth Fault Relays

A simple overcurrent and earth fault relay does not provide good protection for a star connected winding, particularly when the neutral point is earthed through an impedance. Restricted earth fault protection, as shown in Fig. 10.5 provides better protection. This scheme is used for the winding of the transformer connected in star where the neutral point is either solidly earthed or earthed through an impedance. The relay used is of high impedance type to make the scheme stable for external faults.

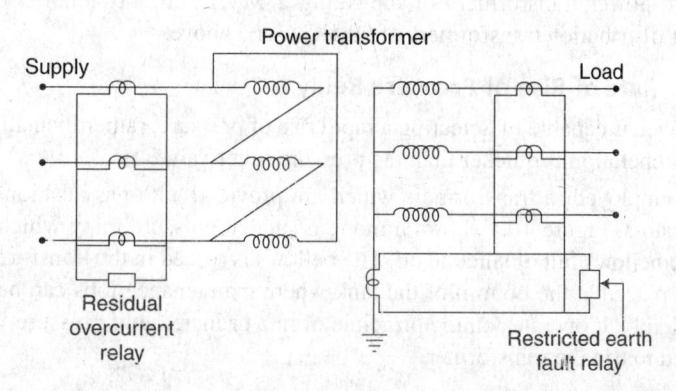

Fig. 10.5 Earth fault protection of a power transformer

For delta connection or ungrounded star winding of the transformer, residual overcurrent relay as shown in Fig. 10.5 is employed. The relay operates only for a ground fault in the transformer.

The differential protection of the transformer is supplemented by restricted earth fault protection in case of a transformer with its neutral grounded through resistance. For such a case only about 40% of the winding is protected with a differential relay pick-up setting as low as 20% of the CT rating.

10.2.10 Overfluxing Protection

The magnetic flux increases when voltage increases. This results in increased iron loss and magnetising current. The core and core bolts get heated and the lamination insulation is affected. Protection against overfluxing is required where overfluxing due to sustained overvoltage can occur. The reduction in frequency also increases the flux density and consequently, it has similar effects as those due to overvoltage. The expression of flux in a transformer is given by

$$\phi = K \frac{E}{f}$$

where, ϕ = flux, f = frequency, E = applied voltage and K = constant.

Therefore, to control flux, the ratio E/f is controlled. When E/f exceeds unity, it has to be detected. Electronic circuits with suitable relays are available to measure the E/f ratio. Usually 10% of overfluxing can be allowed without damage. If E/f exceeds 1.1, overfluxing protections operates. Overfluxing does not requires high speed tripping and hence instantaneous operation is undesirable when momentary disturbances occur. But the transformer should be isolated in one or two minutes at the most if overfluxing persists.

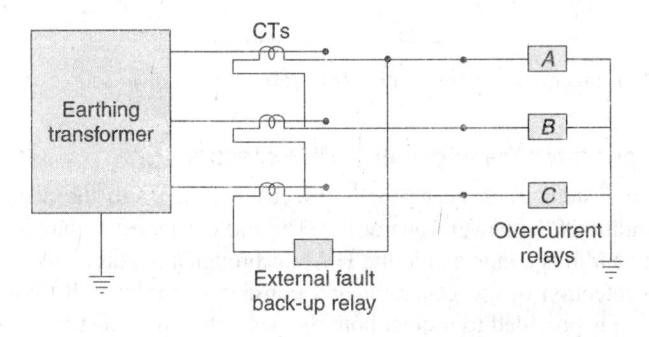

10.2.11 Protection of Earthing Transformer

The function of an earthing transformer is to provide a grounding point for the power system where machines have delta connection. An earthing transformer is connected either in star-delta or zig-zag fashion. When a fault occurs only zero sequence current flows from the earthing transformer to the grounding point. Positive or negative sequence currents can flow only towards the earthing transformer and not away from it. An earthing transformer can be protected by IDMT overcurrent relays fed by delta connected CTs, as shown in Fig. 10.6.

Fig. 10.6 *Protection of earthing transformer*

The CTs are connected in delta and zero sequence currents circulate in it. An overcurrent relay with time delay is inserted in this delta. The time setting of this relay is selected to coordinate with the time setting of the earth fault relays. This relay is used as a back-up relay for external faults.

10.2.12 Protection of Three-Winding Transformer

In a three-winding transformer, one of the three windings is connected to the source of supply. The other two windings feed loads at different voltages. One line diagram is shown in Fig. 10.7 for the protection of a three-winding transformer.

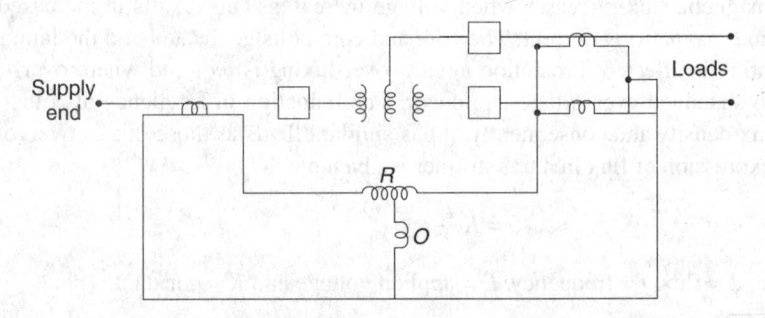

Fig. 10.7 *Protection of three-winding transformer with power source at one end*

When a three-winding transformer is connected to the source of supply at both primary and secondary side, the distribution of current cannot readily be predicted and there is a possibility of current circulation between two sets of paralleled CTs without producing any bias. Figure 10.8 shows protective scheme for such a situation. In this case, the restraint depends on the scalar sum of the currents in the various windings.

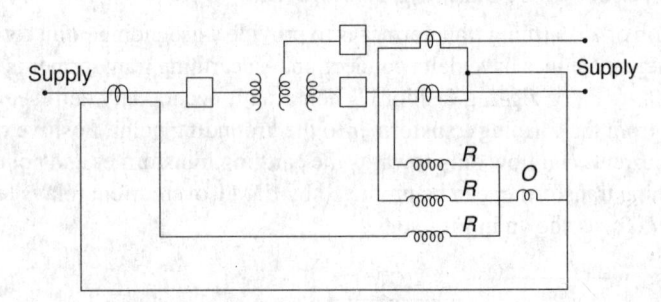

Fig. 10.8 *Protection of three-winding transformer with power source at both ends*

10.2.13 Generator-Transformer Unit Protection

In a modern system, each generator is directly connected to the delta connected primary winding of the power transformer. The star connected secondary winding is HV winding and it is connected to the HV bus through a circuit breaker. In addition to normal protection of the generator and transformer, an overall biased differential protection is provided to protect both the generator and transformer as one unit.

Figure 10.9 shows an overall differential protection. Usually harmonic restraint is not provided because the transformer is only connected to the busbar at full voltage. However, there is a possibility of a small inrush current when a fault near the busbar is cleared, suddenly restoring the voltage.

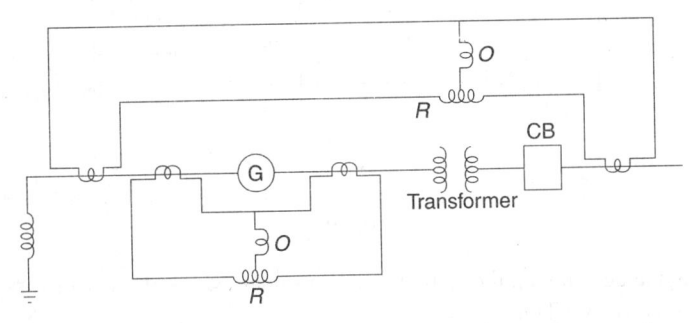

Fig. 10.9 *Differential protection of generator transformer unit*

10.2.14 Miscellaneous

Tank-earth Protection

If the transformer tank is nominally insulated from earth (an insulation resistance of 10 ohms being sufficient), the primary of a CT is connected between the tank and earth. A relay is connected to the secondary of the CT. This protection is similar to the frame earth scheme of protection for busbar discussed in the next section. This is also called Harward protection.

Neutral Displacement

In case of unearthed transformer, an earth fault elsewhere in the system may result in the displacement of the neutral. A neutral displacement detection scheme is shown in Fig. 10.10. The secondary of the potential transformer is connected in open delta. Its output which is applied to the relay is proportional to the zero sequence voltage of the line, i.e. any displacement of the neutral point. For details, see Ref. 5.

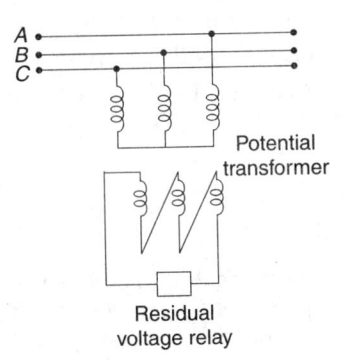

Fig. 10.10 *Neutral displacement detection*

Example 10.1 A three-phase, 11 kV/132 kV, Δ-Y connected power transformer is protected by differential protection. The CT_s on the LV side have a current ratio of 500/5. What must be the current ratio of the CT_s on the HV side and how should they be connected.

Solution: In order that circulating currents in the relay are in phase opposition, the CT_s on the delta connected LV side of the transformer should be connected in star and the CT_s on the star connected HV side of the transformer should be connected in delta. Connections of CT_s on LV and HV sides are shown in Fig. 10.11

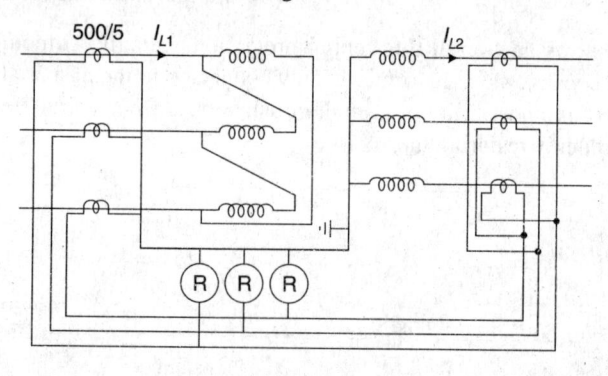

Fig. 10.11

Let the line currents on the primary and secondary sides of the transformer be I_{L1} and I_{L2} respectively. Then,

$$\sqrt{3} \times 11 \times I_{L1} = \sqrt{3} \times 132 \times I_{L2}$$

or

$$I_{L2} = \frac{11}{132} I_{L1}$$

For I_{L1} of 500 A, $I_{L2} = \frac{11}{132} \times 500 = 41.66$ A

Since the CT_s on the LV side are connected in star, the current through the secondary of the CT and the pilot wire will be 5 A. The CT_s on the HV side being delta connected will have a current of $5/\sqrt{3}$ A in the secondary.

Hence CT ratio on HV side

$$= 41.66/5/\sqrt{3}$$
$$= \sqrt{3} \times 41.66/5$$
$$= 72.15/5$$

Example 10.2 A three-phase, 11 kV/33 kV, Y-Δ connected power transformer is protected by differential pratection. The CT_s on the LV side have a current ratio of 400/5. What must be the ratio of CT_s on the HV side. How the CT_s on both the sides of the transformer are connected.

Solution: The connections of the CT_s on both the sides of the transformer are shown in Fig. 10.12.

Let the line currents on the primary and secondary sides of the transformer be I_{L1} and I_{L2} respectively.

Then,

$$\sqrt{3} \times 11 \times I_{L1} = \sqrt{3} \times 33 \times I_{L2}$$

or

$$I_{L2} = \frac{11}{33} I_{L1}$$

For I_{L1} of 400 A, $I_{L2} = \frac{11}{33} \times 400 = 133.3$ A

Fig. 10.12

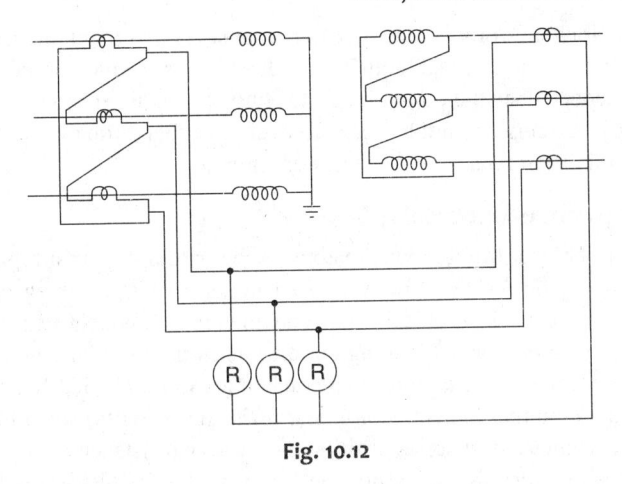

The current through secondary of the CT on the primary side of the transformer will be 5A. Since the CT_s on the primary side are connected in delta, the current through its pilot wire will be $5\sqrt{3}$ A. The CT_s on the secondary side being star connected will have a current of $5\sqrt{3}$ in the secondary.

Hence CT ratio on HV side

$$= 133.3/5\sqrt{3}$$
$$= 76.7/5$$

10.3 BUSZONE PROTECTION

10.3.1 Differential Current Protection

Figure 10.13 shows a scheme of differential current protection of a buszone. The operating principle is based on Kirchhoff's law. The algebraic sum of all the currents entering and leaving the busbar zone must be zero, unless there is a fault therein. The relay is connected to trip all the circuit breakers. In case of a bus fault the algebraic sum of currents will not be zero and relay will operate.

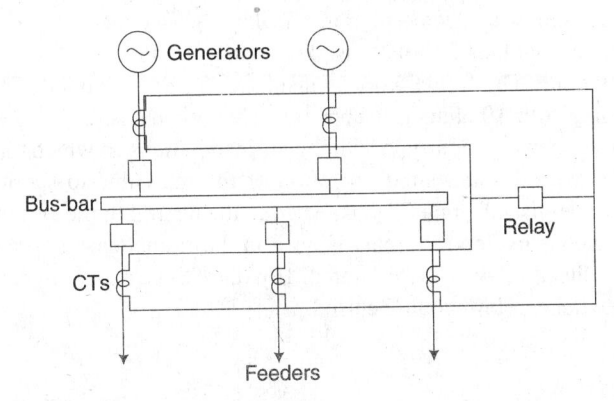

Fig. 10.13 *Differential current protection of bus-zone*

The main drawback of this type of differential scheme is that there may be a false operation in case of an external fault. This is due to the saturation of one of the CT of the faulted feeder. When the CT saturates, the output is reduced and the sum of all the CT secondary currents will not be zero. To overcome this difficulty, high impedance relay or biased differential scheme can be employed.

10.3.2 High Impedance Relay Scheme

Figure 10.14 shows a differential protective scheme employing a high impedance relay. A sensitive dc polarised relay is used in series with a tuning circuit which makes the relay responsive only to the fundamental component of the differential (spill) current of the CTs. The tuning circuit makes the relay insensitive to dc and harmonics, thereby making it more stable on heavy external faults. To prevent excessive voltages on internal faults, a non-linear resistance (thyrite) and a high set over current relay, connected in series with the non-linear resistance are employed. The high set relay provides fast operation on heavy faults. Its pick-up is kept high to prevent operation on external faults.

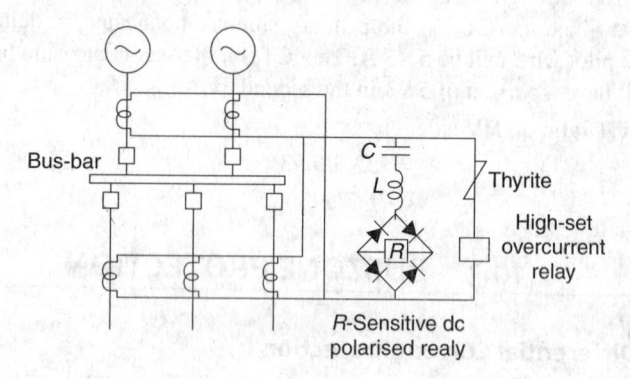

Fig. 10.14 *Differential protection using high impedance relay*

10.4 FRAME LEAKAGE PROTECTION

Figure 10.15 shows a scheme of frame leakage protecting. This is more favoured for indoor than outdoor installations. This is applicable to metal clad type switch gear installations. The frame work is insulated form the ground. The insulation is light, anything over 10 ohms is acceptable. This scheme is most effective in case of isolated-phase construction type switchgear installations in which all faults involve ground. To avoid the undesired operation of the relay due to spurious currents, a check relay energised from a CT connected in the neutral of the system is employed. An instantaneous overcurrent relay is used in the frame leakage protection scheme if a neutral check relay is incorporated. If neutral check relay is not employed, an inverse time delay relay should be used.

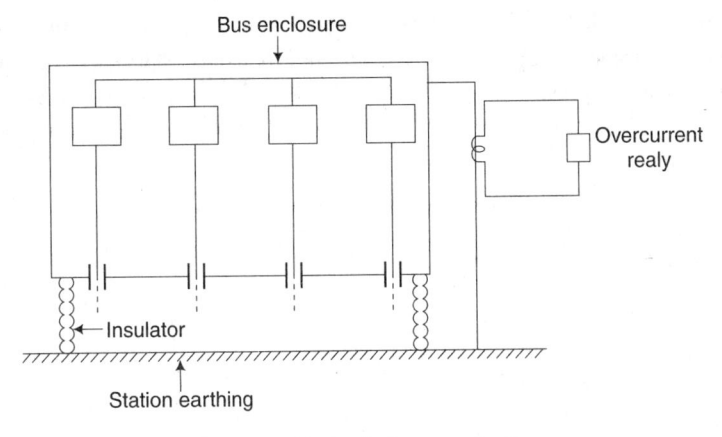

Fig. 10.15 *Frame leakage protection*

EXERCISES

1. Discuss the various types of faults encountered in transformers.
2. What type of protective scheme is employed for the protection of a large power transformer against short-circuits? With neat sketches discuss its working principle.
3. Explain the connection of CT secondaries for differential protection of star-delta connected power transformer.
4. Explain the phenomenon of magnetising inrush. Which harmonie is the most dominant in magnetising inrush current? What are the factors on which the magnitude of the magnetising inrush current depends?
5. What is magnetising inrush current? What measures are taken to distinguish between the fault current and magnetising inrush current? Discuss the protective scheme which protects the transformer against faults but does not operate in case of magnetising inrush current.
6. What is Buchholz relay? Which equipment is protected by it? For what types of faults is it employed? Discuss its working principle.
7. What protective devices other than the differential protection are used for the protection of a large transformer? Briefly describe them.
8. A three-phase, 400 V/11 kV Y-Δ connected transformer is protected by differential protection scheme. The CT_s on the LV side have a current ratio of 1000/5. What must be the ratio of CT_s on the HV side and how should they be connected? (**Ans.** 7/5)
9. A three-phase, 132 kV/33 kV star-delta connected power transformer is protected by differential protection scheme. Determine the ratio of CT_s on the HV side of the transformer, if that on the LV side is 300/5. How are the CT secondaries connected? (130/5)
10. A three-phase, 50 MVA, 132 kV/66 kV Y-Δ transformer is protected by differential protection. Suggest suitable CT ratios and show the connection of the CT_s on either side of the transformer.

11. With a neat sketch, discuss the differential scheme for buszone protection.
12. What is frame leakage protection? Discuss its working principle and field of application.

11 Numerical Protection

11.1 INTRODUCTION

The modern power systems which have grown both in size and complexity require fast, accurate and reliable protective schemes to protect major equipment and to maintain the system stability. The conventional protective relays are either of electromechanical or static type. The electromechanical relays have several drawbacks such as high burden on instrument transformers, high operating time, contact problems, etc. Though static relays have inherent advantages of compactness, lower burden, less maintenance and high speed; but they suffer from a number of disadvantages, e.g. inflexibility, inadaptability to changing system conditions and complexity. Therefore, increasing interest is being shown in the use of on-line digital computers for protection. The concept of numerical protection employing computers which shows much promise in providing improved performance has evolved during the past three decades. In the beginning the numerical protection (previously known as digital protection) philosophy was to use a large computer system for the total protection of the power system. This protection system proved to be very costly and required a large space. If a computer is required to perform other control functions in addition to protection, it can prove to be economical. With tremendous advancement in technology and present downward trend in the cost of computer hardware, powerful and economical computers are available today. New generations of computers tend to make computer relaying a viable alternative to the traditional relaying system. There are many advantages of using computers for power system protection, compared to currently used conventional relays. The computer offers reduced protection costs, better system performance, and more flexibility than conventional approaches. An especially important incentive for advocating the use of computer relaying is that computers are always active, thus permitting constant monitoring and self checking of numerical relays. As a result, the reliability of the numerical protection system and hence that of the power system itself, are greatly enhanced.

With the tremendous developments in VLSI and computer hardware technology, microprocessors that appeared in the seventies have evolved and made remarkable progress in recent years. The latest fascinating innovation in the field of computer technology is the development of microprocessors, mierocontrollers, digital signal processors (DSPs) and Field Programmable Gate Arrays (FPGAs) which are making in-roads in every activity of the mankind. With the development of fast economical powerful and sophisticated microprocessors, microcontrollers, DSPs and FPGAs, there is a growing trend to develop numerical relays based on these devices.

The conventional relays of electromechanical and static types had no significant drawbacks in their protection functions, but the additional features offered by microprocessor technologies encouraged the evolution of numerical relays that introduced many changes in power industry. Economics and additionally, functionality were probably the main factors that forced the power industry to accept and cope with the changes brought by microprocessor microcontroller based numerical relays.

The main component of the numerical protection scheme is the numerical relay. Numerical relays are the latest development in the area of protection. These relays acquire the sequential samples of the ac quantities in numeric (digital) data form through the data acquisition system, and process the data numerically using an algorithm to calculate the fault discriminants and make trip decisions. Numerical relays have been developed because of tremendous advancement in VLSI and computer hardware technology. They are based on numerical (digital) devices, e.g. microprocessors, microcontrollers, digital signal processors (DSPs) etc. At present microprocessor/microcontroller based numerical relays are widely used. The block schematic diagram of a numerical relay is shown in Fig. 11.1. These relays use different relaying algorithms to process the acquired information. Microprocessor/micro controller based relays are called numerical relays specifically if they calculate the algorithm numerically. The term 'digital relay' was originally used to designate a previous generation relay with analog measurement circuits and digital coincidence time measurement (angle measurement), using microprocessors. Nowadays the term *numerical relay* is widely used in place of "digital relay". Sometimes both terms are used in parallel. Similarly, the term *numerical protection* is widely used in place of *digital protection*. Sometimes both these terms are also used in parallel. Thus other nomenclatures for numerical relay are digital relay, computer-based relay or microprocessor-based relay.

The present downward trend in the cost of very large scale integrated (VLSI) circuits has encouraged wide application of numerical relays for the protection of modern complex power networks. Multifunction numerical relays were introduced in the market in the late eighties. These devices reduced the product and installation costs drastically. This trend has continued until now and has converted microprocessor / microcontroller based numerical relays to powerful tools in the modern substations. Intelligent numerical relays using artificial intelligence techniques such as artificial neural networks (ANNs) and fuzzy logic systems are presently under active research and development stage.

The main features of numerical relays are their economy, compactness, flexibility, reliability, self monitoring and self-checking capability, multiple functions, low burden on instrument transformers and improved performance over conventional relays of electromechanical and static types.

While numerical relays have several advantages, they also have a few shortcomings which are not directly offset by specific benefits. Major shortcomings of numerical relays are short life cycle, susceptibility to transients, and setting and testing complexities.

The basic protection principles have remained essentially unchanged throughout the evolution of the numerical relays, the adoption of this technology has provided many benefits, and a few shortcomings, compared to the previous technologies.

Thus, the evolution of numerical protection systems has not been without its challenges. However, the power industry has come to the conclusion that the benefits far outweigh the shortcomings. Therefore, the acceptance of numerical protection systems has gradually increased during the previous three decades.

Modern power system protection devices are built with integrated functions. Multi-functions like protection, control, monitoring and measuring are available today in numerical power system protection devices. Also, the communication capability of these devices facilitates remote control, monitoring and data transfer. Some numerical protection devices also offer adaptable characteristics, which dynamically change the protection characteristic under different system conditions by monitoring the input parameters.

11.2 NUMERICAL RELAY

The numerical relay is the latest development in the area of power system protection and differs from conventional ones both in design and methods of operation. It has been developed because of tremendous advancement in VLSI and computer hardware technology. It is based on numerical (digital) devices e.g. microprocessors, microcontrollers, digital signal processors (DSPs) etc. This relay acquires sequential samples of the ac quantities in numeric (digital) data form through the data acquisition system (DAS), and processes the data numerically using a relaying algorithm to calculate the fault discriminants and make trip decisions. In a numerical relay, the analog current and voltage signals monitored through primary transducers (CTs and VTs) are conditioned, sampled at specified instants of time and converted to digital form for numerical manipulation, analysis, display and recording. This process provides a flexible and very reliable relaying function, thereby enabling the same basic hardware units to be used for almost any kind of relaying scheme. Thus, a numerical relay has an additional entity, the software, which runs in the background and makes the relay functional. Hardware is more or less the same in almost all the numerical relays. The software used in a numerical relay depends upon the processor used and the type of the relay. Hence, with the advent of the numerical relay, the emphasis has shifted from hardware to software. The evolution of the modern numerical relay has thus taken place from a torque balancing device to a programmable information processor.

The block diagram of a typical numerical relay is shown in Fig. 11.1. This relay samples voltages and currents, which, at the power system level, are in the range of hundreds of kilo volts and kilo amperes respectively. The levels of these signals are reduced by voltage and current transformers (transducers). The outputs of the transducers are applied to the signal conditioner (also called 'analog input subsystem'). Signal conditioner is one of the important components of the data acquisition system (DAS). It brings real-world signals into digitizer. In this case, the signal conditioner electrically isolates the relay from the power system, reduces the level of the input voltages, converts currents to equivalent voltages and removes high frequency components from the signals using analog filters. The relay is isolated from the power system by using auxiliary transformers which receive analog signals and reduce their levels to make them suitable for use in the relays. Since the A/D converters accept voltage signals

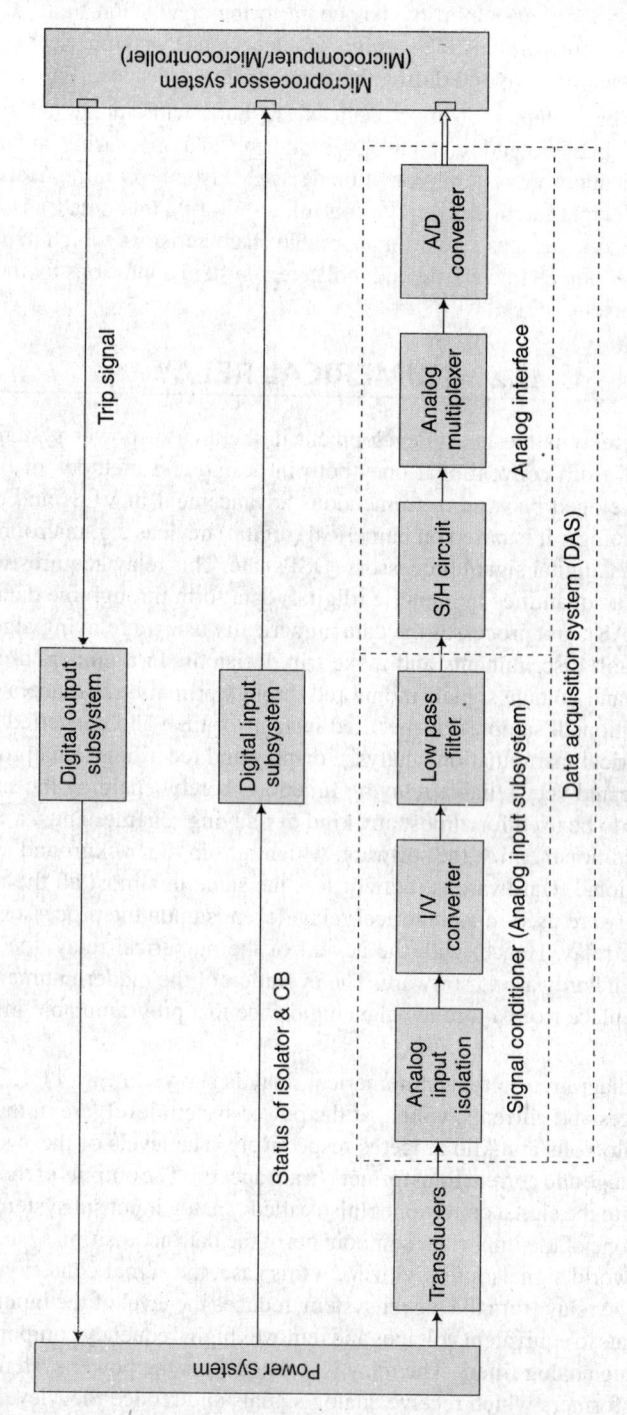

Fig. 11.1 Block diagram of a typical numerical relay

only, the current signals are converted into proportional voltage signals by using I/V converters or by passing through precision shunt resistors. Anti-aliasing filters (which are low-pass filters) are used to prevent aliasing from affecting relaying functions.

The outputs of the signal conditioner (the analog input subsystem) are applied to the analog interface, which includes sample and hold (S/H) circuits, analog multiplexers and analog-to-digital (A/D) converters. These components sample the reduced level signals and convert their analog levels to equivalent numbers that are stored in memory. The status of isolators and circuit breakers in the power system is provided to the relay via the digital input subsystem and are read into the microcomputer/microcontroller memory.

After quantization by the A/D converter, analog electrical signals are represented by discrete values of the samples taken at specified instants of time. The signals in the form of discrete numbers are processed by a relaying algorithm using numerical methods. A relaying algorithm which processes the acquired information is a part of the software. The algorithm uses signal-processing technique to estimate the real and imaginary components of fundamental frequency voltage and current phasors. In some cases, the frequency of the system is also measured. These measurements are used to calculate other quantities, such as impedances. The computed quantities are compared with pre-specified thresholds (settings) to decide whether the power system is experiencing a fault or not. If there is fault in the power system, the relay sends a trip command to circuit breakers for isolating the faulted zone of the power system. The trip output is transmitted to the power system through the digital output subsystem.

Numerical relays nave revolutionized all aspects of protection and control. Modern numerical relays offer several advantages in terms of protection, reliability, trouble shooting and fault information and are capable of providing complete protection with added functions like control and monitoring. Most numerical relays are available today with 'self-check' feature. These relays are capable of periodically checking their hardware and software, and in case a problem is noticed, the relays give an alarm for corrective action. It is extremely easy to service a numerical relay as in most of the cases it requires only certain cards to be replaced. If sufficient spares are maintained at the various locations, the down time of a relay can be substantially reduced.

Systems which use numerical relays have the potential for improved system operation which arises from the increased quantity of information available in the relay. This information can be accessed quickly and effectively by means of remote communication links which result in improved post fault analysis of faults. The data so collected can be used in the planning of maintenance programs more realistically and for interpreting trends and changes of patterns in the data over a period of time. In general, all these will result in a significant improvement in the operation of the power system.

11.2.1 Advantages of Numerical Relays

The advantages of numerical relays are discussed below.

(i) Compactness and Reliability Numerical relays are compact in size and reliable in operation.

(ii) Flexibility   Numerical relays offer flexibility because of programmability. Some general purpose hardware can be used to perform a variety of protection functions with the change of stored program only. Drastic addition and/or alteration of the protection logic hardly require any hardware replacement.

(iii) Adaptive Capability   Numerical relays can adapt themselves to changing system conditions by monitoring the operating quantities from the digital inputs of the relay. The behaviour of the processor can be made to change automatically depending on the external conditions which change with time. The basis of this change can be either local information available to the processor, or an external source such as data link or the central computer system.

(iv) Multiple Functions   Numerical relays provide many functions that were not available in electromechanical or static designs. These features include multiple setting groups, programmable logic, adaptive logic, self-monitoring, self-testing, sequence-of-events recording, oscillography, and ability to communicate with other relays and control computers. While these features make the relays very powerful, they also introduce factors, such as complexity, that were not associated with earlier technologies.

(v) Detailed Logical and Mathematical Capabilities   The programmer in numerical relaying scheme is free to provide almost any characteristic within the limits of his understanding. Specific protection problems can be broken down into fine details and each handled separately. In addition measurement problem can be stated as mathematical equations and directly implemented.

(vi) Economic Benefits   The cost per function of numerical relays is lower compared to the cost of their electromechanical and static counterparts. The reduction in cost is due to the lower cost of components, production equipment and production techniques. Numerical relays provide greater functionality at a reduced price.

(vii) Less Panel Space   Numerical protection systems require significantly less panel space than the space required by electromechanical and static systems that provide similar functions. The reduction in size is a result of the high level of integration of the hardware and the ability of using one physical device for performing multiple protection functions, such as, overcurrent and distance relaying.

(viii) Low Burden on Transducers (Instrument Transformers)   Numerical relays impose significantly less burden on transducers (CTs and VTs). When relays of the previous generations were used, the ability to provide protective functions was limited by the burden that could be placed on transducers. This is not the case when numerical relays are used. In addition, numerical relays can be programmed to detect saturation of instrument transformers for minimizing incorrect operations.

(ix) Self-monitoring and Self-testing   Numerical relays have the ability to perform self-monitoring and self-testing functions. These features reduce the need for routine maintenance because the relays automatically take themselves out of service and alert the operator of the problem when they detect functional abnormalities.

(x) Sequence of Events and Oscillography   Reporting features, including sequence of events recording and oscillography are a natural by product of numerical

protection systems. These features make it possible to analyze the performance of the relays as well as system disturbances in a better way at minimum additional costs.

(xi) Communication Facility Communication makes the relay more intelligent and the operating personnel can set the relay and also download the fault information. It is also possible to upload the revised software to the relay at site without sending the relay back to the manufacturer. The data processed by the relay can be accessed through the relay communication port. A separate high-end communication system is also available for numerical relay data communication and control.

(xii) Metering Facility Numerical relays are provided with metering functions and separate panel mounted meters can be eliminated. Some relays can also provide energy meter function. With the metering facility in the relay, the operating personnel can view different parameters online.

(xiii) Memory Action Since the pre-fault voltage samples are within the computer memory, voltage collapse due to short-circuit at relaying point does not introduce any difficulty when the voltages are used in input quantities.

(xiv) Standardization Numerical protection systems are constructed with a standard hardware feature, and any special feature can be met with modification of the program only. This standardization provides a considerable simplification in production, testing and maintenance.

11.2.2 Disadvantages of Numerical Relays

While numerical relays have several advantages, they also suffer form a few disadvantages. Major disadvantages are follows.

(i) Short Life Cycle All microprocessor-based systems, including numerical protection systems, have short life cycles. Since each generation of microprocessor-based systems increases the functionality as compared with the previous generation, the pace of change makes the numerical protection systems obsolete in shorter times. Because of this, it becomes difficult for the users to maintain expertise in using the latest designs of the equipment. Another variation of this disadvantage is in the from of changes in the software used on the existing hardware platforms.

(ii) Susceptibility to Transients Microprocessor-based numerical protection systems are more susceptible to incorrect operations due to transients because of the nature of the technology compared to the systems built with the electromechanical technology.

(iii) Setting and Testing Complexities Numerical relays, which are designed to replace the functions of electromechanical or static relays, offer programmable functions that increase the application flexibility compared with the fixed function relays. The multifunction numerical relays, therefore, have a number of settings. There may be problems in management of the increased number of settings and in conduction of functional tests. Setting management software is generally available to create, transfer, and track the relay settings. Numerical relays are generally tested by using special testing techniques, specifically the ability to enable and disable selected functions. This increases the possibility that the desired settings may not be invoked after testing. In order to ensure that correct setting and logic are activated after the tests are completed, proper procedures must be followed.

11.2.3 Multifunction Numerical Relays

Multifunction numerical relays consolidate protection functions and reduce the product and installation costs drastically. These devices reduce the cost and complexity of the overall application. In these relays, self-monitoring is much more than the function of testing and monitoring the data acquisition system, hardware, memory and software self-monitoring is also a test to ensure that everything in and about the relay installation is correct. They can provide a diagnosis of its own internal hardware and software and can also monitor the connected system. They can also monitor the health of the instrument transformers (CTs and VTs). In numerical relays, the triggering of the fault record, the duration of the recording, and the type of recording can be specified by the user. In addition to integrating the overall protection control scheme into the relay, there is possibility to extend the benefits of self diagnostics beyond the internal subsystems of the relay. Since the information regarding the health status of the control scheme is provided to the relay using binary inputs, it is possible to use numerical relay programmable logic control functions to detect and respond to external control scheme malfunctions.

11.3 DATA ACQUISITION SYSTEM (DAS)

Data acquisition is the process of sampling of real-world analog signals and conversion of the resulting samples into digital numeric values that can be manipulated by a computer. The system which performs data acquisition is called data acquisition system (DAS). Data acquisition typically involves the conversion of analog signals into digital values for processing. For numerical relaying, the data acquisition system acquires sequential samples of the analog ac quantities (voltages and currents) and converts them into digital numeric values for processing. The voltages and currents at the power system level are in the range of hundreds of kilovolts and kiloamperes respectively. The levels of voltage and current signals are reduced by transducers (voltage current transformers). The output of the transducers are applied to the DAS. DAS also employ various signal conditioning techniques to adequately modify various different electrical signals into voltages that can then be digitized using an analog-to-digital converter (ADC).

The main components of DAS are the signal conditioner (analog input subsystem) and the analog interface.

The signal conditioner makes the signals from the transducers compatible with the analog interface, and the analog interface makes the signals compatible with the processor (e.g. microprocessor, microcontroller, DSP etc.)

11.3.1 Signal Conditioner (Analog Input Subsystem)

Signal conditioner (also called analog input subsystem) is necessary to make the signals from the transducers compatible with the analog interface. Signal conditioning circuitry converts analog input signals into a form that can be converted to digital values. Since the output signals of transducers are often incompatible with data acquisition hardware, the analog signals must be conditioned to make them compatible. Common ways to condition analog signals include the following:

  (i) Analog input isolation and scaling
  (ii) Current to voltage conversion
  (iii) Filtering

The relay is electrically isolated from the power system by using auxiliary transformers which receive analog signals from the transducers and reduce their levels to make them suitable for use in the relays, as shown in Fig. 11.2(a) and (b). Metal Oxide Varistor (MOV) which has a high resistance at low voltages and low resistance at high voltages due to highly nonlinear current-voltage characteristic is used to protect circuits against excessive transient voltages.

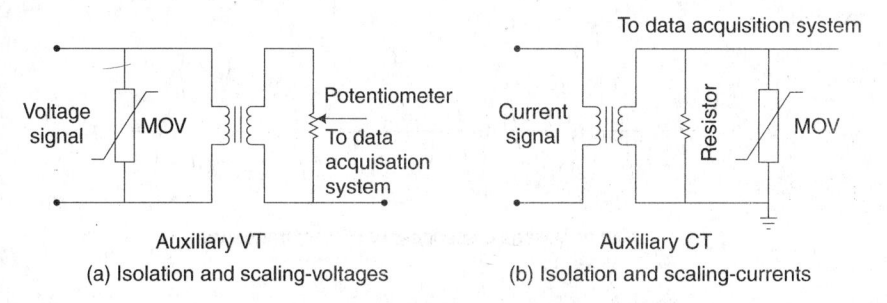

Fig. 11.2 *Auxiliary transformers for isolation and scaling*

Since the A/D converters can handle voltages only, the current are converted to proportional voltages by using current to voltage (I/V) converters or by passing the current through presision shunt resistors.

The phenomenon of appearance of a high frequency signal as a lower frequency signal that distorts the desired signal is called aliasing. To prevent aliasing from affecting the relaying functions, anti-aliasing filters (which are low-pass filters) are used along with analog input isolation block. The block diagram of a typical signal conditioner is shown in Fig. 11.3.

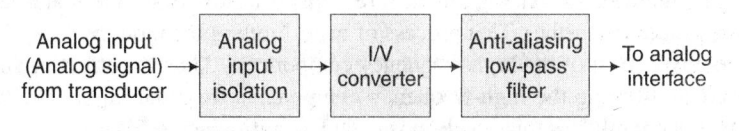

Fig. 11.3 *Block diagram of a typical signal conditioner*

11.3.2 Aliasing

Post-fault power system signals contain dc offset and harmonic components, in addition to the major fundamental frequency component. In order to convert analog signals to sequences of numbers, an appropriate sampling rate should be used, because high-frequency components which might be present in the signal, could be incorrectly interpreted as components of lower frequencies. The mechanism of a high-frequency component in an input signal manifesting itself, as a low-frequency signal is called 'aliasing'. It is the appearance of a high-frequency signal as a lower-frequency signal that distorts the desired signal.

For explaining the phenomenon of aliasing, let us consider a signal of 550 Hz (11th harmonic component) as the high-frequency component, as shown in Fig. 11.4(a). If this signal is sampled 500 times a second, the sampled values at different instants would be as shown in Fig. 11.4(b). The reconstruction of the sampled sequence, and its interpretation by an algorithm, indicates that the signal is of the 50 Hz frequency. This misrepresentation of the high frequency component as a low- frequency component is referred to as aliasing. Therefore, for obtaining a correct estimate of the component of a selected frequency, the sampling rate should be chosen in such a manner that components of higher frequencies do not appear to belong to the frequency of interest.

(a) An ac voltage component of 550-Hz frequency

(b) Samples of 550 Hz signal taken at 500 samples per second

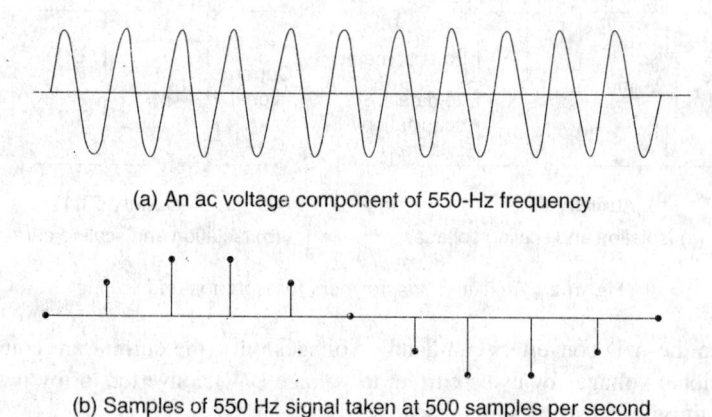

Fig. 11.4 *Sampling of 550 Hz signal at rate of 500 samples per second, i.e, at sampling frequency of 500 Hz*

Since it is not possible to select a sampling frequency (sampling rate) that would prevent the appearance of all high frequency components as components of frequency of interest, the analog signals are applied to low-pass filters and their outputs are processed further. This process of band-limiting the input by using low-pass filter removes most of the high-frequency components. Thus the effect of aliasing is removed by filtering the high-frequency components from the input. The low-pass filter that accomplishes this function is called an anti-aliasing filter.

11.3.3 Sampling

Sampling is the process of converting a continuous time signal, such as a current or voltage, to a discrete time signal. The selected sampling rate should be as high as is practical taking into account the capabilities of the A/D converter and the processor. Modern numerical relays use sampling rate that are as high as 96 samples per fundamental cycle.

In order that the samples represent the analog signal uniquely and contain enough information to recreate the original waveform, a continuous signal is sampled properly. With the unique collection of numerical values of samples, the original waveform can be easily recreated.

The sampling theorem states that in order to preserve the information contained in a signal, it must be sampled at a sampling frequency f_s of at least twice the largest frequency (f_m) present in the sampled information (i.e. $f_s \geq 2f_m$). The sampling theorem is frequently called the 'Shannon sampling theorem', or 'the Nyquist sampling theorem'. Therefore, in order to avoid aliasing error, frequencies above one-half the sampling frequency must be removed by using anti-aliasing filter (low-pass filter).

11.3.4 Analog Interface

The analog interface makes the signal compatible with the processor. The outputs of the signal conditioner are applied to the analog interface which includes sample and hold (S/H) circuits, analog multiplexers and analog to digital (A/D) converters. These components sample the reduced level signals and convert their analog levels to equivalent numbers that are stored in memory for processing. The block diagram of a typical analog interface is shown in Fig. 11.5.

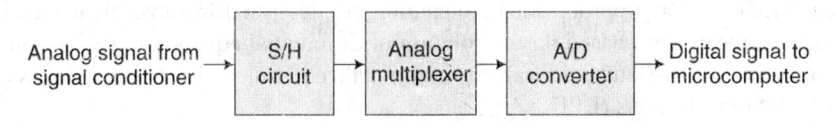

Analog signal from signal conditioner → S/H circuit → Analog multiplexer → A/D converter → Digital signal to microcomputer

Fig. 11.5 *Block diagram of a typical analog interface*

A sample and hold (S/H) circuit is used to acquire the samples of the time verying analog signal and keep the instantaneous sampled values constant during the conversion period of ADC. A S/H circuit has two modes of operation namely Sample mode and Hold mode. When the logic input is high it is in the Sample mode, and the output follows the input with unity gain. When the logic input is low it is the hold mode and the output of the S/H circuit retains the last value it had until the command switches for the sample mode. The S/H circuit is basically an operational amplifier which charges a capacitor during the sample mode and retains the value of the change of the capacitor during the hold mode.

An analog multiplexer has many input channels and only one output. It selects one out of the multiple inputs and transfers it to a common output. Any input channel can be selected by sending proper commands to the multiplexer through the microcomputer.

Analog to digital (A/D) converters take the instantaneous (sampled) values of the continuous time (analog) signal, convert them to equivalent numerical values and provide the numbers as binary outputs that represent the analog signal at the instants of sampling. Thus, after quantization by the A/D converter, analog electrical signals are represented by discrete values of the samples taken at specified instants of time. The signals in the from of discrete numbers are processed by a relaying algorithm using numerical methods. A relaying algorithm which processes the acquired information is a part of the software.

11.4 NUMERICAL RELAYING ALGORITHMS

In a numerical power system relaying scheme, the processor executes different functions, like data acquisition and processing under an elaborate program control. At the core of this program is a signal processing algorithm which processes the incoming digitized relaying data to detect the fault. Recently, strong interest in applying numerical techniques to protective relaying has been indicated by many publication in this field. Most of the work reported in journals concentrates on the development of numerical signal processing algorithms for relaying of EHV lines in general and distance relaying in particular. Numerous algorithms for numerical distance relaying of transmission lines have been derived during the past three decades and a comparative evaluation of different algorithms for line relaying has also been reported.

The aim of the most of these algorithms is to extract the fundamental frequency components from the complex post-fault voltage and current signals containing a transient dc offset component and harmonic frequency components in addition to the power frequency fundamental component. As the microprocessor requires computationally simple and fast algorithm in order to perform the relaying functions, only a few algorithms are suitable for microprocessor implementation. Algorithm which are suitable for microprocessor implantation are described in subsequent sections. These algorithm are based on the solution of differential equation, discrete Fourier transform, Walsh-Handmaid transform, rationalized Haar transform techniques and Block Pulse functions (BPF).

The exponentially decaying dc offset present in the relaying signals gives rise to fairly large errors in the phasor estimates unless the offset terms are removed prior to the execution of the algorithm. Data window is the time span covered by the sample set needed to execute the algorithm. Fourier, Walsh, Haar and BPF algorithms can either be for a full-cycle window or a half-cycle window. The full-cycle data window increases the operating time of the relays to more than one cycle. Since the distance relay's operation within a cycle after the fault inception is desirable, the half cycle data window is preferable. The half cycle window Fourier, Walsh, Haar and BPF algorithms in particular require that the dc offset be removed prior to processing, while the differntial equation algorithms ideally do not require its elimination. Therefore the removal of de offset component is also discussed in Section 11.12.

11.5 MANN–MORRISON TECHNIQUE

This is one of the first techniques proposed for numerical distance relaying. It suggests the use of sine and cosine functions and their derivatives as the orthogonal set of functions. The voltage and current waveforms are assumed to be pure sinusoids. Let the voltage at the relay location be represented by

$$v = V_m \sin(\omega t + \phi_V) \tag{11.1}$$

On differentiation,

$$v' = \frac{dv}{dt} = V_m w \cos(\omega t + \phi_V) \tag{11.2}$$

v' can be approximated by the difference between the two samples on either side of the central sample point. Thus

$$v' = \frac{1}{2T_S}[v_{k+1} - v_{k-1}] \quad (11.3)$$

where T_S is the sampling interval.

From Eqs. (11.1) and (11.2) the peak magnitude (modulus) and phase angle (argument) of the voltage signal in terms of v and v' at an arbitrary sampling instant k are given by

$$V_m = \left[v_k^2 + \left(\frac{v_k'}{\omega}\right)^2\right]^{1/2} \quad (11.4)$$

$$\phi_V = \tan^{-1}\left(\frac{\omega v_k}{v_k'}\right) \quad (11.5)$$

Similarly, if the current is represented by

$$i = I_m \sin(\omega t + \phi_i) \quad (11.6)$$

Then,

$$i' = I_m \omega \cos(\omega t + \phi_i) \quad (11.7)$$

or

$$i' = \frac{1}{2T_s}[i_{k+1} - i_{k-1}] \quad (11.8)$$

The peak magnitude (modulus) and phase angle (argument) of the current signal in terms of i and i' at an arbitrary sampling instant k are given by

$$I_m = \left[i_k^2 + \left(\frac{i_k'}{\omega}\right)^2\right]^{1/2} \quad (11.9)$$

$$\phi_i = \tan^{-1}\left(\frac{\omega i_k}{i_k'}\right)$$

The phase difference ϕ between voltage and current is given by

$$\phi = \phi_V - \phi_i = \tan^{-1}\left(\frac{\omega v_k}{v_k'}\right) - \tan^{-1}\left(\frac{\omega i_k}{i_k'}\right) \quad (11.10)$$

The line impedance from the relay location to the fault point is calculated by

$$Z = \frac{V_m < \phi_v}{I_m < \phi_i} = \left[\frac{v_k^2 + (v_k'/\omega)^2}{i_k^2 + (i_k'/\omega)^2}\right]^{1/2} < (\phi_v - \phi_i) \quad (11.11)$$

This algorithm has a fast response due to a three sample window but its drawback is that it does not recognize the possible existence of an exponentially decaying dc offset or of higher harmonics in the incoming signal. The effect of the dc offset can be reduced by using mimic impedance while the higher harmonics have to be filtered out by using a low-pass filter. The accuracy of this method is not good.

11.6 DIFFERENTIAL EQUATION TECHNIQUE

The differential equation algorithm is based on a model of the system rather than on a model of the signal. This algorithm for the computation of line parameters, i.e. R and X is based on the solution of the differential equation representing the transmission line model. The transmission line is modeled as a series R–L circuit resulting in the following differential equation.

$$V = Ri + L\frac{di}{dt} \tag{11.12}$$

This relationship holds for both steady-state and transient conditions. The solution for R and L is accomplished by numerical differentiation over two successive time periods. Then the solution of the resulting simultaneous linear equations are obtained.

The derivation of the numerical algorithm for the circuit parameters is a straight forward application of difference. Simultaneous samples of both voltage and current signals are taken at a fixed rate, at time t_0, t_1 and t_2. Thus, there are two sample periods, the first from t_0 to t_1 is labelled A; the second from t_1 to t_2 is labelled B. If V_0, V_1, V_2 are the three consecutive samples of voltage signal and i_0, i_1, i_2 the three consecutive samples of current signal, the average values during the periods A and B are as follows:

$$i_A = \frac{i_0 + i_1}{2} \qquad\qquad i_B = \frac{i_1 + i_2}{2}$$

$$V_A = \frac{V_0 + V_1}{2} \qquad\qquad V_B = \frac{V_1 + V_2}{2}$$

$$\frac{di_A}{dt} = \frac{i_1 - i_0}{t_1 - t_0} \qquad\qquad \frac{di_B}{dt} = \frac{i_2 - i_1}{t_2 - t_1}$$

Now for these two periods, the follow differential equations hold.

$$V_A = Ri_A + L\frac{di_A}{dt} \tag{11.13}$$

$$V_B = Ri_B + L\frac{di_B}{dt} \tag{11.14}$$

By substitution of the known values, there remain two algebraic equations with two unknown, R and L. The solution of the simultaneous Eqs. (11.13) and (11.14) gives the expressions for R and L. For the calculation algorithm, the following definitions are made.

$$CA = i_0 + i_1 \qquad\qquad CB = i_1 + i_2$$
$$VA = V_0 + V_1 \qquad\qquad VB = V_1 + V_2$$
$$DA = i_1 - i_0 \qquad\qquad DB = i_2 - i_1$$
$$DT = t_1 - t_0 = t_2 - t_1$$

Than
$$R = \frac{DA * VB - DB * VA}{DA * CB - DB * CA} \tag{11.15}$$

$$L = \frac{DT}{2}\frac{VA * CB - VB * CA}{DA * CB - DB * CA} \tag{11.16}$$

The line reactance $X (= \omega L)$ is proportional to the line inductance L.

Then
$$X = \omega\frac{DT}{2}\frac{VA * CB - VB * CA}{DA * CB - DB * CA} \tag{11.17}$$

Equations (11.15) and (11.17) are programmed to determine R and X from the sampled values of voltage and current.

As this algorithm requires a very short data window of three-samples only, it is very fast and is most suitable for microprocessor implementation. The program flowchart for the computation of R and X using this algorithm is shown in Fig. (11.6)

11.7 DISCRETE FOURIER TRANSFORM TECHNIQUE

In this technique, the algorithm for extracting the fundamental frequency components from the complex post-fault relaying signals is based on discrete Fourier transform. The discrete Fourier transform (DFT) of a data sequence is used to evaluate the Fourier coefficients. In this approach, the fundamental Fourier sine and cosine coefficients are obtained by correlating the incoming data samples with the stored samples of reference fundamental sine and cosine waves, respectively. The fundamental Fourier sine and cosine coefficients are respectively equal to the real and imaginary parts of the fundamental frequency component. Using the DFT, the real and imaginary components of the fundamental frequency voltage and current phasors are calculated. The real and reactive components (R and X) of the apparent impendence of the line is then calculated from these four quantities.

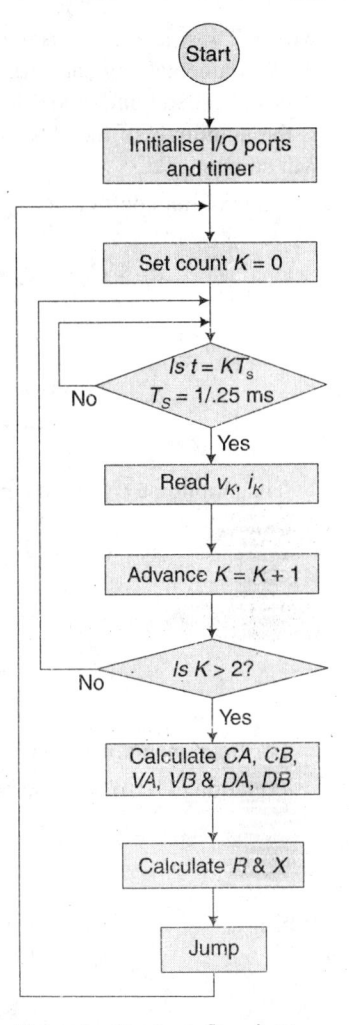

Fig. 11.6 Program flowchart for computation of R and X

11.7.1 Fourier Representation of Signals

Fourier series are used to decompose periodic signals into the sum of sinusoidal components of appropriate amplitude. If the periodic signal has a period of T (seconds), then the frequencies of the sinusoidal components in the Fourier series are integral multiples of the fundamental frequency (1/T).

(i) Fourier Series with Real Coefficients

The Fourier series expansion of any periodic signal $x(t)$ with period T is given by

$$x(t) = \frac{a_0}{2} + \sum_{k=1}^{\infty} (a_K \cos kwt + b_K \sin kwt)$$

$$= \frac{a_0}{2} + \sum_{k=1}^{\infty} C_K \sin(kwt + \phi_K)$$

$$= \frac{a_0}{2} + \sum_{k=1}^{\infty} C_K \cos(k\omega t - \psi_K) \quad (11.18)$$

where $K = 0, 1, 2, \ldots, \infty$ is the integral number of cycle in period T, and $\omega = 2\pi/T$ is the fundamental angular frequency (rad/s). The coefficients a_0, a_K, b_K are real numbers and called Fourier coefficients. They are also called real Fourier coefficients. C_K is the amplitude of the Kth component, and $\psi_K = 90 - \phi_K$ is the phase angle of the Kth components.

The Fourier coefficients a_K and b_K are given by

$$a_K = C_K \sin \phi_K = C_K \cos \psi_K = \frac{2}{T} \int_0^T x(t) \cos k\omega t \, dt,$$

$$K = 0, 1, 2, 3, \ldots, \infty$$

$$b_K = C_K \cos \phi_K = C_K \sin \psi_K = \frac{2}{T} \int_0^T x(t) \sin k\omega t \, dt,$$

$$K = 0, 1, 2, 3, \ldots, \infty \quad (11.19)$$

The amplitude C_K and the phase angles ψ_K and ϕ_K are given by

$$C_K = \sqrt{a_K^2 + b_K^2}$$

$$\psi_K = 90 - \phi_K = \tan^{-1} \frac{a_K}{b_K}$$

$$\phi_K = 90 - \psi_K = \tan^{-1} \frac{b_K}{a_K} \quad (11.20)$$

For fundamental frequency component, $K = 1$.

Equation (11.18) can also be expressed in an alternative form as:

$$x(t) = F_0 + \sqrt{2} F_1 \sin \omega t + \sqrt{2} F_2 \cos \omega t + \sqrt{2} F_3 \sin 2\omega t + \sqrt{2} F_4 \cos 2\omega t + \ldots \quad (11.21)$$

where

$$F_0, \sqrt{2} F_1, \sqrt{2} F_2, \sqrt{2} F_3, \sqrt{2} F_4, \ldots \text{ are equal to}$$

$$\frac{a_0}{2}, b_1, a_1, b_2, a_2, \ldots \text{ respectively.}$$

The $\sqrt{2}$ in Eq. (11.21) is due to the fact that the peak value of the magnitude (amplitude) of a phasor is $\sqrt{2}$ times the rms value of a sinusoid. By convention, the magnitude of a phasor is the rms value of a sinusoid.

The Fourier coefficients associated with the fundamental frequency sine and cosine waves are called fundamental Fourier sine and cosine coefficients. The fundamental Fourier coefficients F_1 and F_2 in Eq. (11.21) are the rms values of the amplitudes of the fundamental Fourier frequency sine and cosine waves, respectively, whereas the fundamental Fourier coefficients b_1 and a_1 in Eq. (11.18) are the peak values of the amplitudes of the fundamental frequency sine and cosine waves, respectively. By comparison of the Fourier coefficients of Eq. (11.21) with that of Eq. (11.18) the Fourier coefficients F_1 and F_2 are expressed in terms b_1 and a_1, respectively, as follows.

$$F_1 = \frac{b_1}{\sqrt{2}}, F_2 = \frac{a_1}{\sqrt{2}} \quad (11.22)$$

If the fundamental frequency component of the signal $x(t)$ is expressed as
$$x_1(t) = C_1 \sin(\omega t + \phi_1)$$
Then from Eqs. (11.19) and (11.22)
$$b_1 = \sqrt{2}\, F_1 = C_1 \cos \phi_1$$
is the peak value of its real component.
$$a_1 = \sqrt{2}\, F_2 = C_1 \sin \phi_1$$
is the peak value of its imaginary component.

$$C_1 = \sqrt{a_1^2 + b_1^2} = \sqrt{2}\,\sqrt{F_1^2 + F_2^2} \quad \text{is its amplitude} \quad (11.23)$$

and $\quad \phi_1 = \tan^{-1} \dfrac{a_1}{b_1} = \tan^{-1} \dfrac{F_2}{F_1} \quad$ is its phase angle $\quad (11.24)$

The phasor representation of the fundamental frequency component in complex form is then given by
$$X_1 = F_1 + jF_2$$
where, F_1 and F_2 are the real and imaginary components, respectively.

If the fundamental frequency component of the signal $x(t)$ is expressed in an alternative form as:
$$x_1(t) = C_1 \cos(\omega t - \psi_1) \quad (11.25)$$
where $\psi_1 = 90 - \phi_1$, then

$a_1 = \sqrt{2}\, F_2 = C_1 \cos \psi_1$ is the peak value of its real component

$b_1 = \sqrt{2}\, F_1 = C_1 \sin \psi_1$ is the peak value of its imaginary component

$C_1 = \sqrt{2}\,\sqrt{F_1^2 + F_2^2}$ is its amplitude

and $\quad \psi_1 = \tan^{-1} \dfrac{F_1}{F_2} \quad (11.26)$

In this case, the phasor representation of the fundamental frequency component in complex form is given by
$$X_1 = F_2 + jF_1 \quad (11.27)$$
where, F_2 and F_1 are the real and imaginary components, respectively.

(ii) Fourier Series, with Complex Coefficients

Equation (11.18) is known as the trigonometric form of the Fourier series in which the Fourier coefficients a_0, a_K and b_K are real numbers. This is also known as Fourier series with real coefficients.

Both since and cosine functions in Eq. (11.18) can be expressed in terms of exponential functions with imaginary exponents. Therefore,

$$\cos k\omega t = \frac{1}{2}(e^{jk\omega t} + e^{-jk\omega t}) \quad (11.28)$$

$$\sin k\omega t = \frac{1}{2j}(e^{jk\omega t} - e^{-jk\omega t}) \quad (11.29)$$

Substituting Eqs. (11.28) and (11.29) in Eq. (11.18) we have

$$x(t) = \frac{a_0}{2} + \frac{1}{2}\sum_{k=1}^{\infty}\left[a_k(e^{jk\omega t}+e^{-jk\omega t}) + \frac{b_k}{j}(e^{jk\omega t}-e^{-jk\omega t})\right]$$

$$= \frac{a_0}{2} + \sum_{k=1}^{\infty}\left[\frac{1}{2}(a_k - jb_k)e^{jk\omega t} + \frac{1}{2}(a_k + jb_k)e^{-jk\omega t}\right]$$

$$= C_{f0} + \sum_{k=1}^{\infty}(C_{fk}\,e^{jk\omega t} + C_{f(-k)}\,e^{-jk\omega t}) \tag{11.30}$$

where the new Fourier coefficients

$$C_{fk} = \frac{1}{2}(a_k - jb_k) \tag{11.31}$$

and

$$C_{f(-k)} = \frac{1}{2}(a_k + jb_k) \tag{11.32}$$

are now complex numbers.

$$C_{f0} = a_0/2 \tag{11.33}$$

is a real number and is consistent with Eqs. (11.31) and (11.32) when k is set equal to zero. C_{fk} is the complex conjugate of $C_{f(-k)}$.

The second exponential series in Eq. (11.30) can be written in a slightly different form by substituting $-k$ for k and changing the limits of the summation properly.

$$\therefore \quad \sum_{k=1}^{\infty} C_{f(-k)}\,e^{-jk\omega t} = \sum_{k=-1}^{-\infty} C_{fk}\,e^{jk\omega t} \tag{11.34}$$

Hence the right side of Eq. (11.30) can be very simply combined into a single summation and the equation can be written compactly as

$$x(t) = \sum_{k=-\infty}^{\infty} C_{fk}\,e^{jk\omega t} \tag{11.35}$$

where C_{fk} is the complex Fourier coefficient.

Equation (11.35) is the complex form of Fourier series. It is also called the exponential form of Fourier series.

The complex Fourier coefficient C_{fk} is obtained by substituting Eq. (11.19) for a_k and b_k in Eq. (11.31) as follows.

$$C_{fk} = \frac{1}{T}\int_0^T x(t)\,e^{-jk\omega t}\,dt \tag{11.36}$$

The amplitude of the kth component is given by

$$2|C_{fk}| = C_k = \sqrt{a_k^2 + b_k^2} \tag{11.37}$$

11.7.2 Discrete Fourier Transform (DFT)

The discrete Fourier transform (DFT) is used to evaluate the Fourier coefficients from N samples of $x(t)$ taken at time $t = 0, T_s, 2T_s, \ldots(N-1)T_s$, where $T_s = T/N$ is the sampling interval. Therefore, the input to the DFT is a sequence of samples (numbers) rather than a continuous function of time $x(t)$. The sequence of samples (i.e. data sequence) which results from periodically sampling the continuous signal $x(t)$ at intervals of T_s is referred to as a discrete-time signal. A system with both continuous and discrete-time signals is called a sampled-data system. A system with only discrete-time signals is called a discrete-time system.

The DFT may be regarded as a discrete-time signal processing technique for the evaluation of the Fourier coefficients. The DFT equation is obtained from Eq. (11.36) by replacing the continuous functions by discrete-time values and the integration by a summation. If the periodic function $x(t)$ is sampled N times per period at the sampling interval of T_s, the N samples represent the period T, so $T = NT_s$. These N samples of $x(t)$ form the data sequence x_m, $m = 0, 1, 2, \ldots(N-1)$.

Therefore, the DFT of a data sequence x_m, $m = 0, 1, 2, \ldots(N-1)$ is defined as:

$$C_{fk} = \frac{1}{N} \sum_{m=0}^{N-1} x_m e^{-j2\pi km/N}, \quad k = 0, 1, 2, \ldots(N-1) \qquad (11.38)$$

The Fourier series of Eq. (11.18) allows k to be any integer. The DFT uses N data samples $x_0, x_1, x_2, \ldots x_{N-1}$, which allows us to solve for only N unknown coefficients. The transform coefficient number k determines the number of cycles in period T and identifies the frequency as k/T Hz.

The computation of Fourier coefficients by using Eq. (11.38) involves complex arithmetic which makes the computation difficult with a microprocessor. Therefore, for microprocessor implementation of the DFT, separate equations for real and imaginary parts are used, instead of using the DFT equation in complex form.

Equation (11.38) can be written as follows.

$$C_{fk} = \frac{1}{N} \sum_{m=0}^{N-1} x_m \left(\cos \frac{2\pi km}{N} - j \sin \frac{2\pi km}{N} \right) \qquad (11.39)$$

or

$$\frac{1}{2}(a_k - jb_k) = \frac{1}{N} \sum_{m=0}^{N-1} x_m \left(\cos \frac{2\pi km}{N} - j \sin \frac{2\pi km}{N} \right), \quad k = 0, 1, 2, \ldots N-1$$

Therefore,

$$a_k = \frac{2}{N} \sum_{m=0}^{N-1} x_m \cos \frac{2\pi km}{N}, \quad k = 0, 1, 2, \ldots N-1 \qquad (11.40)$$

and

$$b_k = \frac{2}{N} \sum_{m=0}^{N-1} x_m \sin \frac{2\pi km}{N}, \quad k = 0, 1, 2, \ldots N-1 \qquad (11.41)$$

The DFT Eq. (11.40) and (11.41) can also be obtained directly from Eq. (11.19) for a_k and b_k by replacing the continuous functions by discrete-time values and integration by a summation. But this long procedure to obtain Eq. (11.40) and (11.41) has been adopted to explain the discrete Fourier transform (DFT).

The DFT Eqs. (11.40) and (11.41) can easily be implemented on microprocessors in order to obtain the Fourier coefficients corresponding to any frequency component. The values of k identifies the frequency.

DFT Matrix Representation

The input to the DFT is the data sequence contained in the data vector X given by

$$[X] = [x_0, x_1, x_2, \ldots x_{(N-1)}]^T \qquad (11.42)$$

where the superscript T denotes the transpose. The output of the DFT is the transform sequence contained in the vector $[C_f]$, given by

$$[C_f] = [C_{f0}, C_{f1}, C_{f2}, \ldots C_{f(N-1)}]^T \qquad (11.43)$$

Computation of the DFT coefficients, i.e. the Fourier coefficients using Eq. (11.38) requires exponentials in the DFT series. Values of the exponents can be found by generating km tables. The factor $e^{-j2\pi/N}$ is common to all coefficients and is denoted by W.

Then Eq. (11.38) can be written in an alternative form as

$$C_{fk} = \frac{1}{N} \sum_{m=0}^{N-1} x_m W^{km}, \quad k = 0, 1, 2, \ldots N-1 \qquad (11.44)$$

where $W = e^{-j2\pi/N}$ and particular values of $k, m = 0, 1, 2, \ldots N-1$ define W^{km}.

The DFT Eq. (11.44) can be written in matrix form as

$$[C_f] = \frac{1}{N}[W^E][X] \qquad (11.45)$$

where W^E is the DFT matrix with row numbers $k = 0, 1, 2, \ldots N-1$ and column numbers $m = 1, 2, \ldots N-1$ and where the entry $W^{E(k,m)}$ is in row k and column m.

11.7.3 Fast Fourier Transform (FFT)

The fast Fourier transform is a fast algorithm for efficient computation of DFT. This algorithm drastically reduces the number of arithmetic operations and memory required to compute the DFT. It is based on manipulation and factorisation of the DFT matrix W^E. If N is the number of samples per period (of fundamental cycle), the easiest case for this algorithm is when $N = 2^n$, where n is an integer. The FFT for this case is called power-of-2 FFT. The more general case is for N having n integral factors. The FFTs for this case are called mixed radix transforms.

The main advantage of this transform is that it reduces the computation involved. The number of complex operations for $N \times 1$ transform is $N \log_2 N$. The main disadvantage with this transform is that complex arithmetic is involved.

The FFT has accelerated the application of Fourier techniques in digital signal processing in many areas other than digital relaying. In relaying applications, N is small (from 4 to 20 for most algorithms) and only a few of the C_{fk} are wanted. Generally, only the fundamental frequency component ($K = 1$) is used in distance relaying while the fundamental and a few harmonics, i.e. the second ($K = 2$) and the fifth ($K = 5$) are used in algorithms for transformer relaying. Hence the FFT has found little application in digital relaying.

11.7.4 Extraction of the Fundamental Frequency Components

The real and imaginary components of the fundamental frequency phasor are computed by using the DFT Eqs. (11.41) and (11.40) for $k = 1$. Using the relations of Eq. (11.24), the equations for the fundamental frequency components are as follows:

$$b_1 = \sqrt{2} F_1 = \frac{2}{N} \sum_{m=0}^{N-1} x_m \sin \frac{2\pi m}{N} \qquad (11.46)$$

$$a_1 = \sqrt{2}\, F_2 = \frac{2}{N} \sum_{m=0}^{N-1} x_m \cos \frac{2\pi m}{N} \qquad (11.47)$$

These components can be computed by using either full-cycle window or half-cycle window. The real and imaginary components of fundamental frequency voltage and current phasors are computed by using the sampled values of voltage and current. The fundamental Fourier coefficients (F_1 and F_2) for the voltage signal and current signal are denoted as $F_{1(v)}$, $F_{2(v)}$ and $F_{1(i)}$ and $F_{2(i)}$ respectively. If the real and imaginary components of the fundamental frequency voltage and current phasors are denoted by V_s, V_c and I_s, I_c, respectively, then

$$F_{1(v)} = V_s,\; F_{2(v)} = V_c$$

and
$$F_{1(i)} = I_s,\; F_{2(i)} = I_c \qquad (11.48)$$

Full-cycle Data Window DFT Algorithm

In this case, a sequence of N uniformly spaced data samples obtained over the full cycle data window is used to compute the real and imaginary components of the fundamental frequency phasor. Both voltage and current signals are sampled simultaneously to acquire N samples for each signal, and then the real and imaginary components of the fundamental frequency voltage and current phasors are obtained by using Eqs. (11.46) to (11.48). If v_m and i_m be the sampled values of voltage and current signals, respectively, the expressions for V_s, V_c, I_s, and I_c, as obtained from Eqs. (11.46) and (11.47), are as follows.

$$F_{1(v)} = V_s = \frac{\sqrt{2}}{N} \sum_{m=0}^{N-1} v_m \sin \frac{2\pi m}{N} \qquad (11.49)$$

$$F_{2(v)} = V_c = \frac{\sqrt{2}}{N} \sum_{m=0}^{N-1} v_m \cos \frac{2\pi m}{N} \qquad (11.50)$$

$$F_{1(i)} = I_s = \frac{\sqrt{2}}{N} \sum_{m=0}^{N-1} i_m \sin \frac{2\pi m}{N} \qquad (11.51)$$

$$F_{2(i)} = I_c = \frac{\sqrt{2}}{N} \sum_{m=0}^{N-1} i_m \cos \frac{2\pi m}{N} \qquad (11.52)$$

Half-cycle Data Window DFT Algorithm

A half-cycle data window algorithm is used on the basis of the assumption that the relaying signals, i.e. voltage and current signals contain only odd harmonics and the transient dc offset component is filtered out from the incoming raw data samples prior to they are being processed. In this case, only $N/2$ samples acquired over the half-cycle data window are used for the computation of the real and imaginary components.

The exponentially decaying dc offset component in the fault current signal is filtered out by employing a digital replica impedance filtering technique. After filtering out of the decaying dc offset component, the fault voltage and current waves contain only odd harmonics and have half-wave symmetry, such that $x(t) = -x(t + T/2)$. For waves having half wave symmetry,

$$x_{l+N/2} = -x_l \text{ where } l = 0, 1, 2, \ldots(N/2-1)$$

In this approach, the data samples $v_{N/2}, v_{N/2+1}, v_{N/2+2}, \ldots v_{N-1}$ are substituted by $-v_0, -v_1, -v_2, \ldots -v_{N/2-1}$, respectively in Eqs. (11.49) and (11.50) for computation of real and imaginary components of the voltage signal; and the data samples $i_{N/2}, i_{N/2+1}, i_{N/2+2}, \ldots i_{N-1}$ are substituted by $-i_0, -i_1, -i_2, \ldots i_{N/2-1}$ respectively in Eqs. (11.51) and (11.52) for the computation of the real and imaginary components of the current signal. As a result of these substitutions, the equations for V_s, V_c, I_s and I_c are obtained as follows.

$$V_s = \frac{2\sqrt{2}}{N} \sum_{m=0}^{N/2-1} v_m \sin \frac{2\pi m}{N} \tag{11.53}$$

$$V_c = \frac{2\sqrt{2}}{N} \sum_{m=0}^{N/2-1} v_m \cos \frac{2\pi m}{N} \tag{11.54}$$

$$I_s = \frac{2\sqrt{2}}{N} \sum_{m=0}^{N/2-1} i_m \sin \frac{2\pi m}{N} \tag{11.55}$$

$$I_c = \frac{2\sqrt{2}}{N} \sum_{m=0}^{N/2-1} i_m \cos \frac{2\pi m}{N} \tag{11.56}$$

where N is the number of samples per fundamental cycle.

11.7.5 Computation of the Apparent Impedance

The main objective of the digital distance relaying of transmission lines is to determine the phasor representations of the voltage and the current signals from their sampled values, and thereafter to calculate the apparent impedance of the line from the relay location to the fault point in order to determine whether the fault lies within the relay's protective zone or not. Since impedance of the linear system is defined in terms of the fundamental frequency voltage and current sinusoidal waves, it is necessary to extract the fundamental frequency components of voltage and current signals from the complex post fault voltage and current signals.

The real and imaginary components of the fundamental frequency voltage phasor, i.e. V_s and V_c, and the real and imaginary components of the fundamental frequency current phasor, i.e. I_s and I_c are obtained by using Eqs. (11.49) to (11.52) for the full-cycle window algorithm or Eq. (11.53) to (11.56) for half-cycle window algorithm. For distance relaying, the half-cycle data window algorithm is preferable.

Knowing V_s, V_c, I_s and I_c, the phasor representations of the fundamental frequency components of voltage and current signals are expressed in complex form as:

$$V = V_s + jV_c \tag{11.57}$$

$$I = I_s + jI_c \tag{11.58}$$

The magnitudes (rms values) and phase angles of the fundamental frequency voltage and current phasor are then given by

$$V = \sqrt{V_s^2 + V_c^2} \tag{11.59}$$

$$\phi_v = \tan^{-1}\frac{V_c}{V_s} \quad (11.60)$$

$$I = \sqrt{I_s^2 + I_c^2} \quad (11.61)$$

$$\phi_i = \tan^{-1}\frac{I_c}{I_s} \quad (11.62)$$

The phase difference ϕ between voltage and current is given by

$$\phi = \phi_v - \phi_i \quad (11.63)$$

When the signal pair is selected properly for a given fault, the ratio of the voltage to current gives the apparent impedance of line.

The voltage and current are expressed in polar from as

$$V = |V| \angle \phi_v \quad (11.64)$$

$$I = I \angle \phi_i \quad (11.65)$$

The apparent impedance is then given by

$$Z = \frac{V}{I} = \frac{|V| \angle \phi_v}{|I| \angle \phi_i}$$

$$= \frac{\sqrt{V_s^2 + V_c^2}}{\sqrt{I_s^2 + I_c^2}} < (\phi_V - \phi_i) = Z < \phi \quad (11.66)$$

Computation of apparent impedance through Eq. (11.66) involves the time consuming operations of squaring and square-rooting. Therefore, in order to avoid the square-rooting operation, the real and reactive components (R and X) of the apparent impedance are calculated as follows.

$$Z = \frac{V_s + jV_c}{I_s + jI_c} = \frac{V_s I_s + V_c I_c}{I_s^2 + I_c^2} + j\frac{V_c I_s - V_s I_c}{I_s^2 + I_c^2}$$

$$= R + jX \quad (11.67)$$

where

$$R = \frac{V_s I_s + V_c I_c}{I_s^2 + I_c^2} \quad (11.68)$$

and

$$X = \frac{V_c I_s - V_s I_c}{I_s^2 + I_c^2} \quad (11.69)$$

Equations (11.68) and (11.69) are programmed to determine R and X from the real and imaginary components of the fundamental frequency voltage and current phasor i.e. V_s, V_c, I_s and I_c.

The program flowchart for the computation of R and X using half-cycle data window DFT algorithm is shown in Fig. 11.7.

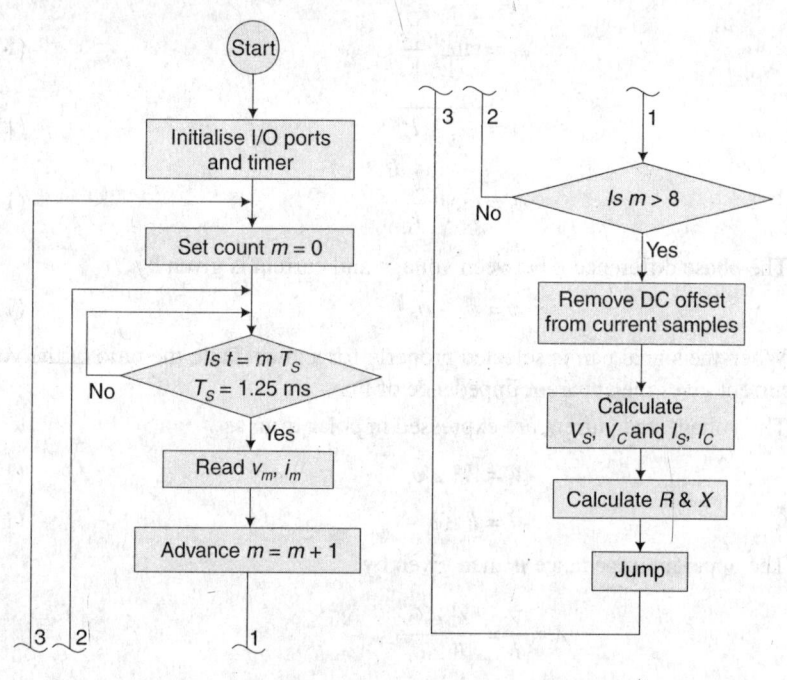

Fig. 11.7 *Program flowchart for computation of R and X*

11.8 WALSH–HADAMARD TRANSFORM TECHNIQUE

In this technique, the algorithm for extracting the fundamental frequency components from the complex post-fault relaying signals is based on Walsh–Hadamard transform (WHT). The Walsh coefficients are obtained by using the Walsh–Hadamard transformation on the incoming data samples, i.e. voltage and current samples acquired over a full-cycle or half-cycle data window. These coefficients are obtained by mere addition and subtraction of data samples. A fast algorithm known as the Fast Walsh–Hadamard transform (FWHT) is available to compute the Walsh coefficients efficiently. Using Fourier–Walsh theory, the Fourier coefficients are expressed in terms of Walsh coefficients. The fundamental Fourier sine and cosine coefficients which are equal to the real and imaginary components of the fundamental frequency phasor, respectively are computed by addition, subtraction and shift operations of the Walsh coefficients. Using the real and imaginary components of the fundamental frequency voltage and current phasors, the real and reactive components (R and X) of the apparent impedance of the line, as seen from the relay location are then calculated. The use of FWHT eliminates the time consuming operation of squaring and square-rooting, and minimizes the multiply and divide operations. Thus rendering this algorithm suitable for microprocessor implementation. This section deals with the Walsh–functions, relationship between Fourier and Walsh coefficients, Walsh–Hadamard transform (WHT), Fast Walsh–Hadamard transform (FWHT) and extraction of fundamental frequency components.

11.8.1 Walsh Functions

The Walsh functions are the basic functions for the Walsh–Hadamard transform (WHT). They are a complete orthonormal set of functions (see Appendix B) which form an ordered set of rectangular waveforms taking only two amplitude values, namely +1 and −1. They are defined over a limited interval T, known as the time base which is required to be known if quantitative values are to be assigned to a functions. Two arguments are required to completely define these functions. These are, time period t, usually normalised to the time base (T) as t/T and an ordering number K, related to the number of zero crossing found within the time base. The functions is written as Wal (k,t).

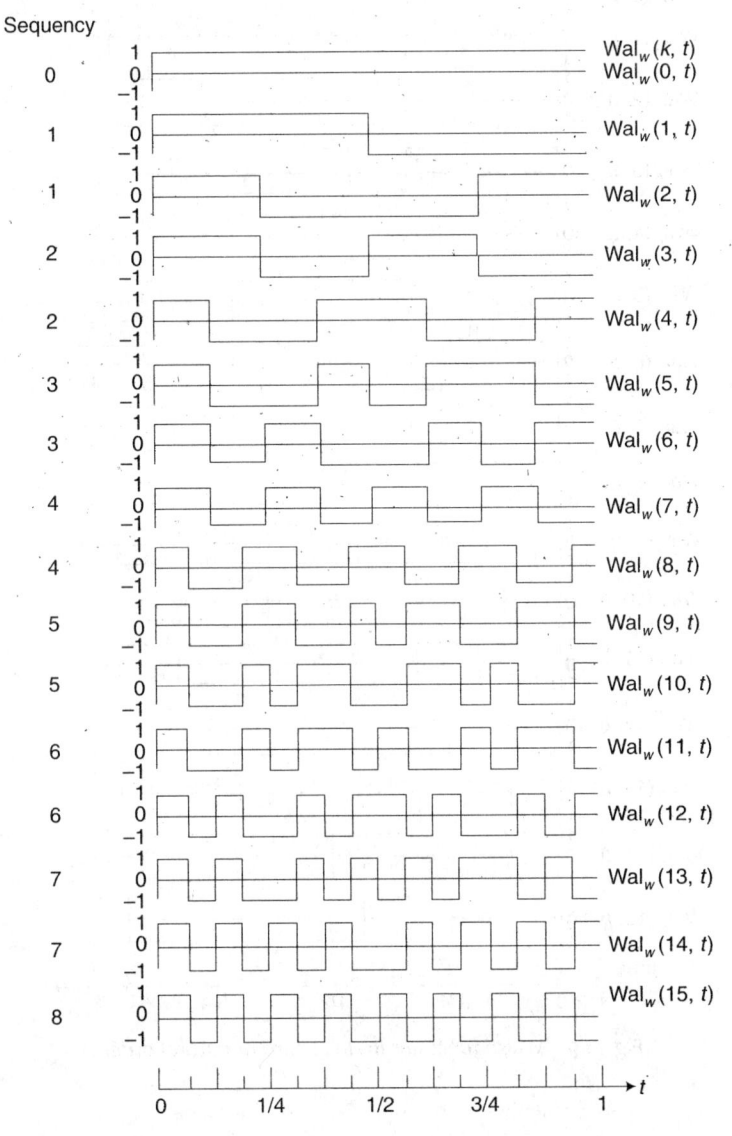

Fig. 11.8 Walsh functions in Walsh or sequency order

The generalised frequency of such functions is known as sequency and is applied to distinguish functions whose zero crossings are not uniformly spaced over an interval and which are not necessarily periodic. For sinusoidal functions, the definition of sequency coincides with that of frequency. The sequency of a periodic function equals one-half the number of sign changes per period while the sequency of an aperiodic function equals one-half the number of sign changes per unit time, if this limit exists. Sequency can be expressed in terms of the number of zero crossing ($n_{z.c.}$) per unit interval in the following manner.

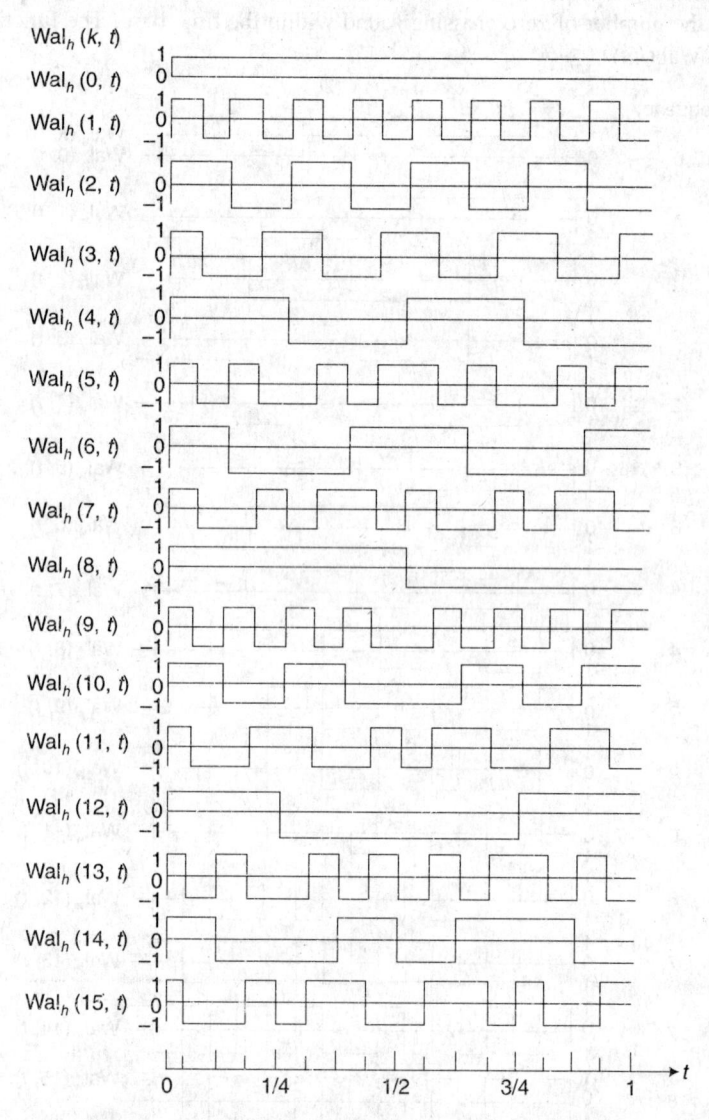

Fig. 11.9 Walsh functions in Hadamard or natural order

$$\text{Sequency} = \begin{cases} 0 & \text{for } n_{zc} = 0 \\ n_{zc}/2 & \text{for even } n_{zc} \\ (n_{zc} + 1)/2 & \text{for odd } n_{zc} \end{cases}$$

where $n_{z.c.}$ is the number of zero crossings per unit interval.

The set of Walsh functions can be generally rearranged in three ways to form three ordering schemes, such as Walsh or sequency order, Hadamard or natural order and Paley or dyadic order. Techniques of transformation from one ordering to another exist. For developing the digital relaying algorithm, Walsh functions of only two orderings, i.e. Walsh or sequency order and Hadamard or natural order have been used. The Walsh functions of these two types of ordering are described below.

(i) Walsh or Sequency Ordering

The set of Walsh functions belonging to this type are denoted by

$$S_w = \{\text{Wal}_w(k,t), \quad k = 0, 1, 2, \ldots(N-1)\} \tag{11.70}$$

where $N = 2^n$, $n = 1, 2, 3, \ldots$, the subscript 'w' denotes Walsh ordering, and k denotes the kth member of S_w. The first 16 Walsh functions in Walsh or sequency order are shown in Fig. 11.8. It is observed that such functions are arranged in ascending values of the number of zero crossing found within the time base.

(ii) Hadamard or Natural Ordering

This set of Walsh functions is denoted by

$$S_h = \{\text{Wal}_h(k,t), \quad k = 0, 1, 2, \ldots(N-1)\} \tag{11.71}$$

where, the subscript 'h' denotes the Hadamard ordering and k denotes the kth member of S_h. The functions belonging to S_h are related to the Walsh order functions by the relation

$$\text{Wal}_h(k,t) = \text{Wal}_w[b(<k>), t] \tag{11.72}$$

where $<k>$ is obtained by the bit reversal of k, and $b(<k>)$ is the Gray code to binary conversion of $<k>$ (see Appendix C). Table (11.1) shown the relationship between the Hadamard-ordered and Walsh-ordered Walsh functions.

The first sixteen Walsh functions in Hadamard order are shown in Fig. 11.9.

11.8.2 Walsh–Hadamard Matrices

In the discrete case, uniform sampling of the set of N Walsh functions of any ordering at N equidistant points results in the ($N \times N$) Walsh–Hadamard matrices of corresponding order. The ($N \times N$) Walsh-Hadamard matrices in Walsh and Hadamard orders are denoted by $[H_w(n)]$ and $[H_h(n)]$ respectively, where $n = \log_2 N$. Each row of the Walsh-Hadamard matrix is a discrete Walsh function. The Walsh–Hadamard matrices $[H_w(4)]$ and $[H_h(4)]$ for $N = 16$ are given in Eqs. (11.73) and (11.74), respectively. $[H_w(4)]$ is obtained by periodic sampling of the set of 16 Walsh functions shown in Fig. 11.8 at 16 equidistant points. Similarly, $[H_h(4)]$ is obtained by periodic sampling of the set of 16 Walsh functions shown in Fig. 11.9 at 16 equidistant points.

Table 11.1 Relationship between Hadamard-ordered and Walsh-ordered Walsh functions

$K_{decimal}$	K_{binary}	$<K>_{binary}$	$b(<K>)_{binary}$	$b(<K>)_{decimal}$	Equation 8.77
0	0000	0000	0000	0	$Wal_h(0, t) = Wal_w(0, t)$
1	0001	1000	1111	15	$Wal_h(1, t) = Wal_w(15, t)$
2	0010	0100	0111	7	$Wal_h(2, t) = Wal_w(7, t)$
3	0011	1100	1000	8	$Wal_h(3, t) = Wal_w(8, t)$
4	0100	0010	0011	3	$Wal_h(4, t) = Wal_w(3, t)$
5	0101	1010	1100	12	$Wal_h(5, t) = Wal_w(12, t)$
6	0110	0110	0100	4	$Wal_h(6, t) = Wal_w(4, t)$
7	0111	1110	1011	11	$Wal_h(7, t) = Wal_w(11, t)$
8	1000	0001	0001	1	$Wal_h(8, t) = Wal_w(1, t)$
9	1001	1001	1110	14	$Wal_h(9, t) = Wal_w(14, t)$
10	1010	0101	0110	6	$Wal_h(10, t) = Wal_w(6, t)$
11	1011	1101	1001	9	$Wal_h(11, t) = Wal_w(9, t)$
12	1100	0011	0010	2	$Wal_h(12, t) = Wal_w(2, t)$
13	1101	1011	1101	13	$Wal_h(13, t) = Wal_w(13, t)$
14	1110	0111	0101	5	$Wal_h(14, t) = Wal_w(5, t)$
15	1111	1111	1010	10	$Wal_h(15, t) = Wal_w(10, t)$

$$[H_w(4)] = \begin{bmatrix} 1 & 1 & 1 & 1 & 1 & 1 & 1 & 1 & 1 & 1 & 1 & 1 & 1 & 1 & 1 & 1 \\ 1 & 1 & 1 & 1 & 1 & 1 & 1 & 1 & -1 & -1 & -1 & -1 & -1 & -1 & -1 & -1 \\ 1 & 1 & 1 & 1 & -1 & -1 & -1 & -1 & -1 & -1 & -1 & -1 & 1 & 1 & 1 & 1 \\ 1 & 1 & 1 & 1 & -1 & -1 & -1 & -1 & 1 & 1 & 1 & 1 & -1 & -1 & -1 & -1 \\ 1 & 1 & -1 & -1 & -1 & -1 & 1 & 1 & 1 & 1 & -1 & -1 & -1 & -1 & 1 & 1 \\ 1 & 1 & -1 & -1 & -1 & -1 & 1 & 1 & -1 & -1 & 1 & 1 & 1 & 1 & -1 & -1 \\ 1 & 1 & -1 & -1 & 1 & 1 & -1 & -1 & -1 & -1 & 1 & 1 & -1 & -1 & 1 & 1 \\ 1 & 1 & -1 & -1 & 1 & 1 & -1 & -1 & 1 & 1 & -1 & -1 & 1 & 1 & -1 & -1 \\ 1 & -1 & -1 & 1 & 1 & -1 & -1 & 1 & 1 & -1 & -1 & 1 & 1 & -1 & -1 & 1 \\ 1 & -1 & -1 & 1 & 1 & -1 & -1 & 1 & -1 & 1 & 1 & -1 & -1 & 1 & 1 & -1 \\ 1 & -1 & -1 & 1 & -1 & 1 & 1 & -1 & -1 & 1 & 1 & -1 & 1 & -1 & -1 & 1 \\ 1 & -1 & -1 & 1 & -1 & 1 & 1 & -1 & 1 & -1 & -1 & 1 & -1 & 1 & 1 & -1 \\ 1 & -1 & 1 & -1 & -1 & 1 & -1 & 1 & 1 & -1 & 1 & -1 & -1 & 1 & -1 & 1 \\ 1 & -1 & 1 & -1 & -1 & 1 & -1 & 1 & -1 & 1 & -1 & 1 & 1 & -1 & 1 & -1 \\ 1 & -1 & 1 & -1 & 1 & -1 & 1 & -1 & -1 & 1 & -1 & 1 & -1 & 1 & -1 & 1 \\ 1 & -1 & 1 & -1 & 1 & -1 & 1 & -1 & 1 & -1 & 1 & -1 & 1 & -1 & 1 & -1 \end{bmatrix}$$

(11.73)

$$[H_h(4)] = \begin{bmatrix}
1 & 1 & 1 & 1 & 1 & 1 & 1 & 1 & 1 & 1 & 1 & 1 & 1 & 1 & 1 & 1 \\
1 & -1 & 1 & -1 & 1 & -1 & 1 & -1 & 1 & -1 & 1 & -1 & 1 & -1 & 1 & -1 \\
1 & 1 & -1 & -1 & 1 & 1 & -1 & -1 & 1 & 1 & -1 & -1 & 1 & 1 & -1 & -1 \\
1 & -1 & -1 & 1 & 1 & -1 & -1 & 1 & 1 & -1 & -1 & 1 & 1 & -1 & -1 & 1 \\
1 & 1 & 1 & 1 & -1 & -1 & -1 & -1 & 1 & 1 & 1 & 1 & -1 & -1 & -1 & -1 \\
1 & -1 & 1 & -1 & -1 & 1 & -1 & 1 & 1 & -1 & 1 & -1 & -1 & 1 & -1 & 1 \\
1 & 1 & -1 & -1 & -1 & -1 & 1 & 1 & 1 & 1 & -1 & -1 & -1 & -1 & 1 & 1 \\
1 & -1 & -1 & 1 & -1 & 1 & 1 & -1 & 1 & -1 & -1 & 1 & -1 & 1 & 1 & -1 \\
1 & 1 & 1 & 1 & 1 & 1 & 1 & 1 & -1 & -1 & -1 & -1 & -1 & -1 & -1 & -1 \\
1 & -1 & 1 & -1 & 1 & -1 & 1 & -1 & -1 & 1 & -1 & 1 & -1 & 1 & -1 & 1 \\
1 & 1 & -1 & -1 & 1 & 1 & -1 & -1 & -1 & -1 & 1 & 1 & -1 & -1 & 1 & 1 \\
1 & -1 & -1 & 1 & 1 & -1 & -1 & 1 & -1 & 1 & 1 & -1 & -1 & 1 & 1 & -1 \\
1 & 1 & 1 & 1 & -1 & -1 & -1 & -1 & -1 & -1 & -1 & -1 & 1 & 1 & 1 & 1 \\
1 & -1 & 1 & -1 & -1 & 1 & -1 & 1 & -1 & 1 & -1 & 1 & 1 & -1 & 1 & -1 \\
1 & 1 & -1 & -1 & -1 & -1 & 1 & 1 & -1 & -1 & 1 & 1 & 1 & 1 & -1 & -1 \\
1 & -1 & -1 & 1 & -1 & 1 & 1 & -1 & -1 & 1 & 1 & -1 & 1 & -1 & -1 & 1
\end{bmatrix}$$

(11.74)

11.8.3 Fourier–Walsh Theory

The Fourier analysis decomposes a signal into a set of sinusoidal waves whereas the Walsh analysis decomposes a signal into a set of rectangular waves. Therefore, the Walsh representation of signals is analogous to the Fourier representation. The Fourier–Walsh theory is used to establish the relationship between the Fourier coefficients and the Walsh coefficients.

The Fourier expansion of the continuous periodic signal $x(t)$ in the interval $(0, T)$ is defined as

$$x(t) = F_0 + \sqrt{2}\, F_1 \sin \frac{2\pi t}{T} + \sqrt{2}\, F_2 \cos \frac{2\pi t}{T}$$
$$+ \sqrt{2}\, F_3 \sin \frac{4\pi t}{T} + \sqrt{2}\, F_4 \cos \frac{4\pi t}{T} + \ldots \quad (11.75)$$

where the Fourier coefficients F_0, F_1, F_2 are given by

$$F_0 = \frac{1}{T} \int_0^T x(t)\, dt$$

$$F_1 = \frac{\sqrt{2}}{T} \int_0^T x(t) \sin \frac{2\pi t}{T}\, dt$$

$$F_2 = \frac{\sqrt{2}}{T} \int_0^T x(t) \cos \frac{2\pi t}{T}\, dt \quad (11.76)$$

$$[A] = \begin{bmatrix}
1 & 0 & 0 & 0 & 0 & 0 & 0 & 0 & 0 & 0 & 0 & 0 & 0 & 0 \\
0 & 0.9 & 0 & 0 & 0 & 0 & 0 & 0 & 0 & 0 & 0 & 0 & 0 & 0 \\
0 & 0 & 0.9 & 0 & 0 & 0 & 0 & 0 & 0 & 0 & 0 & 0 & 0 & 0 \\
0 & 0 & 0 & 0.9 & 0 & 0.30 & -0.30 & 0 & 0 & 0 & 0 & 0 & 0 & 0 \\
0 & -0.373 & 0 & 0 & 0 & 0 & 0 & 0 & 0.180 & 0 & 0 & 0.30 & 0 & 0 \\
0 & 0 & 0.373 & 0 & 0 & 0 & 0 & 0.180 & 0 & 0 & 0.30 & 0 & 0.129 & 0 \\
0 & 0 & 0 & 0 & 0 & 0.724 & 0 & 0 & 0 & 0 & 0 & 0 & 0 & 0 \\
0 & 0 & 0 & 0 & 0 & 0 & 0.724 & 0 & -0.435 & 0.435 & 0 & -0.30 & 0 & -0.053 \\
0 & 0 & 0 & 0 & 0.9 & 0 & 0 & 0 & 0 & 0 & 0 & 0 & 0.053 & 0 \\
0 & -0.074 & 0 & 0 & 0 & -0.484 & 0.484 & 0 & 0 & 0 & 0 & 0 & -0.053 & 0 \\
0 & 0 & -0.074 & 0 & 0 & 0 & 0 & 0 & 0 & 0.65 & 0 & 0 & 0 & -0.69 \\
0 & 0 & 0 & -0.373 & 0 & 0 & 0 & 0.65 & 0 & 0 & 0.724 & 0 & 0.269 & 0 \\
0 & 0 & 0 & 0.373 & 0 & 0 & 0 & 0 & -0.269 & 0.269 & 0.724 & 0 & 0 & 0 \\
0 & 0 & 0 & 0 & 0 & 0.2005 & 0.2005 & 0 & 0 & 0 & 0 & 0 & 0 & 0 \\
0 & -0.179 & 0.179 & 0 & 0 & 0 & 0 & 0 & 0 & 0 & 0 & 0 & 0.65 & 0.65 \\
0 & 0 & 0 & 0 & 0 & 0 & 0 & 0 & 0 & 0 & 0 & 0 & 0 & 0.9
\end{bmatrix} \quad (11.82)$$

The Walsh expansion of the signal $x(t)$ is defined as

$$x(t) = \sum_{k=0}^{\infty} W_{wk} \, \text{Wal}_w\left(k, \frac{t}{T}\right) \qquad (11.77)$$

where, the Walsh coefficients W_{wk} are given by

$$W_{wk} = \frac{1}{T} \int_0^T x(t) \, \text{Wal}_w\left(k, \frac{t}{T}\right) dt \qquad (11.78)$$

A finite series expansion which consists of N terms, where N is of the form $N = 2^n$, is obtained by truncating the above series of Eqs. (11.75) and (11.77) and by reducing the number of terms from ∞ to N.

The Fourier and Walsh coefficients vectors are given by

$$[F] = [F_0, F_1, F_2, \ldots F_{N-1}]^T \qquad (11.79)$$

$$[W_w] = [W_{w0}, W_{w1}, W_{w2}, \ldots W_{w(N-1)}]^T \qquad (11.80)$$

The Fourier coefficients (F_k, $k = 0, 1, 2, \ldots, N-1$) are the components of vector $[F]$, representing $x(t)$ over the interval $(0, T)$. The square of the length of this vector, $[F]^T[F] = \sum_{k=0}^{N-1} F_k^2$, is equal to the mean square value of $x(t)$ over this interval. Similarly, the Walsh coefficients (W_{wk}, $k = 0, 1, 2, \ldots, N-1$) are the components of another vector $[W_w]$ that has the same length and represents $x(t)$ in terms of different basis. By substituting Eq. (11.75) in Eq. (11.78), an orthogonal matrix $[A]$ relating the two vectors $[W_w]$ and $[F]$ is obtained, resulting in the following equation.

$$[W_w] = [A][F] \qquad (11.81)$$

The orthogonal matrix $[A]$, for $N = 16$ is given in Eq. (11.82).

Since $[A]$ is an orthogonal matrix, $[A]^{-1} = [A]^T$, where $[A]^{-1}$ is the inverse of $[A]$ and $[A]^T$ is the transpose of $[A]$.

Therefore, from Eq. (11.81)

$$[F] = [A]^T [W_w] \qquad (11.83)$$

On expansion of Eq. (11.83), the following relations for the fundamental Fourier coefficients F_1 and F_2 can be obtained.

$$F_1 = 0.9 W_{w1} - 0.373 W_{w5} - 0.074 W_{w9} - 0.179 W_{w13}$$
$$F_2 = 0.9 W_{w2} + 0.373 W_{w6} - 0.074 W_{w10} + 0.179 W_{w14} \qquad (11.84)$$

11.8.4 Computation of Walsh Coefficients

The fundamental Fourier coefficients F_1 and F_2 can be computed from the Walsh coefficients by using Eq. (11.84). The Walsh coefficients are computed by using the Walsh–Hadamard transformation on the incoming data samples, i.e. voltage and current samples acquired over full-cycle or half-cycle data window. Walsh–Hadamard transform (WHT) coefficients are called Walsh coefficients. A fast algorithm known as Fast Walsh–Hadamard transform (FWHT) is available for computing the Walsh coefficients efficiently. In the Eq. (11.84) the Walsh coefficients are the Walsh ordered

Walsh–Hadamard Transform (WHT)$_w$ coefficient. (WHT)$_w$ coefficients are denoted by W_{wk}, $k = 0, 1, 2, \ldots(N-1)$. The Walsh coefficients W_{wk} can be efficiently computed by using the Hadamard-ordered Fast Walsh–Hadamard Transform (FWHT)$_h$.

11.8.5 Walsh-Ordered Walsh–Hadamard Transform (WHT)$_w$

The (WHT)$_w$ of a data sequence x_m, $m = 0, 1, 2, \ldots(N-1)$ is defined as

$$[W_w] = \frac{1}{N}[H_w(n)][X] \tag{11.85}$$

where N is the number of data samples per cycle. $[W_w]$ is the (WHT)$_w$ coefficients (or Walsh coefficient) vector which is given by Eq. (11.80).

$[H_w(n)]$ is the $(N \times N)$ Walsh-ordered Walsh–Hadamard matrix.

$[H_w(4)]$ for $N = 16$ is given in Eq. (11.73)

and $[X]$ is the data vector which contains the data sequence x_m, $m = 1, 2, \ldots(N-1)$ as its components and is given by

$$[X] = [x_0, x_1, x_2, \ldots x_{N-1}]^T \tag{11.86}$$

11.8.6 Hadamard Ordered Walsh–Hadamard Transform (WHT)$_h$

The (WHT)$_h$ of a data sequence x_m, $m = 0, 1, 2, \ldots(N-1)$ is defined as

$$[W_h] = \frac{1}{N}[H_h(n)][X] \tag{11.87}$$

where $[W_h]$ is the (WHT)$_h$ coefficient vector which is given by

$$[W_h] = [W_{h0}, W_{h1}, W_{h2}, \ldots W_{h(N-1)}]^T \tag{11.88}$$

and $[H_h(n)]$ is the $(N \times N)$ Hadamard ordered Walsh–Hadamard matrix.

$[H_h(4)]$ for $N = 16$ is given in Eq. (11.74).

11.8.7 Fast Walsh–Hadamard Transform (FWHT)

The Fast Walsh–Hadamard Transform (FWHT) is an algorithm to compute the WHT coefficients efficiently. It is based on the matrix factoring or matrix partitioning technique. Direct computation of WHT coefficients using Eqs. (11.85) and (11.87) requires N^2 additions and subtractions. Whereas the FWHT reduces the computation to $N \log_2 N$ additions and subtraction. The fast algorithm for computing the (WHT)$_w$ coefficients efficiently is called the fast Walsh-ordered Walsh–Hadamard transform (FWHT)$_w$. Similarly the fast algorithm for computing the (WHT)$_h$ coefficients efficiently is called fast Hadamard-ordered Walsh–Hadamard transform (FWHT)$_h$. As there exists a relationship between (WHT)$_w$ and (WHT)$_h$ coefficients, the (WHT)$_w$ coefficients can also be efficiently computed by using the (FWHT)$_h$.

11.8.8 Computation of Walsh Coefficients Using (FWHT)$_w$

The (FWHT)$_w$ is based on *spares* matrix factoring or matrix partitioning techniques of $H_w(n)$. Meanwhile, $H_w(4)$ for $N = 16$, as given in Eq. (11.73) can be factored into *sparse* matrices as:

$[H_w(4)]$

$$= \left\{ \text{diag} \left[\begin{bmatrix} 1 & 1 \\ 1 & -1 \end{bmatrix}, \begin{bmatrix} 1 & -1 \\ 1 & 1 \end{bmatrix}, \begin{bmatrix} 1 & 1 \\ 1 & -1 \end{bmatrix}, \begin{bmatrix} 1 & -1 \\ 1 & 1 \end{bmatrix}, \begin{bmatrix} 1 & 1 \\ 1 & -1 \end{bmatrix}, \begin{bmatrix} 1 & -1 \\ 1 & 1 \end{bmatrix}, \begin{bmatrix} 1 & 1 \\ 1 & -1 \end{bmatrix}, \begin{bmatrix} 1 & -1 \\ 1 & 1 \end{bmatrix} \right] \right\}$$

$$\times \left\{ \text{diag} \left[\begin{bmatrix} U_2 & U_2 \\ U_2 & -U_2 \end{bmatrix}, \begin{bmatrix} U_2 & -U_2 \\ U_2 & U_2 \end{bmatrix}, \begin{bmatrix} U_2 & U_2 \\ U_2 & -U_2 \end{bmatrix}, \begin{bmatrix} U_2 & -U_2 \\ U_2 & U_2 \end{bmatrix} \right] \right\}$$

$$\times \left\{ \text{diag} \left[\begin{bmatrix} U_4 & U_4 \\ U_4 & -U_4 \end{bmatrix}, \begin{bmatrix} U_4 & -U_4 \\ U_4 & U_4 \end{bmatrix} \right] \right\} \begin{bmatrix} U_8 & U_8 \\ U_8 & -U_8 \end{bmatrix} \left[U_{16}^{BRO} \right] \quad (11.89)$$

where U_2, U_4, U_8 and U_{16} are unit (identity) matrices of orders 2, 4, 8 and 16, respectively, and $[U_{16}^{BRO}]$ results from rearranging the columns of U_{16} in bit reversed order (BRO).

The signal flowgraph based on these sparse matrix factors for an efficient computation of Walsh coefficients is shown in Fig. 11.10. Based on Fig. 11.10 and Eq. (11.89), the following observations can be made.

(i) The number of matrix factors is $\log_2 N$. This is equal to the number of stages in the signal flow graph.

(ii) Any row of these matrix factors contains only two non-zero elements (± 1), which correspond to an addition/ subtraction.

(iii) Each sparse matrix factor corresponds to a stage in the signal flow graph in the reverse order that is, the last matrix factor is equivalent to the first stage; the one before the last matrix factor is equivalent to the second stage, and so, the first matrix factor is equivalent to the last stage.

(iv) The total number of arithmetic operations (additions/subtractions) required to compute all the Walsh coefficients is $N \log_2 N$.

11.8.9 Computation of Walsh Coefficients Using (FWHT)$_h$

The Walsh coefficients W_{wk}, $k = 0, 1, 2, \ldots(N-1)$ can also be efficiently computed by using Hadamard-ordered Fast Walsh–Hadamard transform (FWHT)$_h$. (FWHT)$_h$ is based on spares matrix factoring or matrix partitioning techniques of $H_h(n)$. $[H_h 4)]$ for $N = 16$ as given in Eq. (11.74) can be factored into sparse matrices as:

$[H_h(4)]$
$= (\text{diag } [[H_h(1)], [H_h(1)], [H_h(1)], [H_h(1)], [H_h(1)], [H_h(1)], [H_h(1)], [H_h(1)]])$
$\times (\text{diag } [[H_h(1)] \otimes U_2, [H_h(1)] \otimes U_2, [H_h(1)] \otimes U_2, [H_h(1)] \otimes U_2])$
$\times (\text{diag } [[H_h(1)] \otimes U_4, [H_h(1)] \otimes U_4]) \, [[H_h(1)] \otimes U_8] \quad (11.90)$

where diag indicates the diagonal elements of the matrix.

$$[H_h(1)] = \begin{bmatrix} 1 & 1 \\ 1 & -1 \end{bmatrix}$$

$\otimes$ indicates kronecker or direct product of matrices (see Appendix D) and U_2, U_4 and U_8 are unit matrices of orders 2, 4, and 8, respectively.

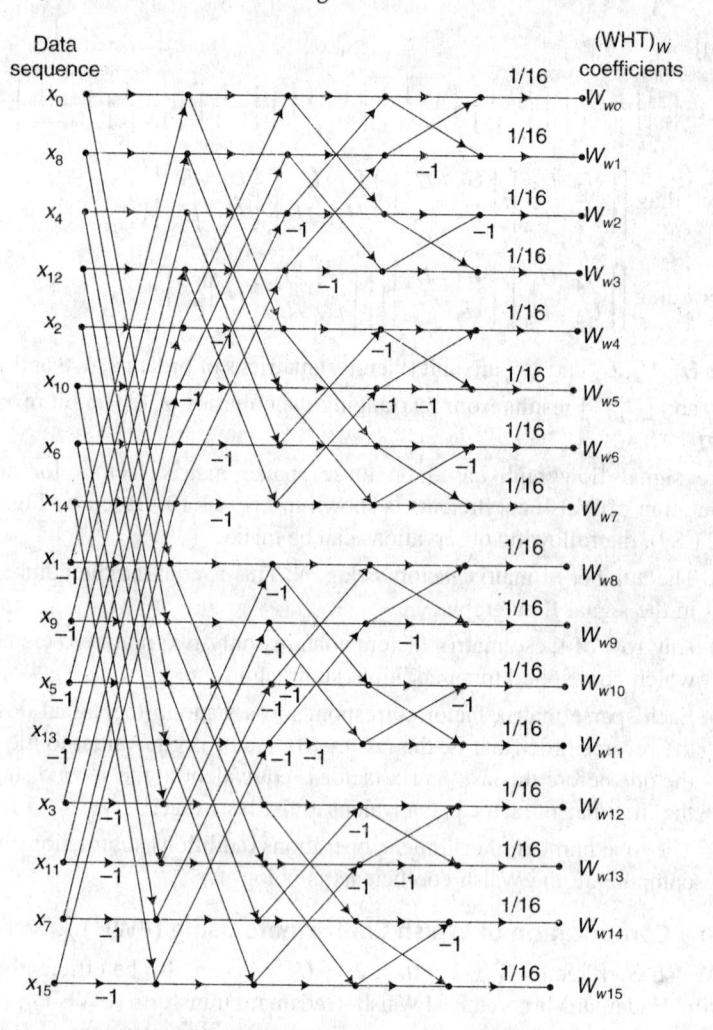

Fig. 11.10 $(FWHT)_w$ signal flow graph, N = 16

The signal flow graph based on these sparse matrix factors for an efficient computation of Walsh coefficients W_{hk} and W_{wk}, is shown in Fig. 11.11. All the properties observed for the $(FWHT)_w$ are also valid for $(FWHT)_h$. The relation between the $(WHT)_w$ and $(WHT)_h$ coefficients can be established in a straight forward manner. From the definitions of these transforms and from Eq. (11.72), it follows that

$$W_{hk} = W_{w[b(<k>)]} \quad k = 0, 1, 2, \ldots (N-1) \qquad (11.91)$$

where $<k>$ denotes the bit-reversal of k, and $b(<k>)$ is the Gray code to binary conversion of k.

For $N = 16$, Eq. (11.91) yields (see Table 11.1)

$W_{h0} = W_{w0},$ $\quad W_{h1} = W_{w15},$ $\quad W_{h2} = W_{w7},$ $\quad W_{h3} = W_{w8}$

$W_{h4} = W_{w3},$ $\quad W_{h5} = W_{w12},$ $\quad W_{h6} = W_{w4},$ $\quad W_{h7} = W_{w11}$

[Figure: (FWHT)$_h$ signal flow graph with data sequence x_0 through x_{15} on the left, butterfly computations with -1 multipliers, $1/16$ scaling, and outputs on the right:]

$W_{h0} = W_{w0}$
$W_{h1} = W_{w15}$
$W_{h2} = W_{w7}$
$W_{h3} = W_{w8}$
$W_{h4} = W_{w3}$
$W_{h5} = W_{w12}$
$W_{h6} = W_{w4}$
$W_{h7} = W_{w11}$
$W_{h8} = W_{w1}$
$W_{h9} = W_{w14}$
$W_{h10} = W_{w6}$
$W_{h11} = W_{w9}$
$W_{h12} = W_{w2}$
$W_{h13} = W_{w13}$
$W_{h14} = W_{w5}$
$W_{h15} = W_{w10}$

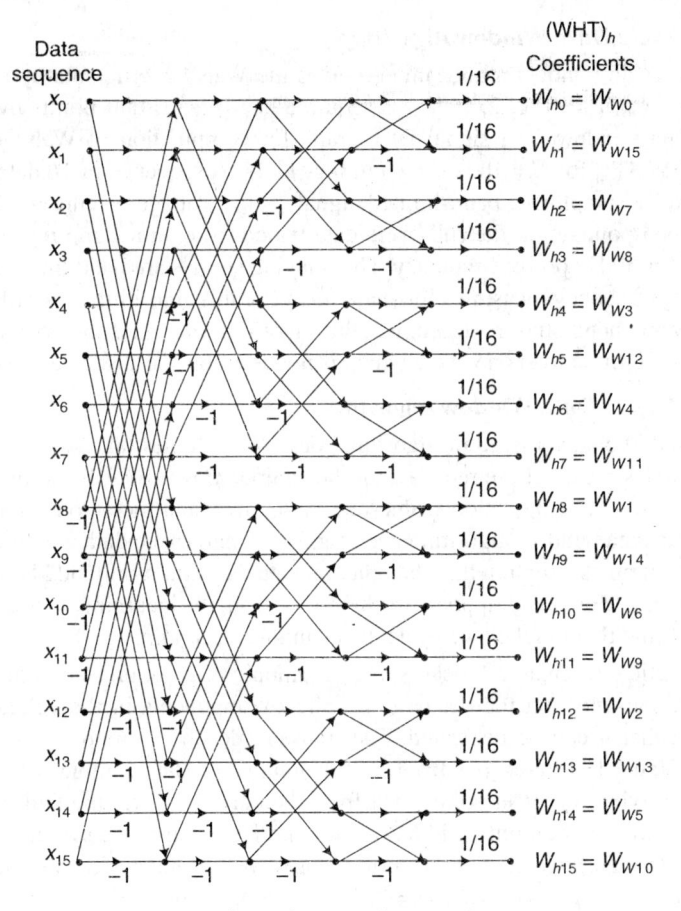

Fig. 11.11 (FWHT)$_h$ signal flow graph, $N = 16$

$W_{h8} = W_{w1}$, $W_{h9} = W_{w14}$, $W_{h10} = W_{w6}$, $W_{h11} = W_{w9}$

$W_{h12} = W_{w2}$, $W_{h13} = W_{w13}$, $W_{h14} = W_{w5}$, $W_{h15} = W_{w10}$

11.8.10 Computation of Walsh Coefficients Required for Computation of Fundamental Fourier Coefficients

The Walsh coefficients required for the computation of the fundamental Fourier coefficients F_1 and F_2 using Eq. (11.89) are W_{w1}, W_{w2}, W_{w5}, W_{w6}, W_{w9}, W_{w10}, W_{w13} and W_{w14}. From the (FWHT)$_h$ signal flowgraph of Fig. 11.11. It is seen that the Walsh coefficients W_{w1}, W_{w2}, W_{w5}, W_{w6}, W_{w9}, W_{w10}, W_{w13} and W_{w14} appear in the last eight terms of the signal flowgraph and are equal to W_{h8}, W_{h12}, W_{h14}, W_{h10}, W_{h11}, W_{h15}, W_{h13} and W_{h9} respectively. Since all the Walsh coefficients required for Eq. (11.84) appear in the lower portion of the signal flowgraph of Fig. 11.11 after the first stage only, the second block of the second stage is needed to be computed further. This reduces a large number of mathematical operations. Therefore, all the Walsh coefficients required for computation of the fundamental Fourier coefficients can be efficiently computed by using (FWHT)$_h$.

(i) Full-cycle Data Window Algorithm

The full-cycle window algorithm computes the Walsh coefficients by using a data sequence x_m; $m = 0, 1, 2, \ldots N - 1$ acquired at N equidistant points over the full-cycle data window. The signal flow graph for computation of Walsh coefficients using $(FWHT)_h$ for $N = 16$ is shown in Fig. 11.11. A sequence of 16 data samples is obtained by sampling a band-limited signal (either voltage or current signal) at 16 equispaced points over the full-cycle data window, i.e. at a sampling frequency of 800 Hz for 50 Hz power frequency. The relaying signals are band limited to half the sampling frequency by using a low-pass filter. For this algorithm, prior filtering of dc offset component is not required. But this algorithm increases the operating time of the relays to more than one cycle which is not desirable for distance relays.

(ii) Half-cycle Data Window Algorithm

In the half-cycle data window algorithm, only $N/2$ samples taken at a sampling rate of N samples pre cycle are required for the computation of Walsh coefficients. Since the distance relay's operation within one cycle after fault inception is desirable, the half-cycle data window algorithm is preferable. A half-cycle data window is used on the basis of an assumption that the relaying signals contain only odd harmonics and the transient dc offset component is filtered out from the incoming data samples prior to processing them in the half-cycle data window algorithm.

In practice, when a fault occurs, even harmonics are not present in the fault voltages and currents and the decaying dc offset components in the voltage signal is so small that it can be neglected. The transient dc offset components in the fault current signal is filtered out from the incoming raw data samples. After filtering out of decaying dc offset component from the current signal, the fault voltage and current waves contain only odd harmonics and have half-wave symmetry such that $x(t) = -x(t + T/2)$, where T is the period. For waves having half-wave symmetry, the data sample $x_{l + N/2} = -x_l$, where $l = 0, 1, 2, \ldots N/2 - 1$. For $N = 16$, $x_{l+8} = -x_l$, where $l = 0, 1, 2, \ldots 7$ and hence the Walsh coefficients can easily be computed by using eight data samples (x_0 to x_7) taken over a half-cycle data window at a sampling rate of 16 samples per cycle. In this approach, the data samples $x_8, x_9, \ldots x_{15}$ are substituted by $-x_0, -x_1, \ldots -x_7$, respectively in Eq. (11.87) and the signal flow graph of Fig. 11.11.

Therefore, for the half-cycle data window algorithm, the equations for various Walsh coefficients obtained from Eq. (11.87) and the signal flow graph of Fig. 11.11 are given by:

$$W_{w1} = W_{h8} = \frac{1}{16}(x_0 + x_1 + x_2 + x_3 + x_4 + x_5 + x_6 + x_7$$
$$- x_8 - x_9 - x_{10} - x_{11} - x_{12} - x_{13} - x_{14} - x_{15})$$
$$= \frac{1}{8}(x_0 + x_1 + x_2 + x_3 + x_4 + x_5 + x_6 + x_7)$$

$$W_{w2} = W_{h12} = \frac{1}{16}(x_0 + x_1 + x_2 + x_3 - x_4 - x_5 - x_6 - x_7$$
$$- x_8 - x_9 - x_{10} - x_{11} + x_{12} + x_{13} + x_{14} + x_{15})$$

$$= \frac{1}{8}(x_0 + x_1 + x_2 + x_3 - x_4 - x_5 - x_6 - x_7)$$

Similarly,

$$W_{w5} = W_{h14} = \frac{1}{8}(x_0 + x_1 - x_2 - x_3 - x_4 - x_5 + x_6 + x_7)$$

$$W_{w6} = W_{h10} = \frac{1}{8}(x_0 + x_1 - x_2 - x_3 + x_4 + x_5 - x_6 - x_7)$$

$$W_{w9} = W_{h11} = \frac{1}{8}(x_0 - x_1 - x_2 + x_3 + x_4 - x_5 - x_6 + x_7)$$

$$W_{w10} = W_{h15} = \frac{1}{8}(x_0 - x_1 - x_2 + x_3 - x_4 + x_5 + x_6 - x_7)$$

$$W_{w13} = W_{h13} = \frac{1}{8}(x_0 - x_1 + x_2 - x_3 - x_4 + x_5 - x_6 + x_7)$$

$$W_{w14} = W_{h9} = \frac{1}{8}(x_0 - x_1 + x_2 - x_3 + x_4 - x_5 + x_6 - x_7) \qquad (11.92)$$

The remaining Walsh coefficients W_{w0} (= W_{h0}), W_{w3} (= W_{h4}) W_{w4} (= W_{h6}), W_{w7} (= W_{h2}), W_{w8} (= W_{h3}), W_{w11} (= W_{h7}), W_{w12} (= W_{h5}), and W_{w15} (= W_{h1}) are zero.

The modified (FWHT)$_h$ signal flowgraph for efficient computation of Walsh coefficients using half-cycle data window algorithm is shown in Fig. 11.12. It is similar to (FWHT)$_h$ signal flowgraph for $N = 8$. The data sequence x_0 to x_7 is obtained over a half-cycle data window at a sampling rate of 16 samples per cycle.

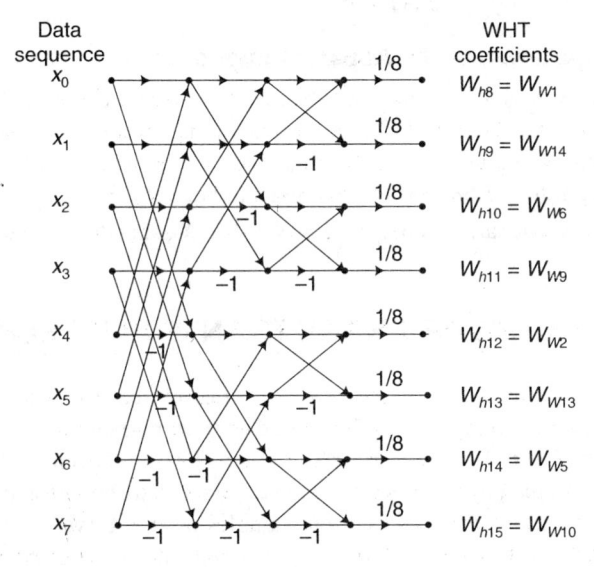

Fig. 11.12 Modified (FWHT)$_h$ signal flowgraph for efficient computation of Walsh coefficients using half-cycle data window

11.8.11 Extraction of the Fundamental Frequency Components

The fundamental Fourier sine and cosine coefficients (F_1 and F_2) are respectively equal to the real and imaginary components of the fundamental frequency phasor.

These coefficients are computed from the Walsh coefficients by using Eq. (11.84). The Walsh coefficients are computed from the data sequence by using $(\text{FWHT})_h$ as described in Sec. 11.8.10. Both voltage and current signals are sampled simultaneously at a sampling rate of 16 samples per cycle in order to acquire either a sequence of 16 samples over a full-cycle window or 8 samples over a half-cycle window of both voltage and current signals. Then the Walsh coefficients required for Eq. (11.84) are computed for both voltage and current signals. The real and imaginary component of the fundamental frequency voltage and current phasors are computed from the Walsh coefficients by using Eq. (11.84). The fundamental Fourier sine and cosine coefficients for both voltage and current signals are denoted as $F_{1(v)}$, $F_{2(v)}$ and $F_{1(i)}$, $F_{2(i)}$, respectively. If V_s and V_c are respectively the real and imaginary components of the fundamental frequency voltage signal, and I_s and I_c are respectively the real and imaginary components of the fundamental frequency current phasor, then

$$F_{1(v)} = V_s,$$
$$F_{2(v)} = V_c$$
$$F_{1(i)} = I_s,$$
$$F_{2(i)} = I_c$$

Knowing V_s, V_c, I_s and I_c, the voltage and current are expressed in complex form as

$$V = V_s + jV_c$$
$$I = I_s + jI_c \tag{11.93}$$

11.8.12 Computation of the Apparent Impedance

The ratio of the voltage to current gives the apparent impedance of the line. The real and reactive components (R and X) of the apparent impedance is computed from the real and imaginary components of the fundamental frequency voltage and current phasor by using Eqs. (11.68) and (11.69) given in Sec. 11.7.5.

The program flowchart for computation of R and X using this algorithm is shown in Fig. 11.13.

11.9 RATIONALISED HAAR TRANSFORM TECHNIQUE

In this case, the algorithm for extracting the fundamental frequency components from the distorted post-fault relaying signals is based on what is known as the Rationalized Haar Transform. The Rationalised Haar Transform (RHT) coefficients C_{rhk}, $k = 0, 1, 2, \ldots N-1$ are obtained by using the rationalised Haar transformation on the incoming data samples, i.e. voltage and current samples acquired over a full-cycle data window or a half-cycle data window at a sampling rate of 16 samples per cycle. These coefficients are obtained by mere addition and subtraction of data samples. A half-cycle data window is used on the basis of the assumption that the relaying signals contain only odd harmonics and the transient dc offset component is filtered out from the incoming raw data samples before they are processed.

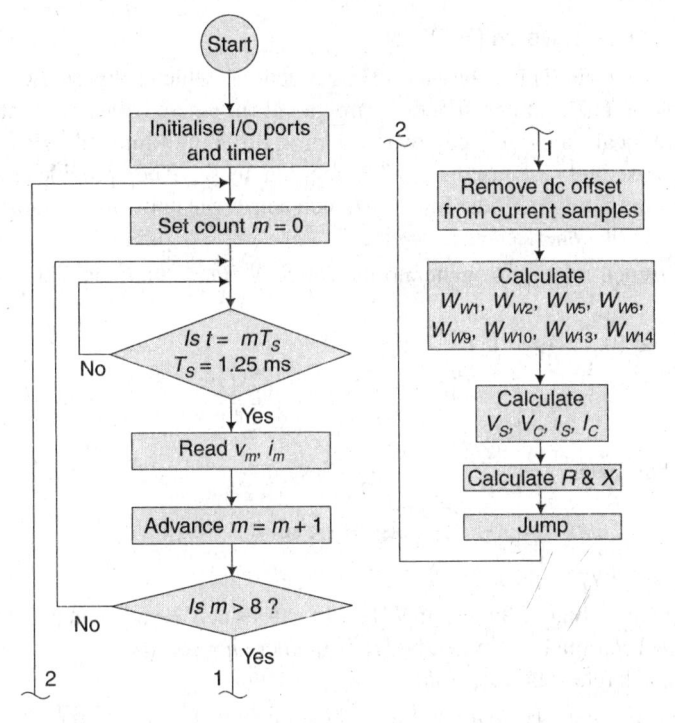

Fig. 11.13 *Program flowchart for computation of R and X*

Relationships to express fundamental Fourier sine and cosine coefficients in terms of the RHT coefficients have been derived. The fundamental Fourier sine and cosine coefficients which are respectively equal to the real and imaginary components of the fundamental frequency phasor can be computed by addition, subtraction and shift operations of the RHT coefficients. Using real and imaginary components of the fundamental frequency voltage and current phasors, the real and reactive components (R and X) of the apparent impedance of the line as seen from the relay location are then calculated. This algorithm has been developed with a sampling rate of 16 samples per cycle, i.e. a sampling frequency of 800 Hz for the 50 Hz power frequency.

Using the discrete Fourier transform, the phasor representation of the voltage and current in rectangular form can also be obtained directly by correlating the incoming data samples with the stored samples of reference fundamental sine and cosine waves. But this method involves time consuming multiplications. Therefore, instead of obtaining the real and imaginary components of voltage and current phasors directly, if the data samples are used to calculate the RHT coefficients first, and thereby the corresponding Fourier sine and cosine coefficients are calculated, the overall computational complexity will be drastically reduced. Introduction to Haar and rationalised Haar transform, algorithm for fast computation of RHT coefficients, relationship between Fourier and RHT coefficients and extraction of the fundamental frequency components are described in this section.

11.9.1 Haar Transform (HT)

The Haar transform (HT) is based on Haar functions, which are periodic, orthogonal and complete. The first two functions are global (non-zero over a unit interval) and the rest are local (non-zero only over a portion of the unit interval). Haar functions become increasingly localised as their number increases. They provide an expansion of a given continuous periodic function which converges uniformly and rapidly. Haar functions can be generated recursively.

A recurrence relation to generate the set of N Haar functions $\{\text{har}(r, m, t)\}$ is given by

$$\text{har}(0, 0, t) = 1 \quad 0 \le t < 1$$

$$\text{har}(r, m, t) = \begin{cases} 2^{r/2} & \dfrac{m-1}{2^r} \le t < \dfrac{m-1/2}{2^r} \\ -2^{r/2} & \dfrac{m-1/2}{2^r} \le t < \dfrac{m}{2^r} \\ 0 & \text{elsewhere for } 0 \le t < 1 \end{cases} \quad (11.94)$$

where $0 \le r \le \log_2 N$ and $1 \le m \le 2^r$.

Uniform sampling of the set of N Haar functions at N equidistant points results in the $(N \times N)$ Haar matrix denoted by $[H_a(n)]$ where $n = \log_2 N$. Each row of the Haar matrix is a discrete Haar function.

The set of first 16 Haar functions is shown in Fig. 11.14, and the corresponding Haar matrix $[H_a(4)]$ for $N = 16$ is given in Eq. (11.95)

The Haar transform and its inverse are respectively defined as

$$[C_h] = \frac{1}{N} [H_a(n)] [X] \quad (11.96)$$

and

$$[X] = [H_a(n)]^T [C_h] \quad (11.97)$$

where N is the number of data samples per cycle, and $[X]$ is the data vector given by

$$[X] = [x_0, x_1, x_2, \ldots x_{(N-1)}]^T \quad (11.98)$$

$[C_h]$ is the HT coefficient vector given by

$$[C_h] = [C_{h0}, C_{h1}, C_{h2}, \ldots C_{h(N-1)}]^T \quad (11.99)$$

and $[H_a(n)]$ is the $(N \times N)$ Haar matrix which is orthogonal,

i.e. $\qquad [H_a(n)] [H_a(n)]^T = N U_N$

where U_N is the matrix of order N.

Haar matrices can be factored into sparse matrices which lead to fast algorithm for efficient computation of *HT*.

11.9.2 Rationalised Haar Transform (RHT)

The rationalised Haar transform (RHT) is the rationalised version of Haar transform (HT). As Haar matrices and their sparse matrix factors contain irrational numbers (power of $\sqrt{2}$), the Haar transform has been rationalised by deleting the irrational

$$H_a(4) = \begin{bmatrix}
1 & 1 & \sqrt{2} & 0 & 2 & 0 & 0 & 0 & 2\sqrt{2} & 0 & 0 & 0 & 0 & 0 & 0 & 0 \\
1 & 1 & \sqrt{2} & 0 & 2 & 0 & 0 & 0 & -2\sqrt{2} & 0 & 0 & 0 & 0 & 0 & 0 & 0 \\
1 & 1 & \sqrt{2} & 0 & -2 & 0 & 0 & 0 & 0 & 2\sqrt{2} & 0 & 0 & 0 & 0 & 0 & 0 \\
1 & 1 & \sqrt{2} & 0 & -2 & 0 & 0 & 0 & 0 & -2\sqrt{2} & 0 & 0 & 0 & 0 & 0 & 0 \\
1 & 1 & -\sqrt{2} & 0 & 0 & 2 & 0 & 0 & 0 & 0 & 2\sqrt{2} & 0 & 0 & 0 & 0 & 0 \\
1 & 1 & -\sqrt{2} & 0 & 0 & 2 & 0 & 0 & 0 & 0 & -2\sqrt{2} & 0 & 0 & 0 & 0 & 0 \\
1 & 1 & -\sqrt{2} & 0 & 0 & -2 & 0 & 0 & 0 & 0 & 0 & 2\sqrt{2} & 0 & 0 & 0 & 0 \\
1 & 1 & -\sqrt{2} & 0 & 0 & -2 & 0 & 0 & 0 & 0 & 0 & -2\sqrt{2} & 0 & 0 & 0 & 0 \\
1 & -1 & 0 & \sqrt{2} & 0 & 0 & 2 & 0 & 0 & 0 & 0 & 0 & 2\sqrt{2} & 0 & 0 & 0 \\
1 & -1 & 0 & \sqrt{2} & 0 & 0 & 2 & 0 & 0 & 0 & 0 & 0 & -2\sqrt{2} & 0 & 0 & 0 \\
1 & -1 & 0 & \sqrt{2} & 0 & 0 & -2 & 0 & 0 & 0 & 0 & 0 & 0 & 2\sqrt{2} & 0 & 0 \\
1 & -1 & 0 & \sqrt{2} & 0 & 0 & -2 & 0 & 0 & 0 & 0 & 0 & 0 & -2\sqrt{2} & 0 & 0 \\
1 & -1 & 0 & -\sqrt{2} & 0 & 0 & 0 & 2 & 0 & 0 & 0 & 0 & 0 & 0 & 2\sqrt{2} & 0 \\
1 & -1 & 0 & -\sqrt{2} & 0 & 0 & 0 & 2 & 0 & 0 & 0 & 0 & 0 & 0 & -2\sqrt{2} & 0 \\
1 & -1 & 0 & -\sqrt{2} & 0 & 0 & 0 & -2 & 0 & 0 & 0 & 0 & 0 & 0 & 0 & 2\sqrt{2} \\
1 & -1 & 0 & -\sqrt{2} & 0 & 0 & 0 & -2 & 0 & 0 & 0 & 0 & 0 & 0 & 0 & -2\sqrt{2}
\end{bmatrix} \quad (11.95)$$

numbers and introducing integer powers of 2. The rationalised Haar transform preserves all the properties of the Haar transform and can be efficiently implemented. The RHT and its inverse are defined as

$$[C_{rh}] = [R_h(n)] [X] \tag{11.100}$$

and

$$[X] = [R_h(n)]^T [P(n)] [C_{rh}] \tag{11.101}$$

where $[X]$ is the data vector, $[C_{rh}]$ is the RHT coefficient vector which contains the RHT coefficient C_{rhk}, $k = 0, 1, 2, \ldots N-1$, as its elements and is given by

$$[C_{rh}] = [C_{rh0}, C_{rh1}, C_{rh2}, \ldots C_{rh(N-1)}]^T \tag{11.102}$$

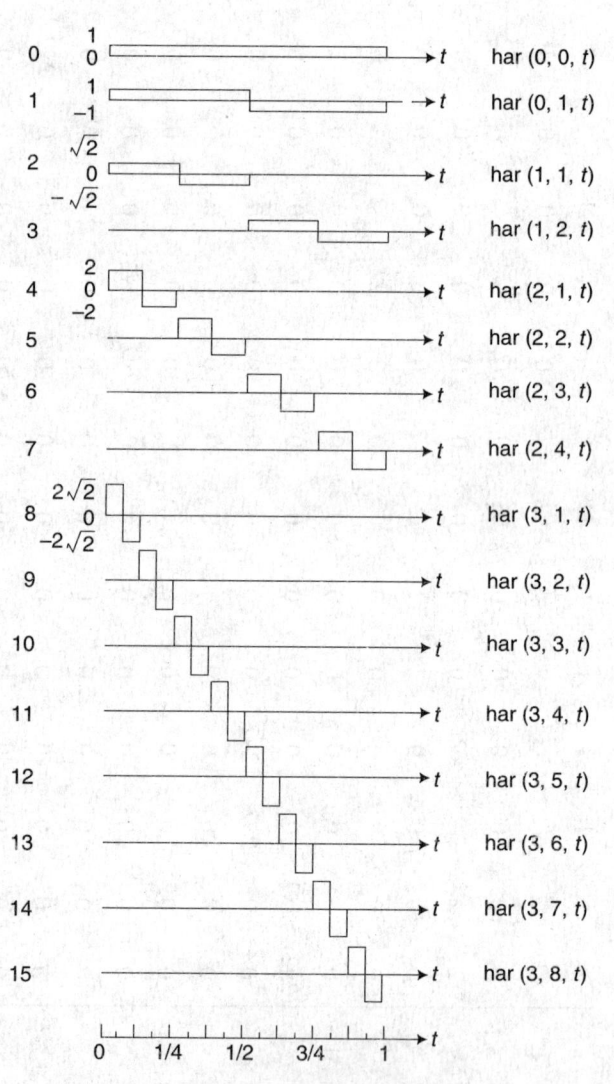

$$n = \log_2 N \quad \text{or} \quad N = 2^n$$

Fig. 11.14 *Haar functions for N = 16*

Numerical Protection

$[R_h(n)]$ is the $(N \times N)$ RHT matrix which is obtained by deleting the irrational numbers (power of $\sqrt{2}$) form the Haar matrix $H_a(n)$, $[P(n)]$ is a diagonal matrix whose diagonal elements are negative integer powers of 2, and

$$[R_h(n)]^T [P(n)] = [R_h(n)]^{-1} \tag{11.103}$$

The RHT matrix $[R_h(4)]$ for $N = 16$, is given in Eq. (11.104) and the diagonal matrix $[P(4)]$ is given in Eq. (11.105).

$$[R_h(4)] = \begin{bmatrix} 1 & 1 & 1 & 1 & 1 & 1 & 1 & 1 & 1 & 1 & 1 & 1 & 1 & 1 & 1 & 1 \\ 1 & 1 & 1 & 1 & 1 & 1 & 1 & 1 & -1 & -1 & -1 & -1 & -1 & -1 & -1 & -1 \\ 1 & 1 & 1 & 1 & -1 & -1 & -1 & -1 & 0 & 0 & 0 & 0 & 0 & 0 & 0 & 0 \\ 0 & 0 & 0 & 0 & 0 & 0 & 0 & 0 & 1 & 1 & 1 & 1 & -1 & -1 & -1 & -1 \\ 1 & 1 & -1 & -1 & 0 & 0 & 0 & 0 & 0 & 0 & 0 & 0 & 0 & 0 & 0 & 0 \\ 0 & 0 & 0 & 0 & 1 & 1 & -1 & -1 & 0 & 0 & 0 & 0 & 0 & 0 & 0 & 0 \\ 0 & 0 & 0 & 0 & 0 & 0 & 0 & 0 & 1 & 1 & -1 & -1 & 0 & 0 & 0 & 0 \\ 0 & 0 & 0 & 0 & 0 & 0 & 0 & 0 & 0 & 0 & 0 & 0 & 1 & 1 & -1 & -1 \\ 1 & -1 & 0 & 0 & 0 & 0 & 0 & 0 & 0 & 0 & 0 & 0 & 0 & 0 & 0 & 0 \\ 0 & 0 & 1 & -1 & 0 & 0 & 0 & 0 & 0 & 0 & 0 & 0 & 0 & 0 & 0 & 0 \\ 0 & 0 & 0 & 0 & 1 & -1 & 0 & 0 & 0 & 0 & 0 & 0 & 0 & 0 & 0 & 0 \\ 0 & 0 & 0 & 0 & 0 & 0 & 1 & -1 & 0 & 0 & 0 & 0 & 0 & 0 & 0 & 0 \\ 0 & 0 & 0 & 0 & 0 & 0 & 0 & 0 & 1 & -1 & 0 & 0 & 0 & 0 & 0 & 0 \\ 0 & 0 & 0 & 0 & 0 & 0 & 0 & 0 & 0 & 0 & 1 & -1 & 0 & 0 & 0 & 0 \\ 0 & 0 & 0 & 0 & 0 & 0 & 0 & 0 & 0 & 0 & 0 & 0 & 1 & -1 & 0 & 0 \\ 0 & 0 & 0 & 0 & 0 & 0 & 0 & 0 & 0 & 0 & 0 & 0 & 0 & 0 & 1 & -1 \end{bmatrix}$$

$$\tag{11.104}$$

$$P(4) = \text{diag}\,[2^{-4}, 2^{-4}, 2^{-3}, 2^{-3}, 2^{-2}, 2^{-2}, 2^{-2}, 2^{-2}, 2^{-1}, 2^{-1}, 2^{-1}, 2^{-1}, 2^{-1}, 2^{-1}, 2^{-1}, 2^{-1}] \tag{11.105}$$

Here, 'diag' indicates diagonal elements of matrix.

11.9.3 Computation of RHT Coefficients

A fast algorithm, based on the matrix factoring technique for efficient computation of RHT coefficients results in reduced computational and memory requirements. This algorithm involves factoring the RHT matrix into sparse matrices. Based on this algorithm, the RHT can be implemented in $2(N - 1)$ additions or subtractions. Transformation time is therefore linearly proportional to the number of terms N, in contrast to Walsh or Fourier techniques where it is proportional to $N \log_2 N$.

The algorithm for efficient computation of RHT coefficients is developed for $N = 16$, i.e. 16 data samples per cycle. The RHT matrix $[R_h(4)]$ for $N = 16$, as given in Eq. (11.104) can be factored into spares matrices as follows.

$$[R_h(4)] = \left(\text{diag}\left[\begin{bmatrix} 1 & 1 \\ 1 & -1 \end{bmatrix}, U_{14}\right]\right)\left(\text{diag}\left[\begin{bmatrix} U_2 & \otimes & (1 \quad 1) \\ U_2 & \otimes & (1 \quad -1) \end{bmatrix}, U_{12}\right]\right)$$

$$\times \left(\text{diag}\left[\begin{bmatrix} U_4 & \otimes & (1 \quad 1) \\ U_4 & \otimes & (1 \quad -1) \end{bmatrix}, U_8\right]\right)\begin{bmatrix} U_8 & \otimes & (1 \quad 1) \\ U_8 & \otimes & (1 \quad -1) \end{bmatrix} \quad (11.106)$$

where U_2, U_4, U_8, U_{12} and U_{14} are unit (identity) matrices of orders 2, 4, 8, 12 and 14, respectively, and $\otimes$ indicates Kronecker product of matrices.

The signal flow graph based on these sparse matrix factors for efficient computation of RHT coefficients is shown in Fig. 11.15.

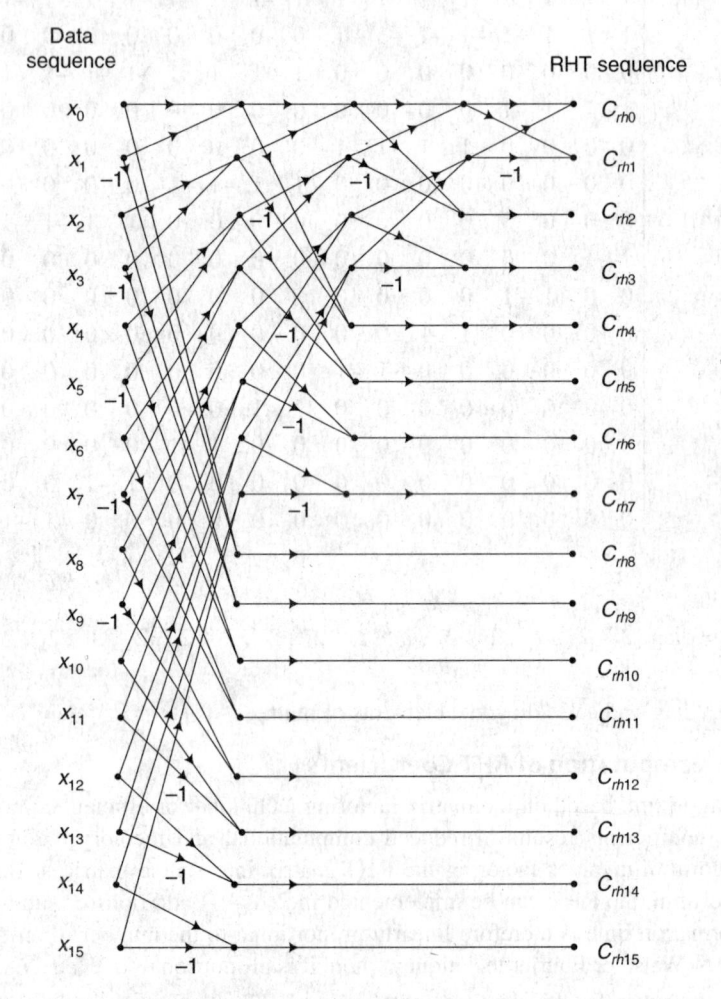

Fig. 11.15 *Signal flowgraph for efficient computation of RHT Coefficients*

(i) Full-cycle Data Window Algorithm

In this case, a sequence of 16 data samples ($x_m = 0, 1, 2, \ldots 15$) obtained over a full-cycle data window is used to compute RHT coefficients. This algorithm does not require prior filtering of dc offset component from the incoming raw data samples. But it increases the operating time of the relays to more than one cycle. The RHT coefficients are computed by using the signal flowgraph shown in Fig. 11.15. Sequences of 16 data samples for both voltage and current signals are obtained by sampling band limited voltage and current signals at 16 equidistant points over a full-cycle data window, i.e. at a sampling frequency (f_s) of 800 Hz for 50 Hz power frequency. The sampling rate in sampled-data system is an important parameter to be considered. Theoretical considerations dictate that the sampling frequency (f_s) must be at least twice the largest frequency (f_m) present in the sampled information to avoid aliasing. Since the sampling frequency used for obtaining the data sequence is 800 Hz, the relaying signals must be band limited to 400 Hz by using a low pass filter.

(ii) Half-cycle Data Window Algorithm

The RHT coefficients can also be computed by using a half-cycle data window. Since the distance relay's operation within a cycle after fault inception is desirable, the half-cycle data window is preferable. A half-cycle data window is used on the basis of assumption that the relaying signals contain only odd harmonics and the transient dc offset component is filtered out from the incoming raw data samples prior to their being processed. In the half-cycle data window algorithm, only eight samples (x_0 to x_7), taken at a sampling rate of 16 samples per cycle are required for the computation of RHT coefficients.

The exponentially decaying dc offset component in the fault current signal is filtered out by employing a digital replica impedance filtering technique. After filtering out of the decaying dc offset component from the current signal, the fault voltage and current waves contain only odd harmonics and have half-wave symmetry, such that $x(t) = -x(t + T/2)$. For waves having half-wave symmetry, the data samples $x_{l+8} = -x_l$, where $l = 0, 1, \ldots 7$ for $N = 16$. Therefore, the RHT coefficient can easily be computed by using eight data samples (x_0 to x_7) taken over a half-cycle data window at a $x_9, \ldots x_{15}$ are substituted by $-x_0, -x_1, \ldots -x_7$, respectively in the signal flow graph of Fig. 11.15 for computation of RHT coefficients.

11.9.4 Relationship Between Fourier and RHT Coefficients

The Fourier expansion of the periodic signal $x(t)$ in the interval $(0, T)$ is defined as

$$x(t) = F_0 + \sqrt{2}\, F_1 \sin \omega t + \sqrt{2}\, F_2 \cos \omega t + \sqrt{2}\, F_3 \sin 2\omega t + \sqrt{2}\, F_4 \cos 2\omega t + \ldots \qquad (11.107)$$

where ω is the fundamental angular frequency which is given by $\omega = 2\pi f = 2\pi/T$, and $F_0, F_1, F_2, F_3, F_4 \ldots$ are Fourier coefficients.

A finite series expansion is obtained by truncating the above series of Eq. (11.107) and reducing the number of terms from ∞ to N where N is of the form $N = 2^n$.

The Fourier coefficients ($F_0, F_1, F_2, \ldots F_{N-1}$) are defined in discrete form as follows.

$$F_0 = \frac{1}{N} \sum_{m=0}^{N-1} x_m$$

$$F_1 = \frac{\sqrt{2}}{N} \sum_{m=0}^{N-1} x_m \sin \frac{2\pi m}{N}$$

$$F_2 = \frac{\sqrt{2}}{N} \sum_{m=0}^{N-1} x_m \cos \frac{2\pi m}{N}$$

...

...

...

$$F_{N-1} = \frac{\sqrt{2}}{N} \sum_{m=0}^{N-1} x_m \sin \pi m \quad (11.108)$$

The above equations can be written in matrix form as follows.

$$[F] = [S][X] \quad (11.109)$$

where $[F]$ is a vector whose components are the Fourier coefficients, and is defined by

$$[F] = [F_0, F_1, F_2, \ldots F_{N-1}]^T \quad (11.110)$$

$[X]$ is the data vector, and $[S]$ is a $(N \times N)$ matrix formed by values of sine and cosine functions at different sampling points, along with the factor, $\sqrt{2}/N$.

Considering N to be even and of the form $N = 2^n$, the elements of matrix S can be obtained as follows.

$$S_{km} = \begin{cases} 1/N & \text{for } k = 0 \\ \sqrt{2}N \sin \dfrac{(K+1)m\pi}{N}, & \text{for odd } k, \text{ i. e. } k = 1, 3, \ldots, (N-1) \\ \dfrac{\sqrt{2}}{N} \cos \dfrac{km\pi}{N}, & \text{for even } k, \text{ i. e. } k = 2, 4, \ldots, (N-2) \end{cases} \quad (11.111)$$

where, $m = 0, 1, 2, \ldots (N-1)$, and S_{km} is the element of matrix $[S]$ in row k and column m.

Thus all the elements of $[S]$ for $N = 16$ can be obtained easily. The Fourier coefficients are the components of vector $[F]$ representing $x(t)$ over the interval $(0, T)$. The square of the length of this vector, $[F]^T [F] = \sum_{k=0}^{N-1} F_k^2$, is equal to the mean square value of $x(t)$ over this interval. Similarly the RHT coefficients are the components of another vector $[C_{rh}]$ that has the same length and represents $x(t)$ in terms of a different basis. By substituting Eq. (11.101) in Eq. (11.109), a matrix $[B]$ relating the two vectors $[F]$ and $[C_{rh}]$ is obtained resulting in the following equation

$$[F] = [B][C_{rh}] \quad (11.112)$$

where

$$[B] = [S] R_h(n)]^T [P(n)] \quad (11.113)$$

For $N = 16$,

$$[B] = [S][R_h(4)]^T [P(4)] \quad (11.114)$$

The matrix $[S]$ (for $N = 16$) is obtained by using Eq. (11.111). By substitution of $[S]$, $[R_h(4)]^T$ being obtained from equation (11.104) and $[P(4)]$ from Eq. (11.105) in Eq. (11.114) and performing multiplication of matrices, matrix $[B]$ is obtained. The matrix $[B]$, in part is given in Eq. (11.115).

$$[B] = \begin{bmatrix}
0.0625 & 0 & 0 & 0 & 0 & 0 & 0 & 0 & 0 & 0 & 0 & 0 & 0 & 0 & 0 & 0 \\
0 & 0.0555 & -0.011 & 0.011 & -0.0227 & 0.0184 & 0.0227 & -0.0184 & 0 & -0.0169 & 0 & 0.0143 & 0.0169 & 0 & 0.0096 & 0 \\
0 & 0.011 & 0.0555 & -0.0555 & 0.0184 & 0.0277 & -0.0184 & -0.0277 & -0.0169 & 0.0033 & -0.0096 & 0.0096 & -0.0033 & -0.0143 & -0.0033 & -0.0143 \\
0 & 0 & 0.0533 & 0.0533 & -0.022 & -0.022 & -0.022 & -0.022 & 0.0033 & 0.0169 & 0.0143 & -0.0129 & 0.0169 & 0.0129 & -0.0169 & -0.0096 \\
0 & 0.022 & 0.533 & -0.0533 & 0.0533 & -0.0533 & 0.0129 & 0.0312 & -0.0312 & 0.0312 & 0.0129 & 0.0312 & -0.0312 & 0.0312 & 0.0312 & -0.0129 \\
0 & 0.01655 & 0.011 & -0.011 & 0.0132 & 0.022 & -0.0132 & 0.066 & -0.0129 & 0.0129 & 0.0312 & -0.0096 & 0.0129 & 0.0312 & -0.0129 & -0.0312 \\
0 & 0.011 & -0.0165 & 0.0165 & 0.066 & 0.0132 & -0.066 & -0.0132 & -0.040 & -0.027 & 0.048 & -0.048 & 0.040 & -0.048 & 0.027 & 0.096 \\
0 & 0 & 0 & 0 & 0.044 & 0.044 & 0.044 & 0.044 & 0.027 & -0.040 & 0.0096 & -0.0096 & -0.027 & -0.0096 & 0.040 & -0.048 \\
0 & 0 & 0 & 0 & 0.044 & 0.044 & 0.044 & 0.044 & -0.044 & -0.044 & 0.044 & 0.044 & -0.044 & 0.044 & -0.044 & 0.044 \\
0 & 0.0074 & -0.011 & 0.011 & 0.044 & 0.088 & -0.044 & -0.0088 & -0.044 & -0.044 & -0.044 & -0.044 & 0.044 & -0.044 & 0.044 & -0.044 \\
0 & 0.011 & 0.0074 & -0.0074 & 0.0088 & -0.044 & -0.0088 & 0.044 & -0.040 & 0.061 & -0.014 & 0.014 & 0.040 & 0.014 & -0.061 & 0.072 \\
0 & 0 & 0.0092 & 0.0092 & 0.022 & -0.022 & 0.022 & -0.022 & 0.061 & 0.040 & -0.072 & 0.072 & -0.061 & 0.072 & -0.040 & -0.014 \\
0 & 0 & 0.022 & -0.022 & -0.0092 & 0.0092 & -0.0092 & 0.0092 & -0.031 & 0.031 & -0.075 & -0.075 & 0.031 & -0.075 & 0.031 & 0.075 \\
0 & 0.0022 & 0.011 & -0.011 & 0.0037 & 0.0055 & -0.0037 & -0.0055 & 0.075 & -0.017 & -0.031 & 0.031 & 0.075 & -0.031 & -0.075 & 0.031 \\
0 & 0.011 & -0.0022 & 0.0022 & -0.0055 & 0.0037 & 0.0055 & -0.0037 & -0.017 & -0.085 & -0.072 & 0.072 & 0.017 & 0.072 & 0.085 & 0.048 \\
0 & 0 & 0 & 0 & 0 & 0 & 0 & 0 & 0.085 & -0.017 & 0.048 & -0.048 & -0.085 & -0.048 & 0.017 & 0.072 \\
0 & 0 & 0 & 0 & 0 & 0 & 0 & 0 & 0 & 0 & 0 & 0 & 0 & 0 & 0 & 0
\end{bmatrix} \quad (11.115)$$

Using Eqs. (11.112), (11.115) and neglecting the higher order RHT coefficients, the fundamental Fourier coefficients F_1 and F_2 can be written in terms of the RHT coefficients as:

$$F_1 = 0.0555 C_{rh1} - 0.011 C_{rh2} + 0.011 C_{rh3} - 0.0277 C_{rh4}$$
$$+ 0.0184 C_{rh5} + 0.0277 C_{rh6} - 0.0184 C_{rh7} \quad (11.116)$$

$$F_2 = 0.011 C_{rh1} + 0.0555 C_{rh2} - 0.0555 C_{rh3} + 0.0184 C_{rh4}$$
$$+ 0.0277 C_{rh5} - 0.0184 C_{rh6} - 0.0277 C_{rh7} \quad (11.117)$$

For voltage and current waves having half-wave symmetry, the data samples x_8, x_9, ...x_{15} are equal to $-x_0$, $-x_1$, ... $-x_7$, respectively. Substitution of this relationship in Eq. (11.100) and signal flow graph of Fig. (11.15) results in the following equations.

$$C_{rh1} = 2(x_0 + x_1 + x_2 + x_3 + x_4 + x_5 + x_6 + x_7) \quad (11.118)$$

$$C_{rh2} = (x_0 + x_1 + x_2 + x_3) - (x_4 + x_5 + x_6 + x_7) \quad (11.119)$$

$$C_{rh3} = -C_{rh2} \quad (11.120)$$

$$C_{rh4} = (x_0 + x_1) - (x_2 + x_3) \quad (11.121)$$

$$C_{rh5} = (x_4 + x_5) - (x_6 + x_7) \quad (11.122)$$

$$C_{rh6} = -C_{rh4} \quad (11.123)$$

$$C_{rh7} = -C_{rh5} \quad (11.124)$$

Substitution of Eqs. (11.120), (11.123) and (11.124) in Eqs. (11.116) and (11.117) results in following equations:

$$F_1 = 0.0555 (C_{rh1} - C_{rh4}) - 0.022 C_{rh2} + 0.0368 C_{rh5} \quad (11.125)$$

$$F_2 = 0.011 C_{rh1} + 0.0555(2 C_{rh2} + C_{rh5}) + 0.0368 C_{rh4} \quad (11.126)$$

Thus, only four RHT coefficients C_{rh1}, C_{rh2}, C_{rh4} and C_{rh5} are required to be calculated for the calculation of fundamental Fourier coefficients F_1 and F_2. The fundamental Fourier sine and cosine coefficients (F_1 and F_2) are respectively, equal to the real and imaginary components of the fundamental frequency phasor.

The multipliers in Eqs. (11.125) and (11.126) are generated by addition, subtraction and shift operations rather than hardware multiplication/division to improve the computational speed.

11.9.5 Extraction of Fundamental Frequency Components

The fundamental Fourier sine and cosine coefficients for both voltage and current signals are computed as $F_1(v)$, $F_2(v)$ and $F_1(i)$, $F_2(i)$ respectively by using Eqs. (11.115) and (11.116). $F_1(v)$ and $F_2(v)$ are respectively equal to the real and imaginary components of the fundamental frequency voltage phasor, namely V_s and V_c; and $F_1(i)$ and $F_2(i)$ are respectively the real and imaginary components of the fundamental frequency current, namely I_s and I_c.

Knowing V_s, V_c, I_s and I_c, the fundamental frequency voltage and current are expressed in complex form as

$V = V_s + jV_c$ and $I = I_s + jI_c$

11.9.6 Computation of the Apparent Impedance

The real and reactive components of the apparent impedance of the line is computed from the real and imaginary components of the fundamental frequency voltage and current phasors by using Eqs. (11.68) and (11.69) given in Sec. 11.7.5.

The program flowchart for computation of R and X using this algorithm is shown in Fig. 11.16.

11.10 BLOCK PULSE FUNCTIONS TECHNIQUE

In this technique the numerical filtering algorithm for extracting the fundamental frequency components from the corrupted post-fault relaying signals is based on block pulse functions, abbreviated as BPF. The Block Pulse Functions (BPF) are defined in a unit, interval of time and as the name suggests the BPF is a set of rectangular pulses, having magnitude unity, which comes one after the other as a block of pulses. The fundamental Fourier sine and cosine coefficients (F_1 and F_2), which are respectively equal to the real and imaginary components of the fundamental frequency phasor are computed for both voltage and current signals by using the BPF coefficients. Relationship to express fundamental Fourier coefficients in terms of the BPF coefficients have been derived. The BPF coefficients are equal to the digitized samples of voltage and current. Using the real and imaginary components of the fundamental frequency voltage and current phasors, the real and reactive components (R and X) of the apparent impendence of the line, as seen from the relay location are then calculated.

The computational simplicity, fastness and accuracy of BPF algorithm renders it most suitable for microprocessor and microcontroller implementation.

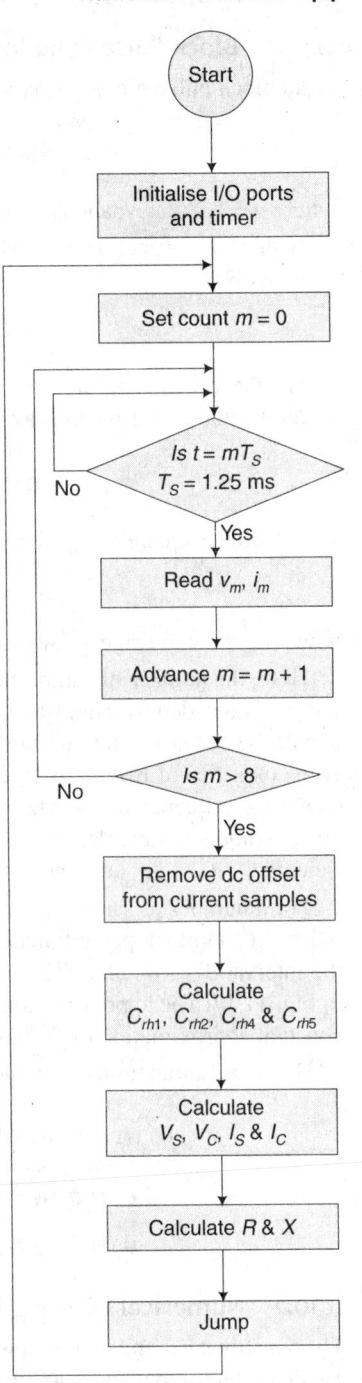

Fig. 11.16 *Program flowchart for computation of R and X*

```
Start
  ↓
Initialise I/O ports and timer
  ↓
Set count m = 0
  ↓
Is t = mT_S
T_S = 1.25 ms  — No →
  ↓ Yes
Read v_m, i_m
  ↓
Advance m = m + 1
  ↓
Is m > 8  — No →
  ↓ Yes
Remove dc offset from current samples
  ↓
Calculate C_rh1, C_rh2, C_rh4 & C_rh5
  ↓
Calculate V_S, V_C, I_S & I_C
  ↓
Calculate R & X
  ↓
Jump
```

11.10.1 Block Pulse Functions (BPF)

A set block pulse functions on a unit time interval (0,1) is defined as

$$\varphi_n = \begin{cases} 1 & (n-1)/N < t < n/N \\ 0 & \text{otherwise} \end{cases} \quad (11.127)$$

where $n = 1, 2, \ldots, N$; and N is the number of samples chosen in one cycle.

If there is a function $f(t)$, which is integrable in $(0, 1)$, it can be approximated using BPF as

$$f(t) \approx \sum_{n=1}^{N} a_n \varphi_n(t) \quad (11.128)$$

where the coefficients a_n are BPF coefficients determined so that the following integral square error is minimized.

$$\varepsilon = \int_0^1 \left(f(t) - \sum_{n=1}^{N} a_n \varphi_n(t) \right)^2 dt \quad (11.129)$$

For such a square fit a_n is given by

$$a_n = \int_{(n-1)/N}^{n/N} f(t)\, dt \quad (11.130)$$

where a_n is the average value of $f(t)$ in the interval $(n-1)/N < t < n/N$.

Here, the number of samples chosen is 12 samples per cycle. Number of samples chosen is decided by sampling theorem, which dictates that the sampling frequency should be at least twice the largest frequency content in the sampled information. If the presence of fifth harmonic is considered as the largest frequency content, the sampling frequency of 600 Hz, *i.e.*, a sampling rate of 12 samples acquired over one cycle window is suitable. From the above equations, it is obvious that the BPF coefficients $a_1, a_2, \ldots, a_{12}$ are the sampled values of function $f(t)$.

For a number of samples $N = 12$, it is clear from Eq. (11.127) that one unit time interval (0,1) block pulse functions (BPFs) are twelve block of magnitude unity for the interval (0 to 1/12), (1/12 to 2/12), (2/12 to 3/12), ..., (11/12 to 12/12) as shown in Fig. 11.17. The approximation in Eq. (11.128) gives the best N segment piecewise constant approximation of $f(t)$ and is unique.

Block pulse functions have the following useful properties:

$$\int_0^1 \phi_n(t)\, \phi_m(t)\, dt = 0 \quad \text{for all } n \neq m \text{ (orthogonality)}$$

$$\phi_n(t)\, \phi_m(t) = 0 \quad \text{for all } n \neq m \text{ (disjoint property)}$$

$$\phi_n(t)\, \phi_m(t) = \phi_n(t) \quad \text{for all } n = m \quad (11.131)$$

11.10.2 Numerical Filtering Using Block Pulse Functions (BPF)

The functions $i(t)$ and $v(t)$ are the post-fault current and voltage signals that contain the dc offset, fundamental and harmonic frequency components, and the BPF coefficients (a_n) will be equal to the sampled values of current and voltage signals. The fundamental Fourier sine and cosine coefficients (F_1 and F_2) are, respectively, equal to the real and imaginary components of the fundamental frequency phasor. F_1 and

F_2 for both current and voltage signals are computed by using the BPF coefficients. The relationships to express fundamental Fourier sine and cosine coefficients in terms of the BPF coefficients are derived as follows.

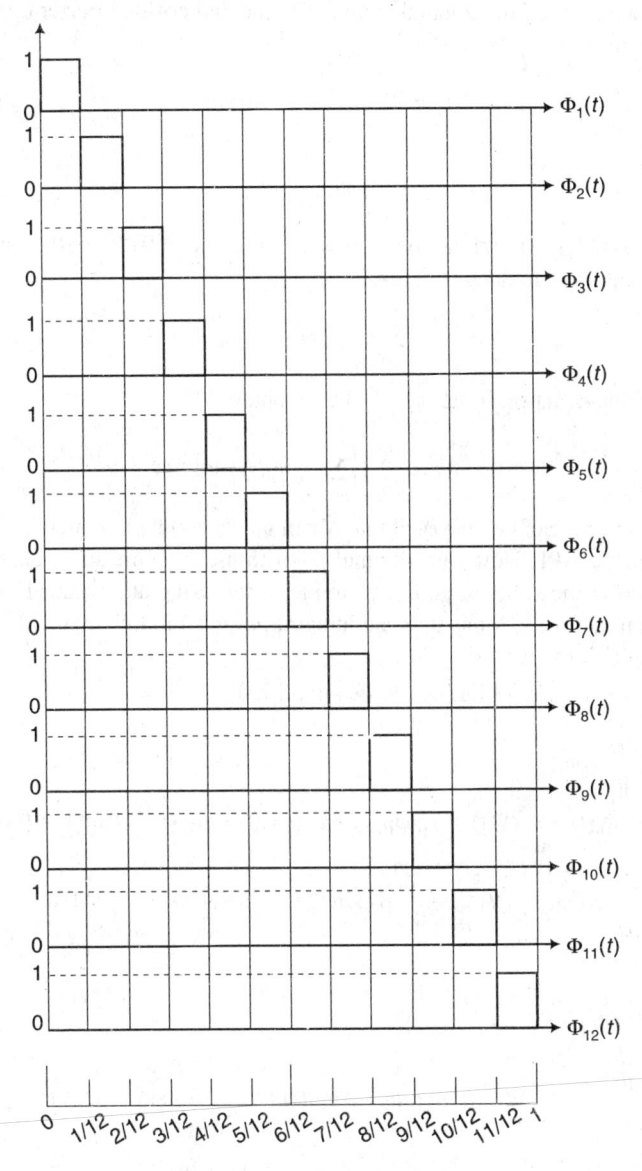

Fig. 11.17 Block pulse functions

Relationship Between Fourier and BPF Coefficients

Taking the fundamental period as 1, current $i(t)$, which is given by time function can be expressed in terms of Fourier coefficient as

$$i(t) = A_o + \sqrt{2} A_1 \sin(2\pi t) + \sqrt{2} B_1 \cos(2\pi t) + \sqrt{2} A_2 \sin(4\pi t)$$
$$+ \sqrt{2} B_2 \cos(4\pi t) + \ldots + \sqrt{2} A_5 \sin(10\pi t) + \sqrt{2} B_5 \cos(10\pi t) \quad (11.132)$$

where A_1 and B_1 are fundamental frequency sine and cosine Fourier coefficients. A_1 and B_1 are given by

$$A_1 = \sqrt{2} \int_0^1 i(t) \sin(2\pi t) \, dt \quad (11.133)$$

$$B_1 = \sqrt{2} \int_0^1 i(t) \cos(2\pi t) \, dt \quad (11.134)$$

From Eq. (11.128), $i(t)$ can be expressed in terms of BPF coefficients a_n, which are simply the current data samples.

$$i(t) \approx \sum_{n=1}^{N} a_n \varphi_n(t) \quad (11.135)$$

Putting this equation in Eq. (11.133), we obtain

$$A_1 = \sqrt{2} \int_0^1 \left(\sum_{n=1}^{N} a_n \varphi_n(t) \right) \sin(2\pi t) \, dt \quad (11.136)$$

At this stage it can be observed that N can take any integral value in Eq. (11.138), while using the BPF. However, in other algorithms, there are some hard restrictions while selecting the value of N. For example, when using algorithms based on Haar and Walsh transforms, N has to be an integral power of 2, i.e, 8, 16, 32 etc. So this is a beneficial point when using BPF.

Taking $N = 12$, Eq. (11.136) can be simplified as follows:

$$A_1 = \sqrt{2} \int_0^1 |a_1 \varphi_1(t) + a_2 \varphi_2(t) + a_3 \varphi_3(t) + a_4 \varphi_4(t) + a_5 \varphi_5(t) + a_6 \varphi_6(t) + a_7 \varphi_7(t)$$
$$+ a_8 \varphi_8(t) + a_9 \varphi_9(t) + a_{10} \varphi_{10}(t) + a_{11} \varphi_{11}(t) + a_{12} \varphi_{12}(t)| \sin(2\pi t) \, dt$$

$$A_1 = \sqrt{2} \int_0^{1/12} a_1 \sin(2\pi t) \, dt + \int_{1/12}^{2/12} a_2 \sin(2\pi t) \, dt + \int_{2/12}^{3/12} a_3 \sin(2\pi t) \, dt$$
$$+ \int_{3/12}^{4/12} a_4 \sin(2\pi t) \, dt + \int_{4/12}^{5/12} a_5 \sin(2\pi t) \, dt + \int_{5/12}^{6/12} a_6 \sin(2\pi t) \, dt$$
$$+ \int_{6/12}^{7/12} a_7 \sin(2\pi t) \, dt + \int_{7/12}^{8/12} a_8 \sin(2\pi t) \, dt + \int_{8/12}^{9/12} a_9 \sin(2\pi t) \, dt$$
$$+ \int_{9/12}^{10/12} a_{10} \sin(2\pi t) \, dt + \int_{10/12}^{11/12} a_{11} \sin(2\pi t) \, dt + \int_{11/12}^{12/12} a_{12} \sin(2\pi t) \, dt]$$

Therefore, after simplification, we get the following equation:

$$A_1 = 0.0302 (a_1 + a_6 - a_7 - a_{12}) + 0.0824 (a_2 + a_5 - a_8 - a_{11})$$
$$+ 0.1125 (a_3 + a_4 - a_9 - a_{10}) \quad (11.137)$$

Similarly from Eq. (11.134) and Eq. (11.135), we obtain

$$B_1 = \sqrt{2} \int_0^1 i(t) \cos(2\pi t)\, dt$$

$$B_1 = \sqrt{2} \int_0^1 \left(\sum_{n=0}^{N} a_n \varphi_n(t)\right) \cos(2\pi t)\, dt$$

$$= \sqrt{2} \int_0^1 |a_1 \varphi_1(t) + a_2 \varphi_2(t) + a_3 \varphi_3(t) + a_4 \varphi_4(t) + a_5 \varphi_5(t) + a_6 \varphi_6(t) + a_7 \varphi_7(t)$$
$$+ a_8 \varphi_8(t) + a_9 \varphi_9(t) + a_{10} \varphi_{10}(t) + a_{11} \varphi_{11}(t) + a_{12} \varphi_{12}(t)| \cos(2\pi t)\, dt$$

$$B_1 = \sqrt{2} \int_0^{1/12} a_1 \cos(2\pi t)\, dt + \int_{1/12}^{2/12} a_2 \cos(2\pi t)\, dt + \int_{2/12}^{3/12} a_3 \cos(2\pi t)\, dt$$
$$+ \int_{3/12}^{4/12} a_4 \cos(2\pi t)\, dt + \int_{4/12}^{5/12} a_5 \cos(2\pi t)\, dt + \int_{5/12}^{6/12} a_6 \cos(2\pi t)\, dt$$
$$+ \int_{6/12}^{7/12} a_7 \cos(2\pi t)\, dt + \int_{7/12}^{8/12} a_8 \cos(2\pi t)\, dt + \int_{8/12}^{9/12} a_9 \cos(2\pi t)\, dt$$
$$+ \int_{9/12}^{10/12} a_{10} \cos(2\pi t)\, dt + \int_{10/12}^{11/12} a_{11} \cos(2\pi t)\, dt + \int_{11/12}^{12/12} a_{12} \cos(2\pi t)\, dt$$

$$B_1 = 0.1125(a_1 - a_6 - a_7 + a_{12}) + 0.0824(a_2 - a_5 - a_8 + a_{11})$$
$$+ 0.0302(a_3 - a_4 - a_9 + a_{10}) \qquad (11.138)$$

The amplitude of the fundamental frequency component of the current is given by

$$I_{m1} = \sqrt{2}\sqrt{(A_1^2 + B_1^2)}$$
$$I_{rms} = \sqrt{(A_1^2 + B_1^2)} \qquad (11.139)$$

In this way all Fourier coefficient are expressed in terms of BPF coefficients. If the Fourier coefficients are determined directly, then it requires a lot of complex computational process. But with the help of BPF coefficients, the Fourier coefficients are determined with simple computations. The fundamental Fourier sine and cosine coefficients (F_1 and F_2), which are respectively equal to the real and imaginary components of the fundamental frequency phasor are equal to A_1 and B_1 given by equations (11.139) and (11.140) respectively. Thus F_1 and F_2 for both current and voltage can be computed by using the BPF coefficients a_n which are simply the digitized samples of current and voltage acquired over one cycle data window at a sampling rate of 12 samples per cycle.

A half-cycle data window is used on the basis of assumption that the relaying signals contain only odd harmonics, and the transient dc offset component is filtered out form the incoming raw data samples prior to processing them. After removal of the dc offset component, the current and voltage waves will have half-wave symmetry. For voltage and current waves having half-wave symmetry, the data samples $a_7, a_8, a_9, a_{10}, a_{11}, a_{12}$ are respectively equal to $-a_1, -a_2, -a_3, -a_4, -a_5, -a_6$.

Substitution of this relationship among data samples in Eqs. (11.137) and (11.138) results in the following equations for the fundamental frequency sine and cosine coefficients (F_1 and F_2).

$$F_1 = A_1 = 0.0604\,(a_1 + a_6) + 0.1648\,(a_2 + a_5)$$
$$+ 0.225\,(a_3 + a_4) \qquad (11.140)$$

$$F_2 = B_1 = 0.0225\,(a_1 - a_6) + 0.1648\,(a_2 - a_5)$$
$$+ 0.0604\,(a_3 - a_4) \qquad (11.141)$$

Thus, in the half-cycle data window algorithm, only six samples (a_1 to a_6) taken at a sampling rate of 12 samples per cycle are required for the computation of real and imaginary components of the fundamental frequency voltage and current signals. The dc offset component in the fault current signal is filtered out by employing digital replica impedance filtering technique described in section 11.12.

11.10.3 Extraction of Fundamental Frequency Components

The fundamental Fourier sine and cosine coefficients for both voltage and current signals are computed as $F_{1(v)}$, $F_{2(v)}$ and $F_{1(i)}$, $F_{2(i)}$ respectively by using Eqs. (11.140) and (11.141) $F_{1(v)}$ and $F_{2(v)}$ are respectively equal to the real and imaginary components of the fundamental frequency voltage phasor, namely V_s and V_c; and $F_{1(i)}$ and $F_{2(i)}$ are respectively the real and imaginary components of the fundamental frequency current, namely I_s and I_c.

Knowing V_s, V_c, I_s and I_c, the fundamental frequency components of voltage and current are expressed in complex form as

$$V = V_s + jV_c \quad \text{and} \quad I = I_s + jI_c$$

11.10.4 Computation of the Apparent Impedance

The real and reactive components of the apparent impedance of the line are computed from the real and imaginary components of the fundamental frequency voltage and current phasors by using Eqs. (11.68) and (11.69) given in Sec. 11.7.5.

The program flowchart for computation of R and X using this algorithm is shown in Fig. 11.18 and

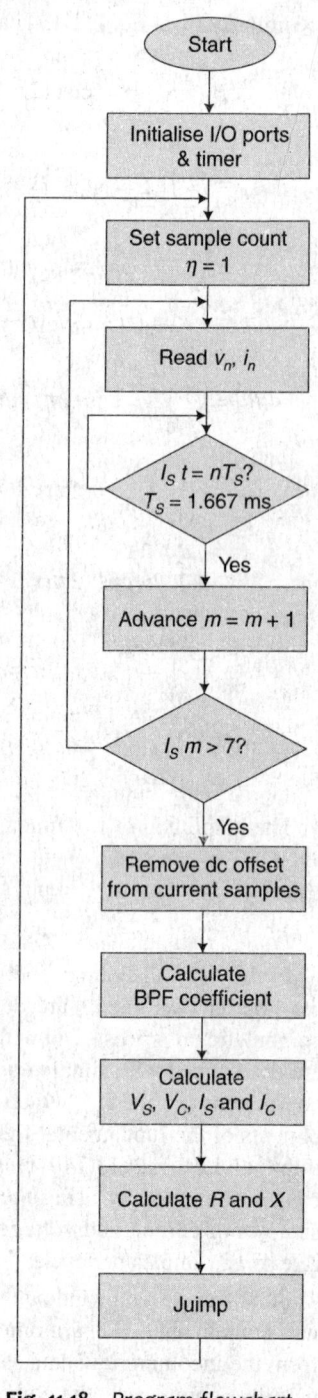

Fig. 11.18 Program flowchart for computation of R and X.

the microprocessor implementation of numerical relaying algorithms is discussed in Chapter 12.

11.11 WAVELET TRANSFORM TECHNIQUE

Wavelet theory is the mathematics associated with building a model for a non-stationary signal, with a set of components that are small waves, called wavelets. There are some conditions that must be met for a function to qualify as a wavelet. They must be oscillatory and have amplitudes that quickly decay t zero at both ends and have average value also equal to zero. The required oscillatory condition and damping lead to damped sinusoids as the building blocks. The product of an oscillatory function with a decay function yields the wavelet. A wavelet is given by Eq. (11.142).

$$g(t) = e^{-\alpha t^2} e^{j\omega t} \qquad (11.142)$$

Wavelets are mainpulated in two ways as shown in Fig. 11.19.

(i) Translation It is the change of central position of the wavelet along the time axis.

(ii) Scaling It is the change of frequency.

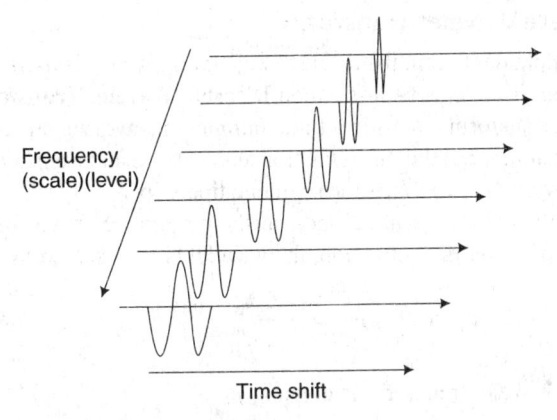

Fig. 11.19 *Change in scale (also called level) of wavelets.*

A number of different wavelets are used to approximate any given function with each wavelet generated from one original wavelet, called a mother wavelet. The new elements, called daughter wavelets are nothing but scaled and translated mother wavelets. Scaling implies that the mother wavelet is either dilated or compressed and translation implies shifting of the mother wavelet in the time domain.

The energy of the scaled daughter wavelet is normalized to keep the energy the same as the energy in the mother wavelet. Equation (11.143) describes the equation of the mother wavelet, where a is the scaling factor and b the translation factor.

$$\psi_{a,b}(t) = \frac{1}{\sqrt{|a|}} \psi\left(\frac{t-b}{a}\right) \qquad (11.143)$$

In the wavelet transformation process, each of the daughter wavelets has the same number of oscillations or cycles as the mother wavelet. Since the wavelets can be of any function provided they satisfy the two conditions-oscillatory and sharp decaying to zero, any suitable wavelet can be selected depending on a particular application.

11.11.1 Continuous Wavelet Transform

The wavelet transform can be accomplished in two different ways depending on what information is required out of this transformation process. The first method is a **Continuous Wavelet Transform (CWT)** where the transformation is done in smooth and continuous fashion and one obtains a surface of wavelet coefficients, $W(a, b)$, for different values of scaling factor a and translation b. The original signal $f(t)$ has one independent variable t, but the wavelets have two independent variables a and b. Therefore a one dimensional signal has been transformed to a two dimensional function of scale (a) and translation (b). The magnitude of the wavelet coefficients provides the information on how close the scaled the translated wavelet is to the original signal. We define continuous wavelet transform as

$$W(a, b) = \int f(t) \frac{1}{\sqrt{|a|}} \psi\left(\frac{t-b}{a}\right) dt \qquad (11.144)$$

For a scaling parameter a, we translate the wavelet by varying the parameter b.

11.11.2 Discrete Wavelet Transform

For computer implementation, the discrete wavelet transform is used. If the transformation is done in discrete steps it is called **Discrete Wavelet Transform (DWT)**. A discrete wavelet transform results in a finite number of wavelet coefficient depending upon the integer number of the discretization step in scale and translation, denoted by m and n respectively. If a_0 and b_0 is the segmentation step sizes for the scale and translation respectively, then the scaling and translating parameter will be $a = a_0^m$ and $b = nb_0 a_0^m$. Therefore after discretization, the wavelet is represented by

$$\psi_{m,n}(t) = \frac{1}{\sqrt{a_0^m}} \psi\left(\frac{t - nb_0 a_0^m}{a_0^m}\right) \qquad (11.145)$$

And discrete wavelet transform is given by

$$\text{DWT}_{(m,n)} = a_0^{-m/2} \int_{-\infty}^{\infty} f(t) \, \psi(a_0^{-m} t - nb_0) \, dt \qquad (11.146)$$

where, $\psi[n]$ is the mother wavelet, and the scaling and translation parameters a and b of Eq. (11.148) are functions of an integer parameter m, $a = a_0^m$ and $b = nb_0 a_0^m$. The result is geometric scaling, i.e. by 1, $1/a$, $1/a^2$ ... and translation by 0, n, $2n$ ... This scaling gives the DWT logarithmic frequency coverage.

Since the purpose of the discretization process is to eliminate the redundancy of the continuous form and to ensure inversion, the choice of a_0 and b_0 must be made so that mother wavelets form an orthonormal basis. This condition is satisfied for instance, if $a_0 = 2$ and $b_0 = 1$. And this choice leads to multi resolution analysis (MRA). The DWT output can be represented in a two dimensional grid with different divisions in time and frequency for the signal as shown in Fig. 11.20. The rectangles in the figure have equal area or constant time bandwidth product such that they narrow at the lower scales (higher frequencies) and widen at the higher scales (lower frequencies) and are shaped proportionally to the magnitude of the DWT output for the input signal.

The DWT isolates the transient component in the top frequency band at precisely the quarter-cycle of its occurrence while the 50 Hz component is represented as a continuous magnitude. This illustrates how the multi-resolution properties of the wavelet transform are well suited to transient signals superimposed on a continuous fundamental.

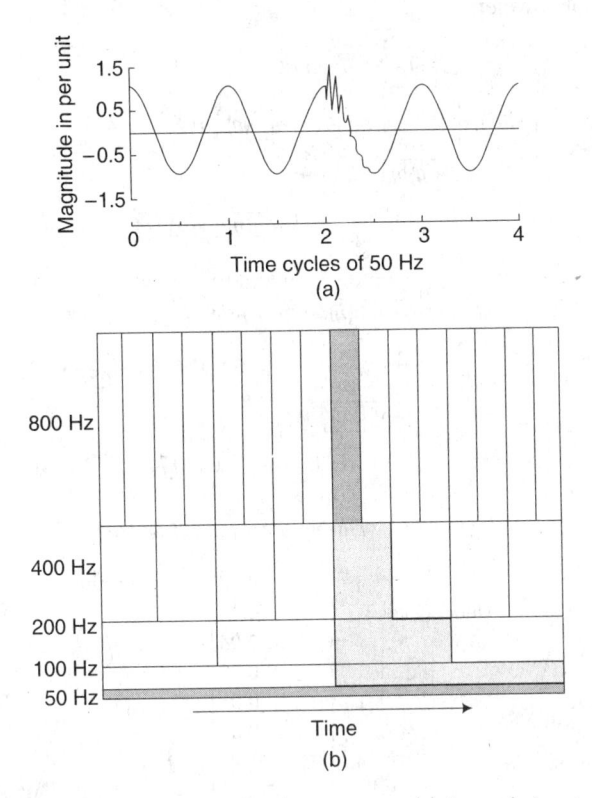

Fig. 11.20 Discrete wavelet transform output: (a) Example input signal (b) DWT on input signal

The inverse discrete wavelet transform is given by

$$f(t) = k \sum_{m=0}^{\infty} \sum_{n=0}^{\infty} W_g f(m,n) a_0^{m/2} g(a_0^{-m} t - nb_0) \qquad (11.147)$$

where K is a constant that depends upon the redundancy of the mother wavelet.

Multistage filter bank implementation of DWT also known as wavelet decomposition tree, is shown in Fig. 11.21, where $x(n)$ is the original signal, and $h(n)$ and $g(n)$ is high-pass and low-pass filters, respectively. $2\downarrow$ Represents down sampling of the input signal by two for the next stage, and $a1$ and $d1$ represent approximations and details at first scale, $a2$ and $d2$ represent approximations and details at second scale, and so on.

Three are several different families of wavelets. Within each family of wavelets (such as the Daubechies family) are wavelet subclasses distinguished by the number of coefficient and by the level of iteration. Wavelets are classified within a family most often by the number of vanishing moments. This is an extra set of mathematical relationships for the coefficients that must be satisfied, and is directly related to the number of coefficients. Figure 11.22 shows examples of different wavelets. The number next to the wavelet name represents the number of vanishing moments (a stringent mathematical definition related to the number of wavelet coefficients) for the subclass of wavelet.

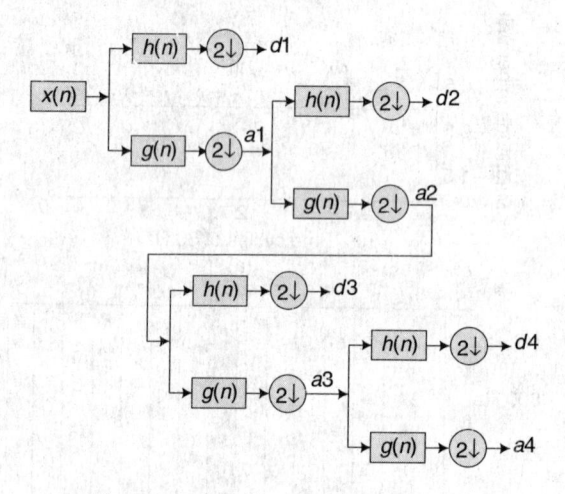

Fig. 11.21 *Wavelet decomposition tree*

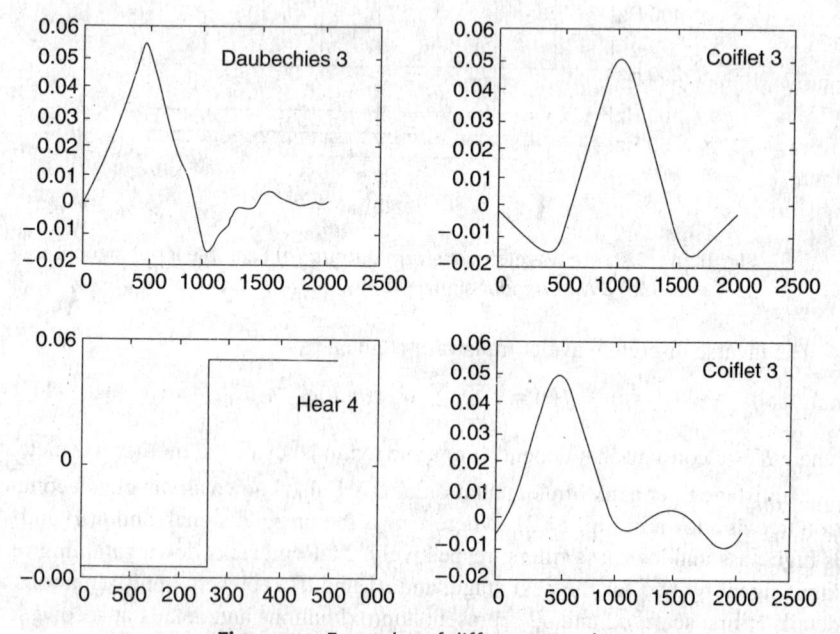

Fig. 11.22 *Examples of different wavelets*

11.11.3 Wavelet Transform Versus Fourier Transform

Similarities Between Fourier and Wavelet Transform

The fast Fourier transform (FFT) and the discrete wavelet transform (DWT) are both linear operations that generate a data structure that contains $\log_2 n$ segments of various lengths, usually filling and transforming it into a different data vector of length 2^n. The mathematical properties of the matrices involved in the transforms are similar as well. The inverse transform matrix for both the FFT and the DWT is the transpose of the original. As a result, both transforms can be viewed as a rotation in function space to a different domain. For the FFT, this new domain contains basis functions that are sines and cosines. For the wavelet transform, this new domain contains more complicated basis functions called wavelets, mother wavelets, or analyzing wavelets.

Both transforms have another similarity. The basis functions are localized in frequency, making mathematical tools such as power spectra (how much power is contained in a frequency interval) and scale grams useful at picking out frequencies and calculating power distributions.

Dissimilarities Between Fourier and Wavelet Transforms

The most interesting dissimilarity between these two kinds of transforms is that individual wavelet functions are localized in space. Fourier sine and cosine functions are not. This localization feature, along with wavelets' localization of frequency, makes many functions and operators using wavelets sparse" when transformed into the wavelet domain. This sparseness, in turn, results in a number of useful applications such as data compression, detecting features in images, and removing noise from time series.

One way to see the time-frequency resolution differences between the Fourier transform and the wavelet transform is to look at the basis function coverage of the time-frequency plane. Figure 11.23 shows a windowed Fourier transform, where the window is simply a square wave. The square wave window truncates the sine or cosine function to fit a window of a particular width. Because a single window is used for all frequencies in the WFT, the resolution of the analysis is the same at all locations in the time-frequency plane.

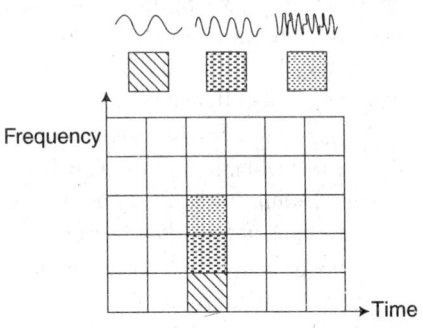

Fig. 11.23 *Fourier basis functions, time-frequency tiles, and coverage of the time-frequency plane.*

An advantage of wavelet transforms is that the windows vary. In order to isolate signal discontinuities, one would like to have some very short basis functions. At the same time, in order to obtain detailed frequency analysis, one would like to have some very long basis functions. A way to achieve this is to have short high-frequency basis functions and long low-frequency ones. This happy medium is exactly what you get with wavelet transforms. Figure 11.24 shows the coverage in the time-frequency plane with one wavelet function, the Daubechies wavelet.

The wavelet transforms do not have a single set of basis functions like the Fourier transform, which utilizes just the sine and cosine functions. Instead, wavelet transforms have an infinite set of possible basis functions. Thus wavelet analysis provides immediate access to information that can be obscured by other time-frequency methods such as Fourier analysis.

The wavelet transform is well suited to wideband signals that may not be periodic and may contain both sinusoidal and impulse components as is typical of fast power system transients. In particular, the ability of wavelets to focus on short time intervals for high-frequency components and long intervals for low-frequency components improves the analysis of signals with localized impulse and oscillations, particularly in the presence of a fundamental and low-order harmonics.

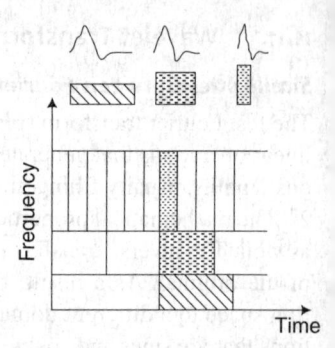

Fig. 11.24 *Daubechies wavelet basis functions, time-frequency tiles, and coverage of the time-frequency plane*

11.11.4 Application of Wavelet Transform to Power System Protection

Power System Disturbances

A power quality problem can be seen as any occurrence manifested in voltage, current, or frequency deviation that results in failure or misoperation of electric equipment. Without determining the existing levels of power quality, electric utilities cannot adopt suitable strategies to provide a better services. Therefore an efficient approach of justifying these electric power quality disturbances is motivated. The various disturbances such as voltage sag, voltage swell, momentary interruption, oscillatory transient and flat top wave shape can be detected and classified using discrete wavelet transform.

Voltage sag is often caused by the fault current, the switching of heavy loads or the starting of large motors. The disturbance describes a 10-90% of the rated system voltage lasting for half a cycle to 1 min. When the voltage drops 30% or more, the system status is considered severe.

Voltage swell often appears on the unfaulted phases of a three phase circuit, where a single phase short circuit occurs. It is a small increases of the rated voltage. This short fall stress the delicate equipment components to premature failure.

A momentary interruption can be seen as a momentary loss of voltage on a power system. Such disturbances describe a drop of 90–100% of rated system voltage lasting for half a cycle to 1 min.

Oscillatory transient is a disturbance whose duration is shorter than sags and swells can be categorized as transient. When a transient exhibits impulsive characteristics it can be called an impulsive transient, which is often caused by lightening or the load switching. When a transient displays oscillatory properties, it can be called an oscillatory transient. Such a case may happen when utility capacitor banks are customarily switched into service only in the morning in anticipation of a higher power demand.

Flat top wave shape outlines a waveform where flattened shapes occur near the peaks periodically. This disturbance may be attributed to a load that compares the dissimilarity between consecutive cycles for classification; they often fail to classify this disturbance because this shortfall periodically shows in each cycle. This inspires the wavelet based method to be an alternative for such applications.

With the world wide proliferation of sensitive power electronic equipment in applications nowadays, any short fall in power system may bring unexpected impacts to industries and utilities. An approach that serves the role of pre-warning for the power quality is very much required. The discrete wavelet transform approach not only detects the shortfalls of the power quality but also classifies the different kinds of disturbances.

Estimation of Harmonic Groups

The discrete wavelet transform can be used to estimate the harmonic groups in voltage and current waveforms. It permits decomposing a power system waveform into uniform frequency bands with an adequate selection of the sampling frequency and the wavelet decomposition tree. The harmonic frequencies can be selected to be in the centre of each band in order to avoid the spectral leakage associated with the imperfect frequency response of the filter bank employed.

Active Power Calculation

When the voltage and current waveform are analyzed using wavelets, the active power can be calculated by simply adding the products of wavelet coefficients without having to reconstruct the signals back to time domain first and then use the traditional integration. Where there is a timeliness requirement, simple and accurate algorithms can calculate electrical quantities directly in the wavelet domain which is essential for real-time application of wavelets and beneficial for compressing electric data of large volume.

Transmission Lines Protection

Transmission lines relaying must be able to detect, locate, estimate and classify disturbances on the supply lines to safeguard the power systems. Therefore, it must be supported by suitable measurement and fast acting methods. The main target of the relay is to calculate the impedance at the fundamental frequency between the relay and the fault point. According to the calculated impedance, the fault is identified as internal or external to the protection zone of the transmission line. The impedance is calculated from the measured voltage and current signals at the relay location. In addition to the fundamental frequency, the signals usually contain some system harmonics and the DC component, which has to be filtered out for the accuracy of the operation of the relay. To apply the wavelet transform for the transmission line relaying the selection of the appropriate wavelet function is very important. The wavelet with desirable frequency is first chosen as the mother wavelet and then the shifted and dilated versions are used to perform analysis. The relay identifies the type of fault that has occurred on the transmission line (i.e., L-G, L-L, L-L-G, L-L-L, L-L-L-G) and its location. In addition it also identifies any other abnormal state than a faulty one, i.e. load jumps, transients, etc. The discrete wavelet analysis when applied to fault detection and classification is found to be most suitable. The algorithm based on Discrete Wavelet Transform has been developed which is independent of fault location, fault inception angle and fault impedance, hence it is simple, robust and generalized. It can be used for the identification of high impedance faults as well as for transmission lines at any voltage level.

Transformer Protection

Transformers are essential and important elements of electric power system and their protection are critical. Traditionally, transformer protection was based on the differential protection schemes but now the studies have focused on discrimination between internal short circuit faults and inrush and over excitation currents to reduce the mal operation of the relays.

The application of discrete wavelet transform on the differential current reveals that each waveform has distinct features and therefore using these features in the waveform signature, automated recognition can be accomplished, i.e. to differentiate between internal fault and magnetizing inrush and over excitation current. The algorithms have been developed which successfully differentiates between magnetizing inrush and fault condition in less than half power frequency cycle.

The incipient faults in equipment containing insulation material such as transformers, cable, etc., are also very important. Detection of these types of faults can provide information to predict failures ahead of time so that the necessary corrective actions are taken to prevent outages and reduce down times. Hence a powerful method based on discrete wavelet transform can be used to discriminate between normal and abnormal operating cases that occur in distribution system related to transformers such as external faults, internal faults, magnetizing inrush, over excitation, load changes, aging, arcing, etc.

Synchronous Generator Protection

Synchronous generators play a very important role in the stability of the power system. Due to its huge size and great cost, the various faults may cause huge loss to the system and severe damage to the machine. The differential protection is used for the protection of synchronous generators. The wideband high frequency signals in a generator created by a fault are usually outside the bandwidth of most current transformers. Hence the differential current which be measured by the protective device contain significant transient component and have much more fault information than the steady state component. As wavelets are well suited to the analysis of the non-stationary signals, one will be able to extract important information from the distorted signals. This information can be used to detect the internal faults.

The stator ground fault is the most common fault to which a generator is subjected. The discrete wavelet transform result of the third harmonic voltages at the generator terminals and neutral has the ability to discriminate the generator stator ground fault from the other distorted signals which may be due to over excitation or sudden reduction of load. Hence it can be used in detecting the stator ground fault with high sensitivity and selectivity.

Conclusion

The basic features of discrete wavelet transform have been presented. Since the analyzing function is localized in time and frequency, it offers important advantages over Fourier methods for power signals containing high frequency transients. The property of multiresolution in time and frequency provided by wavelets allows accurate time location of transient components while simultaneously retaining information about the fundamental frequency and its lower order harmonics. This property of the discrete wavelet transform facilities the detection of physically relevant features in transient signals to characterize the source of the transient or the state of the post-disturbance system. The application of discrete wavelet transform reveals that each waveform has distinct features and using these features in the waveform signature, automated recognition can be accomplished.

11.12 REMOVAL OF THE DC OFFSET

The exponentially decaying dc offset present in some of the relaying signals gives rise to fairly large errors in the phasor estimates unless the offset terms are removed prior to the execution of the algorithms. Fourier, Walsh and Haar algorithms can either be of full-cycle data window or half-cycle data window. The full cycle data window algorithms raise the operating time of the relays to more then one cycle. Since the distance relays operation within a cycle after fault inception is desirable, the half-cycle data window algorithms are preferable. The half-cycle window Fourier, Walsh and Haar algorithms in particular require that the dc offset be removed prior to processing, while the differential equation algorithms ideally do not require its elimination.

In practice, the decaying dc offset component in the voltage signal is so small that it can be neglected. The dc offset component in the fault current signal is filtered out by employing the replica impedance filtering technique.

Replica impedance filtering prior to signal processing improves the accuracy obtainable from half-cycle window algorithms and its use has therefore been advocated. The conventional method of eliminating the dc offset component is with the use of a mimic (or replica) impendence across the CT secondary terminals. The mimic impedance used in the current transformer secondary plays the role of an analog filter in as much as it filters out dc offset in the current signal. The use of mimic impedance eliminates the exponentially decaying dc offset when the x/r ratio of the mimic impedance matches with the X/R ratio of the fault circuit (or primary circuit). Complete suppression of the dc offset by using mimic (or replica) impedance is, however impossible, since exact matching of the X/R ratios of the primary and secondary circuits is difficult owing to the fact that the X/R ratio of the primary circuit up to the point of fault is variable because of circuit switching or fault arc resistance. In addition, any noise (or extraneous high frequency components) present in the current tends to be amplified by the mimic circuit. In case of microprocessor-based relays, the low-pass antialiasing filter attenuates a significant part of the noise.

As microcomputer capabilities have improved, the dc offset component can easily be removed by employing a digital replica impedance filtering technique. In this approach, the dc offset is removed with the help of software.

The fault currents often contain an exponentially decaying dc offset term and odd harmonics. A general expression for such signal is

$$i(t) = I_0 e^{-t/\tau} + I_{m1} \sin(\omega t + \theta_1) + I_{m3} \sin(3\omega t + \theta_3) + \ldots \qquad (11.148)$$

where $\tau = L/R$ is the time constant of the faulted circuit.

If T_s be the sampling interval and N be the number of samples per cycle, the instantaneous value of the current at any Kth sampling instant can be given by

$$i_K = I_0 e^{-KTs/\tau} + I_{m1} \sin(\omega KT_s + \theta_1) + I_{m3} \sin(3\omega KT_s + \theta_3) + \ldots \quad (11.149)$$

and the instantaneous value of the current at $(K + M)$ th sampling instant where, $M = N/2$ can be given by

$$i_{K+M} = I_0 e^{-(K+M)Ts/\tau} + I_{m1} \sin\{\omega(K+M)T_s + \theta_1\}$$

$$+ I_{m3} \sin\{3\omega(K+M)T_s + \theta_3\} + \ldots \quad (11.150)$$

Now $\quad M\omega T_s = 180°$

Therefore, from Eqs. (11.149) and (11.150)

$$i_K + i_{K+M} = I_0 e^{-KTs/\tau} + I_0 e^{-(K+M)Ts/\tau}$$

$$= I_0 e^{-KTs/\tau}[1 + e^{-MTs/\tau}]$$

$$= I_0 \beta^K [1 + B^M] \quad (11.151)$$

where $\beta = e^{-Ts/\tau}$ is the decrement ratio for the dc offset between consecutive samples and $I_0 \beta^K$ is the dc offset component in sample K.

Similarly,

$$i_{K+1} + i_{(K+1)+M} = I_0 \beta^{K+1}[1 + B^M] \quad (11.152)$$

From Eqs. (11.151) and (11.152),

$$\beta = \frac{i_{K+1} + i_{(K+1)+M}}{i_K + i_{K+M}} \quad (11.153)$$

Once β is known, $I_0 \beta^K$ can be calculated from Eq. (11.153). This approach requires $(N/2 + 2)$ samples after the onset of the fault.

The new sample set I'_K of the current signal is given by

$$i'_K = i_K - I_0 \beta^K = i_K - (dc)_K \quad (11.154)$$

Therefore, i'_K does not contain any dc offset component. i'_K is processed for the extraction of the fundamental frequency components and calculation of impedance.

If β is assumed to be determined solely by the X/R ratio of the transmission line, it is supposed to be a known parameter. The dc offset component in sample K can be estimated by using Eq. (11.151). This approach requires $(N/2 + 1)$ samples after the onset of the fault. The multiplication of the dc offset components of the previous current sample by β yields the dc offset component in the next sample. The new current samples which do not contain any dc component are then evaluated by using Eq. (11.154).

11.13 NUMERICAL OVERCURRENT PROTECTION

Overcurrent protection is the simplest form of power system protection. It is widely used for the protection of distribution lines, motors and power equipment. It incorporates overcurrent relays for the protection of an element of the power system. An overcurrent relay operates when the current in any circuit exceeds a certain predetermined value. The value of the predetermined current above which the overcurrent relay operates is known as its pick-up value $I_{pick-up}$.

A protection scheme which incorporates numerical overcurrent relays for the protection of an element of a power system, is known as a numerical overcurrent protection scheme or numerical overcurrent protection. A straight forward application of the numerical relay is in numerical overcurrent protection. A numerical overcurrent relay acquires sequential samples of the current in numeric (digital) data form through the Data Acquisition System (DAS), and processes the data numerically using a numerical filtering algorithm to extract the fundamental frequency component of the current and make trip decision. In order to make the trip decision, the relay compares the fundamental frequency component of the current (I) with the pick-up setting and computes the plug setting multiplier (PSM), given by ($I/I_{pick-up}$) at which the relay has to operate. If the fundamental frequency component of the fault current (I) exceeds the pick-up $I_{pick-up}$ (i.e., PSM > 1), the relay issues a trip signal to the circuit breaker. The time delay required for the operation of the relay depends on the type of overcurrent characteristic to be realised. In case of instantaneous overcurrent relay there is no intentional time delay. For definite time overcurrent relay, the trip signal is issued after a predetermined time delay. In orders to obtain inverse-time characteristics, the relay either computes the operating time corresponding to the fault current or selects the same from the look-up table.

The numerical filtering algorithms based on DFT, FWHT, RHT and BPF which are discussed in Sections 11.7, 11.8, 11.9 and 11.10 respectively, can be used for extraction of the fundamental frequency component of the fault current I. The fundamental Fourier sine and cosine coefficients (F_1 and F_2) are respectively equal to real and imaginary components (I_S and I_C) of the fundamental frequency current phasor I. I in complex form is given by

$$I = I_S + jI_C$$

and
$$|I| = \sqrt{I_S^2 + I_C^2} \tag{11.155}$$

The operating time t can be computed by using the following expression for time-current characteristic.

$$t = \frac{k}{I^n - 1} \tag{11.156}$$

The values of k and n for various characteristics are as follows:

Characteristic	k	n
(i) IDMT	0.14	0.02
(ii) Very inverse	13.50	1.00
(iii) Extremely Inverse	80.00	2.00
(iv) Definite-Time	Any value based on design	0

The block diagram of a typical numerical overcurrent relay is shown in Fig. 11.25. The current derived from the Current Transformer (CT) is applied to the signal conditioner for electrical isolation of the relay from the power system, conversion of current signal into proportional voltage signal and removal of high frequency components from the signals using analog low-pass filter. The output of the signal conditioner is applied to the analog interface which includes S/H circuit, analog multiplexer and A/D converter (ADC).

After quantization by the A/D convertor, along current (i.e., voltage proportional to current) is represented by discrete values of the samples taken at specified instants of time. The current in the form of discrete numbers is processed by a numerical filtering algorithm which is a part of the software. The algorithm uses signal-processing technique to estimate the real and imaginary components of the fundamental frequency current phasor. The measured value of the current is compared with the pick-up value to decide whether there is a fault or not. If there is a fault in any element of the power system, the relay sends a trip command to circuit breaker for isolating the faulty element.

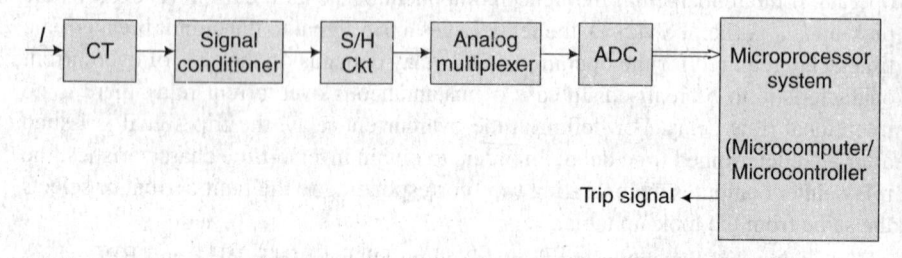

Fig. 11.25 Block diagram of a typical numerical overcurrent relay

11.14 NUMERICAL DISTANCE PROTECTION

Distance protection is a widely used protective scheme for the protection of transmission and sub-transmission lines. It employs a number of distance relays which measure the impedance or some components of the line impedance at the relay location. Since the measured quantity is proportional to the distance (line-length) between the relay location and the fault point, the measuring relay is called a distance relay.

A distance protection scheme which incorporates numerical distance relays for the protection of lines is known as a numerical distance protection scheme or numerical distance protection. In a numerical distance relay, the analog voltage and current signals monitored through primary transducers (VTs and CTs) are conditioned, sampled at specified instants of time and converted to digital form for numerical manipulation, analysis display and recording. The voltage and current signals in the form of discrete numbers are processed by a numerical filtering algorithm to extract the fundamental frequency components of the voltage and current signals and make trip decisions.

The extraction of the fundamental frequency components from the complex post-fault voltage and current signals that contain transient dc offset component and harmonic frequency components, in addition to the power frequency fundamental components, is essential because the impedance of a linear system is defined in terms of the fundamental frequency voltage and current sinusoidal waves. The numerical filtering algorithms based on DFT, FWHT, RHT and BPF which are discussed in Sections 11.7, 11.8, 11.9 and 11.10 respectively can be used for extraction of the fundamental frequency components of the voltage and current signals. The fundamental Fourier sine and cosine coefficients for both voltage and current signals are computed as $F_{1(v)}$, $F_{2(v)}$ and $F_{1(i)}$, $F_{2(i)}$ respectively by using any numerical filtering algorithm. $F_{1(v)}$ and $F_{2(v)}$ are respectively equal to the real and imaginary components of the fundamental frequency voltage phasor, namely V_S and V_C; and $F_{1(i)}$ and $F_{2(i)}$ are respectively the real and imaginary components of the fundamental frequency current phasor, namely I_S and I_C.

Knowing V_S, V_C, I_S and I_C, the fundamental frequency voltage and current are expressed in complex form as

$$V = V_S + jV_C \quad \text{and} \quad I = I_S + jI_C$$

The real and reactive components (R and X) of the apparent impedance of the line as seen by the relay are computed from V_S, V_C, I_S and I_C by using Eqs. (11.68) and (11.69) given in Sec. 11.7.5.

Using the computed values of R and X, the relay examines whether the fault point lies within the defined protective zone or not. If the fault point lies in the protective zone of the relay, the relay issues a trip signal to the circuit breaker.

The block diagram of a typical numerical distance relay is shown in Fig. 11.26.

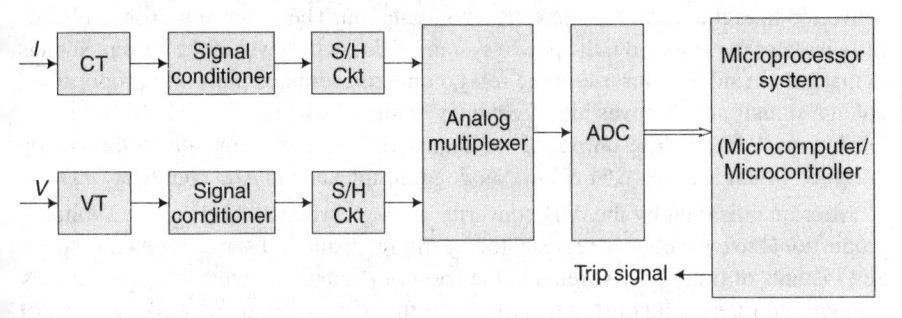

Fig. 11.26 Block diagram of a typical numerical distance relay

11.15 NUMERICAL DIFFERENTIAL PROTECTION

Differential protection is one of the most sensitive and effective methods of protection of electrical equipment (e.g., generators, transformers, large motors, bus zones, etc.) against internal faults. It is based on current comparison. It makes use of the fact that any internal fault in an electrical equipment would cause the current entering it, to be different from that leaving it. The percentage differential protection is widely used for the protection of electrical equipment against internal faults. The main component of a percentage differential protection scheme is the percentage differential relay. Percentage differential relay provides high sensitivity to light internal faults with high security (high restraint) for external faults and makes differential protection scheme more stable.

The operating condition of the percentage differential relay is given by the following expression:

$$(I_{1S} - I_{2S}) \geq K \frac{(I_{1S} + I_{2S})}{2} \qquad (11.157)$$

where, I_{1S} and I_{2S} are the secondary currents of CT_1 and CT_2 respectively, i.e., CT_S at the two ends of the protected equipment (element).

$(I_{1S} - I_{2S}) = I_d$ is the differential operating current.

$(I_{1S} + I_{2S})/2 = I_r$ is the restraining current or through current

and $\qquad K$ = slope or Bias

At the threshold of operation of the relay, the differential operating current (I_d) is equal to a fixed percentage of the restraining current (I_r); and for operation of the relay the differential operating current must be greater than this fixed percentage of the restraining (through fault) current.

A differential protection scheme which incorporates numerical differential relays for the protection of electrical equipment is known as numerical differential protection scheme or numerical differential protection. In a numerical differential relay, the current signals I_{1S} and I_{2S} monitored through current transformers CT_1 and CT_2 are conditioned by the signal conditioner, sampled at specified instants of time and converted to digital form for numerical manipulation. The signal conditioner electrically isolates the relay from the power system, reduces the level of the current signals to make them suitable for use in the relay, converts current signals into proportional voltage signals and removes high frequency components from the signals using analog low-pass filters. The output of the signal conditioner is applied to the analog interface which includes S/H circuit, analog multiplexer and A/D converter (ADC).

After quantization by the A/D converter analog current I_{1S} and I_{2S} (i.e., voltages proportional to currents) are represented by discrete values of samples taken at specified instants of time. The currents in the form of discrete numbers are processed by a numerical filtering algorithm to extract the fundamental frequency components of the current signals I_{1S} and I_{2S}, and make trip decisions.

The numerical filtering algorithms based on DFT, FWHT, RHT and BPF which are discussed in Sections 11.7, 11.8, 11.9 and 11.10 respectively can be used for extraction of the fundamental frequency components of the current I_{1S} and I_{2S}. Using the fundamental frequency components of I_{1S} and I_{2S}, the relay examines whether the operating condition of the relay given by Eq. (11.157) is satisfied or not. If the operating condition is satisfied, the relay issues a trip signal to the circuit breaker.

In case of numerical differential protection of power transformer, second and fifth harmonic components of I_{1S} and I_{2S} are also required to be extracted for examining the harmonic restraint conditions of the relay.

The block diagram of a typical numerical differential relay is shown in Fig. 11.27 and the operating characteristic of the relay is shown in Fig. 11.28.

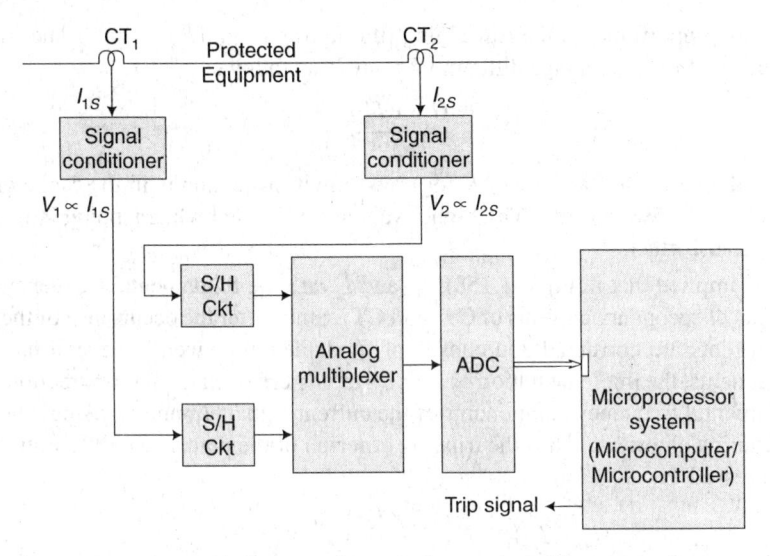

Figure 11.27 Block diagram of a typical numerical differential relay

11.15.1 Numerical Differential Protection of Generator

The tripping criterion of the numerical perecentage (biased) differential relay used for numerical protection of synchronous generator is as follows:

$$(I_{1S} - I_{2S}) \geq \frac{K(I_{1S} + I_{2S})}{2}$$

For generator protection, the value of K lies in the range of 0.05 to 0.15 (i.e. 5% to 15%)

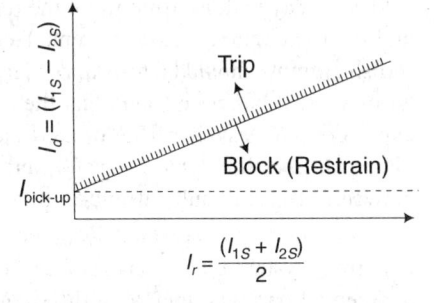

Fig. 11.28 Operating characteristic of percentage differential relay

In other words, if the differential operating current $(I_{1S} - I_{2S})$ exceeds K times the restraining (through) current, the relay issues trip signal to the circuit breaker. The differential operating current is the phasor difference of CTs secondary currents (i.e. $I_{1S} - I_{2S}$) and the restraining current is the mean through current, i.e., $(I_{1S} + I_{2S})/2$. In this protection scheme, the current I_1 entering the protected equipment (i.e. stator winding of the generator) and current I_2 leaving the protected equipment are monitored through CTs connected at each end of the winding. The secondary currents of both CTs (i.e., I_{1S} and I_{2S}) are conditioned and converted into proportional voltage signals. The currents (i.e., voltages proportional to currents) in the form of discrete numbers are processed by a numerical filtering algorithm to extract the fundamental frequency components of the current signals I_{1S} and I_{2S}, and make trip decision.

11.15.2 Numerical Differential Protection of Power Transformer

In percentage differential protection scheme for power transformer, the operating signal is proportional to the phasor difference of currents I_{1S} and I_{2S} and the restraining signal is proportional to the restraining (through) current $(I_{1S} + I_{2S})/2$. The tripping condition of the percentage differential relay is as follows:

$$(I_{1S} - I_{2S}) \geq K \frac{(I_{1S} + I_{2S})}{2} \quad (11.158)$$

where the value of bias (slope) K for transformer protection is in the range of 0.15 to 0.60 (i.e., 15% to 60%). The higher value of K is used when tapping is included in the transformer.

It is implied that in Eq. (11.158), I_{1S} and I_{2S} are the fundamental frequency components of secondary currents of CT_1 and CT_2. Since after the occurrence of the fault, the currents are corrupted and consist of fundamental as well as several harmonic components, the implementation of above tripping criterion requires extraction of the fundamental frequency components of the current signals I_{1S} and I_{2S} using a numerical filtering algorithm. Thus the tripping criterion of the percentage differential relay is expressed as

$$(I_{1S} - I_{2S})_1 \geq K \left(\frac{I_{1S} + I_{2S}}{2} \right)_1 \quad (11.159)$$

where 1 refers to the fundamental frequency components of I_{1S} and I_{2S}.

There may be false tripping of the relay due to magnetizing inrush current (MIC) and over-excitation inrush current. Hence the operation of the relay due to these inrush currents should be restrained. The magnetizing inrush current is rich in second harmonics. It has been found that the second harmonic content of the MIC is never expected to be less than 15% of the fundamental frequency component for both single-phase and three-phase transformers. This observation is used for distinguishing between MIC and fault current for relaying purpose.

Short duration overexcitation (over voltage) which may arise occasionally during abnormal system conditions may cause saturation of transformer resulting in high differential currents. Increased differential current due to overexcitation is for short duration and may resemble internal fault current causing erroneous operation of percentage differential relay.

Over-excitation causes an increase in inrush current with a large increase in the odd harmonics components, especially the third and the fifth harmonic components. Since the third harmonic component gets trapped inside a delta-connection of the transformer, it is not possible to monitor the third harmonic component of the over-excitation inrush current. Hence the only option left is to monitor the fifth harmonic component. It has been found that the fifth harmonic content of the over-excitation in rush current not less than 8% of the fundamental frequency component. Thus the existence of fifth harmonic component is excess of 8% of the fundamental frequency component is an indication of overexcitation (over voltage) inrush current for which the relay must not operate.

On the basis of above discussions, the magnetizing inrush and overexcitation inrush conditions for which the relay must not operate are expressed by the following expressions.

Magnetizing inrush condition exists when,

$$(I_{1S} - I_{2S})_2 \geq 0.15 \, (I_{1S} - I_{2S})_1 \quad (11.160)$$

and over-excitation inrush condition exists when

$$(I_{1S} - I_{2S})_5 \geq 0.08 \, (I_{1S} - I_{2S})_1 \quad (11.161)$$

In the above expression, 2 and 5 refer to second and fifth harmonic components respectively.

Combining Eqs. (11.159) to (11.161), the final operating (tripping) criterion is given by

$$(I_{1S} - I_{2S})_1 \geq K \left(\frac{I_{1S} + I_{2S}}{2} \right)_1 \quad (11.162)$$

and the final restraining (blocking) criterion is given by

$$(I_{1S} - I_{2S})_2 \geq 0.15 \, (I_{1S} - I_{2S})_1$$

$$(I_{1S} - I_{2S})_5 \geq 0.08 \, (I_{1S} - I_{2S})_1 \quad (11.163)$$

The current signals I_{1S} and I_{2S} monitored through current transformers CT_1 and CT_2 are conditioned by the signal conditioner, sampled at specified instants of time and converted to digital form for numerical manipulation. The currents in the form of discrete numbers are processed by numerical filtering algorithm to extract the fundamental frequency, second harmonic and fifth harmonic components of currents I_{1S} and I_{2S}. Fundamental frequency components of the currents are used to examine the operating condition given by eq. (11.162) where as second and fifth harmonic components are used to examine the blocking conditions given by Eq. (11.163). The relay issues a trip signal if the criterion given by Eq. (11.162) is satified.

The numerical filtering algorithms for extraction of the fundamental frequency components of the currents are already discussed in Sections 11.7, 11.8, 11.9 and 11.10. Second and fifth harmonic components of the current can be extracted by using the equations as given below.

Algorithm Based on DFT

The Fourier coefficients F_3 and F_4 are the real and imaginary components of the second harmonic phasor. F_3 and F_4 are obtained from Eqs. (11.41) and (11.40) by substituting $K = 2$, for second harmonic component.

$$F_3 = \frac{b_2}{\sqrt{2}} = \frac{\sqrt{2}}{N} \sum_{m=0}^{N-1} i_m \sin \frac{4\pi m}{N} \quad (11.164)$$

$$F_4 = \frac{a_2}{\sqrt{2}} = \frac{\sqrt{2}}{N} \sum_{m=0}^{N-1} i_m \cos \frac{4\pi m}{N} \quad (11.165)$$

Similarly for fifth harmonic component $K = 5$, and the Fourier coefficient F_9 and F_{10} are the real and imaginary components of the fifth harmonic phasor.

F_9 and F_{10} are given by

$$F_9 = \frac{b_5}{\sqrt{2}} = \frac{\sqrt{2}}{N} \sum_{m=0}^{N-1} i_m \sin \frac{10\pi m}{N} \quad (11.166)$$

$$F_{10} = \frac{a_5}{\sqrt{2}} = \frac{\sqrt{2}}{N} \sum_{m=0}^{N-1} i_m \cos \frac{10\pi m}{N} \quad (11.167)$$

Algorithm Based on FWHT

On expansion of Eq. (11.83) the following expressions for F_3, F_4, F_9 and F_{10} can be obtained.

$$F_3 = 0.9\ W_{W3} - 0.373\ W_{W11} \tag{11.168}$$

$$F_4 = 0.9\ W_{W4} + 0.373\ W_{W12} \tag{11.169}$$

$$F_9 = 0.180\ W_{W1} + 0.435\ W_{W5} + 0.65\ W_{W9} - 0.269\ W_{W13} \tag{11.170}$$

$$F_{10} = 0.180\ W_{W2} - 0.435\ W_{W6} + 0.65\ W_{W10} + 0.269\ W_{W14} \tag{11.171}$$

Algorithm Based on RHT

Using Eqs. (11.112) and (11.115), the Fourier coefficients F_3, F_4, F_9 and F_{10} can be written in terms of the RHT coefficients as

$$\begin{aligned} F_3 = &\ 0.0533\ (C_{rh2} + C_{rh3}) + 0.022\ (-C_{rh4} + C_{rh5} - C_{rh6} + C_{rh7}) \\ &+ 0.0312\ (-C_{rh8} + C_{rh10} - C_{rh12} + C_{rh14}) \\ &+ 0.0129\ (C_{rh9} - C_{rh11} + C_{rh13} - C_{rh15}) \end{aligned} \tag{11.172}$$

$$\begin{aligned} F_4 = &\ 0.022\ (C_{rh2} + C_{rh3}) + 0.0533(C_{rh4} - C_{rh5} + C_{rh6} - C_{rh7}) \\ &+ 0.0129\ (C_{rh8} - C_{rh10} + C_{rh12} - C_{rh14}) \\ &+ 0.0312\ (C_{rh9} - C_{rh11} + C_{rh13} - C_{rh15}) \end{aligned} \tag{11.173}$$

$$\begin{aligned} F_9 = &\ 0.0074\ C_{rh1} + 0.011\ (-C_{rh2} + C_{rh3}) + 0.044\ (C_{rh4} - C_{rh6}) \\ &+ 0.0088\ (C_{rh5} - C_{rh7}) + 0.40\ (-C_{rh8} + C_{rh12}) \\ &+ 0.014\ (-C_{rh9} + C_{rh13}) + 0.061\ (C_{rh10} + C_{rh14}) \\ &+ 0.072\ (-C_{rh11} + C_{rh15}) \end{aligned} \tag{11.174}$$

$$\begin{aligned} F_{10} = &\ 0.11\ C_{rh1} + 0.0074\ (C_{rh2} - C_{rh3}) + 0.088\ (C_{rh4} - C_{rh6}) \\ &+ 0.044\ (-C_{rh5} + C_{rh7}) + 0.061\ (C_{rh8} - C_{rh12}) \\ &+ 0.072\ (-C_{rh9} + C_{rh13}) + 0.040\ (C_{rh10} - C_{rh14}) \\ &+ 0.014\ (C_{rh11} - C_{rh15}) \end{aligned} \tag{11.175}$$

Equations (11.164), (11.165), (11.168), (11.169), (11.172) and (11.173) can be used for extraction of second harmonic components and Eqs. (11.166), (11.167), (11.170), (11.171), (11.174) and (11.175) can be used for extraction of fifth harmonic components.

EXERCISES

1. Describe the principle of numerical protection. How is this method of protection different from conventional methods?
2. With the help of block diagram, discuss the operation of the numerical relay. What is a multifunction numerical relay?
3. What are the advantages of numerical relays over conventional relays? Is there any disadvantage of numerical relays? If so, discuss the disadvantages.
4. What is a data acquisition system? Discuss the functions of various components of the data acquisition system (DAS).

5. What is the role of signal conditioner in a data acquisition system? Discuss the functions of various components of the signal conditioner.
6. What do you mean by aliasing? How can aliasing be removed? State and explain Shannon's sampling theorem.
7. Explain the sample and derivative method of estimating the rms value and phase angle of a signal. State the underlying assumptions.
8. How can the apparent impedance of the line from relay location to the fault point be calculated by using an algorithm based on the solution of the differential equation representing the transmission line model?
9. Why is it necessary to extract the fundamental frequency components from the complex post-fault relaying signals?
10. What do you mean by data window? Why is the half-cycle data window preferred over the full-cycle data window for numerical distance relaying?
11. How can R and X of the line as seen by the relay be calculated by using an algorithm based on the discrete Fourier transform (DFT).
12. What do you mean by FWHT? How can R and X of the line be calculated by an algorithm based on FWHT.
13. What do you mean by RHT? How can R and X of the line, upto the fault point, be calculated by using the algorithm based on RHT.
14. What do you mean by Block Pulse Functions (BPF)? How can the fundament frequency components be extracted from the complex post-fault relaying signals using the algorithm based on BPF?
15. What do you mean by continuous and discrete wavelet transforms? What are the similarities and dissimilarities between Fourier and wavelet transforms?
16. Discuss the application of wavelet transform to power system protection.
17. Discuss the numerical overcurrent protection of power system.
18. Describe a digital technique for the removal of the DC offset component from the current signal.
19. Discuss the operation of a numerical distance relay with the help of block diagram.
20. Discuss the numerical differential protection of electrical equipment. Draw the block diagram of a numerical differential relay.
21. What is the tripping criterion of the numerical differential relay used for percentage differential protection of synchronous generator. How can this tripping criterion be implemented.
22. Discuss the tripping and blocking criterion of the harmonic restraint numerical differential relay for the percentage differential protection of power transformer.
23. Draw the block diagram of a numerical percentage differential relay for the protection of power transformer and discuss its operation.
24. How can the second and fifth harmonic components of the CT secondary currents be extracted for examining the blocking criterion of the percentage differential relay during magnetizing and overexcitation inrush conditions of the power transformer?

12 Microprocessor-Based Numerical Protective Relays

12.1 INTRODUCTION

Computer hardware technology has tremendously advanced since early 1970s and new generations of computers tend to make digital computer relaying a viable and better alternative to the traditional relaying systems. The advent of microprocessors in the 1970s initiated a revolution in the design and development of numerical protection schemes. Microprocessors have afforded us in protective relaying the remarkable capability of sampling voltages and currents at very high speed, manipulating the data to accomplish a distance or overcurrent measurement, retaining fault information and performing self-checking functions. With the development of economical, powerful and sophisticated microprocessors, there is a growing interest in developing micorprocessor-based numerical protective relays which are more flexible because of being programmable and are superior to conventional electromechanical and static relays. The main features which have encouraged the design and development of microprocessor-based numerical protective relays are their economy, compactness, reliability, flexibility, adaptive capability, self-checking ability and improved performance over conventional relays. A number of desired relaying characteristics, such as overcurrent, directional, impedance, reactance, mho, quadrilateral, elliptical, etc. can be obtained using the same interface. Using a multiplexer, a microprocessor can get the desired signals to obtain a particular relaying characteristic. Different programs are used to obtain different relaying characteristics using the same interfacing circuitry.

A microprocessor by itself cannot perform a given task, but must be programmed and connected to a set of additional system devices which include memory elements and input/output devices. In general, a set of system devices, including the microprocessor which acts as CPU, memory, and input/output devices, interconnected for the purpose of performing some well-defined task is called a microcomputer or a microprocessor-based system. The single-chip microcomputer is called a microcontroller. The interconnection of the different components, which is a primary concern in the design of a microprocessor-based system, must take into account the nature and timing of the signals that appear at the interfaces between components. For the purpose of achieving compatibility of signals, it is generally necessary to select appropriate components and design supplimentary circuits. Therefore, to connect an input/output device to a microprocessor, an I/O interface circuit is typically interposed between the device and the system bus which is a communication path between the microprocessor and the I/O devices (peripherals). This circuit serves to match the signal formats and timing characteristics of the microprocessor interface to those of I/O interface. The overall task of connecting I/O devices and microprocessors is termed "interfacing".

12.2 IC ELEMENTS AND CIRCUITS FOR INTERFACES

Various IC elements and circuits are available today for developing interfaces for microprocessor-based systems. This section discusses a few IC circuits used for interfaces. Some other important IC elements such as A/D converter, multiplexer and Sample and Hold have been described in detail in Section 12.3.

12.2.1 Operational Amplifier

The operational amplifier popularly known as Op-Amp is a direct-coupled high gain amplifier to which feedback is added to control its overall response characteristics. It is used to perform a wide variety of linear and non-linear functions and is available in IC form. The operational amplifier has gained wide acceptance as a versatile, predictable, and economic system building block.

The schematic diagram of the Op-Amp is shown in Fig. 12.1(a). It has two inputs, namely the inverting input and non-inverting input and a single output. Figure 12.1(b) shows a basic inverting amplifier. The output voltage V_o is given by

$$V_o = -\frac{R_2}{R_1} V_i \qquad (12.1)$$

where $-(R_2/R_1)$ is the voltage gain, and V_o and V_i are output and input voltages respectively.

The voltage gain can be adjusted by selecting suitable values of R_1 and R_2, and the input voltage signal V_i can be amplified to the desired value.

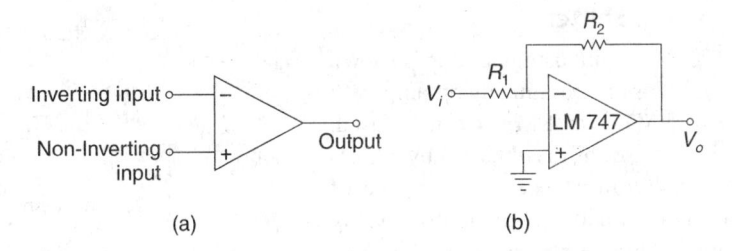

Fig. 12.1 *(a) Symbol of Op-Amp (b) Inverting Op-Amp*

When $R_1 = R_2$, the gain is equal to -1 and the circuit acts as an *inverter*, i.e. $V_o = -V_i$. The operational amplifiers LM741, LM747 and LM324 can be used for the purpose. LM741, LM747 and LM324 consist of one, two and four opamps, respectively. The selection of the IC depends upon the number of opamps required for the interfacing circuit.

12.2.2 Zero-Crossing Detector

Figure 12.2 shows the circuit of a zero-cross detector using voltage comparator LM311. Op-Amps LM741 or LM747 can also be used for developing the circuit of a zero-cross detector, but LM311 gives better results than LM741 or LM747. The output of the basic zero-cross detector circuit using LM311 or LM747 is a square wave. LM311 gives the square wave output on voltage comparison principle. Pin 4, i.e. the negative input point of LM311 is grounded which acts as a reference. When the input signal is equal to the reference voltage, the output is a square wave. Since the reference voltage is zero, the square wave output is available starting at zero instant. In the case of an operational amplifier, feedback is not applied and hence the square wave is obtained due to saturation. When the input voltage is very small, a trapezoidal output is obtained.

The square wave output can be rectified using a diode in order to obtain only the positive half of the wave. The output wave is differentiated using an R-C circuit in order to obtain a spiked output as shown in Fig. 12.2. The magnitude of the output voltage becomes equal to the supply voltage for LM311. The supply voltage of ± 12 V is generally available in the laboratory. Therefore, the magnitude of the output voltage is 12 V. This voltage is reduced to 5 V for compatibility with I/O port of the microcomputer.

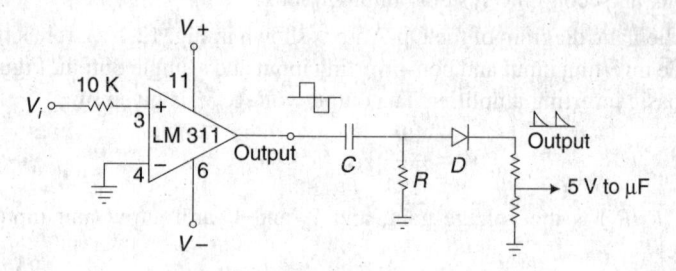

Fig. 12.2 Zero-cross detector

12.2.3 Phase Shifter

A phase shifter circuit using op-amp is shown in Fig. 12.3. For this circuit, any op-amp such as LM741, LM747 or LM324 can be used. The desired phase shift is obtained by adjusting the variable resistance R. The value of R for a phase shift of 90° is approximately 15 K ohms.

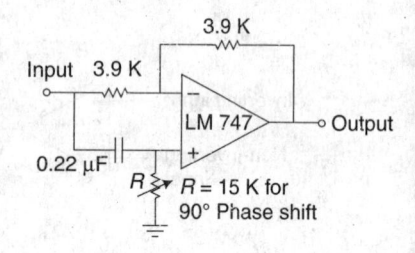

Fig. 12.3 Phase shifter

12.2.4 Current to Voltage Converter

A current to voltage converter is shown in Fig. 12.4(a). The current in R_s is zero due to the virtual ground at the input of the amplifier, and hence the current flows through R_2. The output voltage in this case is given by $V = -i_s R_2$. A current to voltage converter circuit incorporating CT is shown in Fig. 12.4(b). This circuit incorporates R_1 also. If the voltage available at the output terminal of the CT is V_1, then the output V_o of the circuit is given by

$$V_o = V_1 \times \frac{R_2}{R_1}$$

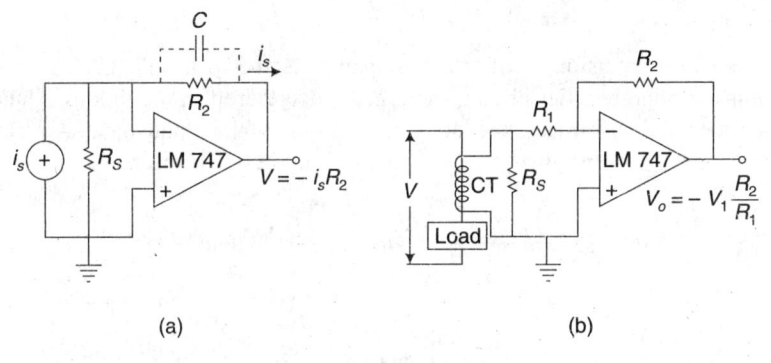

Microprocessor-Based Numerical Protective Relays 455

Fig. 12.4 (a) Current to voltage converter (b) I to V converter with CT

12.2.5 Summing Amplifier

The circuit of a summing amplifier is shown in Fig. 12.5. The output voltage V_o is given by

$$V_o = -\left(\frac{R_f}{R_1}V_1 + \frac{R_f}{R_2}V_2 + \ldots + \frac{R_f}{R_n}V_n\right) \quad (12.2)$$

12.2.6 Differential Amplifier

The circuit of differential amplifier amplifier is shown in Fig. 12.6(a). The output voltage V_o is given by

$$V_o = \frac{R_2}{R_1}(V_2 - V_1), \quad \text{if } \frac{R_2}{R_1} = \frac{R_4}{R_3} \quad (12.3)$$

where V_1 and V_2 are input voltages.

The circuit of a differential amplifier to obtain $(V_2 - V_1)$ using summing circuit is shown in Fig. 12.6(b). This circuit gives more accurate result.

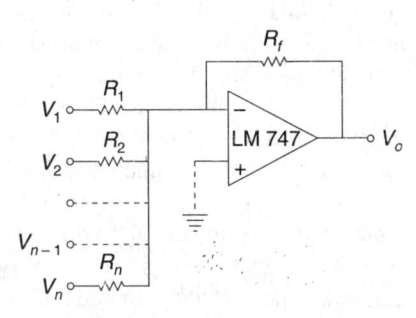

Fig. 12.5 Summing amplifier

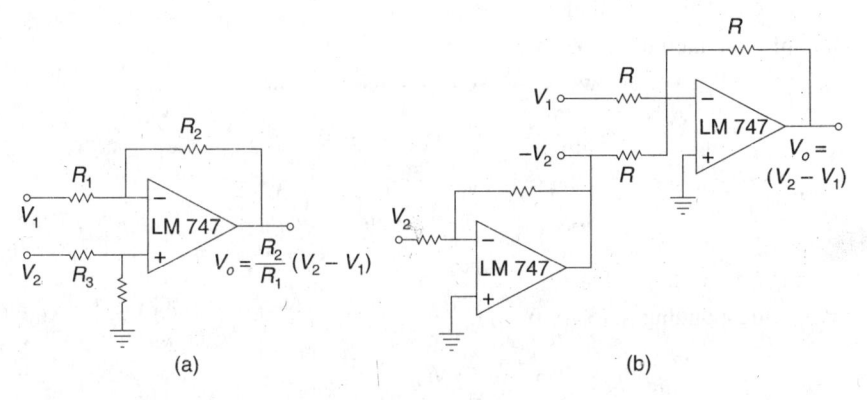

Fig. 12.6 (a) Differential amplifier (b) Differential amplifier using summing amplifier

12.2.7 Precision Rectifier

The circuit of a precision rectifier using opamps is shown in Fig. 12.7. In case of an ordinary bridge rectifier circuit, there is a voltage drop in the diodes. Thus this rectifier does not give an output exactly proportional to the input signal. A precision rectifier using Op-Amps gives a very accurate result in this respect.

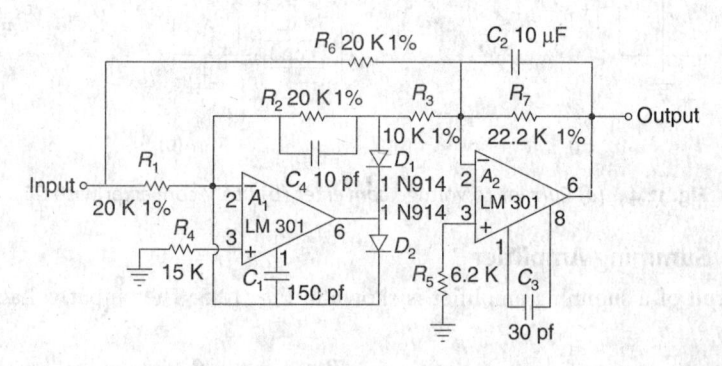

Fig. 12.7 Precision rectifier

12.2.8 Overvoltage Protection

An overvoltage protection circuit is shown in Fig. 12.8. An IC whose voltage is to be kept within a specified limit can be protected by using this circuit. The selection of suitable ratings of the zener diodes and resistance R depends upon the specified voltage limit. A sample calculation for the value of R to keep the specified voltage within 5 volts is illustrated. In the circuit as shown in Fig. 12.8, two zener diodes are connected back to back, i.e. one is forward biased and the other is reverse biased. If the peak voltage is to be limited to 5 volts and the voltage drop in the forward biased zener is 0.6 volt, the reverse biased zener will take $(5 - 0.6) = 4.4$ volts. Selecting a zener diode of 4.9 V rating the resistance R is calculated. Any other rating between 4 and 5 volts for the zener diode may be selected.

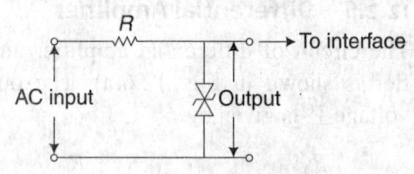

Fig. 12.8 Overvoltage protection circuit

Sample calculation for the value of R:

$$\text{Power} = 0.5 \text{ W} = 500 \text{ mW}$$

$$V_{peak} = 4.9 \text{ V}, \quad V_{rms} = \frac{4.9}{\sqrt{2}} = 3.46 \text{ V}$$

Hence

$$I_{rms} = \frac{500}{3.46} = 144.5 \text{ mA}$$

Taking 50% of this value, $I_{rms} = 72.25$ mA

V_{rms} corresponding to $(5 - 0.6)$

$$V = \frac{4.4}{\sqrt{2}} = 3 \text{ V}$$

Therefore, $$R = \frac{3}{72.25 \times 10^{-3}} \, \Omega = \frac{3000}{72.25} \, \Omega$$
$$= 41.5 \, \Omega$$

A resistance of 47 Ω may be used.

12.2.9 Active Low-pass Filter

The sampling rate in a sampled-data system is an important parameter to be considered. Theoretical considerations dictate that the sampling frequency (f_s) must be at least twice the largest frequency (f_m) present in the sampled information to avoid aliasing errors. A low-pass filter is used to avoid aliasing error in a sampled-data system. The low-pass filter can be passive, consisting of resistance and capacitance exclusively, or active ones utilising operational amplifiers. An active low-pass filter design leads to smaller component sizes, and hence it is preferable. Figure 12.9 shows the circuit of an active low-pass filter. The cut-off frequency f_c of this filter is given by

$$f_c = \frac{1}{2\pi \sqrt{R_1 R_2 C_1 C_2}} \qquad (12.4)$$

The relaying signals must be band limited to 400 Hz by using a low-pass filter of cut-off frequency 400 Hz in order to avoid aliasing errors. If the values of R_1, R_2 and C_1 are known, the value of C_2 can easily be calculated for cut-off frequency of 400 Hz by using Equation (12.4).

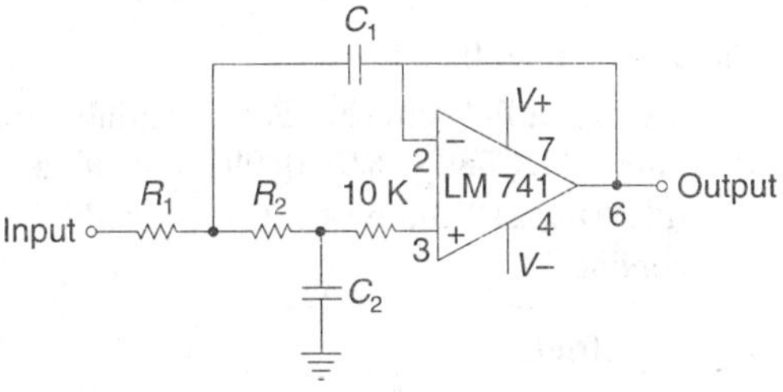

Fig. 12.9 Active low-pass filter

If the selected values of R_1, R_2 and C_1 are as follows:

$$R_1 = R_2 = 10 \, \text{K}$$
$$C_1 = 0.1 \, \mu\text{F},$$

then the value of C_2 for cut-off frequency of 400 Hz, as calculated from Equation (12.4) is 0.0155 μF.

12.3 A/D CONVERTER, ANALOG MULTIPLEXER, S/H CIRCUIT

The microprocessor accepts signals in digital form. Therefore analog signals must be converted into digital form before feeding them to the microprocessor for processing. Both voltage and current are analog quantities. As the microprocessor accepts only voltage signal in digital form, the current signal is first converted into proportional voltage signal, and then the voltage signal is converted into digital form for applying to the microprocessor. An A/D converter is used to convert analog signals into digital forms. If more than one analog quantity is to be converted into digital form by using only one A/D converter, analog multiplexers are used to select any one analog quantity at a time for A/D conversion. For time-varying voltages such as ac voltage, a sample and hold circuit is used to keep the desired instantaneous voltage constant during conversion period. This section discusses A/D converters, analog multiplexer, S/H circuits and their interfacing to the microprocessor.

12.3.1 Analog to Digital Converter

An A/D converter is used to convert analog signals into digital forms. The digital output of the ADC is fed to the microprocessor for further processing. The successive approximation method is the most popular method for analog to digital conversion. This method has an excellent compromise between accuracy and speed. In this method, an unknown analog voltage V_{in} is compared with a fraction of reference voltage V_r. The comparison is made n times with different fractions of V_r for n-bits digital output. The value of a particular bit is set to 1, if V_{in} is greater than the set fraction of V_r and is set to 0, if V_{in} is less than the set fraction of V_r. This fraction is given by

$$\sum_{i=1}^{n} b_i 2^{-i} V_r$$

where b_i is either 0 or 1.

A variety of A/D converters are available today from different makes in IC form. ADC0800, ADC0808, ADC0809, ADC0816 and ADC0817 are 8-bit A/D converters; and ADC1210 and ADC1211 are 12-bit A/D converters available from National Semiconductor.

12.3.2 Clock

Any A/D converter requires a suitable clock frequency to perform the conversion task. The clock frequency for ADC0800 lies in the range of 50 kHz, to 800 kHz, for ADC0808/0809 in the range of 10 kHz to 1280 kHz and for ADC1210/1211 in the range of 65 kHz to 130 kHz. The clock frequency available at 8085 based microprocessor kit is about 3 MHz. The clock frequencies available from 8284 clock Generator/Driver at 8086 based microprocessor kit are 4.9 MHz and 2.45 MHz. A suitable clock frequency for A/D converters can be obtained by dividing the clock frequencies available on microprocessor kits by using IC packages 7490 and 7474. 7490 is a monolithic decade counter which contains four master-slave flip-flops. The pin configuration is shown in Fig. 12.10(a) and the connection diagrams for divide by 10 and divide by 4 are shown in Fig. 12.10(b) and 12.10(c), respectively. The unit A divides the input frequency by 2. If the output is taken at Q_A, the frequency will be half of the input frequency. The units B, C and D combined together divide the input frequency by 5. When input is applied at input B and output is taken at Q_D, the frequency will be divided by 5. When all the four units (A, B, C and D) are connected as shown in Fig. 12.10(b) the frequency will be divided by 10. In this case, $R_o(1)$, $R_o(2)$, $R_g(1)$ and $R_g(2)$ are grounded. The frequency can also be divided by 2 by using 7474 Dual D type flip-flop. The circuit for divide by two using 7474 is shown in Fig. 12.11. As the speed of conversion of the A/D converter depends upon the clock frequency, it should be kept as high as possible within the range.

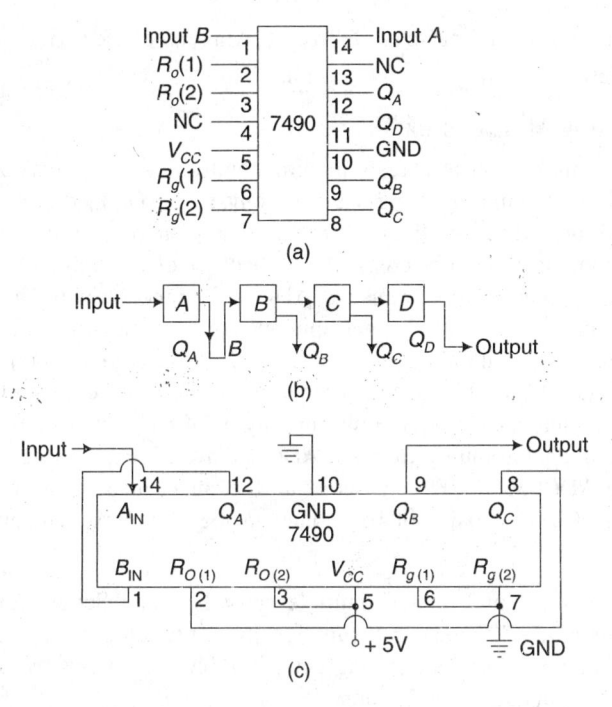

Fig. 12.10 (a) Pin configuration of 7490 (b) Connection diagram of 7490 for divide by ten (c) Connection diagram of 7490 for divide by four

Alternatively, a suitable clock for the A/D converter can also be obtained by using Intel 8253 timer/counter which is usually available on the board of the microprocessor kit. A clock of the desired frequency suitable for A/D converter can be obtained by using 8253 as a square wave generator in MODE3.

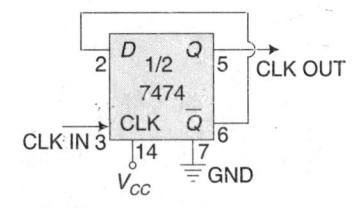

Fig. 12.11 Circuit for divide by two using 7474

12.3.3 ADC0800

ADC0800 is an 8-bit monolithic A/D converter from National Semiconductors fabricated by using P-channel ion-implanted MOS technology and available in an 18-pin DIP. It contains a high input impedance comparator, 256 series resistors called 256 R-network and analog switches, control logic and output latches. Conversion is performed using the successive approximation technique where the unknown analog voltage is compared to the resistor tie-points using analog switches. When the appropriate tie-point voltage matches the unknown voltage, conversion is complete and digital outputs contain an 8-bit complementary binary word corresponding to the unknown analog voltage. Following are the main features of ADC0800.

(i) Supply voltages   + 5 V_{DC} and − 12 V_{DC}
(ii) Clock frequency range   50 to 800 kHz
(iii) Conversion time   40 clock periods
(iv) Input range   ± 5 V, 10 V

(v) Start of conversion pulse   Min: 1, Max 3.5 clock period
(vi) Resolution   8 bits

12.3.4 Analog Multiplexer

Many applications of microprocessor require monitoring of several analog signals. A system can be built either by using one A/D converter (ADC) for each of the analog inputs or only one ADC for all analog inputs. The system using one ADC for each of the analog inputs will not be cost-effective because of the high cost of ADCs. The cost effective approach will be to use an ADC in conjunction with an analog multiplexer which has many input channels and only one output. An analog multiplexer selects one out of the multiple inputs and transfers it to a common output. Any input channel can be selected by sending proper commands to the multiplexer through the microcomputer. An analog multiplexer may be of 8 channels or 16 channels. At present both an analog multiplexer and ADC are available on a single IC package. For example ADC0808/0809 consists of an 8-channel analog multiplexer and an 8-bit ADC. ADC0816 consists of 16-channel analog multiplexer and an 8-bit ADC.

AM3705

AM3705 is an 8-channel analog multiplexer of National Semiconductors which operates at standard +5 V and –15 V supplies. Its input analog voltage range is ± 5 V. Figure 12.12 shows its schematic diagram and Table 12.1 shows its logic inputs to switch on any particular desired channel.

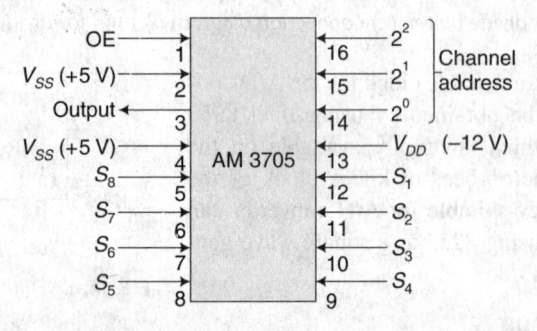

Fig. 12.12   Schematic diagram of AM3705

Table 12.1 Logic for AM3705

Logic Inputs			Output Enable	Channel
2^2	2^1	2^0	OE	ON
L	L	L	H	S_1
L	L	H	H	S_2
L	H	L	H	S_3
L	H	H	H	S_4
H	L	L	H	S_5
H	L	H	H	S_6
H	H	L	H	S_7
H	H	H	H	S_8

12.3.5 Sample and Hold Circuit

An ADC takes a finite time, known as conversion time (t_c) to convert an analog signal into digital form. If the input analog signal is not constant during the conversion period, the digital output of ADC will not correspond to the starting point analog input. A sample and hold (S/H) circuit is used to keep the instantaneous value of the rapidly varying analog signal (such as ac signal) constant during the conversion period. As the name implies a S/H circuit has two modes of operation namely Sample mode and Hold mode. When the logic input is high it is in the sample mode, and the output follows the input with unity gain. When the logic input is low it is in the hold mode and the output of the S/H circuit retains the last value it had until the command switches for the sample mode. The S/H circuit is basically an operational amplifier which charges a capacitor during the sample mode and retains the value of the charge of the capacitor during the hold mode.

LF398

The LF398 is an monolithic sample and hold circuit from National semiconductors which utilises BI-FET technology to obtain high accuracy with fast acquisition of signal and low droop rate. The rate at which the output of S/H circuit decreases is called the droop rate. This rate should should be very small so that the output remains practically constant. LF398 operates from ± 5 V to ± 18 V supplies. The only component which is to be connected externally is the hold capacitor. A low droop rate of about 5 mV/min is obtained with a hold capacitor of 1 μF.

Figure 12.13(a) shows the schematic diagram of LF398. C_h is a hold capacitor of a suitable value which is to be connected externally by the user. It is connected between pin 6 and ground. The logic input which is a pulse 5 V high is applied to pin 8. The width of this pulse should be equal to the acquisition time, depending on the value of the hold capacitor C_h. The acquisition time corresponding to the value of the hold capacitor is determined from the graph, shown in Fig. 12.13(b). When the logic input is made high, the input analog voltage is applied to the hold capacitor which gets charged up to the instantaneous value of the input voltage. When the logic input goes low the input voltage is switched off. An Op-Amp incorporated in the S/H circuit isolates the hold capacitor from any load. Thus the hold capacitor holds the instantaneous value of the applied input voltage.

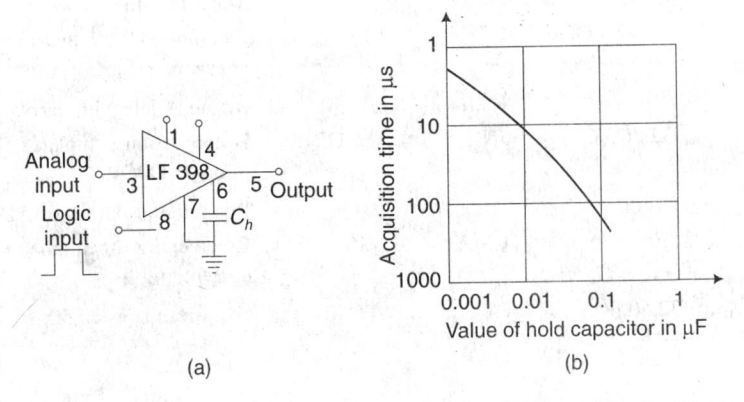

Fig. 12.13 (a) Schematic diagram of LF398 (b) Acquisition time for LF398

Acquisition Time

The acquisition time of the S/H circuit is the time required for the capacitor to charge up to the value of the input signal after the sample command is given, i.e. the logic input is made high.

Aperture Time

The aperture time is the delay between the sample state and the hold state. It is very small in nanoseconds, the typical value of this time being 10 μs.

Selection of Hold Capacitor

A significant source of error in sample and hold circuit is the dielectric absorption in the hold capacitor. Therefore, the hold capacitor selected for S/H circuit should be made of dielectric with very low hysteresis such as polystyrene, polypropylene and teflon.

12.3.6 Interfacing of A/D Converter ADC0800

The circuit for interfacing of ADC0800 to the 8085 microprocessor through programmable I/O port 8255 is shown in Fig. 12.14. A simple program to read the digital output corresponding to a single dc input voltage is developed. As the input dc voltage remains constant, the sample and hold circuit is not required in this case. The supply voltage of ± 5 V_{dc} should be very accurate as it is used as reference voltage. This is obtained using op-amp circuit as shown in Fig. 12.14. 8255-1 indicates that the first unit of the I/O ports provided on the kit has been used. The ports have been defined as follows.

Program I

Memory Address	Machine Code	Labels	Mnemonics	Operands	Comments
FC00	3E,98		MVI	A,98H	Obtain control word.
FC02	D3,03		OUT	03	Initialise I/O ports.
FC04	3E,08		MVI	A,08	For S/C pulse.
FC06	D3,02		OUT	02	Make PC_3 high.
FC08	3E,00		MVI	A,00	To make PC_3 low.
FC0A	D3,02		OUT	02	Make PC_3 low.
FC0C	DB,02	LOOP	IN	02	Examine PC_7 for end of conversion.
FC0E	17		RAL		Rotate A left with carry.
FC0F	D2,0C,FC		JNC	LOOP	Is conversion complete? If No, jump to LOOP.
FC12	DB,00		IN	00	Read digital output of ADC.
FC14	2F		CMA		Complement Accumulator content.
FC15	32,50,FC		STA	FC50	Store result in FC 50.
FC18	76		HLT		

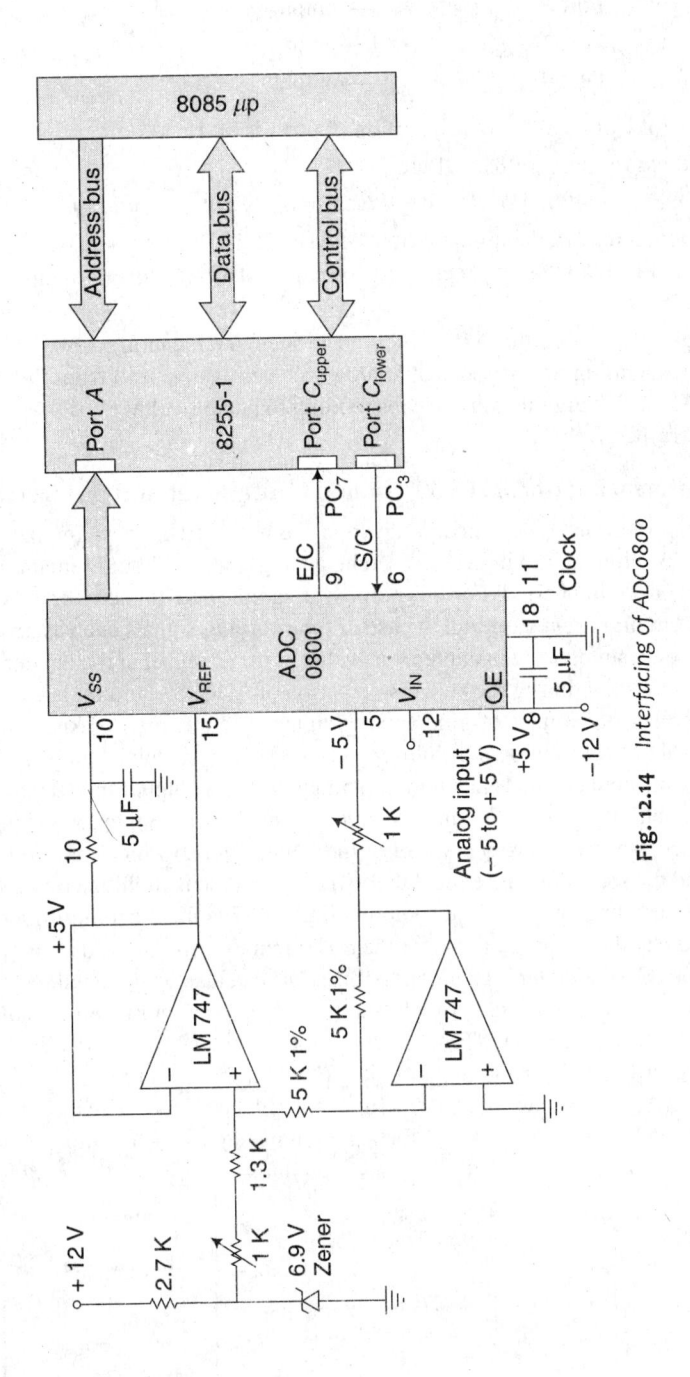

Fig. 12.14 Interfacing of ADC0800

Port A	— input
Port B	— output
Port C_{upper}	— input
Port C_{lower}	— output

The control word for this configuration of ports is 98 H.

The address for ports of 8255.1 are:

Port A – 00, Port B – 01, Port C – 02, Control word – 0.3.

The program to read the digital output of ADC CMA is given here. CMA instruction is used to obtain the correct result because ADC0800 gives complementary output.

The digital output of the ADC is stored in memory location FC50. The digital values corresponding to a number of analog voltages can be read from the memory location FC50. The digital values corresponding to analog voltages of 0 and 5 V are 80 and FF, respectively.

12.3.7 Interfacing of ADC0800, Analog Multiplexer and S/H Circuits

ADC takes a definite time, depending upon the clock frequency, for the conversion of analog signals into digital form. The analog voltage should remain constant during the conversion period. For this purpose, a sample and hold circuit is used. The use of S/H circuit is very essential in case of time varying signals such as ac voltage. The S/H circuit samples the instantaneous value of the ac signal at the desired instant and holds it constant during the conversion period of ADC. S/H circuit is also used to obtain a number of samples of time varying signals at some interval known as sampling interval (T_s). An analog multiplexer is used to select signals of any one channel from the multi-input channels and transfer this signal to its output for A/D conversion. In order to obtain a number of simultaneous samples of more than one signal, S/H circuit is incorporated in every channel of the multiplexer to which input signals are applied, and the sampling pulse to all the S/H circuits is sent simultaneously through the microprocessor at the required sampling interval. The 8253 counter/timer can be used to generate the sampling pulse for the S/H circuit. It can also be used to generate clock for the A/D converter. The clock for the ADC can also be obtained by reducing the clock of 3 MHz or 1.5 MHz available at the microprocessor kit to a suitable value by using the 7490 IC package. The circuit diagram of interfacing of ADC0800, analog multiplexer (AM3705) and S/H circuit (LF398) is shown in Fig. 12.15. Channels S_3 and S_5 of AM3705 have been used for input signals 1 and 2, respectively. The program to read the digital values of the analog input signals is as follows.

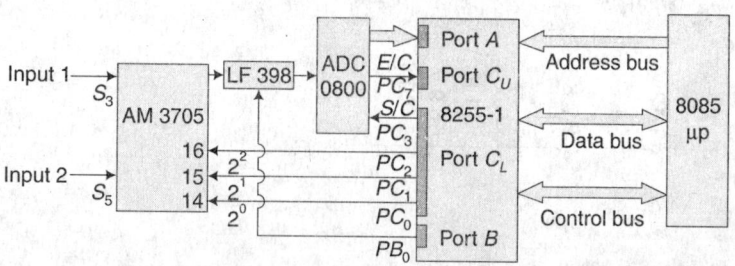

Fig. 12.15 Interfacing of ADC0800, analog multiplexer and S/H circuit

Program 2

Memory Address	Machine Code	Labels	Mnemonics	Operands	Comments
FC00	3E,98		MVI	A,98	Initialise I/O ports of 8255-1.
FC02	D3,03		OUT	03	
FC04	3E,02		MVI	A,02	Switch on multiplexer's channel S_3 for input 1.
FC06	D3,02		OUT	02	
FC08	3E,01		MVI	A,01	Logic high to S/H.
FC0A	D3,01		OUT	01	
FC0C	06,05		MVI	B,05	Delay for acquisition time of S/H.
FC0E	05	BACK	DCR	B	
FC0F	C2,0E,FC		JNZ	BACK	
FC12	3E,00		MVI	A,00	Logic low to S/H.
FC14	D3,01		OUT	01	
FC16	3E,0A		MVI	A,0A	Start of conversion
FC18	D3,02		OUT	02	Pulse to ADC without affecting multiplexer channel S_3.
FC1A	3E,02		MVI	A,02	
FC1C	D3,02		OUT	02	
FC1E	DB,02	LOOP	IN	02	
FC20	17		RAL		Is conversion over?
FC21	D2,1E,FC		JNC	LOOP	
FC24	DB,00		IN	00	Read digital value of input 1.
FC26	2F		CMA		
FC27	D6,80		SUI	80	
FC29	32,00,FD		STA	FD00	
FC2C	3E,04		MVI	A,04	Switch on multiplexer's channel S_5 for Input 2.
FC2E	D3,02		OUT	02	
FC30	3E,01		MVI	A,01	Logic high to S/H
FC32	D3,01		OUT	01	
FC34	06,05		MVI	B,05	Delay for acquisition time of S/H.
FC36	05	ABOVE	DCR	B	
FC37	C2,36,FC		JNZ	ABOVE	
FC3A	3E,00		MVI	A,00	Logic low to S/H.
FC3C	D3,01		OUT	01	
FC3E	3E,0C		MVI	A,0C	S/C pulse to ADC without affecting multiplexer's channel S_5.

Pgm contd...

Pgm contd...

FC40	D3,02	OUT	02	
FC42	3E,04	MVI	A,04	
FC44	D3,02	OUT	02	
FC46	DB,02 READ	IN	02	
FC48	17	RAL		
FC49	D2,46,FC	JNC	READ	
FC4C	DB,00	IN	00	Read digital value of input 2.
FC4E	2F	CMA		
FC4F	D6,80	SUI	80	
FC51	32,01,FD	STA	FD01	
FC54	76	HLT		

The digital values of input 1 and input 2 are stored in the memory locations FD00 and FD01, respectively.

12.3.8 ADC0808/ADC0809

The ADC0808 or ADC0809 is a monolithic CMOS device from National Semiconductors which contains an 8-bit A/D converter and an 8-channel multiplexer. It uses the successive approximation technique for conversion. It operates with a single +5 V supply and its clock frequency range is 10 to 1280 kHz. The conversion time at clock frequency 640 kHz is 100 μs. Its maximum analog voltage input is +5 V. Figure 12.16 shows the schematic diagram of ADC0808/ADC0809 and Table 12.2 shows the logic for the multiplexer channel.

Table 12.2 Logic for multiplexer-channels of ADC0808/0809

Analog Channel	Address		
	C	B	A
IN0	0	0	0
IN1	0	0	1
IN2	0	1	0
IN3	0	1	1
IN4	1	0	0
IN5	1	0	1
IN6	1	1	0
IN7	1	1	1

12.3.9 Interfacing of ADC0808/ADC0809

Figure 12.17 shows the interfacing of ADC0808/ADC0809 to Intel 8085 microprocessor through the programmable I/O port 8255. An analog voltage is connected to IN1. The +5 V reference voltage applied to pin 12 should be very accurate. It should not be given from the stabilised power supply units which are generally available in the laboratory. A circuit for obtaining +5 V reference Voltage is shown in Fig. 12.18.

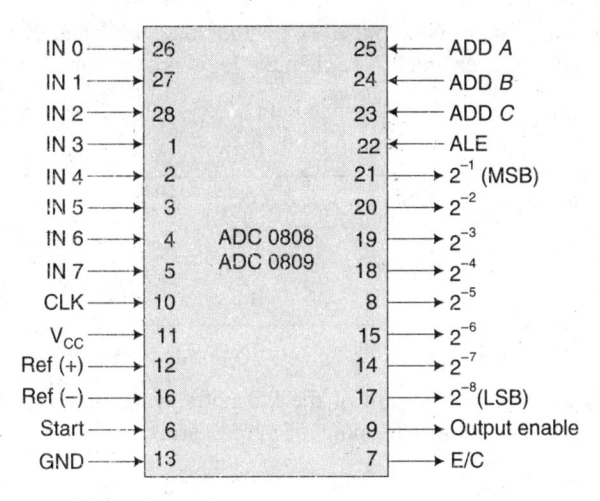

Microprocessor-Based Numerical Protective Relays **467**

```
IN 0  → 26        25 ← ADD A
IN 1  → 27        24 ← ADD B
IN 2  → 28        23 ← ADD C
IN 3  →  1        22 ← ALE
IN 4  →  2        21 → 2⁻¹ (MSB)
IN 5  →  3        20 → 2⁻²
IN 6  →  4   ADC 0808   19 → 2⁻³
IN 7  →  5   ADC 0809   18 → 2⁻⁴
CLK   → 10         8 → 2⁻⁵
V_CC  → 11        15 → 2⁻⁶
Ref(+)→ 12        14 → 2⁻⁷
Ref(−)→ 16        17 → 2⁻⁸ (LSB)
Start →  6         9 → Output enable
GND   → 13         7 → E/C
```

Fig. 12.16 *Schematic diagram of ADC0808/ADC0809*

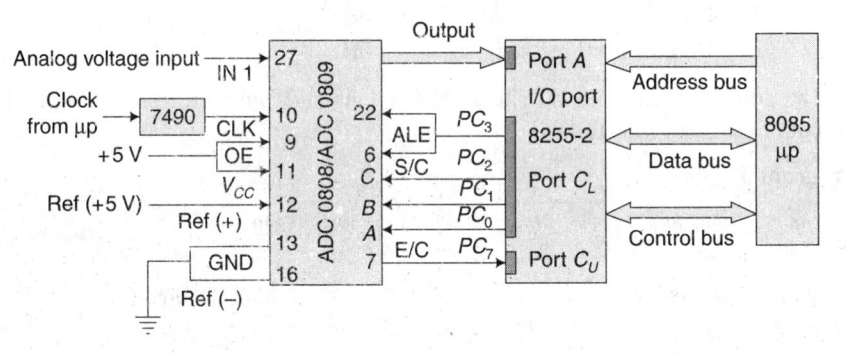

Fig. 12.17 *Interfacing of ADC0808/0809 to 8085 μp*

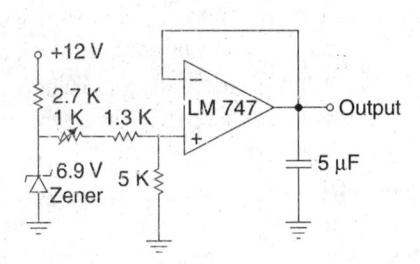

Fig. 12.18 *Circuit for +5 V reference voltage*

The logic to switch on IN1 is as follows.

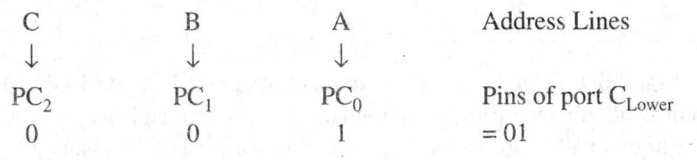

C	B	A	Address Lines
↓	↓	↓	
PC_2	PC_1	PC_0	Pins of port C_{Lower}
0	0	1	= 01

To send the start of conversion pulse, PC_3 is first made high and then low, without affecting the multiplexer channel IN1. Then the logic is as follows.

$$\underbrace{\underset{1}{PC_3}}_{\text{for S/C}} \quad \underbrace{\underset{0}{PC_2} \quad \underset{0}{PC_1} \quad \underset{1}{PC_0}}_{\text{for IN1}} \quad = 09\ H$$

Again

$$\underbrace{\underset{0}{}}_{\text{for S/C}} \quad \underbrace{\underset{0}{} \quad \underset{0}{} \quad \underset{1}{}}_{\text{for IN1}} \quad = 01\ H$$

8255-2 indicates the second unit of the I/O ports available on the board of the microprocessor kit. The address for ports of 8255-2 are:

    Port A   — 08
    Port B   — 09
    Port C   — 0A
    Control word register   — 0B

The control word is 98 H. The assembly language program to read the digital output of the ADC is as follows.

Program 3

Memory Address	Machine Code	Labels	Mnemonics	Operands	Comments
FD00	3E,98		MVI	A,98	Initialise I/O ports.
FD02	D3,0B		OUT	0B	Ports of 8253-2.
FD04	3E,01		MVI	A,01	Switch on Multiplexer's channel INI.
FD06	D3,0A		OUT	0A	S/C pulse to ADC
FD08	3E,09		MVI	A,09	without affecting
FD0A	D3,0A		OUT	0A	multiplexer's channel.
FD0C	3E,01		MVI	A,01	
FD0E	D3,0A		OUT	0A	
FD10	DB,0A	LOOP	IN	0A	Read E/C signal.
FD12	17		RAL		
FD13	D2,10,FD		JNC	LOOP	Is conversion over? If No, jump to Loop.
FD16	DB,08		IN	08	Read digital output of ADC.
FD18	32,50,FD		STA	FD50H	Store results.
FD1B	76		HLT		

The digital output of the ADC is stored in memory location FD50. The digital values corresponding to different analog voltages can be read from this memory location. The digital values corresponding to analog voltages of 0, 2.5 and 5 V are 00, 80 and FF, respectively.

12.3.10 Bipolar to Unipolar Converter

The interface shown in Fig. 12.17 is suitable for converting the instantaneous values of ac voltage into digital form. This will give the correct result only for positive ac voltages having maximum value of 5 V. In order to consider the negative values of ac voltage, the interface circuit has to be modified. Figure 12.19 shows a circuit employing Op-Amps which takes into account the negative values of the input signal having a maximum negative value of –5 V. This circuit is called bipolar to unipolar converter because it converts both positive and negative values (maximum ±5 V) to only positive values having maximum values of +5 V. The first part of the circuit is a summing circuit and the second part is an inverter.

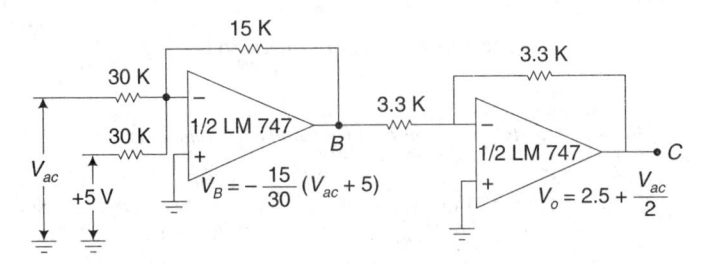

Fig. 12.19 Bipolar to unipolar converter

The output of the summing circuit is given by

$$V_B = -\frac{15}{30}(V_{ac} + 5) = -\left(2.5 + \frac{V_{ac}}{2}\right) \qquad (12.5)$$

The output of the summing circuit is fed to the inverter. The output of the inverter is given by

$$V_o = 2.5 + \frac{V_{ac}}{2}$$

The range of the ac input signal is kept ± 5 V peak. For the variation of input signal from – 5 V to + 5 V, the output varies from 0 to +5 V.

12.3.11 ADC1210 and ADC1211

ADC1210 and ADC1211 are 12-bit CMOS A/D converters from National Semiconductors. They can operate over a wide range of voltage supply from single +5 V to ±15 V, and convert both ac and dc analog voltages into digital form. They use the successive approximation technique for the conversion. They offer 12-bit resolution with 10-bit accuracy.

Figure 12.20 shows the interfacing of ADC1211 using a single supply (+5 V) operating configuration to the 8085 microprocessor through 8255. This gives a TTL compatible output. The clock frequency range of ADC1211 is 65 kHz to 130 kHz. It gives a complementary output. A clock frequency of 1.535 MHz is available on the 8085 microprocessor kit. This clock frequency is divided by 12 by using 8253 Timer/Counter in MODE 3. The frequency at the output of the 8253 is 1.535 MHz/12 = 128 kHz, which is suitable for the ADC1211. The interfacing of ADC1211 to 8086 microprocessor is discussed in Section 12.13 The 8085 assembly language program to read the digital output of ADC1211 is as follows.

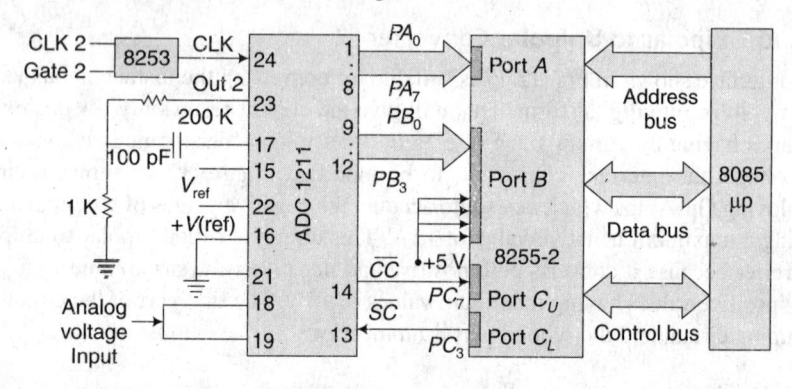

Fig. 12.20 Interfacing of ADC1211 to 8085 µp

Program 4

Memory Address	Machine Code	Labels	Mnemonics	Operands	Comments
FC00	3E,96		MVI	A,96	Control word for Mode3, counter 2 of 8253.
FC02	D3,13		OUT	13	
FC04	3E,0C		MVI	A,0C	For division by 12.
FC06	D3,12		OUT	12	
FC08	3E,9A		MVI	A,9A	Control word for 8255-2.
FC0A	D3,0B		OUT	0B	
FC0C	21,00,FD		LXI	H,FD00	Memory address for storing result.
FC0F	3E,00		MVI	A,00	Logic low to $\overline{SC}$.
FC11	D3,0A		OUT	0A	
FC13	3E,08		MVI	A,08	To make PC_3 high.
FC15	D3,0A		OUT	0A	Logic high to $\overline{SC}$.
FC17	DB,0A	LOOP	IN	0A	Read $\overline{CC}$ to check end of conversion.
FC19	17		RAL		
FC1A	DA,17,FC		JC	LOOP	Is conversion over? If No, jump to Loop.
FC1D	DB,08		IN	08	
FC1F	2F		CMA		Complement accumulator.
FC20	77		MOV	M,A	Store low order 8-bit in FD00.
FC21	23		INX	H	
FC22	DB,09		IN	09	
FC24	2F		CMA		
FC25	77		MOV	M,A	Store high order 4-bit in memory (FD01).
FC26	76		HLT		
			12-bit digital output of ADC1211		
		FD00	— 8-bit (Least significant byte)		
		FD01	— 4-bit (Most significant high order bits)		

12.4 OVERCURRENT RELAYS

An overcurrent relay is the simplest form of protective relay which operates when the current in any circuit exceeds a certain predetermined value, i.e. the pick-up value. It is extensively used for the protection of distribution lines, industrial motors and equipment. Using a multiplexer, the mircoprocessor can sense the fault currents of a number of circuits. If the fault current in any circuit exceeds the pick-up value, the microprocessor sends a tripping signal to the circuit breaker of the faulty circuit breaker of the faulty circuit. As the microprocessor accepts signals in voltage form, the current signal derived from the current transformer is converted into a proportional voltage signal using a current to voltage converter. The ac voltage proportional to the load current is converted into dc using a precision rectifier. Thus, the microprocessor accepts dc voltage proportional to load current.

The block schematic diagram of the relay is shown in Fig. 12.21(a). The output of the rectifier is fed to the multiplexer. The microcomputer sends a command to switch on the desired channel of the multiplexer to obtain the rectified voltage proportional to the current in a particular circuit. The output of the multiplexer is fed to the A/D converter to obtain the signal in digital form. The A/D converter ADC0800 has been used for this purpose. The microcomputer sends a signal to the ADC for starting the conversion. The microcomputer reads the end of conversion signal to examine whether the conversion is over or not. As soon as the conversion is over, the microcomputer reads the current signal in digital form and then compares it with the pick-up value.

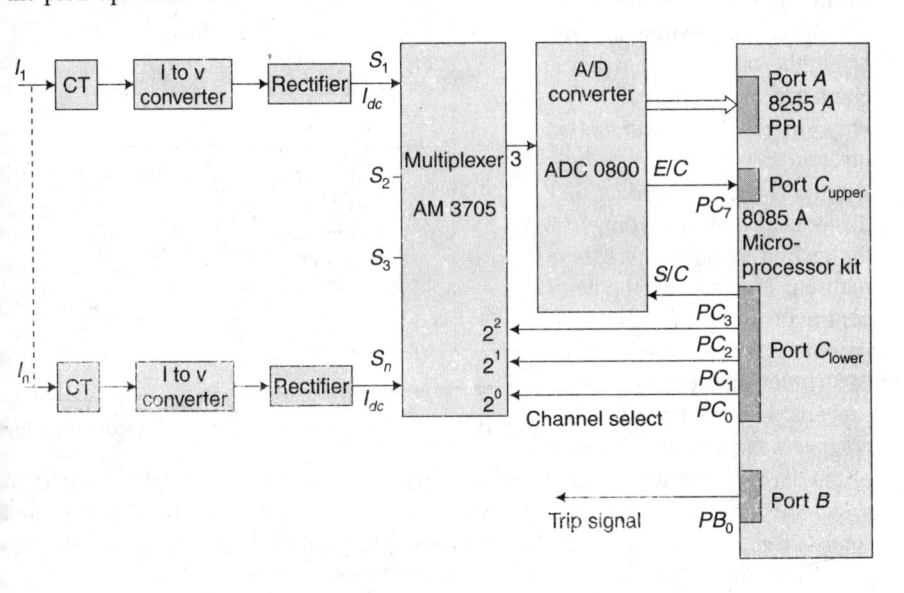

Fig. 12.21 (a) Block schematic diagram of overcurrent relay

In the case of a definite time overcurrent relay, the microcomputer sends the tripping signal to the circuit breaker after a predetermined time delay if the fault current exceeds the pick-up value. In case of instantaneous overcurrent relay there is no intentional time delay. In order to obtain inverse-time characteristics, the operating times for different vales of currents are noted for a particular characteristic. These values are stored in the memory in tabular form. The microcomputer first determines the magnitude of the fault current and then selects the corresponding time of operation from the look-up table. A delay subroutine is started and the trip signal is sent after the desired delay. Using the same program, any characteristic such as IDMT, very inverse or extremely inverse can be realised by simply changing the data of the look-up table according to the desired characteristic to be realised. The microcomputer continuously measures the current and moves in a loop and if the measured current exceeds the pick-up value, it compares the measured value of the current with the digital values of current give in the look-up table in order to select the corresponding count for a time delay. Then it goes in delay subroutine and sends a trip signal to the circuit breaker after the predetermined time delay. The program flowchart is shown in Fig. 12.21(b) and the program is as follows.

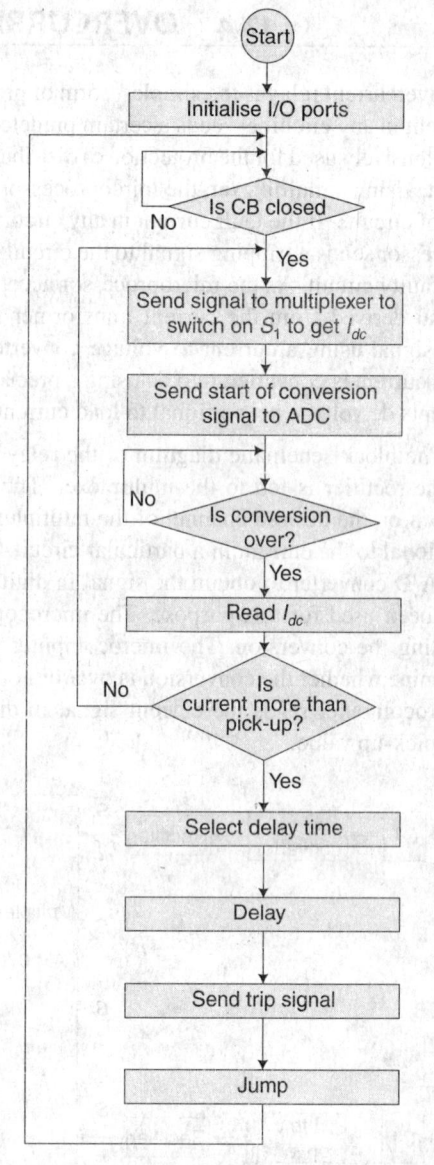

Fig. 12.21 (b) Program flowchart for overcurrent relay

Microprocessor-Based Numerical Protective Relays 473

Program 5

Memory Address	Machine Codes	Label	Mnemonics	Operands	Comments
2000	3E,98		MVI	A,98	Initialise I/O ports.
2002	D3,03		OUT	03	
2004	3E,00	MESH	MVI	A,00	Signal to multiplexer to switch on channel S_1.
2006	D3,02		OUT	02	
2008	3E,08		MVI	A,08	Start of conversion signal.
200A	D3,02		OUT	02	
200C	3E,00		MVI	A,00	
200E	D3,02		OUT	02	
2010	DB,02	CHECK	IN	02	Check whether conversion is over.
2012	17		RAL		
2013	D2,10,20		JNC	CHECK	
2016	DB,00		IN	00	Read I_{dc}.
2018	2F		CMA		
2019	D6,80		SUI	80	
201B	32,00,24		STA	2400	Store I_{dc} in memory.
201E	21,00,22		LXI	H,2200	
2021	46		MOV	B,M	Count for lookup table in reg. B.
2022	23	SEEK	INX	H	
2023	BE		CMP	M	
2024	D2,2E,20		JNC	AHEAD	
2027	05		DCR	B	
2028	C2,22,20		JNZ	SEEK	
202B	CA,04,20		JZ	MESH	
202E	24	AHEAD	INR	H	
202F	46		MOV	B,M	Count for delay in B.
2030	0E,FF	BEHIND	MVI	C,FF	
2032	16,FF	CHAIN	MVI	D,FF	
2034	15	MOVE	DCR	D	
2035	C2,34,20		JNZ	MOVE	
2038	0D		DCR	C	
2039	C2,32,20		JNZ	CHAIN	
203C	05		DCR	B	
203D	C2,30,20		JNZ	BEHIND	
2040	3E,01		MVI	A,01	
2042	D3,01		OUT	01	Send trip signal.
2044	76		HLT		

Look-up Table

Memory Address	Digital Values of Current	Memory Address	Count for Delay in Register B	Delay Time in s
2200	0C (COUNT)	—	—	—
2201	7F	2301	3	0.19
2202	7A	2302	5	0.32
2203	6D	2303	9	0.57
2204	66	2304	0B	0.70
2205	60	2305	0D	0.83
2206	5A	2306	10	1.00
2207	53	2307	16	1.40
2208	4D	2308	1E	1.90
2209	46	2309	2D	2.90
220A	40	230A	50	5.10
220B	3A	230B	7D	8.00
220C	33	230C	DB	14.00

In order to avoid false tripping of an overcurrent relay due to transients the program can be modified slightly. When the fault current exceeds the pick-up value, the fault current is measured once again by the microprocessor to confirm whether it is a fault current or transient. In case of any transient of short duration, the measured current above pick-up value will not appear in the second measurement. But if there is an actual fault, it will again appear in the second measurement also, and then the microprocessor will issue a tripping signal to disconnect the faulty part of the system. The program flowchart is shown in Fig. 12.22 and the program is given below.

Program6

Memory Address	Machine Codes	Label	Mnemonics	Operands	Comments
2100	3E,98		MVI	A,98	Initialise I/O ports.
2102	D3,03		OUT	03	
2104	0E,02	BEGIN	MVI	C,02	Count for transient.
2106	3E,00		MVI	A,00	Signal to multiplexer to switch on S_1.
2108	D3,02		OUT	02	
210A	3E,08	BEHIND	MVI	A,08	Start of conversion pulse to ADC.
210C	D3,02		OUT	02	
210E	3E,00		MVI	A,00	
2110	D3,02		OUT	02	
2112	DB,02	CHECK	IN	02	Check whether conversion is over.

Pgm contd...

Pgm contd...

2114	17		RAL	
2115	D2,12,21		JNC	CHECK
2118	DB,00		IN	00	Read I_{dc}.
211A	2F		CMA	
211B	D6,80		SUI	80
211D	32,00,25		STA	2500	Store I_{dc} to check program if required.
2120	21,00,22		LXI	H,2200	Count for look-up table
2123	46		MOV	B,M
2124	23	SEEK	INX	H	1st point of look-up table.
2125	BE		CMP	M
2126	D2,30,21		JNC	ABOVE	If fault go to AHEAD.
2129	05		DCR	B
212A	C2,24,21		JNZ	SEEK	Search other points of the table.
212D	CA,04,21		JZ	BEGIN	Go to START. If I measured does not correspond to any point of table.
2130	0D	ABOVE	DCR	C
2131	CA,3D,21		JZ	AHEAD
2134	11,6A,15		LXI	D,156A
2137	CD,A0,03		CALL	03A0	Small delay for transient.
213A	C3,0A,21		JMP	BEHIND
213D	24	AHEAD	INR	H
213E	46		MOV	B,M
213F	0E,FF	MESH	MVI	C,FF
2141	16,33	CHAIN	MVI	D,33
2143	15	MOVE	DCR	D
2144	C2,43,21		JNZ	MOVE
2147	0D		DCR	C
2148	C2,41,21		JNZ	CHAIN
214B	05		DCR	B
214C	C2,3F,21		JNZ	MESH
214F	3E,01		MVI	A,01	Trip signal.
2151	D3,01		OUT	01
2153	C3,04,21		JMP	BEGIN

Look-up Table: Same as that for the previous program.

12.5 IMPEDANCE RELAY

The characteristic of an impedance relay is realised by comparing voltage and current at the relay location. The ratio of voltage (V) to current (I) gives the impedance

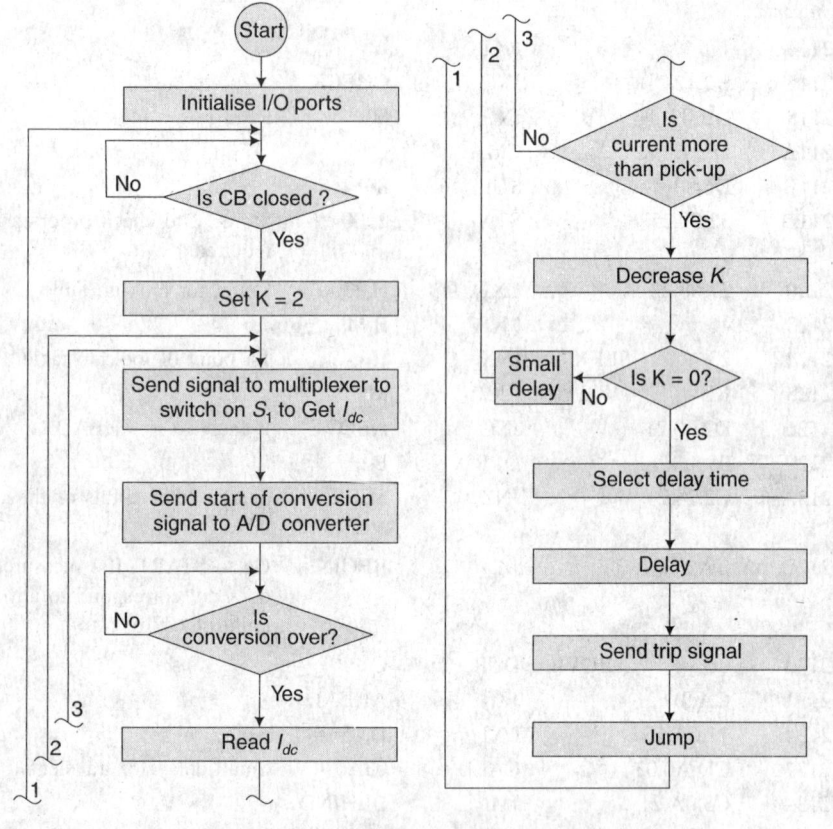

Fig. 12.22 *Program flowchart for overcurrent relay*

of the line section between the relay location and the fault point. The rectified voltage (V_{dc}) and rectified current (I_{dc}) are proportional to V and I, respectively. Therefore, for comparison V_{dc} and I_{dc} are used. The following condition should be satisfied for the operation of the relay.

$$K_1 V_{dc} < K_2 I_{dc} \quad \text{or} \quad \frac{V_{dc}}{I_{dc}} < \frac{K_2}{K_1}$$

or
$$\frac{V}{I} < K$$

or
$$Z < K \tag{12.6}$$

where K_1, K_2 and K are constants.

The values of K for different zones of protection are calculated and stored in the memory as data to obtain the desired characteristic.

The characterisitcs of the three zone impedance relays are shown in Fig. 6.4. Z_1, Z_2 and Z_3 are the values of impedances for I, II and III zones of protection, respectively. t_1, t_2 and t_3 are the operating times for I, II, and III zones of protection, respectively. As the impedance relay is non-directional, a directional unit is also incorporated to give a directional feature so that the relay can operate for the fault in forward direction only.

The block schematic diagram of the interface for the impedance relay is shown in Fig. 12.23. The levels of voltage and current signals are stepped down to the electronic level by using potential and current transformers. The current signal derived from the current transformer is converted into proportional voltage signal using a current to voltage converter. The voltage and current signals are then rectified using precision rectifiers to convert them into dc. The rectified voltage and current signals (V_{dc} and I_{dc} respectively) are fed to two different channels of the multiplexer which are switched on sequentially by proper commands from the microcomputer. The output of the multiplexer is fed to the A/D converter through a sample and hold circuit. The multiplexer (AM3705), sample and hold (LF398) and 8-bit A/D converter ADC0800 form the data acquisition system (DAS). The data acquisition system (DAS) is interfaced to the microprocessor using 8255A programmable peripheral interface. A clock of 300 kHz for ADC0800 is obtained by dividing the 3 MHz clock of the microprocessor by ten using the IC package 7490. Suitable clock for ADC can also be obtained by using the programmable timer/counter, Intel 8253, the controls for analog multiplexer, sample and hold, and ADC are all generated by the microcomputer under program control.

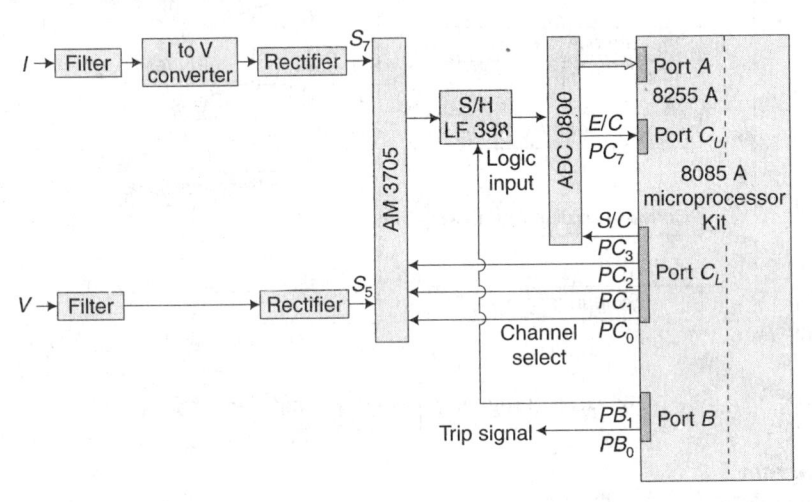

Fig. 12.23 Block schematic diagram of interface for impedance relay

The microcomputer reads V_{dc} and I_{dc}, calculates the impedance Z seen by the relay and then compares Z with Z_1, i.e. the predetermined value of impedance for the first zone of protection. If Z is less than Z_1, the microcomputer sends a tripping signal to the trip coil of the circuit breaker instantaneously. If Z is greater than Z_1, the comparison is made with Z_2, i.e. the value of impedance for the second zone of protection. If Z is less than Z_2, the microcomputer takes up a delay subroutine and sends the trapping signal to the trip coil after a predetermined delay. If Z is greater than Z_2 but less than Z_3, a greater delay is provided before the tripping signal is sent. If Z is more than Z_3, the microcomputer again reads the voltage and current signals and proceeds according to its program. The assembly language program for realisation of impedance relay characteristic is given below and the program flowchart is shown in Fig. 12.24.

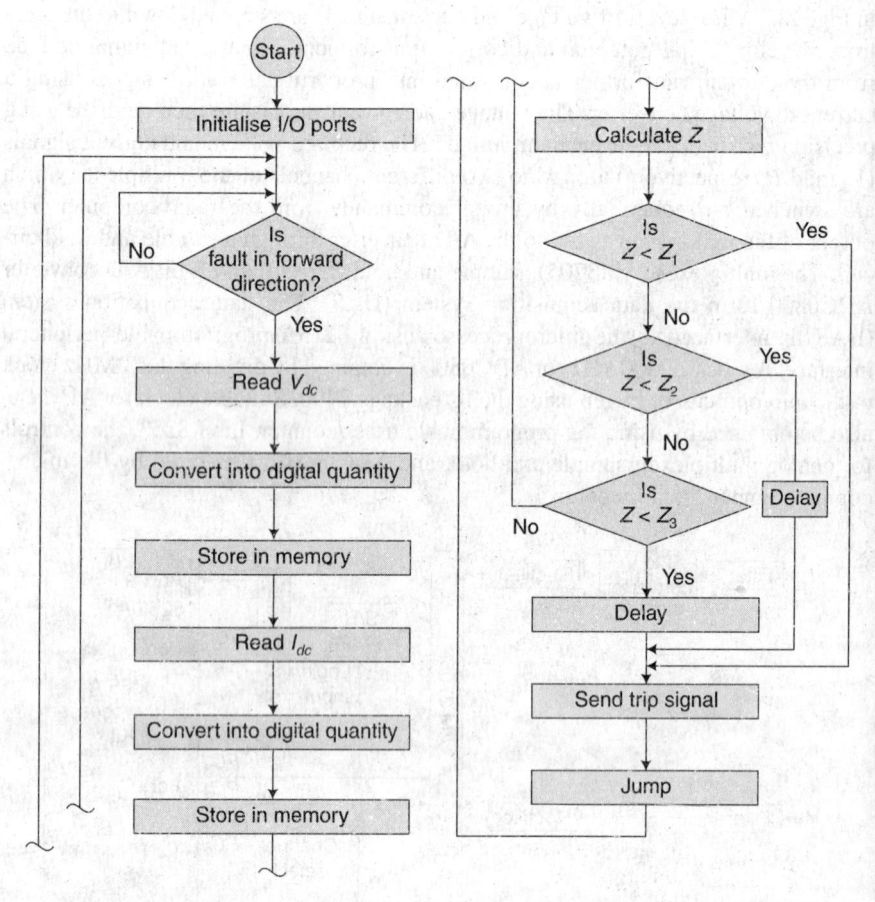

Fig. 12.24 Program flowchart for impedance relay

Program 7

Memory Address	Machine Code	Label	Mnemonics	Operands	Comments
2100	3E,98		MVI	A,98	Initialise I/O ports.
2102	D3,03		OUT	03	
2104	3E,04	BEGIN	MVI	A,04	Switch on S_5 for V_{dc}.
2106	D3,02		OUT	02	
2108	3E,02		MVI	A,02	Logic high to S/H.
210A	D3,01		OUT	01	
210C	16,03		MVI	D,03	Delay for acquisition.
210E	15	BEHIND	DCR	D	

Pgm contd...

Pgm contd...

210F	C2,0E,21		JNZ	BEHIND	
2112	3E,00		MVI	A,00	
2114	D3,01		OUT	01	
2116	3E,0C		MVI	A,0C	S/C pulse to ADC without affecting the channel address of S5.
211A	3E,04		MVI	A,04	
211C	D3,02		OUT	02	
211E	DB,02	CHECK	IN	02	Is conversion over?
2120	17		RAL		
2121	D2,1E,21		JNC	CHECK	
2124	DB,00		IN	00	Read V_{dc} at port A.
2126	2F		CMA		
2127	D6,80		SUI	80	
2129	32,50,24		STA	2450	
212C	3E,06		MVI	A,06	Switch on multiplexer channel S_7 for I_{dc}.
212E	D3,02		OUT	02	
2130	3E,02		MVI	A,02	Logic high to S/H.
2132	D3,01		OUT	01	
2134	16,03		MVI	D,03	Delay for acquisition.
2136	15	MOVE	DCR	D	
2137	C2,36,21		JNZ	MOVE	
213A	3E,00		MVI	A,00	
213C	D3,01		OUT	01	
213E	3E,0E		MVI	A,0E	S/C pulse to ADC without affecting S.
2140	D3,02		OUT	02	
2142	3E,06		MVI	A,06	
2144	D3,02		OUT	02	
2146	DB,02	LOOK	IN	02	Is conversion over?
2148	17		RAL		
2149	D2,46,21		JNC	LOOK	
214C	DB,00		In	00	Read I_{dc}
214E	2F		CMA		
214F	D6,80		SUI	80	
2151	32,52,25		STA	2552	
2154	CD,00,23		CALL	2300	Multiply V_{dc} by a constant.
2157	CD,00,25		CALL	2500	Divide by I_{dc}.
215A	21,53,25		LXI	H,2553	Z in 2553.
215D	46		MOV	B,M	
215E	23		INX	H	

Pgm contd...

Address	Code	Label	Mnemonics	Operands	Comments
215F	23		INX	H	Z_1 in 2555.
2160	7E		MOV	A,M	Z_1 in accumulator.
2161	B8		CMP	B	
2162	D2,17,24		JNC	TRIP 1	Zone 1 trip.
2165	23		INX	H	Z_2 in 2556.
2166	7E		MOV	A,M	Z_2 in accumulator.
2157	B8		CMP	B	
2168	D2,05,24		JNC	TRIP 2	Zone 2 trip.
216B	23		INX	H	Z_3 in 2557.
216C	7E		MOV	A,M	Z_3 in accumulator.
216D	B8		CMP	B	
216E	D2,00,24		JNC	TRIP 3	Zone-3 trip.
2171	C3,04,21		JMP	BEGIN	

Data

2555 – Value of Z_1
2556 – Value of Z_2
2557 – Value of Z_3

Multiplication Subroutine

Memory Address	Machine Code	Label	Mnemonics	Operands	Comments
2300	21,50,24		LXI	H,2450	
2303	5E		MOV	E,M	Get multiplicand from 2450.
2304	16,00		MVI	D,00	
2306	23		INX	H	
2307	7E		MOV	A,M	Get multiplier from 2451, constant.
2308	21,00,00		LXI	H,0000	
230B	06,08		MVI	B,08	
230D	29	HEAD	DAD	H	
230E	17		RAL		
230F	D2,13,23		JNC	SEE	
2312	05	SEE	DCR	B	
2314	C2,0D,23		JNZ	HEAD	
2317	22,50,25		SHLD	2550	Result in 2550 and 2551.
231A	C9		RET.		

Delay Subroutine

Memory Address	Machine Code	Label	Mnemonics	Operands	Comments
2400	0E,06	TRIP 3	MVI	C,06	For 0,4 sec delay
2402	C3,07,24		JMP		

Pgm contd...

Pgm contd...

Address	Code	Label	Mnemonics	Operands	Comments
2405	0E,10	TRIP 2	MVI	C,03	For 0.2 sec delay
2407	16,FF		MVI	D,FF	
2409	1E,33		MVI	E,33	
240B	1D	SEND	DCR	E	
240C	C2,0B,24		JNZ	FIRST	
240F	15		DCR	D	
2410	C2,09,24		JNZ	SEND	
2413	0D		DCR	C	
2414	C2,07,24		JNZ	THID	
2417	3E,01	TRIP 1	MVI	A,01	
2419	D3,01		OUT	01	
241B	C3,04,20		JMP	BEGIN	

Division Subroutine

Memory Address	Machine Code	Label	Mnemonics	Operands	Comments
2500	2A,50,25		LHLD	2550	Get dividend.
2503	3A,52,25		LDA	2552	Get divisor.
2506	4F		MOV	C,A	
2507	06,08		MVI	B,08	
2509	29	DIV	DAD	H	
250A	7C		MOV	A,H	
250B	91		SUB	C	
250C	DA,11,25		JC	AHEAD	
250F	67		MOV	H,A	
2510	2C		INR	L	
2511	05	AHEAD	DCR	B	
2512	C2,09,25		JNZ	DIV	
2515	22,53,25		SHLD	2553	Quotient in 2553.
2518	C9		RET.		

12.6 DIRECTIONAL RELAY

A directional relay senses the direction of power flow. The polarity of the instantaneous value of the current at the moment of the voltage peak is examined to judge the direction of power flow. The program developed for this relay is able to judge whether the fault point is in the forward or reverse direction with respect to relay location as in Fig. 12.25(a). The instantaneous value of the current at the moment of voltage peak is $I_m \cos \phi$, as shown in Fig. 12.25(b). For a fault point lying in the forward direction, $I_m \cos \phi$ is positive for ϕ lying within $\pm 90°$. For a fault lying in the reverse direction, $I_m \cos \phi$ becomes negative, as shown in Fig. 12.25(c).

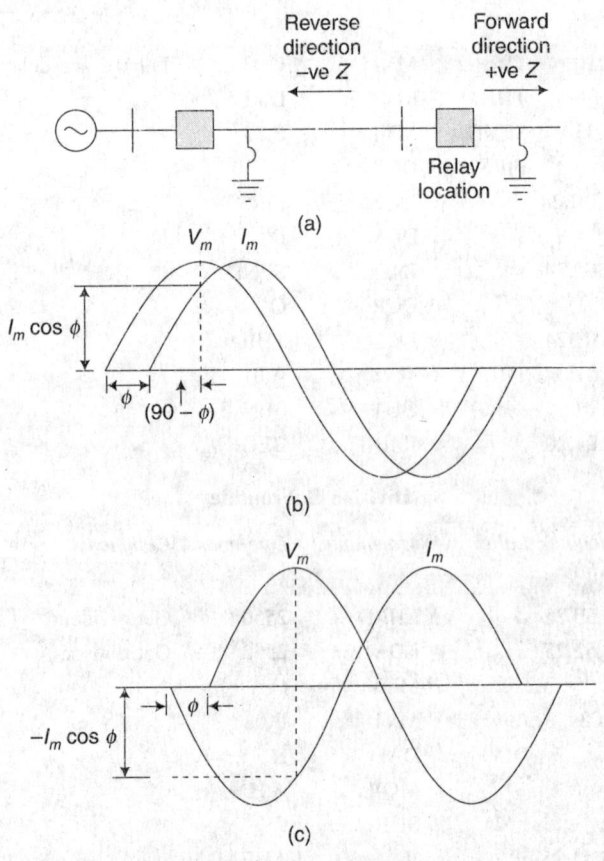

Fig. 12.25 (a) Location of directional relay (b) Instantaneous value of current at peak voltage (c) $I_m \cos \phi$ for fault in reverse direction

The block schematic diagram of the interface for this relay is shown in Fig. 12.26. A phase-shifter and a zero-crossing detector are used to obtain a pulse at the moment of voltage peak. The voltage signal is fed to the phase-shifter to get a phase-shift of 90°. Then the output of the phase-shifter is fed to a zero-crossing detector to obtain the required pulse. The microcomputer reads the output of the zero-crossing detector to examine whether the voltages has crossed its peak value. After receiving the pulse the microcomputer sends a command to the multiplexer to switch on the channel S_2 to obtain the instantaneous value of the current at the moment of voltage peak. The microcomputer reads this value of the current through the A/D converter and examines whether it is positive.

This type of relay can be used to sense the reversal of power flow where it is required. When the power flow is reversed, $I_m \cos \phi$ becomes negative and the microcomputer sends a tripping signal to the circuit breaker. The program flowchart is shown in Fig. 12.27. This relay can also be used in conjunction with overcurrent relays and impedance relays to provide directional features. When it is used as a directional relay, it energises other relays only when the fault lies in the forward direction. It can also be used as a directional relay in conjunction with overcurrent relays for the protection of parallel lines.

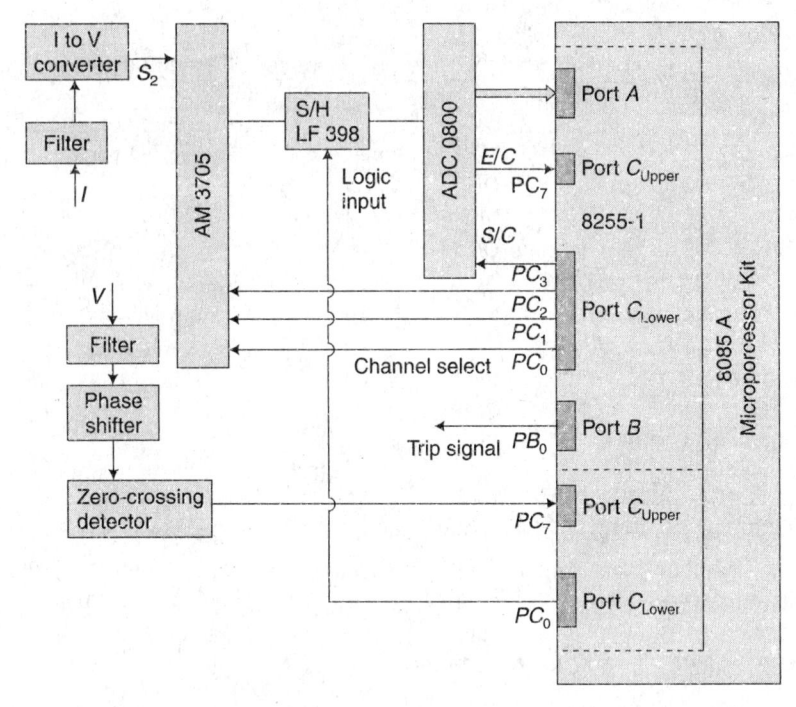

Fig. 12.26 Block schematic diagram of interface for directional relay

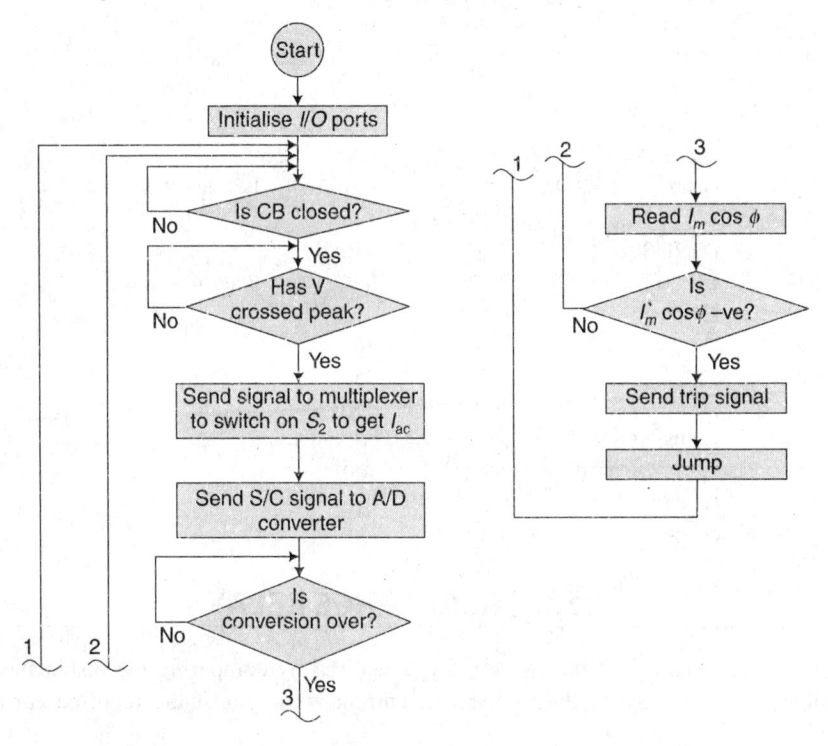

Fig. 12.27 Program flowchart for reverse power relay

484 Power System Protection and Switchgear

Program 8

Memory Address	Machine Code	Label	Mnemonics	Operands	Comments
2000	3E,98		MVI	A,98	Initialise I/O ports.
2002	D3,03		OUT	03	
2004	3E,9A		MVI	A,9A	
2006	D3,0B		OUT	0B	
2008	DB,0A	START	IN	0A	
200A	47		MOV	B,A	Check whether V 90 crossed zero?
200B	DB,0A		IN	0A	
200D	B8		CMP	B	
200E	CA,08,20		JZ	START	
2011	DA,08,20		JC	START	
2014	3E,01		MVI	A,01	Switch on S_2 for $I_m \cos \phi$.
2016	D3,02		OUT	02	
2018	3E,01		MVI	A,01	
201A	D3,0A		OUT	0A	Logic high to S/H.
201C	16,03		MVI	D,03	Time for acquisition.
201E	15	BACK	DCR	D	
201F	C2,1E,20		JNZ	BACK	
2022	3E,00		MVI	A,00	
2024	D3,0A		OUT	0A	
2026	3E,09		MVI	A,09	S/C pulse to ADC.
2028	D3,02		OUT	02	
202A	3E,01		MVI	A,01	
202C	D3,02		OUT	02	
202E	DB,02	READ	IN	02	Is conversion over?
2030	17		RAL		
2031	D2,2E,20		JNC	READ	
2034	DB,00		IN	00	Read $I_m \cos \phi$.
2036	2F		CMA		
2037	D6,80		SUI	80	
2039	32,02,25		STA	2502	
203C	DA,42,20		JC	TRIP	
203F	D2,08,20		JNC	START	
2042	3E,01	TRIP	MVI	A,01	
2044	D3,01		OUT	01	
2046	C3,08,20		JMP	START	

12.7 REACTANCE RELAY

The characteristic of a reactance relay is realised by comparing the instantaneous value of the voltage at the moment of current zero against the rectified current.

The instantaneous value of voltage at the moment of current zero is $V_m \sin \phi$, as shown in Fig. 12.28. For the operation of the relay, the conditions to be satisfied is as follows.

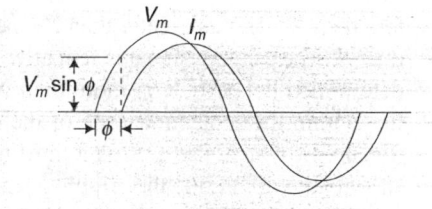

Fig. 12.28 Instantaneous values of voltage at current zero

$$V_m \sin \phi < K_1 I_{dc}$$

or

$$\frac{V_m \sin \phi}{I_{dc}} < K_1$$

or

$$\frac{V \sin \phi}{I} < K$$

(as V_m and I_{dc} are proportional to rms values V and I, respectively)

or

$$Z \sin \phi < K$$

or

$$X < K \quad (12.7)$$

The block schematic diagram of the interface for the realisation of reactance relay characteristic is shown in Fig. 12.29. The microcomputer reads the output of the zero crossing detector to examine whether the current has crossed its zero point. As soon as the current crosses its zero point, the microcomputer sends a command to the multiplexer to switch on channel S_4, and gets the instantaneous value of the voltage, i.e. $V_m \sin \phi$ through the A/D converter. Then the microcomputer sends a command to the multiplexer to switch on the channel S_7 to get the rectified current. Thereafter, the microcomputer calculates X, the reactance as seen by the relay and compares it with X_1, the predetermined value of the reactance for the first zone of protection. The microcomputer sends a tripping signal instantaneously, if the measured value of X is

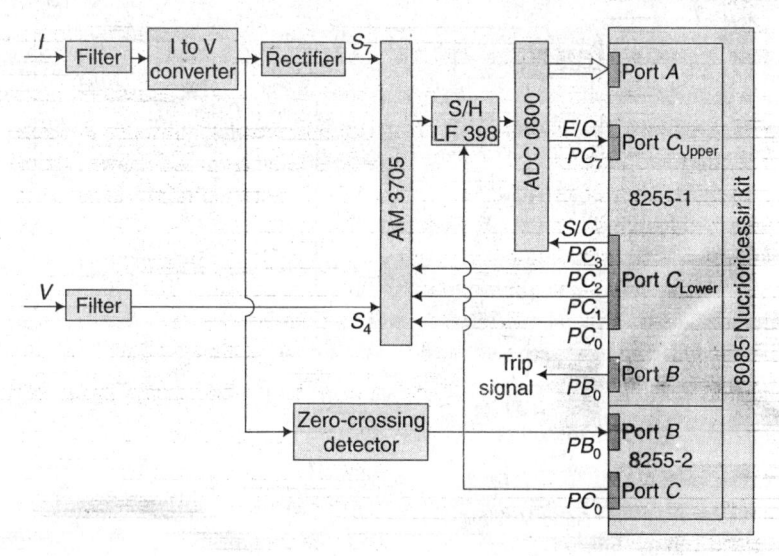

Fig. 12.29 Block schematic diagram of interface for reactance relay

less then X_1. If X is greater than X_1, but less than X_2, the tripping signal is sent after a predetermined delay. If X is more than X_2 but lies within the protection zone of the directional unit which also acts as a third unit as shown in Fig. 12.30, the tripping signal is sent after a greater predetermined delay.

As the reactance relay is a non-directional relaying unit, a directional relay is used in conjunction with it to provide directional features. The directional unit also serves the purpose of the third unit. The directional unit used for reactance relays has the characteristic of mho relay passing through the origin. The program of the directional unit is incorporated in the main program of the reactance relaying protective scheme. If the fault

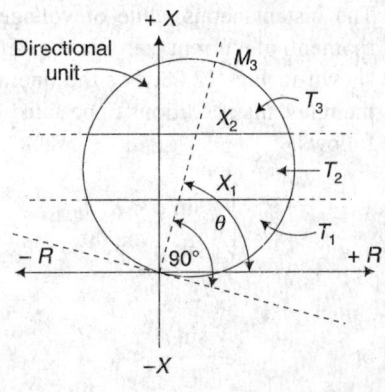

Fig. 12.30 *Reactance relay characteristics with directional unit*

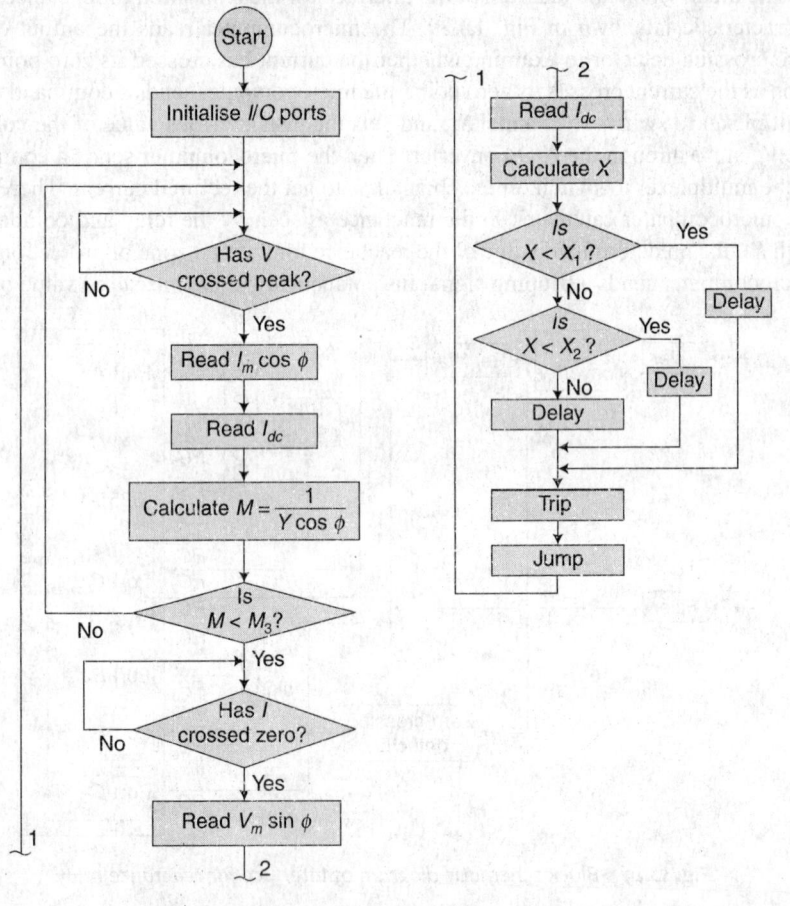

Fig. 12.31 *Program flowchart for reactance relay*

point lies within the protection zone of the directional unit then only the reactance relay program is taken up to check the position of the fault point, i.e. whether it lies in the I, II or III zone of protection. Depending upon the zone of protection, the tripping signal is sent with or without delay. The program flowchart is shown in Fig. 12.31.

12.8 GENERALISED MATHEMATICAL EXPRESSION FOR DISTANCE RELAYS

A generalised mathematical expression for the operating conditions of mho, offset mho and impedance relays can be derived as follows. The derivation of this expression is based on the operating condition of the offset mho relay having a positive offset.

Figure 12.32 (a) shows the characteristic of an offset mho relay on the $R\text{-}X$ diagram. The radius of this circle is

$$r = \frac{Z_r - Z_o}{2} \quad (12.8)$$

where $Z_r = R_r + jX_r$ = Impedance of the protected line section.
and $Z_o = R_o + jX_o$ = Impedance by which the mho circle is offset.

The centre of the mho circle is offset from the origin by

$$C = \frac{Z_r + Z_o}{2} \quad (12.9)$$

If $Z (= R + jX)$ is the impedance seen by the relay, then the operating condition for the offset mho relay, shown in Fig. 12.32 (a), is given by

$$Z - C \leq r \quad (12.10)$$

or $$\left| Z - \frac{(Z_r + Z_o)}{2} \right| \leq \left| \frac{(Z_r - Z_o)}{2} \right|$$

or $$\left| (R + jX) - \frac{(R_r + jX_r) + (R_o + jX_o)}{2} \right| \leq \left| \frac{(R_r + jX_r) - (R_o + jX_o)}{2} \right|$$

or $$\left| \left(R - \frac{(R_r + R_o)}{2} \right) + j \left(X - \frac{(X_r + X_o)}{2} \right) \right| \leq \left| \frac{(R_r - R_o)}{2} + j \frac{(X_r - X_o)}{2} \right|$$

or $$\left[\left(R - \frac{(R_r + R_o)}{2} \right)^2 + \left(X - \frac{(X_r + X_o)}{2} \right)^2 \right] \leq \left[\left(\frac{(R_r - R_o)}{2} \right)^2 + \left(\frac{(X_r - X_o)}{2} \right)^2 \right] \quad (12.11)$$

R_r, R_o, X_r and X_o are constants for a particular characteristic and hence the above expression can be written in the generalise form as

$$(R - K_1)^2 + (X - K_2)^2 \leq K_3 \quad (12.12)$$

where, the constants K_1, K_2 and K_3 are given by

$$K_1 = \frac{(R_r + R_o)}{2} \quad (12.13)$$

$$K_2 = \frac{(X_r + X_o)}{2}$$

$$K_3 = \left[\left(\frac{R_r - R_o}{2}\right)^2 + \left(\frac{X_r - X_o}{2}\right)^2\right]$$

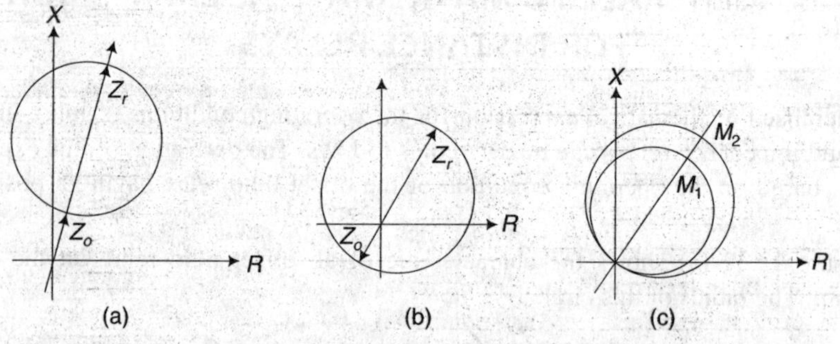

Fig. 12.32 (a) Offset mho relay with positive offset
(b) Offset mho relay with negative offset
(c) Mho relay

Equation (12.12) is the generalised expression for the offset mho, mho and impedance relays. The values of K_1, K_2 and K_3 in the generalised expression are constant for a particular characteristic and different for different characteristics. Substituting the proper values of constants K_1, K_2 and K_3 the desired mho, offset mho or impedance characteristic can be realised. The values of K_1, K_2 and K_3 for different characteristics are computed as follows.

(i) For offset mho characteristic having negative offset as shown in Fig. 12.32(b), the negative values of R_o and X_o are used in computation of K_1, K_2 and K_3 using Eq. (12.13).

$$K_1 = \frac{(R_r - R_o)}{2} \qquad (12.14)$$

$$K_2 = \frac{(X_r - X_o)}{2}$$

and $$K_3 = \left[\left(\frac{(R_r + R_o)}{2}\right)^2 + \left(\frac{(X_r + X_o)}{2}\right)^2\right]$$

(ii) For mho characteristic as shown in Fig. 12.32(c), R_o and X_o are reduced to zero, and consequently the values of K_1, K_2 and K_3 obtained from Eq. (12.13) are as follows.

$$K_1 = \frac{R_r}{2}, \quad K_2 = \frac{X_r}{2} \quad \text{and} \quad K_3 = \left(\frac{R_r}{2}\right)^2 + \left(\frac{X_r}{2}\right)^2$$

(iii) For impedance characteristic, the displacement of the centre of the circle from the origin is zero

i.e. $$C = \frac{Z_o + Z_r}{2} = 0 \quad \text{or} \quad Z_o = -Z_r$$

or $\quad (R_o + jX_o) = -(R_r + jX_r)$ (12.15)

Therefore, $R_o = -R_r$ and $X_o = -X_r$.

Then from Eq. (12.13), the values of K_1, K_2 and K_3 for impedance characteristic are as follows.

$$K_1 = K_2 = 0 \quad \text{and} \quad K_3 = R_r^2 + X_r^2$$

Consequently, the operating conditions for the impedance relay becomes

$$R^2 + X^2 \leq R_r^2 + X_r^2 \quad (12.16)$$

or $\quad R^2 + X^2 \leq r^2,$

where r is the radius of the circle. Therefore, a offset mho, mho or impedance characteristic can easily be realised by using the generalised Eq. (12.12) and substituting the appropriate values of K_1, K_2 and K_3 for the desired characteristic.

12.9 MEASUREMENT OF R AND X

In microprocessor-based distance relaying, the microprocessor calculates the active and reactive components (R and X) of the apparent impedance (Z) of the line from the relay location to the fault point from the ratios of the appropriate voltages and currents, and then compares the calculated values of R and X to the pickup value of the relay to be realised in order to determine whether the fault occurs within the protective zone of the relay or not. The active and reactive components of the apparent impedance (Z) are resistance (R) and reactance (X), respectively. The algorithm presented in this section for the calculation of R and X assumes that the waveforms of voltage and current presented to the relay are pure fundamental frequency sinusoids. But the post-fault voltage and current waveforms fail to be pure fundamental frequency sinusoids due to the presence of harmonics and dc offset components resulting from the fault. Therefore, analog band-pass filtering of the voltage and current signals is necessary in order to eliminate harmonics and dc offset components and to obtain pure fundamental frequency sinusoidal signals. An active band-pass filter having a band-pass of 40 to 60 Hz is employed to filter harmonics and dc offset. Filtering of harmonics and dc offset from the post-fault voltage and current signals can also be achieved by employing digital filters. A number of digital algorithms, based on the digital filtering technique have been proposed for the calculation of R and X. A few digital algorithms which are suitable for microprocessor-based protective relaying are discussed in the subsequent sections.

12.9.1 Measurement of Resistance

The resistance as seen by the relay from the relay location to fault point is given by

$$R = Z \cos \phi = \frac{V}{I} \cos \phi$$

$$= \frac{V_m \cos \phi}{\sqrt{2} \, A_1 I_{dc}} \quad \text{(as } I \propto I_{dc}, \text{ or } I = A_1 I_{dc} \text{ where } A_1 \text{ is a constant}$$

$$= A \frac{V_m \cos \phi}{I_{dc}} \quad (12.17)$$

where A is a constant, and equal to $1/\sqrt{2} \, A_1$. $V_m \cos \phi$ is the instantaneous value of the voltage at the moment of peak current, as shown in Fig. 12.33(a). Therefore, the resistance is proportional to the ratio of voltage at the instant of current peak to the rectified current I_{dc}. To obtain a pulse at the moment of peak current, a phase shifting circuit and a zero-crossing detector have been used. The current signal is fed to the phase shifter to get a phase shift of 90°. Then the output of the phase shifter is fed to the zero crossing detector, as shown in Fig. 12.33(b) to obtain the required pulse. The microcomputer reads the output of the zero-crossing detector, in order to examine whether the current signal has reached its peak. As soon as the current crosses its peak, the microcomputer sends a command to the multiplexer to switch on channel S_4 to obtain the instantaneous value of the voltage at the moment of peak current, which is equal to $V_m \cos \phi$. The microcomputer reads this instantaneous value of the voltage through the A/D converter, and stores the digital voltage in the memory. The current signal is converted to dc using a precision rectifier. The microcomputer gets the rectified current I_{dc} through the multiplexer channel S_7 and A/D converter. After getting the values of $V_m \cos \phi$ and I_{dc}, the microcomputer calculates the value of resistance.

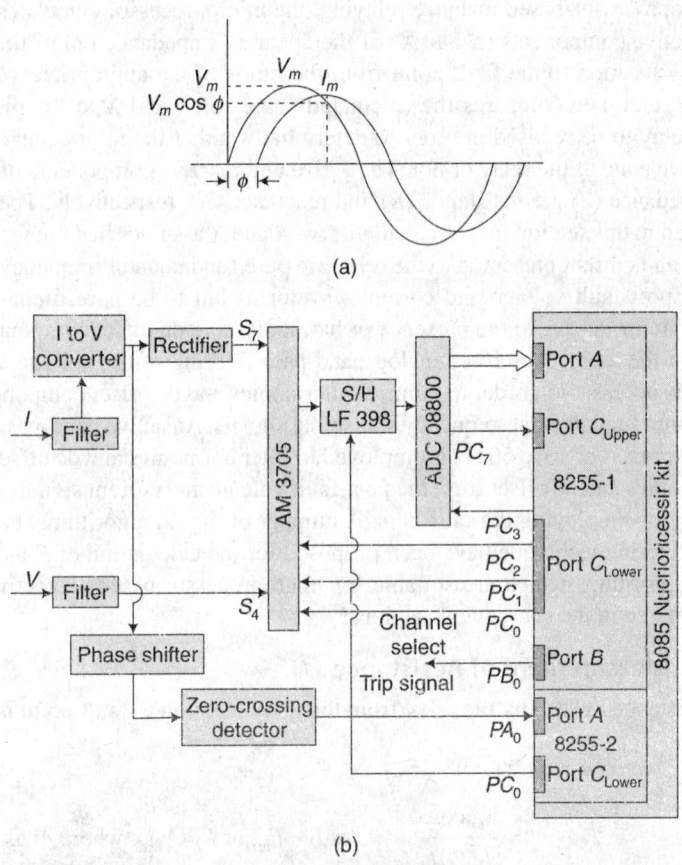

Fig. 12.33 (a) Instantaneous value of voltage at the instant of peak current
(b) Block schematic diagram of interface for resistance measurement

12.9.2 Measurement of Reactance

The reactance seen by the relay is given by

$$X = Z \sin \phi = \frac{V}{I} \sin \phi$$

$$= \frac{V_m \sin \phi}{\sqrt{2} \, A_1 I_{dc}} = A \frac{V_m \sin \phi}{I_{dc}} \qquad (12.18)$$

The instantaneous value of the voltage at the moment of zero current is $V_m \sin \phi$, as shown in Fig. 12.28. Therefore, the reactance is proportional to the ratio of voltage at the moment of current zero to the rectified current. The measurement of reactance using this technique has already been discussed in Section 12.7, while describing the reactance relay. The interface for the measurement of reactance will be the same as that for the reactance relay, i.e. as shown in Fig. 12.11. The microcomputer reads the output of the zero-crossing detector and examines whether the current has reached its zero instant. As soon as the current crosses its zero, the microcomputer reads the instantaneous value of the voltage through the multiplexer channel S_4 and A/D converter. Then the microcomputer reads the rectified current I_{dc} through the multiplexer channel S_7 and A/D converter. After receiving the values of $V_m \sin \phi$ and I_{dc} the microcomputer computes the reactance X. The program flowchart for the measurement of R and X is shown in Fig. 12.34.

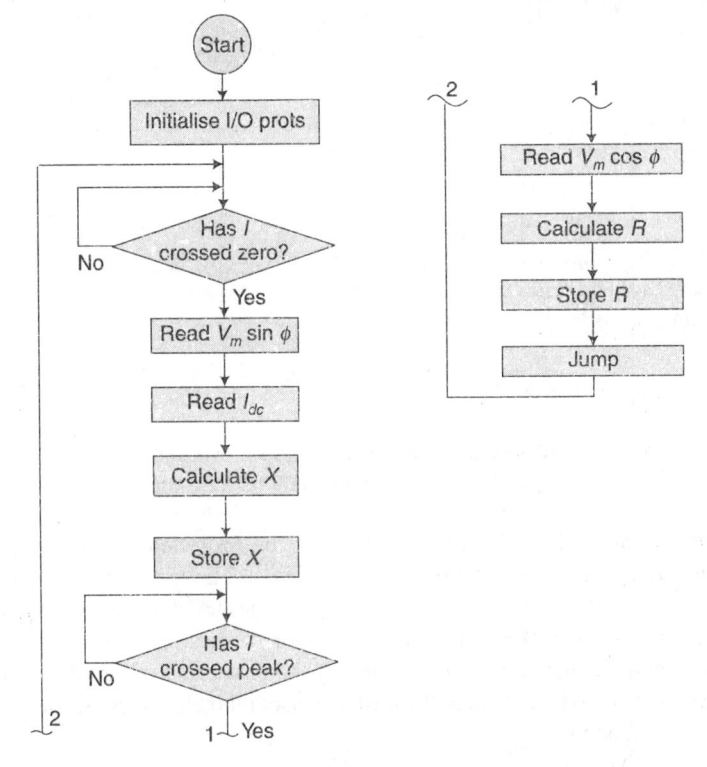

Fig. 12.34 *Program flowchart for measurement of R and X*

12.10 Mho AND OFFSET Mho RELAYS

The characteristic of a mho relay on the impedance $(R - X)$ diagram is a circle passing through the origin, as shown in Fig. 12.32(c). With such a characteristic, the mho relay is inherently directional as it detects faults in the forward direction only. The mho characteristic, occupying the least area on the $R - X$ diagram is least affected by power surges which remain for longer periods in case of long lines. Therefore, a mho relay is best suited for the protection of long transmission lines against phase faults. Through this relay is affected by arc resistances more than any other type of the relay, the value of arc resistance in comparison to the impedances of the long lines is much less. Thus, an arc resistance does not affect much the performance of a mho relay in case of long lines. For very long lines, the tripping area of mho relays can be further reduced by blinders.

Offset mho relays, whose characteristics are shown in Fig. 12.32(a) and (b) are used for zone III in a three-zone protective scheme employing mho relays for zones I and II for the protection of power transmission lines against phase faults. The scheme is shown in Fig. 12.35(a). For long transmission lines, mho and offset mho characteristics, as shown in Fig. 12.35(b) are used to discriminate between loads and faults. An offset mho characteristic with negative offset has more tolerance to arc resistance. By offsetting the III zone mho characteristic to overlap the origin, it can also be used for power swing blocking.

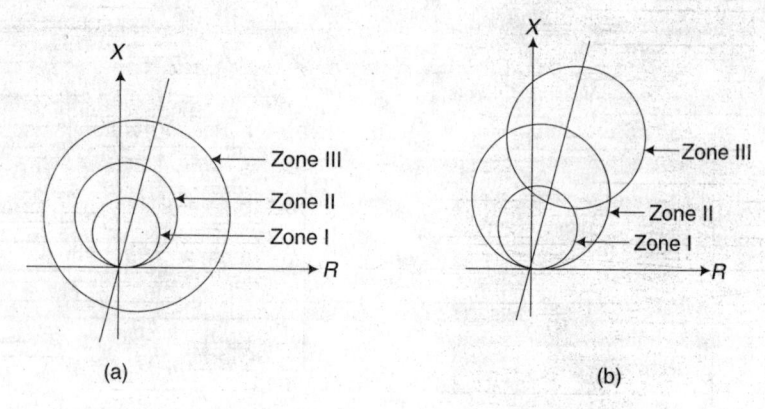

Fig. 12.35 *(a) Mho and offset mho relays for transmission lines*
(b) Mho and offset mho relays for long lines

The tripping area of a mho characteristic can be restricted by two overlapping mho characteristics as shown in Fig. 12.36(a) and (b). The tripping area is the common area between two mho characteristics. This types of characteristic occupies a very small area on the R–X diagram and hence it is least affected by power surges. A restricted mho characteristic can be used for the protection of very long lines. To take care of the resistive faults near the busbar, offset mho characteristic, as shown in Fig. 12.36(b) can be used.

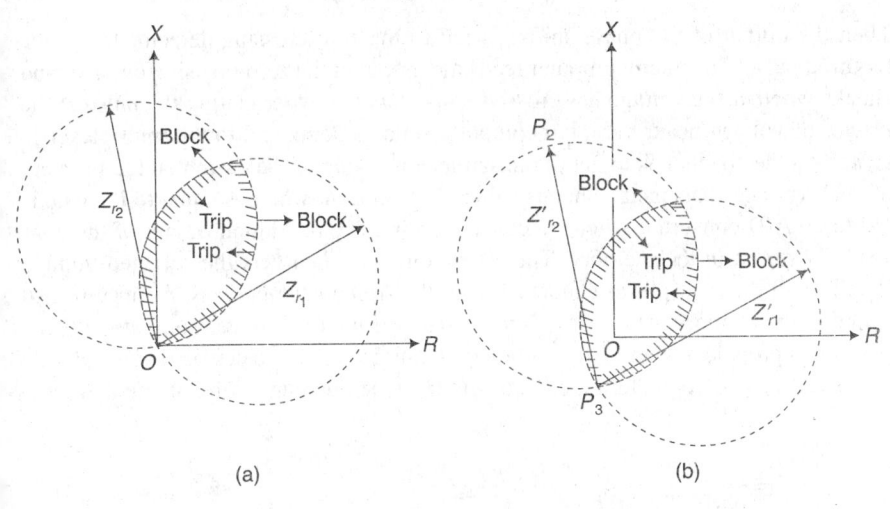

Fig. 12.36 (a) Restricted mho passing through origin (b) Restricted offset mho relay

12.10.1 Realisation of Mho Characteristic

The characteristic of a mho relay can be realised using three techniques i.e. (i) instantaneous amplitude comparison technique, (ii) phase comparison technique and (iii) generalised mathematical expression.

(i) Instantaneous Amplitude Comparison Technique

The characteristic of a mho relay is realised by comparing the instantaneous value of current at the moment of voltage peak against the rectified voltage. The instantaneous value of current at the moment of voltage peak is $I_m \cos \phi$, as shown in Fig. 12.25(b). The condition to be satisfied for the operation of the relay is as follows.

$$I_m \cos \phi > K_1 V_{dc} \quad \text{or} \quad I_m/V_{dc} \cos \phi > K_1$$

or
$$I/V \cos \phi > K_2 \quad \text{(as } I_m \text{ and } V_{dc} \text{ are proportional to the rms values of } I \text{ and } V \text{ respectively)}$$

or
$$Y \cos \phi > K_2 \quad \text{or} \quad \frac{1}{Y \cos \phi} < \frac{1}{K_2}$$

or
$$M < K \tag{12.19}$$

where K_1, K_2 and K are constants.

If the design angle θ is introduced while feeding the voltage and current signals to the relay, the above expression is modified and is given by

$$\frac{1}{Y \cos (\phi - \theta)} < K \tag{12.20}$$

By changing θ a mho characteristic can be shifted towards the R-axis to increase its tolerance to arc resistance, as shown in Fig. 12.37. The value of θ is not kept more than 75° to have a reasonable tolerance for arc resistances when a fault occurs near the bus.

The block schematic diagram of the interface for realisation of mho relay is shown in Fig. 12.38. The voltage signal is fed to the phase-shifter to get phase-shift of 90°.

Then the output of the phase shifter is fed to the zero-crossing detector to get the required pulse. The microcomputer reads the output of the zero-crossing detector and checks whether the voltage has crossed its peak value. After getting the pulse at the instant of voltage peak, the microcomputer sends a command to the multiplexer to switch on the channel S_2 to get the instantaneous value of the current at the moment of peak voltage. This instantaneous value of the current which is equal to $I_m \cos\phi$ is fed to an A/D converter. After the conversion is over, the digital output of the converter is stored in the memory. The microcomputer then gets the rectified voltage V_{dc} through the multiplexer channel S_5 and the A/D converter. This quantity is also stored in the memory. After obtaining this data the microcomputer computes $V/I \cos\phi$ which is proportional to $V_{dc}/I_m \cos\phi$ and is equal to M. The calculated value of M is compared with the predetermined value of M_1 which remains stored in the memory.

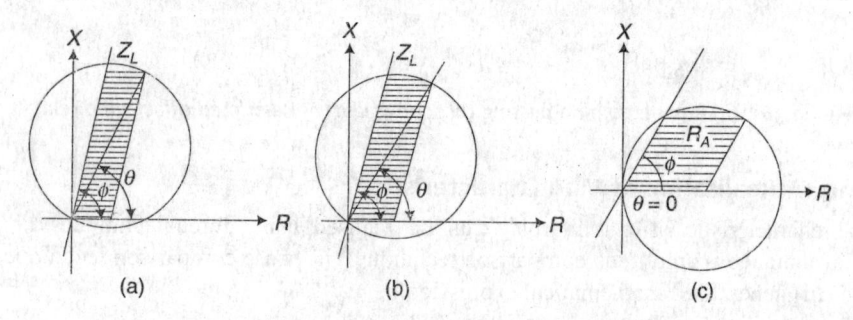

Fig. 12.37 Mho relay's characteristics with different values of θ

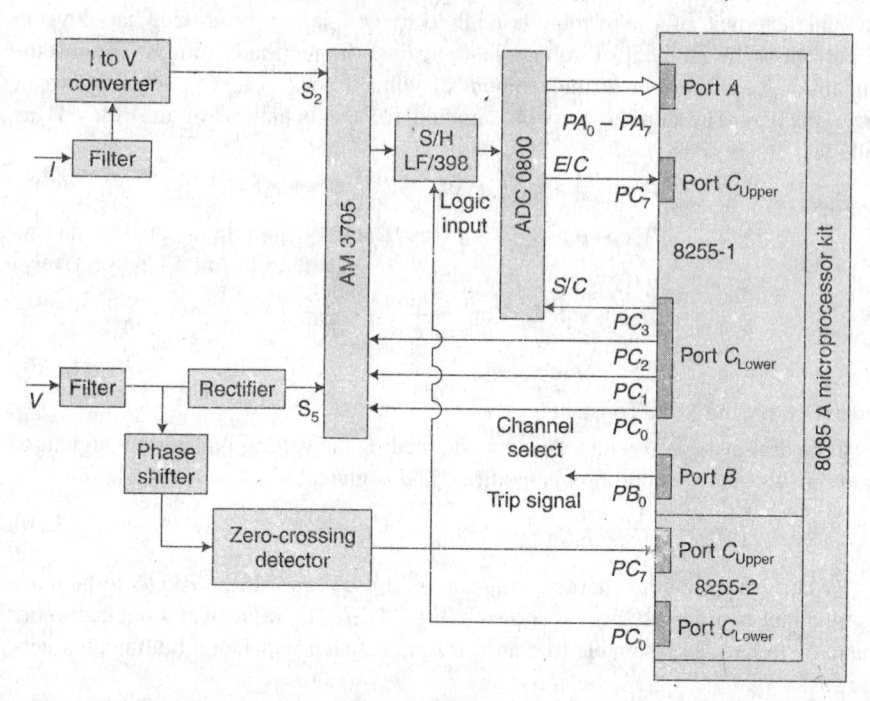

Fig. 12.38 Block schematic diagram of interface for mho relay

M_1, M_2 and M_3 are the predetermined values of M for I, II, and III zones of protection, respectively. If M is less than M_1, the tripping signal is sent instantaneously. If M is greater than M_1 but less than M_2, the tripping signal is sent after a predetermined delay. If M is greater than M_2, the tripping signal is sent after a predetermined delay. If M is greater than M_2 but less than M_3, a greater delay is provided to send the tripping signal. If M is greater than M_3, the microcomputer goes back to the starting point, starts reading the voltage and current signals and again proceeds according to the program. The program flowchart is shown in Fig. 12.39.

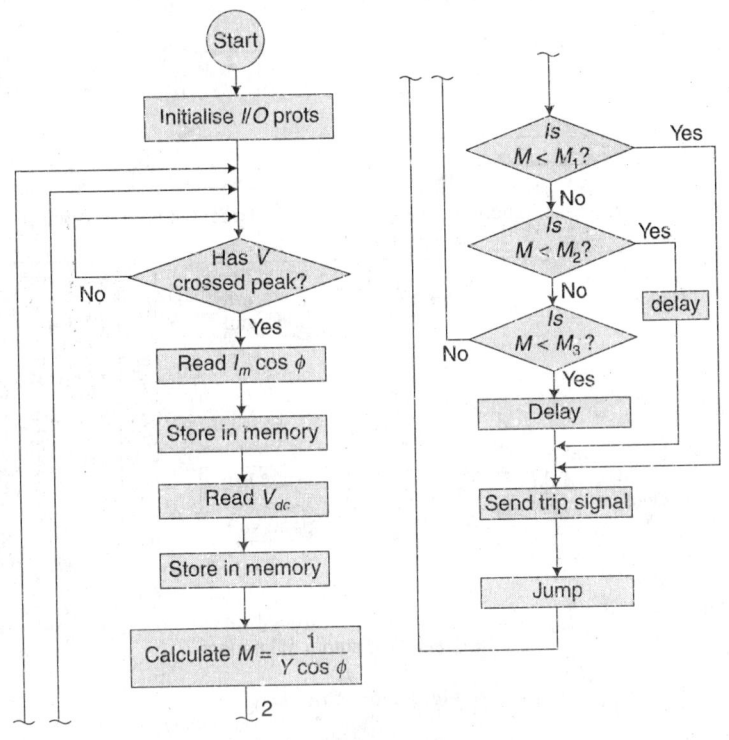

Fig. 12.39 *Program flowchart for mho relay*

(ii) Phase Comparison Technique

For realisation of a mho characteristic, the phase angle between $(IZ_r - V)$ and V is measured. Figure 12.40(a) shows a phasor diagram showing V, IR and IZ_r. The phasors of this diagram are divided by I and the resulting phasor diagram is obtained in the Z-plane as shown in Fig. 12.40(b). The diameter of the mho circle is Z_r. For any point lying within the circle, the phase angle between the phasors $(Z_r - Z)$ and Z is less than $\pm 90°$. For points lying on the right hand side of the diameter, the phase angle is positive and for points lying on the left hand side, it is negative.

The block schematic diagram of the interface for the measurement of phase angle between $(IZ_r - V)$ and V is shown in Fig. 12.40(c). The ac input signals are converted into square waves. The phase angle is measured between the positive going zeros of the square waves. The phasor ($IZ_r - V$) is taken as reference. For measurement of phase angle between ($IZ_r - V$) and V, the microcomputer measures the time between positive going zero points of their corresponding square waves. The phase angle is proportional to the measured time. The microcomputer then compares the measured phase angle with 90°, in order to determine whether the fault point lies within the mho circle or not. If the measured phase angle is less than 90°, the microcomputer issue a tripping signal.

(a) Phasor diagram

(b) Phasor diagram in Z-plane

(c) Schematic diagram of interface

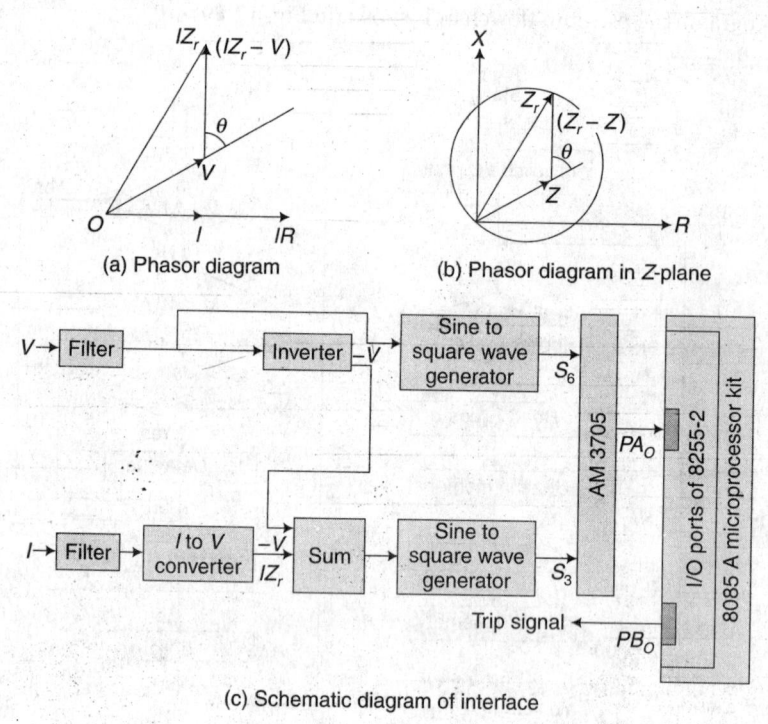

Fig. 12.40 *Mho relay*

(iii) Generalised Expression Technique

The mho characteristic is realised with the help of a generalised expression (12.12). For this characteristic, the values of constants K_1, K_2 and K_3 are respectively equal to $R_r/2$, $X_r/2$ and $(R_r/2)^2 + (X_r/2)^2$. The constants are predetermined and stored in the memory. The microcomputer first of all calculates the resistance and reactance (R and X) as seen by the relay and then examines whether the operating condition as expressed by Eq. (12.12) is satisfied or not. If this condition is satisfied, the microcomputer sends a tripping signal. The block diagram of the interface is shown in Fig. 12.41.

12.10.2 Realisation of Offset Mho Characteristic

The offset mho characteristic is also realised using the generalised expression or Eq. (12.12). The constants K_1, K_2 and K_3 for the offset mho characteristic are predetermined and stored in the memory. The microcomputer measures R and X at the relay location, makes calculations for different terms of the equation (12.12), and then checks whether the operating condition is satisfied. If the operating condition is satisfied, the microcomputer sends the trip signal. The block diagram of the interface is shown in Fig. 12.41 and the program flowchart in Fig. 12.41.

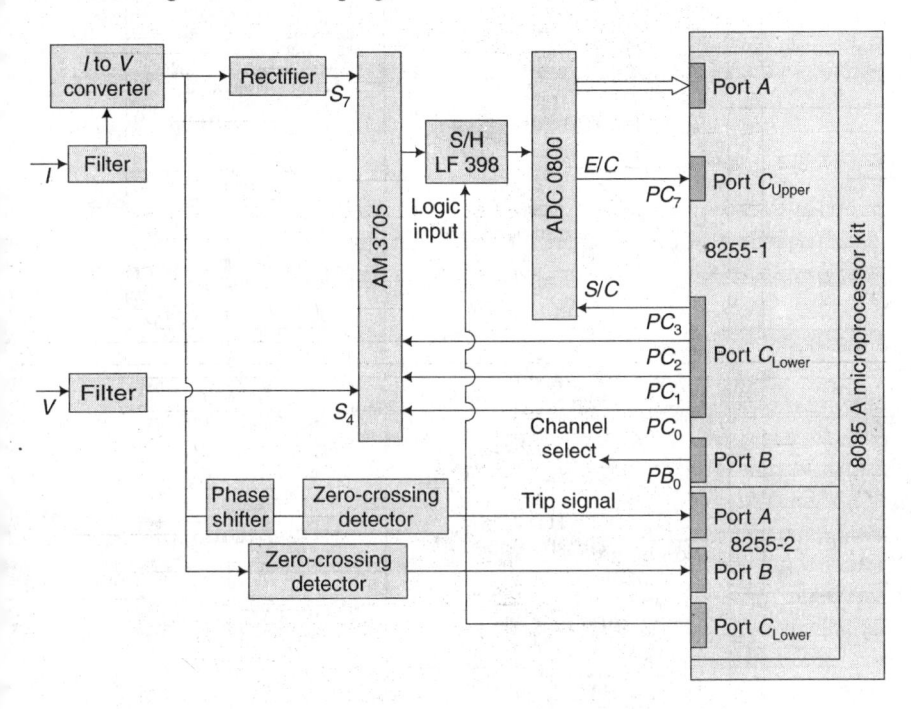

Fig. 12.41 Block schematic diagram of interface for Mho, offset Mho, restricted Mho and quadrilateral relays

12.10.3 Realisation of Restricted Mho Characteristic

Restricted mho and restricted offset mho characteristics are shown in Fig. 12.36 (a) and (b). The block schematic diagram of the interface for realisation of these characteristics is shown in Fig. 12.41. For these characteristics, the mho or offset mho circle, which is near the X-axis has a negative value for the constant K_1. Therefore, the term $(R + K_1)^2$ replaces the term $(R - K_1)^2$ in the expression of Eq. (12.12). The microcomputer first of all measures R and X and then examines whether the fault point lies in the I mho characteristic which is near the R-axis. If the fault points does not lie in this circle, it goes to the starting point and measures R and X again. If the fault point lies in the I circle, the microcomputer then checks whether it also lies within the II circle which is near the X-axis. If the fault point lies within both circles, i.e. in the common area of the circles, the microcomputer sends a tripping signal to the circuit breaker. The program flowchart is shown in Fig. 12.43.

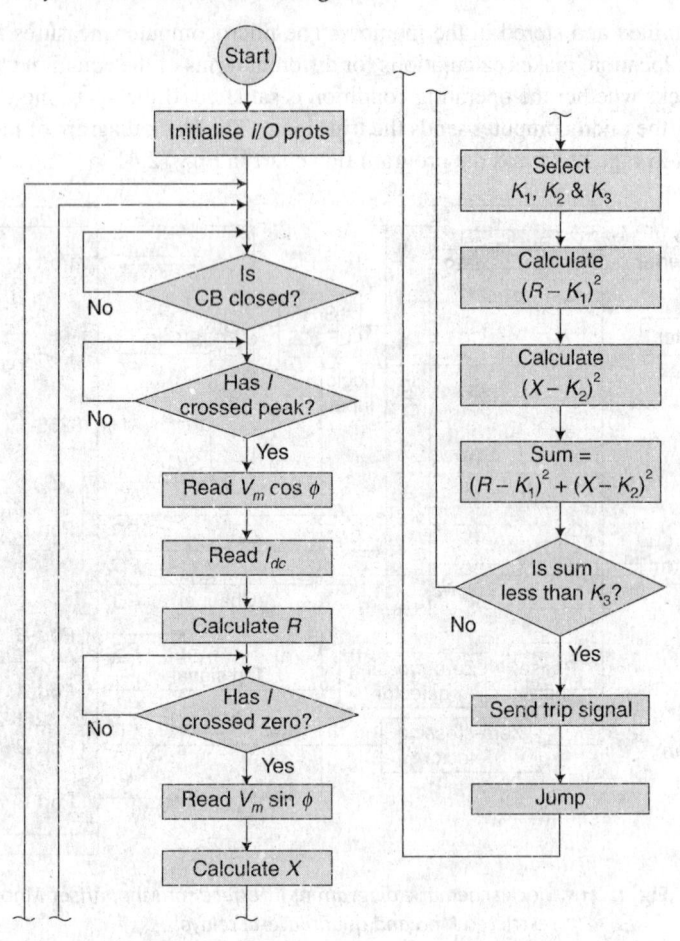

Fig. 12.42 *Program flowchart for offset MHO relay*

12.11 QUADRILATERAL RELAY

The quadrilateral relay is best suited for the protection of EHV/UHV and ELD transmission lines as it possesses the valuable property of possessing the least tendency for mal-operation due to heavy power swings, fault resistance and overloads. It is also suitable for short and medium line. Its characteristic can be designed to just enclose the fault area of the line to be protected. The electromagnetic version of this relay requires four units, each corresponding to one side of the quadrilateral, or a mho relay with two blinders. A static relay employing a multi-input comparator gives a better quadrilateral characteristic than the electromagnetic relays. But this relays does not possess the flexibility needed for obtaining different characteristic. A microprocessor-based scheme can easily obtain a quadrilateral characteristic using the same interface which is used for other types of distance relays. It has the flexibility to obtain a desired quadrilateral characteristics by simply providing the proper data. Quadrilateral characteristics are shown in Fig. 12.44(a) and (b).

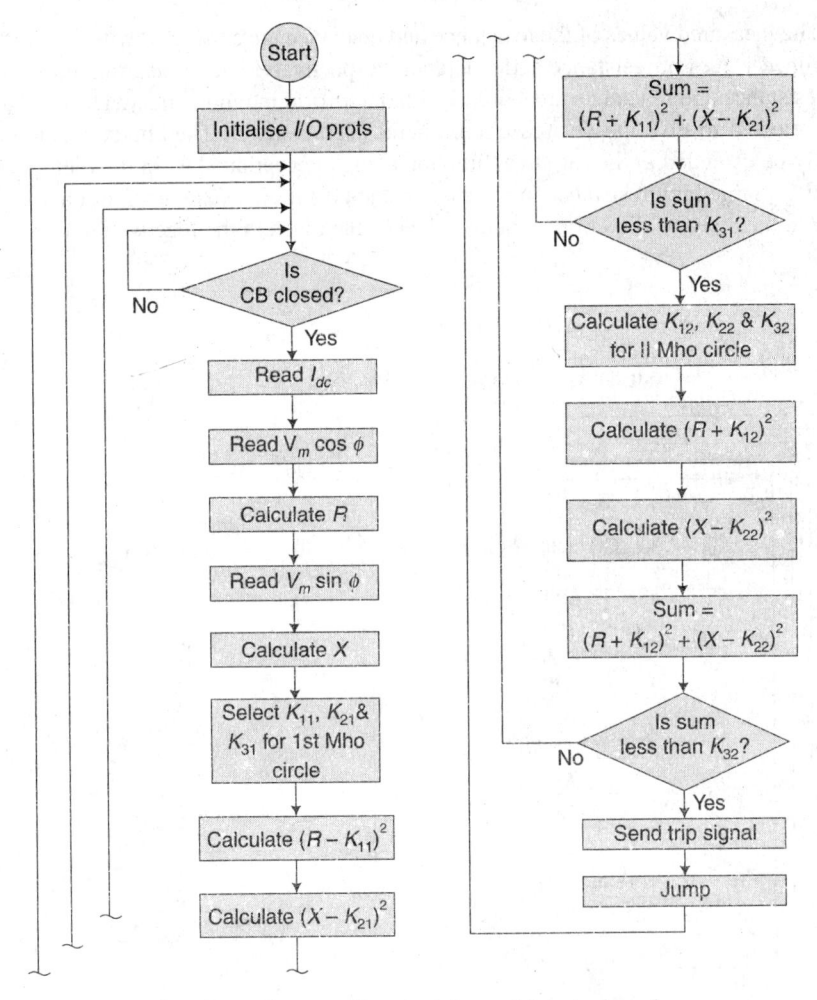

Fig. 12.43 Program flowchart for restricted mho relay

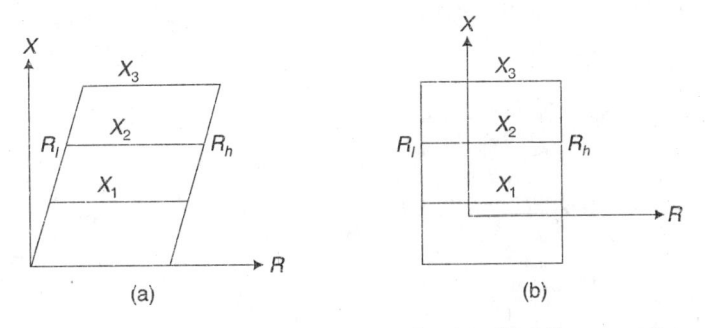

Fig. 12.44 (a) Quadrilateral relay passing through origin (b) Offset quadrilateral relay

The block schematic diagram of the interface for realisation of quadrilateral characteristic is shown in Fig. 12.41. The microcomputer measures the resistance and reactance at the relay location using the technique as described in Section 12.9.

The measured values of the resistance and reactance are compared with the predetermined values of resistance and reactance, respectively. The predetermined values of resistance and reactance are stored in a tabular form in the memory. These values are selected to give the desired quadrilateral characteristic. The characteristic can easily be extended in the resistive direction using appropriate data in the memory to cater for large fault resistance in the case of a short line. The extension of characteristic in the resistive direction is independent of the reach in the reactive direction.

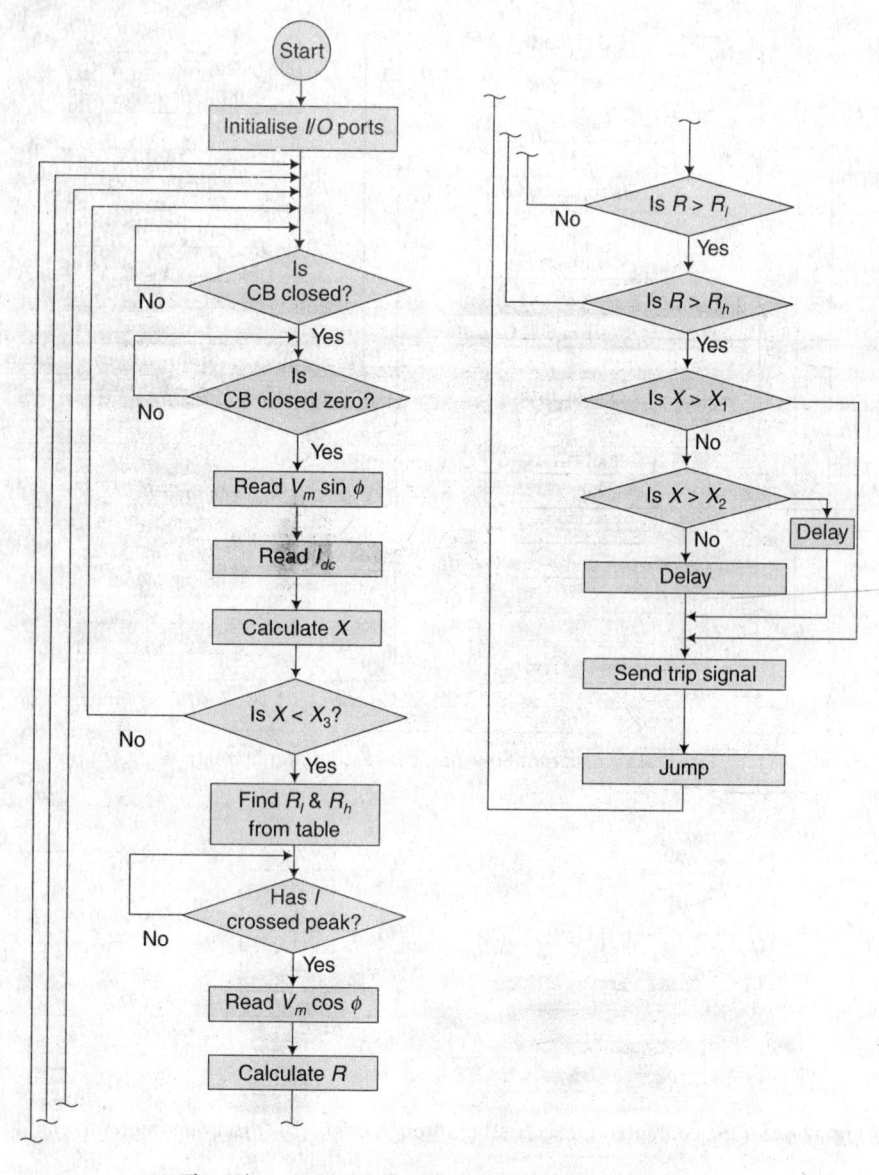

Fig. 12.45 *Program flowchart for quadrilateral relay*

Corresponding to a particular value of the line reactance, R_l and R_h are the lower and higher limits of resistances on the characteristic curve shown in Fig. 12.44(a). To obtain the desired quadrilateral characteristic, R_l and R_h for different values of line reactances are determined and stored in the memory. The microcomputer first measures the line reactance X and compares it with X_3 the predetermined value of the reactance for the third zone of protection. If X is greater than X_3, it goes back to the starting point and measures the same again.

If X is less than X_3, it proceeds further to measure R, the resistance seen by the relay. Corresponding to X, the measured values of the reactance, the microcomputer selects the values of R_l and R_h from the look-up table. For the operation of the relay, the condition to be satisfied is $R_l < R < R_h$. If X is less than X_1, a tripping signal is sent to the circuit breaker instantaneously. If X is greater than X_1 but less than X_2, the tripping signal is sent after a predetermined delay. If X is greater than X_2 but less than X_3, a greater delay is provided. The program flowchart is shown in Fig. 12.45.

12.12 GENERALISED INTERFACE FOR DISTANCE RELAYS

The schematic diagram of the generalised interface for realisation of different distance relay characteristics is shown in Fig. 12.46. Any desired characteristic can be realised using the same hardware and by changing the software only.

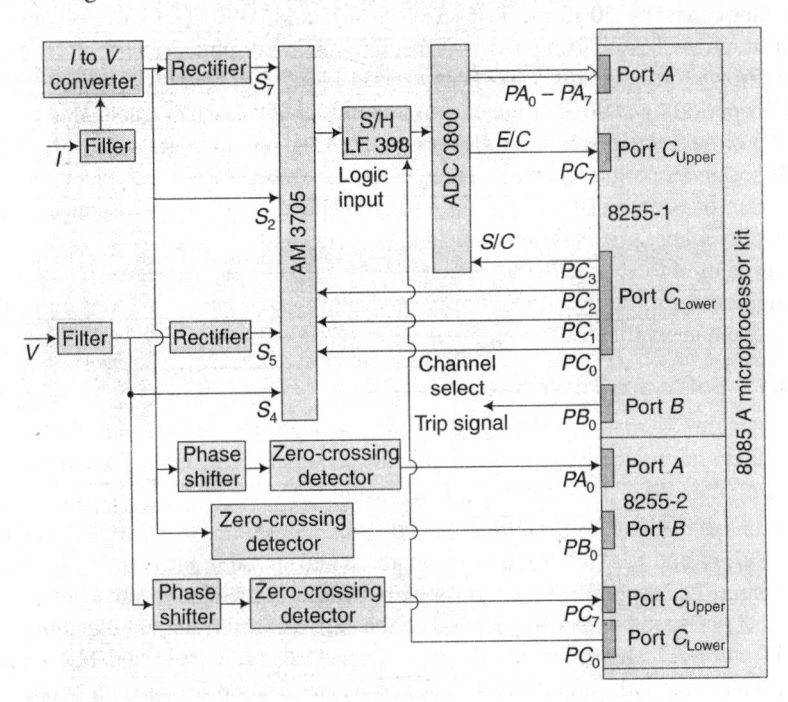

Fig. 12.46 Block schematic diagram of generalised interface for distance relays

12.13 MICROPROCESSOR IMPLEMENTATION OF DIGITAL DISTANCE RELAYING ALGORITHMS

Algorithms described in Sections 11.6, 11.7, 11.8, 11.9 and 11.10 can easily be implemented on the 8086 microprocessor. The hardware and software descriptions for the microprocessor implementation of digital distance relaying algorithms based on solution of the differential equation, DFT, FWHT, RHT and BPF are as follows.

12.13.1 Hardware Description

The block schematic diagram of the 8086 microprocessor based digital distance relays is shown in Fig. 12.47. The levels of voltage and current signals are stepped down to the electronic level by using auxiliary voltage and current transformers. The current signal derived from the current transformer is converted into proportional voltage signal. As the sampling frequency (f_s) used for data acquisition is 800 Hz, an active low-pass filter of cut-off frequency 400 Hz is used to avoid aliasing error. The data acquisition system (DAS) consists of signal conditioner, active low-pass filter, bipolar to unipolar converter, sample and hold circuits (LF 398), analog multiplexer (AM 3705) and 12-bit A/D converter (ADC 1211). ADC 1211 has been used in single supply (+5 V) operating configuration. A clock of 122.5 kHz for ADC 1211 is obtained by dividing 2.45 MHz peripheral clock (PCLK) from 8284 clock generator/driver by 20 using IC packages 7474 and 7490. The DAS is interfaced to the microprocessor using 8255 A programmable peripheral interface. The 8253 programmable timer/counter has been used in Mode 3 to generate a square wave of 1.25 ms period i.e., 800 Hz frequency for sampling of relaying signals. The controls for S/H circuit, the analog multiplexer and the ADC are all generated by the microprocessor under program control. Three LEDs of different colours are used for visual indication of occurrence of faults in three protective zones of the distance relay. A slave relay whose contacts are connected in series with the trip circuit of the circuit breaker is used to actuate the trip circuit after receiving the trip signal output by the microprocessor through an I/O port line on occurrence of fault in any one of the three protective zones of the relay.

12.13.2 Software Description

The software for implementing the algorithms is developed in the assembly language of the 8086 microprocessor. Voltage and current signals are sampled simultaneously at a sampling rate of 16 samples per cycle using DAS.

The 3 sampled values for differential equation algorithm and 9 sampled values for DFT, FWHT and RHT algorithms are acquired and stored in the memory for further processing. The program flowchart for computation of real and reactive components (R and X) of the apparent impedance of the line using differential equation, DFT, FWHT and RHT algorithms are shown in Figs. 11.6, 11.7, 11.13 and 11.16 respectively. Using the calculated values of R and X, any desired distance relay operating characteristic can be realised. A generalised program flowchart for realisation of any distance relay characteristic is shown in Fig. 12.48.

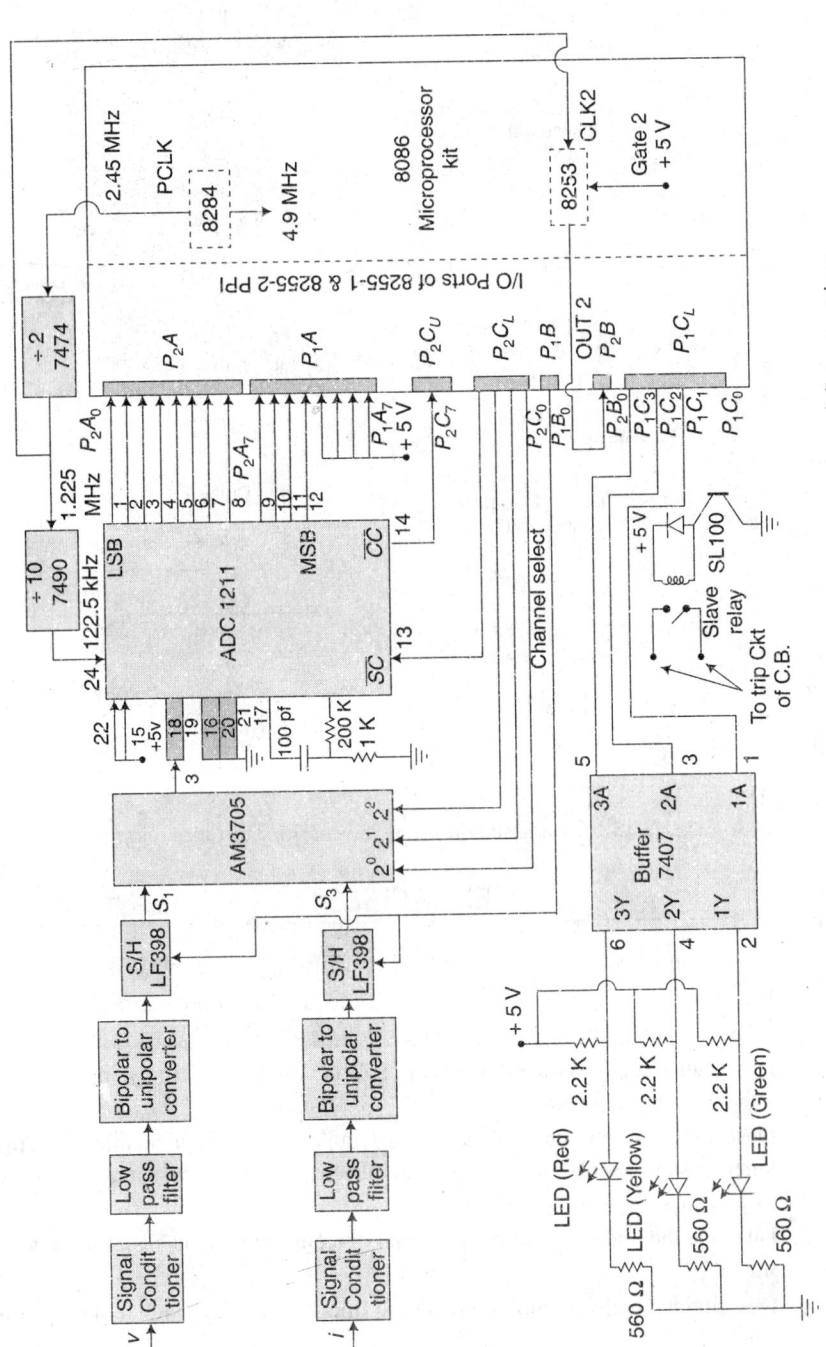

Fig. 12.47 Block schematic diagram of the 8086 microprocessor based distance relays

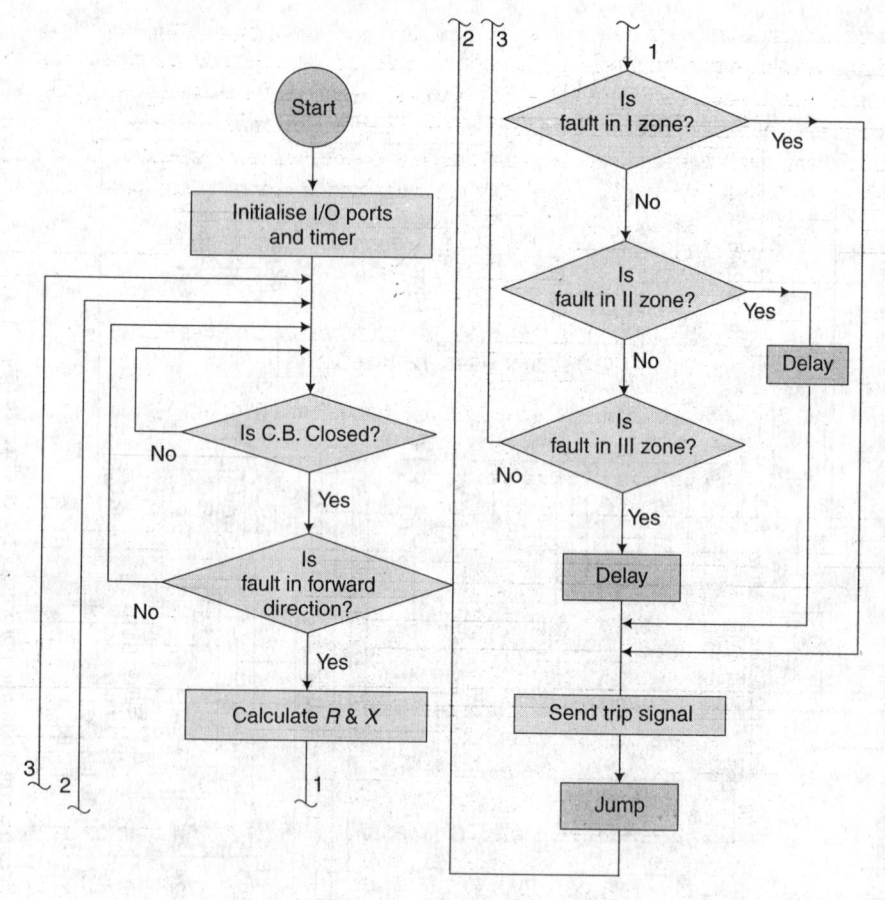

Fig. 12.48 *Generalised program flowchart for distance relays*

EXERCISES

1. Design an active low-pass filter for cut-off frequency of 300 Hz.
2. Describe the functions of analog multiplexer and sample and Hold circuit. How can any channel of the analog multiplexer AM 3705 be selected?
3. How can the A/D converter ADC 0809 be used to read any instantaneous value of the a.c. voltage?
4. Describe a microprocessor-based data acquisition system to acquire the simultaneous samples of both voltage and current signals at the sampling interval of 1.25 ms.
5. Describe the realisation of a directional overcurrent relay using a microprocessor.
6. Describe the realisation of a directional impedance relay using a microprocessor.

7. Derive a generalised mathematical model of distance relays for numerical protection.
8. Describe the realisation of Mho, offset Mho, and restricted Mho relays by using a generalized mathematical model and a microprocessor.
9. How can a quadrilateral distance relay be realised using a microprocessor?
10. Derive the mathematical model of an elliptical distance relay and describe its realisation using a microprocessor.
11. How can numerical distance relaying algorithms be implemented on the 8086 microprocessor? Is it possible to implement these algorithms on the 8085?

13 Artificial Intelligence Based Numerical Protection

13.1 INTRODUCTION

Artificial Intelligence (AI) is a branch of computer science that attempts to approximate the results of human reasoning by organizing and manipulating factual and heuristic knowledge. This generally involves borrowing characteristics from human intelligence, and applying them as algorithms in a computer friendly way. A more or less flexible or efficient approach can be taken depending on the requirements established, which influences how artificial the intelligent behaviour appears. Artificial Intelligence is a multidisciplinary research area. Computer scientists, mathematicians, electrical and electronic engineers are involved in it because they want to build more powerful and intelligent systems. Psychologists, neurophysiologists, linguists and philosophers are involved in AI because they wish to understand (using computer models) the principles which make intelligence possible. The applications of AI is in the field of knowledge representation, reasoning and control, learning (knowledge acquisition), handling uncertainty, plan formation, vision (object recognition, motion and image sequence analysis, scene understanding), and speech (speech recognition and synthesis, story comprehension), etc. Area of artificial intelligence related to power system protection includes the following:

Artificial Neural Networks (ANNs) These are the networks that simulate intelligence by attempting to reproduce the types of physical connections that occur in human brains. They are capable of representing highly nonlinear functions and performing multi-input, multi-output mapping. ANNs have been applied in fault diagnosis, performing one or more of the following tasks:

(i) Pattern recognition, parameter estimation, and nonlinear mapping applied to condition monitoring
(ii) Training based on both time and frequency domain signals obtained via simulation and/or experimental results
(iii) Real time, online supervised diagnosis
(iv) Dynamic updating of the structure with no need to retrain the whole network
(v) Filtering out transients, disturbance and noise
(vi) Fault protection in incipient stages due to operation anomalies
(vii) Operating conditions clustering based on fault types

Fuzzy Systems These are the systems that deal with imprecision, ambiguity, and vagueness in information. Adaptive fuzzy systems utilize the learning capabilities of ANNs or the optimization strength of genetic algorithms to adjust the system parameter set in order to enhance the intelligent system performance based on a prior knowledge. Its application to fault diagnosis include

   (i) Evaluating performance indices using linguistic variables
  (ii) Predicting abnormal operation and locating faulty element
 (iii) Utilizing human expertise reflected to fuzzy if-then rules
  (iv) System modeling, nonlinear mapping, and optimizing diagnostic system parameters through adaptive fuzzy systems
   (v) Fault classification and prognosis

Expert Systems These are the systems which emulate the human thought through knowledge representation and inference mechanisms. The application includes

   (i) Emulating and implementing human expertise
  (ii) Building and online updating system knowledge bases
 (iii) Signal filtering, information search, and feature extraction
  (iv) Data managing and information coding in knowledge bases
   (v) Employment of user interactive sessions
  (vi) Fault classification, diagnosis, and location
 (vii) Knowledge base building through simulation and/or experimentation

Genetic Algorithm It is based on the theory of natural selection and evolution. This search algorithm balances the need for exploitation and exploration. When selection and cross over tend to converge on a good but sub-optimal solution it is known as exploitation. When selection and mutation create a parallel, noise tolerant, hill climbing algorithm, preventing a premature convergence it is known as exploration.

The application includes

   (i) Economics dispatch
  (ii) Power system planning
 (iii) Reliability planning
  (iv) Reactive power allocation
   (v) Load flow problem

13.2 ARTIFICIAL NEURAL NETWORK (ANN)

A neural network is a massively parallel distributed processor that has a natural propensity for storing experiential knowledge and making it available for use. It resembles the brain in two respects:

1. Knowledge is acquired by the network through a learning process.
2. Inter-neuron connection strengths known as synaptic weights are used to store the knowledge.

A neural network derives its computing power through, first, its massively parallel distributed structure and, second, its ability to learn and therefore generalize. Generalization refers to the neural network producing reasonable outputs for inputs not encountered during training (learning). These two information-processing capabilities make it possible for neural networks to solve complex (large-scale) problems that are currently intractable.

13.2.1 Biological Neural Networks

Artificial neural networks are biologically inspired mathematical models (simple to complex) to solve the same kind of difficult, complex problems in a similar manner as we think the human brain might have done. The human brain consists of a large number of neural cells that process information. Each cell works as simple processor and only the massive interaction between all cells and their parallel processing makes the brain abilities possible. These basic cells are called *neurons*.

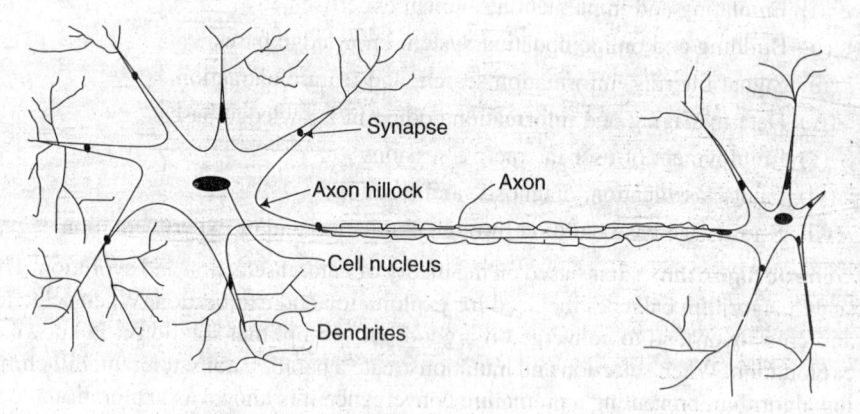

Fig. 13.1 Structure of a biological neuron

Figure 13.1 shows the schematic diagram of the structure of a biological neuron. It consists of a cell body called soma where the cell nucleus is located. Several spine-like nerve fibres are connected to the cell body. They are called *dendrites*, which receive signals from the other neurons. The single long fibre extending from the cells body, called the *axon*, branches into strands and substrands connecting many other neurons at the synaptic junctions or *synapse*. A synapse is an elementary and functional unit between two neurons. A biological neural network is superior to its artificial counterpart for pattern recognition due to its robustness and fault tolerance, flexibility, collective computation and its ability to deal with a variety of data situations.

13.2.2 Neuron Modeling

The artificial counterpart of the biological neural network consists of inputs as dendrites, synaptic weights as synapse, summation and activation function as internal biological activation and output as axon. Artificial neural networks transform a set of inputs into a set of outputs through a network of neurons, each of which generates one output as a function of its inputs. The inputs and outputs are usually normalized, and the output is a nonlinear function of the inputs that is controlled by weights on the inputs. A network learns these weights during training, which can be supervised or unsupervised. A wide variety of network connections and training techniques exist. Neural networks can recognize, classify, convert, and learn patterns. A pattern is a qualitative or quantitative description of an object or concept or event. A pattern class is a set of patterns sharing some common properties. Pattern recognition refers to the categorization of input data into identifiable classes by recognizing significant features or attributes of the data. In the neural network approach, a pattern is represented by a set of nodes along with their activation levels.

A neuron is an information processing unit that is fundamental to the operation of a neural network. A simple model of artificial neural network is given in Figure 13.2.

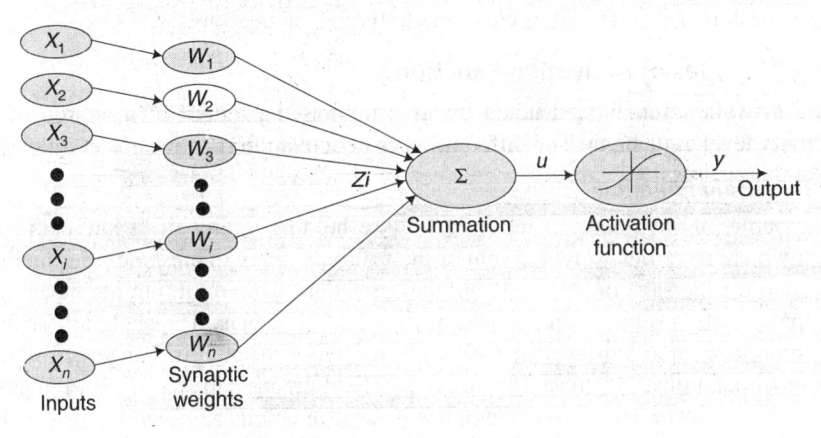

Fig. 13.2 *Model of artificial neural network*

Artificial neural networks involve three important processes:

(i) A set of synapses or connecting links, each of which is characterized by a weight or strength of its own. A dot product provides a confluence between a n-dimensional input vector, X, and n-dimensional synaptic vector, W. Thus, the resulting vector, Z, can be expressed as,

$$Z_i = W_i X_i, \quad \text{for } i = 1, 2, \ldots n \tag{13.1}$$

where

$W = [W_1, W_2, W_3, \ldots W_i, \ldots W_n]^T$ is the vector of synaptic weights,

$X = [X_1, X_2, X_3, \ldots X_i, \ldots X_n]^T$ is the vector of neural inputs and

$Z = [Z_1, Z_2, Z_3, \ldots Z_i, \ldots Z_n]^T$ is the vector of weighted neural inputs.

Thus, the synaptic operation assigns a relative significance to each incoming input signal, X_i, according to the past experience stored in W_i.

(ii) Forming a unit net activation by combining the inputs in different ways e.g. additive, weighted additive, multiplication, etc. This provides a linear mapping from n-dimensional neural input space, X, to the one-dimensional space, U.

$$U = \sum_{i=1}^{n} Z_i = \sum_{i=1}^{n} W_i X_i, \qquad (13.2)$$

where U is the intermediate accumulated sum.

(iii) Mapping this activation value into the artificial unit output using appropriate activation function. A nonlinear mapping operation on U yields a neural output, Y, given by,

$$Y = \varphi[U - W_0], \qquad (13.3)$$

where $\varphi(.)$ is a non-linear function and
W_0 is the threshold.

The commonly used activation functions are linear, linear threshold, threshold, step, ramp, sigmoid and Gaussian. The sigmoid or logistic activation function is most popular binary signal function due to its computational simplicity, semilinearity, squashing property and its closeness to biological neurons.

13.2.3 Types of Activation Function

The activation function, denoted by $\varphi(.)$, defines the output of a neuron of the activity level at its input. The different types of activation function are as follows:

1. Threshold Function

The output of any neuron is the result of thresholding, if any, of its internal activation which, in turn is weighted sum of the neuron's inputs. Thresholding sometimes is done for the sake of scaling down the activation mapping it into the meaningful output for the problem, and sometimes for adding a bias. Thresholding (scaling) is important for multi-layer networks to preserve a meaningful range across each layer's operations. The most often used threshold function ids the sigmoid function. A step function or a ramp function or just a linear function can be used, when the bias is simply added to the activation. The sigmoid function accomplishes mapping the activation into the interval [0, 1]. This function can be expressed mathematically as:

$$\varphi(v) = \frac{1}{(1 + e(-av))} \qquad (13.4)$$

correspondingly the output of the neuron k employing such a threshold function is expressed as

$$y_k = \begin{cases} 1 & \text{if } v_k \geq 0 \\ 0 & \text{if } v_k < 0 \end{cases} \qquad (13.5)$$

where v_k is the internal activity level of the neuron.

2. Piecewise Linear Function

The piecewise linear function is denoted by

$$\varphi(v) = \begin{cases} 1 & v \geq 1/2 \\ v & -1/2 < v < 1/2 \\ 0 & v \leq -1/2 \end{cases} \qquad (13.6)$$

3. Sigmoid Function

More than one function goes by the same sigmoid function. They differ in their formulas and their ranges. They all have a graph similar to a stretched letter s.

This function is by far the most common form of the activation function based on the construction of artificial neural network. An example of sigmoid function is logistic function, defined by

$$\varphi(v) = \frac{1}{(1 + e(-av))} \tag{13.7}$$

where a is the slope of the parameter of the sigmoid function. By varying the parameter a, the sigmoid function of different slope can be obtained. A sigmoid function assumes a continuous range of value from 0 to 1.

It is to be noted that the sigmoid function is differential, whereas the threshold function is not.

The sigmoid is so important and popular because of the following reasons:
(i) It squashes.
(ii) It is semilinear (nondecreasing and differentiable every where). This influences its use with certain training approaches.
(iii) It is expressible in close form.
(iv) The derivative of sigmoid with respect to av is very easy to form.

The threshold function is commonly referred to as the signum function.

$$\varphi(v) = \frac{(1 - e(-v))}{(1 + e(-v))} \tag{13.8}$$

4. Step Function
The step function is also frequently used as a threshold function. The function is 0 to start with and remains so to the left of some threshold value B. A jump to 1 occurs for the value of the function to the right B and the function then remains at the level 1. In general a step function can have a finite number of points at which jumps of equal or unequal size occur. When the jumps are equal and at many points, the graph will resemble a stair case.

5. Ramp Function
To describe the ramp function simply, a step function that makes a jump from 0 to 1 at some point is considered. Instead of letting it take a sudden jump like that at one point, it is let to gradually gain in value, along a straight line (looks like a ramp), over a finite interval reaching from an initial 0 to a final 1. Thus a ramp function is obtained. A ramp function can be thought as a piecewise linear approximation of a sigmoid.

6. Linear Function
A linear function is a sample one given by an equation of the form

$$\varphi(v) = av + B \tag{13.9}$$

when $a = 1$, the application of the threshold function amounts to simply adding a bias equal to B to the sum of the inputs.

13.2.4 Types of Neural-Network Models
Although a single neuron-processing unit can handle simple pattern-classification problems, the strength of neural computation comes from the neurons connected in a network. An artificial neural network consists of several processing units interconnected in a predetermined manner to accomplish a desired pattern recognition

task. The arrangement of the processing units, connections, and pattern input/output is known as the topology the network. The networks are organized into layers of processing units having the same activation dynamics and output function. Connections can be made from the units of one layer to the units of another layerar among the units of the same layer or both. The connections across the layers and among the units within a layer can be either in a feed forward manner or in a feedback manner.

The rules governing the behaviour of the network are specified in the activation and the synaptic dynamics. The adjustment of the synaptic weights is represented by a set of learning equations which helps to learn the pattern information in the input samples. The learning law should lead to convergence of weights and the training time should be a small as possible. The learning process is nothing but the adjustments of the weights to produce the desired output.

If the weight adjustment is determined based on the deviation of the desired output from the actual output then it is called supervised learning. If a given set of pattern updates the weights based on local information and no desired output is specified it is called unsupervised learning.

We can classify the neural-network models depending on whether they learn with or without supervision and whether they contain open or closed synaptic loops. Supervised feedforward models provide the most tractable and most applied neural models. Unsupervised feedback models are biologically most plausible, but mathematically most complicated. Unsupervised feedforward neural models tend to converge to locally sampled pattern class centroids.

The method and capacity to retrieve information can also classify the network as being autoassociative or heteroassociative. *Autoassociation* is the phenomenon of associating an input vector with itself as the output, whereas heteroassociation is that of recalling a related vector given an input vector.

However artificial neural networks can be broadly classified as
- Feedback neural networks
- Feedforward neural networks
- Laterally connected neural networks
- Recurrent networks
- Hybrid networks

13.2.5 Neural Network Learning

The neural network learning can be classified as
1. Supervised or error influenced learning based network
2. Unsupervised or response influenced learning based network

The multilayer perceptron is an example of supervised and kohonen is unsupervised but radial basis function network has both supervised and unsupervised learning.

Supervised Learning

In supervised learning, the network can learn when training is used. The learning assumes the availability of the teacher or supervisor who classifies the training examples into classes. This form of learning involves cross-correlation between error signal and the input.

Supervised learning algorithms utilize the information on the class membership of each training instance. This information allows supervised learning algorithms to detect pattern misclassification as a feedback to themselves. Error information contributes to the learning process by rewarding accurate classifications and/or punishing misclassification—a process known as credit and blame assignment. It also helps eliminate implausible hypothesis.

Supervised learning usually refers to the search in the space of all possible weight combination along error gradients. The supervisor uses class membership information to define a numerical error signal or vector, which guides the gradient search. The performance of the neural network will deteriorate if over trained. Figure 13.3 shows the block diagram of supervised learning.

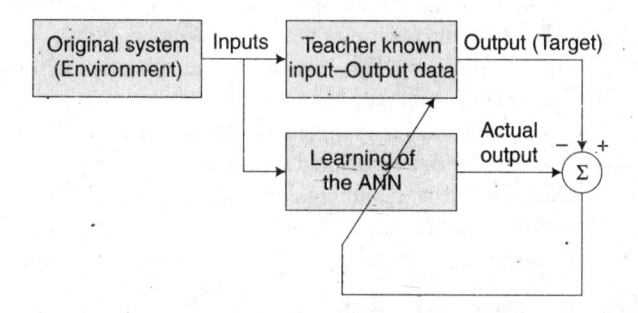

Fig. 13.3 *Supervised adaptive learning block diagram*

Supervised learning can be performed in an offline or online manner. In the offline case, a separate computational facility is used to design the supervised learning system. Once the desired performance is accomplished, the design is 'frozen', which means that the neural network operates in a static manner.

Supervised learning can be performed in an offline or online manner. In the offline case, a separate computational facility is used to design the supervised learning system. Once the desired performance is accomplished, the design is 'frozen', which means that the neural network operates in a static manner. On the other hand, in online learning, the learning procedure is implemented solely within the system itself, not requiring a separate computational facility. In other words, learning is accomplished in real time, with the result that the neural network is dynamic. Therefore, the requirement of online training has a more severe requirement on a supervised learning procedure than the offline learning.

A disadvantage of supervised learning, regardless of whether it is performed offline or on-line, is the fact that without a supervisor, a neural network cannot learn new strategies for particular situations that are not covered by the set of examples used to train the network.

Unsupervised Learning

In unsupervised or self-organized learning there is no external supervisor or critic to oversee the learning process and it involves cross-correlation between input and output. In other words, there are no specific examples of the function to be learned by the network. A provision is made for a task independent measure of quality of representation that the network is required to learn, and the free parameters of the network are optimized with respect to that measure. Once the network has become tuned to the statistical regularities of the input data, it develops the ability to form internal representations for encoding features of the input and thereby create new classes automatically. Figure 13.4 shows the block diagram of unsupervised learning.

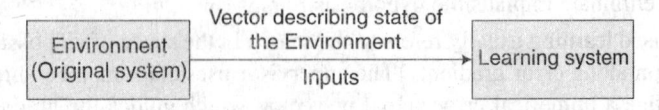

```
                       Vector describing state of
  ┌──────────────┐         the Environment      ┌────────────────┐
  │ Environment  │ ──────────────────────────→  │ Learning system│
  │(Original system)│            Inputs         └────────────────┘
  └──────────────┘
```

Fig. 13.4 *Block diagram of unsupervised learning*

Unsupervised learning refers to how neural networks modify their parameters in biological plausible ways. In this learning mode, the neural network does not use the class membership of training instances. Instead, it uses information associated with a group of neurons to modify local parameters. Unsupervised learning systems adaptively cluster instances into cluster or decision classes by competitively selecting 'winning' neurons and modifying associated weights. The fan-in vectors tend to estimate pattern class statistics. It is the amalgamation of input-output pairs, and hence the disappearance of the external supervisor providing target outputs, that gives its name. This kind of learning is sometimes referred to as self-organization.

Supervised Versus Unsupervised Learning

Most of the supervised and unsupervised algorithms contemplate step learning, in which weights are updated, after each and every training pattern is presented to the network. In batch learning, the error is summed overall the training patterns in training set to produce the difference weight vector for updating the weights.

Unsupervised learning algorithms use unlabeled instances and blindly follow and process them. They often have less computational complexity and less accuracy than supervised learning algorithms. Unsupervised learning can be designed to learn rapidly. This makes unsupervised learning more practical in high-speed, real-time environments, where there is not enough time and information to apply supervised techniques. Unsupervised learning has also been used for scientific discovery.

Reinforcement learning is somewhere between supervised learning and unsupervised learning. It receives a feedback that tells the system whether its output response is right or wrong, but no information on what the right output should be is provided then the system employ some random search strategy so that the space of plausible and rational choices is searched until a correct answer is found.

Back-Propagation Algorithm

Among the algorithm used to perform supervised learning, the back-propagation algorithm has emerged as the most widely used and successful algorithm for the design of multilayer feedforward networks. There are two distinct phases to the operation of back-propagation learning is the forward phase and the backward phase. In the forward phase, the input signals propagate through the network layer, eventually producing some response at the output of the network. The actual response so produced is compared with a desired (target) response, generating error signals that are then propagated in a backward direction through the network. In the backward phase of operation, the free parameters of the network are adjusted so as to minimize the sum of squared errors. Back-propagation learning has been applied successfully to solve some difficult problems such as speech recognition from text, handwritten digit recognition and adaptive control. Unfortunately, back-propagation and other supervised learning algorithms may be limited by their poor scaling behaviour. The hybrid use of unsupervised and supervised learning procedures may provide a more acceptable solution than supervised learning alone, particularly if the size of the problem is large.

13.2.6 Multilayer Feedforward Neural Networks

In multilayer feedforward neural networks, neurons are connected in a layered structure and neurons in a given layer receive inputs from neurons in the layer immediately below it and send their outputs to neurons in the layer immediately above it. Their outputs are a function of only the current inputs and do not depend upon inputs and/or outputs. The name *feedforward* in MFNN implies that all the information in the network flows in the forward direction and during normal processing; there is no feedback from the outputs to the inputs. Multilayer feedforward neural networks have proved successful as pattern classifiers and function approximations. A multilayer feedforward network with one hidden layer is shown in Fig. 13.5. Each node represents a single neural processing unit. Neurons in a given layer receive inputs from neurons in the layer immediately below it and send their output to neurons in the layer immediately above it. Nodes in the input layer function like hyperplanes that effectively partition n-dimensional space into various regions. Each node in the hidden layer represents a cluster of points that belong to the same class. Each node in this layer can combine hyperplanes to create convex decision regions. The nodes in the output layer represent the number of classes. Each node in the output layer can combine convex region to form concave regions. Thus, it is possible to form any arbitrary regions with sufficient layers and sufficient hidden units. It can ideal with non-linear classification problems because it can form more complex decision regions.

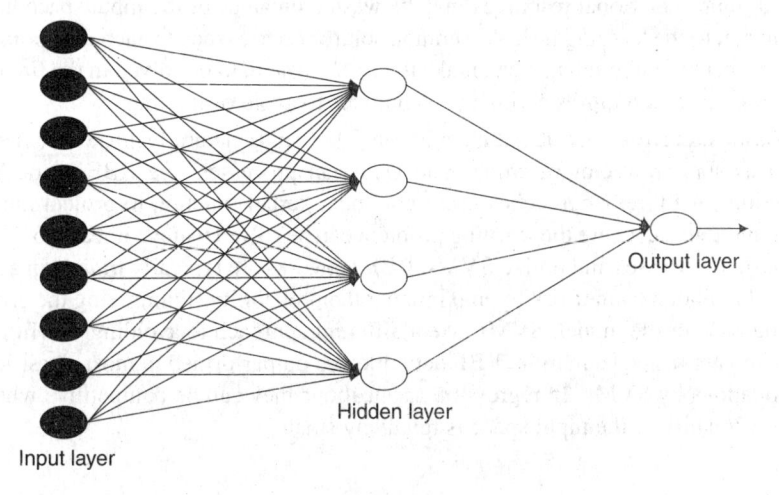

Fig. 13.5 *Multilayer feedforward neural network*

The learning of the multilayer feedforward neural networks can be carried out using the back-propagation rule. It carries out a minimization of the mean square error between the obtained outputs corresponding to the input state and the desired target state.

13.2.7 Radial Basis Function Neural Network

Radial basis functions are powerful techniques for interpolation in multidimensional space. A RBF is a function which has built into a distance criterion with respect to a centre. Radial basis functions have been applied in the area of neural networks where they may be used as a replacement for the sigmoidal hidden layer transfer characteristic in multi-layer perceptrons. RBF network have two layers of processing: In the first, input is mapped onto each RBF in the 'hidden' layer. The RBF chosen is usually a Gaussian. In regression problems the output layer is then a linear combination of hidden layer values representing mean predicted output. The interpretation of this output layer value is the same as a regression model in statistics. In classification problems the output layer is typically a sigmoid function of a linear combination of hidden layer values, representing a posterior probability. Performance in both cases is often improved by shrinkage techniques, known as ridge regression in classical statistics and known to correspond to a prior belief in small parameter values (and therefore smooth output functions) in a Bayesian framework.

RBF networks have the advantage of not suffering from local minima in the same way as multi-layer perceptrons. This is because the only parameters that are adjusted in the learning process are the linear mapping from hidden layer to output layer. Linearity ensures that the error surface is quadratic and therefore has a single easily found minimum. In regression problems this can be found in one matrix operation. In classification problems the fixed nonlinearity introduced by the sigmoid output function is most efficiently dealt with using least squares.

RBF networks have the disadvantages of requiring good coverage of the input space by radial basis functions. RBF centres are determined with reference to the distribution of the input data, but without reference to the prediction task. As a result, representational resources may be wasted on areas of the input space that are irrelevant to the learning task. A common solution is to associate each data point with its own centre, although this can make the linear system to be solved in the final layer rather large, and requires shrinkage techniques to avoid overfitting.

Associating each input datum with an RBF leads naturally to kernel methods such as support vector machines and Gaussian processes (the RBF is the kernel function). All three approaches use a nonlinear kernel function to project the input data into a space where the learning problem can be solved using a linear model. Like Gaussian processes and unlike SVMs, RBF networks are typically trained in a maximum likelihood framework by maximizing the probability (minimizing the error) of the data under the model. SVMs take a different approach to avoiding overfitting by maximizing instead a margin. RBF networks are outperformed in most classification applications by SVMs. In regression applications they can be competitive when the dimensionality of the input space is relatively small.

13.2.8 Artificial Neural Network Design Procedure and Consideration

The design process for artificial neural network consists of the following steps.
  (i) Preparation of suitable training data
  (ii) Selection of a suitable ANN architecture
  (iii) Training of the ANN
  (iv) Testing of ANN

The flowchart in Fig. 13.6 explains the working of finding a solution to the given problem with the help of artificial neural network. The design process is iterative. The network architecture and the training algorithm are chosen according to the requirement of the problem with the help of the Table 13.1. When we train the network with the help of examples along with input and corresponding output and the network is supposed to organize itself according to input information. This is called supervised learning. And when we train the network with the help of examples of only inputs and the network is supposed to organize itself according to input information, it is known as unsupervised learning. The neural networks synaptic weights are adjusted to produce the desired output in the learning process. Thus the learning is nothing but updating the weights in subsequent presentations. After training is complete, we test

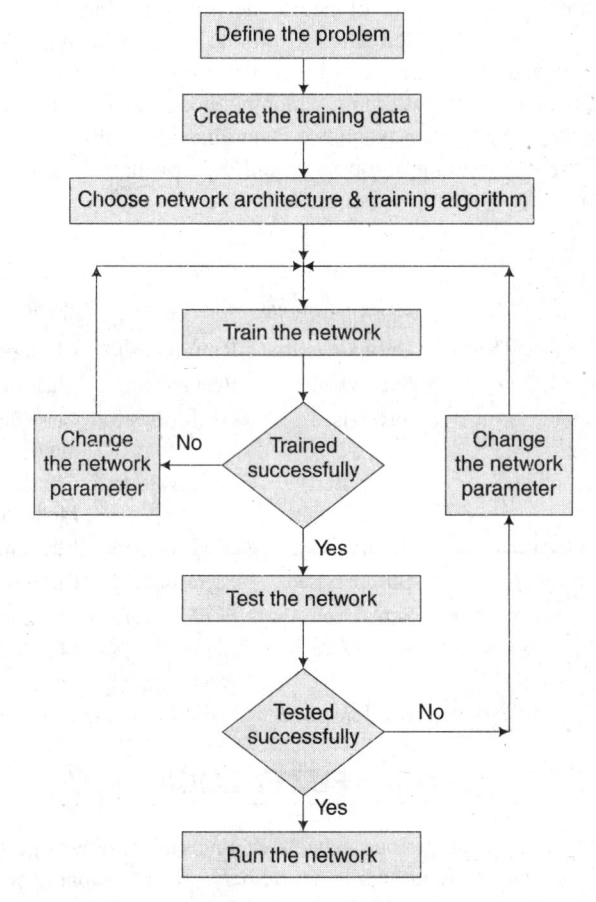

Fig. 13.6 *Flowchart of ANN application design*

the network and then finally run the network. If the network does not perform satisfactorily then the network structure and parameters must be changed, the network retrained and then tested.

Selection and manipulation of the suitable training patterns is of prime importance. Training patterns should contain all the necessary information to generalize the problem. Preprocessing of the data used for training makes it easier for the network to learn.

Architecture of a neural network is application specific. Architectural differences might come on account of the number and types of inputs, number and types of outputs and complexity of the application which would then govern the number of layers and the number of neurons in different layers. These parameters of the network are decided by the experimentation of training and testing a number of network configurations.

The nonlinear mapping function or transfer function introduces non-linearity that enhances the networks ability to model complex functions. It also influences the shape of the decision boundary. A sigmoidal function can produce an arbitrary decision boundary with smooth curves and edges. Choosing the number of nodes involve the training and testing of neural networks having different number of nodes. The process is terminated when a suitable network with satisfactory performance is established. For learning to be successful, it is essential for the network to perform correctly on the pattern on which network was not trained. Training and testing is generally interleaved to achieve generalization. Table 13.1 shows different models, their training paradigm, topology and their primary functions according to which they are being currently being used.

Table 13.1 Neural network models and their functions

Model	Training paradigm	Topology	Primary functions
Adaptive Resonance Theory	Unsupervised	Recurrent	Clustering
ARTMAP	Supervised	Recurrent	Classification
Back propagation	Supervised	Feed-forward	Classification, modeling, time-series
Radial basis function networks	Supervised	Feed-forward	Classification, Modeling, time-series
Probabilistic neural networks	Supervised	Feed-forward	Classification
Kohonen feature map	Unsupervised	Feed-forward	Clustering
Learning vector quantization	Supervised	Feed-forward	Classification
Recurrent back propagation	Supervised	Limited recurrent	Modeling, time-series
Temporal difference learning	Reinforcement	Feed-forward	Time-series

13.3 FUZZY LOGIC

Fuzzy sets are efficient at various aspects of uncertain knowledge representation and are subjective and heuristic, while neural networks are capable of learning from examples, but have the shortcoming of implicit knowledge representation.

A fuzzy-logic system is inflected in three basic elements: fuzzification, fuzzy inference, and defuzzification. Degrees of membership in the fuzzifier layer are calculated according to IF-THEN rules. They base their decisions on inputs in the form of linguistic variable derived from membership functions which are formulas used to determine the fuzzy set to which a value belongs and the degree of membership in that set. The variables are then matched with the specific linguistic IF -THEN rules and the response of each rule is obtained through fuzzy implication. To perform compositional rule of inference, the response of each rule is weighted according to the impedance or degree of membership of its inputs and the centroid of the response is calculated to generate the appropriate output. Table 13.2 shows an example of fuzzy variables to represent the different fault types along with their equivalent fuzzy fault code.

Table 13.2 Fuzzy variables to represent fault type with equivalent fault code

Fault Type	b_3	b_2	b_1	b_0	Equivalent decimal number	Triplets A	B	C
a-g	1	0	0	1	9	8.5	9.0	9.5
b-g	0	1	0	1	5	4.5	5.0	5.5
c-g	0	0	1	1	3	2.5	3.0	3.5
a-b	1	1	0	0	12	11.5	12.0	12.5
b-c	0	1	1	0	6	5.5	6.0	6.5
c-a	1	0	1	0	10	9.5	10.0	10.5
a-b-g	1	1	0	1	13	12.5	13.0	13.5
b-c-g	0	1	1	1	7	6.5	7.0	7.5
c-a-g	1	0	1	1	11	10.5	11	11.5
a-b-c	1	1	1	0	14	13.5	14	14.5
a-b-c-g	1	1	1	1	15	14.5	15.0	15.5

This approach needs sequence components to solve the classification problem and the training is carried out using the information from both designer's experiences and sample data sets. The other drawback of this approach is that the number of fuzzy rules increases exponentially with respect to inputs.

The process performed in a fuzzy-set approach is described by the block diagram in Fig. 13.7.

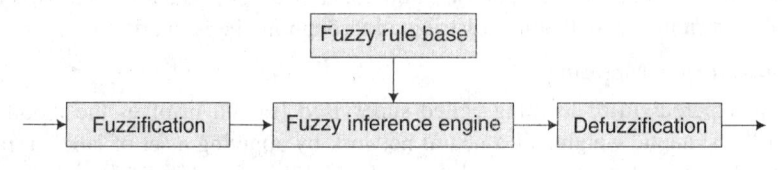

Fig. 13.7 *Fuzzy logic scheme for fault classification*

The process uses a collection of fuzzy membership functions and rules, instead of Boolean logic, to reason about data. Crisp values are first transformed into fuzzy values to be able to use them to apply rules formulated by linguistic expressions. Then the fuzzy system transforms the linguistic conclusion back to a crisp value. These steps are described as follows:

Fuzzification Crisp input values, and are transformed into fuzzy sets to be able to use them for computing the truth values of the premise of each rule in the rule base.

Inference The truth value for the premise of each logic rule is computed and applied to the conclusion part of each rule. This results in one fuzzy set to be assigned to each output variable for each rule. All of the fuzzy sets assigned to each output variable are combined together to form a single fuzzy set for each output variable.

Defuzzification The single fuzzy sets are converted back to crisp values.

13.4 APPLICATION OF ARTIFICIAL INTELLIGENCE TO POWER SYSTEM PROTECTION

Power system problems regarding classification or the encoding of an unspecified non-linear function are well-suited for artificial neural networks. ANNs can be especially useful for problems that need quick results, such as those in real time operation, because of the ability of quickly generating results after receiving a set of inputs.

The ability to scale ANN applications to realistic dimensions for power system problems is a major issue. Component-related applications generally have a limited number of inputs, but realistic power systems have potentially tens of thousands of inputs at the system level. Training times are usually non linear with the problem size. A related concern is the size of the training set. Because of the complex, non-linear behaviour of power systems, ANNs can require large training sets to obtain sufficient accuracy. The trade off between training set size and solution accuracy is difficult to control analytically. It is quite possible to add cases to the training set that do not appreciably improve the accuracy of the ANN solutions.

ANNs are a clear first choice for most classification problems, and for fast approximations to the results of complex numerical analyses. So, they are suitable for many applications in real-time control, operations, and operations planning.

The use of neural network offers the following useful properties and capabilities:

1. Nonlinearity

A neuron is basically a nonlinear device. Consequently, a neural network, made up of an interconnections of neurons, is itself non linear. Moreover, the nonlinearity is of a special kind in the sense that it is distributed throughout the network.

2. Input-Output Mapping

A popular paradigm of learning called supervised learned involves the modification of the synaptic weights of a neural network by applying a set of labeled training samples, or task examples. Each example consists of a unique input signal and the corresponding desired response. The network is presented an example picked at random from the set, and the synaptic weights (free parameters) of the network are modified so as to minimize the difference between the desired response and the actual response of the network produced by the input signal in accordance with an appropriate statistical criterion. The training of the network is repeated for many examples in the set until the network reaches a steady state. Where there are no further significant changes in the synaptic weights; the previously applied training examples may be reapplied during the training session but in a different order. Thus the network learns from the examples by constructing an input–output mapping for the problem at hand.

3. Adaptivity

Neural networks have a built-in capability to adapt their synaptic weights to changes in the surrounding environment. In particular, a neural network trained to operate in a specific environment can be easily retrained to deal with minor changes in the operating environmental conditions. Moreover, when it is operating in a non stationary environment (i.e., one whose statistics change with time), a neural network can be designed to change its synaptic weights in real time. The natural architecture of a neural network for pattern classification, signal processing, and control applications, coupled with the adaptive capability of the network, make it an ideal tool for use in adaptive pattern classification, adaptive signal processing, and adaptive control.

4. Evidential Response

In the context of pattern classification a neural network can be designed to provide information not only about which particular pattern to select, but also about the confidence in the decision made. This latter information may be used to reject ambiguous pattern, should they arise, and thereby improve the classification performance of the network.

5. Contextual Information

Knowledge is represented by the very structure and activation state of a neural network. Every neuron in the network is potentially affected by the global activity of all other neurons in the network. Consequently, contextual information is dealt with naturally by a neural network.

6. Fault Tolerance

A neural network, implemented in hardware form, has the potential to be inherently fault tolerant in the sense that its performance is degraded gracefully under adverse operating conditions.

7. VLSI Implementability

The massively parallel nature of a neural network makes it potentially fast for the computation of certain tasks. This same feature makes a neural network ideally suited for implementations using very-large-scale-integrated (VLSI) technology. The particular virtue of VLSI is that it provides a means of capturing truly complex benaviour in a highly hierarchical fashion, which makes it possible to use a neural network as a tool for real-time applications involving pattern recognition, signal processing, and control.

8. Uniformity of Analysis and Design

Basically, neural networks enjoy universality information processors, in the sense that the same notation is used in all the domains involving the application of neural networks. This feature manifests itself in different ways:

(1) Neurons, in one form or another, represent an ingredient common to all neural networks.

(2) This commonality makes it possible to share theories and learning algorithms in different applications of neural networks.

(3) Modular networks can be built through a seamless integration of modules.

9. Neurobiological Analogy The design of neural network is motivated by analogy with the brain, which is a living proof that fault-tolerant parallel processing is not only physically possibly but also fast and powerful. Neurobiologists look to (artificial) neural networks as a research tool for the interpretation of neurobiological phenomena. The neurobiological analogy is useful in an important way: It provides a hope and belief that physical understanding of neurobiological structures could indeed influence the art of electronics and thus VLSI.

13.5 APPLICATION OF ANN TO OVERCURRENT PROTECTION

Overcurrent protection is used for the protection of generators, power transformers, motors and transmission and distribution lines. Therefore, overcurrent relays are amongst the most widely used relays in a power system. The overcurrent relay senses the fault current and operates only when the current in the circuit exceeds the pick-up value. The various time–current characteristics for the overcurrent relay can be realized using artificial intelligence approaches to power system protection.

For realization of the overcurrent relay, the magnitude of the fundamental frequency component of the current signal can be estimated using appropriate digital filtering algorithm. Using data acquisition system 16 samples of the current signal is acquired and fed as input to the input layer of the ANN. The output layer of overcurrent relay has only one neuron, and is used for estimation of the magnitude of the fundamental component of the current signal thereafter the operating (tripping) time corresponding to a particular value of current is computed using the time–current characteristics.

13.6 APPLICATION OF ANN TO TRANSMISSION LINE PROTECTION

Identification and classification of faults on a transmission line are essential for relaying decision and auto re-closing requirements. The implementation of a pattern recognizer for power system diagnosis has provided great advances in the protection field. An artificial neural network can be used as a pattern classifier for distance relay operation. The magnitudes of three-phase voltage (V_a, V_b, V_c) and current (I_a, I_b, I_c) signals of the transmission line are measured using current and voltage transformers and surge filters. The waveforms are sampled and digitized using analog to digital converters. After the data acquisition, the signals are fed to a pattern recognizer. The ANN recognizer module then verifies for the fault, and if it exists, then issues a trip signal. This ANN module can be either trained online or offline and the trip decision depends on how the module was trained. The advantage of ANN based relays is that it has the ability to learn aspects related to the fault condition and network configuration. Figure 13.8 shows the block diagram of the ANN based distance relays. The ANN approach works as a pattern classifier being able to recognize the changing power system conditions and consequently improve the performance of ordinary relays based on digital principle.

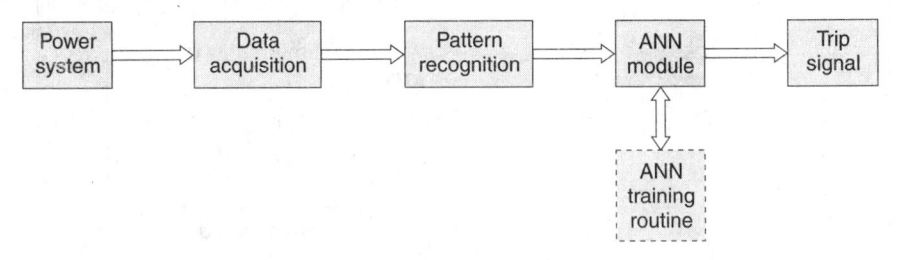

```
Power  →  Data         →  Pattern      →  ANN     →  Trip
system    acquisition     recognition     module     signal
                                            ⇅
                                          ANN
                                          training
                                          routine
```

Fig. 13.8 *Block diagram of the ANN based distance relay*

13.7 NEURAL NETWORK BASED DIRECTIONAL RELAY

With varieties of neural network based relays being designed for transmission line protection, a single neural network can be made to function as directional relay using current and voltage samples. The direction identification is formulated as a three-class problem, i.e., forward fault, reverse fault, and no fault. The modular neural network is preferred due to model complexity reduction, higher learning capability, more suitable incorporation of new training data sets, robustness, and easier insights into the networks.

In this approach, a module discriminates forward fault only from the other states and another module identifies reverse fault only, and hence the task is subdivided. Therefore, this approach requires two modules of neural network to take care of each phase. Figure 13.9 shows the modular structure of neural for the directional relay.

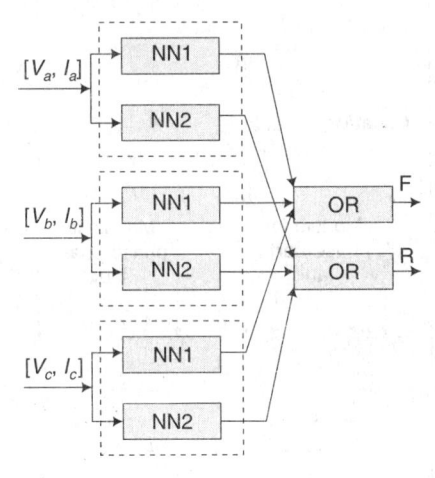

Fig. 13.9 *Overall structure of NN module for the directional*

13.8 ANN MODULAR APPROACH FOR FAULT DETECTION, CLASSIFICATION AND LOCATION

The ANN module shown in the block diagram of the distance relay for the protection of transmission lines can have separate modules for detection, classification and location is as shown in Fig. 13.10.

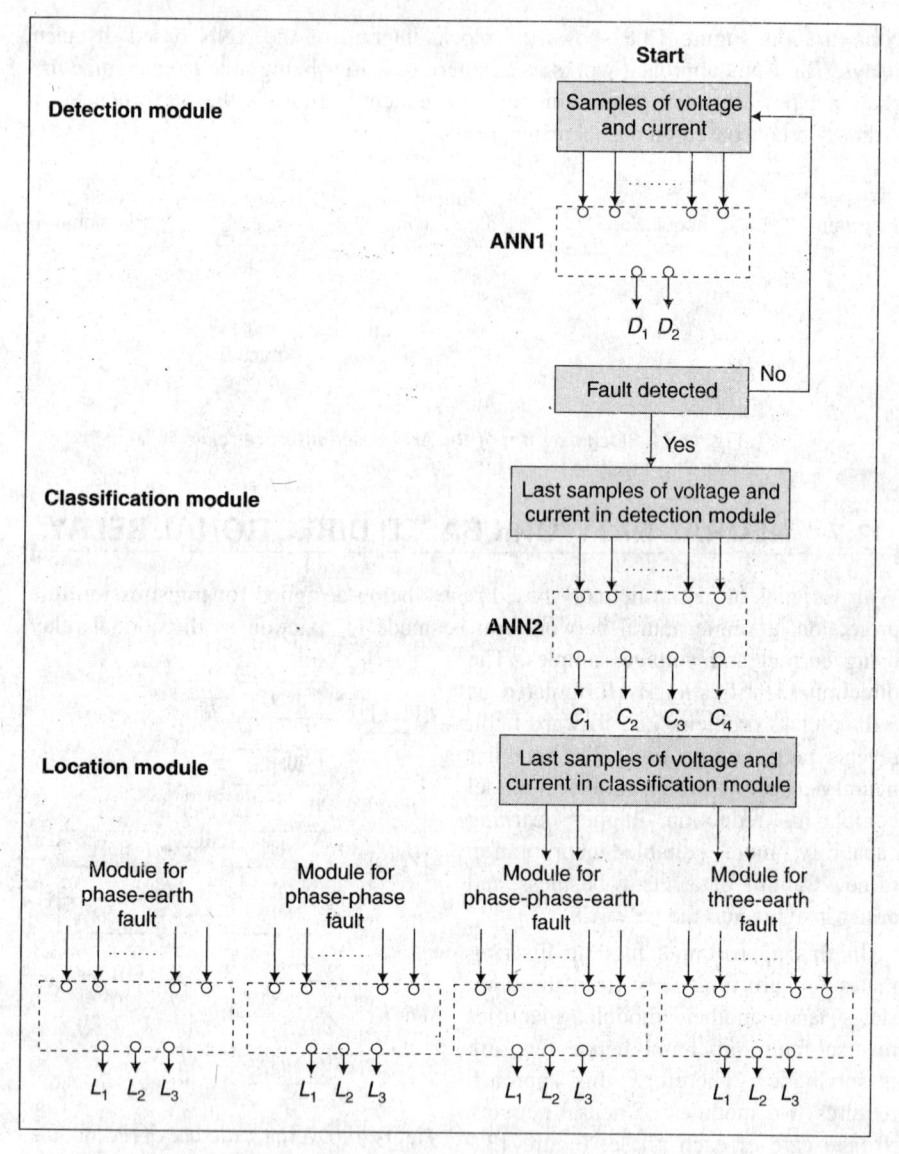

Fig. 13.10 *ANN modular approach to fault detection, classification and location*

The fault detector module detects the fault inception and issues a trip signal indicating this condition. The inception of fault introduces abrupt changes of amplitude and phase in voltage and current signals. Fault signals can contain different transient components, such as exponentially decaying dc-offset (mainly in current signals) and high frequency damped oscillations (mainly in voltage signals), along with other components. These changes of amplitude and phase, and the appearance of transient components, can be used to detect the inception of a fault. The expected outputs of the ANN module for detection are as shown in Table 13.4. The two outputs give the information of the fault situation (forward or reverse) and a normal network situation. When a fault situation is detected the last acquired samples of voltage and current are fed to the classification module. The expected output for classification module is shown in Table 13.5. This tells us about the type of fault which had made the system to trip. After classification the location module tells where the fault had occurred, i.e., in which zone of protection. Different modules can be used for different types of fault to locate the zone of protection in which the fault had taken place. Table 13.6 shows the output of location module.

Table 13.4 Detection module

Situation	D_1	D_2
Normal	0	0
Reverse fault	1	0
Forward fault	0	1

Table 13.5 Classification module

Fault Situation	C_1	C_2	C_3	C_4
a-g	1	0	0	1
b-g	0	1	0	1
c-g	0	0	1	1
a-b	1	1	0	0
b-c	0	1	1	0
c-a	1	0	1	0
a-b-g	1	1	0	1
b-c-g	0	1	1	1
c-a-g	1	1	0	1
a-b-c	1	1	1	0
a-b-c-g	1	1	1	1

Table 13.6 Location module

Fault Situation	L_1	L_2	L_3
Zone 1	1	0	1
Zone 2	0	1	0
Zone 3	0	0	1

13.9 WAVELET FUZZY COMBINED APPROACH FOR FAULT CLASSIFICATION

The samples of voltages and currents are fed in as inputs to the per-processing stage. The pre-processing stage is used to perform feature extraction from the sampled signals with the help of wavelet transform. Wavelet decomposition is achieved by a combination of high and low pass filters. The obtained coefficients, instead of the original waveforms, are fed as pattern inputs to the processing stage to perform pattern recognition and provide an output associated with the pattern. The ANN is trained to obtain the most appropriate weight values based on training data. The fuzzy block rule based post processing improves the accuracy. Fuzzy logic allows one to take into account qualitative information provided by human experts, while ANNs are fault tolerant and present generalization capability, responding well for unseen patterns. Figure 13.11 shows the flow chart of the Wavelet-ANN-Fuzzy combined approach for fault identification and classification scheme.

13.10 APPLICATION OF ANN TO POWER TRANSFORMER PROTECTION

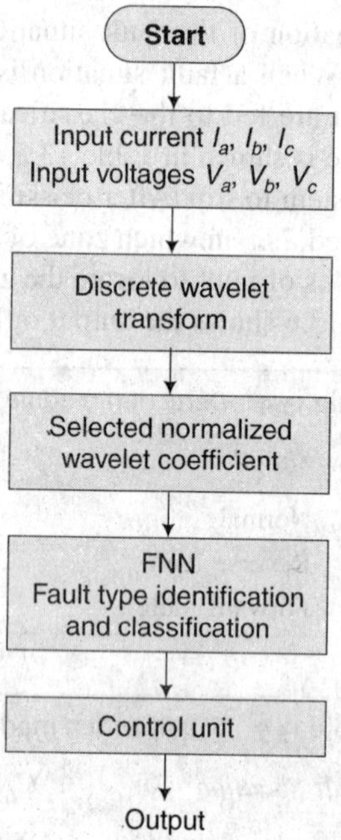

Fig. 13.11 Flowchart of the wavelet-'ANN'-fuzzy combined approach for fault identification and classification scheme

```
Start
  ↓
Input current $I_a$, $I_b$, $I_c$
Input voltages $V_a$, $V_b$, $V_c$
  ↓
Discrete wavelet transform
  ↓
Selected normalized wavelet coefficient
  ↓
FNN
Fault type identification and classification
  ↓
Control unit
  ↓
Output
```

A power transformer is an essential and important element of a power system and its protection is one of the most challenging problems in the power system relaying area. Differential protection has been recognized as the principal basis for the protection of power transformer against internal faults. The differential protection concepts are based on the assumption that during an internal fault, the fundamental component of the differential current becomes substantially higher than the current during normal operating conditions. A differential relay which is used in the differential protection scheme converts the primary and secondary currents of the transformer to a common base and compares them. The nonzero value of the differential current indicates an internal fault.

The flowchart of the algorithm is shown as in Fig. 13.12. The direct and differential currents are calculated using the sampled quantities. If there is a significant differential current, the ANN is used to discriminate the internal fault conditions with other operating conditions.

The differential relaying principle in the case of power transformer shows certain limitations. Detection of a differential current does not provide a clear distinction between internal faults and other conditions. Magnetizing inrush currents, over-excitation of the transformer core, external faults combined with current transformers (CTs) saturation and/or CTs and protected transformer ratio mismatch are the most relevant phenomenon which may upset the current balance causing the relay to maloperate. The conventional approach to mitigate those problems is to apply percentage (biased) differential characteristics along with second and fifth harmonic restraints for inrush and over-excitation conditions, respectively.

Since power transformer differential protection problem of distinguishing among magnetizing inrush, over-excitation internal fault and external fault currents of power transformer can be considered as a current waveform recognition or pattern recognition problem, the use of ANN seems to be an excellent choice.

Example of Multilayer Feedforward Neural Network for Power Transformer Protection

The differential current or both the primary and secondary currents can be fed to a multilayer feedforward neural network for distinguishing among normal, magnetizing inrush, over-excitation, internal fault and external fault currents. If the differential current is taken as input then the number of neurons in the input layer can be 16 i.e., the number of samples acquired of the differential current. If the primary and secondary currents are fed as input for pattern recognition then the number of neurons in the input layer can be 32, i.e., 16 samples of the primary current and 16 samples of the secondary current. The number of neurons in the hidden layer and the number of hidden layers are decided by the experimentation of training and testing a number of network configurations.

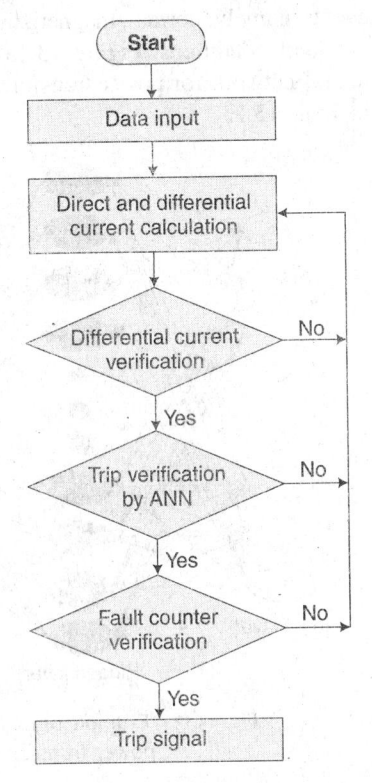

Fig. 13.12 Flowchart of the power transformer protection based on neural network

Most ANN systems can be made to be very accurate (based on the training data) by increasing the number of hidden layers and nodes in these layers. In term of ANNs, numbers of hidden layers and hidden layer nodes may make the system very accurate for the training data, but sometimes the variations that will be present in subsequent data (not involved in training process) may produce large deviations from the expected output. In general, it is up to the analyst to choose an appropriate compromise between the number of layers and nodes and the degree of accuracy obtainable with the training data. The utility of the hidden layers is to add sets of synaptic connections, and thus reinforces the neural interactions that contribute strongly to the success of the learning process, particularly when the size of the input layer is large. Multiple hidden layers may be considered in accordance with the system complexity and the experimentation.

The numbers of output layer neurons are decided as per the number of classification desired at the output. It gives the overall response of the ANN. In this thesis, there are 5 neurons in the output layer, corresponding to 5 different transformer events, namely: normal, magnetizing inrush, over-excitation, internal fault and external fault conditions. Figure 13.13 shows an example of multi-layer feedforward neural network for power transformer protection with the desired output as shown in Table 13.7.

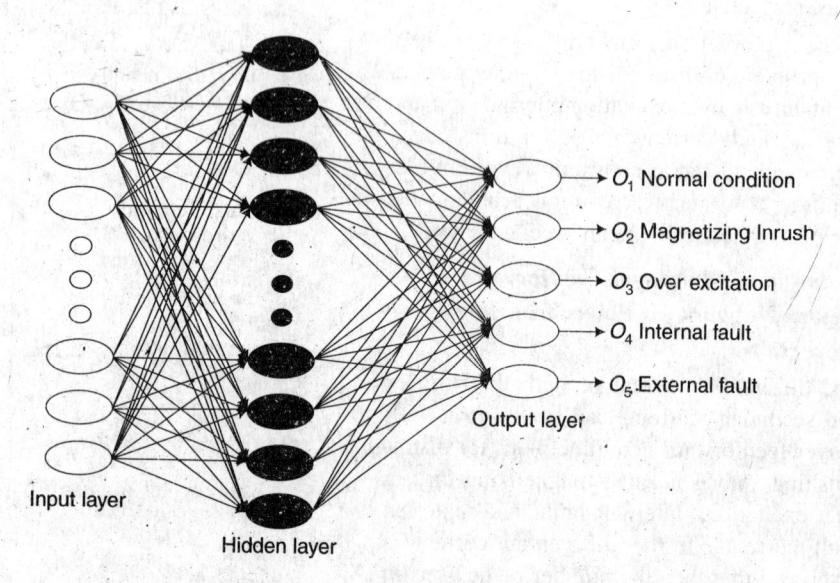

Fig. 13.13 *Example of multilayer feedforward neural network for power transformer protection*

Table 13.7 Desired output of MFNN for different operating conditions of power transformer

Output Neurons	O_1	O_2	O_3	O_4	O_5
Normal Conditions	1	0	0	0	0
Magnetizing Inrush	0	1	0	0	0
Over Excitation	0	0	1	0	0
Internal Fault	0	0	0	1	0
External Fault	0	0	0	0	1

13.11 POWER TRANSFORMER PROTECTION BASED ON NEURAL NETWORK AND FUZZY LOGIC

The internal and external fault with respect to the protection zone are considered as different patterns and the FNN algorithms can be used to recognize these patterns. Utilization of the FNN to pattern recognition of power transformers makes it enable to be robust against specific phenomenon related to the better response and consequently increases the relays performance in comparison with the traditional approaches.

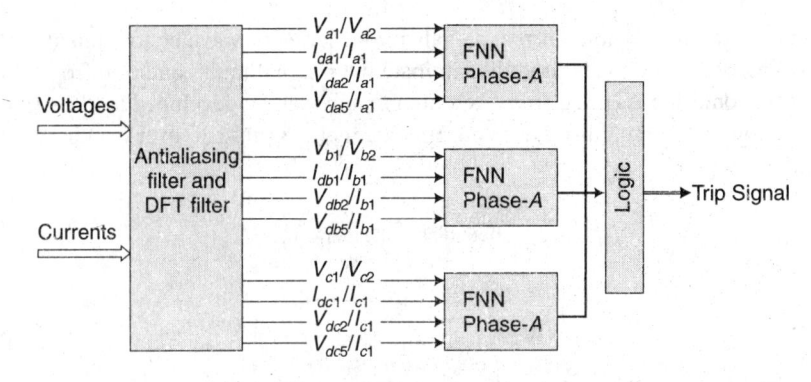

Fig. 13.14 *Structure of the fuzzy-neuro differential protection for power transformer*

Figure 13.14 shows the structure of fuzzy-ANN approach to differential protection of power transformer. First primary and secondary three phase voltage and current signals are processed by anti-aliasing filters and then magnitudes of harmonics of voltage and current be obtained by any filtering algorithm. The fundamental component of primary to secondary voltage ratio of each phase, the fundamental component of the differential current to primary current, 2nd and 5th harmonic component of the differential current to primary current is fed as the input to the FNN module. When the 2nd harmonic component of the differential current is more than 16% of the fundamental frequency component of the differential current, the magnetizing inrush condition exists. The over excitation exists, when the 5th harmonic component of the differential current is more than 8% of the fundamental frequency component of the differential current. Taking these constraints into consideration we can classify the different conditions of the power transformer.

13.12 POWER TRANSFORMER PROTECTION BASED UPON COMBINED WAVELET TRANSFORM AND NEURAL NETWORK

Combining the wavelet transform with neural network the internal fault can be discriminated from other operating conditions of the power transformer. The wavelet transform is firstly applied to decompose the differential current signals of power transformer system into a series of wavelet components, each of which covers a specific frequency band. Thus the time and frequency domain features of the transient signals are extracted. Then the neural network is designed to classify internal fault from all other operating conditions of the power transformer. The extracted features from the wavelet transforms are used as inputs to these ANN architectures.

The flow chart in Fig. 13.15 explains that the wavelet transform is applied to differential current signals for extracting several features and the neural network is used for distinguishing internal fault from the other operating conditions of the power transformer.

The differential current signals are obtained from the power transformer modeling with various operating conditions. Then the wavelet transform is applied to decompose the differential current signals into a series of wavelet coefficients. The first ANN architecture is built to discriminate the internal fault condition from other operating conditions. The second ANN architecture can be developed to classify the different operating conditions, i.e., normal, magnetizing inrush, over excitation and CT saturation.

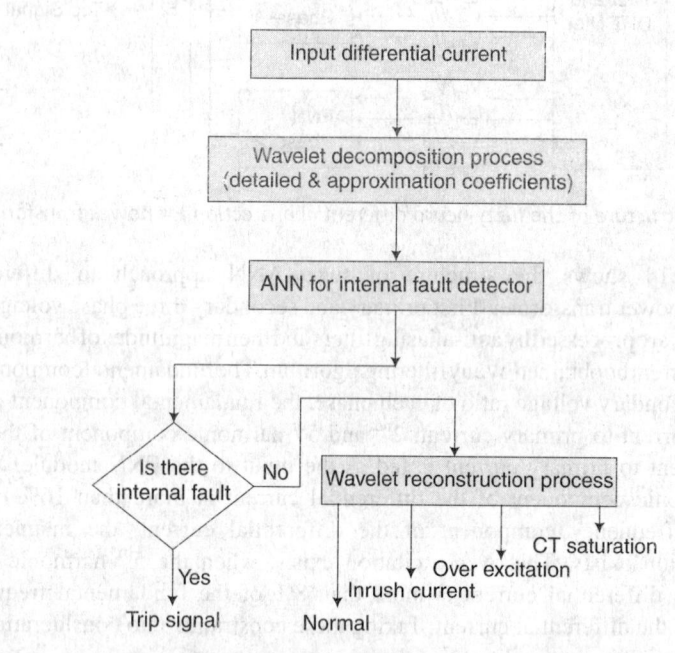

Fig. 13.15 *Flowchart for the power transformer protection based upon combined wavelet transform and neural network*

13.13 APPLICATION OF ANN TO GENERATOR PROTECTION

Synchronous generators are a major component in a power system and any faults will hamper the available power to a system. Therefore, protecting these vital units against abnormal operating conditions and faults, while at the same time keeping protection schemes simple, reliable and fast in operation, has always posed a challenge to the power system protection engineer.

Stator winding faults of synchronous generator are considered serious problems because of the damage associated with high fault currents and high cost of maintenance. A high-speed bias differential relay is normally used to detect three phase, phase-phase and double phase to ground faults. Detection of single line to ground faults depends on the generator grounding type which can be classified into low and high impedance grounding. In case of low impedance grounding, a differential relay can detect and provide protection of only about, 95% of the windings. However, for high impedance grounding, ground faults are not normally detectable by the differential relay because the fault current is, usually, less than the sensitivity of the relay.

Figure 13.16 shows an ANN based digital differential protection scheme for generator stator winding protection. The waveforms of the generator differential current are sampled and digitized using analog to digital converters. After the data acquisition, the signals are fed to a pattern recognizer. Differential current patterns of the generator are used to train the ANN at different loading conditions and different types of internal and external faults. The ANN based fault detector module is used to discriminate between three generator states, i.e., the normal operation state, external fault state and internal fault state. In the event of an internal fault the trip logic module issues a trip signal and activates the ANN based fault classifier module, which identifies the faulted phase.

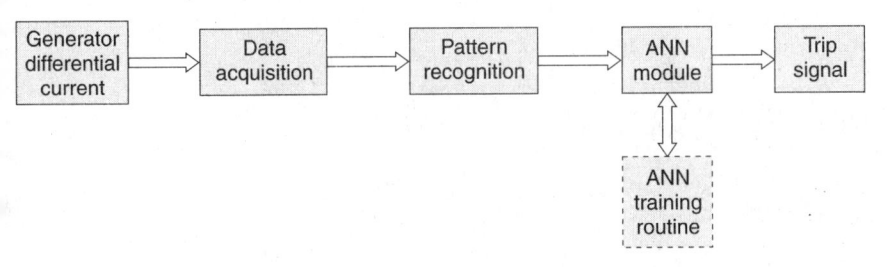

Fig. 13.16 Block diagram of ANN based generator differential protection

EXERCISES

1. What are the various areas of Artificial Intelligence? How can they be applied to power system protection. Discuss the advantages of AI based numerical protection.
2. What is an ANN? Discuss the useful properties and capabilities of ANNs for their application to power system protection.
3. Briefly describe the various types of neural-network models.
4. What do you mean by activation function? What are the different types of activation function.
5. What is neural network learning? Differentiate between supervised and unsupervised learning.
6. Differentiate between multilayer feed forward neural network and radial basis function neural network. Give examples of both the neural networks.
7. Discuss the steps to be followed for the design of ANNs for their application to power system protection.
8. What is fuzzy logic? How can it be applied to power system protection?
9. Discuss the application of ANN to overcurrent protection of the power system?
10. Discuss the application of ANN to protection of transmission lines.
11. Discuss the ANN based differential protection of generators.

12. Discuss the ANN based harmonic restraint differential protection of power transformers.
13. Discuss the ANN modular approach to fault detection, classification and location.
14. Discuss the protection of power transformer based upon combined wavelet transform and neural network.
15. Discuss the fuzzy-neuro differential protection for power transformer.

14 Circuit Breakers

14.1 INTRODUCTION

Circuit breaker is a mechanical switching device capable of making, carrying and breaking currents under normal circuit conditions and also making, carrying for a specified time, and automatically breaking currents under specified abnormal circuit conditions such as those of short-circuits (faults). The insulating medium in which circuit interruption is performed is designated by suitable prefix, such as oil circuit breaker, air-break circuit breaker, air blast circuit breaker, sulphur hexafluorid (SF_6) circuit breaker, vacuum circuit breaker, etc.

The function of a circuit breaker is to isolate the faulty part of the power system in case of abnormal conditions such as faults. A protective relay detects abnormal conditions and sends a tripping signal to the circuit breaker. After receiving the trip command from the relay, the circuit breaker isolates the faulty part of the power system. The simplified diagram of circuit breaker control for opening operation is shown in Fig. 14.1. When a fault occurs in the protected circuit (i.e. the line in this case), the relay connected to the CT and VT detects the fault, actuates and closes its contacts to complete the trip circuit. Current flows from the battery in the trip circuit. As the trip coil of the circuit breaker is energized, the circuit breaker operating mechanism is actuated and it operates for the opening operation to disconnect the faulty element.

A circuit breaker has two contacts – a fixed contact and a moving contact. The contacts are placed in a closed chamber containing a fluid insulating medium (either liquid or gas) which quenches (extinguishes) the arc formed between the contacts. Under normal conditions the contacts remain in closed position. When the circuit breaker is required to isolate the faulty part, the moving contact moves to interrupt the circuit. On the separation of the contacts, an arc is formed between them and the current continues to flow from one contact to the other through the arc as shown in Fig. 14.2. The circuit is interrupted (isolated) when the arc is finally extinguished.

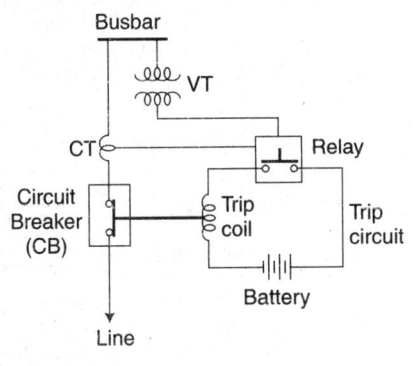

Fig. 14.1 *Simplified diagram of circuit breaker control for opening operation*

When the moving contact starts separating, the insulating fluid between the two contacts experiences a vary high electric stress. The electric stress is almost inversely proportional to the distance between two electrodes, i.e., between two contacts. Thus, at the instant of contact separation, the insulating medium between the contacts is subjected to extremely high electric stress leading to breakdown. The breakdown of the insulating medium between the contacts results in formation of a conducting channel or an arc between the contacts. As the moving contact moves further away, the arc is also drawn along with it. The arc which is in the plasma form continues to carry the pre-opening current. In spite of physical separation of the two contacts, current continues to flow through the arc and thus, the interruption is not effective. The interruption of the current takes place only when the arc is finally quenched (extinguished) and ceases to exist. Thus, circuit breaking may be called an art of arc quenching.

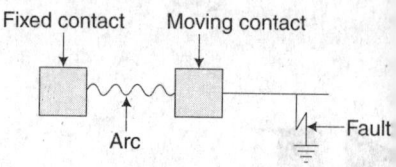

Fig. 14.2 *Separation of the contacts of circuit breaker*

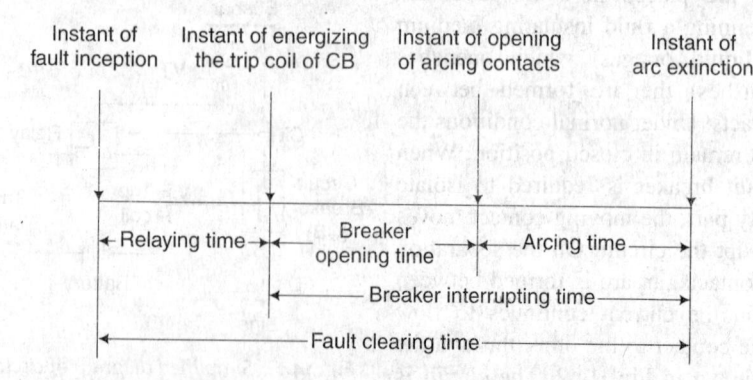

14.2 FAULT CLEARING TIME OF A CIRCUIT BREAKER

A circuit breaker is required in the power system to give rapid fault clearance, in order to avoid overcurrent damage to equipment and loss of system stability. The fault tripping signal to the circuit breaker is derived from the protective relay via the trip circuit. After fault inception, the relay senses the fault and closes its contacts to complete the trip circuit as shown in Fig. 14.1. The relay takes some time to close its contacts. After closing of the contacts of the trip circuit, the trip coil of the circuit breaker is energized and the operating mechanism of the breaker comes into operation. The contacts of the circuit breaker start separating to clear the fault. On the separation of the contacts, an arc is formed between them and the current continues to flow through the arc. The fault is cleared when the arc is finally extinguished.

Figure 14.3 shows the various components of fault clearing time of a circuit breaker. The fault clearing time is the sum of relaying time and breaker interrupting time.

Fig. 14.3 *Fault clearing time of a circuit breaker*

The various components of the fault clearing time of the circuit breaker are defined as follows:

Relaying time = Time from fault inception to the closure of trip circuit of CB.

Breaker opening time = Time from closure of the trip circuit to the opening of the contacts of the circuit breaker.

Arcing time = Time from opening of the contacts of CB to final arc extinction.

Breaker interrupting time = Breaker opening time + arcing time

Fault clearing time = Relaying time + breaker interrupting time

The relaying time for electromechanical relays can vary from about one cycle to five cycles. Static relays are faster than electromechanical relays. Numerical relays give very fast operation and their relaying time is within one cycle. The contact opening time of the circuit breaker may be between about one and three cycles. The arcing time is now generally between one and two half-cycles, depending upon the instant in the current half-cycles at which the contacts part. Therefore, fast relays and modern circuit breakers make it possible to achieve fault clearance in as little as about three cycles of 50 Hz current, but the time varies considerably from system to system and in some cases in different parts of one system.

14.3 ARC VOLTAGE

Figure 14.4 shows the wave shape of arc voltage. The voltage drop across the arc is called arc voltage. As the arc path is purely resistive, the arc voltage is in phase with the arc current. The magnitude of the arc voltage is very low, amounting to only a few per cent of the rated voltage. A typical value may be about 3 per cent of the rated voltage.

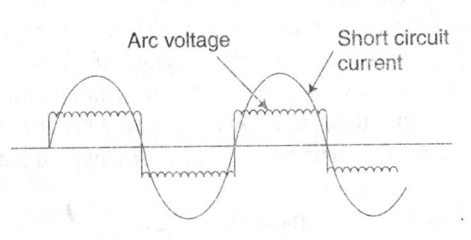

Fig. 14.4 *Short circuit current and arc voltage*

14.4 ARC INTERRUPTION

There are two methods of arc interruption:
  (i) High Resistance Interruption
  (ii) Current Zero Interruption

14.4.1 High Resistance Interruption

In this method of arc interruption, its resistance is increased so as to reduce the current to a value insufficient to maintain the arc. The arc resistance can be increased by cooling, lengthening, constraining and splitting the arc. When current is interrupted the energy associated with its magnetic field appears in the form of electrostatic energy. A high voltage appears across the contacts of the circuit breaker. If this voltage is very high and more than the withstanding capacity of the gap between the contacts, the arc will strike again. Therefore, this method is not suitable for a large-current interruption. This can be employed for low power ac and dc circuit breakers.

14.4.2 Current Zero Interruption

This method is applicable only in case of ac circuit breakers. In case of ac supply, the current wave passes through a zero point, 100 times per second at the supply frequency of 50 Hz. This feature of ac is utilised for arc interruption. The current is not interrupted at any point other than the zero current instant, otherwise a high transient voltage will occur across the contact gap. The current is not allowed to rise again after a zero current occurs. There are two theories to explain the zero current interruption of the arc.

  (i) Recovery rate theory (Slepain's Theory)
  (ii) Energy balance theory (Cassie's Theory)

Recovery Rate Theory

The arc is a column of ionised gases. To extinguish the arc, the electrons and ions are to be removed from the gap immediately after the current reaches a natural zero. Ions and electrons can be removed either by recombining them into neutral molecules or by sweeping them away by inserting insulating medium (gas or liquid) into the gap. The arc is interrupted if ions are removed from the gap at a rate faster than the rate of ionisation. In this method, the rate at which the gap recovers its dielectric strength is compared with the rate at which the gap recovers its dielectric strength is compared with the rate at which the restriking voltage (transient voltage) across the gap rises. If the dielectric strength increases more rapidly than the restriking voltage, the arc is extinguished. If the restriking voltage rises more rapidly than the dielectric strength, the ionisation persists and breakdown of the gap occurs, resulting in an arc for another half cycle. Figure 14.5 explains the principle of recovery rate theory.

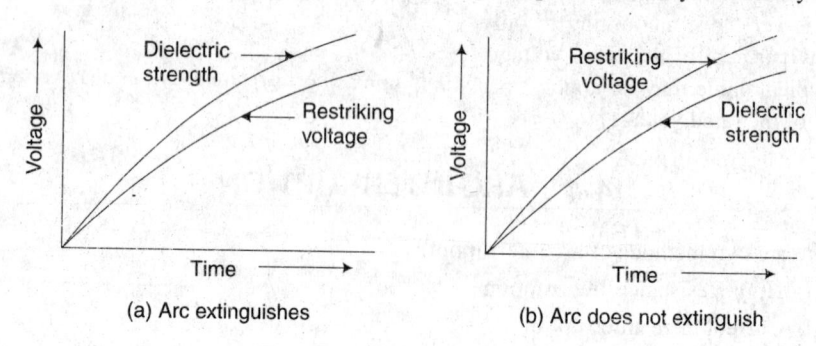

Fig. 14.5 *Recovery rate theory*

(a) Arc extinguishes   (b) Arc does not extinguish

Energy Balance Theory

The space between the contacts contains some ionised gas immediately after current zero and hence, it has a finite post-zero resistance. At the current zero moment, power is zero because restricking voltage is zero. When the arc is finally extinguished, the power again becomes zero, the gap is fully de-ionised and its resistance is infinitely high. In between these two limits, first the power increases, reaches a maximum value, then decreases and finally reaches zero value as shown in Fig. 14.6. Due to the rise of restriking voltage and associated current, energy is generated in the space between the contacts. The energy appears in the form of heat. The circuit breaker is designed to remove this generated heat as early as possible by cooling the gap, giving a blast of air or flow of oil at high velocity and pressure. If the rate of removal of heat is faster than the rate of heat generation the arc is extinguished. If the rate of heat generation is more than the rate of heat dissipation, the space breaks down again resulting in an arc for another half cycle.

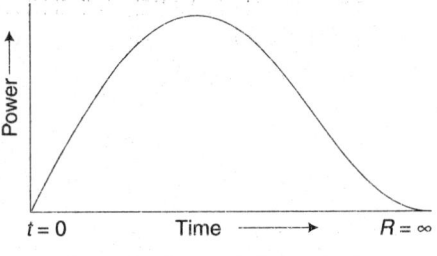

Fig. 14.6 Energy balance theory

14.5 RESTRIKING VOLTAGE AND RECOVERY VOLTAGE

The voltage across the contacts of the circuit breaker is arc voltage when the arc persists. This voltage becomes the system voltage when the arc is extinguished. The arc is extinguished at the instant of current zero. After the arc has been extinguished, the voltage across the breaker terminals does not normalise instantaneously but it oscillates and there is a transient condition. The transient voltage which appears across the breaker contacts at the instant of arc being extinguished is known as *restriking voltage*. The power frequency rms voltage, which appears across the breaker contacts after the arc is finally extinguished and transient oscillations die out is called *recovery voltage*. Figure 14.7 shows the restriking and recovery voltage.

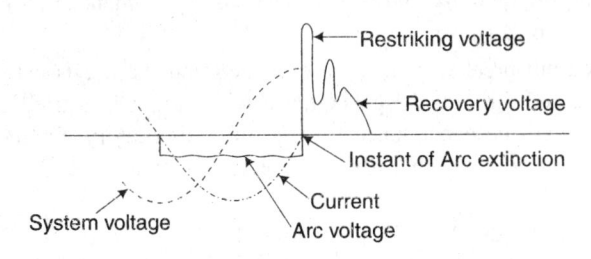

Fig. 14.7 Restriking and recovery voltage

14.5.1 Expression for Restriking Voltage and RRRV

Figure 14.8(a) shows a short circuit (fault) on a feeder beyond the location of the circuit breaker. Figure 14.8(b) shows an equivalent electrical circuit where L and C are the inductance and capacitance per phase of the system up to the point of circuit breaker location, respectively. The resistance of the circuit has been neglected. During the time of fault a heavy fault current flows in the circuit. When the circuit breaker is closed, the fault current flows through L and the contacts of the circuit breaker, the capacitance C being short-circuited by the fault. Hence, the circuit of Fig. 14.8(b) becomes completely reactive and the fault current is limited entirely by inductance of the system.

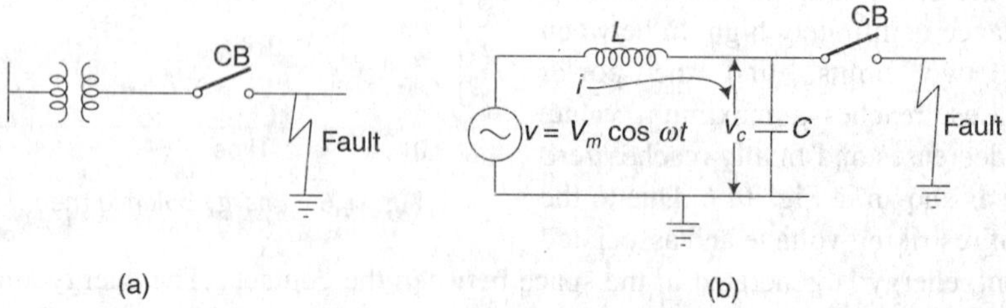

Fig. 14.8 (a) Fault on a feeder near circuit breaker (b) Equivalent electrical circuit for analysis of restriking voltage

The fault is cleared by the opening of the circuit breaker contacts. The parting of the circuit breaker contacts does not in itself interrupt the current because an arc is established between the parting contacts and the current continues to flow through the arc. Successful interruption depends upon controlling and finally extinguishing the arc. Extinction of the arc takes place at the instant when the current passes through zero.

Since the circuit of Fig. 14.8(b) is completely reactive, the voltage at the instant of current zero will be at its peak. The voltage across the circuit breaker contacts and therefore across the capacitor C, is the arc voltage. In high-voltage circuits it is usually only a small percentage of the system voltage. Hence, the arc voltage may be assumed to be negligible.

For the analysis of this circuit, the time is measured form the instant of interruption (arc extinction), when the fault current comes to zero. Since the voltage is a sinusoidally varying quantity and is at its peak at the moment of current zero, it is expressed as '$V_m \cos \omega t$'.

When the circuit-breaker contacts are opened and the arc is extinguished, the current i is diverted through the capacitance C, resulting in a transient condition. The inductance and the capacitance form a series oscillatory circuit. The voltage across the capacitance which is restriking voltage, rises and oscillates, as shown in Fig. 14.7.

The natural frequency of oscillation is given by

$$f_n = \frac{1}{2\pi} \frac{1}{\sqrt{LC}} \qquad (14.1)$$

and the natural angular frequency is

$$\omega_n = \frac{1}{\sqrt{LC}} \qquad (14.2)$$

The voltage across the capacitance which is the voltage across the contacts of the circuit breaker can be calculated in terms of L, C, f_n and system voltage.

Circuit Breakers

The mathematical expression for the transient condition is as follows:

$$L\frac{di}{dt} + \frac{1}{C}\int i\,dt = V_m \cos \omega t \qquad (14.3)$$

Immediately after the instant of arc extinction, the voltage across the capacitance v_c) which is the restriking voltage, oscillates at the natural frequency given by Eq. 14.1). Since the natural frequency oscillation is a fast phenomenon, it persists for only a small period of time. During this short period which is of interest, the change in the power frequency term is very little and, hence negligible, because $\cos \omega t \approx 1$. Hence the sinusoidally varying voltage $V_m \cos \omega t$ in Eq. (14.3) can be assumed to remain constant at V_m during this short interval of time i.e., the transient period.

Substituting $V_m \cos \omega t \approx V_m$, the Eq. (14.3) can be written as

$$L\frac{di}{dt} + \frac{1}{C}\int i\,dt = V_m \qquad (14.4)$$

$$i = \frac{dq}{dt} = \frac{d(cv_c)}{dt} \qquad (14.5)$$

where, v_c = voltage across the capacitor = Restriking voltage

Therefore,
$$\frac{di}{dt} = \frac{d^2(cv_c)}{dt^2} = c\frac{d^2v_c}{dt^2} \qquad (14.6)$$

$$\frac{1}{C}\int i\,dt = \frac{q}{C} = v_c \qquad (14.7)$$

Substituting these values in Eq. (14.4), we get

$$LC\frac{d^2v_c}{dt^2} + v_c = V_m \qquad (14.8)$$

Taking Laplace Transform of both sides of Eq. (14.8), we get

$$LCS^2 v_c(s) + v_c(s) = \frac{V_m}{s}$$

where, $v_c(s)$ is the Laplace Transform of v_c.

Other terms are zero as initially $q = 0$ at $t = 0$

or
$$v_c(s)[LCS^2 + 1] = \frac{V_m}{s}$$

or
$$v_c(s) = \frac{V_m}{s(LCS^2 + 1)} = \frac{V_m}{LCS\left(s^2 + \frac{1}{LC}\right)}$$

$$w_n = \frac{1}{\sqrt{LC}}, \text{ therefore, } \frac{1}{LC} = w_n^2$$

or
$$v_c(s) = \frac{w_n^2 V_m}{s(s^2 + w_n^2)} = \frac{w_n V_m}{s}\left(\frac{w_n}{s^2 + w_n^2}\right) \qquad (14.9)$$

Taking the inverse Laplace of Eq. (14.9), we get

$$v_c(t) = w_n V_m \int_0^t \sin \omega_n t$$

$$= w_n V_m \left[\frac{-\cos \omega_n t}{w_n} \right]_0^t$$

As $v_c(t) = 0$ at $t = 0$, constant = 0.

or
$$v_c(t) = V_m (1 - \cos \omega_n t) \tag{14.10}$$

This is the expression for the restriking voltage.

The maximum value of the restriking voltage occurs at $t = \frac{\pi}{\omega_n} = \pi \sqrt{LC}$

Hence, the maximum value of restriking voltage = $2V_m$

= 2 × peak value of the system voltage

The amplitude factor of the restriking voltage is defined as the ratio of the peak of the transient voltage to the peak value of the system frequency voltage. If losses are ignored, this factor becomes 2.

The Rate of Rise of Restriking Voltage (RRRV)

$$= \frac{d}{dt} [V_m (1 - \cos \omega_n t)]$$

or
$$\text{RRRV} = V_m w_n \sin \omega_n t \tag{14.11}$$

The maximum value of RRRV occurs when $\omega_n t = \pi/2$ i.e., when $t = \pi/2\omega_n$.

Hence, the maximum value of RRRV = $V_m \omega_n$

Example 14.1 | For a 132 kV system, the reactance and capacitance up to the location of the circuit breaker is 3 ohms and 0.015 μF, respectively. Calculate the following:

(a) The frequency of transient oscillation
(b) The maximum value of restriking voltage across the contacts of the circuit breaker
(c) The maximum value of RRRV

Solution:

(a) The frequency of transient oscillation

$$L = \frac{3}{2\pi 50}, \quad f = 50, \text{ the system frequency}$$

$$= \frac{3}{100 \pi} = 0.00954 \text{ H}$$

$$f_n = \frac{1}{2\pi \sqrt{LC}}$$

$$= \frac{1}{2\pi \sqrt{0.00954 \times 0.015 \times 10^{-6}}}$$

$$= \frac{10^5}{2\pi \times 1.1962} = \frac{10^5}{7.5241} = 13.291 \text{ kHz}$$

(b) The restriking voltage

$$v_c = V_m [1 - \cos \omega_n t]$$

The maximum value of the restriking voltage $= 2V_m$

$$= 2 \times \frac{132}{\sqrt{3}} \sqrt{2} = 215.56 \text{ kV}$$

(c) The maximum value of RRRV $= \omega_n V_m$

$$= 2\pi f_n \times \frac{132}{\sqrt{3}} \times \sqrt{2} \times 1000$$

$$= 2\pi \times 13.291 \times 1000 \times \frac{132}{\sqrt{3}} \times \sqrt{2} \times 1000 \text{ V/s}$$

$$= 9010.45 \times 10^6 \text{ V/s} = 9.01045 \text{ kV}/\mu\text{s}$$

14.6 RESISTANCE SWITCHING

To reduce the restriking voltage, RRRV and severity of the transient oscillations, a resistance is connected across the contacts of the circuit breaker. This is known as resistance switching. The resistance is in parallel with the arc. A part of the arc current flows through this resistance resulting in a decrease in the arc current and increase in the deionisation of the arc path and resistance of the arc. This process continues and the current through the shunt resistance increases and arc current decreases. Due to the decrease in the arc current, restriking voltage and RRRV are reduced. The resistance may be automatically switched in with the help of a sphere gap as shown in Fig. 14.9. The resistance switching is of great help in switching out capacitive current or low inductive current.

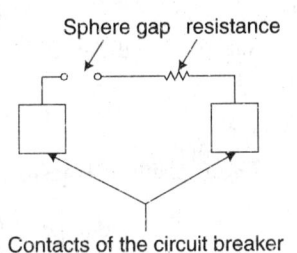

Fig. 14.9 Resistance switching

The analysis of resistance switching can be made to find out the critical value of the shunt resistance to obtain complete damping of transient oscillations. Figure 14.10 shows the equivalent electrical circuit for such an analysis.

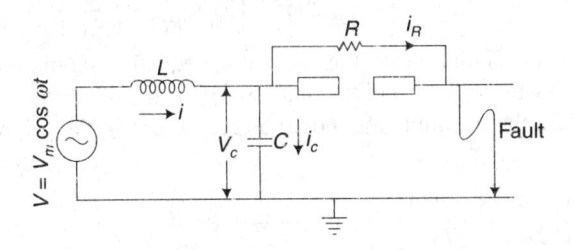

Fig. 14.10 Circuit for analysis of resistance switching

As the period of transient oscillations is very small, the change in the power frequency term during this short period is very little and hence negligible, because $\cos \omega t \approx 1$. Hence, the sinusoidally varying voltage $V_m \cos \omega t$ can be assumed to remain constant at V_m during the transient periods, i.e., $V_m \cos \omega t = V_m$.

Hence, the voltage equation is given by

$$L\frac{di}{dt} + \frac{1}{C}\int i_C\, dt = V_m \quad \text{and} \quad i = i_c + i_R$$

Therefore, the above equation becomes

$$L\frac{d(i_c + i_R)}{dt} + v_c = V_m$$

or

$$L\frac{di_c}{dt} + L\frac{di_R}{dt} + v_c = V_m \tag{14.12}$$

$$i_c = \frac{dq}{dt} = \frac{d(Cv_c)}{dt}$$

Therefore,

$$\frac{di_c}{dt} = \frac{d^2(Cv_c)}{dt^2} = C\frac{d^2v_c}{dt^2} \tag{14.13}$$

$$\frac{di_R}{dt} = \frac{d(v_c/R)}{dt} = \frac{1}{R}\frac{dv_c}{dt} \tag{14.14}$$

Substituting these values in Eq. (14.12), we get

$$LC\frac{d^2v_c}{dt^2} + \frac{L}{R}\frac{dv_c}{dt} + v_c = V_m \tag{14.15}$$

Taking Laplace Transform, of both sides of Eq. (14.15), we get

$$LCS^2 v_c(S) + \frac{L}{R} S\, v_c(S) + v_c(S) = \frac{V_m}{S}$$

Other terms are zero, as $v_c = 0$ at $t = 0$

or

$$LCv_c(S)\left[S^2 + \frac{1}{RC}S + \frac{1}{LC}\right] = \frac{V_m}{S}$$

or

$$v_c(S) = \frac{V_m}{SLC\left[S^2 + \frac{1}{RC}S + \frac{1}{LC}\right]} \tag{14.16}$$

For no transient oscillation, all the roots of the equation should be real. One root is zero, i.e. $S = 0$ which is real. For the other two roots to be real, the roots of the quadratic equation in the denominator should be real. For this, the following condition should be satisfied.

$$\left[\left(\frac{1}{2RC}\right)^2 - \frac{1}{LC}\right] \geq 0 \quad \text{or} \quad \frac{1}{4R^2C^2} \geq \frac{1}{LC}$$

or

$$\frac{4}{LC} \leq \frac{1}{R^2C^2} \quad \text{or} \quad R^2 \leq \frac{LC}{4C^2}$$

or

$$R^2 \leq \frac{1}{4}\cdot\frac{L}{C} \quad \text{or} \quad R \leq \frac{1}{2}\sqrt{\frac{L}{C}} \tag{14.17}$$

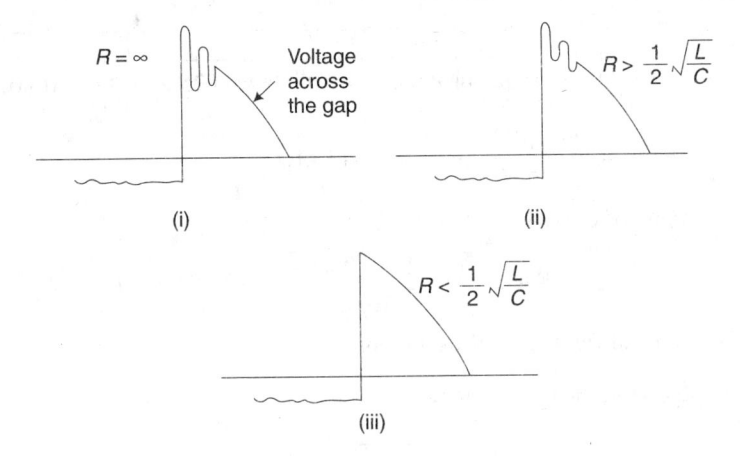

Fig. 14.11 Transient oscillations for different values of R

Therefore, if the value of the resistance connected across the contacts of the circuit breaker is equal to or less than $\frac{1}{2}\sqrt{L/C}$ there will be no transient oscillation. If $R > \frac{1}{2}\sqrt{L/C}$, there will be oscillation. $R = \frac{1}{2}\sqrt{L/C}$ is known as critical resistance. Figure 14.11 shows the transient conditions for three different values of R. The frequency of damped oscillation is given by

$$f = \frac{1}{2\pi}\sqrt{\frac{1}{LC} - \frac{1}{4C^2R^2}} \qquad (14.18)$$

Example 14.2 In a 220 kV system, the reactance and capacitance up to the location of circuit breaker is 8 Ω and 0.025 μF, respectively. A resistance of 600 ohms is connected across the contacts of the circuit breaker. Determine the following:
(a) Natural frequency of oscillation
(b) Damped frequency of oscillation
(c) Critical value of resistance which will give no transient oscillation
(d) The value of resistance which will give damped frequency of oscillation, one-fourth of the natural frequency of oscillation

Solution:

$$L = \frac{8}{2\pi 50} = \frac{8}{100\pi} = 0.02544 \text{ H}$$

(i) Natural frequency of oscillation $= \dfrac{1}{2\pi}\sqrt{\dfrac{1}{LC}}$

$$= \frac{1}{2\pi}\sqrt{\frac{1}{0.02544 \times 0.025 \times 10^{-6}}}$$

$$= 6.304 \text{ kHz}$$

(ii) Frequency of damped oscillation is given by

$$f = \frac{1}{2\pi}\sqrt{\frac{1}{LC} - \frac{1}{4C^2R^2}}$$

$$= \frac{1}{2\pi}\sqrt{\frac{1}{0.02544 \times 0.025 \times 10^{-6}} - \frac{1}{4(0.025 \times 10^{-6})^2 \times (600)^2}}$$

$$= \frac{1}{2\pi}\sqrt{\frac{10^{10}}{6.36} - \frac{10^{10}}{9}} = 3.413 \text{ kHz}$$

(iii) The value of critical resistance

$$R = \frac{1}{2}\sqrt{\frac{L}{C}} = \frac{1}{2}\sqrt{\frac{0.02544}{0.025 \times 10^{-6}}} = 504.35 \text{ }\Omega$$

(iv) The damped frequency of oscillation is

$$\frac{1}{4} \times 6.304 \text{ kHz} = 1576 \text{ Hz}$$

$$1576 = \frac{1}{2\pi}\sqrt{\frac{1}{LC} - \frac{1}{4C^2R^2}}$$

$$= \frac{1}{2\pi}\sqrt{\frac{1}{0.02544 \times 0.025 \times 10^{-6}} - \frac{1}{4(0.025 \times 10^{-6})^2 \times R^2}}$$

or

$$1576 = \frac{1}{2\pi}\sqrt{\frac{10^{10}}{6.36} - \frac{10^{16}}{25R^2}}$$

Therefore, $R = 520.8 \text{ }\Omega$.

14.7 CURRENT CHOPPING

When low inductive current is being interrupted and the arc quenching force of the circuit breaker is more than necessary to interrupt a low magnitude of current, the current will be interrupted before its natural zero instant. In such a situation, the energy stored in the magnetic field appears in the form of high voltage across the stray capacitance, which will cause restriking of the arc. The energy stored in the magnetic field is $\frac{1}{2} L i^2$, if i is the instantaneous value of the current which is interrupted. This will appear in the form of electrostatic energy equal to $\frac{1}{2} C v^2$. As these two energies are equal, they can be related as follows.

$$\frac{1}{2} L i^2 = \frac{1}{2} C v^2$$

$$\therefore \quad v = i \sqrt{L/C} \quad (14.19)$$

Figure 14.12 shows the current chopping phenomenon. If the value of v is more than the withstanding capacity of the gap between the contacts, the arc appears again. Since the quenching force is more, the current is again chopped. This phenomenon continues till the value of v becomes less than the withstanding capacity of the gap. The theoretical value of v is called the prospective value of the voltage.

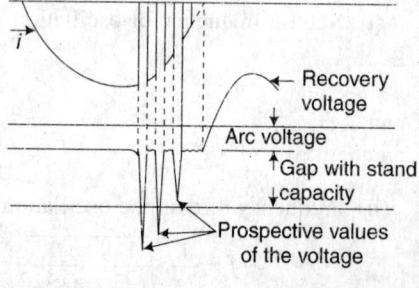

Fig. 14.12 *Current chopping*

Example 14.3 | A circuit breaker interrupts the magnetising current of a 100 MVA transformer at 220 kV. The magnetising current of the transformer is 5% of the full load current. Determine the maximum voltage which may appear across the gap of the breaker when the magnetising current is interrupted at 53% of its peak value. The stray capacitance is 2500 μF. The inductance is 30 H.

Solution: The full load current of the transformer

$$= \frac{100 \times 10^6}{\sqrt{3} \times 220 \times 10^3} = 262.44 \text{ A}$$

Magnetising current $= \dfrac{5}{100} \times 262.44 = 34.44$ A

Current chopping occurs at $0.53 \times 34.44 \sqrt{2} = 25.83$ A

$$\frac{1}{2} Li^2 = \frac{1}{2} Cv^2$$

$\therefore \qquad \dfrac{1}{2} \times 30 \times (25.83)^2 = \dfrac{1}{2} \times 2500 \times 10^{-6} v^2$

$\therefore \qquad v = 2829$ kV

14.8 INTERRUPTION OF CAPACITIVE CURRENT

The interruption of capacitive current produces high voltage transients across the gap of the circuit breaker. This occurs when an unloaded long transmission line or a capacitor bank is switched off. Figure 14.13(a) shows an equivalent electrical circuit of a simple power system. C represents stray capacitance of the circuit breaker. C_L represents line capacitance. The value of C_L is much more than C. Figure 14.13(b) shows transient voltage across the gap of the circuit breaker when capacitive current is interrupted.

At the instance M, the capacitive current is zero and the system voltage is maximum. If an interruption occurs, the capacitor C_L remains charged at the maximum value of the system voltage. After instant M, the voltage across the breaker gap is the difference of V_C and V_{CL}. At instant N, i.e. half-cycle from A, the voltage across the gap is twice the maximum value of V_C. At this moment, the breaker may restrike. If the arc restrikes, the voltage across the gap becomes practically zero. Thus, the voltage across the gap falls from $2V_{C\max}$ to zero. A severe high frequency oscillation occurs. The voltage oscillates about point S between R and N, i.e. between $-3V_{\max}$ and $V_{\max}$. When restriking current reaches zero, the arc may be interrupted again. At this stage, the capacitor C_L remains charged at the voltage $-3V_{\max}$. At instant P, the system voltage reaches its positive maximum shown by the point T in the figure, and at this moment the voltage across the gap becomes $4V_{\max}$. The capacitive current reaches zero again and there may be an interruption. If the interruption occurs at this moment, the transient voltage oscillates between $P(-3V_{\max})$ and $Q(+5V_{\max})$. In this way, the voltage across the gap goes on increasing. But in practice, it is limited to 4 times the peak value of the system voltage. Thus, it is seen that there is a problem of high transient voltage while interrupting a capacitive current.

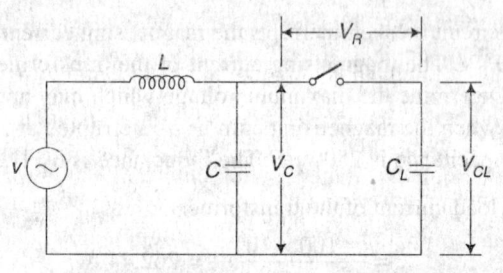

(a) Electrical circuit of a simple power system

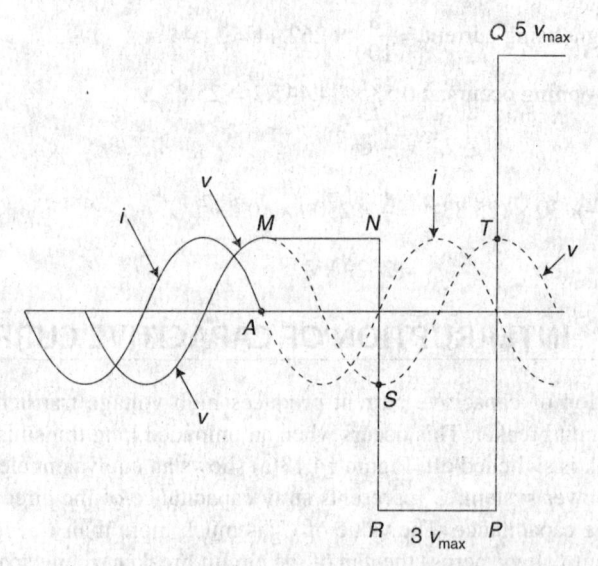

(b) Transient voltage across the gap of the circuit breaker

Fig. 14.13 *Interruption of capacitive current*

14.9 CLASSIFICATION OF CIRCUIT BREAKERS

Circuit breakers can be classified using the different criteria such as, intended voltage application, location of installation, their external design characteristics, insulating medium used for arc quenching, etc.

14.9.1 Classification Based on Voltage

Circuit breakers can be classified into the following categories depending on the intended voltage application.

  (i) Low Voltage Circuit Breaker (less than 1 kV)
 (ii) Medium Voltage Circuit Breaker (1 kV to 52 kV)
(iii) High Voltage Circuit Breakers (66 kV to 220 kV)
(iv) Extra High Voltage (EHV) Circuit Breaker (300 kV to 765 kV)
 (v) Ultra High Voltage (UHV) Circuit Breaker (above 765 kV)

14.9.2 Classification Based on Location

Circuit breakers based on their location are classified as

(i) Indoor type

(ii) Outdoor type

Low and medium voltage switchgears, and high voltage Gas Insulated Switchgears (GIS) are categorised as Indoor Switchgears, whereas the Switchgears which have air as an external insulating medium, i.e. Air-Insulated Switchgear (AIS), are categorised as outdoor Switchgears.

14.9.3 Classification Based on External Design

Circuit breakers can be classified into following categories depending on their external design.

(i) Dead tank type

(ii) Live-tank type

This clasification is for outdoor circuit breakers from the point of view of their physical structural design.

14.9.4 Classification Based on Medium Used for Arc Quenching

Out of the various ways of classification of circuit breakers, the general way and the most important method of classification is on the basis of medium used for insulating and arc quenching. Depending on the arc quenching medium employed, the following are important types of circuit breakers

(i) Air-break circuit breakers:

(ii) Oil circuit breakers

(iii) Air blast circuit breakers

(iv) Sulphur hexafluoride (SF_6) circuit breakers

(v) Vacuum circuit breakers

The development of circuit breakers outlined above has taken place chronologically in order to meet two important requirements of the power system which has progressively grown in size. Firstly, higher and higher fault currents need to be interrupted, i.e., breakers need to have larger and larger breaking capacity. Secondly, the fault interruption time needs to be smaller and smaller for maintaining system stability.

14.10 AIR-BREAK CIRCUIT BREAKERS

Air-break circuit breakers are quite suitable for high current interruption at low voltage. In this type of a circuit breaker, air at atmospheric pressure is used as an arc extinguishing medium. Figure 14.14 shows an air-break circuit breaker. It employs two pairs of contacts—main contacts and arcing contacts. The main contacts carry current when the breaker is in closed position. They have low contact resistance. When contacts are opened, the main contacts separate first, the arcing contacts still remain closed. Therefore, the current is shifted from the main contacts to the arcing contacts. The arcing contacts separate later on and the arc is drawn between them.

In air-break circuit breakers, the principle of high resistance is employed for arc interruption. The arc resistance is increased by lengthening, splitting and cooling the arc. The arc length is rapidly increased employing arc runners and arc chutes. The arc moves upward by both electromagnetic and thermal effects. It moves along the arc runner and then it is forced into a chute. It is split by arc splitters. A blow-out coil is employed to provide magnetic field to speed up arc movement and to direct the arc into arc splitters. The blow-out coil is not connected in the circuit permanently. It comes in the circuit by the arc automatically during the breaking process. The arc interruption is assisted by current zero in case of ac air break circuit breakers. High resistance is obtained near current zero.

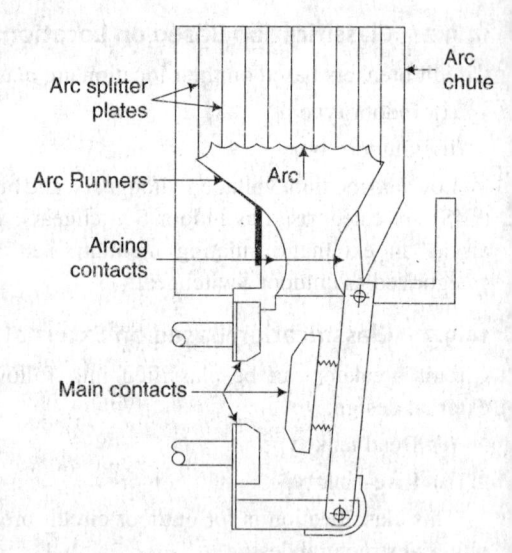

Fig. 14.14 Air-break circuit breaker

AC air-break circuit breakers are available in the voltage range 400 to 12 kV. They are widely used in low and medium voltage system. They are extensively used with electric furnaces, with large motors requiring frequent starting, in a place where chances of fire hazard exist, etc. Air-break circuit breakers are also used in dc circuit up to 12 kV.

14.11 OIL CIRCUIT BREAKERS

Mineral oil has better insulating properties than air. Due to this very reason it is employed in many electrical equipment including circuit breakers. Oil has also good cooling property. In a circuit breaker when arc is formed, it decomposes oil into gases. Hence, the arc energy is absorbed in decomposing the oil. The main disadvantage of oil is that it is inflammable and may pose a fire hazard. Other disadvantages included the possibility of forming explosive mixture with air and the production of carbon particles in the oil due to heating, which reduces its dielectric strength. Hence, oil circuit breakers are not suitable for heavy current interruption at low voltages due to carbonisation of oil. There are various types of oil circuit breakers developed for use in different situations. Some important types are discussed below.

14.11.1 Plain-break Oil Circuit Breakers

In a plain-break oil circuit breaker there is a fixed and a moving contact immersed in oil. The metal tank is strong, weather tight and earthed. Figure 14.15 shows a double break plain oil circuit breaker. When contacts separate there is a severe arc which decomposes the oil into gases. The gas obtained from the oil is mainly hydrogen. The volume of gases produced is about one thousand times that of the oil decomposed. Hence, the oil is pushed away from arc and the gaseous medium surrounds the arc.

The arc quenching factors are as follows.

(i) Elongation of the arc.
(ii) Formation of gaseous medium in between the fixed and moving contacts. This has a high heat conductivity and high dielectric strength.
(iii) Turbulent motion of the oil, resulting from the gases passing through it.

A large gaseous pressure is developed because a large amount of energy is dissipated within the tank. Therefore, the tank of the circuit breaker is made strong to withstand such a large pressure. When gas is formed around the arc, the oil is displaced. To accommodate the displaced oil, an air cushion between the oil surface and the tank is essential. The air cushion also absorbs the mechanical shock produced due to upward oil movement. It is necessary to provide some form of vent fitted in the tank cover for the gas outlet. A sufficient level of oil above the contacts is required to provide substantial oil pressure at the arc.

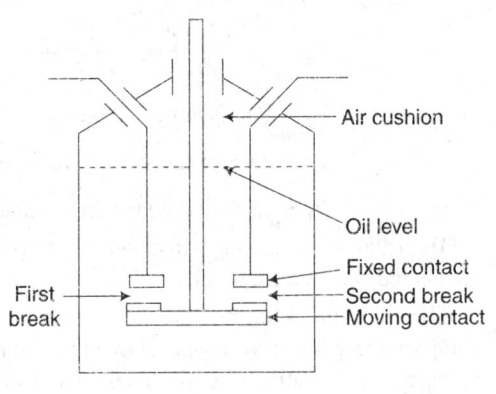

Fig. 14.15 *Plain-break oil circuit breaker*

Certain gap between the contacts must be created before the arc interruption occurs. To achieve this, the speed of the break should be as high as possible. The two breaks in series provide rapid arc elongation without the need for a specially fast contact. The double break also provides ample gap distance before arc interruption. But this arrangement has the disadvantage of unequal voltage distribution across the breaks.

Figure 14.16 shows voltage distribution across breaks. C is the capacitance between the fixed contact and moving contact, C' is the capacitance between the moving contact and earth. V_1 is the voltage across the first contact and V_2 across the second contact. Suppose the fault current is i. The voltages V_1 and V_2 will be expressed as follows.

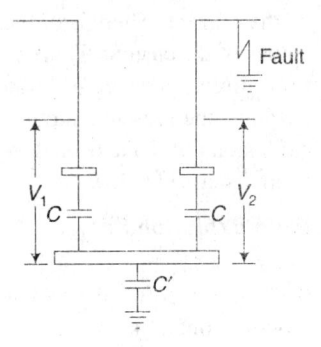

Fig. 14.16 *Voltage distribution across breaks*

$$V_1 = \frac{i}{wC} \quad (14.20) \quad \text{and} \quad V_2 = \frac{i}{w(C+C')} \quad (14.21)$$

The capacitance between the moving contact and earth C' is in parallel with C, the capacitance of the second break.

$$\frac{V_1}{V_2} = \frac{C+C'}{C} \quad (14.22)$$

Taking $C = 8$ pF and $C' = 16$ pF, we get

$$\frac{V_1}{V_2} = \frac{C+C'}{C} = \frac{8+16}{8} = \frac{24}{8} = 3$$

$$\frac{V_1 + V_2}{V_2} = \frac{3+1}{1} = 4$$

$V_1 + V_2 = V =$ system voltage

$V_2 = 25\%$ of the system voltage

$V_1 = 75\%$ of the system voltage

To equalise the voltage distribution across the breaks, non-linear resistors are connected across each break.

The plain-break circuit breakers are employed for breaking of low current at comparatively lower voltages. They are used on low voltage dc circuits and on low voltage ac distribution circuits. Their size becomes unduly large for higher voltages. Also, they require large amount of transformer oil. They are not suitable for autoreclosing. Their speed is slow. They can be used up to 11 kV with an interrupting capacity up to 250 MVA.

14.11.2 Self-generated Pressure Oil Circuit-breaker

In this type of circuit-breakers, arc energy is utilised to generate a high pressure in a chamber known as explosion pot or pressure chamber or arc controlling device. The contacts are enclosed within the pot. The pot is made of insulating material and it is placed in the tank. Such breakers have high interrupting capacity. The arcing time is reduced.

Since the pressure is developed by the arc itself, it depends upon the magnitude of the current. Therefore, the pressure will be low at low current and high at high values of the current. This creates a problem in designing a suitable explosion pot. At low current, pressure generated should be sufficient to extinguish the arc. At heavy currents, the pressure should not be too high so as to burst the pot. Various types of explosion pots have been developed to suit various requirements. A few of them have been discussed below.

Plain Explosion Pot

Figure 14.17 shows a plain explosion pot. This is the simplest form of an explosion pot. When the moving contact separates a severs arc is formed. The oil is decomposed and gas is produced. It generates a high pressure within the pot because there is a close fitting throat at the lower end of the pot. The high pressure developed causes turbulent flow of streams of the gas into the arc resulting in arc-extinction. If the arc extinction does not occur within the pot, it occurs immediately after the moving contact leaves the pot, due to the high velocity axial blast of the gas which is released through the throat. Since the arc extinction in the plain explosion pot is performed axially, it is also known as an axial extinction pot. This type of a pot is not suitable for breaking of heavy currents. The pot may burst due to very

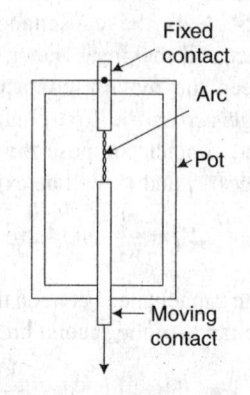

Fig. 14.17 Plain explosion pot

large pressure. At low currents, the arcing time is more. Hence, this type of an explosion pot is suitable for the interruption of currents of medium range.

Cross-jet Explosion Pot

Figure 14.18 shows a cross-jet explosion pot. It is suitable for high current interruptions. Arc splitters are used to obtain an increased arc length for a given amount of contact travel. When the moving contact is separated from the fixed contact, an arc is formed, as shown in Fig. 14.18(a). The arc is pushed into the arc splitters as shown in Fig. 14.18(b), and finally it is extinguished, as in Fig. 14.18(c). In this type of a pot, the oil blast is across the arc and hence it is known as a cross-jet explosion pot.

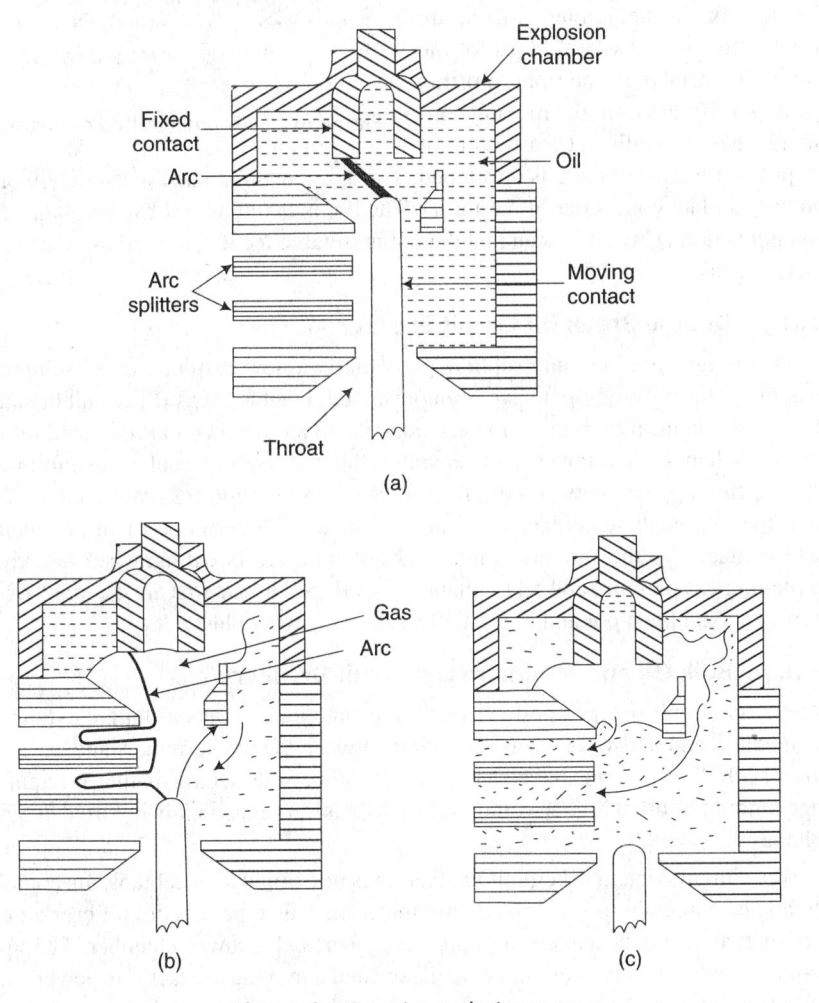

Fig. 14.18 *Cross-jet explosion pot*

Self-compensated Explosion Pot

This type of a pot is a combination of a cross-jet explosion pot and a plain explosion pot. Figure 14.19 shows a self-compensated explosion pot. Its upper portion is a cross-explosion pot, and the lower portion a plain explosion pot. On heavy currents the rate of gas generation is very high and consequently, the pressure produced is also very high. The arc extinction takes place when the first or second lateral orifice of the arc splitter is uncovered by the moving contact. The pot operates as a cross-jet explosion pot. When the current is low, the pressure is also low in the beginning. So the arc is not extinguished when the tip of the moving contact is in the upper portion of the pot. By the time the moving contact reaches the orifice at the bottom of the pot, sufficient pressure is developed. The arc is extinguished when the tip of the moving contact comes out of the throat. The arc is extinguished by the plain explosion pot action. Thus, it is seen that the pot is suitable for low as well as high current interruptions.

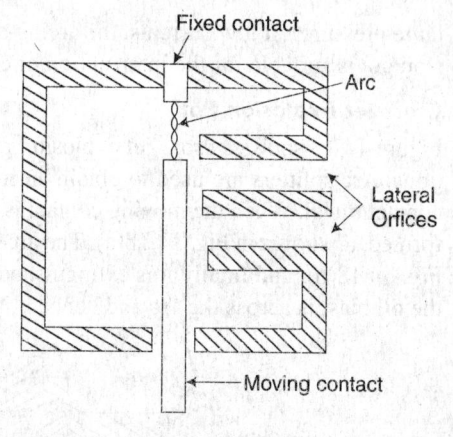

Fig. 14.19 Self-compensated explosion pot

14.11.3 Double Break Oil Circuit Breaker

To obtain high speed arc interruption, particularly at low currents, various improved designs of the explosion pot have been presented. Double break oil circuit breaker is also one of them. It employs an intermediate contact between the fixed and moving contact. When the moving contact separates, the intermediate contact also follows it. The arc first appears between the fixed contact and the intermediate contact. Soon after, the intermediate contact stops and a second arc appears between the intermediate contact and the moving contact. The second arc is extinguished quickly by employing gas pressure and oil momentum developed by the first arc. Figure 14.20(a) shows an axial blast pot and Fig. 14.20(b) shows a cross blast pot.

14.11.4 Bulk Oil and Minimum Oil Circuit Breakers

In bulk oil circuit breakers, oil performs two functions. It acts as an arc extinguishing medium and it also serves as insulation between the live terminals and earth. The tank of a bulk oil circuit breaker is earthed. Its main drawback is that it requires a huge amount of oil at higher voltages. Due to this very reason it is not used at higher voltages.

A minimum content oil circuit breaker does not employ a steel tank. Its container is made of porcelain or other insulating material. This type of a circuit breaker consists of two sections, namely an upper chamber and a lower chamber. The upper chamber contains an arc control device, fixed and a moving contact. The lower chamber acts as an insulating support and it contains the operating mechanism. These two chambers are filled with oil but they are physically separated form each other. The arc control device is placed in a resin bounded glass fiber cylinder (or backelised paper encloser). This cylinder is also filled with oil. The fiber glass cylinder is then placed in a porcelain cylinder. The annular space between the fiber-glass cylinder and the porcelain insulator is also filled with oil.

Minimum oil circuit breakers are available in the voltage range of 3.3 kV to 420 kV. Nowadays they are superseded by SF_6 circuit breakers.

One of the important advantages that the bulk oil circuit breaker has over both the low content oil circuit breakers and air blast circuit breakers is that the protective current transformers can be accommodated on the bushings instead of being supplied as a separated piece of apparatus.

The number of interrupter units contained in a tank depends upon the fault current to be interrupted, and the system voltage. Up to 11 kV voltage, the minimum oil circuit breakers generally employ a single interrupter per phase. The typical figures for higher voltages are, two per phase at 132 kV, and six per phase at 275 kV. Each interrupter has a provision of resistance switching (a typical value being 1200 Ω, linear wire wound resistor) to damp restriking voltage.

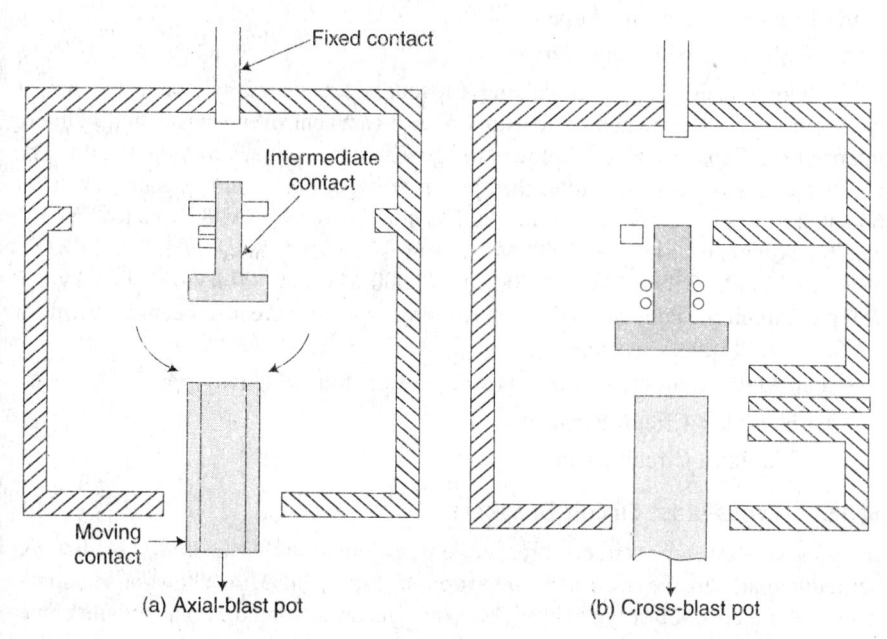

Fig. 14.20 *Double break oil circuit breakers*

14.12 AIR BLAST CIRCUIT BREAKERS

In air blast circuit breakers, compressed air at a pressure of 20-30 kg/cm^2 is employed as an arc quenching medium. Air blast circuit breakers are suitable for operating voltage of 132 kV and above. They have also been used in 11 kV–33 kV range for certain applications. At present, SF_6 circuit breakers are preferred for 132 kV and above. Vacuum circuit breakers are preferred for 11 kV–33 kV range. Therefore, the air blast circuit breakers are becoming obsolete.

The advantages of air blast circuit breakers over oil circuit breakers are:
 (i) Cheapness and free availability of the interrupting medium, chemical stability and inertness of air

(ii) High speed operation
(iii) Elimination of fire hazard
(iv) Short and consistent arcing time and therefore, less burning of contacts
(v) Less maintenance
(vi) Suitability for frequent operation
(vii) Facility for high speed reclosure

The disadvantages of an air blast circuit breaker are as follows
(i) An air compressor plant has to be installed and maintained
(ii) Upon arc interruption the air blast circuit breaker produces a high-level noise when air is discharged to the open atmosphere. In residential areas, silencers need to be provided to reduce the noise level to an acceptable level
(iii) Problem of current chopping
(iv) Problem of restriking voltage

Switching resistors and equalising capacitors are generally connected across the interrupters. The switching resistors reduce transient overvoltages and help arc interruption. Capacitors are employed to equalise the voltage across the breaks. The number of breaks depends upon the system voltage. For example, there are 2 for 66 kV, 2 to 4 for 132 kV, 2 to 6 for 220 kV, 4 to 12 for 400 kV, 8 to 12 for 750 kV. The breaking capacities are, 5000 MVA at 66 kV, 10,000 MVA at 132 kV, 20,000 MVA at 220 kV; 35000 MVA at 400 kV, 40,000 MVA at 500 kV; 60,000 MVA at 750 kV. Circuit breakers for higher interrupting capacity have also been designed for 1000 kV and 1100 kV systems.

An air-blast circuit breaker may be either of the following two types.
(i) Cross-blast Circuit Breakers
(ii) Axial-blast Circuit Breakers

14.12.1 Cross-blast Circuit Breakers

In a cross-blast type circuit breaker, a high-pressure blast of air is directed perpendicularly to the arc for its interruption. Figure 14.21(a) shows a schematic diagram of a cross-blast type circuit breaker. The arc is forced into a suitable chute. Sufficient lengthening of the arc is obtained, resulting in the introduction of appreciable resistance in the arc itself. Therefore, resistance switching is not common in this type of circuit breakers. Cross-blast circuit breakers are suitable for interrupting high current (up to 100 kA) at comparatively lower voltages.

14.12.2 Axial-blast Circuit Breakers

In an axial-blast type circuit breaker, a high-pressure blast of air is directed longitudinally, i.e. in line with the arc. Figure 14.21(b) and (c) show axial-blast type circuit breakers. Figure 14.2.1(b) shows a single blast type. Whereas Fig. 14.21(c) shows a double blast type or radial blast type. Axial blast circuit breakers are suitable for EHV and super high voltage application. This is because interrupting chambers can be fully enclosed in porcelain tubes. Resistance switching is employed to reduce the transient overvoltages. The number of breaks depends upon the system voltage, for example, 4 at 220 kV and 8 at 750 kV. Air-blast circuit breakers have also been commissioned for 1100 kV system.

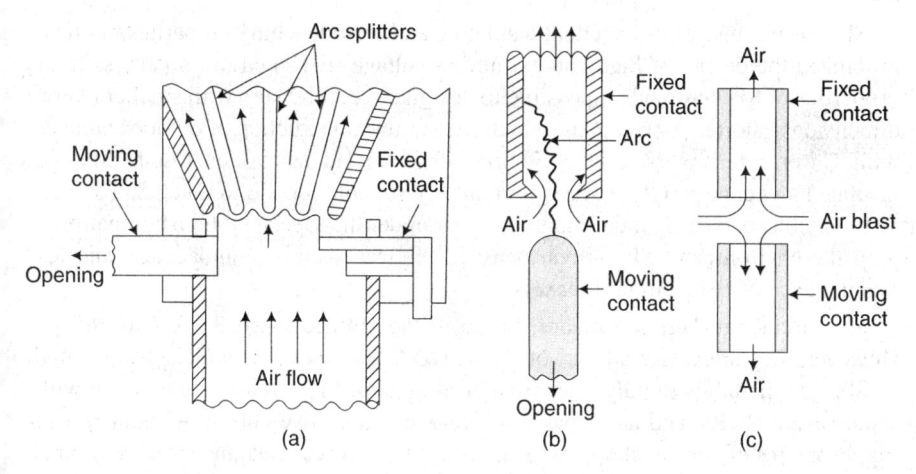

Fig. 14.21 (a) Cross-blast circuit breaker (b) Single blast type axial-blast circuit breaker (c) Double blast type (or radial-blast type) axial-blast circuit breaker

14.13 SF$_6$ CIRCUIT BREAKERS

Sulphur hexafluoride (SF$_6$) has good dielectric strength and excellent arc quenching property. It is an inert, nontoxic, nonflammable and heavy gas. At atmospheric pressure, its dielectric strength is about 2.35 times that of air. At 3 atmospheric pressure its dielectric strength is more than that of transformer oil. It is an electronegative gas, i.e. it has high affinity for electrons. When a free electron comes in collision with a neutral gas molecule, the electron is absorbed by the neutral gas molecule and a negative ion is formed. As the negative ions so formed are heavy they do not attain sufficient energy to contribute to ionisation of the gas. This property gives a good dielectric property. Besides good dielectric strength, the gas has an excellent property of recombination after the removal of the source which energizes the arc. This gives an excellent arc quenching property. The gas has also an excellent heat transfer property. Its thermal time constant is about 1000 times shorter than that of air.

Under normal conditions, SF$_6$ is chemically inert and it does not attack metals or glass. However, it decomposes to SF$_4$, SF$_2$, S$_2$, F$_2$, S and F at temperatures of the order of 1000°C. After arc extinction, the products of decomposition recombine in a short time, within about 1 microsecond. In the presence of moisture, the decomposition products can attack contacts, metal parts and rubber sealings in SF$_6$ circuit breakers. Therefore, the gas in the breaker must be moisture-free. To absorb decomposition products, a mixture soda lime (NaOH + CaO) and activated alumina can be placed in the arcing chamber.

One major disadvantage of SF$_6$ is its condensation at low temperature. The temperature at which SF$_6$ changes to liquid depends on the pressure. At 15 atm. pressure, the gas liquefies at a temperature of about 10°C. Hence, SF$_6$ breakers are equipped with thermostatically controlled heaters wherever such low ambient temperatures are encountered.

SF_6 gas because of its excellent insulating and arc-quenching properties has revolutionized the design of high and extra high voltage (EHV) circuit breakers. These properties of SF_6 has made it possible to design circuit breakers with smaller overall dimensions, shorter contact gaps, which help in the construction of outdoor breakers with fewer interrupters, and evolution of metalclad (metal enclosed) SF_6 gas insulated switchgear (GIS). SF_6 is particularly suitable for use in metalclad switchgear which is becoming increasingly popular under the aspects of high compatibility with the environment. SF_6 offers many advantages such as compactness and less maintenance of EHV circuit breakers.

SF_6 circuit breakers are manufactured in the voltage range 3.3 kV to 765 kV. However, they are preferred for votlages 132 kV and above. The dielectric strength of SF_6 gas increases rapidly after final current zero. SF_6 circuit breakers can withstand severe RRRV and are capable of breaking capacitive current without restriking. Problems of current chopping is minimised. Electrical clearances are very much reduced due to high dielectric strength of SF_6.

14.13.1 Properties of SF_6 Gas

The properties of SF_6 gas can be divided as
  (i) Physical properties
  (ii) Chemical properties
  (iii) Electrical properties

1. Physical Properties of SF_6 Gas

The physical properties of SF_6 gas are as follows:
  (i) It is a colourless, odourless, non-toxic and non-inflammable gas.
  (ii) Pure gas is not harmful to health.
  (iii) It is in gas state at normal temperature and pressure.
  (iv) It is heavy gas having density 5 times that of air at 20°C and atmospheric pressure.
  (v) The gas starts liquifying at certain low temperatures. The temperature of liquification depends on pressure. At 15 atm. pressure, the gas liquifies at a temperature of about 10°C.
  (vi) It has an excellent heat transfer property. The heat transfer capability of SF_6 is 2 to 2.5 times that of air at same pressure.
  (vii) The heat content property is much higher than air. This property of SF_6 assists cooling of arc space after current zero.

2. Chemical Properties of SF_6 Gas

  (i) It is chemically stable at atmospheric pressure and at temperatures up to 500°C.
  (ii) It is a chemically inert gas.
     The property of chemical inertness of this gas is advantageous in switchgear. Because of this property, it has exceptionally low reactivity and does not attack metals, glass, plastics, etc. The life of contacts and other metallic parts is longer in SF_6 gas. The components do not get oxidised or deteriorated. Hence the maintenance requirements are reduced.

(iii) Moisture is very harmful to the properties of this gas. In the presence of moisture, hydrogen fluoride is formed during arcing which can attack the metallic and insulating parts of the circuit breaker.

(iv) It is non-corrosive on all metals at ambient temperatures.

(v) It is an electronegative gas. The ability of an atom to attract and hold electrons is designated as 'electronegativity'. Because of electronegativity, the arc-time constant (the time between current zero and the instant at which the conductance of contact spaces reaches zero value) of SF_6 gas is very low (≤ 1 μs) and the rate of rise of dielectric strength is high. Hence SF_6 circuit breakers are suitable for switching condition involving high RRRV.

(vi) The products of decomposition of SF_6 recombine in a short time after arc extinction. During arc extinction process, SF_6 decomposes to some extent into SF_4, SF_2, S_2, F_2, S and F at temperatures of the order of 1000°C. The products of decomposition recombine to form the original gas in a short time upon cooling. The remainder of the decomposition products is absorbed by a mixture of soda lime (NaOH + CaO) and activated alumina.

3. Electrical Properties of SF_6 Gas

(i) Dielectric properties

Its dielectric strength at atmospheric pressure is 2.35 times that of air and 30% less than that of dielectric oil used in oil circuit breakers. The excellent dielectric strength of SF_6 gas is because of electronegativity (electron attachment) property of SF_6 molecules. In the attachment process, free electrons collide with the neutral gas molecules to form negative ions by the following processes.

Direct attachment process: $SF_6 + e \rightarrow SF_6^-$

Dissociative attachment process: $SF_6 + e \rightarrow SF_5^- + F$

The negative ions formed are heavier and immobile as compared to the free electrons and therefore under a given electric field the ions do not accumulate sufficient energy to lead to cumulative ionization in the gas. This process is an effective way of removing electrons. Otherwise, this would have contributed to the cumulative ionisation, to the current growth and to the sparkover of the gas. Therefore, this property gives rise to very high dielectric strength for SF_6 gas.

The dielectric strength of the gas increases with pressure. At three times the atmospheric pressure, its dielectric strength is more than that of dielectric oil used in oil circuit breakers. The dielectric strengths of SF_6, air and dielectric oil as a function of pressure for a particular electrode configuration and a gap of 1 cm is shown in Fig. 14.22.

SF_6 gas maintains high dielectric strength even after mixing with air. A mixture of 30% SF_6 and 70% air by volume has a dielectric strength twice that of air at the same pressure. Below 30% of SF_6 by volume, the dielectric strength falls sharply.

Because of electronegativity of the gas, its arc-time constant is very low (≤ 1 μs). The arc-time constant is defined as the time required for the medium to regain its dielectric strength after final current zero.

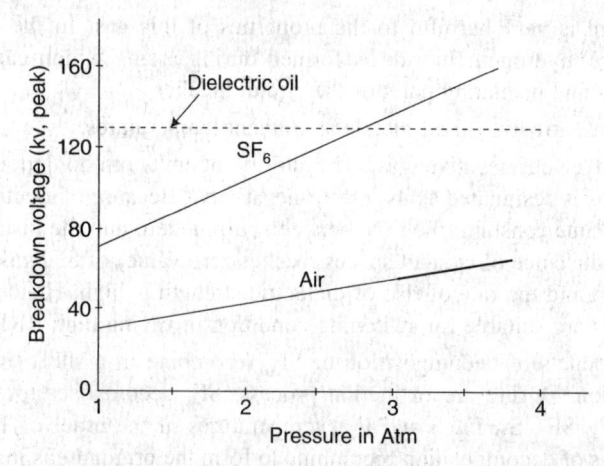

Fig. 14.22 *Breakdown voltages of SF_6, air and dielectric oil as a function of pressure*

(ii) Corona inception voltage

Corona inception voltage for SF_6 in a non-uniform electric field is also considerably higher than that for air.

(iii) Dielectric constant

Because of being non-polar (i.e., dipole moment is zero), the dielectric constant of SF_6 is independent of the frequency of the applied voltage. Further, the dielectric constant changes by only 7% over a pressure range of 0 to 22 atmospheres.

(iv) Arc-interrupting capacity

Besides possessing a high dielectric strength, the molecules of SF_6 when dissociated due to sparkovers, recombine rapidly after removal of the source which energizes the spark (arc) and the gas recovers its dielectric strength. This makes SF_6 very effective in quenching the arc. SF_6 is approximately 100 times as effective as air in quenching arc. Figure 14.23 compares the arc quenching ability of SF_6 and air.

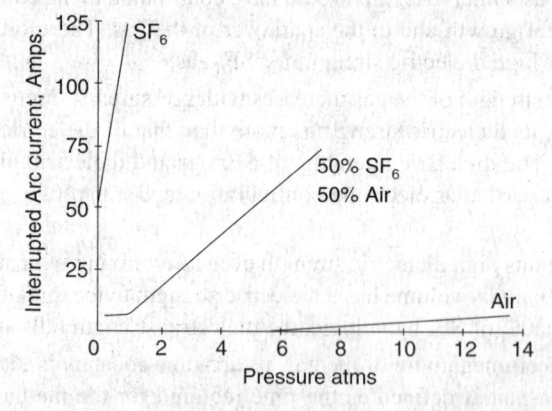

Fig. 14.23 *Current interrupting capacity of SF_6, air and a mixture of both gases*

The arc quenching property of SF_6 can be due to several factors, especially, its large electron attachment rate. If the free electrons in an electric field can be absorbed before they attain sufficient energy to create additional electrons by collision, the breakdown mechanism can be delayed or even stopped.

The arc temperature in SF_6 is lower than that in air for the same arc current and radius. The high heat transfer capability and low arc temperature also provide an excellent arc quenching (extinguishing) capacity to SF_6.

14.13.2 Arc Extinction in SF_6 Circuit Breakers

The final extinction of the arc requires a rapid increase of the dielectric strength between the contacts of the circuit breaker, which can be achieved either by the deionisation of the arc path or by the replacement of the ionised gas by cool fresh gas. The various deionisation processes are high pressure, cooling by conduction, forced convection and turbulence.

The basic requirement in arc extinction is not only the dielectric strength, but also, high rate of recovery of dielectric strength. In SF_6 gas, the dielectric strength is quickly regained because electrons get attached with the molecules to become ions. This is due to electronegativity property of SF_6 gas.

In SF_6 circuit breakers, SF_6 gas is blown axially along the arc during the arcing period. The gas removes the heat from the arc by axial convection and radial dissipation. As a result of this, the diameter of the arc is reduced during the decreasing node of the current wave. The arc diameter becomes very small at current zero. In order to extinguish the arc, the turbulent flow of the gas is introduced.

From the properties of SF_6, it is clear that it is a remarkable medium for arc extinction. The arc extinguishing properties of SF_6 are improved by moderate rates of forced gas flow through the arc space.

The SF_6 gas at atmospheric pressure can interrupt currents of the order of 100 times the value of those that can be interrupted in air with a plain breaker interrupter.

In SF_6 circuit breakers, the gas is made to flow from a high pressure zone to a low pressure zone through a convergent-divergent nozzle. The mass flow depends on nozzle-throat diameter, the pressure ratio and the time of flow. The location of the nozzle is such that the gas flows axially over the arc length. In the divergent portion of the nozzle, the gas flow attains almost supersonic speed and thereby the gas takes away the heat from the periphery of the arc, causing reduction in the diameter of the arc. The arc diameter finally becomes almost zero at current zero and the arc is extinguished. After filling the contact space with fresh SF_6 gas, the dielectric strength of the contact space is rapidly recovered due to electronegativity of the gas and turbulent flow of the fresh gas in the contact space.

14.13.3 Types of SF_6 Circuit Breakers

The following are two principal types of SF_6 circuit breakers:

(i) Double Pressure Type SF_6 Circuit Breaker This type of circuit breaker employs a double pressure system in which the gas from a high-pressure compartment is released into the low-pressure compartment through a nozzle during the arc extinction process. This type of SF_6 circuit breaker has become obsolete.

(ii) Puffer-type (Single-pressure Type) SF$_6$ Circuit Breaker In this type of circuit breaker the SF$_6$ gas is compressed by the moving cylinder system and is released through a nozzle during arc extinction. This is the most popular design of SF$_6$ circuit breaker over wide range of voltages from 3.3 kV to 765 kV.

Double pressure type SF$_6$ circuit breaker

This is the early design of SF$_6$ circuit breakers which employed a double pressure system. In this system, SF$_6$ gas at high pressure of 14 to 18 atmospheric pressure stored in a separate tank is released into the arcing zone to cool the arc and build up the dielectric strength of the contact gap after arc extinction. Its operating principle is similar to that of air-blast circuit breakers. Because of their complicated design and construction due to the requirement of various auxiliaries such as gas compressors, high pressure storage tank, filters and gas monitoring and controlling devices, this type of circuit breakers have become obsolete.

Puffer-type SF$_6$ circuit breaker

This type of circuit breakers are also sometimes called single-pressure or impulse type SF$_6$ circuit breakers. In this type of breakers, gas is compressed by a moving cylinder system and is released through a nozzle to quench the arc. This type is available in the voltage range 3.3 kV to 765 kV.

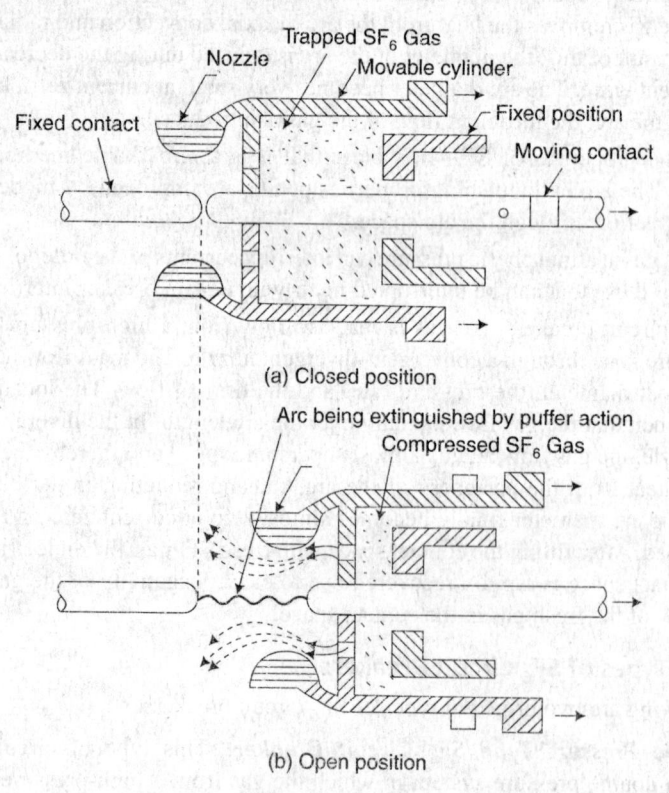

Fig. 14.24 Puffer-type SF$_6$ circuit breaker

Figure 14.24(a) shows a puffer-type breaker in closed position. The moving cylinder and the moving contact are coupled together. When the contacts separate and the moving cylinder moves, the trapped gas is compressed. The trapped gas is released through a nozzle and flows axially to quench the arc as shown in Fig. 14.24(b). There are two types of tank designs. Live tank design and dead tank design. In live tank design, interrupters are supported on porcelain insulators. In the dead tank design, interrupters are placed in SF_6 filled-tank which is at earth potential. Live tank design is preferred for outdoor substations.

A number of interrupters (connected in series) on insulating supports are employed for EHV systems up to 765 kV. Two interrupters are used in a 420 kV system. Breaking time of 2 to 3 cycles can be achieved. In the circuit breaker the steady pressure of the gas is kept at 5 kg/cm^2. The gas pressure in the interrupter compartment increases rapidly to a level much above its steady value to quench the arc.

14.13.4 Advantages of SF_6 Circuit Breakers

(i) Low gas velocities and pressures employed in the SF_6 circuit breakers prevent current chopping and capacitive currents are interrupted without restriking.

(ii) These circuit breakers are compact, and have smaller overall dimensions and shorter contact gaps. They have less number of interrupters and require less mantenance.

(iii) Since the gas is non-inflammable, and chemically stable and the products of decomposition are not explosive, there is no danger of fire or explosion.

(iv) Since the same gas is recirculated in the circuit, the requirement of SF_6 gas is small.

(v) The operation of the circuit breaker is noiseless because there is no exhaust to atmosphere as in case of air blast circuit breakers

(vi) Because of excellent arc quenching properties of SF_6, the arcing time is very short and hence the contact erosion is less. The contacts can be run at higher temperatures without deterioration.

(vii) Because of inertness of the SF_6 gas, the contact corrosion is very small. Hence contacts do not suffer oxidation.

(viii) The sealed construction of the circuit breaker avoids the contamination by moisture, dust, sand etc. Hence the performance of the circuit breaker is not affected by the atmospheric conditions.

(ix) Tracking or insulation breakdown is eliminated, because there are no carbon deposits following an arcing inside the system.

(x) Because of the excellent insulating properties of the SF_6, contact gap is drastically reduced.

(xi) As these circuit breakers are totally enclosed and sealed from atmosphere, they are particularly suitable for use in such environments where explosion hazards exist.

14.13.5 Disadvantages of SF$_6$ Circuit Breakers

(i) Problems of perfect sealing. There may be leakage of SF$_6$ gas because of imperfect joints.
(ii) SF$_6$ gas is suffocating to some extent. In case of leakage in the breaker tank, SF$_6$ gas may lead to suffocation of the operating personnel.
(iii) Arced SF$_6$ gas is poisonous and should not be inhaled or let out.
(iv) Influx of moisture in the breaker is very harmful to SF$_6$ circuit breaker. There are several cases of failures because of it.
(v) There is necessity of mechanism of higher energy level for puffer-types SF$_6$ circuit breakers. Lower speeds due to friction, misalignment can cause failure of the breaker.
(vi) Internal parts should be cleaned thoroughly during periodic maintenance under clean and dry environment.
(vii) Special facilities are required for transporting the gas, transferring the gas and maintaining the quality of the gas. The performance and reliability of the SF$_6$ circuit breaker is affected due to deterioration of quality of the gas.

14.14 VACUUM CIRCUIT BREAKERS

The dielectric strength and arc interrupting ability of high vacuum is superior to those of porcelain, oil, air and SF$_6$ at atmospheric pressure. SF$_6$ at 7 atm. pressure and air at 25 atm. pressure have dielectric strengths higher than that of high vacuum. The pressure of 10^{-5} mm of mercury and below is considered to be high vacuum. Low pressures are generally measured in terms of torr; 1 torr being equal to 1 mm of mercury. It has now become possible to achieve pressures as low as 10^{-8} torr.

When contacts separate in a gas, arc is formed due to the ionised molecules of the gas. The mean free path of the gas molecules is small and the ionisation process multiplies the number of electrons to form an electron avalanche. In high vacuum, of the order of 10^{-5} mm of mercury, the mean free path of the residual gas molecules becomes very large. It is of the order of a few metres. Therefore, when contacts are separated by a few mm in high vacuum, an electron travels in the gap without collision. The formation of arc in high vacuum is not possible due to the formation of electron avalanche. In vacuum arc electrons and ions do not come from the medium in which the arc is drawn but they come from the electrodes due to the evaporation of their surface material. The breakdown strenght is independent of gas density. It depends only on the gap length and surface condition and the material of the electrode. The breakdown strength of highly polished and thoroughly degassed electrodes is higher. Copper-bismuth, silver-bismuth, silver-lead and copper-lead are good materials for making contacts of the breaker.

When contacts are separated in high vacuum, an arc is drawn between them. The arc does not take place on the entire surface of the contacts but only on a few spots. The contact surface is not perfectly smooth. It has certain microprojections. At the time of contact separation, these projections form the last points of separation. The current flows through these points of separation resulting in the formation of a few hot spots. These hot spots emit electrons and act as cathode spots. In addition to thermal emission, electrons emission may be due to field emission and secondary emission.

Figure 14.25 shows the schematic diagram of a vacuum circuit breaker. Its enclosure is made of insulating material such as glass, porcelain or glass fibre reinforced plastic. The vapour condensing shield is made of synthetic resin. This shield is provided to prevent the metal vapour reaching the insulating envelope. As the interrupter has a sealed construction a stainless metallic bellows is used to allow the movement of the lower contact. One of its ends is welded to the moving contact. Its other end is welded to the lower end flange. Its contacts have large disc-shaped faces. These faces contain spiral segments so that the arc current produces axial magnetic field. This geometry helps the arc to move over the contact surface. The movement of arc over the contact surface minimises metal evaporation, and hence erosion of the contact due to arc. Two metal end flanges are provided. They support the fixed contact, outer insulating enclosure, vapour condensing shield and the metallic bellows. The sealing technique is similar to that used in electronic valves.

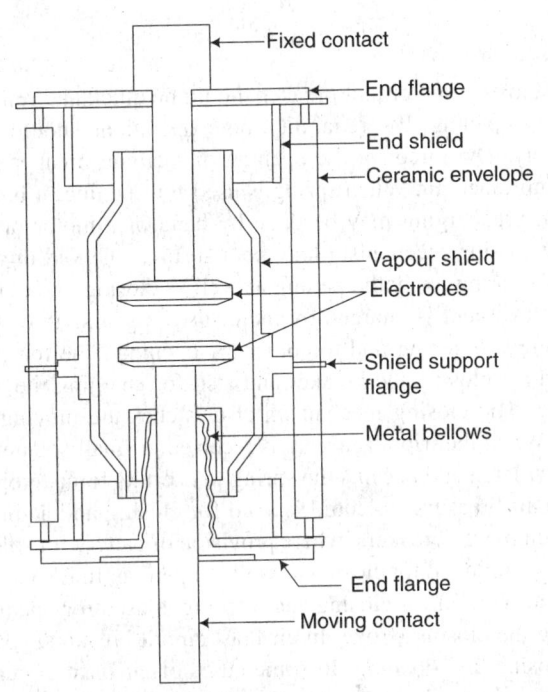

Fig. 14.25 *Vacuum circuit breaker*

The vacuum circuit breaker is very simple in construction compared to other types of circuit breaker. The contact separation is about 1 cm which is adequate for current interruption in vacuum. As the breaker is very compact, power required to close and open its contacts is much less compared to other types of breaker. It is capable of interrupting capacitive and small inductive currents, without producing excessive transient overvotlages. Vacuum circuit breakers have other advantages like suitability for repeated operations, least maintenance, silent operation, long life, high speed of dielectric recovery, less weight of moving parts, etc. The vapour emission depends on the arc current. In ac, when the current decreases, vapour emission decreases. Near current zero, the rate of vapour emission tends to zero. Immediately after current zero, the remaining vapour condenses and the dielectric strength increases rapidly. At current zero, cathode spots extinguish within 10^{-8} second. The rate of dielectric recovery is many times higher than that obtained in other types of circuit breakers. Its typical value may be as high as 20 kV/μs.

Vacuum circuit breakers have now become popular for voltage ratings up to 36 kV. Up to 36 kV they employ a single interrupter.

14.15 OPERATING MECHANISM

To open and close the contacts of a circuit breaker, one of the following mechanisms is employed.

(1) Spring
(2) Solenoid
(3) Compressed air

Opening and closing of a circuit breaker should be quick and reliable. Closing is slow compared to opening. For reliability, both operations should be independent of the main supply. The force for the opening of a circuit breaker may be applied by spring or compressed air. When spring is used for opening, it is precharged, i.e. compressed. The precharging may be done by hand or a motor or by the closing mechanism. The closing mechanism may be a spring, solenoid or compressed air. If the system uses spring for both opening as well as closing, the closing spring has a higher energy level and is charged by motor-driven gears. Two separate springs are employed, one for closing and the other for opening. The force of the closing spring is utilized for closing the breaker and also for charging the spring which is used for opening. The closing mechanism also latches the moving contact in the closed position. When the tripping signal is received, a small solenoid is energised, which releases the latch and permits the spring to exert its force to open the contact. For large oil circuit breakers, solenoid is used for closing and a spring is used for opening. For medium size breakers, where provision of battery for solenoid is uneconomical, spring is employed for closing as well as opening. In EHV oil circuit breakers, compressed air is used for closing and a spring is used for opening. The spring is charged during the closing stroke. In air blast circuit breakers, compressed air is used for both closing and opening. In some cases of air blast circuit breakers, the moving contact is held in closed position by a spring. When compressed air enters the arc chamber and its pressure exceeds the spring force, the contacts are opened. The contacts automatically come in the closed position by the spring action when the supply of compressed air is stopped. Therefore, the supply of compressed air must be maintained till the auxiliary circuit breaker switch is opened.

In SF_6 circuit breaker, compressed air may be used for closing and a spring for opening or compressed air for both closing as well as opening. In vacuum circuit breakers, a solenoid or spring mechanism is fixed to the lower end to move a metallic bellows upward or downward inside the chamber during closing or opening the contacts. The closing mechanism must be tripfree, i.e. if a tripping signal comes during the closing operation, the circuit breaker must trip immediately.

Springs are very good for opening as their force is large in the beginning and gradually decreases as the distance of travel of the moving contact increases. But it is not suitable for closing because the closing operation must be a smart one. However, in circuit breakers of small capacity, it is also used for closing. The spring which is used for closing is charged by hand or a motor. In small circuit breakers, the spring can be charged by hand. For large ones, it is charged by means of a motor. The spring which is used for opening is charged by the closing mechanism.

Solenoids are very good for closing. The force of attraction increases when the distance between the contacts decreases. They are not suitable for opening because they are slow in action. Compressed air is suitable for both closing as well as opening as its force remains almost constant even when the distance between the contacts increases or decreases. Figure 14.26 shows the characteristics of the opening and closing mechanisms.

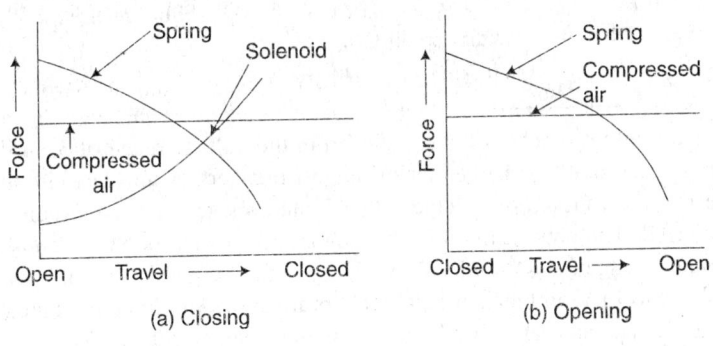

Fig. 14.26 *Characteristics of operating machanism*

14.16 SELECTION OF CIRCUIT BREAKERS

Table 14.1 shows the summary of various types of circuit breakers, their voltage ranges and arc quenching medium they employ. Table 14.2 shows the choice of circuit breakers for various voltage ranges.

Table 14.1 Types of circuit breakers

Type	Arc Quenching Medium	Voltage Range and Breaking Capacity
Miniature circuit breakers	Air at atmospheric pressure	400-600V; for small current rating
Air-break circuit breaker	Air at atmospheric pressure	400 V-11 kV; 5-750 MVA
Minimum oil breakers	Transformer oil circuit	3.3 kV-220 kV; 150-25000 MVA

Contd...

Contd...

Vacuum circuit breakers	Vacuum	3.3 kV-33 kV; 250-2000 MVA
SF_6 circuit breakers	SF_6 at 5 kg/cm^2 pressure	3.3-765 kV; 1000-50,000 MVA
Air blast circuit breakers	Compressed air at high pressure (20-30 kg/cm^2)	66 kV-1100 kV; 2500-60,000 MVA

Table 14.2 Selection of circuit breakers

Rated Voltage	Choice of circuit Breakers	Remark
Below 1 kV	Air-break CB	
3.3 kV-33 kV	Vacuum CB, SF_6 CB, minimum oil CB	Vacuum preferred
132 kV-220 kV	SF_6 CB, air blast CB, minimum oil CB	SF_6 preferred
400 kV-760 kV	SF_6 CB, air blast CB	SF_6 is preferred

Earlier oil circuit breakers were preferred in the voltage range of 3.3 kV-66 kV. Between 132 kV and 220 kV, either oil circuit breakers or air blast circuit breakers were recommended. For voltages 400 kV and above, air blast circuit breakers were preferred. The present trend is to recommend vacuum or SF_6 circuit breakers in the voltage range 3.3 kV-33 kV. For 132 kV and above, SF_6 circuit breakers are preferred. Up to 1 kV, air break circuit breakers are used. Air blast circuit breakers are becoming obsolete and oil circuit breakers are being superseded by SF_6 and vacuum circuit breakers.

14.17 HIGH VOLTAGE DC (HVDC) CIRCUIT BREAKERS

At present, HVDC transmission lines are used for point to point transmission of large power over long distances. Such lines have many advantages over ac transmission lines such as lower cost, less stability problems, less corona loss and less radio interference, etc. HVDC circuit breakers are not essential for single HVDC transmission lines which are used for point to point transmission. The current in HVDC lines is controlled by controlling the firing circuits of the thyristors employed in rectifiers and inverters. Switching operations are performed from ac side with the help of ac circuit breakers. If HVDC circuit breakers are available, parallel HVDC lines, HVDC lines with a tap-off line, closed loop circuit, etc. can also be planned and designed.

In ac circuits, current passes through natural current zeros, and hence it is possible to design ac circuit breakers to interrupt large currents. This feature is not available in dc. If a high current is suppressed abruptly in dc, a very high transient voltage appears across the contacts of the circuit breakers. Therefore, in dc circuit breakers, some external circuits have to be provided to bring down the current from full value to zero, smoothly without suppressing it abruptly. The additional circuit creates artificial current zeros which are utilised for arc interruption as shown in Fig. 14.27.

Figure 14.28 shows the schematic diagram of a HVDC circuit breaker. It consists of a main circuit breaker MCB and a circuit to produce artificial current zero and to suppress transient voltage. The main circuit breaker MCB may either be an SF_6 or vacuum circuit breaker. R and C are connected in parallel with the main circuit breaker to reduce dv/dt after the final current zero. L is a saturable reactor in series with the main circuit breaker. It is used to reduce dI/dt before current zero. C_p and L_p are connected in parallel to produce artificial current zero after the separation of the contacts in the main circuit breaker MCB. A non-linear resistor is used to suppress the transient overvoltage which may be produced across the contacts of the main circuit breaker.

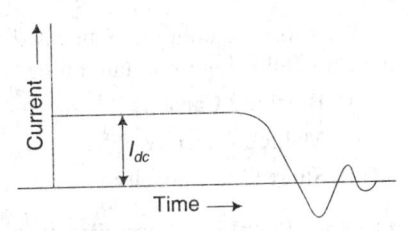

Fig. 14.27 Artificial current zeros in dc

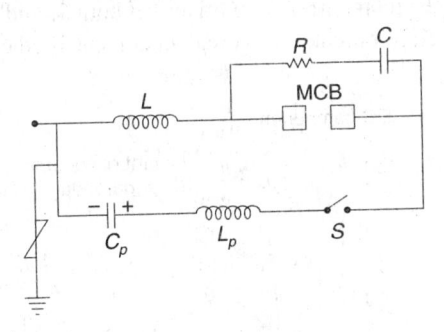

Fig. 14.28 HVDC circuit breaker

Switch S, which is a triggered vacuum gap, is switched immediately after the opening of the contacts of the main circuit breaker. The capacitor C_p is precharged in the direction as shown in the figure. When S is closed, the precharged capacitor C_p discharges through the main circuit breaker and sends a current in opposition to the main circuit current. This will force the main circuit current to become zero with a few oscillations. The arc is interrupted at a current zero.

14.18 RATING OF CIRCUIT BREAKERS

A circuit breaker has to perform the following major duties under short-circuit conditions.
   (i) To open the contacts to clear the fault
   (ii) To close the contacts onto a fault
   (iii) To carry fault current for a short time while another circuit breaker is clearing the fault.

Therefore, in addition to the rated voltage, current and frequency, circuit breakers have the following important ratings.
(i) Breaking Capacity
(ii) Making Capacity
(iii) Short-time Capacity.

14.18.1 Breaking Capacity

The breaking capacity of a circuit breaker is of two types.
(i) Symmetrical breaking capacity
(ii) Asymmetrical breaking capacity

Symmetrical Breaking Capacity

It is the rms value of the ac component of the fault current that the circuit breaker is capable of breaking under specified conditions of recovery voltage.

Asymmetrical Breaking Capacity

It is the rms value of the total current comprising of both ac and dc components of the fault current that the circuit breaker can break under specified conditions of recovery voltage.

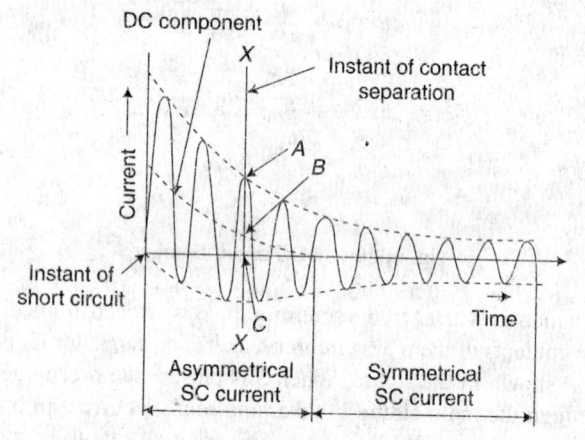

Fig. 14.29 *Short-circuit current waveform*

Figure 14.29 shows a short-circuit current wave. The short-circuit current contains a dc component which dies out gradually as shown in the figure. In the beginning, the short-circuit current is asymmetrical due to the dc component. When dc dies out completely, the short-current becomes symmetrical.

The line X-X indicates the instant of contact separation. AB is the peak value of the ac component of the current at this instant. Therefore, the symmetrical breaking current which is the rms value of the ac component of the current at the instant of contact separation is equal to current $AB/\sqrt{2}$. The section BC is the dc component of the short-circuit current at this instant. Therefore, asymmetrical breaking current is given by

$$I_{asym.} = \left(\frac{AB}{\sqrt{2}}\right)^2 + (BC)^2 \qquad (14.23)$$

The breaking capacity of a circuit breaker is generally expressed in MVA. For a three-phase circuit breaker, it is given by

$$\text{Breaking capacity} = \sqrt{3} \times \text{rated voltage in kV} \times \text{rated current in kA}.$$

The breaking capacity will be symmetrical if the rated current in the above expression is symmetrical (British practice). The breaking capacity will be asymmetrical if the rated current is asymmetrical (American practice). The rated asymmetrical breaking current is taken by designer as 1.6 times the rated symmetrical current.

14.18.2 Making Capacity

The possiblity of a circuit breaker to be closed on short-circuit is also considered. The rated making current is defined as the peak value of the current (including the dc component) in the first cycle at which a circuit breaker can be closed onto a short-circuit. I_p in Fig. 14.29 is the making current. The capacity of a circuit breaker to be closed onto a short-circuit depends upon its ability to withstand the effects of electromagnetic forces.

$$\text{Making current} = \sqrt{2} \times 1.8 \times \text{symmetrical breaking current}.$$

The multiplication by $\sqrt{2}$ is to obtain the peak value and again by 1.8 to take the dc component into account.

$$\text{Making capacity} = \sqrt{2} \times 1.8 \times \text{symmetrical breaking capacity}$$
$$= 2.55 \times \text{symmetrical breaking capacity}.$$

14.18.3 Short-time Current Rating

The short-time current rating is based on thermal and mechanical limitations. The circuit breaker must be capable of carrying short-circuit current for a short period while another circuit breaker (in series) is clearing the fault. The rated short-time current is the rms value (total current, both ac and dc components) of the current that the circuit breaker can carry safely for a specified short period. According to British standard, the time is 3 seconds if the ratio of symmetrical breaking current to rated normal current is equal to or less than 40 and 1 second if this ratio is more than 40. According to ASA there are two short-time ratings, one is the current which the circuit breaker can withstand for 1 second or less. Another is rated 4-second current which is the current that the circuit breaker can withstand for a period longer than 1 second but not more than 4 seconds.

14.18.4 Rated Voltage, Current and Frequency

In a power system, the voltage level at all points is not the same. It varies, depending upon the system operating conditions. Due to this reason manufacturers have specified a rated maximum voltage at which the operation of the circuit breaker is guaranteed. The specified voltage is somewhat higher than the rated nominal voltage.

The rated current is the rms value of the current that a circuit breaker can carry continuously without any temperature rise in excess of its specified limit.

The rated frequency is also mentioned by the manufacture. It is the frequency at which the circuit breaker has been designed to operate. The standard frequency is 50 Hz. If a circuit breaker is to be used at a frequency other than its rated frequency, its effects should be taken into consideration.

14.18.5 Rated Operating Duty

The operating duty of a circuit breaker prescribes its operations which can be performed at stated time intervals. For the circuit breakers which are not meant for autoreclosing, there are two alternative operating duties as given below:

(i) $O - t - CO - t' - CO$

(ii) $O - t'' - CO$

where O denotes opening operation, CO denotes closing operation followed by opening without any intentional time lag, and t, t' and t'' are time intervals between successive operations. According to IEC, the value of t and t' is 3 minutes and t'' is 15 seconds.

For circuit breakers with auto-reclosing, the operating duty is as follows.

$O - Dt - CO$, where Dt is the dead time of the circuit breaker, which is expressed in cycle.

According to B.S.S. there is only one operating duty for the circuit breakers not intended for auto-reclosing. It is written as follows.

$B - 3 - MB - 3 - MB$, where B denotes breaking and MB denotes making followed by breaking without any intentional time delay. Three is the time interval in minutes.

For circuit breakers with auto-reclosing, the operating duty is written as

$$B - Dt - MB$$

Dt is the dead time and it is expressed in cycles.

14.19 TESTING OF CIRCUIT BREAKERS

There are two types of tests of circuit breakers namely routine tests and type tests. Routine tests are performed on every piece of circuit breaker in the premises of the manufacturer. The purpose of the routine tests is to confirm the proper functioning of a circuit breaker. Type tests are performed in a high voltage laboratory, such tests are performed on sample pieces of circuit breakers of each type to confirm their characteristics and rated capacities according to their design. These tests are not performed on every piece of circuit breaker. All routine and type tests are performed according to Indian Standard (IS) codes or International Electromechanical Commission (IEC) codes or British Standard (BS) codes.

A few important type tests, such as breaking capacity, making capacity, short-time current rating tests will be discussed here. These tests come in the category of short circuit testing of circuit breakers. For circuit breakers of smaller capacity, these tests are carried out by direct testing techniques. Circuit breakers of large capacities are tested by the synthetic testing method. In addition to short circuit tests, mechanical tests, thermal tests, dielectric tests (power frequency tests, impulse tests), capacitive charging-current breaking test, small inductive breaking current test, etc. are also performed. For details see the relevant IS codes.

14.19.1 Short-circuit Testing Stations

There are two types of short-circuit testing stations.

(i) Field type testing station

(ii) Laboratory type testing station

In a field type testing station, the power required for testing is derived from a large power system. The circuit breaker to be tested is connected to the power system. Large amount of power is easily available for testing. Hence, this method of testing is economical for testing of circuit breakers, particularly high voltage circuit breakers. But it lacks flexibility. Its drawbacks are:

(i) For research and development work, tests cannot be repeated again and again without disturbing the power system

(ii) The power available for testing varies, depending upon the loading conditions of the system.

(iii) It is very difficult to control the transient recovery voltage, RRRV, etc.

In a laboratory type testing station, the power required for testing is taken form specially designed generators which are installed in the laboratory for such testing. Its advantages are:

(i) For research and development work, tests can be carried out again and again to confirm the designed characteristics and capacity.

(ii) Current, voltage, restriking voltage, RRRV, etc. can be controlled conveniently.

(iii) Tests for circuit breakers of large capacity can be carried out using synthetic testing.

The drawbacks of laboratory type testing stations are:

(i) High cost of installation.

(ii) Availability of limited power for testing of circuit breakers

Short-circuit Generator

In a laboratory, short-circuit generators are used to provide power for testing. The design of such generators is different from a conventional generator. These are specially designed to have very low reactance to give the maximum short-circuit output. To withstand high electromagnetic forces their windings are specially braced and made rugged. They are provided with a flywheel to supply kinetic energy during short circuits. This also helps in speed regulation of the set. The generator is driven by a three-phase induction motor.

The short-circuit current has a demagnetising effect on the field of generator. This results in reducing the field current. Consequently, the generator's emf is reduced. Impulse excitation or super excitation is employed to counteract the demagnetisation effect of armature reaction. At the time of short-circuit, the field current is increased to about 8 to 10 times its normal value.

Short-circuit Transformer

Such a transformer has a low reactance and it is designed to withstand repeated short-circuits. To get different voltage for tests, its windings are arranged in sections. By series and parallel combinations of these sections, the desired test voltage is obtained. To get lower voltage than the generated voltage, a three-phase transformer is generally used. For voltage higher than the generated voltages, normally banks of single phase transformers are employed. As the transformers remain in the circuit for a short time, they do not poses any cooling problem.

Master Circuit Breaker

It is used as a backup circuit breaker. If the circuit breaker under test fails to operate, the master circuit breakers opens. The master circuit breaker is set to operate at a predetermined time after the initiation of the short-circuit. After every test, it isolates the circuit breaker under test form the supply source. Its capacity is more than the capacity of the circuit breaker under test.

Making Switch

This switch is used to apply short-circuit current at the desired moment during the test. The making switch is closed after closing the master circuit breaker and the test circuit breaker. It must be bounce-free to avoid its burning or contact welding. To achieve this, a high pressure is used in the chamber. Its speed is also kept high.

Capacitors

Capacitor are used to control RRRV. They are used in synthetic testing which will be discussed while describing such testing. Capacitors are also used for voltage measurement purpose.

Reactors and Resistors

Resistors and reactors are used to control short-circuit test current. They also control power factor. The resistor controls the rate of decay of the dc component of the current. They control the transient recovery voltage.

14.19.2 Testing Procedure

During the short-circuit test, several switching operations are performed in a sequence in a very short time. For example, the sequence of switching operations for breaking capacity test is as follows.

(i) After running the driving motor of the short-circuit generator to a certain speed it is switched off.
(ii) Impulse excitation is switched on.
(iii) Master circuit breaker is closed.
(iv) Oscillograph is switched on.
(v) Making switch is closed.
(vi) Circuit breaker under test is opened.
(vii) Master circuit breaker is opened.
(viii) Exciter circuit of the short-circuit generator is switched off.

It is not possible to perform this sequence of operations manually. There is an automatic control for the purpose. The time of operation for the above sequence is of the order of 0.2 second.

14.19.3 Direct Testing

In direct testing, the circuit breaker is tested under the conditions which actually exist on power systems. It is subjected to restriking voltage which is expected in practical situations. Figure 14.30 shows an arrangement for direct testing. The reactor X is to control short-circuit current. C, R_1 and R_2 are to adjust transient restriking voltage. Short-circuit tests to be performed are as follows.

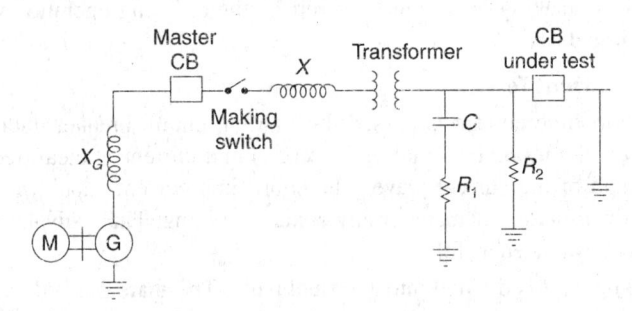

Fig. 14.30 *Direct testing of circuit breaker*

Test for Breaking Capacity

First of all, the master circuit breaker and the circuit breaker under test are closed. Then the short-circuit current is passed by closing the making switch. The short-circuit current is interrupted by opening the breaker under test at the desired moment. The following measurements are taken.

(i) Symmetrical breaking current
(ii) Asymmetrical breaking current
(iii) Recovery voltage
(iv) Frequency of oscillation and RRRV

The circuit breaker must be capable of breaking all currents up to its rated capacity. As it is not possible to test at all values of current, tests are performed at 10%, 30%, 60% and 100% of its rated breaking current.

Test for Making Capacity

The master circuit breaker and the making switch are closed first, then the short-circuit is initiated by closing the circuit breaker under test. The rated making current, i.e. the peak value of the first major loop of the short-circuit current wave is measured.

Duty Cycle Test

The following duty cycle tests are performed.

(i) B – 3 – B – 3 – B tests are performed at 10%, 30%, and 60% of the rated symmetrical breaking capacity.

(ii) B – 3 – MB – 3 – MB tests are performed (a) at not less than 100% of the rated symmetrical breaking capacity and (b) at not less than 100% of the rated making capacity.

This test can also be performed as two separate tests.
  (a) M – 3 – M make test
  (b) B – 3 – B – 3 – B break test
(iii) B – 3 – B – 3 – B tests are performed at not less than 100% of the rated asymmetrical breaking capacity.

Here B denotes the breaking operation and M denotes the making operation, while MB denotes the making operation followed by the breaking operation without any intentional time delay.

Short-time Current Test

The rated short-time current is passed through the circuit breaker under test for a specified short duration (1 second or 3 second) and current is measured by taking an oscillograph of the current wave. The short-time current should not cause any mechanical or insulation damage or any contact welding. The equivalent rms short-time current is evaluated as follows.

The time-interval is divided into 10 equal parts. These are marked as $t_0, t_1, t_2, \ldots t_9, t_{10}$. The asymmetrical rms values of the current at these intervals are marked as $I_0, I_1, I_2, \ldots I_9, I_{10}$. The equivalent rms value of the short-time current using Simpson formula is given by

$$I = \sqrt{\frac{1}{3}\left[I_0^2 + 4\left(I_1^2 + I_3^2 + I_5^2 + I_7^2 + I_9^2\right) + 2\left(I_2^2 + I_4^2 + I_6^2 + I_8^2 + I_{10}^2\right)\right]}$$

14.19.4 Indirect Testing

The testing to HV circuit breakers of large capacity also requires very large capacity of the testing station, which is uneconomical. It is also not practical to increase the short-circuit capacity of the testing station. Therefore, indirect methods of testing are used for the testing of large circuit breakers. The important indirect methods of testing are:
  (i) Unit testing
  (ii) Synthetic testing

Unit Testing

Generally, high voltage circuit breakers are designed with several arc interrupter units in series. Each unit can be tested separately. From the test results of one unit, the capacity of the complete breaker can be determined. This type of testing is known as unit testing.

Synthetic Testing

In this method of testing, there are two sources of power supply for the testing, a current source and a voltage source. The current source is a high current, low voltage source. It supplies short-circuit current during the test. The voltage source is a high voltage, low current source. It provides restriking and recovery voltage.

There are two methods of synthetic testing—parallel current injection method and series current injection method. Parallel injection method is widely used as it is capable of providing RRRV and recovery voltage as required by various standards.

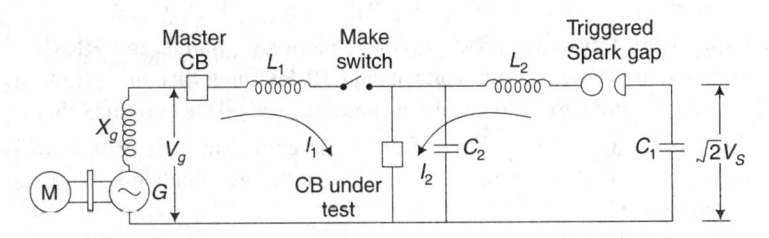

Fig. 14.31 *Synthetic testing of circuit breaker*

Figure 14.31 shows a circuit for synthetic testing. It is a circuit for parallel current injection method of synthetic testing. The high current source is a motor driven generator. It injects a high short-circuit current I_1 into the circuit breaker under test at a relatively reduced voltage, V_g. The inductance L_1 is to control the short-circuit current. The master circuit breaker and the circuit breaker under test are tripped before current I_1 reaches its natural zero. These circuit breakers are fully opened by the time t_0. The capacitor C_1 is a high voltage source to provide recovery voltage. It is charged prior to the test, to a voltage $\sqrt{2}\, V_s$. This voltage is equal to the peak power frequency voltage which will appear across the contacts at the moment the circuit breaker under test interrupts the current. L_2 and C_2 control transient recovery voltage and RRRV. The triggered spark gap is fired at t_1, slightly before the short-circuit current I_1 reaches its natural zero. It is done to properly simulate the pre-current zero zone during the test. There is a control circuit to fire the triggered spark gap at the appropriate moment. Figure 14.32 shows waveforms during synthetic testing.

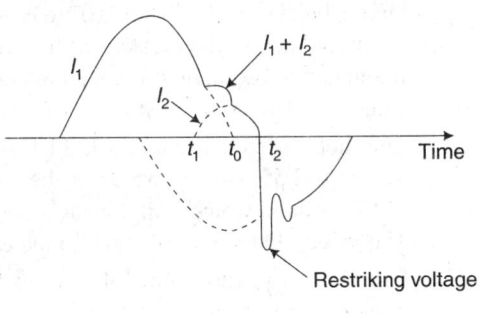

Fig. 14.32 *Waveform during synthetic testing*

EXERCISES

1. What are the different types of circuit breaker when the arc quenching medium is the criterion? Mention the voltage range for which a particular types of circuit breaker is recommended.
2. Discuss the recovery rate theory and energy balance theory of arc interruption in a circuit breaker.
3. What is the function of an explosion pot in an oil circuit breaker? What are the different types of explosion pot? Describe their construction, working principle and application.

4. Describe with neat sketches the working of the cross-jet explosion pot of an oil circuit breaker. Compare its merits and demerits with other types of arc control devices.

5. What are the advantages of an air blast circuit breaker over the oil circuit breaker?

6. Explain the terms: restriking voltage, recovery voltage and RRRV. Derive expressions for restriking voltage and RRRV in terms of system voltage, inductance and capacitance. What measures are taken to reduce them?

7. What is resistance switching? Derive the expression for critical resistance in terms of system inductance and capacitance, which gives no transient oscillation.

8. In a 132 kV system, the inductance and capacitance up to the location of the circuit-breaker are 0.4 H and 0.015 μF, respectively. Determine (a) the maximum value of the restriking voltage across the contacts of the circuit breaker, (b) frequency of transient oscillation and the maximum value of RRRV. [**Ans.** (a) 216 kV, (b) 2.06 kHz (c) 1.39 kV/μs]

9. In a 132 kV system, the reactance per phase up to the location of the circuit breaker is 5 Ω and capacitance to earth is 0.03 μF. Calculate (a) the maximum value of restriking voltage, (b) the maximum value of RRRV, and (c) the frequency of transient oscillation.
[**Ans.** (a) 216 kV, (b) 4.94 kV/μs, (c) 729 Hz]

10. In a 132 kV system, reactance and capacitance up to the location of the circuit breaker is 5 Ω and 0.02 μF, respectively. A resistance of 500 Ω is connected across the breaker of the circuit breaker. Determine the (a) natural frequency of oscillation, (b) damped frequency of oscillation, and (c) critical value of resistance. [**Ans.** (a) 8.7 kHz, (b) 3.9 kHz, (c) 445 Ω]

11. The short-circuit current of a 132 kV system is 8000 A. The current chopping occurs at 2.5% of peak value of the current. Calculate the prospective value of the voltage which will appear across the contacts of the circuit breakers. The value of stray capacitance to the earth is 100 pF. [**Ans.** 4.9 MV]

12. Explain the phenomenon of current chopping in a circuit breaker. What measures are taken to reduce it?

13. Discuss the problem associated with the interruption of (a) low inductive current, (b) capacitive current, and (c) fault current if the fault is very near the substation.

14. What are the different types of air blast circuit breaker? Discuss their operating principle and area of applications. Which type is less affected by current chopping?

15. With a neat sketch, describe the working principle of an axial air blast type circuit breaker. Explain why resistance switching is used with this type of circuit breaker.

16. In a multi-break circuit breaker, what measures are taken to equalise the voltage distribution across the breaks?

17. Discuss the properties of SF_6 which make it most suitable to be used in circuit breakers.
18. Briefly describe the various types of SF_6 circuit breakers.
19. Discuss the arc extinction phenomenon in SF_6 circuit breakers.
20. Discuss the operating principle of SF_6 circuit breaker. What are its advantages over other types of circuit breakers? For what voltage range is it recommended?
21. Describe the construction, operating principle and application of vacuum circuit breaker. What are its advantages over conventional type circuit breakers? For what voltage ranges is it recommended?
22. Briefly describe miniature circuit breaker (MCB) and moulded case circuit breaker (MCCB) what are their advantages over conventional breakers and fuse-switch units.
23. What are the various types of operating mechanisms which are used for opening and closing of the contacts of a circuit breaker? Discuss their merits and demerits.
24. Discuss the selection of circuit breakers for different ranges of the system voltage.
25. Enumerate various types of ratings of a circuit breaker. Discuss symmetrical and asymmetrical breaking capacity, making capacity and short-time current rating.
26. What are the main parts of equipment which are used for the testing of a circuit breaker in a laboratory type testing station?
27. Discuss how breaking capacity and making capacity of a circuit breaker are tested in a laboratory type testing station.
28. What are the different methods of testing of circuit breakers? Discuss their merits and demerits. Which method is more suitable for testing the circuit breakers of large capacity?
29. What is the difficulty in the development of HVDC circuit breaker? Describe its construction and operating principle.
30. In a 132 kV system, the inductance and capacitance per phase up to the location of the circuit breaker is 10 H and 0.02 μF, respectively. If the circuit breaker interrupts a magnetising current of 20 A (instantaneous), current chopping occurs. Determine the voltage which will appear across the contacts of the circuit breaker. Also calculate the value of the resistacne which should be connected across the contacts to eliminate the transient restriking voltage.

[**Ans.** 447.2 kV, 11.2 kΩ]

15 Fuses

15.1 INTRODUCTION

A fuse in its simplest form, is a small piece of thin metal wire (or strip) connected in between two terminals mounted on an insulated base and forms the weakest link in series with the circuit. The fuse is the cheapest, simplest and oldest protective device and is used as current interrupting device under overload and/or short circuit conditions. It is designed so that it carries the working current safely without overheating under normal conditions and melts due to sufficient i^2R heating when the current exceeds a certain predetermined value in abnormal conditions, caused by overloads or short circuits and thus interrupts the current. A fuse, being a thermal device, possesses inverse time-current characteristic, i.e. the operating time decreases as the fault current increases. Fuses are used to protect cables, electrical equipment and semiconductor devices against damage from excessive currents due to overloads and/or short circuits in low-voltage and medium voltage circuits. They are relatively economical, they do not require relays or instrument transformers and they are reliable. They are available in a large range of sizes and can be designed "one-shot" or as reusable devices with replaceable links. Modern HRC (High Rupturing Capacity) fuses provide a reliable discrimination and accurate characteristics. Recently, HRC fuses of voltage ratings up to 66 kV have been developed for application to distribution systems. The initial cost of a fuse is small but its installation, maintenance (against corrosion and deterioration) and replacement (periodically and after blowing) can be quite expensive (for example, on remote rural distribution networks).

15.2 DEFINITIONS

(i) Fuse

A fuse is a protective device used for protecting cables and electrical equipment against overloads and/or short-circuits. It breaks the circuit by fusing (melting) the fuse element (or fuse wire) when the current flowing in the circuit exceeds a certain predetermined value. The term fuse in general refers to all parts of the device.

(ii) Fuse Element (or Fuse Wire)

It is that part of the fuse which melts when the current flowing in the circuit exceeds a certain predetermined value and thus breaks the circuit.

Materials commonly used for fuse element (or fuse wires) are tin, lead, zinc, silver, copper, aluminium, etc. Practical experience has shown that the materials of low melting point such as tin, lead and zinc are the most suitable materials for the fuse element. But these materials of low melting point have high specific resistance as can be seen from Table 15.1. For a given current rating, the fuse element made from low melting point metal of high specific resistance will have greater diameter and hence, greater mass of the metal than those made from high melting point metal of low specific resistance. Thus, the use of low melting point metals for fuse elements introduces the problem of handling excessive mass of vaporized metal released on fusion. For small values of current, an alloy of lead and tin containing 37 per cent lead and 63 per cent tin is used. Either lead-tin alloy or copper is mostly used as ordinary fuse wire for low range current circuits.

For currents above 15A, lead-tin alloy is not found suitable as the diameter of the fuse wire will be large and after fusing, the vaporized metal released will be excessive. It has been found practically that silver is quite a suitable and satisfactory material for fuse elements, because it is not subjected to oxidation and its oxide is unstable and there is no deterioration of the material. The only drawback is that it is a costlier material. Despite being costlier, the present trend is to use silver as the material for fuses meant for reliable protection of costly and precious equipment.

Table 15.1 *Melting points and specific resistances of certain metals*

Metal	Melting point in °C	Specific resistance in microhm-cm
Tin	230	11.2
Lead	328	21.5
Zinc	419	6.1
Aluminium	670	2.85
Silver	960	1.64
Copper	1090	1.72

(iii) Rated Current
The rated current of a fuse is the current it can carry indefinitely without fusing.

(iv) Minimum Fusing Current
It is the minimum current (rms value) at which the fuse element will melt.

The minimum fusing current depends upon various factors, such as the material length, shape and area of cross-section of the fuse element, size and location of the terminals, the type of enclosure employed and the number of strands in the stranded fuse wire. The minimum fusing current for a stranded fuse wire will be less than the product of the minimum fusing current for one strand and the number of strands. For example, the typical values of fusing current for stranded fuse wires having 2, 3, 4 and 7 strands will be 1.66, 2.25, 2.75 and 4 times respectively of the fusing current for one strand.

An approximate value of the minimum fusing current for a round fuse wire is given by

$$I = K d^{3/2} \tag{15.1}$$

where K is a constant depending upon the material of the fuse wire, and d is the diameter of the wire. Preece has given the value of the constant K for different materials as shown in Table 15.2.

Table 15.2 Values of constant K for certain materials

Material	Values of constant K for d in cm
Tin	405.5
Lead	340.5
Iron	777
Aluminium	1870
Copper	2530

Table 15.3 gives the values of the rated current and approximate fusing current for tinned copper fuse wires of different S.W.G. and diameter.

Table 15.3 Rated and approximate fusing current for tinned copper fuse wires

S.W.G	Diameter in cm	Rated current in A	Approximate fusing current in A
20	0.09	34	70
25	0.05	15	30
30	0.031	8.5	13
35	0.021	5.0	8
40	0.012	1.5	3

(v) Fusing Factor

It is defined as the ratio of the minimum fusing current to rated current, i.e.

$$\text{Fusing factor} = \frac{\text{Minimum fusing current}}{\text{Rated current}}$$

This factor is always more than unity.

(vi) Prospective Current

Figure 15.1 shows the first major loop of the fault current. The prospective current (shown dotted) is the current which would have flown in the circuit if the fuse had been absent, i.e. it had been replaced by a link of negligible resistance. It is measured in terms of the rms value of the a.c. (symmetrical) component of the fault current in the first major loop. In Fig. 15.1, I_p is the peak value of the prospective current.

(vii) Cut-off Current

The current at which the fuse element melts is called the cut-off current (i_c in Fig. 15.1) and is measured as an instantaneous value. It is thus possible for a prospective current (rms value) to be numerically less than the cut-off current. Since the fault current normally having a large first loop generates sufficient heat energy, the fuse element melts well before the peak of the prospective current is reached.

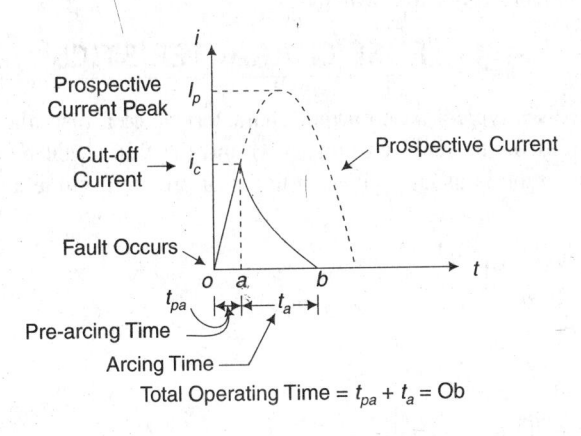

Fig. 15.1 Cut-off characteristic

Figure 15.1 Illustrates the cut-off action. On occurrence of a fault, the current starts increasing. It would have reached the peak of the prospective current (I_p), if no fuse were there to protect. But the cut-off action of the fuse does not allow the current to reach I_p. Because of the sufficient heat energy generated by the fault current, the fuse element is cut-off at i_c and an arc is initiated. After a brief arcing time, the current is interrupted. The cut-off current depends upon the rated current, the prospective current and the asymmetry of the fault current waveform. The cut-off property has a great advantage that the fault current does not reach the prospective peak.

(viii) Pre-arcing Time or Melting Time
This is the time taken from the instant of the commencement of the current which causes cut-off to the instant of cut-off and arc initiation. In Fig. 15.1, *oa* is the pre-arcing time (t_{pa}).

(ix) Arcing Time
This is the time taken from the instant of cut-off (arc initiation) to the instant of arc being extinguished or the current finally becoming zero. In Fig. 15.1, *ab* is the arcing time (t_a).

(x) Total Operating Time
It is the sum of the pre-arcing time and the arcing time. In Fig. 15.1, *ob* is the total operating time, i.e. $t_{pa} + t_a$.

(xi) Rupturing Capacity (or Breaking Capacity)
It is the MVA rating of the fuse corresponding to the largest prospective current which the fuse is capable of breaking (rupturing) at the system voltage. A fuse is never required to pass an actual current equivalent to its rupturing or breaking capacity. When a particular rupturing capacity is specified, it is expected that the fuse will successfully operate in a circuit having prospective current equivalent to its rupturing capacity; but the fuse melts much earlier due to cut-off action. Hence a fuse never allows to pass a current equivalent to its rupturing capacity.

15.3 FUSE CHARACTERISTICS

Figure 15.2 shows a typical time-current characteristic of a fuse, the current scale being in multiples of the rated current (= 1) and the time scale being logarithm. In practice, this graph is usually given in terms of pre-arcing time and prospective current.

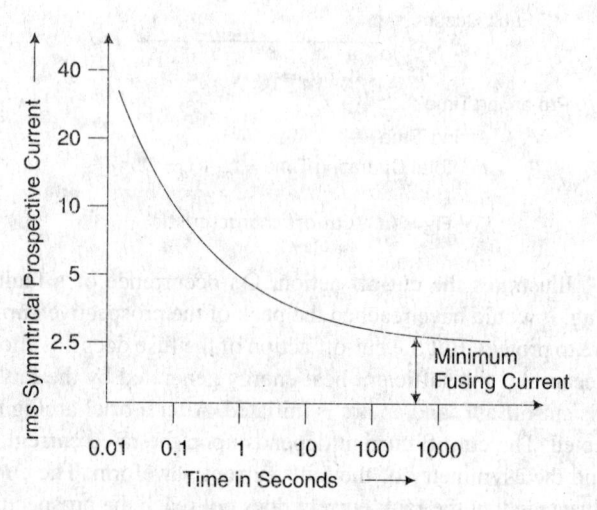

Fig. 15.2 *Typical time-current characteristic of a fuse*

By observation of the characteristic, it is clear that as the prospective current increases, the pre-arcing time decreases. It is also clear that the characteristic becomes asymptotic and there is a minimum current below which the fuse does not operate. This current is called the minimum fusing current. The operating time of the fuse for currents near the minimum fusing current is long.

15.4 TYPES OF FUSES

Following are two main types of fuses:

(i) Rewirable Type

This type of fuse is rewirable, i.e. the blown-out fuse element can be replaced by a new one. The fuse element can be either open or semi-enclosed. Therefore, this type of fuse is further of two types.

(a) Open type, and
(b) Semi-enclosed type.

(ii) Totally Enclosed or Cartridge Fuse

The fuse clement of this type of fuse is enclosed in a totally enclosed container and is provided with metal contacts on both sides. This type of fuse is also of two types.

(a) D-type cartridge fuse, and
(b) Link type cartridge fuse or high rupturing capacity (HRC) cartridge fuse.

Different types of fuses are described in the following sections.

15.4.1 Open Fuse

An open fuse element is a thin piece of wire of tin, lead or copper inserted directly in a circuit. It is the simplest and cheapest form of protection but due to fire hazard and unreliable operation caused by oxidation, it is seldom used any longer.

15.4.2 Semi-enclosed Rewirable Fuse

This fuse is most commonly used in house wiring and small current circuits. Figure 15.3 shows various components of this fuse. The fuse wire is fitted on a porcelain 'carrier' which is fitted in the porcelain base. Whenever the fuse wire blows off due to overload or short-circuit, the fuse carrier can be pulled out, the new wire can be placed and service can be restored. The fuse wire may be of lead, tinned copper or an alloy of tin-lead. The fuse wire should be replaced by a wire of correct size and to the specification, otherwise it may prove dangerous with the possibility of the equipment burning out.

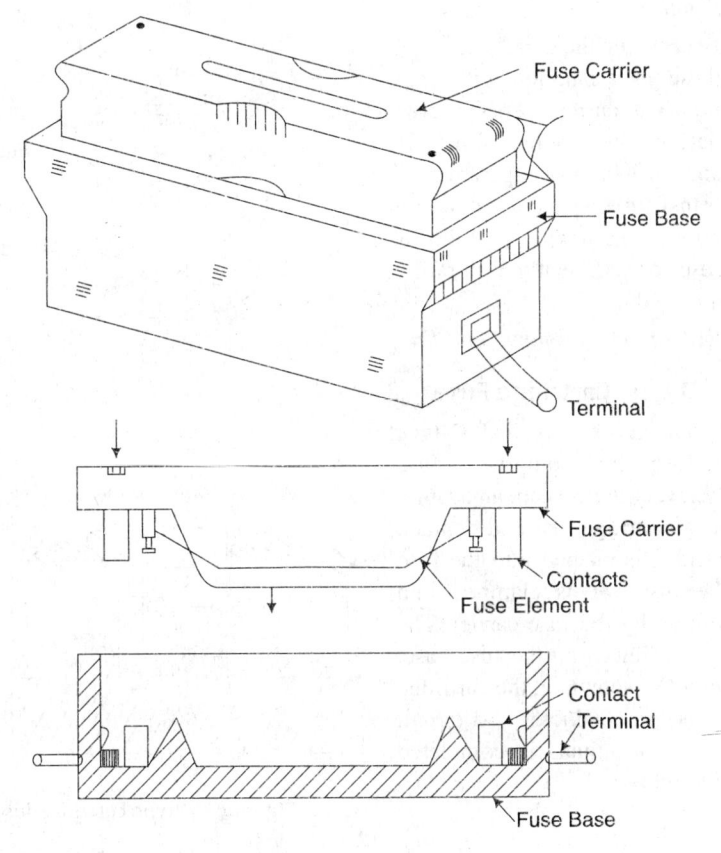

Fig. 15.3 *Semi-enclosed rewirable fuse*

Because of its simplicity and low cost, this type of fuse is extensively used as a protective device on low voltage circuits. But, since the fuse wire is exposed to atmosphere, it is affected by ambient temperature. As the fuse is affected by ambient condition, its time-current characteristic is not uniform and gets deteriorated with time, resulting in unreliable operation and lack of discrimination. Therefore, these fuses are mainly used for domestic and lighting loads. For all important and costly equipment operating at comparatively lower voltages (up to 33 kV), cartridge fuses are used because they provide more reliable protection.

Rewirable fuses suffer from the following disadvantages.

(i) *Unreliable operation*

The operation of these fuses is unreliable because of the following factors.

(a) Since the fuse wire is exposed to the atmosphere, it gets oxidised and deteriorated, resulting in a reduction of the wire section with the passage of time. This increases the resistance, causing operation or the fuse at lower currents

(b) Local heating caused by loose connection, etc.

(ii) *Lack of discrimination*

Proper discrimination cannot be ensured due to unreliable operation.

(iii) *Small time lag*

Because of the small time lag, these fuses can blow with large transient currents encountered during the starting of motors and switching-on operation of transformers, capacitors and fluorescent lamps, etc. unless fuses of sufficiently high rating are used.

(iv) *Risk of external flame and fire*

15.4.3 D-type Cartridge Fuse

Figure 15.4 shows a typical D-type cartridge fuse which comprises a fuse base, a screw type fuse cap, an adapter ring and a cartridge-like end contact. The cartridge is pushed into the fuse cap. The fuse-link is clamped into the fuse base by the fuse carrier. The fuse cap is screwed on the fuse base. After complete screwing, the cartridge tip touches the conductor and circuit between two terminals is completed through the fuse link.

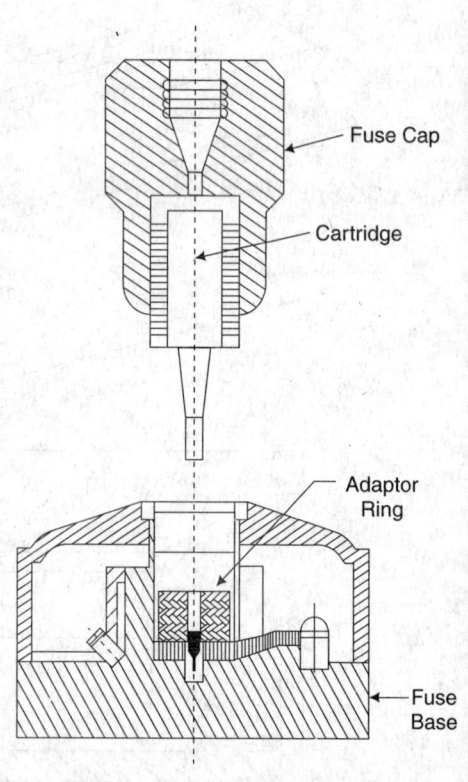

Fig. 15.4 *D-type cartridge fuse*

These fuses have none of the disadvantages of the rewirable fuse. Due to their reliable operation, coordination and discrimination to a reasonable extent can be achieved with them.

15.4.4 High Rupturing Capacity (HRC) Cartridge Fuse

The HRC fuses cope with increasing rupturing capacity on the distribution system and overcome the serious disadvantages suffered by the semi-enclosed rewirable fuses.

In an HRC fuse, the fuse element surrounded by an inert arc quenching medium is completely enclosed in an outer body of ceramic material having good mechanical strength. The unit in which the fuse element is enclosed is called 'fuse link'. The fuse link is replaced when it blows off. In its simplest form (Fig. 15.5), an HRC fuse consists of a cylindrical body of ceramic material usually steatite, pure silver (or bimetalic) element, pure quartz powder, brass end-caps and copper contact blades. The fuse element is fitted inside the ceramic body and the space within the body surrounding the element is completely filled with pure powdered quartz. The ends of the fuse element are connected to the metal end-caps which are screwed to the ceramic body by means of special forged screws. End contacts (contact blades) are welded to the metal end-caps. The contact blades are bolted on the stationary contacts on the panel.

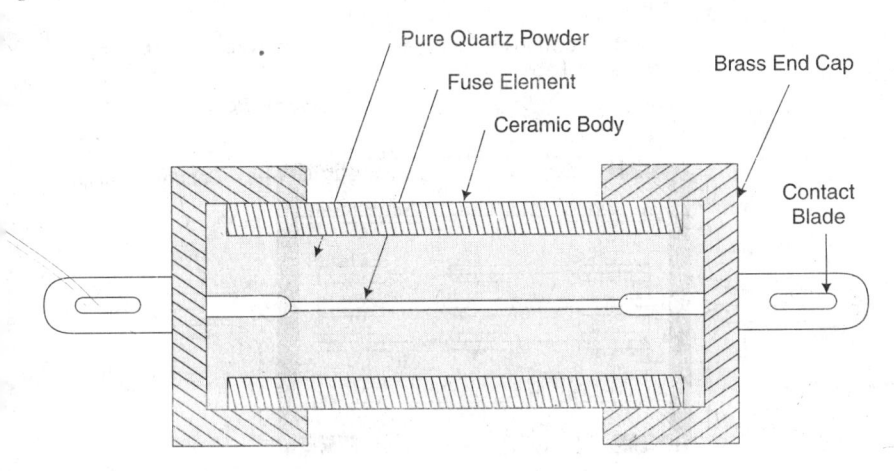

Fig. 15.5 *HRC fuse*

The fuse element is either pure silver or bimetalic in nature. Normally, the fuse element has two or more sections joined together by means of a tin joint. The fuse element in the form of a long cylindrical wire is not used, because after melting, it will form a string of droplets and an arc will be struck between each of the droplets. Later on these droplets will also evaporate and a long arc will be struck. The purpose of the tin joint is to prevent the formation of a long arc. As the melting point of tin is much lower than that of silver, tin will melt first under fault condition and the melting of tin will prevent silver from attaining a high temperature. The shape of the fuse element depends upon the characteristic desired.

When the fuse carries normal rated current, the heat energy generated is not sufficient to melt the fuse element. But when a fault occurs, the fuse element melts before the fault current reaches its first peak. As the element melts, it vaporizes and disperses. During the arcing period, the chemical reaction between the metal vapour and quartz powder forms a high resistance substance which helps in quenching the arc. Thus the current is interrupted.

Figure 15.6 shows an HRC fuse developed by General Electric Company. The cylindrical body made of ceramic material is closed by metal end-caps which carry the copper terminal tags. The brass end-caps and the copper terminal-tags are electro-tinned. The fuse element made of pure silver is surrounded by silica as the arc quenching medium. In order to increase the breaking capacity of the fuse, two or more widely separated fuse elements are used in parallel. The breaking capacity is increased due to a greater surface area of the fuse element in contact with the silica filler. An indicating device consisting of a fine resistance wire connected in parallel with the fuse elements and led through a small quantity of mild explosive held in a pocket in the side of the fuse and covered by a label is also provided. It indicates whether the fuse has blown out or not. When the fuse operates on occurrence of a fault, the fine wire is automatically fused resulting in the combustion of the explosive material. The combustion of the explosive material chars the label, and thus indicates that the fuse has blown out.

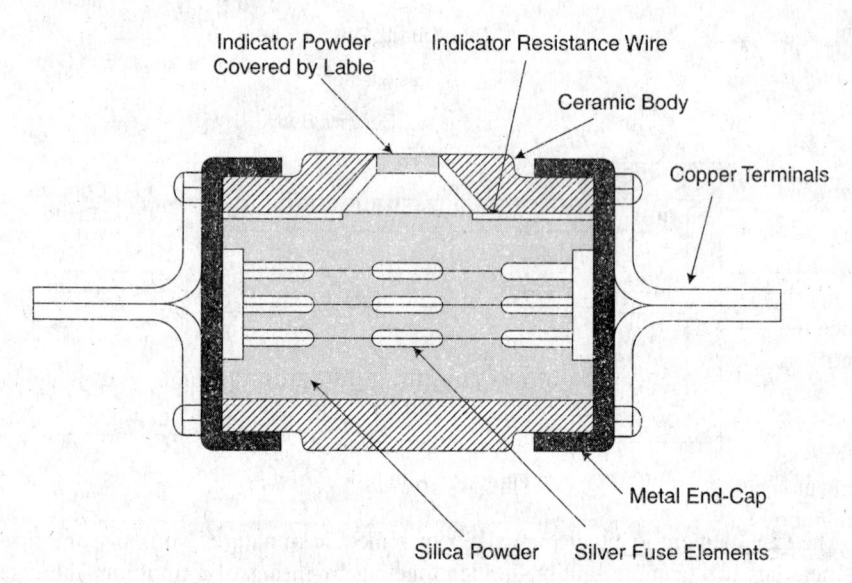

Fig. 15.6 *HRC fuse with indicator*

Advantages of HRC Fuses
The HRC cartridge fuses possess the following advantages.
  (i) Capability of clearing high values of fault currents
  (ii) Fast operation
  (iii) Non-deterioration for long periods

(iv) No maintenance needed
(v) Reliable discrimination
(vi) Consistent in performance
(vii) Cheaper than other circuit interrupting devices
(viii) Current limitation by cut-off action
(ix) Inverse time-current characteristic

Disadvantages of HRC fuses

The following are the main disadvantages of these fuses.
(i) It requires replacement after each operation.
(ii) Inter-locking is not possible.
(iii) It produces overheating of the adjacent contacts.

15.4.5 Expulsion Type High-Voltage Fuse

In an expulsion type fuse, the arc caused by operation of the fuse is extinguished by expulsion of gases produced by the arc.

An expulsion fuse contains a hollow tube made of synthetic-resin-bonded paper in which the fuse element is placed and the ends of the element are connected to suitable fittings at each end. The length of the tube is generally longer than the conventional enclosed fuses. On the occurrence of a fault, the arc produced by fusing of the fuse element causes decomposition of the inner coating of the tube resulting in the formation of gases which assist arc extinction. Such fuses are developed for 11 kV, 250 MVA and are commonly used for protection of distribution transformers, overhead lines and cables terminating with overhead lines.

15.4.6 Drop-out Fuse

This is also an expulsion type high voltage fuse with one pole in closed position (Fig. 15.7). When the fuse element gets fused on the occurrence of a fault, the fuse element-carrying tube drops down under gravity about its lower hinged support, and hence the operation of the fuse can be readily spotted from a distance. As the fuse element-carrying tube drops down on operation of the fuse, it provides additional isolation. The blown fuse element is replaced by lifting the complete tube from the hinge and bringing it down by means of a special insulated rod. After replacing the fuse element, the tube is replaced in the hinge and the device is closed in a way similar to closing of isolators. These fuses are extensively used for protection of outdoor transformers, and the fuse-isolator combination is generally pole mounted.

Fig. 15.7 Expulsion type drop-out fuse with one pole in closed position

15.5 APPLICATIONS OF HRC FUSES

The applications of HRC fuses are enormous but some very important applications are as follows.

(i) Protection of low voltage distribution systems against overloads and short-circuits
(ii) Protection of cables
(iii) Protection of busbars
(iv) Protection of motors
(v) Protection of semiconductor devices
(vi) Back up protection to circuit breakers.

15.6 SELECTION OF FUSES

The following points should be taken into account while selecting a fuse.

(i) It should be able to withstand momentary over-current due to starting a motor and transient current surges due to switching on transformers, capacitors and fluorescent lighting, etc.
(ii) Its operation must be ensured when sustained overload or short-circuit occurs.
(iii) It should provide proper discrimination with the other protective devices.
(iv) Its selection should depend upon the load circuit. For the purpose of protection, electric circuits are broadly classified as steady load circuits and fluctuating load circuits.

    (a) **Steady load circuits:** In these circuits, the load does not fluctuate much from its normal value (e.g. circuit of heating devices). If both overload and short-circuit protection of these circuits is required, fuses having a current-rating equal to or a little higher than the anticipated steady-load current are selected. However, if the fuse has to provide protection only against short-circuit, then a fuse of much greater rating than the normal load-current can be selected.

    (b) **Fluctuating load circuits:** In these circuits there are wide fluctuations of load and peaks of a comparatively short duration occur during starting or switching on. Motors, transformers, capacitors and fluorescent lighting are examples of this category of load. The main criterion for selection of a fuse for fluctuating loads is that the fuse should not blow under transient overloads. Hence the time-current characteristics of the fuse must be above the transient current characteristic of the load, with a sufficient margin.

15.6.1 Fuses for Motors

The fuses for protection of motors are selected keeping in view the starting current, its duration and frequency. When the starting-current of the motor is known, it is assumed that the starting-current surge will persist for about 20 seconds and a suitable cartridge fuse which will withstand the starting current for this time is selected.

When the starting current is not known, it is assumed to be about five times the full-load current.

15.6.2 Fuses for Capacitors

Protection of capacitors is a particularly difficult task due to the presence of transient current surges during switching operations. Therefore, cartridge fuse links having rated current about 50% greater than the rated currents of the capacitors are selected.

15.6.3 Fuses for Transformers and Fluorescent Lighting

The selected fuse should be capable of withstanding the transient current surges during switching on. Generally fuses having rated current about 25 to 50% greater than the normal full load current of the apparatus are selected.

15.7 DISCRIMINATION

When two or more protective devices, e.g. two or more fuses, a fuse and a circuit breaker, etc. are used for the protection of the same circuit, the term discrimination concerns the correct operation of the correct device on occurrence of a fault. For proper discrimination, there should be coordination between the protective devices. In order to obtain proper discrimination between two adjacent fuses carrying the same current, the pre-arcing time of the major fuse (nearer the source) must exceed the total operating time of the minor fuse (far from the source).

15.7.1 Discrimination between Two Fuses

In order to understand this type of discrimination, consider a radial circuit as shown in Fig. 15.8. The power to the radial line is being fed at the left end. The fuse F_1 which is nearer the feeder is called the major fuse, and the fuse F_2 which is far from the feeder is called the minor fuse. If a fault occurs at the far end of the feeder, i.e. at point F, the fault current will flow through both the fuses F_1 and F_2. If the fuses used do not have discriminative character, there is a likelihood that the major fuse F_1 is blown out and thus supply to the whole line will be interrupted, although there is no fault between F_1 and F_2. Therefore, when a fault occurs beyond F_2, only F_2 should operate and F_1 should remain unaffected. This is called proper discrimination. For proper discrimination in this case, the pre-arcing time of the major fuse F_1 must be greater than the total operating time of the minor fuse F_2.

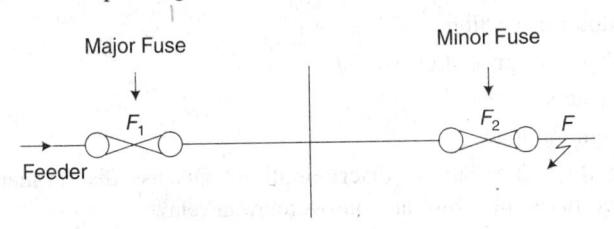

Fig. 15.8 *Discrimination between two fuses*

15.7.2 Discrimination between Fuses and Overcurrent Protective Devices

In motor circuits, fuses provide short-circuit protection and the overcurrent relay provides overcurrent (overload) protection. The characteristics of the fuses and overcurrent relay are coordinated in such a way that the overcurrent relay operates for currents within the breaking capacity of the circuit breaker (or the contactor), and the fuses operate for faults of larger current. For this purpose, the characteristic of the overcurrent protective device should be below the characteristic of the fuse, as shown in the Fig. 15.9. The fuse is so selected that the intersection of the characteristics of these two protective devices must take place at a point (A) corresponding to six times the full-load current, keeping in view that the protective devices do not operate unduly during starting. In this case, the fuse provides back-up protection to the motor and is connected on the supply side.

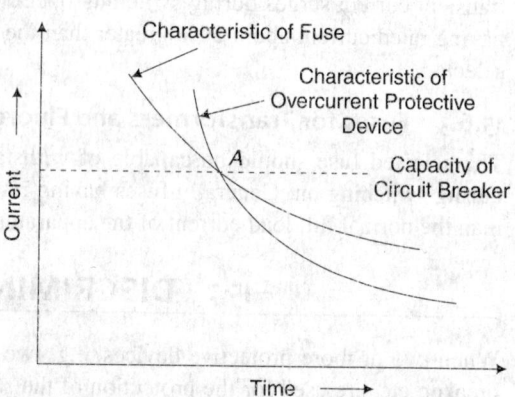

Fig. 15.9 *Discrimination between fuse and overcurrent protective device*

EXERCISES

1. Describe the construction and operation of the HRC cartridge fuse. What are its advantages and disadvantages?
2. Explain the following terms:
   (i) Minimum fusing current
   (ii) Rated current
   (iii) Fusing factor
   (iv) Prospective current
   (v) Cut-off current
3. Explain the points to be considered while selecting a fuse.
4. What are the considerations in selecting a fuse for
   (i) Transformer protection
   (ii) Motor protection
   (iii) Capacitor protection
   (iv) Heaters
   (v) Lighting loads
5. What do you mean by discrimination? Discuss discrimination between (a) two fuses (b) a fuse and an overcurrent relay.

6. Write short notes on
   (i) Semi-enclosed rewirable fuse
   (ii) D-type cartridge fuse
   (iii) Applications of HRC fuses
   (iv) Expulsion type high voltage fuse
   (v) Drop-out fuse

16 Protection Against Overvoltages

16.1 CAUSES OF OVERVOLTAGES

Overvoltages (or surges) on power systems are due to various causes. The normal operating voltages of the system do not stress the insulation severely. But the voltage stresses due to overvoltages can be so high that they may become dangerous to both the lines as well as the connected equipment and may cause damage, unless some protective measures against these overvoltages are taken. Overvoltages arising on a power system can be generally classified into two main categories as follows.

External Overvoltages These overvoltages originate from atmospheric disturbances, mainly due to lightning. These overvoltages take the form of a unidirectional impulse (or surge) whose maximum possible amplitude has no direct relationship with the operating voltage of the system. They may be due to any of the following causes.

 (i) Direct lightning strokes
 (ii) Electromagnetically induced overvoltages due to lightning discharge taking place near the line (commonly known as 'side stroke')
 (iii) Voltages induced due to changing atmospheric conditions along the line length
 (iv) Electrostatically induced overvoltages due to the presence of charge clouds nearby
 (v) Electrostatically induced overvoltages due to the frictional effects of small particles such as dust or dry snow in the atmosphere or due to change in the altitude of the line

Internal Overvoltages These overvoltages are caused by changes in the operating conditions of the network. Internal overvoltages can further be divided into two groups as follows.

 (i) **Switching overvoltages (or transient overvoltages of high frequency)**

 These overvoltages are caused by the transient phenomena which appear when the state of the network is changed by a switching operation or fault condition. These overvoltages are generally oscillatory and take the form of a damped sinusoid. The frequency of these overvoltages may vary from a few hundred Hz to a few kHz, and it is governed by the inherent capacitances and inductances of the circuit. For example, switching on and off of equipment, such as switching of high voltage reactors and switching of a transformer at

no load causes overvoltages of transient nature. Other examples of this type of overvoltage are—if a fault occurs on any phase, the voltage with respect to ground of the other two healthy phases can exceed the normal value until the fault is cleared; when the contacts of a circuit breaker open in order to clear the fault, the final extinction of current is followed by the appearance of a restriking voltage across the contacts whose characteristics are a high amplitude which may reach twice that of the system voltage, and a relatively high frequency.

(ii) **Temporary overvoltages (or steady-state overvoltages of power frequency)**

These overvoltages are the steady-state voltages of power-system frequency which may result from the disconnection of load, particularly in the case of long transmission lines.

Transient overvoltages arising on the power system are assessed by an overvoltage factor which is defined as the ratio of the peak overvoltage to the rated peak system-frequency phase voltage. This ratio is also referred to as the amplitude factor.

The examination of overvoltages on the power system includes a study of their magnitudes, shapes, duration, and frequency of occurrence. This study should be performed not only at the point where an overvoltage originates but also at all other points along the transmission network through which the surges may travel. It is also essential to know the causes and effects of these overvoltages, so that suitable protective measures can be taken.

16.2 LIGHTNING PHENOMENA

The discharge of the charged cloud to the ground is called lightning phenomenon. A lightning discharge through air occurs when a cloud is raised to such a high potential with respect to the ground (or to a nearby cloud) that the air breaks down and the insulating property of the surrounding air is destroyed. This raising of potential is caused by frictional effects due to atmospheric disturbances (e.g., thunderstorms) acting on the particles forming the cloud. The cloud and the ground form two plates of a gigantic capacitor whose dielectric medium is air. During thunderstorms, positive and negative charges are separated by the movement of air currents forming ice crystals in the upper layer of cloud and rain in the lower part. The cloud becomes negatively charged and has a larger layer of positive charge at its top. As the separation of charge proceeds in the cloud, the potential difference between concentrations of charges increases and the vertical electric field along the cloud also increases. The total potential difference between the two main charge centres may vary from 100 to 1000 MV. Only a part of the total charge is released to the ground by lightning, and the rest is consumed in inter-cloud discharges.

As the lower part of the cloud is negatively charged, the ground gets positively charged by induction. Lighting discharge requires breakdown of the dielectric medium, i.e. air between the negatively charged cloud and the positively charged ground. Though the electric field (potential gradient) required for the breakdown of air at STP is 30 kV/cm peak in a cloud where the moisture content in the air is large and also because of the high altitude (i.e., lower pressure) the electric field required for breakdown of air is only about 10 kV/cm. The lightning discharge mechanism is well explained with the help of Fig. 16.1.

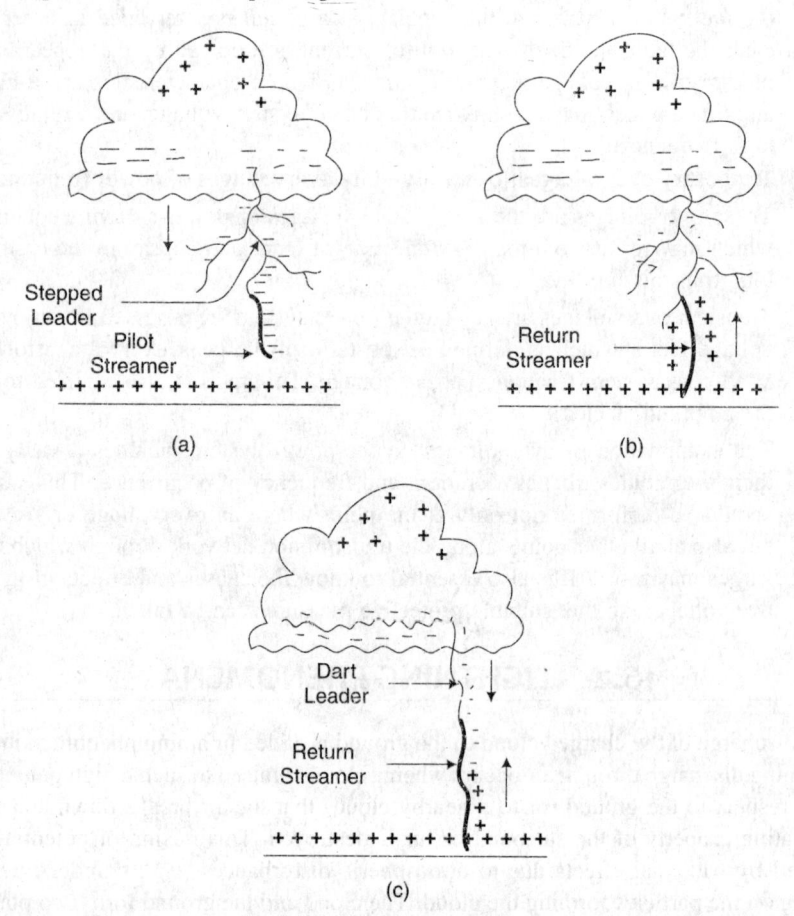

Fig. 16.1 *Lightning mechanism*

When a potential gradient of about 10 kV/cm is set up in the cloud, the air immediately surrounding the cloud gets ionised and the first process of the actual lightning discharge starts. At this instant, a streamer called a 'pilot streamer' starts from the cloud towards the ground which is not visible. The current associated with this streamer is of the order of 100 amperes, and the most frequent velocity of propagation of the streamer is about 0.15 m/μs (i.e. 0.05 per cent of the velocity of light). Depending upon the state of ionisation of the air surrounding the pilot streamer, it is branched into several paths, a stepped leader is formed, as shown in Fig. 16.1(a). It is called a stepped leader because of its zig-zag shape. It contains a series of steps about 50 m in length. The velocity of propagation of these steps should be more than 16.6 per cent that of light. A portion of the charge in the centre from which the strike originated is lowered and distributed over this entire system of temporary conductors. This process continues until one of the leaders strikes the ground. When one of the stepped leaders strikes the ground, an extremely bright return streamer, as shown in Fig. 16.1(b) propagates upward from the ground to the cloud following the same path as the main channel of the downward leader. The charge distributed along the leaders is thus discharged progressively to the ground, giving rise to the very large currents associated with lightning discharges. The conventional directions of the current in the stepped leader and return streamer are the same. The current varies between 1 kA and 200 kA and the velocity of propagation of the streamer is about 10 per cent that of light. It is here that the negative charge of the cloud is being neutralized by the positive induced charge on the ground. It is this instant which gives rise to the lightning flash which is visible with our naked eye.

After the neutralization of the most of the negative charge on the cloud any further discharge from the cloud may have to originate from another charge centre within the cloud near the already neutralized charge centre. Such a discharge from another charge centre will, however, make use of the already ionised path, and consequently it will have a single branch and will be associated with a high current. This streamer of discharge is called a dart leader as shown in Fig. 16.1(c). The velocity of propagation of the dart leader is about 3 per cent that of light. The dart leader can cause more severe damage than the return stroke.

Though the discharge current in the return streamer is relatively large but since it continues only for a few microseconds, it contains less energy and hence this streamer is called a cold lightning stroke. The dart leader is called a hot lightning stroke because even though the current in this leader is relatively smaller, it contains relatively more energy since it continues for some milliseconds.

It has been observed that every thundercloud may consist of as many as 40 separate charged centres due to which a heavy lightning stroke may occur. This lightning stroke is called multiple (or repetitive) stroke.

When a lightning stroke takes place to a point adjacent to the line, the current associated with it induces a voltage on the line which is in the form of a travelling wave. A lightning discharge may have currents in the range of 10 kA to about 100 kA, and the duration of the lightning flash may be of the order of a few microseconds. The current wave form is generally a unidirectional pulse which rises to a peak value in about 3 μs and decays to a small values in several tens of microseconds. Therefore, the overvoltage due to lightning will have travelling waves of very steep wave front which may be dangerous to the line and equipment of the power system and may cause damage.

16.3 WAVE SHAPE OF VOLTAGE DUE TO LIGHTNING

Lightning sets up steep-fronted, unidirectional voltage waves which can be represented as the difference of two exponentials. Thus

$$v = V(e^{-at} - e^{-bt}) \tag{16.1}$$

where a and b are constants which determine the shape and v is magnitude of the steep voltage and V is equal to the crest (peak) value of the impulse voltage wave. The steep is dependent on whether the surge is induced or is the result of a direct stroke. The wave shape of this type is shown in Fig. 16.2.

The wave shape is generally defined in terms of the times t_1 and t_2 in microseconds. Where t_1 is the time taken by the voltage wavefront to reach its peak value, and t_2 is the time taken for the tail to fall to 50 per cent of the peak value. Both times are measured from the virtual zero time point (i.e., from the start of the wave). The wave is then referred to as t_1/t_2 wave. The standard wave chosen for the testing purpose is a 1/50 wave which implies that t_1 is 1 μs and t_2 is 50 μs.

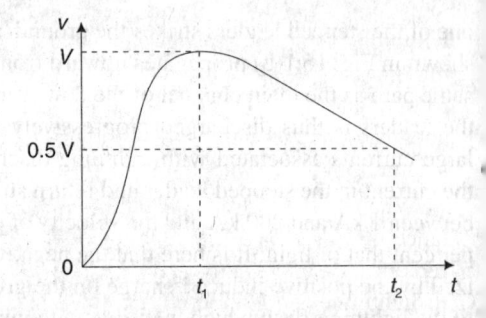

Fig. 16.2 Wave shape of voltage due to lightning

16.4 OVERVOLTAGES DUE TO LIGHTNING

Lightning causes two kinds of voltage surges (overvoltage), one by direct stroke to a line conductor, and the other induced by indirect stroke when bound charges are dissipated following a lightning discharge to an object near the line conductor. A direct stroke to a phase (line) conductor is the most severe lightning stroke as it produces the highest overvoltage for a given stroke current. The current in a direct stroke to a phase conductor is hardly affected by the voltage which it sets up between the conductor and ground, so that a direct lightning stroke approximates to a constant current source. If a current I flows in a direct stroke to a phase conductor, forward and backward travelling currents i_f and i_b, each of magnitude $I/2$ flow in the conductor away from the point of strike in both directions of the line as shown in Fig. 16.3. Therefore, the voltage surge magnitude at the striking point is given by

$$V = \frac{I Z_0}{2} \qquad (16.2)$$

where Z_0 is the surge impedance of the line.

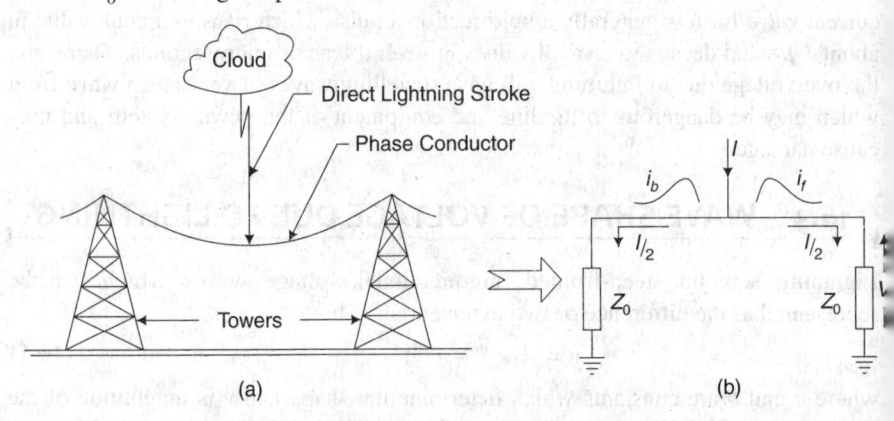

Fig. 16.3 Development of lightning overvoltage

The lightning current magnitude (I) is rarely less than 10 kA. Therefore, for a typical overhead line having a surge impedance Z_0 of 400 Ω, the lightning surge voltage will probably have a magnitude in excess of 2000 kV. Any overvoltage surge appearing on a transmission line due to internal or external disturbances propagates in the form of a travelling wave towards the ends of the line. The lightning surge is rapidly attenuated, partly due to corona as it travels along the line. In this way, corona acts as a natural safety valve to some extent.

When a tower of the transmission system is directly struck by lightning, the resistance offered to the lightning current is that of the tower footing and any ground rods or counterpoise wires in parallel. In calculation of overvoltage developed due to lightning in this case, certain assumptions must be made concerning several of the factors such as wave shape and magnitude of lightning current, surge impedance of lightning stroke channel, shape of potential wave at tower top, effect of surges on tower and footing impedance, etc. All these assumptions can be thrown into the lightning current and tower footing resistance, thus giving

$$V = RI \qquad (16.3)$$

where, V = voltage across insulation at tower, R = tower footing resistance, and I = lightning current in tower.

For very high towers, such as the ones at river crossings, which are subjected to direct strokes of very high rate of change of current, the tower inductance may be taken into account. The voltage (v) which may be sufficient to cause insulator flashover is given by

$$v = L\frac{di}{dt} \qquad (16.4)$$

where, L = tower inductance, and di/dt = rate of change of current due to direct stroke.

In the case of an indirect stroke, a negative charge in a cloud base causes bound positive charges on the conductors of a nearby transmission line, and discharge of the cloud due to a lightning stroke in the neighbourhood causes the conductor potential to rise to v since high insulation resistance prevents rapid dissipation of the charge. The peak value of the induced voltage surge is given by

$$v = Eh \qquad (16.5)$$

where, E is the mean electric field near the ground (earth) surface under a thundercloud which may vary from 0.5 kV/cm to 2.8 kV/cm, and h is the height of the phase conductor above ground.

If i_f be the forward travelling wave or the current and i_b be the backward travelling wave of the current, then in a case where the total conductor current $i = 0$, we have $i_f = i_b$ and $V_f = V_b = v/2$. So that forward and backward travelling waves of voltage and current set-off in opposite directions along the transmission line from the place nearest to the cloud discharge.

The amplitudes of voltages induced on lines indirectly by lightning strokes on a tower, ground wire or nearby ground or object, are however normally much less than those caused by direct strokes to lines. Such strokes, however, are of real concern on low voltage lines (33 kV and below) supported on small insulators. They are of little importance on high-voltage lines whose insulators can withstand hundreds of kilovolts without a flashover.

16.5 KLYDONOGRAPH AND MAGNETIC LINK

16.5.1 Klydonograph

The klydonograph is an instrument for the measurement of surge voltage on transmission lines caused by lightning. It measures voltage by means of Lichtenberg figures, when suitably coupled to the line whose surge voltage is to be measured. The klydonograph contains a rounded electrode connected to the line whose surge voltage is to be measured. The electrode rests on the emulsion side of a photographic film or plate, which in turn rests on the smooth surface of an insulating plate made of homogeneous insulating material, backed by a metal plate electrode as shown in Fig. 16.4.

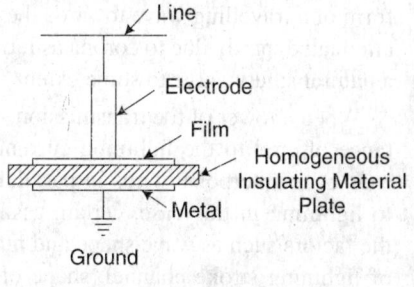

Fig. 16.4 *Klydonograph*

The photographic plate or the film is turned or moved by a clockwork mechanism for bringing in the element of time. Three assemblies are generally placed in the same box, for simultaneously measuring the voltages on the three phases of a transmission line.

With this arrangement, a positive Lichtenberg figure is produced by a positive surge, and a negative Lichtenberg figure by a negative surge, as illustrated in Fig. 16.5. Positive Lichtenberg figures are found to be superior to the negative ones for voltage measurement purpose, since they are much larger than the negative figures for the same voltage, as shown in Fig. 16.5. Diameter of positive Lichtenberg figure is a

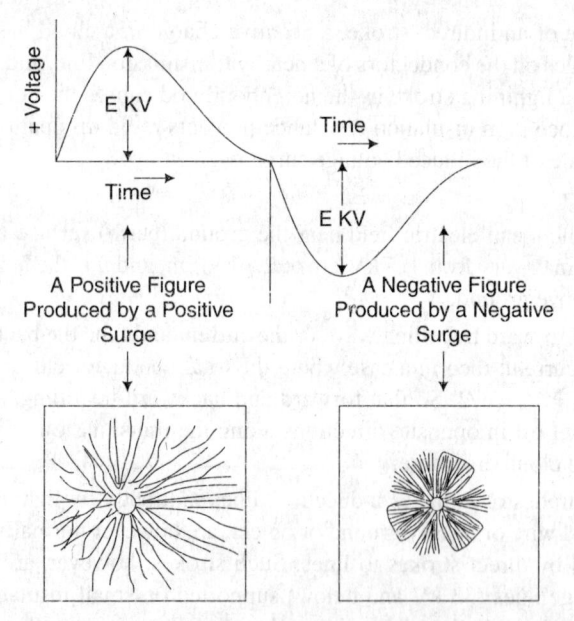

Fig. 16.5 *Positive and negative Lichtenberg figures produces by positive and negative surge voltages of same magnitude and wave shape*

function of the maximum value of the impressed voltage. The shape and configuration of the figure depend on the wave shape of the impressed voltage.

16.5.2 Magnetic Link

The magnetic link is an instrument for the measurement of surge currents due to lightning. It contains a small bundle of laminations made of cobalt-steel inserted in a cylindrical molded plastic container, with the open end sealed in. It is placed in an unmagnetised state in the vicinity of the conductor whose surge current is desired to be measured.

When the current flows through the conductor, a magnetic field is set up around it which is proportional to the current. The magnetising force of this field magnetizes the magnetic link which is placed in the vicinity of the conductor. After the current has passed, the link is left with a residual magnetism which is a function of the magnitude of the current producing it for unidirectional surges. The residual magnetism may be measured by a suitable instrument, and the magnitude of the current which produces it may be determined by an experimentally obtained calibration curve of current versus residual magnetism, or deflection on the instruments.

The principle of operation of the magnetic link is based on the assumption that the surge current is unidirectional in polarity and that surge currents of any duration, no matter how short, leave the same residual magnetism as a direct current of the same magnitude. The magnetic link is extensively used to measure currents in direct strokes, in transmission-line tower legs, ground wires, phase conductors, in counterpoise conductors and in ground leads of arresters.

Magnetic links are generally placed in brackets fastened to the conductor so that their axes coincide with the normal direction of the magnetic lines of force.

16.6 PROTECTION OF TRANSMISSION LINES AGAINST DIRECT LIGHTNING STROKES

Direct lightning strokes are of concern on lines of all voltage classes, as the voltage that may be set up is in most instances limited by the flashover of the path to ground. Increasing the length of insulator strings merely permits a higher voltage before flashover occurs. Therefore protective methods must be adopted to avoid flashover or breakdown of line insulators due to overvoltage caused by direct lightning strokes.

The most generally accepted and effective method of protecting lines against direct strokes is by the use of overhead ground wires. This method of protection is known as shielding method which does not allow an arc path to form between the line conductor and ground.

Ground wires are conductors running parallel to the main conductors of the transmission line, supported on the same towers (or supports) and adequately grounded at every tower or support. They are made of galvanised steel wires or ACSR conductors. They are provided to shield the lines against direct strokes by attracting the lightning strokes to themselves rather than allowing them to strike the lines (phase conductors). When a ground wire is struck by direct lightning stroke, the impedance through which the current flows is very much reduced and a correspondingly higher current is required to cause flashover.

Overhead ground wires protect the lines by intercepting the direct strokes, keeping them off the phase conductors (lines), and by providing multiple paths for conducting the strokes current to ground. The multiple paths for the stroke current result in a reduced voltage drop. When the ground wire is struck in midspan, the current gets divided and flows towards both towers. At the tower, the current again gets divided between the tower and the outgoing ground wire. When the tower is struck by a direct lightning stroke and there is only one overhead ground wire, the current gets divided into three-path, one through the tower and two through the branches of the ground wire. In addition to the above functions, ground wires also reduce the voltages induced on the conductors from nearby strokes by increasing the capacitance between conductors and ground.

In order to provide efficient protection to lines against direct strokes, the ground wires must satisfy the following requirements.

(i) There should be an adequate clearance between the line conductors and the ground or the tower structure.
(ii) There should be an adequate clearance between the ground wires and line conductors, especially at the midspan, in order to prevent flashover to the line conductors up to the protective voltage level used for the line design.
(iii) The tower footing resistance should be as low as economically justifiable.

The ground wires should be placed at such a height above the conductors and so located that they should be well out on towers and should not be exactly over the conductors in order to avoid any possibility of a short circuit occurring in the event of conductors swinging under ice loading, etc.

A few terms relating to protection of transmission lines are explained below.

Protective Ratio

The protective ratio is defined as the ratio of the induced voltage on a conductor with ground wire protection to the induced voltage which would exist on the conductor without ground wire protection. The induced voltage under discussion is the induced voltage between phase conductor and ground due to nearby stroke.

Protective Angle

The protective angle (α) afforded by a ground wire is defined as the angle between a vertical line through the ground wire and a slanting line connecting the ground wire and the phase conductor to be protected. This angle is shown in Fig. 16.6.

The protective angle (α) is in the region of 20° to 45°. Experience with lines and also tests have shown that when the protective angle does not exceed 30° (on level spans), good shielding of the line conductors is obtained and the probability of direct stroke to the conductor is also reduced. But a protective angle of 45° is satisfactory when the tower is on a hill-side. The angle should be decreased by the angle of the slope of the hill.

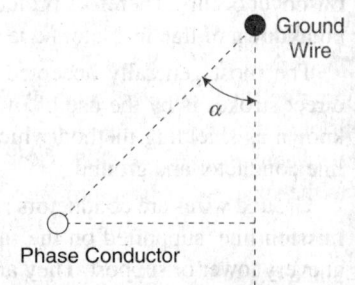

Fig. 16.6 *Protective angle α of an overhead ground wire*

Protective Zone

The protective zone is defined as the volume between the base plane cbc and the slanting planes ac, extending from the ground wire to the plane of the conductors. Figure 16.7(a) shows the cross-section of this volume. The plane ac cuts the base plane at c, at a distance ky from the point b, vertically under the apex a. The ground wire is at a height $ab = y$ above the base plane. The ratio $ky/y = k$ is called the protective ratio. The protective zone is sometimes called the protective wedge or protective shed.

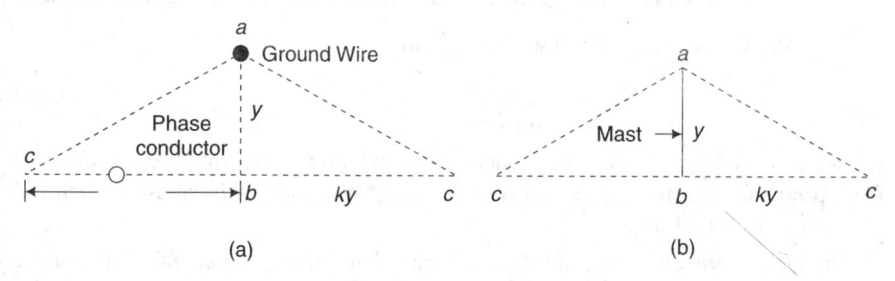

Fig. 16.7 (a) Zone of protection of an overhead ground wire
(b) Cone of protection of a rod or mast

If the line ab in Fig. 16.7(b) represents a rod or mast with a height y, then it is said to be protective cone around the mast. The radius on the base line is equal to ky and the protective ratio is $ky/y = k$.

Height of Ground Wire

For protection against direct strokes, the ground wire should be located at a height at least 10 per cent greater than y, calculated from the following equation.

$$\frac{X}{H} = \sqrt{2\left(\frac{y}{H}\right) - \left(\frac{y}{H}\right)^2} - \sqrt{2\left(\frac{h}{H}\right) - \left(\frac{h}{H}\right)^2} \qquad (16.6)$$

where, X = horizontal spacing between the conductor and ground wire, H = height of cloud, y = height of ground wire, and h = height of conductor.

Though recommendations about the protective angle and protective ratio vary, experience has shown that double-circuit lines will be completely shielded if two ground wires arc used, placed one above each outermost conductor, at a height above the top conductor equal to the vertical spacing between conductors. For horizontally arranged single-circuit lines, two overhead ground wires will give compete shielding if placed above the plane of the conductors at a height about two-thirds the spacing between conductors. The distance between the ground wires should be about equal to the spacing of conductors.

Coupling Factor

Since a voltage of the same polarity as the ground wire voltage has induced on the conductors, the entire ground wire or tower top voltage does not appear across the line insulation. The ratio of this induced voltage on the conductor to the ground wire voltage is known as the coupling factor. Therefore, the voltage across the line insulation is the ground wire voltage times the coupling factor, neglecting the normal frequency voltage.

The electromagnetic and electrostatic coupling factors are equal unless the effective radius of the ground wire is increased by a corona, in which case, the electrostatic coupling increases and the electromagnetic coupling remains unaffected. The coupling factor with the effect of corona considered is calculated by the following equation.

$$\text{Coupling factor } (C) = \sqrt{\text{electrostatic coupling} \times \text{electromagnetic coupling}}$$

Coupling factor is calculated by the equation,

$$C = \frac{\log b/a}{\log 2h/r} \qquad (16.7)$$

where, C = coupling factor, a = distance from conductor to ground wire, b = distance from conductor to image of ground wire, h = height of ground wire above ground and r = radius of ground wire.

The electromagnetic coupling factor is calculated using the actual radius of the ground wire and the electrostatic coupling factor is calculated by the same equation using the effective radius of the ground wire due to corona.

Reduction of Tower Footing Resistance

The effectiveness of a ground wire in controlling the lightning overvoltage on a line to reasonable values depends to a great extent on low tower footing resistance. The tower-footing resistance should be in the region 10-20 Ω, preferably closer to 10 Ω. For medium voltage lines it may be even less than 10 Ω. Therefore, sometimes some special means are employed to reduce the tower footing resistance, since by itself the tower footing may not be able to give a low resistance. The tower footing resistance may be reduced by driving rods near the tower and connecting them to the tower base or by burying conductors (counterpoise wires) in the ground and connecting them to the tower base. Driven rods are usually of galvanised iron, copper weld, or copper-bearing steel. Sometimes driven galvanised iron pipe is also used. Counterpoise wires are usually of copper, aluminium, galvanised steel, stranded cable, or galvanized steel strip.

The method to be used for reducing the tower footing resistance either by driving rods or by counterpoise wires depends upon the soil conditions. Either method is satisfactory if the proper low resistance is obtained. Deep rods may be driven only in soil free from rocks. Deep-driven rods can be successfully used in the sandy soil.

As regards counterpoise wires, the hand-trenching method may be used for small amounts of wire. However, if the amount of wire is considerable, it is usually installed by feeding into a trench. The trench is shallow, usually 30 to 60 cm deep, and does not require extensive filling.

Counterpoise wires are used extensively because of the difficulty encountered in driving rods. They can be arranged either radially (non-continuous) or tower-to-tower (continuous). Typical arrangements of radial counterpoise wires are shown in Fig. 16.8 and tower-to-tower counterpoise wires are shown in Fig. 16.9.

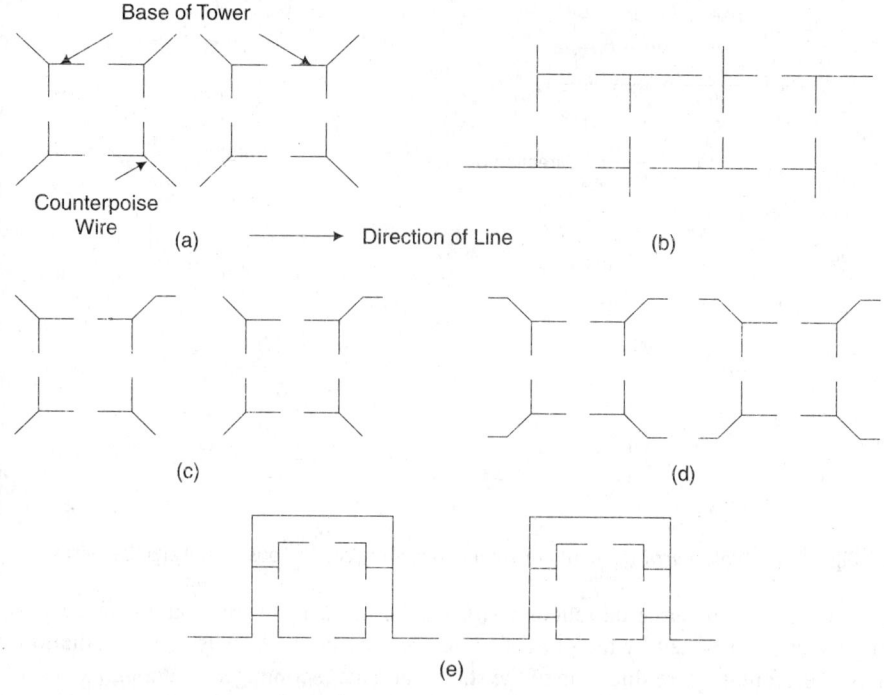

Fig. 16.8 *Typical arrangements of radial (non-continuous) counterpoise wires*

Continuous counterpoise wires are more effective than non-continuous counterpoise wires. Current is drawn into the counterpoise from 1.5 to 2.5 km away from the point of the stroke and fed into the stroke by means of the counterpoise wire and the overhead ground wires. The concentration of the current at any tower is thus reduced. The reduced current at any tower causes reduced voltage drop at the tower base, and thus the tower potential is held to a value below the flashover voltage of the line insulators.

16.7 PROTECTION OF STATIONS AND SUB-STATIONS FROM DIRECT STROKES

The station structure itself is commonly built of steel members on which lines terminate and which support the buses, isolating switches, etc. The power transformers and circuit breakers usually rest on concrete pads on the ground.

The task of protection of such stations can be resolved into two parts. The first being the protection of stations (including sub-stations) from direct lightning strokes, and the second is the protection of electrical equipment of stations from travelling waves coming in over the lines. The protection of stations from direct strokes will be discussed in this section and the protection of equipment from travelling waves will be discussed in the subsequent section.

Wherever it is likely for direct strokes of lightning to strike the line at or near the station, there is a possibility of exceedingly high rates of surge-voltage rise and

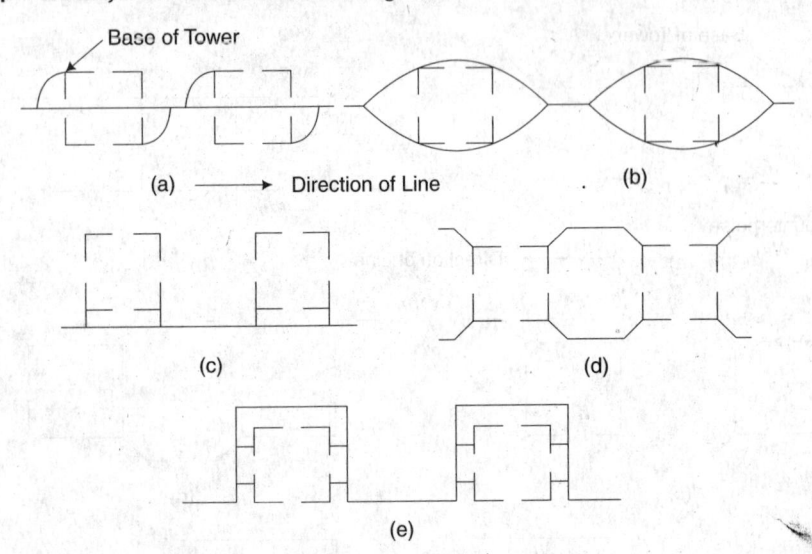

Fig. 16.9 *Typical arrangements of tower-to-tower (continuous) counterpoise wires*

large magnitude of surge-current discharge. If the stroke is severe enough, the margin of protection provided by the protective device may be inadequate. The installation may, therefore, require shielding of the station and the incoming lines enough to limit the severity of surges particularly in the higher voltage class of 66 kV and above. This can be done by properly placed overhead ground wires, masts, or rods.

Therefore protection of stations from direct strokes involves the principles already discussed in Sec. 16.6. If supports are available for overhead ground wires, these may be run over the station in such a way that the station and all apparatus will lie in the protected zone. Sufficient ground wires should be strung so that the entire station is covered, including any apparatus outside the main structure and any apparatus on top of the structure as shown in Fig. 16.10.

For a small station, it is sometimes sufficient to run one or two ground wires across the station from adjacent line towers as shown in Fig. 16.11(a). For a more extensive station, extra ground wire may be used, fanning out from the towers to the station structure, as also over the station if mechanically feasible as shown in Fig. 16.11(b). If there are no isolating switches or other apparatus above the station structure, then the steel of this structure will act as a grounded object to protect the apparatus and buses below it.

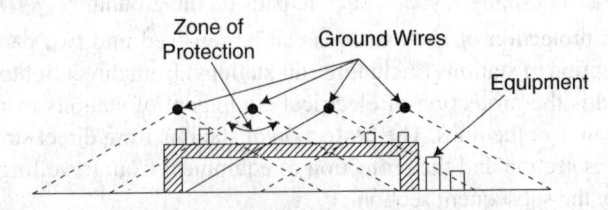

Fig. 16.10 *Protection of a station by overhead ground wires (elevation view)*

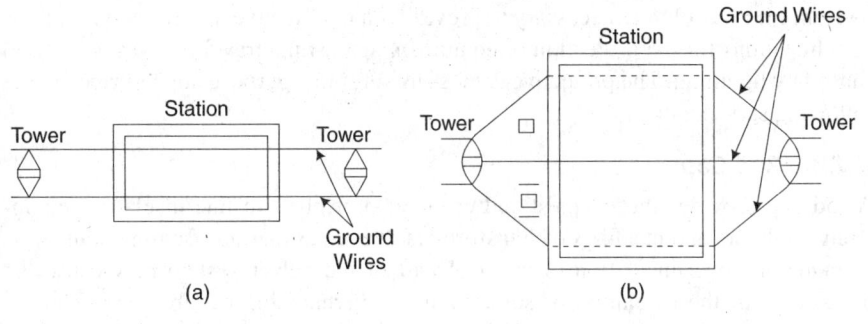

Fig. 16.11 *(a) Overhead ground wires carried over a small station between adjacent towers (plan view) (b) Overhead ground wires carried over a large station between adjacent towers (plan view)*

If it is not feasible to run ground wires over the station, then it may be possible to erect masts or rods at the corners or over vertical columns, so that the buses and apparatus in the station will fall within the cone of protection of the masts or rods as in Fig. 16.7(b). Such a cone may be assumed to have a base radius equal to twice the height of the mast. Figure 16.12 illustrates this method of protection.

16.8 PROTECTION AGAINST TRAVELLING WAVES

Shielding the lines and stations (including sub-stations) by overhead ground wires, masts or rods undoubtedly provides adequate protection against direct lightning strokes and also reduces electrostatically or electromagnetically induced over voltages, but such shielding does not provide protection against travelling waves which may come in over the lines and reach the terminal equipment. Such travelling waves can cause the following damage to electrical equipment.

(i) Internal flashover caused by the high peak voltage of the surge may damage the insulation of the winding.

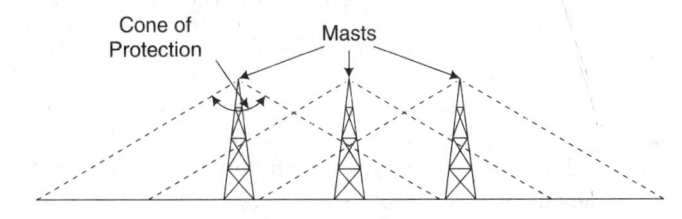

Fig. 16.12 *Method of protecting station by vertical masts (elevation view)*

(ii) Internal flashover between interturns of the transformer may be caused by the steep-fronted voltage wave.

(iii) External flashover between the terminals of the electrical equipment caused by the high peak voltage of the surge may result in damage to the insulators.

(iv) Resonance and high voltages resulting from the steep-fronted wave may cause internal or external flashover of an unpredictable nature causing building up of oscillations in the electrical equipment.

Hence it is absolutely necessary to provide some protective devices at the stations or sub-stations for the protection of equipment against the travelling waves (surges) caused by lightning. The protective devices used for this purpose are described in the following sub-sections.

16.8.1 Rod Gap

A rod gap provides the simplest and cheapest protection to line insulators, equipment insulators and bushings of transformers. It is clear that in case of serious over-voltages, the over-insulation of any one part of the power system may cause the breakdown of the insulation of some vital and perhaps inaccessible part. Thus, it is preferred that it is the line insulators that flashover rather than the bushings of a transformer breaking down. Again, it is preferred that a bushing breakdown before the insulation of the transformer of which it forms a part. In the case of transformers, rod gaps which are also called coordinating gaps are installed to protect the apparatus. Rod gaps provide back-up protection to the bushings of transformers in case of the primary protective devices, i.e. lightning arresters fail.

A rod gap consists of two rods of approximately 1.2 cm diameter or square, which are bent at right angles as shown in Fig. 16.13. One rod is connected to the line while the other rod is connected to ground. In case of a transformer, they are fixed between bushing insulators. In order to avoid cascading across the insulator surface under very steep-fronted waves, the rod gap should be adjusted to breakdown at about 20 per cent below the impulse flashover voltage of the insulation of the equipment to be protected (i.e. bushing insulator of a transformer in Fig. 16.13). Further, the distance between the gap and the insulator should be more than one-third of the gap length $\left(\text{i.e. } L_1 > \dfrac{L}{3}\right)$ in order to prevent the arc from being blown to the insulator. The accurate breakdown value of the rod gap cannot easily be predicted because the breakdown of air depends upon the atmospheric conditions (i.e., humidity, temperature and pressure), as also upon polarity, steepness and the waveshape of the wave.

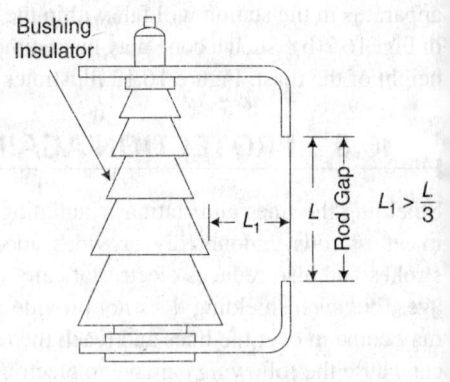

Fig. 16.13 Rod gap

The major disadvantage of the rod gap is that it does not interrupt the power frequency follow current after the surge has disappeared. This means that every operation of the rod gap creates an L-G fault which can only be cleared by the operation of the circuit breaker. Thus, the operation of the rod gap results in circuit outage and interruption of power supply.

16.8.2 Arcing Horn

The damage to line insulators from heavy arcs formed due to overvoltages is a serious maintenance problem. Several protective devices have been developed to keep an insulator string free from the arc. Arcing horn is one of such protective devices. It consists of small horns attached to the clamp of the line insulator string. Horns with a large spread, both at the top of the insulator and at the clamp are required to be effective, as shown in Fig. 16.14. In the case of lightning impulse, the arc formed tends to cascade the string. In order to avoid this cascading, the gap between horns should be considerably less than the length of the string. Protection of line insulators by arcing horns thus results in reduced flashover voltage. In any event, flashover persists as a power arc until the line trips out. The protection of line insulators by arcing horns is especially used in hilly areas.

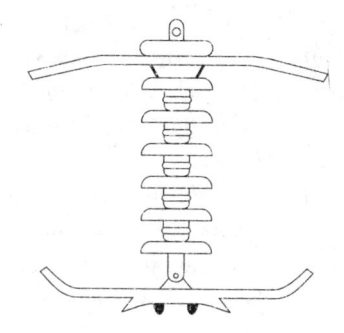

Fig. 16.14 Suspension string with arcing horns

The grading ring when used in conjunction with an arcing horn fixed at the top of the insulator string serves the purpose of an arcing shield. In the event of an arc forming following a flashover caused by some type of overvoltage, the arc will usually take the path between the horn and the shield and the insulator string will remain clear from the arc.

16.8.3 Lightning Arresters or Surge Diverters

Lightning arresters are also known as surge diverters or surge arresters. They are connected between the line and ground at the substation and always act in shunt (parallel) with the equipment to be protected, and perform their protective function by providing a low-impedance path for the surge currents so that the surge arrester's protective level is less than the surge voltage withstanding capacity of the insulation of equipment being protected. A lightning arrester's protective level is the voltage appearing across the terminals of the arrester at sparkover or during the flow of current through the arrester after sparkover. The main purpose of lightning arresters is to divert or discharge the surge to the ground.

The action of the surge diverter can be studied with the help of Fig. 16.15. When the travelling surge reaches the diverter, it sparks over at a certain pre-fixed voltage as shown by point P and provides a relatively low-impedance path to ground for the surge current. The current flowing to ground through the surge impedance of the line limits the amplitude of the overvoltage across the line and ground known as 'residual voltage' (as shown by point Q in Fig. 16.15) to such a value which will protect the insulation of the equipment being protected.

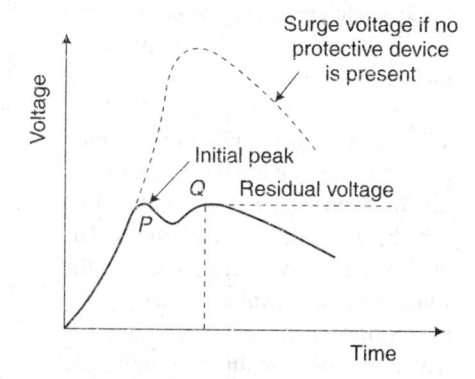

Fig. 16.15 Voltage characteristic of surge diverter

It is however, essential that the low-impedance path to ground must not exist before the overvoltage appears and it must cease to exist immediately after the voltage returns back to its normal value.

An ideal lightning arrester or surge diverter should possess the following characteristics.

  (i) It should not draw any current at normal power frequency voltage, i.e. during the normal operation.
  (ii) It should breakdown very quickly when the abnormal transient voltage above its breakdown value appears, so that a low-impedance path to ground can be provided.
  (iii) The discharge current after breakdown should not be so excessive so as to damage the surge diverter itself.
  (iv) It must be capable of interrupting the power frequency follow-up current after the surge is discharged to ground.

Impulse Ratio of Lightning Arresters

It is defined as the ratio of the breakdown impulse voltage of a wave of specified duration to the breakdown voltage of a power-frequency wave. The impulse ratio of any lightning arrester is therefore a function of time duration of the transient wave.

Types of Lightning Arresters (or Surge Diverters)

The following are the main types of lightning arresters.

  (i) Expulsion Type Lightning Arrester
  (ii) Non-linear Surge Diverter
  (iii) Metaloxide Surge Arrester (MOA)

(i) Expulsion type lightning arrester

This type of arrester is also known as Expulsion Gap or Protector Tube. It consists of a fibre tube with an electrode at each end. The lower electrode is solidly grounded. The upper electrode forms a series gap with the line conductor, as shown in Fig. 16.16. When a surge appears on the conductor, the series gap breaks down, resulting in formation of arc in the fibre tube between the two electrodes. The heat of the arc vaporises some of the fibre of the tube walls resulting in the generation of an inert gas. This gas is expelled violently through the arc so that arc is extinguished and the power frequency current is prevented from flowing after the surge discharge.

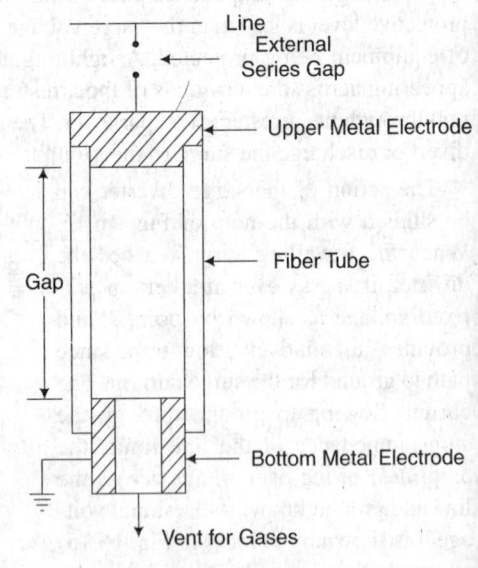

Fig. 16.16 *Expulsion type lightning arrester*

The expulsion arrester derives its name from the fact that gases formed during its operation are expelled from the tube through a vent.

This type of arrester is mainly used to prevent flashover of line insulators, isolators and bus insulators.

(ii) Non-linear surge diverter

This type of diverter is also called valve type lightning arrester, or conventional non-linear type lightning arrester. It consists of a divided spark-gap (i.e. several short gaps in series) in series with non-linear resistor elements. The divided spark-gap and the non-linear resistor elements are placed in leaktight porcelain housing which ensures reliable protection against atmospheric moisture, condensation and humidity (see Fig. 16.17). Figure 16.18 shows a typical valve type lightning arrester.

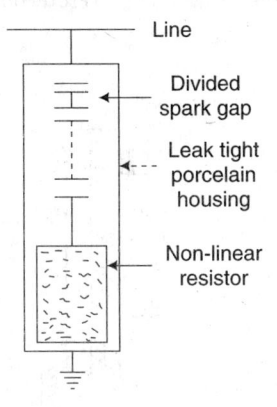

Fig. 16.17 *Value-type lightning arrester (non-linear diverter)*

The functions of the diverter's divided spark-gap are as follows
(a) It prevents the flow of current through the diverter under normal conditions.
(b) It sparks over at a predetermined voltage.
(c) It discharges high-energy surges without any change in sparkover characteristics.
(d) It interrupts the flow of power-frequency follow current from the power system after the surge has been dissipated.

The functions of the non-linear resistors are as follows:
(a) They provide a low-impedance path for the flow of surge current after gap sparks over.
(b) They dissipate surge energy.
(c) They provide a relatively high-resistance path for the flow of power frequency follow current from the power system, thereby assisting the divided gap to interrupt the power frequency current (i.e. reseal against system voltage).

The divided spark-gap consists of a number of short-gaps in series. Each of them has two electrodes across which a grading resistor of high ohmic value is connected. The grading resistors ensure even grading of voltage between the various gaps. The system is similar to that of a number of capacitors connected in series and across each of these capacitors is a high value resistor. The voltage grading by means of resistors also makes it possible to raise the interrupting capacity of the divided spark-gap.

In case of relatively slow variations in voltage, there is no sparkover across the gaps as the influence of parallel grading resistors across these gaps prevails over that of the spark-gap capacitances with regard to one another and to the ground. But when there are large rapid variations in voltage, the even grading of the voltage across the series of gaps no longer remains and the influence of unbalanced capacitances of the gaps prevails over the grading resistors. The surges are mainly concentrated on the upper spark gaps which on sparking over result in the sparkover of the complete arrester.

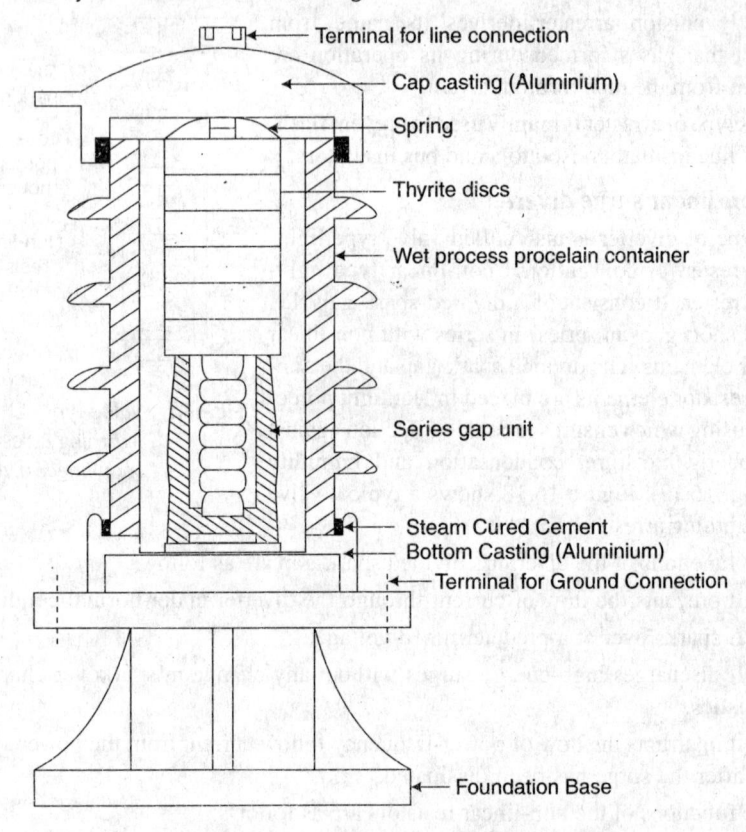

Fig. 16.18 *A typical value-type lightning arrester*

The main function of the diverter is the protection of the insulation against dangerously high overvoltages and for this reason the breakdown voltage of the diverter at system frequency is made greater than 1.8 times the normal value.

The ideal characteristic for the non-linear resistor would be RI = constant, since such a resistor will maintain constant voltage by changing its resistance in inverse proportion to the current. The non-linear resistor is in the form of discs of 9 cm diameter and 2.5 cm thickness. The material used for such resistors, called by trade names of "Thyrite", and "Metrosil" is a hard ceramic substance in the form of cylindrical blocks which consist of small crystals of silicon carbide bound together by means of an inorganic binder, with the whole assembly being subjected to heat treatment. The non-linear characteristic is attributed to the properties of the electrical contacts between the grains of silicon carbide. As the material used for non-linear resistors for the arrester is known by trade name as thyrite, this type of lightning arrester is also sometimes called Thyrite type lightning arrester.

Figure 16.19 shows the volt-ampere characteristic of the non-linear resistor, the dotted curve being the static characteristic, while the solid curve is the dynamic characteristic, corresponding to the application of a voltage surge. If a horizontal line, tangential to the dynamic characteristic is drawn, then its intercept with the voltage axis gives the residual voltage. The residual voltage is defined as the peak (crest) value of the voltage which appears between the terminals of the surge diverter at the instant when the surge current discharges through it. This voltage varies between 3 kV and 6 kV, depending on the type i.e. station type or line type, the discharge current and the rate of change of this current.

A non-linear resistor type lightning arrester is also known as valve type lightning arrester. The valve type arrester derives its name from the fact that the non-linear resistance (similar to valve opening) regulates itself to the flowing current and limits the voltage when high surge (lightning) currents flow through the arrester.

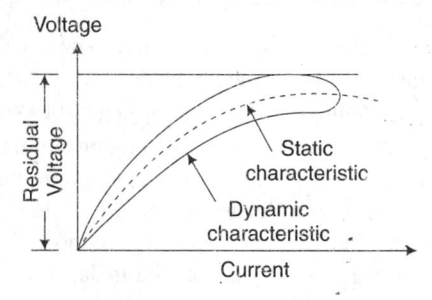

Fig. 16.19 *Volt-ampere characteristics of non-linear diverter*

The valve arrester works as an insulator at normal system voltage, but it becomes a good conductor of low resistance to limit the lightning (surge) voltages to a value lower than the basic impulse level of the equipment being protected. It re-establishes itself as an insulator after the discharge of the surge current.

Comparison of Expulsion and Non-linear Types of Surge Diverters The series gap of the expulsion-type surge diverter (lightning arrester) acts as an isolator under normal conditions. When the diverter is required to operate, the impedance of the series gap is very low and the arc formed in the fibre tube is also very low, with the result that the follow-up current is almost the same as that due to the line to ground (L-G) fault, its value therefore being determined by the factors used in the calculation of fault currents. Thus the current through the expulsion type diverter may become as high as several kilo-amperes (kA). Secondly, as the follow-up current is determined by the network under fault conditions, it is a reactive current and lags the system voltage by 90 degree, i.e. it passes through a zero value when the system voltage is maximum. Since the arc voltage at the instant of breakdown falls rapidly to a very low value, there is no phenomenon which may be analogous to residual voltage. Therefore, the level of protection is fixed only by the value of the impulse breakdown voltage. It gets increasingly difficult for the expulsion gap to interrupt currents of low value, because of the small amount of deionising gas formed by it.

Though the sub-divided series gap of the non-linear surge diverter is also an isolator during normal conditions, it actually plays a role in the performance of the diverter during its operation. As the current through the diverter is limited to a comparatively low value by the non-linear resistors, it is practically in phase with the system voltage. Because of the current being practically in phase with the system voltage, an instant of current zero is also an instant of zero for the system voltage. Thus, the sub-divided gap interrupts the current at the first zero. As the impulse breakdown voltage is usually less than the residual voltage, it is the latter which determines the level of protection provided by the non-linear surge diverter.

(iii) Metal-oxide surge arrester (MOA)

The metal oxide surge arrester abbreviated as MOA is a recently developed ideal surge arrester. It is a revolutionary advanced surge protective device for power systems. It is constructed by a series connection of zinc oxide (ZnO) elements having a highly non-linear resistance. The excellent non-linear characteristic of zinc oxide element has enabled to make surge arresters without series connected spark gaps, i.e. fully solid-state arresters suitable for system protection up to the highest voltages.

As mentioned earlier, the conventional non-linear surge diverters almost exclusively use Silicon Carbide (SiC) non-linear resistors. As this material is not ideal, it is not non-linear enough and thus imposes certain design restrictions. Also, its characteristics call for a large number of spark gaps. A new class of non-linear material was recently developed in Japan by Japan's Matsushita Electric Company. The new ceramic material is basically formed from zinc oxide together with addition of other oxides, such as bismuth and cobalt oxides. As the main constituent of this new ceramic material is zinc oxide, the non-linear resistor made of this material is popularly known as zinc oxide element and surge arresters made of zinc oxide elements are called metal oxide surge arresters. Such material can be used to make resistors with a much higher degree of non-linearity over a large current range. With such resistors, one can design arresters having voltage-current characteristics very close to ideal. Because of the high degree of non-linearity, this material allows considerable simplification in arrester (diverter) design.

The metal oxide surge arrester (MOA) which consists of a series connected stack of discs of zinc oxide elements, operates in a very simple fashion. It is dimensioned so that the peak value of the phase to ground voltage in normal operation never exceeds the sum of the rated voltages of the series-connected discs. The resistive losses in the arrester in normal operation are therefore very small. When an overvoltage occurs, the current will rise with the wavefront according to the characteristics without delay. No breakdown occurs but a rather continuous transition to the conducting state is observed. At the end of the voltage transient, the current is reduced closely following the *I-V* curve (i.e. in contrast to the conventional arrester, there is no follow-up current).

The metal oxide surge arrester has the following marked advantages over conventional arresters.

(i) Series spark-gap is not required.

(ii) It has very simple construction and is a fully solid-state protective device.

(iii) Significant reduction in size.

(iv) Quick response for steep discharge current.

(v) Very small time delay in responding to overvoltages.

(vi) Superior protective performance.

(vii) Outstanding durability for multiple operating duty cycle.

(viii) No abrupt transient such as that occurs at the time of sparkover in a conventional arrester.

(ix) Negligible power follow-up current after a surge operation.

MOA is especially suitable for gas insulated sub-stations (GIS), since it can be installed directly in SF_6.

Figure 16.20 illustrates the difference in the operations of MOA and conventional arresters, responding to an on-coming surge $v(t)$. The voltage wave shape of MOA is smooth and can be expressed as follows

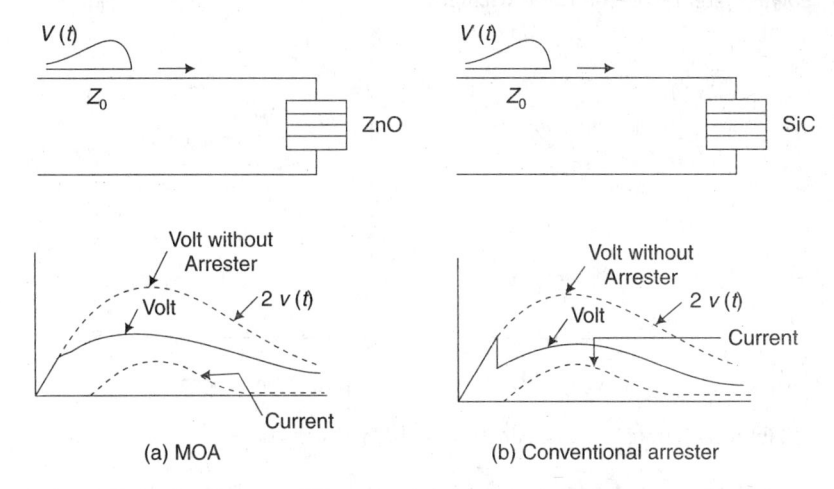

Fig. 16.20 Operation of MOA and conventional arrester responding to an on-coming surge

$$V = 2v(t) - Z_0 i \qquad (16.18)$$

where Z_0 is the surge impedance of the line and i is the instantaneous current of the arrester.

The voltage wave shape of the conventional arrester has a peak value at the time of sparkover of the series gap. The voltage after sparkover or the discharge voltage can be expressed by Eq. (16.18).

As seen from Fig. 16.20, the protection performance of the conventional arrester is controlled by both the sparkover voltage and the discharge voltage. However, the protection performance of MOA is simply related to the discharge voltage, given by Eq. (16.18).

The protection level of MOA is decided by the maximum discharge voltage encountered under normal operating conditions. Though it is difficult to get a clear-cut expression like the V-t curve in a conventional arrester, the individual voltage waveform should be considered in precise discussion of insulation coordination.

For rough estimation of the voltage waveform, it is convenient to assume a fixed functional relation between the voltage and current of MOA, as follows.

$$V = v(i) \qquad (16.19)$$

The voltage waveform or the time dependence of V can be calculated by connecting Eqs. (16.18) and (16.19). A graphical method of solving the equation is shown in Fig. 16.21 (a). The V and I values corresponding to $v(t)$ are represented by P_0 which is the intersection point of two curves corresponding to Eqs. (16.18) and (16.19). In this simple treatment, the point P_0 moves on the $v(i)$ curve shown by a dotted line in Fig. 16.21(a). However, the actual V-I point moves along the thick line which cannot be conformed to a single line. A typical V-I characteristic of conventional SiC arrester is conceptually shown in Fig. 16.21(b) for comparison. In this case, the change from sparkover voltage (SOV) to a lower discharge voltage is abrupt.

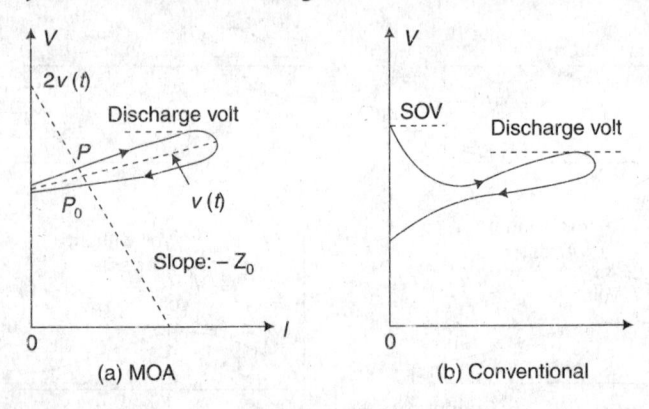

Fig. 16.21 *Transient V-I characteristics of MOA and conventional arrester*

The main drawback of MOA is that the absence of spark-gaps results in a continuous flow of current through the device so that there is theoretically speaking the danger of thermal runway. However, as very stable resistors have been developed, these apprehensions are almost eliminated. Furthermore, the absence of spark-gaps makes the voltage grading system unnecessary. The MOA is inherently self-regulating in that the current flow of 0.5 to 1 mA at normal supply voltage leads to reliable operation even in polluted conditions.

16.8.4 Surge Absorber

A surge absorber is also known as a surge modifier. It is a device which absorbs the energy contained in a travelling wave and reduces the amplitude of the surge and the steepness of its wave front. The energy absorption in the form of corona loss due to corona formation acts as a kind of a natural safety-valve in the case of travelling waves which raise the line voltage above the corona level. There may be some cases in which a length of line adjacent to the terminal station may be working at a lower critical corona voltage than the main length of the line. Another method of lessening the risk of travelling waves is the installation of a length of steel conductor at the terminal end, say 1.5 to 3 kms away. The skin effect of the wire to steep fronted waves being equivalent to an appreciable increase in the high frequency resistance. These methods are compromises which do not provide adequate protection against disturbances originating near the station.

A better method of reducing the risk of travelling waves is to terminate the line at a distance of 0.8 to 1.5 km from the station (or sub-station) and to connect it to the station with an underground cable. This method has the following three advantages.

(a) The cable gets shielded from all electrostatic phenomena, and from accidental grounds (earths) through various causes.

(b) When the travelling waves enter the cable from the overhead line, they are reduced in magnitude to about 20 per cent of their incident value because of the ratio of the natural impedances.

(c) When a travelling wave is transmitted into the cable, it gets attenuated rapidly due to the dielectric losses taking place in the cable at high frequencies.

The most modern surge absorber is the Ferranti surge absorber (Fig. 16.22). It consists of an aircored inductor connected in series with each line and surrounded by a grounded metallic sheet called a dissipator. The inductor is magnetically coupled but electrically isolated from the dissipator. The inductor is insulated from the dissipator by the air.

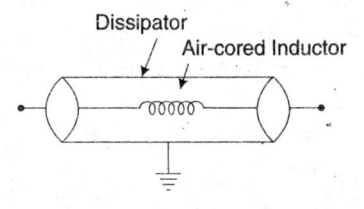

Fig. 16.22 Ferranti surge absorber

This surge absorber acts as an air cored transformer whose primary is the inductor and the dissipator is the short-circuited secondary of a single turn. Whenever the travelling wave is incident on the surge absorber, the energy contained in the wave is dissipated in the form of heat generated in the dissipator; firstly, due to the current set up in it by ordinary transformer action, the secondly, by eddy currents. The steepness of the wavefront is also reduced because of the series inductance.

16.9 PETERSON COIL

The Peterson coil is also known as a ground-fault neutralizer or arc suppression coil. It is an iron cored reactor connected between the neutral point and ground and capable of being tuned with the capacitance of the healthy phases to produce resonance when a line to ground (L-G) fault occurs Fig. 16.23(a). It is mainly used to prevent arcing grounds which lead to overvoltages on systems with an ungrounded (isolated) neutral. The Peterson coil makes arcing ground faults self-extinguishing and in case of sustained ground fault on one of the lines, it reduces the fault current to a very low value so that the healthy phases can be kept in operation even with one line grounded. In order to select the proper value of inductive reactance, depending upon the length of the transmission line and hence, the capacitance to be neutralised. Peterson coil is provided with tappings.

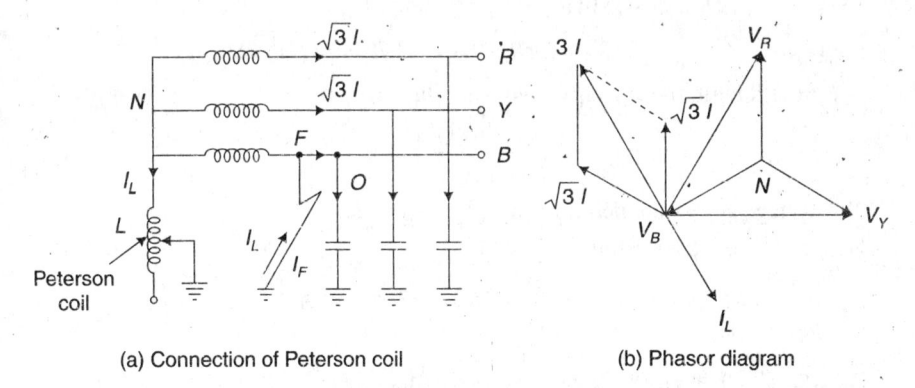

(a) Connection of Peterson coil

(b) Phasor diagram

Fig. 16.23 Resonant grounded 3-phase system

In an undergrounded system, when a ground fault occurs on any one line, the voltage of the healthy phases is increased $\sqrt{3}$ times (i.e. $\sqrt{3}\ V_p$, where V_p is the phase voltage). Hence the charging currents become $\sqrt{3}\ I$ per phase, where I is the charging current of line to ground of one phase. The phasor sum of the charging currents of the healthy phases becomes 3 times the normal line to neutral charging current of one phase, as shown in phasor diagram of Fig. 16.23(b).

Hence, $$I_c = 3I = \frac{3V_P}{1/\omega C} = 3V_P \omega C \qquad (16.10)$$

where I_c is the resultant charging current and I is the charging current of line to ground of one phase.

If L is the inductance of the Peterson coil connected between the neutral and the ground, then

$$I_L = \frac{V_P}{\omega L} \qquad (16.11)$$

In order to obtain satisfactory cancellation (neutralisation) of arcing grounds, the fault current I_L flowing through the Peterson coil should be equal to the resultant charging current I_C.

Therefore, for balance condition

$$I_L = I_C$$

or $$\frac{V_P}{\omega L} = 3V_P \omega C \quad \text{(From Eqs. (16.10) and (16.11))}$$

or $$L = \frac{1}{3\omega^2 C} \qquad (16.12)$$

The inductance L of Peterson coil is calculated from Eq. (16.12).

Example 16.1 | A 33 kV, 3 phase 50 Hz, overhead line 60 km long has a capacitance to ground of each line equal to 0.015 μF per km. Determine the inductance and kVA rating of the Peterson coil.

Solution: $L = \dfrac{1}{3\omega^2 C} = \dfrac{1}{3 \times (314)^2 \times 0.015 \times 60 \times 10^{-6}} = 3.75$ H

$$X_L = 2\pi f L = 314 \times 3.75 = 1177.5 \ \Omega$$

For ground fault, the current in the neutral is given by

$$I_N = I_L = \frac{V_P}{\omega L} = \frac{33 \times 1000}{\sqrt{3} \times 1177.5} = 16.18 \text{ A}$$

The voltage across the neutral is the phase voltage.

Hence, kVA rating = $V_p \times I_N$

$$= \frac{33}{\sqrt{3}} \times 16.18 = 308.3 \text{ kVA}$$

Example 16.2 | A 50 Hz overhead line has the line to ground capacitance of 1.2 μF. It is decided to use a ground-fault neutralizer. Determine the reactance to neutralise the capacitance of (i) 100% of the length of the line, (ii) 95% of the length of the line, and (iii) 80% of the length of the line.

Solution:

(i) The inductive reactance of the coil for 100% neutralisation will be

$$\omega L = \frac{1}{3\omega C} = \frac{1}{3 \times 314 \times 1.2 \times 10^{-6}} = \frac{10^6}{3 \times 314 \times 1.2} = 884.6 \ \Omega$$

(ii) The inductive reactance for neutralizing 95% of the capacitance will be

$$\omega L = \frac{1}{3\omega C} = \frac{10^6}{3 \times 314 \times 1.2 \times 0.95} = 931.2 \ \Omega$$

(iii) The inductive reactance for 80% neutralization will be

$$\omega L = \frac{1}{3\omega C} = \frac{10^6}{3 \times 314 \times 1.2 \times 0.8} = 1105.8 \ \Omega$$

16.10 INSULATION COORDINATION

Insulation coordination is the correlation of the insulation of electrical equipment and lines with the characteristics of protective devices such that the insulation of the whole power system is protected from excessive overvoltages.

The main aim of insulation coordination is the selection of suitable values for the insulation level of the different components in any power system and their arrangement in a reasonable manner so that the whole power system is protected from overvoltages of excessive magnitude. Thus, the insulation strength of various equipment, like transformers, circuit breakers, etc. should be higher than that of the lightning arresters and other surge protective devices. The insulation coordination is thus the matching of the volt-time flashover and breakdown characteristics of equipment and protective devices, in order to obtain maximum protective margin at a reasonable cost. The volt-time curves of equipment to be protected and the protective device are shown in Fig. 16.24. Curve A is the volt-time curve of the protective device and curve B is the volt-time curve of the equipment to be protected. From volt-time curves A and B of Fig. 16.24, it is clear that any insulation having a voltage withstanding strength in excess of the insulation strength of curve B will be protected by the protective device of curve A.

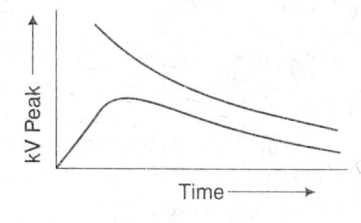

Fig. 16.24 *Volt-time curves of protective device and the equipment to be protected*

16.10.1 Volt-time Curve

The breakdown voltage of any insulation or the flashover voltage of a gap depends upon both the magnitude of the voltage and the time of application of the voltage. The volt-time curve is a graph of the crest flashover voltages plotted against time to flashover for a series of impulse applications of a given wave shape. The construction of the volt-time curve and the terminology associated with impulse testing are shown in Fig. 16.25. The construction of the volt-time curve is based on the application of impulse voltages of the same wave shape but of different peak values to the insulation whose volt-time curve is required. If an impulse voltage of a given wave shape and polarity is adjusted so that the test specimen (i.e. a particular insulation) flashes over on the front of the wave, the value of voltage corresponding to the point on front of the wave at which flashover occurs is called front flashover. If an impulse voltage of the same wave shape is adjusted so that the test specimen flashes over on the tail of the wave at 50 per cent of the applications and fails to flashover on the other 50 per cent of the applications, the crest value of this voltage is called the critical flashover voltage. If an impulse voltage causes flashover of the test specimen exactly at the crest value, then it is called crest flashover. If flashover does not take place, the wave is called a full wave and if flashover does take place, the wave is called a chopped

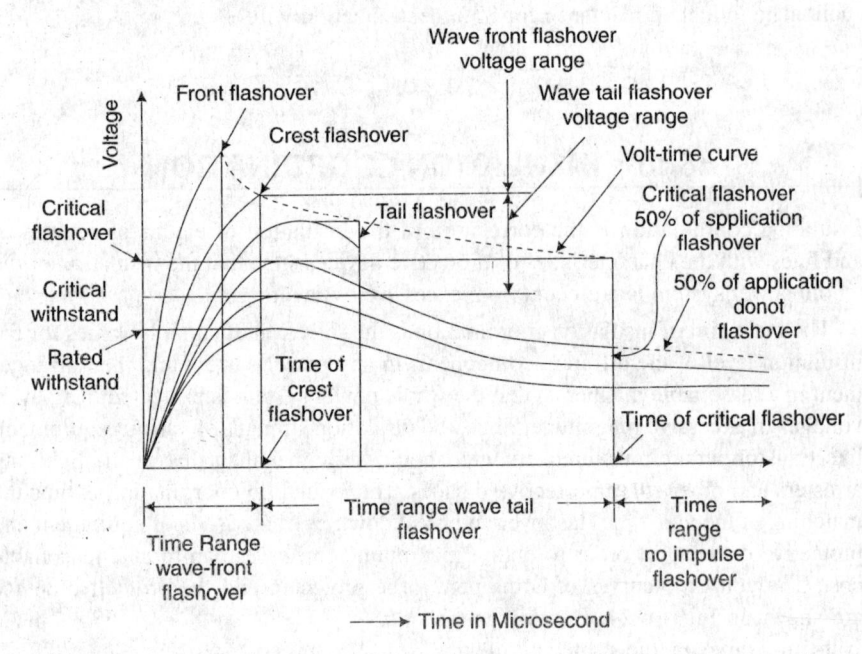

Fig. 16.25 *Construction of volt-time curve and the terminology associated with impluse testing*

wave. The applied impulse voltage reduced to just below the flashover voltage of the test specimen is called the critical withstand voltage. The rated withstand voltage is the crest value of the impulse wave that the test specimen will withstand without disruptive discharge.

Figure 16.25 illustrates for an assumed impulse, its wave front flashover, crest flashover, critical flashover voltage, critical withstand voltage, rated flashover voltage, wavetail flashover and a volt-time curve.

16.11 BASIC IMPULSE INSULATION LEVEL (BIL)

The insulation strength of equipment like transformers, circuit breakers, etc. should be higher than that of the lightning arresters and other surge protective devices. In order to protect the equipment of the power system from overvoltages of excessive magnitude, it is necessary to fix an insulation level for the system to see that any insulation in the system does not breakdown or flashover below this level and to apply protective devices that will give the apparatus effective protection.

Protection Against Overvoltages

Several methods of providing coordination between insulation levels in the station and on the line leading to the station have been offered. The best method is to establish a definite common level for all the insulation in the station and bring all insulation to or above this level. This limits the problem to three fundamental requirements, namely, the selection of a suitable insulation level, the assurance that the breakdown or flashover strength of all insulation in the station will equal or exceed the selected level, and the application of protective devices that will give the apparatus as good protection as can be justified economically.

The common insulation level for all the insulation in the station is known as Basic Impulse Insulation Level (BIL) which have been established in terms of withstanding voltages of apparatus and lines.

Basic impulse insulation level can be defined as reference level expressed in impulse crest voltage with a standard wave not longer than a 1.2/50 microsecond wave, according to Indian standards (1.5/40 μs in USA and 1/50 μs in UK). Apparatus insulation as demonstrated by suitable tests should be equal to or greater than the BIL.

The basic impulse insulation level for a system is selected such that the system could be protected with a suitable lightning protective device, e.g. a lightning arrester. The margin between the BIL and the lightning arrester should be fixed such that it is economical and it also ensures protection to the system. The BIL chosen must be higher than the maximum expected surge voltage across the selected arrester.

EXERCISES

1. What are the causes of overvoltages arising on a power system? Why is it necessary to protect the lines and other equipment of the power system against overvoltages?
2. Describe the phenomenon of lightning and explain the terms pilot streamer, stepped leader, return streamer, dart leader, cold lightning stroke and hot lighting stroke.
3. How can the magnitude of overvoltages due to direct and indirect lightning strokes on overhead lines be calculated?
4. What protective measures are taken against lightning overvoltages?
5. What is a ground wire? What are the requirements to be satisfied by ground wires to provide efficient protection to lines against direct lightning strokes? How do ground wires protect the overhead lines against direct lightning strokes.
6. What is tower-footing resistance? Why is it required to have this resistance as low as economically possible? What are the methods to reduce this resistance?
7. Explain the terms overvoltage factor, protective ratio, protective angle, protective zone and coupling factor.

8. Differentiate between surge diverter and surge absorber. What are the characteristics of an ideal surge diverter?
9. What is the necessity of protecting electrical equipment against travelling waves? Describe in brief the protective devices used for protection of equipment against such waves.
10. Describe the protection of stations and sub-stations against direct lightning strokes.
11. Describe the construction and principle of operation of (i) Expulsion type lightning arrester, and (ii) Valve type lightning arrester.
12. Compare the protection performance of expulsion-type surge diverter with that of non-linear surge diverter.
13. Describe the construction and the operation of metal oxide surge arrester? What are its advantages over conventional arresters? Discuss the difference in protection performance of MOA and conventional non-linear type arrester with the help of suitable figures.
14. What is a Peterson coil? What protective function are performed by this device?
15. A 132 kV, 3-phase, 50 Hz transmission line 200 km long contains three conductors of effective diameter 2.2 cm, arranged in a vertical plane with 4.5 m spacing and regularly transposed. Find the inductance and kVA rating of the Peterson coil in the system.
16. A 50 Hz overhead line has line to ground capacitance of 1.5 μF. It is decided to use ground-fault neutralizer. Determine the reactance to neutralize the capacitance of (i) 85% of the length of the line (ii) 80% of the length of the line, and (iii) 75% of the length of the line.
17. Explain the term insulation coordination. Describe the construction of volt-time curve and the terminology associated with impulse testing.
18. Write short notes on the following.
    (i) Klydonograph and magnetic link
    (ii) Rod gap
    (iii) Arcing horns
    (iv) Ferranti surge absorber
    (v) Basic impulse insulation level

17 Modern Trends in Power System Protection

17.1 INTRODUCTION

The modern power system network formed by inter-connecting several individually controlled ac networks is a highly complicated network. Each individually controlled ac network has its own generating stations, transmission and distribution systems, loads and a load control centre. The increase in the size of generating units of each individually controlled network is also being compelled with the rapid increase in electrical power demand worldwide. The interconnection of different load centres located in different areas is necessitated by increased power demand. The regional load control centre controls the generation in its own geographical region to maintain the system frequency between 50.5 Hz and 49.5 Hz. The exchange of power (Import/Export) between neighbouring ac networks is dictated by the National Load Control Centre. The entire interconnected ac network is called a *national grid*. The interconnection of neighbouring National Grids forms a *super grid*. For proper operation and control of modern interconnected power system grown in both size and complexity, power system engineers have to face a number of challenging tasks including real-time data recording, communication networking and man-machine interfacing. The protection of such power system requires fast and reliable relays to protect major equipment and to maintain system stability.

A fault on a power system is characterized by increase in current magnitude, decrease in voltage magnitude and change in the phase shift between them. The fault also causes transients or noise, such as dc offset harmonics, distortion in the electrostatic and magnetic fields surrounding the power system element, travelling waves, etc. Transients may lead to relay malfunction, when the relay is assumed to be fed only with sinusoidal quantities. Thus, any event or fault-generated noise needs prior filtering from the relaying signals (currents and voltages) before feeding them to the relay. The function of the protective relay is to sense the fault and to issue a trip signal to circuit breaker for disconnecting the faulty element of the power system. There have been rapid developments in relaying technology during the last three decades. Because of advancement in computer technology, numerical relays based on microprocessor or microcontroller have been developed various relay technologies such as electromechanical, static and numerical technologies are available today. Although all the technologies are presently in use, numerical relay technology is gaining popularity over the others because of its communication capabilities which facilitate integrated protection, control and metering features.

The concept of adaptive relaying is used for the protection of complex interconnected power system. The settings of many relays are dependent upon the assumed conditions in the power system and it is not possible to realise these conditions in complex interconnected power system network. Hence it is desirable that the relay settings adapt themselves to the real-time system as and when system conditions change. Such a relaying scheme is called adaptive relaying. Adaptive relaying is defined as the protection system whose settings can be changed automatically so that it is attuned to the prevailing power system conditions.

During the last three decades, software packages have been developed for carrying out the various duties related to supervision, protection and control of power system with the help of this software, several functions can be performed automatically from a central system control centre. Extensive monitoring of network operations, load dispatching, load frequency control, load shedding, optimum loading of various plants, remote back-up protection, etc., are possible from a central grid control centre.

With advancements in computer technology, electromechanical and discrete electronic devices used for power system protection, metering and supervisory control are being replaced by integrated protection and control systems. The need of the day is for modern substations and for installing intelligent electronic devices such as numerical relays, programmable logic controllers (PLCs), etc., at the bay level. These are called bay controllers. An integration system combines the hardware and software components needed to integrate the various bay controllers, providing data collection, control and monitoring from a central work station. This system consists of network interface devices, host computer and printer. The network interface devices provide interface between the bay controller specific protocol and the system's local area network (LAN). Numerical protection, along with integrated system and suitable communication medium, provides a total solution for protection and control of power system. The integrated protection and control system is the latest trend in the world and the need of the day. Supervisory control and data acquisition (SCADA) is one of the features provided by the integrated protection and control system. The SCADA system provides support to the system operators or operating centres to locally or remotely monitor and control equipment.

Field Programmable Gate Arrays (FPGA) is a new and exciting device in the area of computing. It is a programmable logic integrated circuit that contains many identical logic cells viewed as standard components. FPGAs provide designers with new possibilities for flexibility and performance. The re-configurability and re-programmability features of FPGA make it possible to implement many relays on the single FPGA chip with faster and reliable operation. At present, protective relays based on FPGAs are at design and development stage.

Gas Insulated substation/switchgear (GIS) reduces the requirement for space and improves reliability, thus increasing its popularity for installation in cosmopolitan cities, industrial townships and hydrostations where the costs of space and construction are very high. In a GIS, the various equipment such as circuit breakers, busbars, isolators, load break switches current transformers, voltage transformers, earthing switches etc. are housed in separate metal enclosed modules filled with SF_6 gas. SF_6

Modern Trends in Power System Protection

gas provides better insulation and are quenching medium. The used of SF_6 gas having higher dielectric strength helps in manifold reduction in the size of the substation components.

Under stable condition, power system, operates at normal frequency and the total mechanical power input from the prime movers to the generators is equal to the sum of all the connected loads, plus all real power losses in the system. Any significant upset of this balance causes a frequency change. The frequency relay provides protection so that machines and systems may fully operate within their capability. In typical applications, the frequency relays are used to balance the load requirement and available system output. Frequency is a reliable indicator of an overload condition. Frequency sensitive relays are therefore used to disconnect load automatically. Such an arrangement is referred to as a load shedding. Automatic load-shedding, based on underfrequency, is necessary to balance load and generation. Underfrequency relays are usually installed at distribution substations, where loads can be disconnected in steps. The overfrequency relay is used for restoring the load. Loads are also restored in steps to avoid oscillation.

This chapter discusses modern topics such as Gas Insulated Substation (GIS), frequency relays and load shedding, field programmable gate arrays (FPGA) based relays, travelling wave relays, adaptive protection, integrated protection and control, supervisory control and data acquisition (SCADA), relay reliability and advantages of fast fault clearing.

17.2 GAS INSULATED SUBSTATION/SWITCHGEAR (GIS)

Conventional air-insulated transmission and distribution substations are of substantial size due to the poor dielectric strength of air. They also suffer variations in the dielectric capability of air to withstand varying ambient conditions and deterioration of the exposed components due to corrosive and oxidising nature of the environment. In order to enhance the life and reliability of power transmission and distribution substations, it is desirable to protect the substation components from corrosion and oxidisation due to environment. Metal encapsulation of the substation equipment provides a simple and effective solution to the problem of durability of the substations. Since the size of the container/enclosure is a direct function of the dielectric strength of the insulating medium, the container/enclosure sizes are large with medium of poor insulation such as air or nitrogen. The use of sulphur hexafluoride (SF_6) gas with higher dielectric strength instead of air helps in manifold reduction in the size of the substation components. On the other hand, the grounded metal encapsulation (metal-enclosure) makes the substation equipment safe, as the live components are no longer within the reach of the operator. As the metal enclosure of the equipment is solidly grounded, the electric field intensity, at the enclosure surface, is reduced to zero. Using this design philosophy, various substation equipment such as circuit breakers, busbars, earthing switches, isolators, instrument transformers (current and voltage transformers), etc., are housed in separate metal-encapsulated (metal-enclosed) modules field with pressurised SF_6 gas. The assembly of such equipment at a substation is defined as Gas Insulated and Metal Enclosed System (GIMES) Gas Insulated and Metal Enclosed System (GIMES) is popularly known as Gas Insulated

Substation (GIS). The term GIS is also sometimes used to refer to Gas Insulated Switchgear. Thus, the substation in which its various components (equipment), like circuit breakers, busbars, isolators, earthing switches, instrument transformers (CTs and VTs), etc., are housed in separate metal enclosed (metal-encapsulated) modules filled with pressurised SF_6 gas is popularly known as Gas Insulated Substation (GIS).

The various modules of GIS are factory assembled and are filled with SF_6 gas. Therefore, they can be easily taken to site for final assembly. Such substations are compact and can be conveniently installed on any floor of a multi-storied building or in an underground substation. GIS reduces the requirement for space and improves reliability, thus increasing its popularity for application in metro cities, industrial townships, and hydrostations where the costs of space and construction are very high. GIS is also preferred in heavily polluted areas where dust, chemical fumes and salt layers can cause frequent flashovers in conventional outdoor substations.

Designs of GIS equipment using SF_6 gas medium for both insulation and interruption are readily available. Because of metal encapsulation and SF_6 gas insulation of the GIS components, the space requirement of GIS is reduced to one-sixth (~17 per cent), as compared to a conventional air insulated yard substation. For a higher KV class, the size of GIS is reduced to about 8 per cent. GIS covering the complete voltage range from 11 kV to 800 kV are internationally available.

17.2.1 Design Aspects of GIS

The GIS contains the same components as a conventional air insulated outdoor substation. The entire installation in a GIS is divided into many sections/bays. The main components/equipment in a section are: circuit breakers, isolators or disconnectors, earthing switches, current and voltage transformers, surge arrestors, etc. The live parts of various components are enclosed in metal housing filled with SF_6 gas. The enclosures are made of non-magnetic material such as aluminium or stainless steel, are gas tight and are solidly grounded. The live parts are supported on cast resin insulators. Some of the insulators are designed as barriers between neighbouring modules. The modular components are assembled together to form a desired arrangement for a section or bay of GIS. The constituent components are assembled side by side. The porcelains and connections, are required in a yard substation, are totally eliminated in this new configuration. The high-voltage conductors (busbars) are supported on simple disc insulators. The connections to the overhead lines or underground cables are arranged through necessary air-to-gas or gas-to-cable terminations. The gas insulated instrument voltage transformer and the surge arrestor are directly installed on the gas insulated bus, using openings provided in the GIS for this purpose.

17.2.2 Modules/Components of GIS

The following are the principal modules insulated by SF_6 gas in a GIS:
 (i) Busbar
 (ii) Isolator or disconnector
 (iii) Circuit breaker
 (iv) Current transformer
 (v) Earthing switch

The auxiliary gas insulated modules/accessories (excluding control panel) required to complete a GIS are
  (i) Instrument voltage transformer
  (ii) Surge and lightning arrester
  (iii) Terminations

Busbar

The busbar which is one of the most elementary components of the GIS system, is of different lengths to cater to the requirement of circuit or bay formation. Co-axial busbars are commonly used in isolated-phase GIS as this configuration results in an optimal stress distribution. The main high voltage conductor made of aluminium or copper is centrally placed in a tubular metal enclosure, and supported by the disc/post insulator, at a uniform distance to maintain concentricity. Two sections of bus are joined by using plug-in connecting elements.

Isolator or Disconnector

Isolators are placed in series with the circuit breaker to provide additional protection and physical isolation. In a circuit, two isolators are generally used, one on the line side and the other on the feeder side. A pair of fixed contacts and a moving contact form the active part of the isolator. The fixed contacts are separated by an isolating gas gap. Isolators are either on-load type load-break switch or no-current break type. They can be motorised for remote control or driven manually. In GIS system, motorized isolators are preferred. Necessary interlocks are provided between the circuit-breaker isolator and earthing switch.

Circuit Breaker

The circuit breaker is the most critical module of a GIS system. It is metal-clad and utilises SF_6 gas, both for insulation and fault interruption. Puffer type SF_6 circuit breakers are commonly used to accomplish fault current interruption in GIS.

Current Transformer

The current transformers used in GIS are essentially in-line current transformers. Gas insulated current transformers have classical coaxial geometry and consist of the following parts: the tubular primary conductor, an electrostatic shield ribbon–would toroidal core and the gas-tight metal enclosure filled with SF_6 gas. The primary of a current transformer is a tubular metal conductor linking two gas-insulated modules, placed on either side of the current transformer. A ribbon–wound silicon steel core formed in toroidal shape is used for the magnetic circuit of the current transformer. A coaxial electrostatic shield, at ground potential, is placed between the high-voltage primary and the toroidal magnetic core of the current transformer for ensuring zero potential at the secondary of the current transformer.

Earthing Switch

Earthing switches used for GIS system are of two types, namely, maintenance earthing switch and fast earthing switch. The maintenance earthing switch is a slow device used to ground the high voltage conductors during maintenance schedule, in order to ensure the safety of the maintenance staff. On the other hand, the fast earthing switch is used to protect the circuit-connected instrument voltage transformer from core saturation caused by direct current flowing through its primary as a result of remnant charge (stored online during isolation/switching off the line). In such a situation, the use of a fast earthing switch provides a parallel low resistance path to drain the residual static charge quickly, thereby protecting the instrument voltage transformer from the damages which may be caused otherwise.

The earthing switch is the smallest module of a GIS. It is made up of two parts: a fixed contact located at the live bus conductor and a moving contact system mounted on the enclosure of the main module.

17.2.3 Accessories of GIS

Incomer and feeder connections form the main accessories of a substation. The supply is received at the incomer from a higher-level substation or from a ring main. The power is received and delivered through either the overhead lines or the underground cables, at a substation. Air-to-gas and cable-to-gas terminations are employed as an interface to the two media in GIS.

An instrument voltage transformer insulated by SF_6 gas and used for metering and protection forms a part of the GIS. A gas-insulated surge arrester which protects the system from switching surges is a critical accessory required for the substation.

The control panels used in GIS are both local and remote types. The local control panel (LCP) provides an access to the various controls and circuit parameters of an individual GIS section/bay. The LCP facilitates the monitoring of gas pressures, status of the switchgear element and operating fluid pressures, of oil, SF_6 gas and air. A remote control panel (RCP), based on microcontroller (MC) and digital signal process or (DSP), can shrink the RCP to a single unit. A single RCP is a must. The number of RCPs required in a GIS depends on the choice of the utility. A man-machine interface (MMI), featuring bay-wise pages and many other similar functional integrations, is the best techno-commercial solution. Connections to the central load dispatch centre (LDC) or central controls enhances the utility of an RCP.

17.2.4 Classification of GIS

Gas-insulated substations (GIS) are classified according to the type of modules or the configuration. The following configurations of modules are generally used in GIS.

(i) Isolated-phase (segregated-phase) module.
(ii) Three-phase common modules
(iii) Hybrid modules
(iv) Compact modules
(v) Highly integrated systems

The isolated-phase GIS module contains an assembly of individual circuit elements such as a pole of circuit breaker, a single pole isolator, one-phase assembly of a current transformer, etc. A single-phase circuit is formed by using individual components and pressurising the element with SF_6 gas forming a gas-tight circuit. A complete three-phase GIS section/bay is formed by three such circuits, arranged side by side.

In the three-phase common module style, a three-phase bay is assembled using the desired number of three-phase elements. The total number of the enclosures is thus reduced to one-third. By using three-phase modules, it is possible to reduce the floor of the substation upto 70 percent.

Hybrid systems use a suitable combination of isolated-phase and three-phase common elements.

Compact GIS system are essentially three-phase common systems, with more than one functional elements in one enclosure.

Highly integrated systems (HIS) are single unit metal enclosed and SF_6 gas insulated substations which are gaining user appreciation as this equipment provides a total substation solution for outdoor substations.

17.2.5 Advantages and Disadvantages of GIS

Advantages of GIS

(i) **Compactness** GIS are compact in size and the space occupied by them is about 15% of that of a conventional air insulated outdoor substation. The compact size of GIS offers a practical solution to vertically upgrade the existing substation and to meet the ever-increasing power demand in developing countries.

(ii) **Cost-effective, reliable and maintenance-free** GIS offer a cost-effective, reliable and maintenance free alternative to conventional air insulated substations.

(iii) **Protection from pollution** The moisture, pollution, dust etc. have little influence on GIS.

(iv) **Reduced installation time** The modular construction reduces the installation time to a few weeks.

(v) **Superior arc interruption** SF_6 gas used in the circuit-breaker has superior arc quenching property.

(vi) **Reduced switching overvoltages** The overvoltages while closing and opening line, cables, motors, capacitors etc. are low.

(vii) **Increased safety** As the metal enclosures of the various modules are at ground potential, there is no possibility of accidental contact by service personnel to live parts.

Disadvantages of GIS

(i) GIS has high cost as compared to conventional outdoor substation.
(ii) Requirement of cleanliness in GIS are very stringent as dust or moisture can cause internal flashover.
(iii) Since GIS are generally indoor, they need separate building which is generally not required for conventional outdoor substations.
(iv) GIS can have excessive damage in case of internal fault and there may be long outage periods as repair of damaged part may be difficult at site.
(v) Maintenance of adequate stock of SF_6 gas is essential. Procurement of gas and supply of gas is problematic.

17.3 FREQUENCY RELAYS AND LOAD-SHEDDING

Under stable conditions, a power system operates at normal frequency and the total power generated matches the load on the system. A fall in the system frequency is a clear indication of loss of balance between the power generated and the power delivered. The decay in system frequency is proportional to the deficiency in power generation with regard to load requirement. An underfrequency relay can be used to sense power deficiency and open circuit breakers to carry out a preplanned schedule of load-shedding. Thus, the frequency relay provides protection so that machines and systems may fully operate within their capability. Since frequency is a reliable indicator of an overload condition, frequency sensitive relays are used to disconnect load automatically. The generators must be equipped with frequency protection in order to adjust the output according to system demands.

The frequency relays have a predetermined limit set in cycles/second (Hz). Any deviations from this predetermined limit can trigger the frequency protection. Triggering of frequency protection below the predetermined limit is called underfrequency, and the triggering of frequency protection above the predetermined limit is called overfrequency.

17.3.1 Frequency Relays

The main types of the frequency relays are as follows:
  (i) Underfrequency relays
  (ii) Overfrequency relays
  (iii) Rate of change of frequency (df/dt) relays

For the operation of underfrequency and overfrequency relays, the duration of half-cycle period of supply frequency is compared with the time period of half-cycle of standard frequency used. The duration by which the supply frequency deviates from the standard frequency is used to detect the error frequency, e.g. if the frequency deviates from 50 cycles by half a cycle, then the duration of half cycle varies by 0.01 ms for a standard frequency of 50 cycles/second.

For the measurement of a frequency with a given accuracy, a minimum time is required. The measurement thus has an additional error-dynamic error. In case of fall of supply frequency the difference between the set frequency and the, actual frequency at the instant, the relay response is greater when df/dt is greater. In order to reduce the dynamic error, df/dt must be included in the measurement.

Underfrequency occurs when the system is overloaded with demand. This happens when a large load is suddenly brought on-line or the loading requirement of a system gradually increases over time.

Overfrequency normally occurs when the system has excessive generation compared to the demand or load. This can be a drastic or gradual change in the system load. Excessive generation can also happen when the system demand decreases.

17.3.2 Microprocessor Based Frequency and df/dt Relays

Frequency measurement in microprocessor-based relay is done on the principle of measurement of the time period of half-cycle which is inversely proportional to the frequency. The sinusoidal voltage signal is stepped down by using voltage transformer and then is converted to square wave by using a zero crossing detector (ZCD). ZCD is developed by using a voltage converter LM 311 or operational amplifier LM 747 or LM 324. A diode is used to rectify the output signal of the ZCD. A potential divider is used to reduce the magnitude to 5 volts.

The microprocessor senses the zero instant of the rectified square wave with the help of a program. As soon as the zero instant is detected, the microprocessor initiates the number of how many times the loop is executed. The microprocessor reads the magnitude of the square wave again and again, and moves in the loop as long as it remains high. It crosses the loop when the magnitude of the square wave becomes zero. Thus the time for half-cycle is measured. The count can be compared with the stored numbers in a look-up table and the frequency can be measured.

The predetermined threshold values of frequency for underfrequency and overfrequency relays are already stored in the memory. The microprocessor compares the measured frequency with threshold values of frequency. If the measured frequency is below the predetermined limit, microprocessor issues the trip signal to underfrequency relay. If the measured frequency is above the predetermined limit, the microprocessor issues the trip signal to the overfrequency relay. The interfacing circuitry is shown in Fig. 17.1.

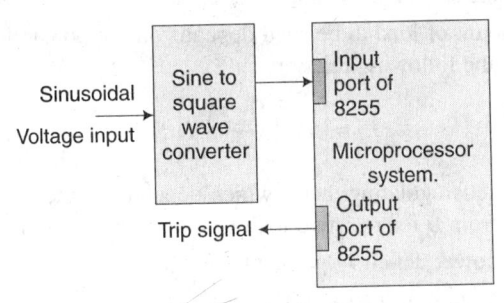

Fig. 17.1 *Interface for frequency measurement*

As soon as the frequency falls below 49.5 Hz, the rate of change of frequency (df/dt) is measured for load shedding. The rate of fall of frequency can be expressed as

$$\frac{df}{dt} = \lim_{T \to 0} \frac{f(t+T) - f(t)}{T} \quad (17.1)$$

where T is the sampling time.

The time period measurement of two consecutive half-cycles are done by using frequency measurement method, and using subtraction and division routines, the df/dt is calculated.

17.3.3 Load-shedding and Restoration

Under normal operating condition of the power system, the amount of electrical power generated is equal to the amount that is consumed. If this state of equilibrium is disturbed due to sudden increase in demand or the demand is more than the generation capacity or by failure of generating or transmission capacity, a power deficit occurs. With small deficit of available generation, the frequency drop may be small and the turbine governors can restore the frequency to normal, provided sufficient spinning reserve is available.

The generation deficiency is characterized by a fall in frequency and therefore it has been a common practice to shed loads with the help of frequency relays until a balance is restored between output and input. Frequency is a reliable indicator of an overload condition. The overload condition causes the generators to slow down and the frequency to drop. In order to halt such a drop, it is necessary to intentionally and automatically disconnect a portion of the load equal to or greater than the overload. After the decline is arrested and the frequency returns to normal, the load may be restored in small increments, allowing the spinning reserve to become active. Frequency-sensitive relays can therefore be used to disconnect load automatically. Such an arrangement is referred to as a load-shedding. Automatic load shedding, based on underfrequency, is necessary since sudden, moderate to severe overloads can plunge a system into a hazardous state much faster than an operator can react. Underfrequency relays are usually installed at distribution substations, where selected loads can be disconnected. The main object of load-shedding is to balance the load and generation. Since the load shed depends on the shortfall the generated power and should before the minimum period possible, the load shedding is done in steps (usually three steps are considered enough).

Since the amount of load to be shed depends on the power deficit, the load to be shed is given by the following relation.

$$L_s = \left[1 - \left(\frac{f_s - f}{f_s}\right) d\right] D \quad (17.2)$$

where, d = self-regulating factor which depends on the type of load and varies from 0 for resistive load to 4 or more for a inductive load.

D = power deficit in per cent

f_s = system normal frequency

f = frequency at any time

L_s = load that must be shed in per cent

Taking an average of $d = 2$ and assuming a 100% overload (i.e., power deficit D of 50%) and maximum frequency fall of 5% (i.e., $f = 95\%$ of normal system frequency f_s), the load to be shedded is 45% i.e., $L_s = 45\%$

This amount of load reduction has to be achieved in suitable steps.

Underfrequency relays will have different frequency setting for load shedding, and the lowest step depends on the critical frequency for the system. The number of setting is usually 2 to 4. Load shedding is utilized to trip predetermined (low-priority) load from the system while keeping the critical load running.

The number of load-shedding steps, the frequency setting for each step, and the amount of load shed at each step should be decided by a system study. The priority must also be decided so that the lowest priority load is taken off first. The same priority used for load shedding should be used when the load is restored.

The relay settings also depend on the rate of fall of frequency (df/dt) as this affects the actual value of system frequency by the time the relay operates. Thus if df/dt is large, the system frequency drops to much lower value than with a smaller value of df/dt.

Another relation for f is

$$f = f_s \sqrt{1 - t\,(D/H)} \qquad (17.3)$$

where, H = inertia constant of the synchronous machines in seconds.

D = deficit in power generation in pu

f = frequency at any time t

Microprocessor or microcontroller can be easily used for automatic load shedding and restoration. The three phase power, frequency and df/dt are required to be measured in microprocessor-based scheme for automatic load-shedding and restoration.

17.4 FIELD PROGRAMMABLE GATE ARRAYS (FPGA) BASED RELAYS

Field Programmable Gate Arrays (FPGA) is a new and exciting device in the area of computing. It is a programmable logic integrated circuit which consists of an array of configurable logic blocks that implement the logical functions of gates. It can be programmed in the field after manufacture. FPGAs are similar in principle to, but have vastly wider application than programmable read only memory (PROM) chips. They are viable alternatives to the two traditional forms of chips, i.e., general-purpose processors and application-specific integrated circuit (ASIC). In FPGAs, the logic functions performed within the logic blocks and sending signals to the chip can alter the connections between the blocks. The configurable logic blocks in FPGAs can be rewired and reprogrammed repeatedly in around a microsecond. FPGAs provide designers with new possibilities for flexibility and performance. By combining high integration with user programmability, they permit designers to rapidly evaluate design alternatives. FPGAs combine the integration of an ASIC with the flexibility of user programmable logic. They present the user with the basic cells and interconnected resources which serve as the building blocks for design implementation. FPGAs allow computer users to tailor microprocessors to meet their own individual needs.

Currently, available relays are either electromechanical or microprocessor based, but all of them are less reliable, slower in operation and have limitations of multitasking. With Field Programmable Gate Arrays (FPGA), we can exploit its re-configurability and re-programmability through software routines to implement many relays on the single FPGA board with faster and reliable operation. FPGA based relays have the following advantages over the conventional relays:

Multitasking With FPGAs, many relays can be implemented on the same chip. Implemented relays may work sequentially, simultaneously, or in a combination of both as per the requirements.

Better Speed As in FPGA, any logic is implemented using a hard-wired approach; thus its speed of operation is better than microprocessors, which in turn uses programming approach.

Use of Less Peripherals Abundance of input/output ports available with an FPGA eliminates the necessity of "Programmable Peripheral Interface (8255)".

Easy Software Handling Design entry is given in the form of schematic diagram or textual hardware description language code (VHDL or VeriLOG).

Flexibility and Re-configurability in Design Re-configurable system combine the advantages of software programming to the performance level of application specific dedicated hardware.

Increased Reliability FPGA provides self checking, fault-detection, fault-tolerant features making whole design more reliable.

17.4.1 Realization of Field Programmable Gate Arrays (FPGA) Based Relays

Field programmable means that the FPGA's function is defined by a user's program rather than by the manufacture of the device. A typical integrated circuit performs a particular function defined at the time of manufacture. In contrast, the FPGA's function is defined by a program written by someone other than the device manufacturer. Depending on the particular device, the program is either burned in permanently or semi-permanently as part of a board assembly process, or is loaded from an external memory each time the device is powered up. This user programmability gives the user access to complex integrated designs without the high engineering costs associated with application-specific integrated circuits.

The realization of a protective relay on FPGA involves many steps. First an algorithm is developed for implementing various logical parts of the design; then a source code is written using Verilog hardware description language. Varilog language is a hardware description language that provides a means of specifying a digital system at various levels of abstraction. The language supports the early conceptual stages of design with its behavioral level of abstraction, and the later implementation stages with its structural abstractions. The language includes hierarchical constructs, allowing the designer to control a description's complexity. The next few steps like generating the net-list, mapping, placement and routing on FPGA can be taken care of by the kit. "Mapping" is the process of assigning a design's logic elements to the specific physical elements, such a CLBs and IOBs, that actually implement logic functions in a device. 'Placing' is the process of assigning physical device locations to the logic in design. 'Routing' is the process of assigning logic nets to physical wire segments in a FPGA that interconnect logic cells.

The FPGA is programmable logic integrated circuit that contains many (64 to over 10,000) identical logic cells that can be viewed as standard components. Each logic cell can independently take on any one of a limited set of personalities. The individual cells and I/O cells are interconnected by a matrix of wires and programmable switches (Interconnect Resources) as shown in Fig. 17.2. A user's design is implemented by specifying the simple logic function for each cell and selectively closing the switches in the interconnect matrix. The arrays of logic cells form a fabric f basic building blocks for logic circuits. Complex designs are created by combining these basic blocks to create the desired circuit. Here, each logic cell combines a few binary inputs (typically between 3 and 10) to one or two outputs according to a Boolean logic function specified in the user program. The cell's combinational logic may be physically implemented as a small look-up table memory (LUT) or as a set of multiplexers and gates.

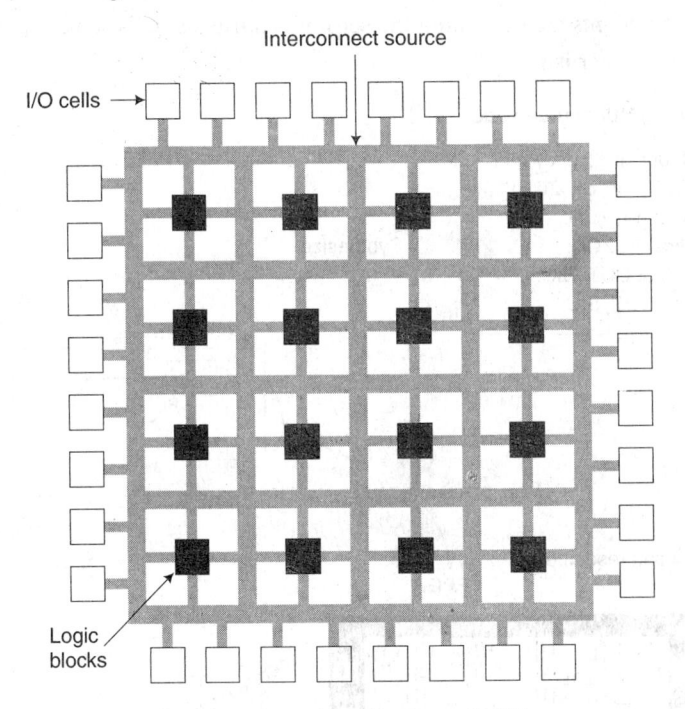

Fig. 17.2 *General architecture of FPGA.*

Individually defining the many switch connections and cell logic functions would be a daunting task. Fortunately, this task is handled by special software. The software translates a user's schematic diagrams or textual hardware description language code then places and routes the translated design. Most of the software packages have hooks to allow the user to influence implementation, placement and routing to obtain better performance and utilization of the device. Libraries of more complex function macros (e.g., adders) further simplify the design process by providing common circuits that are already optimized for speed or area.

17.4.2 Frequency Relays (Under Frequency and Over Frequency)

A frequency sensitive relay is necessary, when due to change in loading, the frequency of line current is violating permissible limits. The algorithm used to build the frequency relay is as follows:

1. The clock counts corresponding to the lower limit of frequency and the upper limit of frequency are defined (in a half-cycle).
2. An external hardware circuit is used for detecting the zero crossing.

3. Total number of clock counts between two consecutive zero crossings is determined. This gives the counts corresponding to the frequency of the input signal. A counter is activated once it detects a zero crossing, and it counts till the next zero crossing is detected.
4. This calculated count is compared with the upper and the lower set frequency limits.
5. If the counts are not within the set limits a trip signal is sent.
6. The counter is reset.

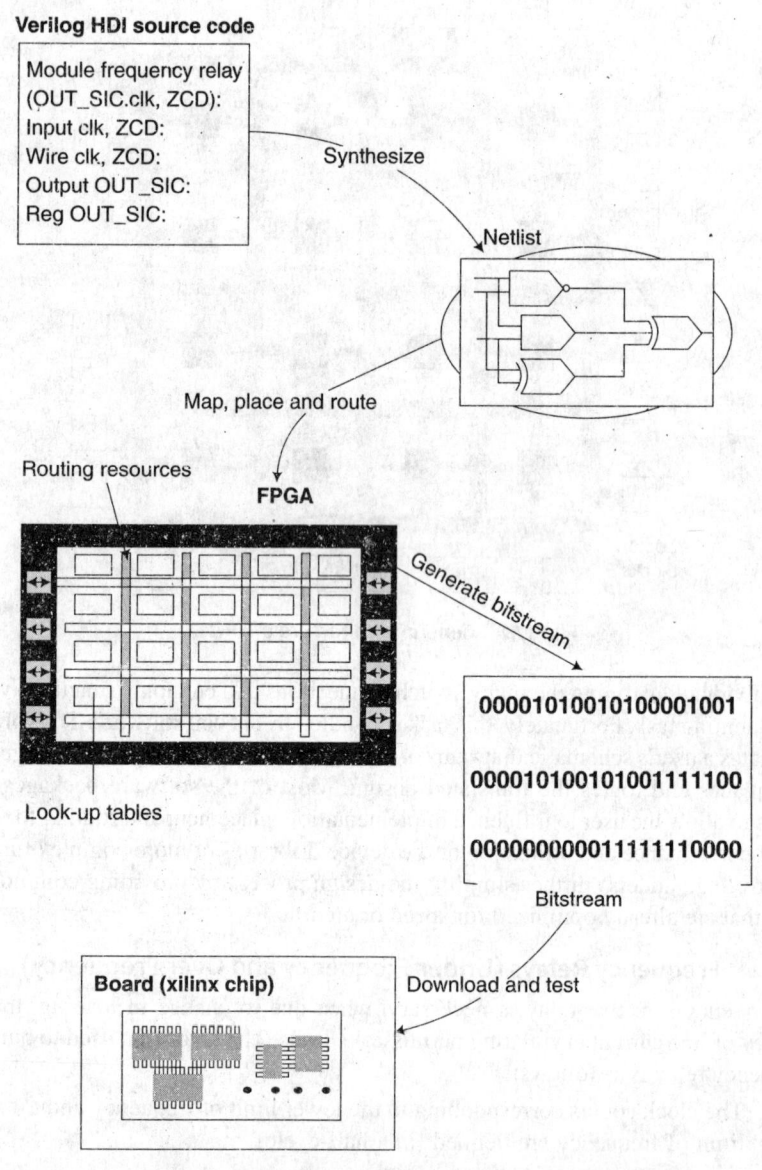

Verilog HDl source code

Module frequency relay
(OUT_SIC.clk, ZCD):
Input clk, ZCD:
Wire clk, ZCD:
Output OUT_SIC:
Reg OUT_SIC:

Synthesize

Netlist

Map, place and route

Routing resources

FPGA

Generate bitstream

0000101001010001001

00001010010100111100

000000000011111110000

Bitstream

Look-up tables

Board (xilinx chip)

Download and test

Fig. 17.3 *Relay implementation using a FPGA kit.*

Modern Trends in Power System Protection 635

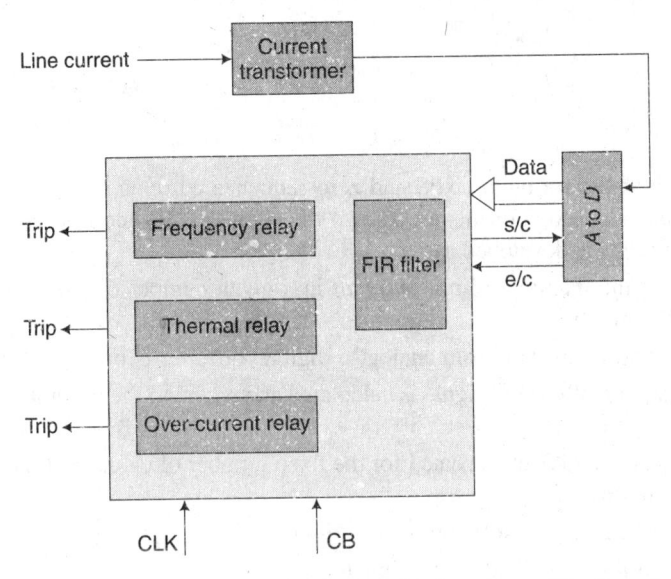

Fig. 17.4 The design of hardware circuit-FPGA chip along with other circuits

17.4.3 Thermal Relay (also for Zero Sequence and Negative Sequence Protection)

The basic algorithms for implementing the Thermal relay and negative sequence protection (NPS) and zero sequence protection are same. For all the three types on integral the form $\int I^2\, dt$ needs to be computed. The only notable difference being, that from the line current- the zero sequence and negative phase sequence components have to be obtained using the following formulae:

$$I_0 = \frac{1}{3}(I_a \angle 0° + I_b \angle \theta_b + I_c \angle \theta_c) \tag{17.4}$$

$$I_2 = \frac{1}{3}(I_a \angle 0° + a^2 I_b \angle \theta_b + a I_c \angle \theta_c) \tag{17.5}$$

where a is an operator equal to $1\angle 120°$

And simple algebraic manipulation yields:

$$I_{r2} = \frac{1}{3}\left(I_{ra} - 0.5 I_{rb} - 0.5 I_{rc} + \frac{\sqrt{3}}{2} I_{xb} - \frac{\sqrt{3}}{2} I_{xc}\right) \tag{17.6}$$

$$I_{x2} = \frac{1}{3}\left(\frac{\sqrt{3}}{2} I_{rc} - 0.5 I_{xc} - 0.5 I_{xb} - \frac{\sqrt{3}}{2} I_{rb}\right) \tag{17.7}$$

$$|I^2| = \sqrt{I_{r2}^2 + I_{x2}^2} \tag{17.8}$$

Where, the subscript 'r' denotes the real pat and 'x' denotes the imaginary part. Similarly, for zero sequence:

$$|I_0| = \frac{1}{3}\sqrt{I_{r0}^2 + I_{x0}^2} \tag{17.9}$$

where,
$$I_{r0} = (I_{ra} + I_{rb} + I_{re}) \tag{17.10}$$
And
$$I_{x0} = (I_{xb} + I_{xc}) \tag{17.11}$$

Using the above formulae, NPS and zero sequence currents are obtained. Once all the symmetrical components are known the algorithm for thermal, NPS or zero sequence currents is developed as:

1. Setting the maximum limit of $\int I^2 dt$ in a given number of clock cycles sets the thermal limit
2. The continuous data from analog to digital converter is obtained for current
3. The square of current signal is calculated and added to the accumulator every clock cycle
4. The step 2 and 3 are repeated for the fixed number of clock cycles as defined in Step 1 above
5. The thermal limit violation is then tested
6. If the limit is violated, a trip signal is sent
7. The accumulator is cleared and again steps 2 to 7 are repeated

17.4.4 Overcurrent Relay (IDMT Relay)

The protective scheme which responds to a rise in current flowing through the protected element over a pre-determined value is called 'over-current protection' and the relays used for this purpose are known as over-current relays. The algorithm used to build the over-current relay is as follows:

1. The inverse time characteristics are defined as to what delay is required at what value of overcurrent.
2. An external hardware circuit (analog to digital converter) is used for getting digital value of current.
3. Initially a start of conversion signal is generated so that conversion is started in A/D converter and digital value of current is sent to the chip.
4. This digital value is compared with the set current limits.
5. If the data is not within limits a delay value is selected, corresponding to the value of overcurrent.
6. The delay is generated before sending the trip signal.
7. Hence inverse-time characteristics have been realized.

17.4.5 Implementation on FPGA Kit

The implementation of the relay on FPGA, as illustrated in Figure 17.3, is managed and processed by the ISE Project Navigator through the following steps in the ISE design flow:

Design Entry Design entry is the first step in the ISE design flow. During design entry, source files based on the design objective are created. The top-level design file is created using a Hardware Description Language (HDL), such a VHDL, Verilog, or ABEL, or using a schematic.

Synthesis After design entry and optional simulation, synthesis is run. During this step, VHDL, Verilog, or mixed language designs becomes netlist files that are accepted as input to the implementation step.

Implementation After synthesis, the design implementation is run, which converts the logical design into a physical file format that can be downloaded to the selected target device. Also, the Project Navigator provides the flexibility to run the implementation process in one step or each of the implementation processes separately.

Verification The functionality of the design can be verified at several points in the design flow. The simulator software can be used to verify the functionality and timing of the design or a portion of the design. The simulator interprets VHDL or Verilog code into circuit functionality and displays logical results of the described HDL to determine correct circuit operation. In-circuit verification can also be run after programming the FPGA.

Device Configuration After generating a programming file, the device can be configured. During configuration, configuration files are generated and the programming files are downloaded from a host computer to a Xilinx device.

All these relays can be integrated and implemented on a single XILINX-FPGA chip. The key features are the following:

- A 'top level' module can be created in which all the relays can be simultaneously implemented.
- The input current signal is common to all the relays. It is phase current value, for which signal conditioning has to be done.
- Then the RTL schematic design is simulated.
- Also the chip area utilized and the percent port utilization can be obtained.

The design of Hardware Circuit-FPGA chip along with other circuits is shown in Fig. 17.4.

The advantages of developing FPGA based relays are numerous.

- By implementing the Filter on same chip, as opposed to, the cost of using external filter is saved.
- The reconfigurability and in-system programmability of FPGA provides the option of producing the most optimal design for any application.
- The development of such single chip solution on FPGA is more cost effective than using any other DSP technique.
- The power requirements of the circuit will be minimal and the response time is superior to any of the solid state relays presently available.

17.5 ADAPTIVE PROTECTION

The concept of adaptive protection is used for the protection of complex interconnected power system. The power system topology changes due to deliberate system switching, e.g. isolation of a part of the equipment for maintenance and unplanned system switching e.g., removal of a fault by a relay from a healthy power system. A change in the power system topology requires updating of the relay settings to adapt to the changed power system condition. The power system topology gets changed but the settings of the protection system remain the same as earlier, that is why the latter may not be able to adequately protect the current system. This is the main basis for using the adaptive protection system.

The settings of the many relays are dependent upon the assumed conditions in the power system and it is not possible to realise these conditions in complex interconnected power system. Hence it is desirable that the relay settings adapt themselves to the real-time system as and when the system conditions change. The protection scheme which uses such a relaying scheme is called adaptive protection.

Because of advances in digital communications, it is possible for the manufacturers of relays to include the communication hardware and protocols in most of their products marketed at this time. This has provided impetus for the protection engineers to investigate and exploit the possibilities of using adaptive protection concepts to reduce the impact of faults and disturbances on power systems and their equipment.

Adaptive protection is a protection philosophy which permits and makes adjustments to protection functions automatically for making them more attuned to the prevailing power system conditions.

Adaptive protection systems are expected to perform the following functions:
 (i) The operation of the relays should be for faults in their respective zone of protection, as defined by the direction that the relay supervises.
 (ii) If a fault is sensed by both the relays in their respective zones of protection, the relay with shorter time delay or the relay which is subjected to the larger fault current should trip first.
 (iii) with change in the power system topology, the relay should automatically switch to the appropriate relay group for proper protection of the power system.

The specific areas which can benefit from incorporating adaptive features in protection include, among others, the following.
 (i) Selecting the most suitable algorithm
 (ii) Balancing security and dependability
 (iii) Changing relay settings as the system configuration modifies
 (iv) Compensating for measurement errors
 (v) Transformer differential protection
 (vi) Protecting circuits connected to a transfer bus
 (vii) Compensating for mutual coupling
 (viii) Compensating for pre-fault load flow
 (ix) Protection of multi-terminal lines
 (x) Backup protection of transmission lines
 (xi) Out of step protection

17.6 INTEGRATED PROTECTION AND CONTROL

With advancements in computer technology, electromechanical and discrete electronic devices used for power system protection, metering and supervisory control are being replaced by integrated protection and control systems. The need of the day is for modern substations and for installing intelligent electronic devices such as numerical relays, programmable logic controllers (PLCs), etc., at the bay level. These are called bay controllers.

With advancements in computer technology, it has now become possible to introduce microcomputers, digital signal processors (DSPs) A/D converters, optical transducers and fibre-optic communication systems to acquire and process electrical power system information in an effective manner and use it in the development of the integrated protection and control system for a substation. Such a system not only provides a cost-effective solution to the problem earlier faced by power utilities while using conventional protection and control equipment but adds the good features of man machine interface (MMI), disturbance records and event recording useful for post-fault analysis.

An integration system combines the hardware and software components needed to integrate the various bay controllers, providing data collection, control and monitoring from a central workstation. This system consists of network interface devices, host computer and printer. The network interface devices provide interface between the bay controller specific protocol and system's local area network (LAN). Numerical protection along with integrated system and suitable communication medium, provides a total solution for protection and control of power system. The integrated protection and control system is the latest trend in the world and need of the day.

The integrated protection and control system provides the following features:
 (i) Power system protection
 (ii) Supervisory control and Data Acquisition (SCADA)
 (iii) Statistical and revenue metering
 (iv) Local control
 (v) Voltage regulator
 (vi) Station Battery monitoring, and
 (vii) Digital fault recording

17.6.1 Power System Protection

A modern numerical relay is most suited to perform the function of the protection and metering of the considered electrical equipment.

The microcomputer subsystem is the heart of the numerical relay. It is responsible for processing the input signals and making protective relaying decisions. The relay software executes a variety of signal processing algorithms and calculates several system parameters. The relay logic software compares these parameters against relay settings to detect fault and abnormal conditions. Communication ports provide remote communication capability to the outside world. The numerical relay also requires a power supply to provide various dc voltages required for its normal operation. The various subsystems of a typical numerical relay are

(i) Analog input unit
(ii) Central processing unit
(iii) Digital input/output unit
(iv) Power supply unit
(v) Communication unit

17.6.2 Supervisory Control and Data Acquisition (SCADA)

Supervisory control and data acquisition (SCADA) is a software package positioned on the top of the hardware modules to which it is interfaced. Programmable Logic Controller (PLC) is one of the popular hardware modules. SCADA software which is based upon a real-time data base (RTDB) is available in different servers. The SCADA systems offer the advantages of functionality, openness, performance and scalability. The SCADA systems perform the critical functions of data acquisition, supervisory control, and alarm display and control.

Most modern day substations and generating stations use some form of SCADA to provide support to the system operators or operating centres to locally or remotely monitor and control equipment. SCADA originates from the control centre in software form and is used for process controlling. SCADA works by gathering data in real time from remote locations in order to control equipment and conditions. From the control centre, SCADA computers are connected to remote locations using various forms of communication equipment and protocols. SCADA system is intended to facilitate the work of the operator (dispatcher) by acquiring and compiling information as well as locating, identifying and reporting faults. On the basis of the information received, the operator makes the necessary decisions via the control system. He can then perform different control operations in substations and power stations or influence the processing of the information acquired.

17.7 RELAY RELIABILITY

The relay reliability is described in the following two ways.

(i) The relay must operate correctly in response to a system problem. This is known as *dependability*.
(ii) The relay must not operate incorrectly during normal system operation. This is known as *security*.

Dependability is defined as the degree of certainty that a relay or relay system will operate correctly. Security is defined as the degree of certainty that a relay or relay system will not operate inadvertently. In more direct terms, dependability is a measure of the ability of the protective relay to perform correctly when required. Security is its ability to avoid unnecessary operation during normal system conditions or in response to system faults outside its zone of protection. Increasing relay dependability tends to reduce relay security, and increasing relay security tends to decrease relay dependability. Numerical relays provide the platform to maximise both dependability and security, because they make informed decisions based on past-recorded historic data.

17.8 ADVANTAGES OF FAST FAULT CLEARING

The main advantages of fast fault clearing with the help of fast acting protective relays and associated circuit breakers are as follows:

(i) The permanent damage to the equipment and components is avoided.
(ii) Chances of fire and other hazards are reduced.
(ii) Risk to the life of personnel is also reduced.
(iv) By quickly isolating the particular faulted section or components, the continuity of power supply is maintained in the remaining health section.
(v) Since the fault arising time is reduced, the power system can be brought back to the normal state sooner.
(vi) The transient state stability limit of the power system is greatly improved.
(vii) The possibility of development of the simplest but the most common fault such as L-G fault into more severe fault, such as 2L-G fault, is avoided.

EXERCISES

1. What is a Gas Insulated Substation? Discuss its advantages and disadvantages as compared to conventional air insulated substation.
2. What are the various components of a GIS? Briefly describe their functions.
3. Discuss the classification of GIS.
4. Briefly describe the various types of frequency relays. How can under frequency and df/dt relays be used for load shedding and restoration?
5. Briefly describe a microprocessor-based scheme for automatic load shedding and restoration.
6. Discuss the realization of FPGA based relays. What are their advantages?
7. Discuss the principle of operation of travelling wave relays. What are the types of these relays? Briefly describe them.
8. What do you mean by adaptive protection? What functions are expected to be performed by an adaptive protection system?
9. What are the specific area which can benefit from incorporating adaptive features in protection?
10. What are the features of integrated protection and control? Discuss the role of SCADA in modern substations.
11. How is relay reliability described? What are the advantages of fast fault clearing?

8086 Assembly Language Programming

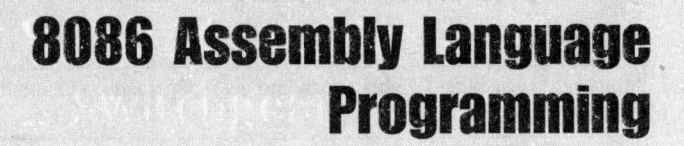

EXAMPLES

(i) 16-bit by 16-bit Unsigned Multiplication

Program

Memory Address	Machine Code	Label	Mnemonics	Operands	Comments
0200	A1,01,05		MOV AX,	[0501]	Bring the 16 bit multiplicand in AX.
0203	BB,01,06		MOV BX,	[0601]	Address of multiplier in BX.
0206	F7,27		MUL AX,	[BX]	Multiply the 16 bit contents of register AX by the 16 bit content of memory whose address is in BX.
0208	BB,03,06		MOV BX,	0603	
020B	89,07		MOV[BX],	AX	Bring low-order word of the 32-bit result in memory locations having addressess 0603 and 0604.
020D	43		INC	BX	
020E	43		INC	BX	
020F	89, 17		MOV [BX],	DX	Bring high-order word of the result in memory locations having addresses 0605 and 0606.
0211	F4		HLT		

For example, let the data be
 Multiplicand = 8754 H
 Multiplier = 6293 H
Data is stored in the following:

Memory Location			Result (product = 341BDD3C)	
Memory Address		Data	Memory Address	Result
0501-LSB of Multiplicand		(54)	0603 – LSB	(3C)
0502-MSB of multiplicand		(87)	0604 –	(DD)
0601-LSB of multiplier		(93)	0605 –	(IB)
0602-MSB of multiplier		(62)	0606 – MSB	(34)

(ii) Multiplication of two 16-bit signed numbers

Program 1

Memory Address	Machine Code	Label	Mnemonics	Operands	Comments
0200	A1,01,05		MOV AX,	[050]	Multiplicand
0203	BB,01,06		MOV BX,	0601	For multiplier.
0206	F7,2F		IMUL AX, [BX]		Multiply the 16-bit signed number in AX by the 16-bit signed number in memory location addressed by BX
0208	BB,03,06		MOV BX,	0603	For result.
020B	89,07		MOV [BX],	AX	Low-order word of the result.
020D	43		INC	BX	
020E	43		INC	BX	
020F	89,17		MOV [BX],	DX	High-order word of the result.
0211	F4		HLT		

Example

Memory Address	Data	Memory Address	Result
0501-LSB of Multiplicand	(FF)	0603-LSB	(00)
0502-MSB of multiplicand	(7F)	0604	(80)
0601-LSB of multiplier	(00)	0605	(00)
0602-MSB of multiplier	(80)	0606-MSB	(C0)

Maximum positive number = 7FFF
Maximum negative number = – 8000
Negative number is represented by 2's complement
2' complement of 8000 = 8000
Result (product) = (+ 7FFF) × (– 8000) = – 3FFF8000
(– 3FFF8000) is represented by 2's complement. The 2's complement of 3FFF8000 is C0008000.

The data and results for signed numbers are stored in the form of 2's complement. The above program is valid for any 16-bit signed numbers.

(iii) Division of unsigned 32-bit number by unsigned 16-bit number

Program 2

Memory Address	Machine Code	Label	Mnemonics	Operands	Comments
0200	BB,01,06		MOV	BX,0601	
0203	8B,07		MOV	AX, [BX]	Bring low-order word of dividend in AX.
0205	43		INC	BX	
0206	43		INC	BX	

Appendix A

0207	8B,17		MOV DX, [BX]		Bring high-order word of dividend in DX
0209	BB,01,05		MOV BX, 0501		Divisor
020C	F7,37		DIV DX, [BX]		
020E	43		INC BX		
020F	43		INC BX		
0210	89, 07		MOV [BX], AX		Quotient
0212	43		INC BX		
0213	43		INC BX		
0214	89, 17		MOV [BX], DX		Remainder
0216	F4		HLT		

Example

Memory Address	Data	Memory Address	Result
for data		for result	
0601 – LSB	(92)	0503-LSB of Quotient	(93)
0602	Dividend (11)	0504-MSB of Quotient	(62)
0603	(IC)	0505-LSB of Remainder	(56)
0604 – MSB	(34)	0506-MSB of Remainder	(34)
0501 – LSB	(54)		
0502 – MSB	Divisor (87)		

Hence, for the above data, Result
Dividend = 341C1192 (32-bit)   Quotient = 6293 H
Divisor   = 8754 (16-bit)   Remainder = 3456 H
The above program is valid for any data.

(iv) Division of 32-bit signed number by 16-bit signed number

Program 3

Memory Address	Machine Code	Label	Mnemonics	Operands	Comments
0200	BB,01,06		MOV BX,	[0601]	
0203	8B,07		MOV AX	[BX]	Low order word of dividend in AX
0205	43		INC	BX	
0206	43		INC	BX	
0207	8B,17		MOV DX,	[BX]	High-order word in DX
0209	BB,01,05		MOV BX,	[0501]	
020C	F7, 3F		IDIV DX,	[BX]	
020E	43		INC BX		
020F	43		INC BX		
0210	89,07		MOV [BX], AX		Quotient
0212	43		INC BX		
0213	43		INC BX		
0214	89,17		MOV [BX], DX		Remainder
0216	F4		HLT		

For example, let dividend = – 341C1192   (2's complement = CBE3EE6E)
Divisor = 7754
Quotient = – 6FCB   (2's complement = 9035)
Remainder = – 05F6   (2's complement = FAOA)

Data		Result	
Memory Address	Data	Memory Address	Data
0601 – LSB	(6E)	0503 – LSB of Quotient	(35)
0602	Dividend (EE)	0504 – MSB of Quotient	(90)
0603	(E3)	0505 – LSB of Remainder	(0A)
0604 – MSB	(CB)	0506 – MSB of Remainder	(FA)
0501 – LSB of Divisor	(54)		
0502 – MSB of Divisor	(77)		

(v) Multibyte addition of two numbers each N bytes long

First number starting from address 0300 H
Second number starting from address 0308 H
Length of each No. = 8 bytes

Example

	MSB							LSB
I No =	8B	FD	BC	49	32	56	98	74
II No =	2A	78	56	DC	ED	32	67	91
Result								
(Addition) =	B6	76	13	26	1F	89	00	05

Result is to be stored in address from 0308 H
Here N = 08 bytes
Words = N/2 = 04 (Word length of numbers)

Program 4

Memory Address	Machine Code	Label	Mnemonics	Operands	Comments
0200	F8		CLC		
0201	B9,04,00		MOV	CX,0004	Word length
0204	BE,00,03		MOV	SI,0300	
0207	8B,04	GO	MOV	AX, [SI]	
0209	11, 44, 08		ADC	[SI + 08], AX	
020C	46		INC	SI	
020D	46		INC	SI	
020E	E0, F7		LOOP NZ,	GO	
0210	F4		HLT.		

Appendix A **647**

I No.		II No.		Result of addition	
Memory	Data	Memory	Data	Memory	Data
0300 -	74 (LSB)	0308 -	91 (LSB)	0308 -	05 (LSB)
0301 -	98	0309 -	67	0309 -	00
0302 -	56	030A -	32	030A -	89
0303 -	32	030B -	ED	030B -	1F
0304 -	49	030C -	DC	030C -	26
0305 -	BC	030D -	56	030D -	13
0306 -	FD	030E -	78	030E -	76
0307 -	8B (MSB)	030F -	2A (MSB)	030F -	B6 (MSB)

(vi) Program to read digital output of ADC 1211 corresponding to analog signals at channels S_1 and S_3 of analog multiplexer

Referring to Fig. 12.47, the 12-bit digital output data is input at two ports having consecutive addresses.

The addresses of the I/O ports of 8255 of the 8086 microprocessor kit of Vinytics make are as follows.

Device name	Port name	Port addresses
8255-1 Port C	Port A (P_1A)	FFF9
	Port B (P_1B)	FFFB
	Port C (P_1C)	FFFD
	Control Word	FFFF

Device name	Port name	Port Addresses
8255-2	Port A (P_2A)	FFF8
	Port B (P_2B)	FFFA
	Port C (P_2C)	FFFC
	Control Word	FFFE

It is clear that ports P_2A and P_1A have consecutive addresses. Therefore, the data is input at these ports. The input port (P_2A) having lower address inputs the low-order 8 bits and the input port (P_1A) having higher address inputs the high-order 4 bits, i.e.

P_2A (Input) - Low-order 8 bits of ADC1211

P_1A (Input) - High-order 4 bits of ADC1211

The program to read the 12-bit digital output of ADC1211 is as follows.

Program 5

Memory Address	Machine Code	Label	Mnemonics	Operands	Comments
0200	BA,FF,FF		MOV DX,	FFFF	Initialise I/O ports of 8255-1 and 8255-2.

0203	B0,98		MOV AL,	98	
0205	EE		OUT [DX]		
0206	4A		DEC DX		
0207	EE		OUT [DX]		
0208	83,EA,03		SUB DX,	03	Address of P_1B in DX
020C	B0,01		MOV AL,	01	Logic high to both S/H.
020D	EE		OUT [DX]		
020E	B9,3F,00		MOV CX,	003F	For delay
0211	E2, FE	HERE	LOOP	HERE	Decrement CX and jump to HERE if CX $\neq$ 0 (DISP = -2 = FE, 2's complement)
0213	B0,00		MOV AL,	00	Logic low to both S/H.
0215	EE		OUT [DX]		
0216	42		INC DX		Address of P_2C in DX
0217	B0,00		MOV AL	00	Switch on channel
0219	EE		OUT [DX]		S_1 of AM3705
021A	B0,00		MOV AL,	00	Logic low to S/C
021C	EE		OUT [DX]		without affecting S_1
21D	B0,08		MOV AL,	08	Logic high to S/C
021F	EE		OUT [DX]		without affecting S_1
0220	EC	READ	IN [DX]		Read $\overline{CC}$ to check end
0221	D0,D0		RCL AL		of conversion and jump
0223	72,FB		JB, READ		on carry to READ $DISP_8 = -05 = FB$
0224	BA, F8, FF		MOV DX	FFF8	Address of P_2A
0227	ED		INW [DX]		Input data.
0228	F7, D0		NOT AX		Complement AX
022A	A3, 00, 05		MOV [0500],	AX	Store content of AX in memory
022D	83, C2, 04		ADD DX,	04	Address of P_2C in DX
0230	B0,02		MOV AL,	02	
0232	EE		OUT [DX]		Switch on S_3
0233	B0, 02		MOV AL,	02	Logic low to S/C
0235	EE		OUT [DX]		without affecting S_3
0236	B0, 0A		MOV AL,	0A	Logic high to S/C
0238	EE		OUT [DX]		
0239	EC	ABOVE	IN [DX]		Read $\overline{CC}$ to check end of conversion.
023A	D0,D0		RCL,	AL	
023C	72,FB		JB,	ABOVE	
023E	83,EA, 04		SUB DX,	04	Address of P_2A in DX.
0241	ED		INW [DX]		Input data
0242	F7, D0		NOT AX		Complement AX
0244	A3, 00, 06		MOV[0600], AX		Store the content of AX in memory
0247	F4		HLT		

Digital values of signal at S_1 are stored at memory addresses 0500 and 0501, and that of signal at S_3 at memory addresses 0600 and 0601.

Addresses for memory locations for storing 12-bit digital values

Analog signal at channel	Addresses of Low-order 8 bits	Addresses of High-order 4 bits
S_1	0500	0501
S_3	0600	0601

B

Orthogonal and Orthonormal Functions

A set of real-valued continuous functions

$\{x_n(t)\} = \{x_0(t), x_1(t), ...\}$ is called orthogonal over a time interval $(0, T)$ if

$$\int_0^T x_m(t)\, x_n(t)\, dt = \begin{cases} K & \text{if } m = n \\ 0 & \text{if } m \neq n \end{cases}$$

where, m and n are integers and K is a constant.

When the constant $K = 1$, $\{x_n(t)\}$ is called an orthonormal set of functions.

Only orthogonal sets of function can be made to completely synthesise any time function to a required degree of accuracy. Further, the characteristics of an orthogonal set are such that recognition of a particular member of the set contained in a given time function can be made using quite simple mathematical operations on the function. Using orthogonal functions, a signal can be expressed as a limited set of coefficients.

Gray-code to Binary Conversion

Let $g_{n-1}, g_{n-2}, \ldots g_2, g_1, g_0$ denote a code word in the n-bit Gray code corresponding to the binary $b_{n-1}, b_{n-2}, \ldots b_2, b_1, b_0$. To convert from Gray code to binary, we start with the leftmost digit and move to the right, making $b_i = g_i$ if the number of 1's preceding g_i even, and making $b_i = \bar{g}_i$ (bar denoting complement) if the number of 1's preceding g_i is odd. During this process, zero 1's is treated as an even number of 1's. For example, the binary number corresponding to the Gray code word 11001101 is found to be 10001001 in the following manner.

g_7	g_6	g_5	g_4	g_3	g_2	g_1	g_0
1	1	0	0	1	1	0	1
↓	↓	↓	↓	↓	↓	↓	↓
1	0	0	0	1	0	0	1
b_7	b_6	b_5	b_4	b_3	b_2	b_1	b_0

Kronecker (or Direct) Product of Matrices

Kronecker (or direct) product of two matrices C and D is given by

$$C \otimes D = \begin{bmatrix} C_{11}D & C_{12}D & \cdots & C_{1n}D \\ C_{21}D & C_{22}D & \cdots & C_{2n}D \\ \vdots & \vdots & & \vdots \\ C_{m1}D & C_{m2}D & \cdots & C_{mn}D \end{bmatrix}$$

If the order of matrix C is $m \times n$ and the order of matrix D is $k \times l$, then the order of $C \otimes D$ is $mk \times nl$.

References

Books

1. A. R. Van C. Warrignton, *Protective Relays–Their Theory and Practice*, Vol. I and II, 3rd ed. Chapman & Hall, London and John Wiley & Sons, New York, 1977.
2. Electricity Council, *Power System Protection*, Vol, I, II and III, Macdonald & Co. Ltd., London, 1st ed. 1969.
3. C.R. Mason, *The Art and Science of Protective Relaying*, John wiley & Sons, 1956.
4. A.E. Guile and W. Paterson, *Electrical Power System*, Vol. I & II, 2nd ed. Pergamon Press, 1977.
5. G.E.C Measurements, *Protective Relays Application Guide*, G.E.C., ed., 1987.
6. D. Robertson, *Power System Protection Reference Manual*, Oriel Press, Stocksfield, London, 1982.
7. B.D. Russel and M.E. Council, *Power System Control and Protection*, Academic Press, 1978.
8. T.S.M. Rao, *Power System Protection: Static Relays with Microprocessor Applications*, 2nd ed., Tata McGraw-Hill, New Delhi, 1989.
9. B. Ravindranath and M. Chander, *Power System Protection and Switchgear*, Wiley Eastern, New Delhi 1977.
10. S.S. Rao, *Switchgear and Protection*, Khanna Publishers, Delhi, 1986.
11. J. Lewis Blackburn and Thomas J. Domain, Protective Relaying : Principles and Applications, 3rd Edition, CRC Press, 2006.
12. Anthony F. Sleva, Protective Relaying Principles, CRC Press, 2009.
13. Stanley H. Horowitz and Arun G. Phadke, Power System Relaying, John wiley & sons, 2008.
14. Walter A. Elmore, Protective Relaying Theory and applications, Marcel Dekker, New York, 1994.
15. Y.G. Paithankar, Transmission Network Protection : Theory and Applications, Marcel Dekker, New York, 1998.
16. Arun G. Phadke and James S. Thorp, Computer Relaying for Power System, second Edition, John wiley & sons, 2009.
17. C. Christopoulos and A. Wright, Electrical Power System Protection, Kluwer Academic Publishers, The Netherlands, 1999.
18. Gerhard Ziegler, Numerical Distance Protection, Wiley-VHC, 2008.
19. Gerhard Ziegler Numerical Differential Protection, : Principles and Applications, Wiley-VHC, 2005.
20. A.T. Johns and S.K. Salman, Digital Protection of Power Systems, IET, 1997.

21. John J. Grainger and William D. Stevenson, Jr. McGraw-Hill, Inc, 1994.
22. D.P. Kothari and I.J. Nagrath, Modern Power System Analysis, Tata McGraw-Hill, 2006.
23. Hadi saadat, Power System Analysis, Tata McGraw-Hill Education, 2002.
24. J.C. Das, Power System Analysis, CRC Press, 2002.
25. Arthur R. Bergen and Vijay Vittal, Power System Analysis, second edition, Prentice-Hall, Inc, New Jersy, 2000.
26. J.D. Glover, M.S. Sarma and T.J. overbye, Power System Analysis and Design, Cengage Learning, 2008.
27. T.K. Nagsarkar and M.S. Sukhija, Power System Analysis, Oxford University Press, New Delhi, 2007.
28. B. Ram, Fundamentals of Microprocessors and microcomputers, 5th edition, Dhanpat Rai & Sons, 2001.
29. D.V. Hall, *Microprocessors and Interfacing–Programming and Software*, McGraw-Hill, 1986.
30. Lance A. Leventhal, *8080/8085 Assembly Language Programming*, Osborne/McGraw-Hill, Berkeley, 1976.
31. National Semiconductor, *Linear Data Book*, Santaclara, California, 1982.
32. Adam Osborne, *An Introduction to Microcomputers, Vol. II : Some Real Products*, Adam Osborne and Associates, Berkeley, 1976.
33. Yu-cheng Liu and Glenn A. Gibson, *Microcomputer System: The 8086/8088 Family: Architecture, Programming and Design*, Prentice-Hall of India, New Delhi, 1988.
34. R. Rector and G. Alexy, *The 8086 Book*, Osborne/McGraw-Hill, 1980.
35. M. Rafiquzzaman, *Microprocessor and Microcomputer Development System: Designing Microprocessor-Based System*, Harper and Row Publishers, New York, 1984.
36. D.F. Elliott and K.R. Rao, "*Fast Transforms: Algorithms, Analysis, Applications*, Academic Press, 1982.
37. N. Ahmad and K.R. Rao, *Orthogonal Transforms for Digital Signal Processing*, Springer-Verlag, Berlin, 1975.
38. D.G. Fink an H.W. Beaty, *Standard Handbook Engineers*, 11th ed., McGraw-Hill Book Company, 1978.
39. E. Kuffel and M. Abdullah, *High Voltage Engineering*, Pergamon Press, 1970.
40. M. Khalifa, *High Voltage Engineering Theory and Practice*, Marcel Dekker, New York and Basel, 1990.
41. W.W. Lewis, *The Protection of Transmission System against Lightning*, John Wiley & Sons, 1950.
42. H. Cotton and H. Barber, *The Transmission and Distribution of Electrical Energy*, B.I. Publications, 1977.
43. V.N. Maller and M.S. Naidu, "Advances in High Voltage Insulation and Arc Interruption in SF_6 and Vacuum, Pergamon Press, Oxford, New York, 1981.

Paper

44. T.U. Patel, "Protective Current Transformers", *Journal of the Institution of Engineers (India)*, Vol. XLIII, No.4, Pt. EL 2, December 1962, pp 87-101.
45. D. Dalasta, F. Free and A.P. Desnoo, "An Improved Static Overcurrent Relay", *IEEE Transactions on Power Apparatus and Systems*, No. 68, October 1963, pp 705-716.

46. S.C. Gupta and Dr. T.S.M. Rao, "Improved Static Overcurrent Relay with IDMT Characteristic", *The Journal of the Institution of Engineers (India)*, Vol. 53, Pt EL3, February 1973, pp 138-140.
47. H.K. Verma and T.S.M. Rao "Inverse Time Overcurrent Relay Using Linear Components", *IEEE Transaction on Power Apparatus and Systems*, Vol. PAS-95, No. 5, Sept./October 1976, pp 1738-1743.
48. J. Rushton and W.D. Humpage, "Power-System Studies for the Determination of Distance Protection Performance", *Proceedings of IEE*, Vol. 119, No. 6, June 1972, pp 677-688.
49. C.H. Griffin, "Principles of Ground Relaying for High Voltage and Extra High Voltage Transmission Lines", *IEEE Transaction on Power Apparatus and System*, Vol PAS-102, No. 2, February 1983, pp 420-432.
50. M. Okamura, F. Andow and S. Suzuki, "Improved Phase-Comparison Relaying with High Performance", *IEEE Transaction on Power Apparatus and Systems*, Vol PAS-99, No. 2, March/April 1980, pp 522-527.
51. J.L. Blackburn, "Ground Fault Relay Protection of Transmission Lines", *IEEE Transaction on Power Apparatus and Systems*, No. 1, August 1952, pp 685-692.
52. W.A. Elmore and J.L. Blackburn, "Negative-Sequence Directional Ground Relaying", *IEEE Transaction on Power Apparatus and Systems*, No. 64, February 1963, pp 913-921.
53. R.A. Larner, "Protective Practices for EHV System", *IEEE Transaction on Power Apparatus and Systems*, No. 64, February 1963, pp 1020-1028.
54. W.D. Humpage and S.P Sabberwal, "Developments in Phase Comparison Techniques for Distance Protection", *Proceedings of IEE*, Vol. 112 (7), 1965, p 1383.
55. J.B. Patrickson, "The Protection of Transmission Lines–Present and Future Trends", Reyrelle Rev., Dec. 1964.
56. J.J. Loving, "Electronic Relay Developments", *A.I.E.E*, Vol. 68, Part 1, 1949, pp 233-242.
57. C. Adamson and L.M. Wedepohl, "Power System Protection with Particular Reference to the application of Junction Transistors to Distance Relays", *Proceedings IEE*, Vol. 1956, p 379.
58. C. Adamson and L.M. Wedpohl, "A Dual Comparator Mho Type Distance Relays Utilizing Transistors", *Proceedings of IEE*, Vol. 103A 1956, p 509.
59. A.T. Johns, "Generalised Phase Comparators Techniques for Distance Protection: Basis for Their Operation and Design", *Proceedings of IEE*, July 1972, p 833.
60. A.T. Johns, "Generalised Phase Comparator Techniques for Distance Protection: Theory and Operation of Multi-Input Devices", *Proceedings of IEE*, November 1972, p 1529.
61. N.M. Anil Kumar, "New Approach to Distance Relays with Quadrilateral Polar Characteristic for EHV Line Protection", *Proceedings of IEE*, October 1970, 1986.
62. H.P. Khincha, K. Parthasarathy, B.S. Ashok Kumar, and C.G. Arun, "New Possibilities in Amplitude and Phase-Comparison Techniques for distance Relays", *Proceedings of IEE*, No. 1970, p 2133.
63. T.S.M. Rao, "Behaviour of Rectifier Bridge Comparator as an Amplitude Comparator", *Journal of the Institution of Engineers (India)*, Vol. XLV, No. 10, Pt ELx5, June 1965, pp220-230.

64. S.P. Patra and S.R. Bandopadhyay, "a Generalised Theory of Protective Relays and its Application to Distance Protection-1", *Journal of the Institution of Engineers (India)*, Vol XLVI, No. 8, Pt. EL4, April 1966. pp 260-269.
65. S.P. Patra and S.R. Bandopadhysy, "A Generalised of Protective Relays and its application to Distance Protection-2", *Journal of the Institution of Engineers (India)*, Vol. XLVI, No. 12, Pt. EL6, August 1966, pp 498-512.
66. G.W. Hampe and B.W. Storer, "Power-Line Carrier for Relaying and joint Usage", *AIEE Power Apparatus and System No.1*, August 1952, pp 661-670.
67. N.M. Anil Kumar and K. Parthasarathy, "Power System Protection-3: Carrier Current Protection of Transmission Lines", Journal of the Institution of Engineers (India), Vol. 49, No. 12, Pt, El6, pp 310-320.
68. C.G. Dewey, C.A. Mathews and W.C. Morris, "Static MHO Distance and Pilot Relaying Principles and Circuits", IEEE Transaction on Power Apparatus and Systems, 1963, Vol. 66, p 391.
69. J. Rushton, "Pilot-Wire Differential Protection Characteristics for Multi-ended Circuits", Proceedings of IEE, November 1965, 2095.
70. K. Parthasarathy, "Pilot Wire Differential Protection", Electrical Times, 10th March 1966, p 320.
71. W.D. Humpage, J. Ruston and P.D. Stevenson, "Differential Pilot Wire Protection using Phase Comparators", *Proceedings of IEE*, July 1966, p 1183.
72. V. Caleca, S.H. Horowiz, A.J McConnel and H.T. Seeley, "Static MHO Distance and Pilot Relaying–Application and Test Results", Proceedings of IEE, 1963, pp 424.
73. J.A. Imlof and others, "Out of Step Relaying for Generators", Working Group Report, *IEEE Transactions on Power Apparatus and Systems*, September/October 1977, pp. 1556-1564.
74. N.M. Anil Kumar and K. Parthasarathy," Power System Protection-1: Generator and Transformer Protection", *The journal of the Institution of Engineering (India)*, Vol. 49, No. 12, Pt. El6, August 1969, pp. 289-300.
75. S. Govindappa, "New Relay for the Protection of Full Generator Winding for Earth Faults", *The Journal of the Institution (India)*, Vol. 52, No. 4, Pt. EL2, December 1971, pp 78-80.
76. B. Ram and B.B. Chakravarty, "Microprocessor-Based Distance Relays", *Journal of Microcomputer Application*, No. 6, 1983 Academic Press Inc. (London).
77. B. Ram, "Modelling of distance Relays for Digital Protection", *Proceedings of International AMSE Conference*, 29-30 October 1987, New Delhi.
78. D.N. Vishwakarma and B. Ram, "Microprocessor-Based Offset MHO Relays", *The Journal of the Institution of Engineers (India)*, Vol. 69, Pt. El2, October 1988, p 78.
79. B. Ram, "Microprocessor Based Overcurrent and Directional Relays", *Journal of the Institution of Engineers (India)*, Vol. 67, Part EL2, October 1986.
80. D. N. Vishwakarma and B. Ram, "Microprocessor-Based Digital Directional Relay", *Proceedings of Fifteenth National Systems Conference*, 13-15 March, 1992, Roorkee.
81. M. Ramamoorthy, "Application of Digital Computers to Power System Protection", *The Journal Institution of Engineers (India)*, Vol 52, No. 10, Pt. EL 5, June 1972, pp 235-238.
82. M. Ramamoorthy," A Static Directional Relay", *The Journal of the Institution of Engineers (India)*, Vol 50, No. 2, Pt. EL1, October 1969, pp 28-30.

83. M. Chamia and S. Liberman," Ultra High Speed Relay EHV/UHV Transmission Lines Development Design and Application", *IEEE Transaction on Power Apparatus and Systems*, Vol. PAS-97, Nov./Dec. 1976, pp 2104-2116.
84. T.F. Gallen, W.D. Breingan and M.M. Chen, "A Digital System for Systems, Comparison Relaying", *IEEE Transaction on Power Apparatus and Systems*, Vol. PAS-98, 3, May/June 1979.
85. W.J. Smolinski, "A Algorithm for Digital Impedance Calculation Using a Single PI-Section Transmission Line Model", *IEEE Transaction on Power Apparatus and Systems*, Vol. PAS-98, No. 5, September/October 1979, pp 1546-1551.
86. M.M. Chen, W.D. Breingan and T.F. Gallen, "Field Experience with a Digital System for Transmission Line Protection", *IEEE Transaction on Power Apparatus and Systems*, Vol PAS-98, No. 5, September/October 1979, pp 1796-1805.
87. G.D. Rockfeller, "Fault Protection with a Digital Computer", *IEEE Transaction on Power Apparatus and Systems*, Vol. PAS-88 No. 4, April 1969.
88. Y. Milki, Y. Sano and J.I. Makino, "Study of High Speed Distance Relay Using Microcomputer", *IEEE Transaction on Power Apparatus and Systems*, Vol. PAS-96, No. 2, March/April 1977, pp 602-613.
89. A.G. Phadke, T. Hlibka and M. Ibrahim, "A Digital Computer System for EHV Substations: Analysis and Field Tests", *IEEE Transactions on Power Apparatus and Systems*, Vol. PAS-95, No. 1, January/February 1976, pp 291-301.
90. A.T. Johns and M.A. Martin, "Fundamental Digital Approach to the Distance Protection of EHV Transmission Lines", *Proceedings of IEE*, Vol. 125, No. 5, May 1978, pp 377-384.
91. P.K. Dash and H.P. Khincha, "New Algorithms for Computer Relaying for Power Transmission Lines", *Electric Machines and Power Systems*, Vol. 14, No. 3-4, 1988, pp 163-178.
92. P.K. Dash, A.K. Panigrahi, H.P. Khincha and A.M. Sharaf, "New Algorithms for Computer Based Power System Measurements for Relaying", *Electric Machines and Power Systems*, vol. 14, No. 3-4, pp 179-193.
93. H.Y. Chu, S.L. Chen and C.L. Haung, "Fault Impedance Calculation Algorithms for Transmission Line Distance Protection", *Electric Power Systems Research*, 10 (1986), pp 69-75.
94. B. Jeyasurya and W.J. Smolinski, "Design and Testing of a Microprocessor-Based Distance Relay", *IEEE Transaction on Power Apparatus and Systems*, Vol. PAS-103, No. 5, May 1984, pp 1104-1110.
95. B. Jeyasurya and M.A. Rahman, "Transmission Line Distance Protection by Spectral Estimation Using Rectangular Wave transforms", *Electric Machines and Power Systems*, Vol. 11, 1986, pp 65-77.
96. "Computer Relaying Tutorial Text", *IEEE Power Society Special Publication No. 79*, EHO148-7-PWR, 1979.
97. B. Jeyasurya and W.J. Smolinski, "Identification of a Best Algorithm for Digital Distance Protection of Transmission Lines", *IEEE Transaction on Power Apparatus and Systems*, Vol. PAS 102, No. 10, October 1983, pp 3358-3369.
98. J.G. Gilbert, E.A. Udren and M. Sackin, "The Development and Selection of Algorithms for Relaying of Transmission Lines by Digital Computer", *Power Systems Operation and Control*, Academic Press, 1978, pp 83-126.

99. A.D. McInness and I.F. Morison, "Real Time Calculation of Resistance and Reactance for Transmission Line Protection by Digital Computer", *IEE Transaction*, Institution of Engineers, Australia, Vol. EE 7, No.1, 1971, pp 16-23.
100. A.M. Ranjbar and B.J. Cory, "An Improved Method for the Digital Protection of High Voltage Transmission Lines", *IEEE Transaction on Power Apparatus and Systems*, Vol. PAS-91, No.2 March/April 1975, pp 544-550.
101. W.D. Breingan, M.M. Chen and T.F. Gellen, "The Laboratory Investigation of a Digital System for the Protection of Transmission Lines", *IEEE Transactions on Power Apparatus and System*, Vol. PAS-98, No. 2, March/April 1979, pp 350-368.
102. H. Kudo, H. Sasaki, K. Seo, M. Takahashi, K. Yoshida and T. Maeda, "Implementation of a Digital Distance Relay Using an Integral Solution of a Differential Equation", *IEEE Transaction on Power Delivery*, Vol. PWRD-3, No. 4, October 1988, pp 1475-1484.
103. B.J. Mann I.F. Morrison, "Digital Calculation of Impedance for Transmission Line Protection", *IEEE Transaction on Power Apparatus and Systems*, Vol. PAS-90, No. 1, January/February, pp 270-279.
104. M.S. Sachdev and M.A. Baribeau, "A New Algorithms for Digital Impedance Relays", *IEEE Transaction on Power Apparatus and Systems*, Vol. PAS-98, No. 6, Nov./Dec./1979, pp 2232-2240.
105. P. Bornard and J.C. Bastide, "A Prototype of Multiprocessor-Based Distance Relay", *IEEE Transaction on Power Apparatus and Systems*, Vol. PAS-101, No. 2, February 1982, pp 491-498.
106. G.R. Slemon, S.D.T. Robertson and M. Ramamoorthy, "High Speed Protection of Power System Based on Improved Power System Models", CIGRE, Paris.
107. M. Ramamoorthy, "A Note Impedance Measurement Using Digital Computers", *IEE-IERE Proceedings*, (India), Vol. 9, No. 6. Nov./Dec. 1971, pp 243-247.
108. P.G. Melaren and M.A. Redfern, "Fourier-Series Technique Applied to Distance Protection", *Proceedings of IEE*, Vol. 122, No. 11, November 1975, pp 1301-1305.
109. D.D. Amoro and A. Ferrero, "A Simplified Algorithm for Digital Distance Protection Based on Fourier Techniques", *IEEE Transaction on Power Delivery*, Vol. PWRD-4, No. 1, January 1989. pp 157-163.
110. J.W. Horton, "The Use of Walsh Function for High Speed Digital Relaying", *IEEE PES Summer Meting*, San Francisco, July 1975, Paper No. A75-582-7, pp 1-9.
111. B.G. Srinivasa Rao, H.P. Khincha and K. Parthasarathy, "Algorithm for Digital Impedance Calculation Using Walsh Function", *Journal of the Institution of Engineers (India)*, Vol. 62, February 1982, pp 184-187.
112. N.M. Balchman, "Sinusoids versus Walsh Functions", *Proceedings of IEEE*, Vol. 62, 1974, pp 346-354.
113. K.K. Islam, S.K. Bose, and L.P. Singh, "On Microprocessor Based Relaying Scheme for EHV/UHV Transmission Line: An Existing 400 KV Line", *Electric Machines and Power Systems*, Vol. 12, 1987, pp 313-324.
114. D.B. Fakruddin, K. Parthasarathy, L. Jenkins and B.W. Hogg, "Application of Haar Function for Transmission Line and Transformer Differential Protection", *International Journal of Electrical Power and Energy Systems*, Vol. 6, No. 3, July 1984, pp 169-180.
115. D.N. Vishwakarma and B. Ram, "Microprocessor-Based Quadrilateral Distance Relay for EHV/UHV Transmission line Protection", *Journal of Microcomputer Applications*, Academic Press (London), Vol. 15, No. 4, October 1992, pp. 347-360.

References

116. W.A. Lewis and L.S. Tippett, "Fundamental Basis for Distance Relaying on 3-Phase Systems", *AIEE Transactions*, Vol. 66, 1947, pp 694-709.

117. B.J. Mann and I.F. Morrison, "Relaying a Three-Phase Line with Digital Computer", *IEEE Transaction on Power Apparatus and Systems*, Vol. PAS-90, No. 2, March/April 1971.

118. A.S. Joglekar and B.N. Karekar, "Contact Materials for the Development of Vacuum Circuit Breakers- A Reviews", *Journal of the Institution of Engineers (India)*, Vol. 53 Pt. EL 6 August 1973.

119. R.B. Shore and V.E. Phillips, "High Voltage Vacuum Circuit Breakers", *IEEE Transaction on Power Apparatus and Systems*, Vol. PAS-94, No. 5, September/October 1975, pp 1821-1830.

120. M. Murano, H. Nishikawa, A. Kobayashi, T. Okazaki and S. Yamashita, "Current Zero Measurement for Circuit Breaking Phenomena", *IEEE Transaction on Power Apparatus and Systems*, Vol. PAS-94, No. 5, September/October 1975, pp 1890-1900.

121. D.G. Tamaskar, "A Circuit for the Synthetic Testing of Vacuum Circuit Breaker-1 & 2", *The Journal of the Institution of Engineers (India)*, Vol. XLVI, No. 8, Pt. EL 4 and No. 12, Pt. EL6.

122. R.A. Larner, "Interrupting Performance of Air-Blast Breakers", *IEEE Transaction on Power Apparatus and Systems*, No. 5, April 1953, pp 303-311.

123. Jan Panek, "Synthetic Methods for Testing of Vacuum Breakers", *IEEE Transaction on Power Apparatus and Systems*, Vol. PAS-97 No. 4, July/August 1978, pp 1328-1326.

124. C.J.O. Garrard, "High Voltage Switchgear", *Proceedings of IEE*, September 1966, p 1523.

125. W. Rieder, "Circuit Breakers-Physical and Engineering Problems I", *IEEE Spectrum*, July 1970, p 35.

126. W. Rieder, "Circuit Breakers-Physical and Engineering Problems II", *IEEE Spectrum*, August 1970, p 90.

127. W. Rieder, "Circuit Breakers-Physical and Engineering Problems III", *IEEE Spectrum*, September 1970, p 80.

128. Siemens Review on SF6 Circuit Breakers, January 1969.

129. A. Selzzer, "Switching Vacuum: A Review", *IEEE Spectrum*, June 1971, p 26.

130. W.M. Leads and Others, "The Use of SF6 for High Power Are Quenching", *AIEE*, Dec. 57, p 906.

131. "New Generation of Oilless Circuit Breakers: 115 KV-345 KV, I, II, III and IV", IEEE Trans, on PAS, MarchApril 1971, p 628.

132. T. Ushio and Other, "Practical Problems on SF_6 Gas Circuit Breakers", *IEEE Trans. on PAS*, September/October 1971, p 2166.

133. R.E. Friedrich and R.N. Yeckley, "A New Concept in Power C.B. Design Utilising SF_6", *AIEE*, October 1959, p 695.

134. Y. Shibuya, "Electrical Breakdown of Long Gaps in SF_6" *IEEE Trans. on PAS*, May/June, p 1065.

135. E. Bolton, D. Birtwhistle, Mrs P. Bownes, M.G. Dwak and G.W. Routledge, "Overhead-line Parameters for Circuit Breakers Application", *Proceedings of IEE*, May 1973, p 561.

136. J.G. Stech and D.T. Switch-Hool. "Statistics of Circuit Breaker Performance", *Proceeding of IEE*, July 1970, p 1337.

137. J.G.P. Anderson, N.S. Ellis, F.O. Mason and Others, "Synthetic Testing of A.C. Circuit Breakers: Part I, Method of Testing and Relative Severety", *Proceeding of IEE*, Vol 113, No. 4, 1966, p 611.

138. J.G.P. Anderson; N.S. Ellis, M.A.S. Hich and Others, "Synthetic Testing of A.C. Circuit Breakers: Part II, Requirements for Circuit Breakers Proving", Proceedings of IEEE,. July 1968, p 996.

139. Casagrande and S. Rovelli, " Synthetic Testing Under Short-circuit Fault Condition", *Proceedings of IEE*, January 1968, p 136.

140. S. Tominga, K. Azumi, Y. shibiwa, M. Imataki, Y. Fujiwara and S. Nishida," Protection Performance of Metaloxide Surge Arreters Based on Dynamic V-I Characteristic", *IEEE Transaction on Power Apparatus and Systems*, Vol. PAS-98, No. 6 November/December 1979, pp. 1860-1871.

141. Castellis and Others, "Lightning Arresters", *IEEE Transaction on PAS*, March/April 1957, p 647.

142. D.N. Vishwakarma and B. Ram, "Microprocessor-Based Digital Impedance Relay", Journal of The Institution of Engineers (India), Vol. 75, May 1994, pp. 5-12.

143. Zahra Moravej, D.N. Vishwakarma and S.P. Singh, "Digital Filtering Algorithms for the Differential Relaying of Power Transformer: An Overview", Electric Machines and Power Systems (USA), Vol. 28, No. 6, June 2000, pp. 485-500.

144. Zahra Moravej, D.N. Vishwakarma and S.P. Singh, "ANN Based Protection Schemes for Power Transformer", Electric Machines and Power Systems (USA), Vol. 28, No. 9 September, 2000, pp. 875-884.

145. Zahra Moravej, D.N. Vishwakarma and S.P. Singh, "Differential Protection of Power Transformer Using ANN", International Journal of Engineering Intelligent Systems for Electrical Engineering and Communication (UK), Vol. 8, No. 4, Dec. 2000, pp. 203-211.

146. Zahra Moravej, D.N. Vishwakarma and S.P. Singh, "Radial Basis Function Neural Network Model for Protection of Power Transformer", Electric Power Components and Systems (USA), Vol.29, No. 4, April 2001, pp. 307-320.

147. Zahra Moravej, D.N. Vishwakarma and S.P. Singh, "Protection and Condition Monitoring of Power Transformer Using ANN", Electric Power Components and Systems (USA), Vol.30, No.3, March, 2002, pp. 217-231.

148. Zahra Moravej, D.N. Vishwakarma and S.P. Singh, "Application of Radial Basis Function (RBF) Neural Network for Differential Relaying of Power Transformer", Computers and Electrical Engineering, Vol.29, Issue 3, May, 2002, pp. 421-434.

149. Zahra Moravej, D.N. Vishwakarma and S.P. Singh, "Intelligent Differential Relay for Power Transformer Protection", Journal of The Institution of Engineers (India), Vol.83, June, 2002, pp. 28-32.

150. Zahra Moravej and D.N. Vishwakarma, "ANN-Based Harmonic Restraint Differential Protection of Power Transformer", Journal of The Institution of Engineers (India), Vol. 84 June, 2003, pp. 1-6.

151. Sumit Ahuja, Love Kothari, D.N. Vishwakarma and S.K. Balasubramanian, "Field Programmable Gate Array Based Overcurrent Relay", Electric Power Components and Systems (USA), Vol.32, No.3, March 2004, pp. 247-255.

152. Kiran Shrivastava and D.N. Vishwakarma, "Microcontroller-Based Quadrilateral Relay for Transmission Line Protection", Electric Power Components and Systems (USA), Vol.35, Issue 12, Dec. 2007, pp. 1301-1315.

153. Kiran Shrivastava and D.N. Vishwakarma, "Numerical Filtering Algorithms for Distance of Transmission Lines", Journal of The Institution of Engineers (India), Vol.89, March 2009, pp. 19-28.
154. Amrita Sinha, D.N. Vishwakarma and R.K. Srivastava, "Modeling and Real Time Simulation of Internal Faults in Turbogenerators", Electric Power Components and Systems (USA), Vol. 37, Issue 9, Sept 2009, pp. 957-969.
155. Amrita Sinha, D.N. Vishwakarma and R.K. Srivastava, "Modeling and Simulation of Internal Faults in Salient-pole Synchronous Generators with Wave Winding", Electric Power Components and Systems (USA), Vol. 38, Issue 1, Jan 2010, pp. 100-114.
156. Amrita Sinha, D.N. Vishwakarma and R.K. Srivastava, "Modeling and Simulation of Faults in Synchronous Generators for Numerical Protection", accepted for publication in International Journal of Power and Energy Conversion (Inderscience Publication).
157. Srivats Shukla and D.N. Vishwakarama, "Development of a Novel single Chip Solution a Motor Protection Relay", Proc. 2nd International Conference of Power engineering, Energy and Electrical Drives 2009 (POWERENG'09)", 18-20 March, 2009, Lisbon, Portugal, pp. 697-702.
158. Srivats Shukla, D.N. Vishwakarma and S.C. Gupta, "Improvement of System Reliability Using SoC Technique", Proc. International Conference on Industrial Technology (ICIT), 2010, 14-17 March, 2010, ViaMar Chile, pp. 229-234.
159. Harish Balaga, D.N. Vishwakarma and Amrita Sinha, "Numerical Differential Protection of Power Transformer using ANN as Pattern Classifier", Proc. International Conference on Power, Control and Embedded Systems (ICPCES-2010), Nov. 28-Dec. 1, 2010, Allahabad.
160. P. Pillay, A. Bhattacharjee, "Application of wavelets to model short-term power system disturbances", IEEE transactions on power systems, Vol. 11, No. 4, pp 2031-2037, Nov 1996.
161. K.P. Soman, K. I. Ramachandran, "Insight into wavelets from theory to practice", Prentice Hall of India, Oct 2005.
162. O. Cherrari, M. Meunier, F. Brouaye, "Wavelets: A new tool for the resonant grounded power distribution system relaying", IEEE transactions on power delivery, Vol. 11, No. 3, pp 1301-1308, July1996.
163. S. Santoso, E. J. Powers, W. M. Grady, P. Hofmann, "Power quality assessment via wavelet transform analysis", IEEE Transaction on power delivery, Vol. 11, No. 2, pp 924-930, Apr 1996.
164. D. C. Robertson, O. I. Camps, J. S. Mayor, W. B. Gish, "Wavelets and electromagnetic power system transients, ", IEEE transaction on power delivery, Vol. 11, No. 2, pp 1050-1058, Apr 1996.
165. S. G. Mallat, "A theory for multiresolution signal decomposition: The wavelet representation", IEEE transaction on pattern analysis and machine intelligence, Vol. 11, No. 7, pp 674-693, July 1989.
166. S. J. Huang, C. T. Hsieh C.L. Huang, "Application of wavelets to classify power system disturbances", Electric power systems research, Vol. 47, pp 87-93, 1998.
167. J. Barros, R. I. Diego, "Application of the wavelet packet transform to the estimation of harmonic groups in current and voltage waveforms", IEEE transaction on power delivery, Vol. 21, No. 1, pp 533-535, Jan 2006.
168. Omar. A. S. Youssef, "A modified wavelet-based fault classification technique", Electric power system research, Vol. 64, pp 165-172, 2003.

169. D. Chanda, N. K. Kishore, A. K. Sinha, "Application of wavelet multiresolution analysis for identification and classification of faults on transmission lines", Electric power system research, Vol. 73, pp 323-333, 2005.
170. K. L. Butler-Purry, M. Bagriyanaik, "Characterization of transients in transformer using discrete wavelet transforms", IEEE transaction on power systems, Vol. 18, No. 2, pp 648-656, May 2003.
171. T. Nengling, H. Zhijian, Y. Xianggen, L. Xiaohua, C. Deshu, "Wavelet based ground fault protection scheme for generator stator winding" Electric power system research, Vol. 62, pp 21-28, 2002.

Index

Accessories of GIS 626
Accuracy Class of CTs 76
Accuracy limit factor or saturation factor 77
Accuracy limit primary current 77
Acquisition Time 462
Active Low-pass Filter 457
Actual Transformation Ratio 43
Adaptive Protection 637
ADC0800 459
ADC0808/ADC0809 466
ADC1210 and ADC1211 469
A/D Converter, Analog Multiplexer, S/H Circuit 457
Advantages of GIS 627
Advantages of SF_6 Circuit Breakers 561
Air Blast Circuit Breakers 553
Air-Break Circuit Breakers 547
Alarm Relays 45
Aliasing 387
AM3705 460
Amplitude Comparator 51
Analog Interface 389
Analog Multiplexer 460
Analog to Digital Converter 458
Angle Impedance (OHM) Relay 267
ANN-based Overcurrent Relay 246
ANN Modular Approach for Fault Detection, Classification and Location 523
Arcing Time 581
Aperture Time 462
Application of ANN to Generator Protection 530
Application of ANN to Overcurrent Protection 522
Application of ANN to Power Transformer Protection 526
Application of ANN to Transmission Lines Protection 522
Application of Artificial Intelligence to Power System Protection 520
Application of Wavelet Transform to Power System Protection 438
Arcing Ground 199
Arcing Horn 606
Arc Interruption 535
Arc Voltage 535
Artificial Intelligence 506
Artificial Neural Network (ANN) 506, 507
Artiicial Neural Network Design Procedure and Consideration 517
Attracted Armature Relays 33
Automatic Reclosing 20
Auto-reclosing 301
Auxiliary CTs 86
Auxiliary Relay 44
Auxiliary Switch 45

B

Back-Propagation Algorithm 514
Backstops 49
Back-up Protection 361
Balanced Beam Relays 35
Balanced (opposed) Voltage Differential Protection 346
Balanced Voltage (or Opposed Voltage) Scheme 312
Ballistic resistance 48

Index

Bar Primary CTs 81
Basic Impulse Insulation Level (BIL) 618
Bearing Overheating Protection 360
Bearings 49
Biological Neural Networks 508
Bipolar to Unipolar Converter 469
Block Pulse Functions (BPF) 428
Block Pulse Functions Technique 427
Breaking Capacity 568
Buchholz Relay 368
Busbar Reactors 139
Bus-zone Protection 375

C

Carrier Acceleration Scheme 322
Carrier Aided Distance Protection 318
Carrier Blocking Scheme 323
Carrier Current Protection 19, 315
Causes of Overvoltages 592
Characteristics of Simple Differential Protection Scheme 336
Chemical Properties of SF_6 Gas 556
Choice of Characteristics for Different Zones of Protection 294
Circuit breaker 18, 533
Circulating Current Scheme 311
Classiication of CTs 80
Classiication of GIS 626
Clock 458
Combination of Current and Time-grading 230
Combined Earth Fault and Phase Fault Protective Scheme 240
Compact GIS system 627
Comparators 51
Compensation for Correct Distance Measurement 295
Computation of RHT Coeficients 421
Computation of the Apparent Impedance 400
Computation of Walsh Coeficients 409
Contacts 49
Continuous Wavelet Transform 434
Counter 48

Counterpoise wires 602, 603
Coupling Capacitor Voltage Transformers
Coupling Factor 601
CT Burden 68
CT Errors 74
Current Chopping 544
Current-graded System 229
Current Limiting Reactors 136
Current Setting 225
Current to Voltage Converter 454
Current Transformers 20, 65
Cut-off current 580

D

Data Acquisition System (DAS) 386
Defuzzification 520
Definite Overcurrent Relay 221
Delayed Auto-reclosing Scheme 303
Design Aspects of GIS 624
Differential Amplifier 455
Differential Current Protection 375
Differential Equation Technique 391
Differential Protection 19, 327
Differential Protection of 3-Phase Circui 346
Differential Relays 327, 328
Directional Earth Fault Relay 240
Directional Overcurrent Relay 236
Directional Relay 481
Directional Relay Connections 235
Direct Testing 573
Direct transfer tripping (under-reaching scheme) 319
Disadvantages of GIS 627
Disadvantages of SF_6 Circuit Breakers 562
Discrete Fourier Transform Technique 393
Discrete Wavelet Transform 434
Discrimination between Fuses and Over-current Protective Devices 590
Discrimination between Two Fuses 589
Distance Protection 19
Double Line-to-Ground (2L-G) Fault 177

Index **665**

uble Pressure Type SF6 Circuit Breaker 559
op-out Fuse 587
P-based Overcurrent Relay 246
type Cartridge Fuse 584
al-input Comparator 17
antages of fast fault clearing 641

rth Fault and Phase Fault Protection 238
rth-fault Compensation 297
rth Fault Protective Schemes 239
rth Fault Relay and Overcurrent Relay 238
rth Fault Relays 370
ective Grounding (Solid Grounding) 206
ectively Grounded System 199
ect of arc Resistance on the Performance of Distance Relays 271
ect of Line Length and Source Impedance on Distance Relays 281
ect of Power Surges (Power Swings) on the Performance of Distance Relays 277
ects of Faults 4
ctrical Properties of SF_6 Gas 557
ctrical Time-delay 47
ctromagnetic CTs 81
ctromagnetic relays 14
ctromagnetic Type VTs 92
ctromechanical Impedance Relay 255
ctromechanical Relays 14, 33
ctronic Time Delay 48
ptical Relay 286
rgy Balance Theory 536
uipment Grounding (Safety Grounding) 212
olution of Protective Relays 6
pert Systems 507
pulsion Type High-Voltage Fuse 587
pulsion Type Lightning Arrester 608
ernal Fault Characteristics 338
ernal Overvoltages 592

Extraction of the Fundamental Frequency Components 398
Extremely Inverse-time Overcurrent Relay 223

F

Fast Fourier Transform (FFT) 398
Fast Operation 12
Fast Walsh-Hadamard Transform (FWHT) 410
Fault 1, 96, 621
Fault clearing Time of a Circuit Breaker 534
Fault impedance 96
Fault Statistics 5
Fault Under Loaded Condition of Synchronous Machine 133
Feeder Reactors 139
Ferranti surge absorber 615
Field Ground-fault Protection 356
Field Programmable Gate Arrays (FPGA) 622
Field Programmable Gate Arrays (FPGA) Based Relays 631
Field Suppression 361
Flag or Target 45
Flux-summing CT 87
Formation of Sequence Networks of a Power System 164
Fourier-Walsh Theory 407
FPGA-based Overcurrent Relay 246
Frame Leakage Protection 376
Frequency Relays 628, 633
Frequency Relays and Load-shedding 628
Full-cycle Data Window DFT algorithm 399
Functions of Substation Grounding System 214
Fuse 578
Fuse Characteristics 582
Fuse element 578
Fuses for Capacitors 589
Fuses for Motors 588
Fuses for Transformers and Fluorescent Lighting 589
Fusing factor 580

Fuzzification 520
Fuzzy Logic 518
Fuzzy Systems 507

G

Gas insulated substation/switchgear (GIS) 622, 623
Generalised Interface for Distance Relays 501
Generalised Mathematical Expression for Distance Relays 487
Generator Reactors 138
Generator-transformer unit protection 352, 372
Genetic algorithm 507
Grounded system 205
Ground fault neutralizer 209
Grounding (Earthing) 198
Grounding through Grounding Transformer 211
Grounding Transformer 200
Ground wires 599

H

Haar functions 418
Haar Transform (HT) 418
Hadamard or Natural Ordering 405
Half-cycle Data Window DFT algorithm 399
Height of Ground Wire 601
High Impedance Relay Scheme 376
Highly integrated systems (HIS) 627
High Rupturing Capacity (HRC) Cartridge Fuse 585
High-speed Relays 16
High-Voltage CTs 87
High Voltage DC (HVDC) Circuit Breakers 566
Hinged Armature-Type Relays 34
Hybrid systems 627

I

Impedance and Reactance Diagrams 103
Impedance Relay 250, 475
Indirect Testing 574
Indoor CTs 81

Induction Cup Relay 41
Induction Disc Relay 38
Induction Relays 37
Inference 520
Input-Output Mapping 520
Input Quantities for Various Types of Distance Relays 269
Instantaneous Overcurrent Relay 222
Instantaneous Relays 16
Instrument transformers 65
Insulation Coordination 617
Integrated protection and control 639
Interfacing 453
Interfacing of ADC0800, Analog Multiple and S/H Circuits 464
Interfacing of A/D Converter ADC0800 4
Internal Fault Characteristics 337
Internal Overvoltages 592
Interruption of Capacitive Current 545
Interturn protection based on zero-seque component 354
Inverse Deinite Minimum Time Overcurr (I.D.M.T) Relay 222
Inverse-time Over-current Relay 222

K

Klydonograph 598

L

LF398 461
Lightning Arresters or Surge Diverters 60
Lightning Phenomena 593
Linear Couplers 80
Linear Function 511
Line-to-Line (L-L) Fault 174
Load-shedding and Restoration 629
Location of Reactors 138
Loss of Excitation 357

M

Magnetic Link 599
Making Capacity 569
Mann-Morrison Technique 390
Measurement of Reactance 491

Index **667**

asurement of Resistance 489
asuring and Protective CTs 66
chanical Time-delay 46
lting Time 581
tal-Oxide Surge Arrester (MOA) 611
IO (Admittance or Angle Admittance) Relay 261
o and Offset Mho Relays 492
IO Relay with Blinders 283
corprocessor-based numerical protective relays 452
crocontroller-based Overcurrent Relays 246
croprocessor Based Frequency and df/dt Relays 628
croprocessor-based Impedance Relay 257
croprocessor-based Overcurrent Relays 245
croprocessor Implementation of Digital Distance Relaying Algorithms 502
nimum fusing current 579
dules/Components of GIS 624
ving Coil Relays 36
ltifunction Numerical Relays 386
lti-input Comparator 17
ltilayer Feedforward Neural Networks 515
lti-shot Auto-reclosing 302
'A Rating of Reactors 140

ional grid 621
gative Sequence Network 158
ural Network Based Directional Relay 523
ural Network Learning 512
aron Modeling 508
atral Displacement 373
atral Grounding (Neutral Earthing) 200
ninal (rated) transformation ratio or Nominal (rated) CT ratio 71
n-effective Grounding 200
nlinearity 520

Non-linear Surge Diverter 609
Numerical Differential Protection 445
Numerical Differential protection of Generator 447
Numerical Differential Protection of Power Transformer 447
Numerical Distance Protection 444
Numerical Filtering using Block Pulse Functions (BPF) 428
Numerical Overcurrent Protection 442
Numerical Overcurrent Relays 245
Numerical protection 379
Numerical relays 7, 15, 62, 380, 381
numerical Relaying Algorithms 390

O

Offset MHO Relay 266
Oil Circuit Breakers 548
Oil Pressure Relief Devices 369
One Conductor Open 197
One-Line Diagram 102
Open Fuse 583
Operating Mechanism 564
Operational Amplifier 453
Optical Fibre Channels 325
Opto-Electronic CTs 82
Opto-Electronic VTs 93
Outdoor CTs 81
Out of Step Tripping and Blocking Relays 280
Overcurrent Protection 19
Overcurrent Protective Schemes 228
Overcurrent relay 221, 370, 471
Overheating Protection 366, 371
Overreach of Distance Relays 275
Overspeed Protection 359, 546
Overvoltage Protection 359
Overvoltages due to Lightning 596

P

Percentage Differential Protection 365
Percentage differential protection for a Y-connected generator with only four leads brought out 352

Index

Percentage or Biased Differential Relay 339
Performance of Protective Relays 12
Permissive over-reach transfer tripping scheme 321
Permissive under-reach transfer tripping scheme 319
Per-unit System 98
Peterson Coil 209, 615
Phase-angle Error 74, 75
Phase Comparator 52
Phase Comparison Carrier Current Protection 316
Phase Fault Protective Scheme 240
Phase-sequence current-segregating Network 94
Phase Shifter 454
Physical Properties of SF6 Gas 556
Piecewise Linear Function 510
Pilot relaying schemes 310
Plug setting multiplier (PSM) 225
Plunger-type Relays 35
Polarised MHO Relay 265
Polarised Moving Iron Relays 36
Positive Sequence Network 157
Power in Terms of Symmetrical Components 151
Power Swing Analysis 278
Precision Rectifier 456
Pre-arcing Time 581
Present Practice in Neutral Grounding 212
Primary and Back-up Protection 10
Principle of Out of Step Tripping Relay 280
Printed Disc Relay 40
Properties of SF_6 Gas 556
Prospective current 580
Protected Zone of Differential Protection Scheme 332
Protection against Auxiliary Failure 360
Protection against Magnetising Inrush Current 367
Protection against Motoring 359
Protection against Pole Slipping 361
Protection against Rotor Overheating because of Unbalanced Three-phase Stator Currents 358

Protection against Stator Interturn Faults
Protection against Travelling Waves 605
Protection against Vibration and Distortion of Rotor 360
Protection against Voltage Regulator Failure 360
Protection of Earthing Transformer 371
Protection of Generators 349
Protection of Parallel Feeders 237
Protection of Ring Mains 237
Protection of Stations and Sub-stations 60
Protection of Three-Winding Transformer 372
Protection of Transmission Lines against Direct Lightning Strokes 599
Protection system 18
Protective angle 600
Protective ratio 600
Protective relay 1, 18, 621
Protective system 1
Protective zone 601
Puffer-type SF_6 Circuit Breakers 560

Q

Quadrilateral Relay 284, 498

R

Radial Basis Function Neural Network 51
Ramp Function 511
Rated Current 579
Rate of Rise of Pressure Relay 369
Rating of Circuit Breakers 567
Ratio Correction Factor 76
Ratio Error (Current Error) 75
Ratio Error (Voltage Error) 91
Rationalised Haar Transform (RHT) 418, 484
Rationalised Haar Transform Technique 4
Reach of Distance Relays 274
Reactance Grounding 209
Reactance Relay 258, 484
Realisation of Mho Characteristic 493
Realisation of Offset Mho Characteristic 4
Realisation of Restricted Mho Characteristic 497

Index **669**

Realization of Field Programmable Gate Arrays (FPGA) Based Relays 632
Recovery Rate Theory 536
Reduction of Measuring Units 298
Reduction of Tower Footing Resistance 602
Reed Relays 37
Relationship Between Fourier and BPF Coefficients 429
Relationship Between Fourier and RHT Coeficients 423
Relay Reliability 640
Reliability 12
Removal of the dc offset 441
Repeat Contactors 45
Resistance Grounding 207
Resistance Switching 541
Resonant Grounding (Arc Suppressing Coil Grounding) 209
Restricted Directional Relay 289
Restricted earth-fault protection by differential system 351
Restricted Impedance Relay 289
Restricted MHO Relay 287
Restricted Reactance Relay 290
Restriking Voltage and Recovery Voltage 537
Reverse Looking Relay with Offset Characteristic 324
Reverse Power or Directional Relay 234
Rod Gap 606
Rogowski Coil 83
Rotor Protection 356

S

Sample and Hold Circuit 461
Sampling 388
Sampling Comparator 270
Saturation current 77
Seal-in Relay 44
Selection of Circuit Breakers 114, 565
Selection of Distance Relays 283
Selection of Fuses 588
Selectivity or Discrimination 11
Semi-enclosed Rewirable Fuse 583

Sensitivity 12
Sequence Impedances and Sequence Networks 154
Sequence Impedances of Power System Elements 155
Sequence Networks of Unloaded Synchronous Machines 157
Series Faults 196
Series-type Faults 169
Settings of Percentage Differential Relay 343
SF_6 Circuit Breakers 555
Short-Circuit Capacity 113
Shunt-type Faults 169
Sigmoid Function 510
Simple (Basic) Differential Protection 329
Simultaneous Faults 4
Single-input Comparator 17
Single Line-to-Ground (L-G) Fault 171
Single-phase (Single-pole) Auto-reclosing 303
Single-shot Auto-reclosing 302
Solid fault 96
Some Other Distance Relay Characteristics 290
Sound-Phase Compensation 298
Special Characteristics 224
Stability 12
Stability Limit and Stability Ratio 338
Static Impedance Relay 256
Static Overcurrent Relays 241
Static Relays 14, 50
Stator-overheating Protection 356
Stator Protection 350
Step Function 511
Step Voltage 213
Sudden Short-circuit of an Unloaded Synchronous Generator 109
Summation Transformer 93
Summing Amplifier 455
Super grid 621
Supervised Learning 512
Supervised Versus Unsupervised Learning 514

Supervisory Control and Data Acquisition (SCADA) 622, 640
Surge Absorber 614
Switched Schemes 299
Swivelling Characteristics 293
Symmetrical (balanced) Faults 97
Symmetrical Components 146
Symmetrical Fault Analysis 107
Symmetrical Faults 3
Synthetic Testing 574

T

Tank-earth Protection 373
Testing of Circuit Breakers 570
The isolated-phase GIS module 626
Thermal Relays 14, 43
Thermal Time-delay 47
Three-phase Auto-reclosing 303
Three-phase common module 626
Threshold Function 510
Time-current characteristics 221
Time-delay Relays 16
Time-graded System 228
Time-lag Relays 44
Time multiplier setting 226
Time SettinG 226
Touch Voltage 214
Transducers 18, 65
Transformer Protection 364
Transient Behaviour of CTs 77
Transley Scheme 313
Transley S Protection 313
Two Conductors Open 197
Types of Amplitude Comparators 53
Types of Faults 3
Types of Percentage Differential Relays 343
Types of Phase Comparators 56

U

Ultra High-speed Relays 16
Under-reach of Distance Relays 274
Ungrounded System 201
Unit Testing 574
Unsupervised Learning 513
Unsymmetrical Fault Analysis 168
Unsymmetrical Faults 3
Unsymmetrical (unbalanced) Faults 97

V

Vacuum Circuit Breakers 562
Very Inverse-time Overcurrent Relay 222
Voltage Transformer Ground-ing 211
Voltage Transformers 22, 89
VT Errors 90

W

Walsh Functions 403
Walsh-Hadamard Matrices 405
Walsh-Hadamard Transform Technique 402
Walsh or Sequency Ordering 405
Wavelet Fuzzy Combined Approach for Fault Classification 525
Wavelet Transform Technique 433
Wavelet Transform Versus Fourier Transform 437
Wave Shape of Voltage due to Lightning 595
Wire Pilot Protection 310
Wound Primary CTs 82

Z

Zero-Crossing Detector 453
Zero-sequence Diagrams of Generators 161
Zero Sequence Network 159, 160
Zero-sequence Networks of 3-phase Loads 160
Zero-sequence Networks of Transformers 162
Zones of Protection 9